# SERIES ON SEMICONDUCTOR SCIENCE AND TECHNOLOGY

*Series Editors*

R. J. Nicholas    University of Oxford
H. Kamimura    University of Tokyo

# SERIES ON SEMICONDUCTOR SCIENCE AND TECHNOLOGY

# Bands and Photons in III–V Semiconductor Quantum Structures

Igor Vurgaftman, Matthew P. Lumb, and Jerry R. Meyer

OXFORD
UNIVERSITY PRESS

# OXFORD
## UNIVERSITY PRESS

Great Clarendon Street, Oxford, OX2 6DP,
United Kingdom

Oxford University Press is a department of the University of Oxford.
It furthers the University's objective of excellence in research, scholarship,
and education by publishing worldwide. Oxford is a registered trade mark of
Oxford University Press in the UK and in certain other countries

First Edition published in 2021

Impression: 1

Published in the United States of America by Oxford University Press
198 Madison Avenue, New York, NY 10016, United States of America

British Library Cataloguing in Publication Data
Data available

Library of Congress Control Number: 2020944521

ISBN 978–0–19–876727–5

DOI: 10.1093/oso/9780198767275.001.0001

Printed and bound by
CPI Group (UK) Ltd, Croydon, CR0 4YY

*ILLVMINAT·OMNEM·STRVCTVRAM·BANDVM*
Inscription on a Roman vase

# Contents

## Part IV Semiconductor Photonic Devices

# Preface

Why do we need yet another book on the physics of semiconductor quantum structures and photonic devices? The writing process has taken enough time from our regular lives (and the purchase may take enough from a potential reader's wallet) that we cannot brush this question aside lightly.

First, we feel that the existing treatments fall a little short of presenting the theory of band structure and optical properties in a fully logical and self-contained manner. As we try to show in this book, this theory is surprisingly accessible at a rather deep level and does not demand much background from the reader, apart from a basic grounding in quantum mechanics and electromagnetism. In some cases, we find that other sources do not make it clear how *minimal* a model can be and still remain adequate for a particular class of devices.

Second, we would like to expand and bring up to date some of the content that can be found elsewhere. We include extended discussions of quantum and interband cascade lasers, superlattice photodetectors, and novel photovoltaic concepts as well as a much-needed update of the comprehensive set of band parameters first reported in a 2001 review by two of the present authors (expanded and corrected in 2003 and 2007 for the wurtzites).

Third, the book tries to shed a ray or two of light on many connections between the physics of III–V semiconductors and newer materials such as graphene and transition metal dichalcogenides. Today's junior researchers may benefit from following these links, sketched in outline, as the field inevitably pivots to the next great thing.

How did we decide which topics should receive the most care? Everyone tries to tell the story they know best, and we are no exception. The book relies on heavily on our published work, and hence emphasizes mid-infrared devices. But this is hardly a compilation, and much of the material appears in the form adopted here for the first time. We also try to take advantage of the extended book format to provide a more granular picture and frame it in a slightly different way that is *fully understandable to ourselves* and, we hope, to any other sufficiently motivated reader willing to go along for the ride. Sometimes that requires dropping rigor and bridging the gaps with dimensional and plausibility arguments. At other times, we use quantitative examples to make the magnitude of the effect clear.

Here is a brief overview of the actual content of the book. Part I lays the foundation for Parts II, III, and IV in describing how the most important regions of the band structure close to the energy gap can be modeled using approaches that range from quite simple to rather involved. The band structure technique of choice is the 8-band $\mathbf{k} \cdot \mathbf{p}$ method and its various simplified forms, although we do not neglect tight-binding approaches, including the effective bond-orbital method, and the empirical pseudopotential method. Chapter 4 discusses the optical properties of bulk III–V semiconductors, starting from the basic concepts and culminating in quantitative analyses of the absorption, gain, and radiative emission in bulk III–V materials. Part II presents a comprehensive tabulated update of the recommended band parameters (here augmented to include dilute bismides, boron nitride, and the optical constants) that were first published by two of us nearly twenty years ago. These parameters are used in the models and simulation results presented in Parts I, III, and IV, and may of course be used for general reference in the reader's own band structure and device modeling calculations. Part III guides the reader into the quantum realm, and its implementation in semiconductor epitaxial layers with atomic-scale thicknesses. It provides tools for computing the band structures and optical properties of quantum wells and superlattices, as well as quantum wires and dots with multidimensional confinement. Chapter 11 discusses the optical properties of quantum heterostructures, focusing on absorption, gain, and radiative emission processes that may result from both interband and intersubband interactions. Finally, Part IV describes the physical principles behind photonic devices based on layered III–V semiconductors and how they are designed using the tools developed in Parts I and III. Lasers, photodetectors, solar cells, and nonlinear devices will be discussed, with particular emphasis on mid-infrared lasers and mid-wave and long-wave superlattice-based photodetectors. Since all of these devices rely on non-equilibrium free carriers, the various radiative and non-radiative recombination processes that can dominate under different conditions will be described.

Everyone dislikes a long and rambling preface. If the reader indulges us for another paragraph, we would like to reiterate the philosophy that has guided the writing of this book. While it comes to you untested in a classroom, we do have vivid memories as well as recent experiences of learning new concepts. Whenever an echo of the familiar resounds in an exotic thicket, we feel a step closer to the destination. For this reason, we avoid plucking a *non-obvious* mathematical result from another reference and sending the reader scurrying off to check its validity (and sometimes despairing of ever doing so!). Instead, we attempt to make every calculation as self-contained as possible in a book of this size. When that stretches the book's scope too far, we outline why the given fact is plausible and how it connects to the accumulated body of physical knowledge. Prizing intelligibility over cleverness, we steer clear of the more obscure theoretical tools, particularly where they are not strictly necessary to understand the material. We also try to paint a continuous narrative rather than serve up a farrago of formulas and calculations. Having given up on citing every relevant paper under the sun, we limit our references to a short list of those articles and books that have been the most useful to us. The exception is the band-parameter update in Part II, which is more catholic and builds on our earlier reviews. Finally, we have aimed to keep the language simple and direct, not peppered with acronyms. The readers can judge for themselves the success (or lack thereof!) in achieving these goals from the text that follows.

# Acknowledgments

We are immensely grateful to the many people who have worked with us or advised us on the topics covered here over the past several decades. These include, first and foremost, the members of our research groups, as well as many world-class scientists, some of whose insights found their way into the book. While we are unable to list everyone who should share at least some credit, special thanks go to Bill Bewley, Chadwick Canedy, Chul Soo Kim, Mijin Kim, Charles Merritt, Stephanie Tomasulo, Jacob Khurgin, Edward Aifer, L. R. Ram-Mohan, Jasprit Singh, and Maria Gonzalez. On a more artistic front, Chul Soo Kim designed the cover image, and Yelena Vurgaftman helped to draw the schematics. It goes without saying that the responsibility for any errors rests with us alone.

We are grateful to Sonke Adlung and Oxford University Press for their interest in the initial idea (that did not flag as deadlines passed ... and passed), and for helping to make the publication process smoother than we could ever imagine. Finally, a word of thanks to our families who have made this book and many other things possible.

# Part I

# Band Structure and Interband Optical Transitions in Bulk Semiconductors

# 1
# Basics of Crystal Structure and Band Structure

III–V semiconductors form crystalline structures with three-dimensional periodic arrangements of the atoms. In this chapter, we will explore the nature of the crystal lattice starting from lower dimensions and progressing to real semiconductor crystals. We also learn why we expect distinct energy bands to form in the solids that crystallize in such lattices. The carrier statistics and occupation of the bands will also be examined.

## 1.1   1D and 2D Crystals

In one dimension, movement is along a single line. Therefore, all periodic 1D structures are alike and characterized uniquely by the lattice spacing $a$. We can also place more than one atom around each periodic lattice point, a concept known as adding a *basis* to the lattice. If the lattice is defined by points with coordinates $na$, where $n$ is an integer, one possible basis is to put two atoms at relative distances $\pm b$ from each lattice point, so that the atomic positions are $na \pm b$, as shown in Fig. 1.1. When $b = a/2$, the lattice with a two-atom basis degenerates into a single-atom basis with half the spacing, unless of course the two atoms are different. This is all we need to know about one-dimensional crystal structures.

Adding another dimension changes everything. We can construct the 2D lattice by tiling the plane with a repeated unit cell and placing the atoms somewhere within each cell. There are five distinct arrangements of the lattice points, and the ones with the highest symmetry tile the plane with squares or equilateral triangles (or, equivalently, regular hexagons).

Symmetry is pleasing to the eye, but which of the possibilities are the most physically important? The real answer will have to wait, but we can start with a guess that the energy is at a minimum when the atoms are packed into the smallest possible volume. To estimate the packing density of each crystal structure, we assume that all the atoms are circular.

*Bands and Photons in III–V Semiconductor Quantum Structures.* Vurgaftman, Lumb, and Meyer,
Oxford University Press (2021). © Vurgaftman, Lumb, and Meyer.
DOI: 10.1093/oso/9780198767275.003.0001

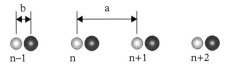

**Figure 1.1** *One-dimensional crystal lattice with period a and a basis of two atoms positioned at relative distances ±b from each lattice point.*

For example, the circles in the square lattice illustrated in Fig. 1.2a just touch when their diameters are equal to the lattice period $a$. Since only one circle fits within each square tile, the fraction of the plane covered by circles is

$$F_{square} = \frac{\text{area of circle}}{\text{area of square}} = \frac{\pi a^2/4}{a^2} = \frac{\pi}{4} \approx 0.785 \tag{1.1}$$

Can we do better? Clearly, a rectangular lattice is a poor choice since the circles would touch only along the shorter side. Then what about the triangular (or hexagonal) lattice shown in Fig. 1.2b? Some of the circles are shared between adjacent hexagonal tiles, so we count only the fraction of each circle that falls within a given hexagon. In this case, one circle lies entirely within each hexagon, and six more are shared equally between three cells. Counting the shared circles as 1/3 each, we obtain a total of three circles contained within each unit cell. The packing ratio is then

$$F_{hex} = \frac{3 \times (\text{area of circle})}{\text{area of hexagon}} = \frac{3\pi a^2/4}{3\sqrt{3}a^2/2} = \frac{\pi}{2\sqrt{3}} \approx 0.907 \tag{1.2}$$

This is impressively high, but can we raise it even closer to 1? The answer is no. In fact, we have run out of basic symmetries to consider. The plane cannot be tiled by regular octagons, dodecagons, etc., because an integer number of internal angles, $(k-2)\pi/k$ each, where $k$ is the number of vertices, does not fit into a

(a)                              (b)

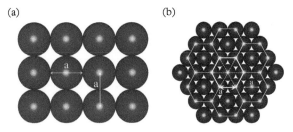

**Figure 1.2** *(a) Square-lattice arrangement of circles, each with diameter a. (b) Hexagonal-lattice arrangement of circles with diameter a.*

$2\pi$ angle at each vertex when $k > 6$. In fact, the so-called *Thue's theorem* states that among all possible periodic *and* aperiodic arrangements, a hexagonal lattice provides the highest planar filling fraction for circles of fixed radius.

We can add a basis to the 2D hexagonal lattice to represent two types of atoms in a compound. The concept will also come in handy when the atoms themselves are identical, but the outer-shell electronic configuration has fewer than six nearest neighbors.

In general, we need two integers $(m,n)$ to describe the position of each atom. The Cartesian coordinates of the atom $(x,y)$ are equal to $(m,n)$ multiplied by the two lattice vectors. For example, if we choose

$$\boldsymbol{a}_1 = a(1,0) \quad \boldsymbol{a}_2 = a\left(\frac{1}{2}, \frac{\sqrt{3}}{2}\right) \tag{1.3}$$

the six nearest neighbors in a hexagonal lattice are $(m,n) = (\pm 1,0), (0,\pm 1), (1,-1)$ and $(-1,1)$. To define the basis we superpose two such lattices, with the second displaced by $(0,-b)$ relative to the first in Cartesian coordinates. If $b = a/\sqrt{3}$, this is the "honeycomb" crystal structure illustrated in Fig. 1.3, in which each atom has three nearest neighbors that occupy sites on the other sublattice. The distance to the six nearest neighbors on the same sublattice is larger by a factor of $\sqrt{3}$. In Cartesian coordinates, the positions of the nearest neighbors with respect to an atom at the origin are

$$\boldsymbol{\delta}_{1,2} = \frac{a}{2\sqrt{3}}\left(\pm\sqrt{3}, 1\right) \quad \boldsymbol{\delta}_3 = \frac{a}{\sqrt{3}}(0,-1) \tag{1.4}$$

The atoms in the two interpenetrating sublattices may or may not be identical. III–V semiconductors do not produce any two-dimensional crystals, but group IV elements potentially can. In particular, *graphene* is a 2D honeycomb crystal of carbon atoms. We can also conceive of *silicene, germanene,* etc. The three nearest

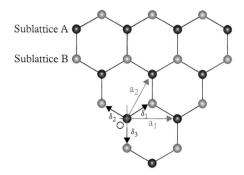

**Figure 1.3** *Two interpenetrating hexagonal lattices that produce a honeycomb crystal structure.*

**Table 1.1** *The 2D lattice structures discussed in the text.*

| Lattice Type | First Lattice Vector | Second Lattice Vector |
|---|---|---|
| Square | $a(1,0)$ | $a(0,1)$ |
| Rectangular | $a(1,0)$ | $b(0,1), a \neq b$ |
| Hexagonal (triangular) | $a(1,0)$ | $a\left(\frac{1}{2}, \frac{\sqrt{3}}{2}\right)$ |

*Note*: The honeycomb pattern is a hexagonal lattice with the two-atom basis described by Eq. (1.4).

neighbors are accommodated by $sp^2$ hybridization of the outer-shell electrons, with the non-hybridized orbitals forming two out-of-plane $\pi$ bonds. Graphene becomes the three-dimensional graphite crystal when atomic planes configured in the honeycomb lattice are stacked vertically and weakly bonded by van der Waals forces. The important 2D lattice structures are summarized in Table 1.1, along with the unit lattice vectors.

## 1.2   3D Crystal Structure

Taking our cue from 2D crystals, we next calculate how densely the atoms—now represented by spheres—can be packed in three dimensions. The simplest arrangement by far is the cubic lattice shown in Fig. 1.4a, in which the atoms occupy the vertices of a cube with side $a$. Any given vertex has six nearest neighbors, separated from the origin by $a$. For example, the atom at $(0,0,0)$ has nearest neighbors at $(\pm 1,0,0)$, $(0,\pm 1,0)$, and $(0,0,\pm 1)$. The number of nearest neighbors in a lattice is referred to as the *coordination number*. A single sphere with diameter $a$ fits into the cube with a packing fraction

$$F_{cube} = \frac{\text{volume of sphere}}{\text{volume of cube}} = \frac{4\pi (a/2)^3/3}{a^3} = \frac{\pi}{6} \approx 0.524 \qquad (1.5)$$

Surely we can do better! What if we place a single atom at the center of the cube $[(x,y,z) = a\left(\frac{1}{2},\frac{1}{2},\frac{1}{2}\right)]$ and position the others at the vertices as shown in Fig. 1.4b? This arrangement is known as *body-centered cubic* (*bcc*). If the sphere at the center just touches the corner spheres, the diagonal of the cube (with length $\sqrt{3}a$) spans two sphere diameters. There are eight spheres in the corners, each shared between eight adjacent cells. Counting the sphere at the center, this means that each cubic cell encloses two spheres of diameter $\sqrt{3}a/2$. The packing fraction is then

$$F_{bcc} = \frac{2 \times (\text{volume of sphere})}{\text{volume of cube}} = \frac{2 \times 4\pi \left(\sqrt{3}a/4\right)^3/3}{a^3} = \frac{\sqrt{3}\pi}{8} \approx 0.680 \qquad (1.6)$$

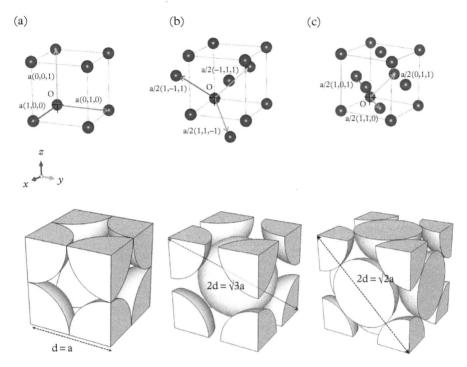

**Figure 1.4** *(a) Cubic lattice, the simplest 3D arrangement, has atoms positioned at the vertices of a cube with side length a. (b) Body-centered cubic lattice, which places a single atom in the center of the cube and others at each vertex. Also shown is the most convenient set of lattice vectors, which are drawn from a corner atom to the center atom in the next cell. (c) Face-centered cubic lattice, which positions additional atoms at the center of each face of the cube. Also shown is a convenient set of lattice vectors that connects a corner atom to the atoms on three adjacent faces. The schematic below each lattice diagram illustrates the space filling by spheres centered at the lattice sites. In the cubic lattice a single sphere of diameter $d = a$ fits into the cube, in the bcc lattice two spheres of diameter $d = \sqrt{3}a/2$ fit into a body diagonal, and in the fcc lattice two spheres of diameter $d = \sqrt{2}a/2$ fit into a face diagonal.*

Naïve though our packing argument may be, the bcc lattice does occur in nature as the crystal structure of a few elementary solids.

We can define the bcc lattice vectors in a number of ways. The most convenient choice is to draw the vectors from a corner atom in one cell symmetrically to the center atoms in the neighboring cells:

$$\boldsymbol{a}_1 = \frac{a}{2}(-1,1,1) \quad \boldsymbol{a}_2 = \frac{a}{2}(1,-1,1) \quad \boldsymbol{a}_3 = \frac{a}{2}(1,1,-1) \tag{1.7}$$

The atom at the cube center is closer to those at the vertices of the same cube (a distance of $\sqrt{3}a/2$) than it is to atoms at the centers of the neighboring cubes (a distance of $a$). Therefore, the coordination number for the bcc lattice is eight.

If denser packing goes together with the higher coordination number, can we find a lattice with a coordination number greater than eight? Try adding more atoms to the cube in the most symmetrical way possible! We can do this by putting the extra atoms at the centers of the six faces rather than at the center of the cube. This *face-centered cubic (fcc)* lattice, illustrated in Fig. 1.4c, looks the same from any given point in any direction. It has a higher coordination number than the bcc lattice, because the nearest neighbors of the atom at $(0,0,0)$ are at the centers of the 12 adjacent faces $a(\pm 1, \pm 1, 0)$, $a(0, \pm 1, \pm 1)$, and $a(\pm 1, 0, \pm 1)$ located on the three intersecting planes. We choose the following symmetric lattice vectors to describe the fcc lattice:

$$\boldsymbol{a_1} = \frac{a}{2}(0,1,1) \quad \boldsymbol{a_2} = \frac{a}{2}(1,0,1) \quad \boldsymbol{a_3} = \frac{a}{2}(1,1,0) \tag{1.8}$$

Does the higher coordination number for the fcc lattice give a higher packing fraction? In the fcc lattice, the atoms at the centers of the six faces are half inside the cube pictured in Fig. 1.4c, while those at the eight vertices are only 1/8 within that cube. This adds up to four spherical volumes inside the cube. Since we can fit two diameters into a face diagonal with length $\sqrt{2}a$, the packing fraction is

$$F_{fcc} = \frac{4 \times (\text{volume of sphere})}{\text{volume of cube}} = \frac{4 \times 4\pi \left(\sqrt{2}a/4\right)^3/3}{a^3} = \frac{\pi}{3\sqrt{2}} \approx 0.740 \tag{1.9}$$

The fraction is indeed higher, but only by $< 10$ percent compared to Eq. (1.6), even though the coordination number is up by 50 percent. This should alert us that we are approaching an upper limit. It also makes intuitive sense that extra open space is unavoidable when atoms are packed in 3D rather than in 2D. Are there any other lattice configurations we have missed? For example, could we have a cubic lattice with additional atoms at the centers of the edges rather than on the faces? No, this "edge-centered lattice" is not in fact a valid lattice, because with gaping holes at the cube centers and six faces it does not look the same from all directions. Of course, if these gaps are filled with atoms, we recover a simple cubic lattice with half the period. We conclude that the potential for improvement within the cubic geometry is at an ebb.

In Section 1.1, a hexagonal geometry gave the tightest fit in 2D. Might we use a similar type of packing in 3D? We can simply borrow the first two lattice vectors from the 2D form in Eq. (1.3) and choose the third to point along the $z$ axis with some length $c$. Now, what value of $c$ (in units of $a$) yields the densest packing arrangement? We start with all the atoms in successive hexagonal planes lying exactly on top of one another, as illustrated in Fig. 1.5a. Here $c$ is simply equal to $a$. As in the analogous 2D case, there are three atoms per unit cell since exactly half of each layer fits inside. The lattice vectors are

$$\boldsymbol{a}_1 = a(1,0,0) \quad \boldsymbol{a}_2 = a\left(\frac{1}{2}, \frac{\sqrt{3}}{2}, 0\right) \quad \boldsymbol{a}_3 = c(0,0,1) \qquad (1.10)$$

The lattice vectors in Eq. (1.10) apply to any hexagonal geometry with $c \neq a$ as well as $c = a$.

The cell volume is the area of the hexagon times $c$ ($= a$), or $3a(\boldsymbol{a}_1 \times \boldsymbol{a}_2) = 3\sqrt{3}a^3/2$. The packing fraction is then

$$F_{hxs} = \frac{3 \times (\text{volume of sphere})}{\text{volume of cell}} = \frac{3 \times 4\pi(a/2)^3/3}{3\sqrt{3}a^3/2} = \frac{\pi}{3\sqrt{3}} \approx 0.605 \qquad (1.11)$$

This is no better than the fcc lattice, but what if we slide the upper layer so that it fits into the interstitials between the atoms of the first layer, as shown in Fig. 1.5b? The third layer is aligned with the first, the fourth with the second etc., as if we were stacking cannonballs (or oranges). The unit cell, which consists of both the first and second layers, now has three extra atoms, for a total of six. The cell volume is again the product of the hexagon's area and $c$, which we would like to calculate in terms of $a$.

The distance between the first and second layers is simply $c/2$ in this geometry. The distance between a given atom and the nearest atoms in the next layer is $a$, so that the two spheres of diameter $a$ are just touching. This is the hypotenuse of the right triangle shown in Fig. 1.5c. The leg in the plane is equal to the distance between one vertex and the center of an equilateral triangle formed by the neighboring vertices, i.e., half the side length divided by the cosine of the half-angle, $a/\left(2\cos 30^\circ\right) = a/\sqrt{3}$. Applying the Pythagorean theorem to the right triangle in Fig. 1.5c, we obtain

$$c = 2\sqrt{\frac{2}{3}}\, a \qquad (1.12)$$

and the packing fraction becomes

$$F_{hcp} = \frac{6 \times (\text{volume of sphere})}{\text{volume of cell}} = \frac{6 \times \frac{4\pi\left(\frac{a}{2}\right)^3}{3}}{\frac{3\sqrt{3}a^2 c}{2}}$$

$$= \frac{6 \times 4\pi(a/2)^3/3}{3\sqrt{2}a^3} = \frac{\pi}{3\sqrt{2}} \approx 0.740 \qquad (1.13)$$

The packing fraction of 74.0 percent for this aptly named *hexagonal close-packed (hcp)* lattice is the same as for the fcc lattice! We have indeed reached a

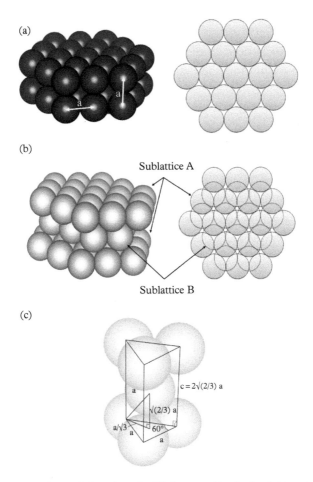

**Figure 1.5** *(a) Projection and plan view of a 3D hexagonal lattice in which each hexagonal layer lies exactly on top of the one below it (c = a); (b) 3D projection and plan view of the hexagonal close-packed (hcp) lattice, in which the atoms of sublattice B fit interstitially between the atoms of sublattice A; (c) out-of-plane lattice constant c for the hcp lattice, as found by applying the Pythagorean theorem to the right triangle composed of the vertex in the first layer, a vertex in the second layer, and the center of the equilateral triangle just below the second-layer vertex.*

fundamental limit, and in fact, there is a surprisingly close connection between the fcc and hcp lattices. When viewed along the [111] direction, the fcc lattice looks very much like the hcp lattice except the third layer is placed in different interstitials of the second layer rather than over the first layer. In the nineteenth century, Gauss gave a simple argument that demonstrated the packing fraction in Eqs. (1.9) and (1.13) is an upper bound when identical spheres are arranged in any kind of lattice. Curiously, the proof that this holds true for irregular as well as regular packing arrangements had to await the end of the twentieth century and could not have been completed without reams of painstaking, computer-assisted analysis. This is the famous *Kepler conjecture* from an influential list of unsolved mathematical problems published by David Hilbert in 1900.

Do actual crystals found in nature take advantage of these mathematical relations? Indeed, many elemental materials, e.g., the noble metals gold, silver, and copper, crystallize in the fcc lattice. Hexagonal lattices with $c$ close to the value in Eq. (1.12) are also quite popular in the natural world (found, e.g., in beryllium, magnesium, and zinc). Even the bcc lattice is not far behind, claiming the alkali metals and some transition metals such as iron and tungsten.

The packing fraction can take us only so far. The lattice structure of group IV, III–V, and II–VI semiconductors is primarily due to the electron configuration in the outer shell. In the simplest case, the outer shells in group IV materials such as C, Si, Ge, and $\alpha$-Sn (the low-temperature form) have two $s$ and two $p$ electrons. A strong covalent bond forms with other atoms in the lattice only when these electrons are shared with up to four other group IV atoms. The favored coordination number of four is even lower than in the simple cubic lattice.

Is there a close-packed lattice structure with a coordination number of four? Perhaps we can find one with a basis. We superpose two spatially displaced copies of one of the lattices we already know. With two interpenetrating fcc lattices, what displacement vector leads to a coordination number of four? If we shift the second lattice by $a/2$ along all axes, we form the rock-salt structure found, e.g., in table salt (NaCl). But Fig. 1.6a illustrates that each atom in the lattice still has six nearest neighbors.

The desired configuration has four atoms equidistant from each other, i.e., forming the vertices of a regular tetrahedron (a triangular pyramid). Can we embed this tetrahedron into the cube that underlies the fcc lattice? This is a well-known geometrical problem in which the vertices of the tetrahedron correspond to the $a/4\,(1,1,1)$, $a/4\,(1,-1,-1)$, $a/4\,(-1,-1,1)$, and $a/4\,(-1,1,-1)$ vertices of the origin-centered cube of side length $a/2$, as shown in Fig. 1.6c. The atoms at the vertices are connected by vectors $\pm a/2\,(1,1,0)$, $\pm a/2\,(1,0,1)$, and $\pm a/2\,(0,1,1)$, and the distance between them is $\sqrt{2}a$. In fact, these vectors are the same as those linking the nearest neighbors in the fcc lattice of Eq. (1.8). All of the atoms at the vertices of the regular tetrahedron are separated from the origin by a distance $\sqrt{3}a/4$. If the first lattice includes an atom at the origin, and the second lattice is displaced by a distance $a/4\,(1,1,1)$ from the first so that it includes atoms at $a/4\,(1,1,1)$, $a/4\,(1,-1,-1)$, $a/4\,(-1,-1,1)$, and $a/4\,(-1,1,-1)$, we have a crystal structure with four nearest neighbors that can accommodate the $sp^3$ covalent bonding found in most bulk semiconductors. This is shown in Fig. 1.6b.

When the atoms of the two interpenetrating sublattices are the same, we have the *diamond* crystal structure, which is found in nature in the diamond form of crystalline C, Si, Ge, and $\alpha$-Sn. When the atoms in the two sublattices are different, as in III–V semiconductors or some group IV materials such as SiC, the covalent

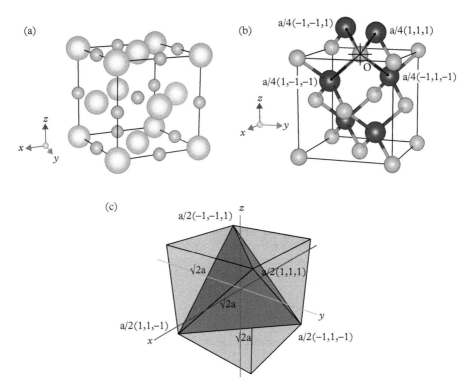

**Figure 1.6** *(a) Rock-salt crystal structure, which superposes two fcc lattices with the second displaced by a/2 along all axes. (b) Diamond crystal lattice formed by two interpenetrating fcc lattices. When different atoms occupy the two sublattices, the structure is called zinc-blende. (c) A regular tetrahedron embedded into a cube to illustrate how a cubic lattice can accommodate the valence of III–V semiconductors.*

bond is partially ionic because some electrons are closer to one ion than the other. The crystal structure is then called *zinc-blende*.

We can also imagine two interpenetrating hexagonal sublattices. Even the hcp lattice itself can be thought of as having a basis that consists of the first and second atomic layers, but we will take it to be a single unit. If two such lattices are superposed, the resulting *wurtzite* crystal is the default for wide-gap III-N semiconductors and many II–VI semiconductors, with the relation between $c$ and $a$ in Eq. (1.12) approximately satisfied. Table 1.2 summarizes the most important types of 3D lattice and crystal structures, along with the unit lattice and basis vectors.

In the diamond and zinc-blende crystals, the second-nearest neighbors are always the same as the atom at the origin and identical to the first-nearest neighbors in the fcc lattice. We will see later that the most important features of the III–V

**Table 1.2** *The most important 3D lattice and crystal structures, with corresponding lattice vectors and displacement vector between the two atoms in the basis, where applicable.*

| Lattice Type | Lattice Vectors | | | Basis Displacement |
|---|---|---|---|---|
| Simple cubic | $\boldsymbol{a}_1 = a(1,0,0)$ | $\boldsymbol{a}_2 = a(0,1,0)$ | $\boldsymbol{a}_3 = a(0,0,1)$ | None |
| Body-centered cubic | $\boldsymbol{a}_1 = \frac{a}{2}(-1,1,1)$ | $\boldsymbol{a}_2 = \frac{a}{2}(1,-1,1)$ | $\boldsymbol{a}_3 = \frac{a}{2}(1,1,-1)$ | None |
| Face-centered cubic | $\boldsymbol{a}_1 = \frac{a}{2}(0,1,1)$ | $\boldsymbol{a}_2 = \frac{a}{2}(1,0,1)$ | $\boldsymbol{a}_3 = \frac{a}{2}(1,1,0)$ | None |
| Hexagonal | $\boldsymbol{a}_1 = a(1,0,0)$ | $\boldsymbol{a}_2 = a\left(\frac{1}{2}, \frac{\sqrt{3}}{2}, 0\right)$ | $\boldsymbol{a}_3 = c(0,0,1)$ | None |
| Hexagonal close-packed | $\boldsymbol{a}_1 = a(1,0,0)$ | $\boldsymbol{a}_2 = a\left(\frac{1}{2}, \frac{\sqrt{3}}{2}, 0\right)$ | $\boldsymbol{a}_3 = 2\sqrt{\frac{2}{3}}\, a(0,0,1)$ | None |
| Diamond/zinc blende | $\boldsymbol{a}_1 = \frac{a}{2}(0,1,1)$ | $\boldsymbol{a}_2 = \frac{a}{2}(1,0,1)$ | $\boldsymbol{a}_3 = \frac{a}{2}(1,1,0)$ | $\boldsymbol{\delta} = a\left(\frac{1}{4}, \frac{1}{4}, \frac{1}{4}\right)$ |
| Rock salt | $\boldsymbol{a}_1 = \frac{a}{2}(0,1,1)$ | $\boldsymbol{a}_2 = \frac{a}{2}(1,0,1)$ | $\boldsymbol{a}_3 = \frac{a}{2}(1,1,0)$ | $\boldsymbol{\delta} = a\left(\frac{1}{2}, \frac{1}{2}, \frac{1}{2}\right)$ |
| Wurtzite | $\boldsymbol{a}_1 = a(1,0,0)$ | $\boldsymbol{a}_2 = a\left(\frac{1}{2}, \frac{\sqrt{3}}{2}, 0\right)$ | $\boldsymbol{a}_3 = c(0,0,1)$ | $\boldsymbol{\delta} = c\left(0,0,\frac{3}{8}\right)$ |

band structure can be described by the underlying fcc lattice, with the fine details emerging from the full zinc-blende symmetry.

## 1.3 Origin of the Bands

With this picture in mind of how the atoms form a lattice, we are ready to tackle the nature of electronic states in a semiconductor crystal. The simplest case is an infinite 1D chain of atoms with periodic spacing $a$, as in Fig. 1.7. As an illustration rather than a practical example, imagine that the chain consists of something like alkali or noble metal atoms with a single $s$ electron in the outer shell.

The electronic state for the $s$ electron in the crystal is described by its wavefunction. The square of the wavefunction is the probability of finding the electron at a particular point along the chain. Since the chain is infinite, and one atom looks exactly like the next, the probability must be the same at all atoms and, more generally, at all points with the same distance from any given atom. In other words, the probability is periodic along the chain, with period $a$. This does not necessarily mean that the wavefunction itself is periodic because it can have an arbitrary position-dependent phase factor $\exp[i\theta(x)]$.

How do we find the spatial variation of the phase $\theta(x)$? Imagine starting at some atom and then moving a few atoms to the right. Because there is nothing to distinguish any particular atom, the wavefunction at our destination cannot depend on the departure point, but solely on the distance traveled. This happens only if the phase $\theta(x)$ is linear in $x$, i.e., $\exp[i\theta(x)] = \exp(ikx)$. Of course, a constant phase "background" has no physical meaning, so the origin is arbitrary. We conclude that spatial invariance along the chain translates into a linear phase variation with distance. The complete wavefunction becomes

$$\Psi_k(x) = e^{ikx} u_k(x) \tag{1.14}$$

where $u_k(x)$ is periodic in $x$ with period $a$. Thus $u_k(x) = u_k(x+a)$ and in general is a function of $k$. The periodic function $u_k(x)$ may tell us, for example, that the

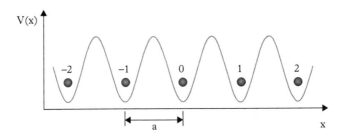

**Figure 1.7** *Schematic illustrating a simple sinusoidal variation of the potential energy along a chain of atoms spaced with period a.*

electrons are more likely to stay between the atoms than at their centers. This form of periodic potential is known as the *Bloch function*. For free carriers moving in the absence of any potential, $u_k(x) = 1$, and the wavefunction in Eq. (1.14) is a pure plane wave with uniform probability density over all space.

From Eq. (1.14), the wavefunctions at neighboring lattice sites are connected via what we can call a recursion relation:

$$\Psi_k(na) = e^{ika}\Psi_k\left[(n-1)\,a\right] \tag{1.15}$$

The symbol $k$ in Eqs. (1.14) and (1.15) is the proportionality factor describing the particle's translation, known as the *wavevector*, or in 1D, as the *wavenumber*. Just as in free space, it determines the period over which the wavefunction oscillates, i.e., the wavelength $\frac{2\pi}{k}$. The Bloch function shows that the wavefunction also oscillates at the atomic scale, but the variation due to $k$ may be much slower, i.e., $\frac{2\pi}{k} \gg a$.

If $k$ varies by more than $2\pi/a$, the wavefunction at a given atomic position $na$ $(n = 0, \pm 1, \pm 2, \ldots)$ repeats itself. This means the range of $-\pi/a$ to $\pi/a$ is a complete representation of the $k$ dependence. To keep things simple, we have chosen a symmetric range of $k$, but could have centered it at any value $k$ without losing (or gaining) any information.

To see this machinery in action, we calculate the states in the sinusoidal potential illustrated in Fig. 1.7:

$$V(x) = 2V_0 \cos\left(\frac{2\pi x}{a}\right) = V_0\left(e^{iGx} + e^{-iGx}\right) \tag{1.16}$$

where for brevity we have introduced $G \equiv 2\pi/a$. Using periodic boundary conditions, the wavefunction can be expanded as the Fourier sum:

$$\Psi(x) = \sum_q a_q e^{iqx} \tag{1.17}$$

In a large crystal, $q$ is essentially continuous, and Eq. (1.17) can be converted to an integral over $q$.

We would like to transform Eq. (1.17) into an expression that resembles Eq. (1.14). If $k$ is constrained to the range $-\pi/a$ to $\pi/a$, we can in general write $q = k + \frac{2\pi l}{a} = k + lG$, where $l$ is an integer. We can substitute our potential in Eq. (1.16) and the trial wavefunction of Eq. (1.17) into the Schrödinger equation and let the chips fall where they may. The kinetic-energy term in that equation is $p^2/2m$, with $p \equiv -i\hbar\partial/\partial x$. For a free particle, the plane-wave wavefunction means that the momentum is proportional to the wavevector $p = \hbar k$. That is, the faster the particle moves, the more rapidly its phase oscillates. In fact, in the pilot-wave interpretation of quantum mechanics proposed by Louis de Broglie and expanded by David Bohm, particles move deterministically with a velocity proportional to the derivative of the phase. We can also loosely associate $k$ with $-i\partial/\partial x$, by saying that it is *dual* to the spatial coordinate $x$ (or to $-ix$). We are interested in the dependence of the energies $E$ and the corresponding wavefunctions on $k$, $G$, and $V_0$ that follow from

$$\left[-\frac{\hbar^2}{2m}\frac{\partial^2}{\partial x^2} + V(x)\right]\Psi(x)$$

$$= \sum_{q'}\left[\frac{\hbar^2 q'^2}{2m}e^{iq'x} + V_0\left(e^{i(q'+G)x} + e^{i(q'-G)x}\right)\right]a_{q'}$$

$$= E\sum_{q'}a_{q'}e^{iq'x} \tag{1.18}$$

We now multiply by $\exp(-iqx)$ and average over the entire crystal. The average is defined as the integral over the crystal divided by its length taken in the infinite limit, i.e., $\lim_{L\to\infty}\frac{1}{L}\int_0^L\dots dx$. When the crystal is very large, the terms with non-zero complex exponentials oscillate so many times that their average is negligible. Therefore, we retain only those $q'$ components for which the exponents vanish. The surviving components of the potential term satisfy $q' \pm G - q = 0$ or $q' = q \mp G$, but in the other terms only the coefficients with $q' = q$ are non-zero. Equation (1.18) becomes

$$\frac{\hbar^2 q^2}{2m}a_q + V_0\left(a_{q+G} + a_{q-G}\right) = Ea_q \tag{1.19}$$

This system of equations spans all values of $q$, but only those separated by $\pm G$ are coupled by the potential terms. This means we can replace $q$ with $k + lG$ ($l = 0, \pm 1, \pm 2, \dots$) and rewrite the wavefunction in Eq. (1.17) in a reduced form:

$$\Psi_k(x) = \sum_l a_{k,l}e^{ikx}e^{ilGx} \tag{1.20}$$

We have inserted the subscript to make it clear that the wavefunction has a value of $k$ in the range $-\pi/a \le k \le \pi/a$. Equation (1.20) has the same form as Eq. (1.14), with the periodic part given by

$$u_k(x) = \sum_l a_{k,l}e^{ilGx} \tag{1.21}$$

Now that we know what $q$ stands for in this case, Eq. (1.19) becomes

$$\frac{\hbar^2(k + lG)^2}{2m}a_{k,l} + V_0\left(a_{k,l+1} + a_{k,l-1}\right) = Ea_{k,l} \tag{1.22}$$

In this form, we recognize that the potential indeed couples the equation for any given $l$ (both positive and negative) to two other equations. How can we go about solving these equations? One way is to truncate the $l$ terms to make the system finite. Is this really justified? We see from Eq. (1.22) that the energies of the $k = 0$ states increase rapidly with $l$ via the $\frac{\hbar^2 (lG)^2}{2m}$ term. So if we are interested only in the lowest energy states, we may discard the terms with large $l$ and incur only a small error. Pushing the concept to its logical limit, we may brashly eliminate all the terms except $l = 0$ and $l = -1$ (or, equivalently, $l = 0$ and $l = 1$). We do not recommend this procedure for any practical solids, but the result below retains the important features of real bands. The admittedly crude approximation rewards us with

$$\begin{bmatrix} \frac{\hbar^2 (k-G)^2}{2m} & V_0 \\ V_0 & \frac{\hbar^2 k^2}{2m} \end{bmatrix} \begin{bmatrix} a_{k,-1} \\ a_{k,0} \end{bmatrix} = E \begin{bmatrix} a_{k,-1} \\ a_{k,0} \end{bmatrix} \tag{1.23}$$

This is something we can solve without much ado! A matrix equation with the general form

$$\begin{bmatrix} \varepsilon_1 & V_0 \\ V_0 & \varepsilon_2 \end{bmatrix} \begin{bmatrix} a_1 \\ a_2 \end{bmatrix} = E \begin{bmatrix} a_1 \\ a_2 \end{bmatrix} \tag{1.24}$$

has the energy eigenvalues

$$\varepsilon_\pm = \frac{\varepsilon_1 + \varepsilon_2}{2} \pm \sqrt{\frac{(\varepsilon_1 - \varepsilon_2)^2}{4} + V_0^2}, \tag{1.25}$$

Substituting the actual values from Eq. (1.23) yields

$$\varepsilon_\pm(k) = \frac{\frac{\hbar^2 (k-G)^2}{2m} + \frac{\hbar^2 k^2}{2m}}{2} \pm \sqrt{\frac{\left(\frac{\hbar^2 (k-G)^2}{2m} - \frac{\hbar^2 k^2}{2m}\right)^2}{4} + V_0^2} \tag{1.26}$$

We see from the plot of $\varepsilon_\pm(k)$ in Fig. 1.8 that the energies are continuous functions of $k$. The two signs lead to two distinct *bands* with different energies at the same $k$. In the lower (upper) band, the energy monotonically increases (decreases) with $k$. We notice that electronic states with different $k$ values generally have different energies.

At $k = \frac{G}{2} = \pi/a$, we have

$$\varepsilon_\pm(k = G/2) = \frac{\hbar^2 G^2}{8m} \pm V_0 \tag{1.27}$$

Here the two bands are separated by an *energy gap* of width $2V_0$, within which no electronic states exist. The monotonic variation of the energies with $k$ means they cannot "overshoot" into the gap.

More generally, the Fourier expansion of the crystal potential contains terms other than $V_G$ and $V_{-G}$ from Eq. (1.16). Provided that $V_G = V_{-G} = V_0$, the

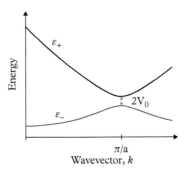

**Figure 1.8** *Wavevector dependence of the energy eigenvalues of Eq. (1.23). At $k = \frac{G}{2}$, the eigenvalues are separated by a gap of $2V_0$.*

energies at $k = G/2$ are still close to the values in Eq. (1.27); i.e., the fundamental bandgap is not affected. Jumping ahead a little, the other Fourier components $V_{lG}$ in the crystal potential induce additional gaps at higher energies.

Stepping back to take stock, our gambit with the minimal model of Eq. (1.23) has paid off beyond all expectations. We found that the electrons in solids reside in continuous bands of states separated by the "no man's land" of the energy gaps. In our three-dimensional world, the dispersions are more difficult to calculate and describe, but there are still bands and gaps. Indeed, we have stolen a glimpse of the true nature of electronic states in solid crystals. We are fortunate that this insight came at very low computational cost.

Now we change perspective and regard the same atomic chain from an entirely different angle. The idea is to simplify the problem by choosing the right basis states. In the case of Eq. (1.17), those are simply plane waves. Might we do better if we start with electron wavefunctions $\phi_l (x - na)$ "similar" to those of the isolated atoms at $x = na$? We do not specialize to $s$ electrons just yet, because the extra work will pay off in Chapter 5. The subscript $l$ refers now to the atom's different electronic states ($s$, $p$, etc.) The crystal potential is a superposition of the atomic potentials at each site:

$$V(x) = \sum_{n=-\infty}^{\infty} V_{at} (x - na) \tag{1.28}$$

Wavefunctions for the electrons in the atomic chain are formed by adding the weighted contributions $\phi_l (x - na)$ for all the atomic states $l$ summed over all the atoms.

We next choose coefficients that satisfy Eq. (1.14). We know from Eq. (1.15) that shifting the wavefunction by $n$ atoms (i.e., by the distance $na$) is equivalent to multiplying by the phase factor $\exp (ikna)$. By pulling that factor out of the overall coefficient, we can incorporate everything we know about the candidate wavefunction into a single formula:

$$\Psi_k(x) = \sum_{l,n} b_{k,l} e^{ikna} \phi_l(x-na) \tag{1.29}$$

This wavefunction depends on the individual atomic states, with the phase factor of magnitude one. The coefficients $b_{k,l}$ do not depend on position, which tells us that the spatial probability density $|\Psi_k(x)|^2$ can be translated by any multiple of $a$ and remain the same. To test whether the wavefunction of Eq. (1.29) is useful in describing electronic states in the crystal, we insert Eqs. (1.28) and (1.29) into the Schrödinger equation:

$$\sum_{l,n} b_{k,l} e^{ikna} \left[ -\frac{\hbar^2}{2m} \frac{\partial^2}{\partial x^2} + V(x) \right] \phi_l(x-na)$$

$$= \sum_{l,n} b_{k,l} e^{ikna} \left[ E_l + \sum_{n' \neq n} V_{at}(x - n'a) \right] \phi_l(x-na)$$

$$= E_k \sum_{l,n} b_{k,l} e^{ikna} \phi_l(x-na) \tag{1.30}$$

To simplify, we follow what we did in the wake of Eq. (1.18). We multiply by the complex conjugate of the atomic wavefunction associated with state $m, \phi_m^*(x)$, where $m$ is not necessarily the same as $l$. The $\phi_m^*(x)$ function plays the role of $\exp(-iqx)$ in the derivation following Eq. (1.18). We again average over the entire crystal by integrating over its length. The integrals coupling the wavefunctions at neighboring atomic sites are assumed to vanish unless the integral includes the potential for the full atomic chain. For $m \neq l$, the surviving terms have the form

$$V_{ml} = \sum_n e^{ikna} \int \phi_m^*(x) V(x) \phi_l(x-na) \tag{1.31}$$

At this point, we restrict to a single band, e.g., the lowest order or $s$-like atomic state, so that only a single *matrix element* appears in Eq. (1.31). We also reason that the coupling between two atomic wavefunctions weakens with increasing distance between the sites at which those wavefunctions reside. In the extreme limit where the sum retains only nearest-neighbor coupling, Eq. (1.31) reduces to

$$V_{ss} = V_{ss}' \left( e^{ika} + e^{-ika} \right) = 2V_{ss}' \cos(ka) \tag{1.32}$$

where $V_{ss}'$ denotes the coupling integral for the nearest-neighbor sites. The Schrödinger equation for the atomic-orbital case is a close analog of Eq. (1.22), although the kinetic energy term is missing, and the expansion coefficients have a different meaning, i.e., given by Eq. (1.29) rather than Eq. (1.21):

$$E_s a_{k,l} + V_{ss} \left( a_{k,l+1} + a_{k,l-1} \right) = E a_{k,l} \tag{1.33}$$

From Eqs. (1.32) and (1.33), we see that the energy of the state corresponding to wavevector $k$ is

$$E(k) = E_s + 2V'_{ss} \cos (ka) \tag{1.34}$$

where $E_s$ is the energy of the $s$ state in an isolated atom.

Starting from the isolated atomic wavefunctions, we have derived the relation between energy and wavevector in a simple single band. In the solid-state literature, this approach is known the *tight-binding* method, since we can imagine the original wavefunctions to be bound to their atoms. Expanding Eq. (1.34) in a Taylor series near $k = 0$, we find that the energy is approximately quadratic in $k$:

$$E (k \to 0) = E_s + 2V'_{ss} - V'_{ss}a^2k^2 \tag{1.35}$$

where for $V'_{ss} < 0$ the curvature in the band is positive. Even though Eq. (1.34) is very different from Eq. (1.26), both are approximately quadratic near $k = 0$. This general feature can traced back to the second-derivative form of the kinetic energy. In the "tight-binding" expression of Eq. (1.35), the free-electron "effective mass" is $m_e = \frac{\hbar^2}{2V'_{ss} a^2}$.

A second characteristic is that both band dispersions are periodic in $2\pi/a$, as anticipated in the discussion that followed Eq. (1.15). In other words, periodicity in real space inevitably imposes periodicity in the *reciprocal space* associated with wavevector $k$. The equivalence of wavevectors separated by $\Delta k = 2\pi/a$ is illustrated in Fig. 1.9. At the boundary wavevectors of $-\pi/a$, $\pi/a$, $3\pi/a$, etc., where $E(k)$ begins to repeat itself, the dispersion from Eq. (1.34) flattens out so as to remain continuous, i.e., $\frac{\partial E}{\partial k}\big|_{k=\pm\pi/a} \to 0$. Have we stumbled onto something general? For a symmetric dispersion around $k = 0$, like that in Eq. (1.34), points with $-k$ are equivalent to those with $+k$. Combined with the $2\pi/a$ periodicity, this requires that the dispersion be continuous and symmetric around $k = \pi/a$ (i.e., $k = \frac{\pi}{a} + \varepsilon$, where $\varepsilon$ is small, is equivalent to both $k = -\frac{\pi}{a} - \varepsilon$ and $k = \frac{\pi}{a} - \varepsilon$). This is only possible when the dispersion's first derivative vanishes at that point, since any non-zero slope would be discontinuous there.

The *group velocity* is proportional to the first derivative of the dispersion $\partial E/\partial k$:

$$v_g = \frac{1}{\hbar} \frac{\partial E}{\partial k} \tag{1.36}$$

We will see that the group velocity at both $k = 0$ and the "zone" boundary in Fig. 1.9 ($k = \pi/a$) vanishes in many crystalline band structures. What is then the physical meaning of the group velocity? One answer is that this is the instantaneous velocity of an isolated electron in state $k$. It is most certainly not the average speed of the electrons observed experimentally, which is forged by the competing effects of accelerating fields and electron collisions with lattice vibrations and defects.

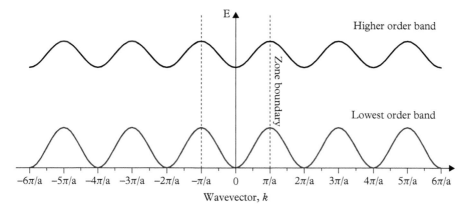

**Figure 1.9** *Periodicity of the band dispersion in reciprocal space, which arises from the periodicity of the wavefunction in real space.*

Even if all the imperfections are removed from the crystal and we operate at absolute zero, the observable velocity depends on how electrons in the other states are moving (possibly in the opposite direction!). In fact, it is not difficult to see that the sum of the group velocities over all the states in a band given by Eq. (1.34) is exactly zero. Put another way, some of the states must remain empty if the electron distribution as a whole is to be non-stationary. We will learn more about the occupation of the electronic states in Section 1.4.

When we include additional atomic orbitals into Eq. (1.31), we multiply the number of bands separated by gaps, much like in the plane-wave model. For example, we can include the higher order $p$-like and $d$-like atomic states or interactions between two dissimilar lattice sites. A similar approach for the zinc-blende lattice will be developed in greater depth in Chapter 5.

In deriving Eq. (1.34), we made little or no reference to the actual states of the isolated atoms. This is a clue that the expansion in Eq. (1.29) can be over an arbitrary set of basis states, which proves to be an important flexibility. Of course, to limit the number of potential matrix elements, it helps to select wavefunctions that are localized around individual atoms. Also, the math is much simpler if the wavefunctions are *orthogonal* to each other, i.e. the integrated product of any two distinct wavefunctions averages to zero:

$$\int_{-\infty}^{\infty} \phi_m^*(x)\phi_l(x)\,dx = \delta_{ml} \tag{1.37}$$

These are just the minimal requirements, with the full discussion deferred until Chapter 5.

## 1.4   Carrier Occupation and Statistics

The electronic states are of little consequence unless electrons occupy them. Of course, some of them must be occupied because the total number of electrons in a crystal is fixed by its constituent atoms (along with the fact that crystals do not remain charged in their environment). Returning to the simple example from Section 1.3, we first assume that each atom in the chain has exactly one *s* electron in its highest atomic orbital. As before, the atoms could be Li or another alkali metal, or a noble metal such as Cu, Ag, or Au as long as we neglect all the core electrons. If the chain is electrically neutral and has $N$ atoms, it also contains $N$ electrons. We define the *electron density* by dividing the number of electrons in the chain by its total length:

$$n = \frac{N}{Na} = \frac{1}{a} \tag{1.38}$$

This is just a long-winded way of stating that there is an average of one electron per period. The same idea applies in three dimensions, where we find the electron density per unit volume by dividing the average number of electrons that occupy a unit cell by its volume.

Another way to obtain the electron density is to count the number of $k$ states occupied by electrons. If the chain is infinite, $k$ is a continuous quantity, but if it has $N$ atoms and fixed boundaries, the wavevector is quantized in units of $2\pi/(Na)$. This is because the phase variation for $k < 2\pi/(Na)$ would otherwise "leak out" into the surrounding areas, which we assume to be impossible. Counting in steps of $2\pi/(Na)$, we obtain the correct total of $N$ when all the $k$ states between $-\pi/a$ and $\pi/a$ are summed. By analogy with Eq. (1.38), we can also define the electron density as the number of occupied $k$ states divided by the extent of the region of interest, i.e., $Na$. For very large $N$, the sum over $k$ states transforms smoothly into an integral. If all the states are occupied as in Eq. (1.38), we obtain

$$n = \frac{1}{2\pi} \int_{-\pi/a}^{\pi/a} dk = \frac{1}{a} \tag{1.39}$$

The integral form allows us to eliminate all the arbitrary lengths and total number of atoms in the chain. All that is left is the factor $\frac{1}{2\pi}$ that appears because we happened to write the spatial variation of the wavefunction as $\exp(ikx)$ rather than $\exp(i2\pi\kappa x)$. We could have eliminated this pesky factor by replacing $k$ with $\kappa \equiv k/(2\pi)$, but we choose to follow the convention enshrined in the literature. With multiple dimensions $d$, the necessary factor is raised to a power $(2\pi)^{-d}$, which is at least easy to remember.

What happens if the bands are less than completely full? It is a fundamental fact of nature that the intrinsic electron property called *spin* leads to two distinct states, with spins $+\frac{1}{2}$ and $-\frac{1}{2}$, for each $k$ vector in a given band. If we assume that the

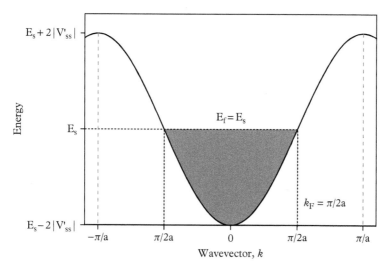

**Figure 1.10** *Electron occupation for an example in which the total number of electrons is half the number of k-states. Each wavevector corresponds to two distinct spin states.*

spin states are *degenerate*, i.e., they have the same energy, the number of available $k$ states is twice that in Eq. (1.39). Obviously, atoms such as Li have only one electron in the outer shell, and these electrons cannot fill the entire band. It follows that the total number of electrons is half the number of available $k$ states, and for a band dispersion given by Eq. (1.34), only states with $-\frac{\pi}{2a} < k < \frac{\pi}{2a}$ are occupied, as shown in Fig. 1.10. The electron density is still $n = \frac{1}{a}$, since each wavevector now corresponds to two distinct spin states. The maximum wavevector (energy) for which the states are occupied is the *Fermi wavevector* $k_F = \frac{\pi}{2a}$ (or *Fermi energy* $E_F = E(k_F)$). Since the energy minimum in the band is $E(k=0) = E_s - 2|V'_{ss}|$, with $V'_{ss} < 0$, we obtain $E_F = E_s$. Of course, for free electrons [$V_0 = 0$ in Eq. (1.26)], $E_F = \frac{\hbar^2 k_F^2}{2m}$.

The transition between occupied and empty states is abrupt only at $T = 0$. At higher temperatures, some of the states above the Fermi energy are occupied at the expense of the states below the Fermi energy. Since we cannot be sure that a particular state is occupied, we can only talk about the occupation probability, which we will now need to calculate.

The energy scale for electronic states lacks a natural absolute reference. It is clearly not the ground energy of the $s$ electrons $E_s$, because there is no observable change if that energy is shifted in either direction. It appears that only energy differences are truly meaningful. The occupation of any given state is independent of all the rest, with the ratio of the occupation probabilities of any two states depending only on the energy difference between them. This requirement is satisfied by the exponential function.

This line of reasoning leads us to the so-called Maxwell–Boltzmann occupation probability $P_k$:

$$P_k \propto \exp\left[-\left(E(k) - \mu\right)/k_B T\right] \tag{1.40}$$

where $\mu$ is the *chemical potential*. The overall minus sign in the exponent enforces the monotonic decrease with energy. We need $\mu$ to maintain the known overall occupation of the band, i.e., the carrier density. Of course, we could have multiplied by a constant factor instead, but there are advantages to giving this parameter units of energy. The Boltzmann constant $k_B$ simply converts from units of energy (joule) to temperature (kelvin). We can define $1/T$ formally as the derivative of the logarithm of the number of accessible states (entropy) with respect to energy, so that temperature $T$ has its intuitive meaning.

We have indicated proportionality rather than equality in Eq. (1.40) because the exponential function has not yet been normalized. Being fermions, electrons occupy electronic states one at a time or not at all; i.e., there are never two electrons in exactly the same state at the same time. If we accept this so-called *Pauli Exclusion Principle* as a fact of nature, the normalization amounts to summing the probabilities of the state being empty and occupied and making sure the total adds up to one. This is equivalent to assuming that each microscopic state is in equilibrium with the surroundings. The normalized probability of finding one electron in a state with energy $E$ becomes

$$
\begin{aligned}
P_k &= \frac{\exp\left[-\frac{1(E-\mu)}{k_B T}\right]}{\exp\left[-\frac{0(E-\mu)}{k_B T}\right] + \exp\left[-\frac{1(E-\mu)}{k_B T}\right]} \\
&= \frac{\exp\left[\frac{\mu-E}{k_B T}\right]}{1 + \exp\left[\frac{\mu-E}{k_B T}\right]} = \frac{1}{1 + \exp\left[\frac{E-\mu}{k_B T}\right]} \equiv f_k(E)
\end{aligned}
\tag{1.41}
$$

where $f_k$ is known as the Fermi–Dirac distribution function.

The denominator of Eq. (1.41) is known as the *partition function*. When the electron energy is large compared to the chemical potential, the exponential form of Eq. (1.40) is approximately correct and we obtain the Maxwell-Boltzmann limit. In a finite crystal with electrical contacts, the chemical potential is equivalent to the *Fermi energy* $(E_F)$ that we already defined. We see from Eq. (1.41) that the Fermi level is the energy at which the occupation probability is exactly 50 percent. Figure 1.11 shows that the occupation probability decreases quickly when $E(k) > E_F$, and approaches unity when $E(k) \ll E_F$ on a scale of $k_B T$.

We can now generalize Eq. (1.39) for $T > 0$ by including the occupation probability

$$n = \frac{1}{2\pi} \int_{-\pi/a}^{\pi/a} s_k f_k \left[E(k)\right] dk \tag{1.42}$$

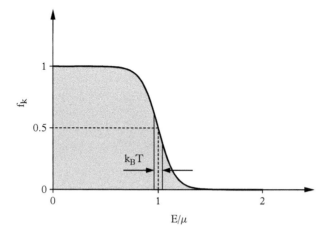

**Figure 1.11** *The Fermi–Dirac distribution function $f_k$.*

Here we have assumed the spin states to be degenerate and replaced the sum over spin by the spin degeneracy factor $s_k = 2$. It is often convenient to perform the integration over energy rather than wavevector using the transformation

$$n = \frac{1}{2\pi} \int_0^\infty \frac{s_k f_k(E)}{dE/dk} dE \qquad (1.43)$$

where $dE/dk$ is expressed as a function of energy rather than $k$ and $E = 0$ is taken to be the bottom of the band. The integral's upper bound should really be the highest energy state in the crystal rather than infinity, but that energy is much higher than the typical Fermi levels that we might as well take it to be infinity. Multiple bands are present in real crystals, and we will want to distinguish the carrier density in a particular band (e.g., one that is mostly empty) from the mostly full bands below the Fermi level that are of little interest. A side benefit of the differential transformation of Eq. (1.43) is that we no longer need to link the energy to a specific $k$ state.

In 3D, Eq. (1.42) becomes

$$n = \frac{1}{(2\pi)^3} \iiint s_k f_k [E(k)] \, d^3k \qquad (1.44)$$

If the electron energy depends only on the magnitude of the wavevector $k$, but not on its direction, Eq. (1.43) becomes

$$n = \frac{1}{(2\pi)^3} \int_0^\infty s_k \left[ \frac{4\pi k^2}{dE/dk} \right] f_k(E) \, dE \qquad (1.45)$$

where $4\pi$ comes from the trivial angular integral, and again the quantity in brackets is expressed in terms of energy rather than $k$. For example, for free carriers with $E(k) = \hbar^2 k^2/2m$:

$$DOS = \frac{1}{(2\pi)^3}\left[\frac{4\pi k^2}{dE/dk}\right] = \frac{1}{2\pi^2}\sqrt{2Em^3}/\hbar^3 \tag{1.46}$$

For the cosine dispersion in Eq. (1.34), we obtain

$$DOS = \frac{1}{(2\pi)^3}\left[\frac{4\pi k^2}{\frac{dE}{dk}}\right] = \frac{k^2}{4\pi^2 a\,|V_{ss}'|\sin(ka)}$$

$$= \frac{1}{4\pi^2 a^3\,|V_{ss}'|}\left[\cos^{-1}\left(\frac{E-E_s}{2V_{ss}'}\right)\right]^2\left(1 - \left(\frac{E-E_s}{2V_{ss}'}\right)^2\right)^{-1/2} \tag{1.47}$$

The quantities in Eqs. (1.46) and (1.47) are known as the *density of states* (DOS) and depend on the dispersion relation for a particular band. Figure 1.12 illustrates how the DOS depends on energy in these two simple examples. In one dimension, the DOS diverges when the dispersion becomes flat and the group velocity vanishes, e.g., at $k=0$ and $k=\pm\pi/a$. The 2D and 3D DOS are better behaved because there are additional directions for the states to "spread into". As $k \to 0$, the density of states in 3D generally goes to zero.

The bottom line is that the DOS always fulfills the same function, namely, to convert a $k$-space summation to an integral over energy. Of course, if the summation over $k$ states (converted to integrals) is straightforward or more

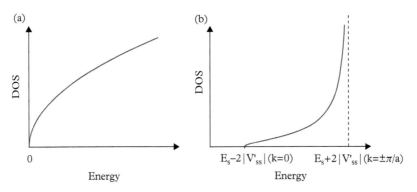

**Figure 1.12** *(a) Density of states for 3D free carriers that follows a $k \propto E^{1/2}$ dependence; (b) density of states for carriers in the 1D band described by Eq. (1.47).*

convenient than the energy integration, we do not need to calculate the DOS and can proceed directly with the $k$ integrations.

We mentioned earlier that when the Fermi level lies well below the bottom of the band ($\frac{E_F}{k_B T} \ll 0$), we can neglect the "1" in the denominator of Eq. (1.41). In this case, the Fermi level typically falls within an energy gap that separates two bands. Most of the states in the upper band are then empty, and the Fermi distribution function becomes

$$f_k = \exp\left(-\frac{E - E_F}{k_B T}\right) \qquad (1.48)$$

For the free-carrier DOS of Eq. (1.46) and accounting for spin degeneracy, we obtain

$$n = \int_0^\infty \frac{1}{\pi^2} \frac{\sqrt{2Em^3}}{\hbar^3} \exp\left(-\frac{E - E_F}{k_B T}\right) dE \qquad (1.49)$$

This integral over energy has the form of the *gamma function* $\Gamma(3/2)$. We can use the properties of this function $[\Gamma(3/2) = \frac{\Gamma\left(\frac{1}{2}\right)}{2} = \frac{\sqrt{\pi}}{2}]$ to evaluate the integral in Eq. (1.49):

$$n = \frac{\sqrt{(k_B T m)^3 / (2\pi^3)}}{\hbar^3} \exp\left(\frac{E_F}{k_B T}\right) \equiv N_c \exp\left(\frac{E_F}{k_B T}\right) \qquad (1.50)$$

where $N_c$ is known as the electron's *effective density of states*.

We see that the carrier density scales exponentially with the Fermi level as long as the electron population is *non-degenerate*, i.e., Maxwell–Boltzmann statistics hold. This is true for Eq. (1.50) written for free electrons, with all the information about the DOS incorporated into $N_c$. For an arbitrary dispersion, we may want to calculate $N_c$ numerically.

In the opposite limit $E_F \gg k_B T$, where the Fermi level lies far above the bottom of the band (*degenerate* statistics), $f_k(E)$ can be approximated as a step function that is unity for $E < E_F$ and zero for $E > E_F$. From Eqs. (1.45) and (1.46), the carrier density for parabolic bands then varies as $E_F^{3/2}$.

When the Fermi level crosses into the band, we can evaluate the more general integral with the Fermi–Dirac distribution numerically. Analytical approximations of the Fermi–Dirac integrals are available, but modern computing power makes them less important.

Of course, the toy band-structure models introduced in this chapter do not accurately describe what happens in a real semiconductor. We will explore more realistic models in Chapters 2, 3, and 5.

## Suggested Reading

- An outline of the road to proving the Kepler conjecture can be found in T. C. Hales, "Cannonballs and honeycombs", *Notices of the AMS* **47**, 440 (2000), but this is not light reading!
- To learn more about lattices and bands, the best option is still the time-honored classic of N. W. Ashcroft and N. D. Mermin, *Solid State Physics* (Saunders College, Philadelphia, 1976).
- A more systematic introductory treatment of classical and quantum statistics is provided in D. V. Schroeder, *Introduction to Thermal Physics* (Addison Wesley, Boston, 2000).

# 2
# Introduction to k·p Theory

The so-called **k·p** theory provides an effective framework for modeling the band structures of III–V semiconductors. This approach starts with a handful of bands that are coupled in accordance with the crystal symmetry discussed in Chapter 1. This chapter describes the **k·p** theory and lays out the assumptions that underlie its most common versions, including Kane's model. We will also learn how the spin–orbit coupling affects the band structure of semiconductors.

## 2.1 Coupled Conduction and Valence Bands

In Chapter 1, we saw that the electronic states in a semiconductor are located in continuous bands separated by energy gaps that are devoid of such states. We also discovered that, loosely speaking, the bands in the crystal may be built from the electronic states of the individual atoms. The outer shells of the elements comprising III–V semiconductors consist only of $s$ and $p$ orbitals. The total electron number (8 for each pair of group III and group V elements) is half the number of orbitals (1 $s$ and 3 $p$ for both group III and group V elements) when both spins are included. What does this imply for the electron occupation? In a perfect crystal at zero temperature, we expect to see a full band that is separated by an energy gap from an empty one above it. The full band is called the *valence band*, and the empty one the *conduction band*. If we could guess which atomic orbitals correspond to the conduction and valence bands at $k = 0$, would we be able to say anything about the dispersions of the two bands?

The **k·p** approximation gives a simple answer. We start with two periodic functions $u_{v0}$ and $u_{c0}$. Both satisfy the relation $u_{n0}(r) = u_{n0}(r + R)$, where $R$ is a linear combination of the lattice vectors $a_1, a_2, a_3$. These are familiar from the Bloch-function form of the wavefunction in Eq. (1.14). They are chosen to satisfy the symmetry characteristics of the conduction- and valence-band states at $k = 0$; otherwise, their functional form within the unit cell is not important. The **k·p** approximation further assumes that the periodic component of the wavefunction in a given band $n$ at arbitrary $k$ can be expanded as a linear combination of $u_{v0}$ and $u_{c0}$ (at $k = 0$):

*Bands and Photons in III–V Semiconductor Quantum Structures.* Vurgaftman, Lumb, and Meyer,
Oxford University Press (2021). © Vurgaftman, Lumb, and Meyer.
DOI: 10.1093/oso/9780198767275.003.0002

$$u_{nk} = a_c u_{c0} + a_v u_{v0} \tag{2.1}$$

Now we substitute the wavefunction of Eq. (1.14), with the Bloch term from Eq. (2.1), into the Schrödinger equation. To avoid writing out the gradients, we define $\boldsymbol{p} \equiv -i\hbar\nabla$. Since $\boldsymbol{p}$ and $\boldsymbol{k}$ are vector quantities, the algebraic products in Chapter 1 are now replaced by vector products (hence the name):

$$\left[\frac{\boldsymbol{p}^2}{2m} + V(\boldsymbol{r})\right] e^{i\boldsymbol{k}\cdot\boldsymbol{r}} u_{nk}(\boldsymbol{r})$$

$$= e^{i\boldsymbol{k}\cdot\boldsymbol{r}} \left[\frac{\boldsymbol{p}^2}{2m} + \frac{\hbar}{m}\boldsymbol{k}\cdot\boldsymbol{p} + \frac{\hbar^2 k^2}{2m} + V(\boldsymbol{r})\right] u_{nk}(\boldsymbol{r})$$

$$= E_{nk} e^{i\boldsymbol{k}\cdot\boldsymbol{r}} u_{nk}(\boldsymbol{r}) \tag{2.2}$$

At $k = 0$, the Schrödinger equation is satisfied by the periodic component of the Bloch function alone. This is because the phase term in Eq. (1.14) reduces to unity:

$$\left[\frac{\boldsymbol{p}^2}{2m} + V(\boldsymbol{r})\right] u_{n0}(\boldsymbol{r}) = E_{n0} u_{n0}(\boldsymbol{r}) \tag{2.3}$$

Using Eq. (2.3), we rewrite Eq. (2.2) as

$$e^{i\boldsymbol{k}\cdot\boldsymbol{r}}\left[\frac{\hbar}{m}\boldsymbol{k}\cdot\boldsymbol{p}\right]\sum_{n=c,v} a_n u_{n0}(\boldsymbol{r}) = \left(E_{nk} - E_{n0} - \frac{\hbar^2 k^2}{2m}\right) e^{i\boldsymbol{k}\cdot\boldsymbol{r}} \sum_{n=c,v} a_n u_{n0}(\boldsymbol{r}) \tag{2.4}$$

where $E_{c0}$ and $E_{v0}$ represent the conduction- and valence-band energies at $k = 0$. Now we multiply Eq. (2.4) by $u_{m0}^*(\boldsymbol{r})$ and average by integrating over the entire crystal. The plane-wave terms on the two sides of the equation cancel, and the $\langle u_{n0}|\boldsymbol{p}|u_{n0}\rangle$ terms vanish for periodic functions because taking the derivative transforms an even function into an odd one and vice versa. Here and below, we use the bracket notation to represent volume integrals, with the function on the left conjugated:

$$\langle u_{v0}|\boldsymbol{p}|u_{c0}\rangle = \frac{1}{V}\int d^3r\, u_{v0}^*\,\boldsymbol{p}u_{c0} \equiv \boldsymbol{p}_{cv} \tag{2.5}$$

We also normalize the Bloch functions to unity when $m = n$ and zero otherwise:

$$\langle u_{m0}|u_{n0}\rangle = \frac{1}{V}\int d^3r\, u_{m0}^*\,u_{n0} = \delta_{mn} \tag{2.6}$$

This leads to a set of two equations representing $n = c$ and $v$:

$$\begin{bmatrix} E_{c0} + \frac{\hbar^2 k^2}{2m} - E_{c,vk} & \frac{\hbar}{m}\boldsymbol{k}\cdot\boldsymbol{p}_{cv} \\ \frac{\hbar}{m}\boldsymbol{k}\cdot\boldsymbol{p}_{vc} & E_{v0} + \frac{\hbar^2 k^2}{2m} - E_{c,vk} \end{bmatrix}\begin{bmatrix} a_c \\ a_v \end{bmatrix} = 0 \tag{2.7}$$

To demonstrate that $\boldsymbol{p}_{cv} = \boldsymbol{p}_{vc}^*$, we take a particular component of the gradient, say $p_x \equiv -i\hbar\partial/\partial x$. Differentiating $\langle u_{v0}|u_{c0}\rangle$, we obtain

$$\frac{\partial}{\partial x}\int d^3 r\, u_{v0}^*\, u_{c0} = \int d^3 r\, \frac{\partial u_{v0}^*}{\partial x}\, u_{c0} + \int d^3 r\, u_{v0}^* \frac{\partial u_{c0}}{\partial x} = 0 \qquad (2.8)$$

To complete the demonstration, take the complex conjugate of the first term and multiply by $-i\hbar$.

We already summarized the solution of a 2-band Hamiltonian in Eqs. (1.24) and (1.25). To simplify, we assume that $\boldsymbol{k}$ and $\boldsymbol{p}_{cv}$ are collinear and denote the energy separating the conduction- and valence-band states at $k = 0$ as $E_g \equiv E_{c0} - E_{v0}$:

$$E_{c,v\boldsymbol{k}} = \frac{E_{c0} + E_{v0}}{2} + \frac{\hbar^2 k^2}{2m} \pm \sqrt{\frac{E_g^2}{4} + \frac{\hbar^2 k^2}{2m} E_p} \qquad (2.9)$$

where we define the Kane energy as $E_P \equiv \frac{2|\boldsymbol{p}_{cv}|^2}{m}$.

Figure 2.1 plots the dispersion relations from Eq. (2.9) for a (a) wide-gap, (b) narrow-gap, and (c) zero-gap semiconductor. In crystals with cubic zinc-blende lattices, there is no preferred direction, and we expect $\boldsymbol{p}_{cv}$ to be isotropic. For the time being, this allows us to sweep under the rug the issue of directionality in a 3D space.

Equation (2.9) closely resembles the well-known solution of the Klein–Gordon equation (and the Dirac equation in 2D) for a relativistic particle:

$$E_\pm = mc^2 \pm \sqrt{m^2 c^4 + \hbar^2 k^2 c^2} \qquad (2.10)$$

Is there a deeper connection? We might liken the carriers in the valence band of a semiconductor to positrons, and the carriers in the conductions band to electrons proper. In fact, the analogy can be taken surprisingly far. The vacuum energy gap of $E_{g0} = 2mc^2$ in Eq. (2.10) corresponds to $E_g$ in a semiconductor. The origin of the energy scale has no absolute meaning. In both cases, the second term under the square root is proportional to $k^2$, with the role of light velocity played by $|\boldsymbol{p}_{cv}|/m$. We will see later that this velocity-like quantity is several hundred times slower than light in a vacuum. Nonetheless, electrons in the coupled two-band model of Eq. (2.9) behave much like relativistic particles, but with the speed of light reduced to the scale of electronic velocities in a solid. When the semiconductor energy gap vanishes, as it does in Fig. 2.1c, the effect on the dispersion is the same as reducing the mass in Eq. (2.10) to zero while keeping the "speed of light" constant. To be sure, Eq. (2.10) lacks the free-electron energy $\hbar^2 k^2/2m$, but it turns out to be unimportant in most III–V semiconductors.

Now we look at a few limiting cases to gain a hands-on understanding of the dispersions in Eq. (2.9) and Fig. 2.1. To begin with, what happens when the energy of a state at nonzero $k$ is very close to $E_{c0}$? For small $k$, the second term under the square root is clearly much smaller than the first, so we can expand in a Taylor series:

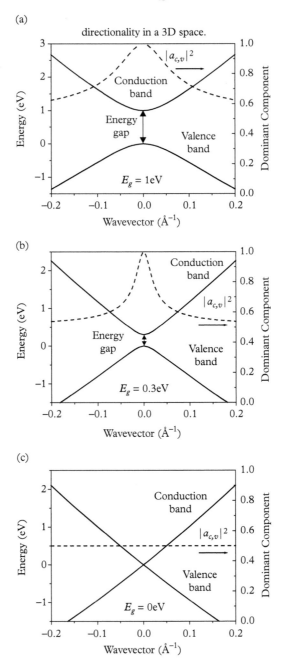

**Figure 2.1** *Coupled conduction and valence bands from the dispersion relations of Eq. (2.9), for $E_P = 25$ eV and $E_g = 1$ (a), 0.3 (b), and 0 (c) eV. The square of the dominant component of the wavefunction is shown on the right-hand vertical scale. Both wavefunction components are equally present for all the states in a zero-gap material of (c).*

$$E_{c\boldsymbol{k}}(k \to 0) \approx \frac{E_{c0} + E_{v0}}{2} + \frac{\hbar^2 k^2}{2m} + \frac{E_g}{2}\left(1 + \frac{\hbar^2 k^2 E_P}{m E_g^2}\right)$$

$$= E_{c0} + \frac{\hbar^2 k^2}{2m}\left(1 + \frac{E_P}{E_g}\right) \equiv E_{c0} + \frac{\hbar^2 k^2}{2m_e} \tag{2.11}$$

The dispersion for $k \to 0$ is quadratic in $k$, just as for a free particle, but the curvature is no longer inversely proportional to the free electron mass. Instead, it scales inversely with the so-called *effective mass*:

$$\frac{1}{m_e} \equiv \frac{1}{m}\left(1 + \frac{E_P}{E_g}\right) \tag{2.12}$$

We easily see the quadratic-dispersion regions in Figs. 2.1a and 2.1b.

What are the typical magnitudes of $E_P$ and the effective mass? Since $\frac{\hbar^2}{2m} = 3.8$ eV $\mathring{A}^2$ and $|\boldsymbol{p}_{cv}|$ captures variation on the atomic scale ($\approx 1\ \mathring{A}$), a reasonable order-of-magnitude guess is $E_P \sim \frac{\hbar^2}{2m}\left(\frac{2}{1\,\mathring{A}}\right)^2 \sim 10$ eV. This is not far from the mark, since Chapter 6 will show that the $E_P$ values in III–V semiconductors fall within the relatively narrow range 22–8 eV. For a typical energy gap of 1 eV, the electron effective mass is $m_e \sim 0.05m$, i.e. much smaller than the free electron mass. From Eq. (2.12), the narrower the gap, the smaller the mass, with values reaching $\sim 0.01m$ in InSb.

Returning to the comparison with Eq. (2.10), the analog of both $E_g$ and $E_P$ is $2mc^2$, which is $10^5$–$10^6$ larger. At face value, this implies a maximum electron velocity of $\sim c/\sqrt{10^5}$, or $\sim 10^8$ cm/s. In real semiconductors, this maximum remains unreachable even in strong electric fields because of the rapid scattering by the vibrating lattice.

The valence-band dispersion is similar to that of the conduction band, except that it slopes down rather than up when $E_P > E_g$:

$$E_{v\boldsymbol{k}}(k \to 0) \approx E_{v0} + \frac{\hbar^2 k^2}{2m}\left(1 - \frac{E_P}{E_g}\right) \equiv E_{v0} - \frac{\hbar^2 k^2}{2m_h} \tag{2.13}$$

This inverted curvature is equivalent to a negative *hole effective mass* $m_h$. If $E_g \ll E_P$, the electron and hole effective masses are both much smaller than the free-particle mass, and have similar values: $m_{e,h}/m \cong E_g/E_P$. If $E_{c0} > E_{v0}$ and the effective mass is much smaller than the free-particle mass, there are no available states between $E_{v0}$ and $E_{c0}$, and $E_g$ is indeed the energy gap. In Sections 2.2 and 2.3 we will find that this picture is incomplete and misses an important hole band with a much heavier mass and weak coupling to the conduction band. Also, some non-III–V materials such as HgTe turn out to be semimetals with $E_{v0} > E_{c0}$. They lack an energy gap, with both electron and hole states occupied at low temperatures in the absence of any doping.

We still need to find the coefficients $a_c$ and $a_v$ that appear in Eq. (2.7). The most straightforward way to estimate them is to neglect the energy terms proportional to $k^2$ near $k = 0$. For simplicity, we also slide the energy scale such that $E_{v0} = 0$ and $E_{c0} = E_g$, and assume that $\mathbf{k} \parallel \mathbf{p}_{cv}$ as before. The conduction-band eigenfunctions then satisfy

$$\frac{\hbar k}{m} p_{vc} a_c - E_g a_v = 0 \tag{2.14}$$

from which we conclude that

$$a_v = \frac{\hbar k p_{cv}^*}{E_g m} a_c \tag{2.15}$$

The conduction-band wavefunction starts out as the original Bloch function $u_{c0}$ at $k = 0$ and then acquires a small valence component proportional to $k$. The same holds for the valence-band wavefunction, except that the (small) conduction-band component is

$$a_c = \frac{\hbar k p_{cv}}{E_g m} a_v \tag{2.16}$$

This is all very well, but what happens when the wavevector is no longer small, i.e., $\frac{\hbar}{m} |\mathbf{k} \cdot \mathbf{p}_{cv}| \gg E_g$? Eventually, the second term under the square root in Eq. (2.9) comes to dominate, and the dispersions become linear over a large fraction of the energy range:

$$E_{c,v\mathbf{k}} \approx \frac{E_{c0} + E_{v0}}{2} \pm \frac{\hbar}{m} |\mathbf{k} \cdot \mathbf{p}_{cv}| \tag{2.17}$$

The dispersions in Fig. 2.1a straighten at large wavevectors for a relatively wide gap. The linear region is even clearer in Fig. 2.1b that plots the dispersions for a narrow-gap material. If the energy gap is exactly zero, as in Fig. 2.1c, the variation is linear even near $k = 0$. The Klein–Gordon relation given by Eq. (2.10) tells us this is true of a massless particle (such as a photon). Apparently, electrons in narrow-gap semiconductors within a certain range of energies behave as if (nearly) stripped of their mass.

What are the eigenfunctions when the gap is zero? In this case, Eq. (2.7) implies that $a_c = \pm a_v$, where the plus sign applies to the conduction band, and the minus sign to the valence band. Instead of mixing in a little of the other component, as in Eqs. (2.15) and (2.16), the conduction and valence states are fully hybridized! The same happens for a material with any gap when $\frac{\hbar^2 k^2}{2m_{e,h}} \gg E_g$, which is exactly where the dispersions become linear in Eq. (2.9). Figure 2.1 shows the magnitudes of the dominant wavefunction components ($a_c$ for the conduction-band states, $a_v$ for the valence band) as a function of wavevector.

We saw that the dispersions remain quadratic in $k$ only near the band edges $E_{c0}$ and $E_{v0}$. If we think of the mass as the inverse curvature of the bands, can we write an expression for the dependence of mass on energy? For electrons, we return to the matrix in Eq. (2.7) and set its determinant to zero:

$$\left(E_{c\boldsymbol{k}} - \frac{\hbar^2 k^2}{2m}\right)\left(E_{c\boldsymbol{k}} - \frac{\hbar^2 k^2}{2m} - E_g\right) - \frac{\hbar^2 k^2}{2m}E_P = 0 \qquad (2.18)$$

As before, we assume the free-electron energy $\frac{\hbar^2 k^2}{2m}$ is small compared to the other terms and discard its square. We define $\Delta E \ll E_g$ as $E_{c\boldsymbol{k}} = E_g + \Delta E$ and calculate its value using Eq. (2.18):

$$\Delta E \approx \frac{\hbar^2 k^2}{2m} \frac{2\Delta E + E_g + E_P}{\Delta E + E_g} \qquad (2.19)$$

Happily, Eq. (2.19) is quadratic in $k$ with the energy-dependent effective mass:

$$m_e(\Delta E) \approx m \frac{\Delta E + E_g}{2\Delta E + E_g + E_P} \qquad (2.20)$$

We can also approximate $m_e(\Delta E)$ using $\Delta E \ll E_P, E_g$, so it can be expressed in terms of the mass at $k = 0$, $m_e(0)$:

$$m_e(\Delta E) \approx m \left(\frac{\Delta E + E_g}{E_g + E_P}\right) = m_e(0)\left(1 + \frac{\Delta E}{E_g}\right) \qquad (2.21)$$

The increase of $\frac{1}{m_e}$ with $\Delta E$ is approximately linear. Figure 2.2 shows examples of the electron mass as a function of energy above the conduction-band edge for materials with three different bandgaps with the clear linear dependence from Eq. (2.21). Note that sometimes it is useful to define the mass as something other than the curvature of the bands, e.g., the "inertial" mass of the carriers in response to an electric field, as we will see in more detail in Parts III and IV.

We have not yet specified which orbitals of the constituent atoms form the conduction- and valence-band states. Based on the identities of the outer-shell electrons ($s^2 p^1$ and $s^2 p^3$), we should have two pairs of $s$- and $p$-like orbitals. Are they associated with particular atoms? It turns out that the orbitals in III–V semiconductors do not center on a specific atom. Instead, they more closely resemble the bonding and antibonding states of a double-quantum-well system, with each well representing the potential of a group III or group V atom. Of course, this is just a crude approximation given the different sizes of the two atoms, but it is sufficient for our qualitative picture.

The bonding configuration in which the wavefunction has the same sign in both wells always has a lower energy than the antibonding state that flips the sign

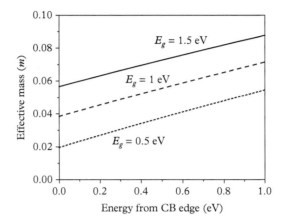

**Figure 2.2** *Dependence of the electron effective mass on energy above the conduction-band edge for materials with three different energy gaps, according to Eq. (2.20) and assuming $E_P = 25$ eV.*

between the wells. For the Hamiltonian in Eqs. (1.24) and (1.25) with $V_0 < 0$ and energies $\varepsilon_1 = \varepsilon_2 = \varepsilon$ for the non-interacting states, we obtain

$$\begin{bmatrix} \varepsilon & -|V_0| \\ -|V_0| & \varepsilon \end{bmatrix} \begin{bmatrix} a_1 \\ a_2 \end{bmatrix} = \varepsilon_\pm \begin{bmatrix} a_1 \\ a_2 \end{bmatrix} \tag{2.22}$$

The energies of the two states are $\varepsilon_\pm = \varepsilon \pm |V_0|$. The minus sign pertains to the bonding state with $a_1 = a_2$.

The upshot of this brief digression is that we expect the bands to have $s$- and $p$-like bonding and antibonding configurations. The $s$-like states are twofold degenerate (including spin), while the $p$-like states are sixfold degenerate. The combined potentials of the ions and core electrons split the $s$- and $p$-like states, with the $s$-like states having lower energy. So a reasonable guess is that the $s$-like bonding band weighs in at the bottom and the $p$-like antibonding band at the top of the energy scale. In turn, this means that the $s$-like antibonding and $p$-like bonding states form the conduction and valence bands. We do not yet have a way to confirm that the conduction electrons are $s$-like, but this is indeed what happens in bulk III–V semiconductors. Figure 2.3 schematically illustrates the relative positions of the various conduction and valence bands, which will prove of use in deriving the band structure models of Section 2.2.

Following much of the literature, we will apply the **k·p** formalism to the band structure near the direct bandgap at $k = 0$. In principle, any point $\mathbf{k_0}$ is suitable for the **k·p** expansion (by setting $\mathbf{k} \to \mathbf{k} - \mathbf{k_0}$ in the equations) as long as we can deduce the identity of the underlying atomic states. We can also add more bands to make the results accurate far away from $\mathbf{k_0}$, although this is rarely the best solution. Other options include various implementations of the tight-binding and pseudopotential methods discussed in detail in Chapter 5.

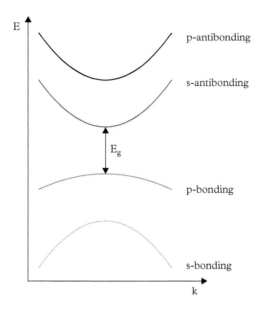

**Figure 2.3** *Schematic ordering of the atomic orbitals comprising the valence and conduction bands closest to the energy gap in a typical III–V semiconductor.*

## 2.2  Spin–Orbit Coupling

What have we accomplished so far? We have derived a simple model for the inter-action of two bands near the conduction- and valence-band edges. We have also found that the band structure of III–V semiconductor may be more complicated than that because the valence band is sixfold degenerate and involves another degree of freedom: orbital angular momentum of one. To make our model more realistic in this case, we need an additional piece, namely, the *spin–orbit coupling*.

How can we understand the spin–orbit interaction? We imagine the electron as moving at a relatively slow velocity compared to the speed of light in the outward-directed electric field created by the nucleus and core electrons of an atom, as illustrated in Fig. 2.4. Screening by the core electrons makes the gradient of this potential relatively strong when compared to the bare Coulomb potential of the ionic charge.

We will ignore the fact that orbital motion involves acceleration in the hope of finding a simple functional form. From the electron's point of view (i.e., in its rest frame), the electric field of the core potential $\boldsymbol{E}$ is transformed into an effective magnetic field:

$$\boldsymbol{B} = -\frac{\boldsymbol{v} \times \boldsymbol{E}}{c^2} \tag{2.23}$$

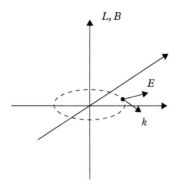

**Figure 2.4** *Conceptual picture of the origin of spin–orbit splitting. The radial electric field arises from the combined potential of the ions and core electrons.*

In effect, the motion of the electron in the electric potential creates current flow that in turn induces a magnetic field according to Ampere's law. More generally, the demarcation between electric and magnetic fields is somewhat arbitrary in the presence of moving charges. For a radial electric field $[\boldsymbol{E} = (E/r)\boldsymbol{r}]$ in vacuum, we can rewrite Eq. (2.23) in terms of orbital angular momentum $\boldsymbol{L} = \boldsymbol{r} \times \boldsymbol{p} = -m\boldsymbol{v} \times \boldsymbol{r}$:

$$\boldsymbol{B} = \frac{\boldsymbol{L}}{mc^2}\frac{E}{r} \tag{2.24}$$

It so happens that the electron itself has a magnetic moment because it has both intrinsic spin angular momentum $\boldsymbol{S}$ and charge $e$:

$$\boldsymbol{\mu} = -\frac{g_e e \boldsymbol{S}}{2m} \tag{2.25}$$

where $g_e \approx 2$ is the electron's *gyromagnetic ratio*. When we measure spin in any chosen direction, the result is either $\hbar/2$ or $-\hbar/2$. The overall energy in a magnetic field is lowered by $\boldsymbol{\mu} \cdot \boldsymbol{B}$, which means we need to add another term to the Hamiltonian:

$$H_{LS} = -\boldsymbol{\mu} \cdot \boldsymbol{B} = \frac{g_e e}{2m}\frac{1}{mc^2}\frac{E}{r}\boldsymbol{L} \cdot \boldsymbol{S} \tag{2.26}$$

How large is this term? Both $\boldsymbol{L}$ and $\boldsymbol{S}$ have units of $\hbar$, and in these units we expect that the angular momenta for the $s$ and $p$ orbitals to be on the order of 1. The factor $eE/r$ is of order 0.1 eV $\mathring{\text{A}}^{-2}$, since the energy scale is $eEr \approx 1$ eV and the atomic dimension is $\approx 3$ $\mathring{\text{A}}$. Following multiplication by $\frac{g_e \hbar^2}{2m}$, the numerator is $\approx 1$ eV$^2$. In the denominator we have the famous electron rest energy $mc^2$, which is equal to half the energy gap between electrons and their positron antiparticles in vacuum $\sim 1$ MeV. Therefore, in vacuum the spin–orbit coupling term is quite weak.

To guesstimate the magnitude of Eq. (2.26) in a semiconductor, we might simply replace $2mc^2$ with $E_g \sim 1$ eV. With this substitution, the spin–orbit coupling is comparable to $E_g$ and definitely not negligible. Rigorous modeling shows that this hunch has the right order of magnitude. However, heavier elements tend to have more screening by core electrons and hence, higher electric fields and stronger spin–orbit coupling.

How does the spin–orbit coupling affect the electronic states near the band edges? For an $s$ state the angular momentum vanishes, and so does $\boldsymbol{L} \cdot \boldsymbol{S}$. This means the spin–orbit interaction has no effect on the conduction band at $k = 0$ because it is comprised of $s$-like antibonding states. On the other hand, the valence band is $p$-like with the orbital angular momentum expectation value $l = 1$. To make progress, we define the total angular momentum $\boldsymbol{J} = \boldsymbol{L} + \boldsymbol{S}$ and calculate the eigenvalues of $\boldsymbol{L} \cdot \boldsymbol{S}$ in terms of $\boldsymbol{J}^2$, $\boldsymbol{L}^2$, and $\boldsymbol{S}^2$:

$$\boldsymbol{L} \cdot \boldsymbol{S} = \frac{1}{2} \left( \boldsymbol{J}^2 - \boldsymbol{L}^2 - \boldsymbol{S}^2 \right) \tag{2.27}$$

Are we justified in deriving Eq. (2.27) as if the angular momenta were simple vectors? Yes, because the orbital and spin parts are unrelated. In the language of quantum mechanics, they commute: $\boldsymbol{L} \cdot \boldsymbol{S} = \boldsymbol{S} \cdot \boldsymbol{L}$.

The angular momentum is calculated as any vector quantity might be. The lengths of $\boldsymbol{J}$, $\boldsymbol{L}$, and $\boldsymbol{S}$ have the values $l = 1$, $s = \frac{1}{2}$, and $j = \frac{3}{2}$ or $j = \frac{1}{2}$ in units of $\hbar$. Of course, if we measure the total angular momentum $\boldsymbol{J}$ along some axis, we can obtain both positive and negative values: $m_j = -\frac{3}{2}, -\frac{1}{2}, \frac{1}{2}, \frac{3}{2}$ for $j = \frac{3}{2}$ and $m_j = -\frac{1}{2}, \frac{1}{2}$ for $j = \frac{1}{2}$. This covers all the possible states of the system; i.e., these $| j, m_j >$ states form a complete set. The dot product has the following eigenvalues when it acts on the $| j, m_j >$ states:

$$\boldsymbol{L} \cdot \boldsymbol{S} \, | j, m_j > = \frac{\hbar^2}{2} \left[ j(j+1) - l(l+1) - s(s+1) \right] | j, m_j > \tag{2.28}$$

The appearance of the eigenvalues, $j(j+1)$ rather than $j^2$, etc., is a distinguishing mark of quantum mechanics. It follows because the angular-momentum operator can modify the states it acts on.

Substituting $j = \frac{3}{2}$ and $j = \frac{1}{2}$, we obtain $j(j+1) - l(l+1) - s(s+1) = 1$ and $-2$, respectively. That this Hamiltonian term has two different values is our clue that the spin–orbit interaction splits the $p$-like bands. The simplest case is $k = 0$. According to Eqs. (2.15) and (2.16) from $\mathbf{k} \cdot \mathbf{p}$ theory, the interaction between the conduction and valence bands vanishes at that point. The ordering of the unmixed valence bands can be deduced from Eq. (2.28).

The one quantity that does not appear in Eq. (2.28) is $m_j$, which means the degeneracy associated with different $j$ (but not $m_j$!) is lifted. The upper state with $j = \frac{3}{2}$ is fourfold degenerate ($m_j = -\frac{3}{2}, -\frac{1}{2}, \frac{1}{2}, \frac{3}{2}$). The lower state with $j = \frac{1}{2}$ is twofold degenerate ($m_j = -\frac{1}{2}, \frac{1}{2}$), which is also true for the conduction band. We will not attempt to calculate the spin–orbit splitting, but instead we assume that $\Delta$

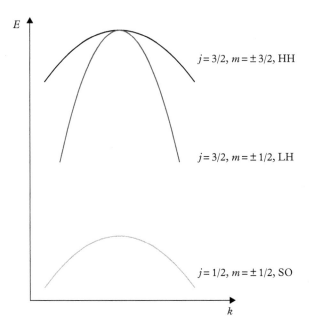

**Figure 2.5** *Schematic ordering of the valence bands in a typical III–V semiconductor.*

is known experimentally and then write the $\boldsymbol{L} \cdot \boldsymbol{S}$ Hamiltonian in the total angular momentum basis $| \, (j, m_j)$. To be concise, we can separate the positive and negative values of $m_j$:

$$H(\boldsymbol{L} \cdot \boldsymbol{S}) \begin{bmatrix} a\left(\left|\frac{1}{2},\pm\frac{1}{2}\right\rangle\right) \\ a\left(\left|\frac{3}{2},\pm\frac{1}{2}\right\rangle\right) \\ a\left(\left|\frac{3}{2},\pm\frac{3}{2}\right\rangle\right) \end{bmatrix} = \Delta \begin{bmatrix} -\frac{2}{3} & 0 & 0 \\ 0 & \frac{1}{3} & 0 \\ 0 & 0 & \frac{1}{3} \end{bmatrix} \begin{bmatrix} a\left(\left|\frac{1}{2},\pm\frac{1}{2}\right\rangle\right) \\ a\left(\left|\frac{3}{2},\pm\frac{1}{2}\right\rangle\right) \\ a\left(\left|\frac{3}{2},\pm\frac{3}{2}\right\rangle\right) \end{bmatrix} \tag{2.29}$$

The ordering of the $p$-like valence-band states predicted by this calculation of the spin–orbit interaction is depicted schematically in Fig. 2.5.

In Eq. (2.29), the coefficients $a\left(|j, m_j\rangle\right)$ multiply the angular-momentum states $|j, m_j\rangle$. We will elide this distinction in what follows to make the equations easier to read. From now on, the reader should mentally convert any vector like

$$\begin{bmatrix} \left|\frac{1}{2},\pm\frac{1}{2}\right\rangle \\ \left|\frac{3}{2},\pm\frac{1}{2}\right\rangle \\ \left|\frac{3}{2},\pm\frac{3}{2}\right\rangle \end{bmatrix} \text{ into } \begin{bmatrix} a\left(\left|\frac{1}{2},\pm\frac{1}{2}\right\rangle\right) \\ a\left(\left|\frac{3}{2},\pm\frac{1}{2}\right\rangle\right) \\ a\left(\left|\frac{3}{2},\pm\frac{3}{2}\right\rangle\right) \end{bmatrix}, \text{ with the full wavefunction given by}$$

$$\Psi(\boldsymbol{r}) = e^{i\boldsymbol{k}\cdot\boldsymbol{r}} \left[ a\left(\left|\tfrac{1}{2},\pm\tfrac{1}{2}\right\rangle\right) \left|\tfrac{1}{2},\pm\tfrac{1}{2}\right\rangle + a\left(\left|\tfrac{3}{2},\pm\tfrac{1}{2}\right\rangle\right) \left|\tfrac{3}{2},\pm\tfrac{1}{2}\right\rangle + a\left(\left|\tfrac{3}{2},\pm\tfrac{3}{2}\right\rangle\right) \left|\tfrac{3}{2},\pm\tfrac{3}{2}\right\rangle \right] \text{ etc.}$$

We already know how to calculate the $\boldsymbol{k}\cdot\boldsymbol{p}$ term using the familiar $|p_x, p_y, p_z\rangle$ basis with spin-1/2 up and down. The spin–orbit splitting emerges naturally from the $|j, m_j\rangle$ states, as shown in Eq. (2.29). This disconnect means that we need to learn how to transform one set into the other! This process will involve multiple steps, as described later. The transformation will be linear, as we have come to expect from the rules of quantum mechanics.

To start deriving the transformation, we choose an arbitrary axis (by convention, the $z$ axis), about which the (orbital) angular momentum is measured. The orbital pointing along this axis ($p_z$) corresponds to a state with zero projection of $\mathbf{L}$ ($m_l = 0$). One analogy is to circularly polarized light, so that the $m_l = 1$ and $m_l = -1$ states are complex-valued combinations of the $p_x$ and $p_y$ orbitals. The overall phase is arbitrary, and the conventional choices are as good as any:

$$|m_l = 1\rangle = -\frac{1}{\sqrt{2}}\left(|p_x\rangle + i|p_y\rangle\right) \tag{2.30a}$$

$$|m_l = -1\rangle = \frac{1}{\sqrt{2}}\left(|p_x\rangle - i|p_y\rangle\right) \tag{2.30b}$$

$$|m_l = 0\rangle = |p_z\rangle \tag{2.30c}$$

Equations (2.30a) and (2.30b) are normalized, and the relative phase follows from the $\langle m_l = 1|m_l = -1\rangle = 0$ orthogonality condition.

The basis in Eq. (2.30) is perfectly acceptable for calculating the $\mathbf{k}\cdot\mathbf{p}$-based dispersion as a function of $k_z$. This is because the $\mathbf{k}\cdot\mathbf{p}$ elements involving $|m_l = 1\rangle$ and $|m_l = -1\rangle$ states vanish, and the $|p_z\rangle$ state is pure. Of course, our random choice of the $z$ axis does not mean that it is now somehow "special". It is simply a bookkeeping device needed to account for the spin–orbit interaction. In fact, any axis chosen along a high-symmetry direction of the zinc-blende lattice is as good as any other.

We now multiply the basis in Eq. (2.30) by the spin-1/2 basis involving the spin-up $|m_s = \frac{1}{2}\rangle$ and spin-down $|m_s = -\frac{1}{2}\rangle$ states. In practice, we simply double the states in Eq. (2.30), calling one half $|m_s = \frac{1}{2}\rangle$ and the other half $|m_s = -\frac{1}{2}\rangle$. This has the unintended consequence of making it tedious to write out the full notation. We adopt the following rules: (1) if the first quantity is an integer and the second is $\frac{1}{2}$ or $-\frac{1}{2}$, the states are $|m_l, m_s\rangle$; (2) if both are half-integers, the states are expressed in the $|j, m_j\rangle$ total angular momentum basis. The remainder of this section deals only with $p$-like states ($l = 1$).

We can add the projections of the angular momenta along the axis we singled out, since they are now just numbers, i.e., the $z$ coordinates of that angular momentum vector. This leads to $m_j = m_l + m_s$. We can next decouple the states with positive and negative $m_j$. For example, the only way to reach $m_j = \frac{3}{2}$ is to add $m_l = 1$ and $m_s = \frac{1}{2}$. On the other hand, there are two ways to obtain $m_j = \frac{1}{2}$: with $m_l = 1$ and $m_s = -\frac{1}{2}$ or with $m_l = 0$ and $m_s = \frac{1}{2}$. Both $j = \frac{3}{2}$ and $j = \frac{1}{2}$ can have $m_j = \frac{1}{2}$.

Do states like $\left|\frac{1}{2},\frac{1}{2}\right\rangle$ and $\left|\frac{3}{2},\frac{1}{2}\right\rangle$ consist of an equal mixture of $\left|0,\frac{1}{2}\right\rangle$ and $\left|1,-\frac{1}{2}\right\rangle$? No, they do not. The clue is that the orbital momentum with $l = 1$ is twice as large as $s = \frac{1}{2}$, so the square of one of the amplitudes should be twice the other. Of course, they must also be orthogonal, i.e., $\langle j', m_{j'} | j, m_j \rangle = 0$, and the coefficients are purely real. We will not attempt to derive the full *Clebsch–Gordan coefficients*, but the following should look at least plausible:

$$\left|\frac{3}{2},\frac{3}{2}\right\rangle = \left|1,\frac{1}{2}\right\rangle \tag{2.31a}$$

$$\left|\frac{3}{2},\frac{1}{2}\right\rangle = \sqrt{\frac{2}{3}}\left|0,\frac{1}{2}\right\rangle + \sqrt{\frac{1}{3}}\left|1,-\frac{1}{2}\right\rangle \tag{2.31b}$$

$$\left|\frac{1}{2},\frac{1}{2}\right\rangle = \sqrt{\frac{1}{3}}\left|0,\frac{1}{2}\right\rangle - \sqrt{\frac{2}{3}}\left|1,-\frac{1}{2}\right\rangle \tag{2.31c}$$

Similar expressions hold for the states with $m_j$ of the opposite sign, except that the signs of $m_l$ and $m_s$ are flipped. The spins are decoupled, so there are no cross terms between $|j, m_j\rangle$ and $|j, -m_j\rangle$. We can also say that $\left|0,\frac{1}{2}\right\rangle$ is a spin-down $p_z$-like state, while $\left|1,-\frac{1}{2}\right\rangle$ is a spin-up linear combination of $p_x$ and $p_y$.

To translate the spin–orbit Hamiltonian of Eq. (2.29) into our new idiom, we rewrite Eq. (2.31) in a matrix form:

$$\begin{bmatrix} \left|\frac{1}{2},\frac{1}{2}\right\rangle \\ \left|\frac{3}{2},\frac{1}{2}\right\rangle \\ \left|\frac{3}{2},\frac{3}{2}\right\rangle \end{bmatrix} = \begin{bmatrix} \sqrt{\frac{2}{3}} & -\sqrt{\frac{1}{3}} & 0 \\ \sqrt{\frac{1}{3}} & \sqrt{\frac{2}{3}} & 0 \\ 0 & 0 & 1 \end{bmatrix} \begin{bmatrix} \left|1,-\frac{1}{2}\right\rangle \\ \left|0,\frac{1}{2}\right\rangle \\ \left|1,\frac{1}{2}\right\rangle \end{bmatrix} = M \begin{bmatrix} \left|1,-\frac{1}{2}\right\rangle \\ \left|0,\frac{1}{2}\right\rangle \\ \left|1,\frac{1}{2}\right\rangle \end{bmatrix} \tag{2.32}$$

Because the last row is decoupled from the other two, $M$ is actually a $2 \times 2$ matrix, plus an identity relation between the $\left|\frac{3}{2},\frac{3}{2}\right\rangle$ and $\left|1,\frac{1}{2}\right\rangle$ coefficients. Equation (2.29) shows that the spin–orbit Hamiltonian contributes $\frac{\Delta}{3}$ when acting on either of these states.

Instead of writing out the wavefunction components, we can refer to them using indices, e.g., $|n_{j,m}\rangle$ stands for a particular $|j, m_j\rangle$, etc. In this more abstract and compact form, Eq. (2.32) becomes

$$|n_{j,m}\rangle = \sum_k \left( |k_{l,s}\rangle \langle k_{l,s} | n_{j,m}\rangle \right) = \sum_k M_{nk} \left( |k_{l,s}\rangle \right) \tag{2.33}$$

What we have are the $\langle n'_{j,m} | H | n_{j,m}\rangle$ matrix elements of the spin–orbit Hamiltonian in the $|j, m_j\rangle$ basis from Eq. (2.29):

$$\left[\left\langle\frac{1}{2},\frac{1}{2}\Big|\left\langle\frac{3}{2},\frac{1}{2}\Big|\left\langle\frac{3}{2},\frac{3}{2}\Big|\right]H_{j,m}\left(\boldsymbol{L}\cdot\boldsymbol{S}\right)\begin{bmatrix}|\frac{1}{2},\frac{1}{2}>\\|\frac{3}{2},\frac{1}{2}>\\|\frac{3}{2},\frac{3}{2}>\end{bmatrix}=\left|n'_{j,m}\right\rangle\left\langle n'_{j,m}\right|H\left|n_{j,m}\right\rangle\left\langle n_{j,m}\right|$$

$$(2.34)$$

What we actually want are the $\langle k'_{l,s}|\boldsymbol{H}|k_{l,s}\rangle$ elements in the $|m_l,m_s\rangle$ basis. To obtain them, we expand $|k_{l,s}\rangle=\sum_n|n_{j,m}\rangle\langle n_{j,m}|k_{l,s}\rangle$. Next we use Eq. (2.33) and the property $\langle k|n\rangle=\langle n|k\rangle^*$ to find that $\langle n_{j,m}|k_{l,s}\rangle=M^*_{nk}$. Putting it all together and finally converting to the index-free notation of Eq. (2.32), we have

$$\langle k'_{l,s}|\boldsymbol{H}|k_{l,s}\rangle=\sum_{n',n}\langle k'_{l,s}|n'_{j,m}\rangle\langle n'_{j,m}|\boldsymbol{H}|n_{j,m}\rangle\langle n_{j,m}|k_{l,s}\rangle$$

$$=\sum_{n,n'}M_{n'k'}H_{n'n}M^*_{nk}=\sum_{n,n'}M^T_{k'n'}H_{n'n}M^*_{nk}$$

$$=\boldsymbol{M}^T\boldsymbol{H}\boldsymbol{M}^* \qquad (2.35)$$

The matrix transpose $\boldsymbol{M}^T$ denotes the substitution of $M_{ba}$ for every element $M_{ab}$ (of course, leaving the diagonal terms unchanged). The transpose is necessary because we should sum over the subscripts that go with those in the Hamiltonian matrix; i.e., we must match rows and columns rather than rows and rows, etc. This is the second-to-last step in Eq. (2.35). In this case, taking a transpose means simply exchanging the signs of the $\sqrt{\frac{1}{3}}$ elements in Eq. (2.32). Of course, complex conjugation has no effect on the real matrix $\boldsymbol{M}$.

Starting from the Hamiltonian of Eq. (2.29) and working through the two required matrix multiplication steps, we obtain the spin–orbit Hamiltonian in the $|m_l,m_s\rangle$ basis. It helps to decouple the $|1,\frac{1}{2}\rangle$ state from the start so the matrix to be transformed is reduced to a $2\times2$ form. The result becomes

$$\boldsymbol{H}_{3\times3,l,s}\left(\boldsymbol{L}\cdot\boldsymbol{S}\right)\begin{bmatrix}|1,-\frac{1}{2}\rangle\\|0,\frac{1}{2}\rangle\\|1,\frac{1}{2}\rangle\end{bmatrix}=\Delta\begin{bmatrix}-\frac{1}{3}&\frac{\sqrt{2}}{3}&0\\\frac{\sqrt{2}}{3}&0&0\\0&0&\frac{1}{3}\end{bmatrix}\begin{bmatrix}|1,-\frac{1}{2}\rangle\\|0,\frac{1}{2}\rangle\\|1,\frac{1}{2}\rangle\end{bmatrix} \qquad (2.36)$$

Do not forget that the full spin–orbit Hamiltonian is actually a $6\times6$ matrix with two blocks, each of which is like Eq. (2.36) in one block and zeroes elsewhere! We show this in the equation form

$$\boldsymbol{H}_{6\times6}\left(\boldsymbol{L}\cdot\boldsymbol{S}\right)\begin{bmatrix}|1,-\frac{1}{2}\rangle\\|0,\frac{1}{2}\rangle\\|1,\frac{1}{2}\rangle\\|-1,\frac{1}{2}\rangle\\|0,-\frac{1}{2}\rangle\\|-1,-\frac{1}{2}\rangle\end{bmatrix}=\begin{bmatrix}H_{3\times3}&0\\0&H_{3\times3}\end{bmatrix}\begin{bmatrix}|1,-\frac{1}{2}\rangle\\|0,\frac{1}{2}\rangle\\|1,\frac{1}{2}\rangle\\|-1,\frac{1}{2}\rangle\\|0,-\frac{1}{2}\rangle\\|-1,-\frac{1}{2}\rangle\end{bmatrix} \qquad (2.37)$$

Solving Eq. (2.36), we obtain the same eigenvalues as Eq. (2.29), which is of course exactly what we expect since the two forms are equivalent. What are the eigenfunctions of the full 6×6 matrix? We can read them off Eqs. (2.30) and (2.31), making sure we do not forget to indicate the spin. For the fourfold-degenerate states with the eigenvalue $\frac{\Delta}{3}$, the wavefunctions are

$$\psi_{HH}^{+} = \left|\frac{3}{2},\frac{3}{2}\right\rangle = \left|1,\frac{1}{2}\right\rangle = -\frac{1}{\sqrt{2}}\left(p_x + ip_y\right)\uparrow \tag{2.38a}$$

$$\psi_{HH}^{-} = \left|-\frac{3}{2},-\frac{3}{2}\right\rangle = \left|-1,-\frac{1}{2}\right\rangle = \frac{1}{\sqrt{2}}\left(p_x - ip_y\right)\downarrow \tag{2.38b}$$

$$\psi_{LH}^{+} = \left|\frac{3}{2},\frac{1}{2}\right\rangle = \frac{1}{\sqrt{3}}\left|1,-\frac{1}{2}\right\rangle + \sqrt{\frac{2}{3}}\left|0,\frac{1}{2}\right\rangle = -\frac{1}{\sqrt{6}}\left(p_x + ip_y\right)\downarrow + \sqrt{\frac{2}{3}}\,p_z\uparrow$$
$$\tag{2.38c}$$

$$\psi_{LH}^{-} = \left|\frac{3}{2},-\frac{1}{2}\right\rangle = \frac{1}{\sqrt{3}}\left|-1,\frac{1}{2}\right\rangle + \sqrt{\frac{2}{3}}\left|0,-\frac{1}{2}\right\rangle = \frac{1}{\sqrt{6}}\left(p_x - ip_y\right)\uparrow + \sqrt{\frac{2}{3}}\,p_z\downarrow$$
$$\tag{2.38d}$$

while for the doubly degenerate states with the eigenvalue $-\frac{2\Delta}{3}$, they are

$$\psi_{SO}^{+} = \left|\frac{1}{2},\frac{1}{2}\right\rangle = -\sqrt{\frac{2}{3}}\left|1,-\frac{1}{2}\right\rangle + \frac{1}{\sqrt{3}}\left|0,\frac{1}{2}\right\rangle = \frac{1}{\sqrt{3}}\left(p_x + ip_y\right)\downarrow + \frac{1}{\sqrt{3}}\,p_z\uparrow$$
$$\tag{2.39a}$$

$$\psi_{SO}^{-} = \left|\frac{1}{2},-\frac{1}{2}\right\rangle = \sqrt{\frac{2}{3}}\left|-1,\frac{1}{2}\right\rangle - \frac{1}{\sqrt{3}}\left|0,-\frac{1}{2}\right\rangle = \frac{1}{\sqrt{3}}\left(p_x - ip_y\right)\downarrow - \frac{1}{\sqrt{3}}\,p_z\uparrow$$
$$\tag{2.39b}$$

As before, Eqs. (2.38) and (2.39) assume that both the electron spin and its wavevector point along the $z$ axis. As we mentioned earlier, we choose this axis arbitrarily, but once it is chosen, we should make sure the directions in $k$-space remain consistent.

We introduced the notations "HH" (heavy-hole) and "LH" (light-hole) in Eqs. (2.38) and (2.39) to stand for the two pairs of states that are degenerate at the top of the valence band, and "SO" (split-off hole) to denote states in the band that lie $\Delta$ lower at $k = 0$. The various states are labeled after the dispersions of the bands they belong to, as shown in Fig. 2.5 and discussed in Section 2.3 using the **k·p** and spin–orbit Hamiltonians of Eqs. (2.7) and (2.36).

Before launching into the detailed analysis, are the states with $m_j = \pm\frac{1}{2}$ ever *not* degenerate? We can reason that the deeper reason for degeneracy is a system's underlying symmetry. For example, in a two-dimensional potential well with equal lateral dimensions (i.e., a quantum wire with square cross section), the first excited energy levels above the ground state correspond to degenerate excitations along the two equivalent orthogonal directions. Since both directions look identical to an electron in the well, rotation by 90° changes precisely nothing.

By analogy, rotating a diamond crystal by 180° about one axis has the same null effect, as is apparent from Fig. 1.6b. The rotation reverses the direction of $k$, but not the angular momentum about the rotation axis. Therefore, the states with $+k$ and $-k$ wavevectors must be degenerate for the diamond lattice and, more generally, for any lattice with *inversion symmetry*. Since there is no change in the angular momentum, the spin-up and spin-down states must also be degenerate for both positive and negative $k$. Another symmetry available to us is to reverse the direction of time flow—moving forward into the past! Then the sign of momentum (and wavevector $k$) flips, since it is proportional to the time derivative of the average position. So does the sign of the angular momentum, which is proportional to momentum. We can visualize the reversal as watching the particle spin in the opposite direction when the movie is played backwards. With both $k$ and spin reversed, spin-up states with wavevector $k$ must be degenerate with spin-down states at $-k$. One consequence is that spin degeneracy is always maintained for states at $k = 0$.

Are we forgetting that any III–V semiconductor crystallizes in a zinc-blende lattice rather than a diamond lattice, and that two distinct atoms in the basis break the inversion symmetry? True, but it turns out not to matter as much as you might think! The asymmetric basis is a small perturbation of the full band-structure, and the spin-up and spin-down states at a given $k$ remain close in energy. In fact, even a diamond crystal no longer has inversion symmetry when we apply an electric field to opposite sides of the crystal, but we can still describe the band-structure of that crystal starting with the degenerate states under inversion symmetry (unless the field is too strong). Another way to break symmetry is to grow a structure in which it is expressly abandoned, e.g., two coupled wells with different widths. Any mechanism that lifts inversion symmetry also removes spin degeneracy at the same $k$, but not the deeper time-inversion degeneracy between the different spin states at $\pm k$. The possible degeneracies are illustrated schematically in Fig. 2.6. Chapter 8 will pick up the discussion of spin splitting.

Given a wavevector $k$ directed along one of the high-symmetry axes of the crystal, we can always find some axis for which the two spin states are directed "up" or "down". Were the wavevector chosen to point along a different, yet equivalent high-symmetry axis, the spin axis would rotate, so that the electron spin does not pick out one equivalent direction over another (in the absence of a magnetic field). The moral is that the "up" or "down" labels should not be interpreted too literally!

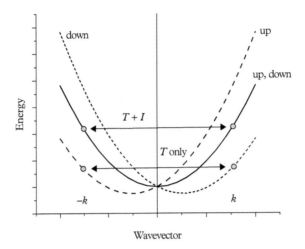

**Figure 2.6** *Schematic dispersion relation for a crystal possessing both inversion (I) and time-reversal (T) symmetries (solid curve, which is degenerate in ±k) as well as for a crystal with broken inversion symmetry (dashed and short-dashed curves, which are no longer symmetric about k = 0).*

## 2.3 Kane's Model

In 1957, Evan Kane introduced a model that marries the basic **k·p** Hamiltonian of Eq. (2.7) with the spin–orbit splitting Hamiltonian of Eqs. (2.36) and (2.37):

$$
H_{\bm{k}\cdot\bm{p}} + H_{6\times6}(\bm{L}\cdot\bm{S}) =
\begin{bmatrix}
E_{c0} + \frac{\hbar^2 k^2}{2m} & \frac{\hbar}{m}\bm{k}\cdot\bm{p}_{cv} & \frac{\hbar}{m}\bm{k}\cdot\bm{p}_{cv} \\
\frac{\hbar}{m}\bm{k}\cdot\bm{p}^*_{cv} & H_{3\times3} + \frac{\hbar^2 k^2}{2m} & 0 \\
\frac{\hbar}{m}\bm{k}\cdot\bm{p}^*_{cv} & 0 & H_{3\times3} + \frac{\hbar^2 k^2}{2m}
\end{bmatrix}
\tag{2.40}
$$

The resulting 8×8 matrix includes in its basis the $s$- and $p$-like orbitals with both spins. We are not interested in spin splitting for the moment, so we take all states to be at least twofold degenerate and halve the size of the matrix. The literature commonly takes the interband matrix element $\langle s|p|p_z\rangle = \langle s|p|p_x\rangle = \langle s|p|p_y\rangle$ to be imaginary. To keep the matrix in Eq. (2.40) real, we pre-multiply the $s$ orbital by $i$: $i|s\rangle\uparrow$ and $i|s\rangle\downarrow$. We could also avoid the imaginary pre-factor and allow the off-diagonal **k·p** elements to become imaginary instead, as we will do in Chapter 3. These choices have no effect on any physical quantity derived from the Hamiltonian.

To keep things as simple as possible, we take the wavevector to point along a single axis, e.g., the $z$ axis, as in Section 2.2. We will re-examine this simplification in Chapter 3, but for now only the $\bm{k}\cdot\bm{p}_{cv} = k_z p_{cv} = k p_{cv}$ term that connects $s$-like and $p_z$-like ($|0,\pm\frac{1}{2}\rangle$) states survives. The interactions of the $s$ and $p_x$, $p_y$ orbitals are proportional to $k_x = k_y = 0$, which leaves behind vanishing matrix elements. For maximum transparency, we choose the energy scale as $E_{v0} = 0$ and $E_{c0} = E_g$ by subtracting $\frac{\Delta}{3}$ from the diagonal elements in Eq. (2.36). This leads to the following $4 \times 4$ matrix equation that keeps the free-electron $\frac{\hbar^2 k^2}{2m}$ terms:

$$
\boldsymbol{H}_{4\times4}
\begin{bmatrix}
i\,|l=0, m_l=0, m_s = \tfrac{1}{2}\rangle \\
|l=1, m_l=1, m_s = -\tfrac{1}{2}\rangle \\
|l=1, m_l=0, m_s = \tfrac{1}{2}\rangle \\
|l=1, m_l=1, m_s = \tfrac{1}{2}\rangle
\end{bmatrix}
$$

$$
=
\begin{bmatrix}
E_g + \frac{\hbar^2 k^2}{2m} & 0 & \frac{\hbar}{m} k p_{cv} & 0 \\
0 & -\frac{2\Delta}{3} + \frac{\hbar^2 k^2}{2m} & \frac{\sqrt{2}\Delta}{3} & 0 \\
\frac{\hbar}{m} k p_{cv} & \frac{\sqrt{2}\Delta}{3} & -\frac{\Delta}{3} + \frac{\hbar^2 k^2}{2m} & 0 \\
0 & 0 & 0 & \frac{\hbar^2 k^2}{2m}
\end{bmatrix}
\begin{bmatrix}
i\,|0,0,\tfrac{1}{2}\rangle \\
|1,1,-\tfrac{1}{2}\rangle \\
|1,0,\tfrac{1}{2}\rangle \\
|1,1,\tfrac{1}{2}\rangle
\end{bmatrix}
\qquad (2.41)
$$

What are the eigenstates of Eq. (2.41) at $k = 0$? We are already familiar with them because we solved the spin–orbit problem in Section 2.2. There are two states with $E = 0$ and one with $E = -\Delta$ (remember we have shifted the origin by $-\frac{\Delta}{3}$) that represent the valence band. There is also the $E = E_g$ state, which is the bottom of the conduction band. We know that the wavefunctions for the valence-band states are given by Eqs. (2.38) and (2.39), and that the $s$ orbital is synonymous with the conduction band at $k = 0$.

The other zeroes in the Hamiltonian of Eq. (2.41) inform us that the heavy-hole states are decoupled from the others at all $k$. Taken at face value, the HH dispersion is free-electron-like rather than hole-like; i.e., it bends upward instead of downward! Does this make any sense? No, because the HH band cannot cross the energy gap. This unphysical behavior arises simply because we have restricted the basis of the Kane Hamiltonian to a single set of $s$- and $p$-like orbitals. We will find in Chapter 3 that the interactions with higher-lying bands, e.g., with the "antibonding" $p$ bands mentioned in Section 2.1, force the HH band to slope gently downward. In other words, the HH band is indeed hole-like, but its effective mass is quite large. Indeed, we anticipated this behavior by calling it a *heavy*-hole band!

What are the dispersions of the other three bands? Since we do not wish to continue carrying the free-electron terms around, we subsume them into the definition of energy: $E \rightarrow E + \frac{\hbar^2 k^2}{2m}$. The eigenvalue equation for the top $3 \times 3$ portion of Eq. (2.41) is then

$$
(E_g - E)\left(E + \frac{2\Delta}{3}\right)\left(E + \frac{\Delta}{3}\right) - (E_g - E)\frac{2\Delta^2}{9} + \frac{\hbar^2}{m^2}k^2 p_{cv}^2\left(E + \frac{2\Delta}{3}\right) = 0
$$

$$
(2.42)
$$

This is a cubic equation in $E$. The second term is cancelled by part of the first, so we can simplify it somewhat:

$$(E - E_g)E(E + \Delta) - \frac{\hbar^2 k^2}{2m}E_P\left(E + \frac{2\Delta}{3}\right) = 0 \qquad (2.43)$$

where the definition of $E_P$ is taken from Eq. (2.9).

Figure 2.7 shows the solutions of Eq. (2.43) for a few representative values of the energy gap and spin–orbit splitting. All three bands are coupled, and the LH and SO bands display an anti-crossing near the split-off energy; i.e., the two bands exchange their dispersion characteristics at larger $k$. The numerical results are all very well, but what we really want is a deeper understanding of what the bands are like and whether we can identify different dispersion regions by analogy to Fig. 2.1.

To this end, we note that at $k = 0$, we recover the three states that were already calculated: $E_{CB}(k = 0) = E_g$, $E_{LH}(k = 0) = 0$, and $E_{SO}(k = 0) = -\Delta$. To reiterate, the first is the conduction band (CB), the second is the light hole (LH) that is degenerate with the omitted HH, and the third is the split-off hole (SO), where the hole wavefunctions are given by Eqs. (2.38) and (2.39). Starting at this marker, we will use Eq. (2.43) to deduce the dispersions at small $k$.

Since the wavevector appears in Eq. (2.43) only as $k^2$, and there are no degenerate states, the expansion can include only terms quadratic in $k$. We are actually repeating the procedure used to derive Eq. (2.11), but with Eq. (2.43) that is cubic in $E$ rather than quadratic. This means that for each band (CB, LH, and SO), we can substitute $E \rightarrow E_i(k = 0) + \frac{\hbar^2 k^2}{2m_i}$, where $m_i$ is the appropriate effective mass, and ignore any terms in $k^4$, $k^6$, etc. This is in fact equivalent to ignoring the $k$ dependence in any term of Eq. (2.43) that does not vanish at $k = 0$ if the band energy $E_i(k = 0)$ is used in place of $E$. We should remember to re-insert the free-electron energy by setting $1/m_i \rightarrow 1/m_i + 1/m$. For the CB states, we obtain

$$E_{CB}(k) \approx E_g + \frac{\hbar^2 k^2}{2m} + \frac{\hbar^2 k^2}{2m}\frac{E_P\left(E_g + \frac{2\Delta}{3}\right)}{E_g(E_g + \Delta)} \qquad (2.44)$$

Therefore, the electron effective mass in a III–V semiconductor is approximately

$$\frac{1}{m_e} \approx \frac{1}{m}\left[1 + \frac{E_P\left(E_g + \frac{2\Delta}{3}\right)}{E_g(E_g + \Delta)}\right] \qquad (2.45)$$

When $\Delta = 0$, the result is the same as in Eq. (2.12). As the spin–orbit splitting increases to $\Delta \sim E_g$, the electrons become a little lighter even if the gap remains unchanged.

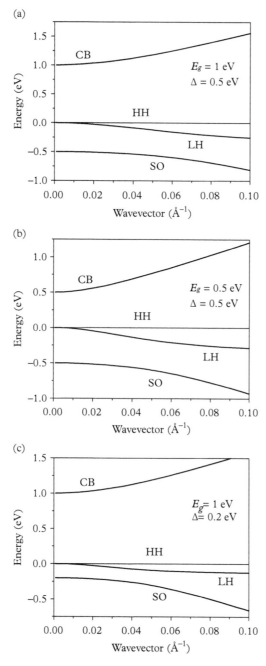

**Figure 2.7** *Coupled conduction- and valence-band solutions of Eq. (2.43), for semiconductors with three different combinations of $E_g$ and $\Delta$ (assuming $E_P = 25\ eV$ in all cases). The uncoupled HH band is indicated as independent of wavevector.*

How good an approximation is Eq. (2.45)? One factor we have not yet included is the potential for interactions with remote bands. By analogy with the HH band, we surmise that the higher-lying states can "press down" on the CB dispersion and make it slope upward more slowly than it would in the absence of remote bands. By this mechanism, $m_e$ may become higher than what we found in Eq. (2.45). We parameterize the difference by adding a $\frac{2F}{m}$ term (with $F < 0$) to the right-hand side.

The LH energy at $k = 0$ approaches zero rather than $E_g$. From Eq. (2.43), we have

$$E_{LH}(k) \approx \frac{\hbar^2 k^2}{2m} - \frac{\hbar^2 k^2}{2m} \frac{2E_P}{3E_g} \tag{2.46}$$

This means the LH mass is independent of the spin–orbit coupling:

$$\frac{1}{m_{LH}} \approx -\frac{1}{m}\left(\frac{2E_P}{3E_g} - 1\right) \tag{2.47}$$

Since $E_P \gg E_g$, the LH mass is indeed negative. Comparing Eqs. (2.45) and (2.47), we find that $m_e < |m_{LH}|$, although the CB and LH masses become quite similar when $E_g \ll \Delta$. Of course, we know from Eq. (2.41) that the CB and LH bands are coupled at nonzero $k$; i.e., the LH acquires some $s$-like character while the CB states have a small $p$-like component. This admixture of "foreign" components is proportional to $k^2$.

Finally, the dispersion and effective mass for the SO states are

$$E_{SO}(k) \approx -\Delta + \frac{\hbar^2 k^2}{2m} - \frac{\hbar^2 k^2}{2m} \frac{E_P}{3(E_g + \Delta)} \tag{2.48}$$

$$\frac{1}{m_{SO}} \approx -\frac{1}{m}\left[\frac{E_P}{3(E_g + \Delta)} - 1\right] \tag{2.49}$$

Of the three hole bands, the LH band is clearly the flyweight. We already saw that the HH band is the undisputed heavyweight, which makes the SO band, if you will, the welterweight. Figure 2.8 plots the relative magnitudes of the electron and hole effective masses as a function of energy gap when $\Delta = E_g/2$ and $\Delta = E_g$.

What happens if the gap shrinks to nothingness, but the spin–orbit splitting $\Delta$ is held fixed? In this case, the CB, HH, and LH states are all degenerate at $k = 0$, while the SO state is lower by $\Delta$. If $\Delta$ is large compared to the CB and LH energies of interest near $k = 0$, we can set $E + \Delta \approx \Delta$ and $E + \frac{2\Delta}{3} \approx \frac{2\Delta}{3}$ in Eq. (2.43). The CB and LH energies are then

$$E_{CB,LH}(k) \approx \frac{\hbar^2 k^2}{2m} \pm \frac{\sqrt{2}\hbar}{\sqrt{3}m} k p_{cv} \tag{2.50}$$

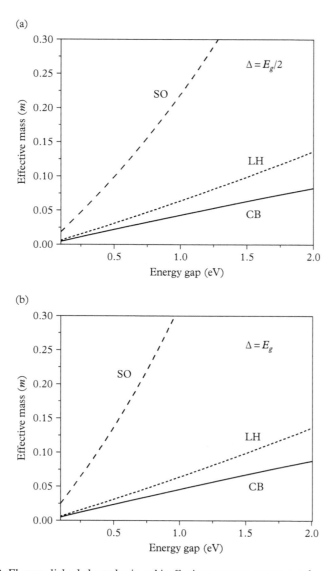

**Figure 2.8** *Electron, light–hole, and spin–orbit effective masses vs energy gap for typical III–V semiconductors with $E_P = 25$ eV and two different values of the spin–orbit splitting relative to the energy gap.*

We see that away from a small region with quadratic dependence near $k = 0$, both dispersions are nearly linear in $k$! This is of course the analog of Eq. (2.17) that differs only by the free-electron energy and a factor of $\sqrt{2/3}$. We conclude that the physical picture of narrow-gap semiconductors described in Section 2.2 carries over to the case of the spin–orbit coupling.

Alternatively, we could have constructed a Hamiltonian equivalent to Eq. (2.41) using a basis in which the spin–orbit interaction is diagonal, i.e., CB, HH, LH, SO rather than $s$ and $p$. Using Eq. (2.32) to transform the matrix elements equal to $\frac{\hbar}{m} k p_{cv}$, the result is

$$
\boldsymbol{H}_{\boldsymbol{k}\cdot\boldsymbol{p}}
\begin{bmatrix}
\left\langle i \,|\, l = 0, j = \tfrac{1}{2}, m_j = \tfrac{1}{2} \right\rangle \\
\left| l = 1, j = \tfrac{3}{2}, m_j = \tfrac{1}{2} \right\rangle \\
\left| l = 1, j = \tfrac{1}{2}, m_j = \tfrac{1}{2} \right\rangle \\
\left| l = 1, j = \tfrac{3}{2}, m_j = \tfrac{3}{2} \right\rangle
\end{bmatrix}
=
$$

$$
\begin{bmatrix}
E_g + \frac{\hbar^2 k^2}{2m} & \sqrt{\frac{2}{3}}\frac{\hbar}{m} k p_{cv} & -\sqrt{\frac{1}{3}}\frac{\hbar}{m} k p_{cv} & 0 \\
\sqrt{\frac{2}{3}}\frac{\hbar}{m} k p_{cv} & \frac{\hbar^2 k^2}{2m} & 0 & 0 \\
-\sqrt{\frac{1}{3}}\frac{\hbar}{m} k p_{cv} & 0 & -\Delta + \frac{\hbar^2 k^2}{2m} & 0 \\
0 & 0 & 0 & \frac{\hbar^2 k^2}{2m}
\end{bmatrix}
\begin{bmatrix}
\left\langle i \,|\, 0, \tfrac{1}{2}, \tfrac{1}{2} \right\rangle \\
\left| 1, \tfrac{3}{2}, \tfrac{1}{2} \right\rangle \\
\left| 1, \tfrac{1}{2}, \tfrac{1}{2} \right\rangle \\
\left| 1, \tfrac{3}{2}, \tfrac{1}{2} \right\rangle
\end{bmatrix}
\tag{2.51}
$$

Now $\Delta$ appears only in one diagonal element. Of course, both Hamiltonians (Eqs. (2.41) and (2.51)) produce the same dispersions, and we can easily transform the eigenfunctions of one into the other. The prescription for that is once again found in Eq. (2.32).

Because we picked the $z$ axis arbitrarily, Kane's model is isotropic, and it does not matter how the wavevector aligns with the zinc-blende crystal axes. This turns out to be a reasonable approximation for the conduction band composed mostly of spherically symmetric $s$ orbitals. However, Chapter 3 will show that some of the valence-band dispersions, particularly for the HH band, can and do depend on the crystal orientation.

Could we have chosen the electron spin to point along an axis other than $z$, so that the spin and the wavevector are not parallel? Yes, but that choice saddles us with a *lot* more algebra, all of it redundant because we already know that the results are independent of wavevector direction.

Kane's model is about as far as we can go with a simple, back-of-the-envelope estimate. In Chapter 3, we will learn how to improve this model.

## Suggested Reading

- For another presentation of Kane's model, see S. L. Chuang, *Physics of Photonic Devices*, 2nd edn (Wiley, New York, 2009).

- A good reference for learning more about how to add angular momenta in quantum mechanics is J. J. Sakurai, *Modern Quantum Mechanics*, rev. edn (Addison Wesley, Boston, 1993).

# 3

# Detailed k·p Theory for Bulk III–V Semiconductors

We now discuss a full implementation of k·p theory for the eight bands near the energy gap in a III–V semiconductor. This model can precisely calculate the band structure for both conduction and valence bands near the energy gap. Even though only eight bands are explicitly included, the terms due to interactions with "remote" bands are present in the complete version of the theory. We will consider the physical meaning of the results and the effect that strain has on the semiconductor band structure. The Hamiltonians for both zinc-blende and wurtzite crystals are introduced.

## 3.1 Valence-Band Hamiltonian with Contributions from Remote Bands

We begin by returning to the basic **k·p** Hamiltonian, but now without constraining the wavevector to point along a particular axis (i.e., it can contain all three components: $k_x$, $k_y$, and $k_z$). At first, we neglect the spin–orbit coupling and use the $\begin{bmatrix} |s\rangle \\ |p_x\rangle \\ |p_y\rangle \\ |p_z\rangle \end{bmatrix}$ basis. Taking all of the states to be doubly degenerate to account for spin, the 4×4 Hamiltonian becomes

$$
\boldsymbol{H}_{4\times4}\begin{bmatrix} |s\rangle \\ |p_x\rangle \\ |p_y\rangle \\ |p_z\rangle \end{bmatrix} = \begin{bmatrix} E_g + \frac{\hbar^2 k^2}{2m} & i\frac{\hbar}{m}k_x p_{cv} & i\frac{\hbar}{m}k_y p_{cv} & i\frac{\hbar}{m}k_z p_{cv} \\ -i\frac{\hbar}{m}k_x p_{cv} & \frac{\hbar^2 k^2}{2m} & 0 & 0 \\ -i\frac{\hbar}{m}k_y p_{cv} & 0 & \frac{\hbar^2 k^2}{2m} & 0 \\ -i\frac{\hbar}{m}k_z p_{cv} & 0 & 0 & \frac{\hbar^2 k^2}{2m} \end{bmatrix}\begin{bmatrix} |s\rangle \\ |p_x\rangle \\ |p_y\rangle \\ |p_z\rangle \end{bmatrix}
\tag{3.1}
$$

*Bands and Photons in III–V Semiconductor Quantum Structures.* Vurgaftman, Lumb, and Meyer,
Oxford University Press (2021). © Vurgaftman, Lumb, and Meyer.
DOI: 10.1093/oso/9780198767275.003.0003

This expression can be streamlined again by including the free-electron terms into the definitions of the band-edge energies: $E_{c,v} \rightarrow E_{c,v} + \frac{\hbar^2 k^2}{2m}$. We have dispensed with the imaginary pre-factor for the $|s\rangle$ state, and the linear-in-$k$ off-diagonal elements have become imaginary. The Hermiticity requirement $H_{ij} = H_{ji}^*$ makes it easy to keep track of them. If we now add the spin–orbit Hamiltonian, expressed in terms of this basis, the problem is effectively formulated for any direction of the wavevector. But first we should fix the unrealistic behavior of the HH mass by taking into account the interactions with remote bands, as mentioned in Chapter 2.

We keep our "fudge" parameter $F$ in the definition of the electron effective mass, as discussed in Chapter 2. It is a simple way to correct for the interactions with the CB antibonding $s$-like states with the antibonding $p$-like states well above them (other states with the same symmetry have a much smaller impact). The dispersion is still parabolic close to the band edge, so an improved version of Eq. (2.45) reads

$$\frac{1}{m_e} \approx \frac{1}{m} \left[ 1 + 2F + \frac{E_P \left( E_g + {}^{2\Delta}\!/_3 \right)}{E_g \left( E_g + \Delta \right)} \right] \tag{3.2}$$

where $F < 0$. If we assume the coupling with remote states to be isotropic, the mass does not depend on wavevector direction.

By contrast, the valence-band states are composed primarily of $p$-like orbitals that are not spherically symmetric, as shown in Fig. 3.1. For this reason, it may not be so easy to compute their accurate (anisotropic) dispersions. As we saw, the valence-band interaction with the $s$-like conduction band does not lead to anisotropy, so we will need to include coupling to the $p$-like and $d$-like remote states to improve on Kane's model. What is the best way to do so?

We would like to avoid increasing the size of the Hamiltonian. After all, its full form with the spin–orbit interaction has already become unwieldy. In fact, we can fold in the information about the interactions in an approximate manner by following the steps that led us from Eq. (2.7) to Eq. (2.11). To recap, we converted a 2×2 matrix equation into a single equation with the conduction-band dispersion quadratic in $k$. The price was that the solution remained accurate only at relatively small energies where the interaction between the bands is weak. Proceeding by analogy, we figure that the remote-band coupling is "weak" and fold its effect into the 8×8 Hamiltonian in Eq. (3.1) plus the spin–orbit interaction. Even though we will treat the remote-band terms within the **k·p** method, the resulting matrix elements are proportional to $k^2$ rather than $k$. This is because the overall term is the product of the original state's **k·p** interaction with the remote state combined with the inverse coupling of the remote state back to the original state. Both diagonal and off-diagonal elements of the Hamiltonian are subject to this interaction. Of course, this is nothing more than second-order perturbation theory, but with a slight twist to account for degenerate states.

To develop these ideas quantitatively, we postulate two separate sets of states: (1) the 8 states comprising the conduction and valence bands of interest (Set $N$); and (2) all the "remote states" (Set $R$), which are important only to the extent they affect Set $N$. The total wavefunction for any given state includes contributions from both sets:

$$\psi = \sum_N C_n \psi_n + \sum_R C_r \psi_r \tag{3.3}$$

where $\psi_{n,r}$ are the wavefunctions of the basis states in the two sets. These are, in fact, the wavefunctions of the unperturbed states at $k = 0$.

We now focus on a particular coefficient $C_m$ from Eq. (3.3) with $m$ that belongs to the 8-band Set $N$. We can cast the Schrödinger equation as a linear set of equations with dimension $N + R$. The $m$th equation in the set is

$$H_{mm} C_m + \sum_{N, n \neq m} H_{mn} C_n + \sum_R H_{mr} C_r = E C_m \tag{3.4}$$

where the Hamiltonian matrix element $H_{mn}$ is defined by analogy with Eq. (2.5):

$$H_{mn} = \langle m | H | n \rangle = \frac{1}{V} \int d^3 r \, \psi_m^* \, H \, \psi_n \tag{3.5}$$

The off-diagonal $H_{mn}$ with $m \neq n$ are linear in $k$, as expected from the **k·p** formulation. Now we can solve Eq. (3.4) for $C_m$:

$$C_m = \frac{\sum\limits_{N, n \neq m} H_{mn} C_n + \sum\limits_R H_{mr} C_r}{E - H_{mm}} \tag{3.6}$$

Even though we obtained Eq. (3.6) for $m$ in Set $N$, the same expression without the restriction to $m \neq n$ is also valid if $m$ is in Set $R$. This is convenient because we can substitute it for $C_r$ in Eq. (3.6) without further ado:

$$C_m = \sum_{N, n \neq m} \frac{H_{mn}}{E - H_{mm}} C_n + \sum_R \frac{H_{mr}}{E - H_{mm}} \left[ \sum_N \frac{H_{rn'}}{E - H_{rr}} C_{n'} + \sum_R \frac{H_{rr'}}{E - H_{rr}} C_{r'} \right] \tag{3.7}$$

Of course, we have no idea what the $C_{r'}$ are. It would be nice to eliminate them, but this cannot be done by simply substituting the equation for $C_r$ (similar to Eq. (3.6)) into Eq. (3.7) again because both sides of that equation contain $C_r$. Instead we argue that the remote states are "remote" because the states from Set $N$ are not coupled to them as strongly as to other states in Set $N$. That is, $C_r$ should be small compared to $C_n$. This being the case, a single substitution may have done the trick, and we can neglect $C_{r'}$ in Eq. (3.7) entirely. This leads to the desired second-order form for $C_m$:

$$C_m = \sum_{N, n \neq m} \frac{H_{mn}}{E - H_{mm}} C_n + \sum_{n' \in N, r \in R} \frac{H_{mr} H_{rn'}}{(E - H_{mm})(E - H_{rr})} C_{n'} \qquad (3.8)$$

Equation (3.8) is quadratic in $k$, since the Hamiltonian matrix elements $H_{mr}$ and $H_{rn'}$ are linear in $k$. To make this expression like the Schrödinger equation, we multiply by $E - H_{mm}$:

$$\sum_{N, n \neq m} H_{mn} C_n + \sum_{n' \in N, r \in R} \frac{H_{mr} H_{rn'}}{(E - H_{rr})} C_{n'} + H_{mm} C_m = E C_m \qquad (3.9)$$

This is still a bit awkward, since the $n$ and $n'$ subscripts can refer to the same states within Set $N$. Fortunately, we can already see that the off-diagonal matrix elements between states $m$ and $n$ within Set $N$ are given by

$$H'_{mn} = H_{mn} + \sum_{r \in R} \frac{H_{mr} H_{rn}}{E - H_{rr}} \qquad (3.10)$$

Can we now substitute the matrix elements in Eq. (3.10) into the full Hamiltonian and solve for the energies and wavefunctions? Not so fast, because we have not yet specified the energy $E$ in the denominator. In fact, this is the energy of the state $E(\mathbf{k})$ at some $k > 0$ that we are seeking to determine in the first place. Whatever it is, its approximate form at small $k$ is $E(\mathbf{k}) = E_{m,n}(k = 0) + O(k^2)$, but we do not know if we should use $E_m$ or $E_n$ here (or if the two are in fact different). In any case, the quadratic-in-$k$ terms in the denominator drop out if we discard all the terms higher than $k^2$ from the final expression, as would be done within second-order perturbation theory. Similarly, we can approximate $H_{rr}$ as $E_r(k = 0)$.

While much ink has been spilled in making the derivations more precise, the somewhat surprising bottom line is that we can evade the question of what energy should be inserted in the denominator! This is because we are not trying to compute the matrix elements from first principles. We just wish to find the functional form in terms of $k_x$, $k_y$, and $k_z$ with a minimum of fitting parameters. From this point of view, lattice symmetry gives rise to the functional form, and experimental results determine the parameter values.

The first term of Eq. (3.10), which comes from first-order $\mathbf{k} \cdot \mathbf{p}$ theory, is already present in Kane's model. The second term produces off-diagonal Hamiltonian elements of the form $H_{mn} = \sum_{ij} A^{ij}_{mn} k_i k_j$, where $i$ and $j$ are not necessarily the same. Our goal is simply to determine which of the coefficients $A^{ij}_{mn}$ do not vanish. We already mentioned that for this purpose we can neglect the $k$ dependence in the denominator.

The more formal version of this approach is known as Löwdin's perturbation theory, named after Per-Olov Löwdin, who described it for the first time in 1951. Another route to the same destination is the method of infinitesimal basis transformations that was developed shortly thereafter. These approaches are equivalent for our purposes, and they apply regardless of whether either Set $N$ or Set $R$ contains any degenerate states. Even though all the gory parts are now swept under the rug, it seems intuitively unlikely that the calculated band energies for states in Set $N$ will be equal to $H_{rr}$. Indeed, this is why these bands are called "remote".

We have discovered that the $H_{mr}H_{rn}$ sum over the remote states $r$ in Set $R$ gives us the correct form for the second-order $\mathbf{k \cdot p}$ matrix elements. The next order of business is to determine what these Hamiltonian matrix elements actually look like. We begin by establishing what kind of symmetry the remote states are likely to have. The easiest cases to explore are the interactions with antibonding $p$-like states and bonding $s$-like states. The $m$ and $n$ subscripts in $H_{mr}$ and $H_{rn}$ correspond to the $s$ and $p$ states from our 8×8 Hamiltonian, which are "purely" $s$-like and $p$-like at $k = 0$. We have already seen how the $H'_{ss}$ matrix element modifies the electron mass. We will not include any corrections to the $H'_{sp}$ matrix elements here because they are dominated by first-order interactions between the conduction and valence bands, but will revisit this point in Chapter 8. Instead, our focus will be on the $H'_{xx}$ and $H'_{xy}$ matrix elements (for compactness, we use subscript $x$ to stand for $p_x$, $y$ for $p_y$, etc.). Recall that these elements were zero in the first-order theory described by Eq. (3.1). The other elements follow because we can permute the axes.

We saw in Chapter 2 that the bonding $s$-like and antibonding $p$-like states are relatively close in energy to the conduction and valence bands. More generally, there will be remote states with $s$, $p$, and $d$ symmetry. What terms can result from coupling the $p_x$ and $p_y$ states to such remote states? Figure 3.1 illustrates that $p$-like orbitals point in a certain direction and hence transform much like the linear functions $x$ and $y$. The $d$-like states are trickier, with a total of five states that transform as the functions $2z^2 - x^2 - y^2$, $\sqrt{3}\left(x^2 - y^2\right)$, $xy$, $yz$, and $xz$. Fortunately, their precise form will be largely irrelevant, so we assume there is non-zero coupling only to the first two, which we call $d_a$ and $d_b$.

What are the relevant momentum matrix elements? We already know from Chapter 2 that non-zero $s$–$p$ elements occur only when the momentum is directed along the same axis as the orbital. Otherwise, the positive part of the orbital cancels the negative part. The surviving momentum elements are $\langle x|p_x|S \rangle$, $\langle y|p_y|S \rangle$, and $\langle z|p_z|S \rangle$. Here we simplify the notation again by omitting the $p$ in the orbital label, so that $p_x$, $p_y$, and $p_z$ refer to the momentum components rather than orbitals. They are easily distinguished because the momentum appears between the vertical lines while the orbitals are in the bra or ket slots. We also use capital letters to refer to remote states with $s$ and $p$ symmetries. Recall that we also have $\langle S|p_x|x \rangle = \langle x|p_x|S \rangle^*$, etc.

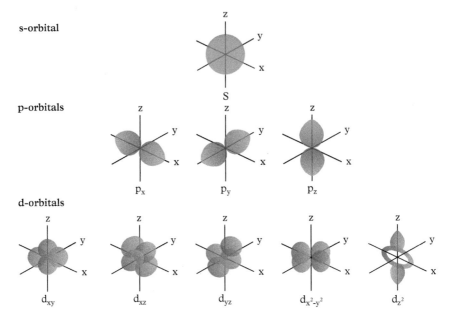

**Figure 3.1** *Schematic illustration of the symmetry for s, p, and d atomic orbitals.*

**Table 3.1** *Non-zero momentum matrix elements connecting s, p, and d-like states.*

| Origin of matrix element | Value |
|---|---|
| **s–p coupling** | $\langle x\|p_x\|S\rangle = \langle y\|p_y\|S\rangle = \langle z\|p_z\|S\rangle$ |
| **p–p coupling** | $\langle x\|p_y\|Z\rangle = \langle z\|p_x\|Y\rangle = \langle y\|p_z\|X\rangle = \langle z\|p_y\|X\rangle = \langle y\|p_x\|Z\rangle$ <br> $= \langle x\|p_z\|Y\rangle$ |
| **p–$d_a$ and p–$d_b$ coupling** | $\langle x\|p_x\|d_a\rangle = -\langle y\|p_y\|d_a\rangle = -\sqrt{3}\,\langle x\|p_x\|d_b\rangle = -\sqrt{3}\,\langle y\|p_y\|d_b\rangle$ |

We should also include the matrix elements between the various $p$-like orbitals, which are non-zero only when all three spatial coordinates appear, i.e., $\langle x\|p_y\|Z\rangle$, $\langle z\|p_x\|Y\rangle$, $\langle y\|p_z\|X\rangle$, etc. As shown in Table 3.1, we can take all of these elements to be the same.

Finally, there are matrix elements between the $p$ and $d$ states. Their values in Table 3.1 are much less intuitive, and their signs relate to the signs of the $x$ and $y$ terms in the corresponding transformation functions. Here we will make yet another approximation and take the coupling strengths with $d_a$ and $d_b$ states to be the same. Of course, different bands may have the same symmetry, but we sum their contributions and retain only a single coefficient for each orbital symmetry. We will see later that this yields enough parameters to accurately describe the valence-band structure of real III–V semiconductors.

We are now ready to form $H'_{xx}$ and $H'_{xy}$ using Table 3.1. The bulk materials are invariant in all directions, so the wavevector components $k_x$, $k_y$, and $k_z$ are just real numbers. To lay the groundwork for extending the treatment to quantum structures in Part III, we assume that they are in fact operators $\hat{k}_x$, $\hat{k}_y$, $\hat{k}_z$, whose order in multiplication is important; i.e., they do not commute. Then within the **k·p** theory we expect that $H'_{xx} \propto \hat{k}_x \hat{k}_x$ and $H'_{xy} \propto \hat{k}_x \hat{k}_y$:

$$H'_{xx} = H_{xx} + \hat{k}_x \langle x|p_x|S\rangle W'_s \langle S|p_x|x\rangle \hat{k}_x + \hat{k}_y \langle x|p_y|Z\rangle W'_p \langle Z|p_y|x\rangle \hat{k}_y$$

$$+ \hat{k}_z \langle x|p_z|Y\rangle W'_p \langle Y|p_z|x\rangle \hat{k}_z + \hat{k}_x \langle x|p_x|d_a\rangle W'_d \langle d_a|p_x|x\rangle \hat{k}_x$$

$$+ \hat{k}_x \langle x|p_x|d_b\rangle W'_d \langle d_b|p_x|x\rangle \hat{k}_x \tag{3.11}$$

$$H'_{xy} = \hat{k}_x \langle x|p_x|S\rangle W'_s \langle S|p_y|y\rangle \hat{k}_y + \hat{k}_y \langle x|p_y|Z\rangle W'_p \langle Z|p_x|y\rangle \hat{k}_x$$

$$+ \hat{k}_x \langle x|p_x|d_a\rangle W'_d \langle d_a|p_y|y\rangle \hat{k}_y + \hat{k}_x \langle x|p_x|d_b\rangle W'_d \langle d_b|p_y|y\rangle \hat{k}_y \tag{3.12}$$

where $W'_s$, $W'_p$, and $W'_d$ are to be determined. These fitting parameters include the unknown information about $(E - H_{rr})^{-1}$ in Eq. (3.10).

Making use of the relationships between the matrix elements in Table 3.1, and defining $W_s = W'_s \langle x|p_x|S\rangle^2$, $W_p = W'_p \langle x|p_y|Z\rangle^2$, $W_d = \frac{2}{3} W'_d \langle x|p_x|d_a\rangle^2$, we compute

$$H'_{xx} = H_{xx} + \hat{k}_x (W_s + 2W_d) \hat{k}_x + \hat{k}_y W_p \hat{k}_y + \hat{k}_z W_p \hat{k}_z \tag{3.13}$$

$$H'_{xy} = \hat{k}_x (W_s - W_d) \hat{k}_y + \hat{k}_y W_p \hat{k}_x \tag{3.14a}$$

Since $x$ and $y$ may be thought of as dummy indices for these purposes, we can easily specify the matrix element $H'_{yx}$ as well:

$$H'_{yx} = \hat{k}_y (W_s - W_d) \hat{k}_x + \hat{k}_x W_p \hat{k}_y \tag{3.14b}$$

$H'_{xy}$ and $H'_{yz}$ are consistent with the Hamiltonian being Hermitian, but the coefficients of $\hat{k}_x \hat{k}_y$ and $\hat{k}_y \hat{k}_x$ are not the same ("symmetric"). We will see in Chapter 8 that the operator ordering in Eqs. (3.13) and (3.14) helps to resolve an ambiguity in defining the Hamiltonians for quantum structures, even though the resolution ultimately proves of little consequence.

At this point, we revert to dealing with bulk materials and stop treating $\hat{k}_x$, $\hat{k}_y$, $\hat{k}_z$ as operators for the duration of Part I. We now insert Eqs. (3.13) and (3.14) and their analogs into Eq. (3.1):

$$H_{4\times4} = \begin{bmatrix} E_g + \frac{\hbar^2 k^2}{2m}(1+2F) & i\frac{\hbar}{m}k_x p_{cv} & i\frac{\hbar}{m}k_y p_{cv} & i\frac{\hbar}{m}k_z p_{cv} \\ -i\frac{\hbar}{m}k_x p_{cv} & -Lk_x^2 - M\left(k_y^2 + k_z^2\right) & -Nk_x k_y & -Nk_x k_z \\ -i\frac{\hbar}{m}k_y p_{cv} & -Nk_x k_y & -Lk_y^2 - M\left(k_x^2 + k_z^2\right) & -Nk_y k_z \\ -i\frac{\hbar}{m}k_z p_{cv} & -Nk_x k_z & -Nk_y k_z & -Lk_z^2 - M\left(k_x^2 + k_y^2\right) \end{bmatrix} \tag{3.15}$$

The definitions $L \equiv -W_s - 2W_d$, $M \equiv -W_p$, $N \equiv -W_s - W_p + W_d$ describe three standard valence-band parameters that have resulted from the three couplings included in Eqs. (3.11) and (3.12). The largest contribution is often from the $p$-like antibonding states above the conduction band, but generally we cannot neglect the $s$-like and $d$-like couplings. Chapter 6 will show that $|W_p|, |W_s| \gg |W_d|$ for most materials based on the recommended valence-band effective-mass parameters. If we neglect any of the couplings, the three coefficients in Eq. (3.15) are no longer independent.

For future reference, we invert the definitions in Eq. (3.15) to obtain the actual coupling coefficients in terms of the more compact parameters $L, M, N$:

$$W_s = \frac{-L + 2(M - N)}{3} \tag{3.16a}$$

$$W_p = -M \tag{3.16b}$$

$$W_d = \frac{-L - M + N}{3} \tag{3.16c}$$

For simplicity, we also neglect the free-electron terms in the lower $3 \times 3$ block of Eq. (3.15). Later we will outline the relation of $L$, $M$, and $N$ to measurable quantities in the band structure, but first we complete the development of the full $8 \times 8$ Hamiltonian that will form the basis of many calculations in the rest of the book.

We double the number of orbitals to 8, to account for up and down spins, and to add the spin–orbit coupling as shown in Chapter 2. The missing ingredient is the spin–orbit Hamiltonian that can be added to the spin-doubled Hamiltonian of Eq. (3.15), i.e., $H(\boldsymbol{L} \cdot \boldsymbol{S})$ in the the $|s, p_x, p_y, p_z\rangle$ basis. We can get there in several steps: (1) convert between the $|s, p_x, p_y, p_z\rangle$ and $|m_l, m_s\rangle$ bases; (2) convert between the $|m_l, m_s\rangle$ and $|j, m_j\rangle$ bases; (3) compound the two transformations to convert between the $|s, p_x, p_y, p_z\rangle$ and $|j, m_j\rangle$ bases; (4) transform the spin–orbit Hamiltonian into the $|s, p_x, p_y, p_z\rangle$ basis using the prescription in Eq. (2.35); and (5) add the resulting spin–orbit Hamiltonian to the spin-doubled version of Eq. (3.15).

The first step of connecting the $|s, p_x, p_y, p_z\rangle$ and $|m_l, m_s\rangle$ bases can be done using Eq. (2.30). The only change is to make the spin-doubled nature of the Hamiltonian explicit and add the $s$-like states. To distinguish between the $s$-like and $p$-like states, we specify the $l$ quantum number: $l = 0$ for $s$ and $l = 1$ for $p$, which gives the conversion formula

$$
\begin{bmatrix}
\,|\,l=0,m_l=0,m_s=\tfrac{1}{2}\rangle\, \\
\,|\,l=1,m_l=1,m_s=\tfrac{1}{2}\rangle\, \\
\,|\,l=1,m_l=0,m_s=\tfrac{1}{2}\rangle\, \\
\,|\,l=1,m_l=-1,m_s=\tfrac{1}{2}\rangle\, \\
\,|\,l=0,m_l=0,m_s=-\tfrac{1}{2}\rangle\, \\
\,|\,l=1,m_l=1,m_s=-\tfrac{1}{2}\rangle\, \\
\,|\,l=1,m_l=0,m_s=-\tfrac{1}{2}\rangle\, \\
\,|\,l=1,m_l=-1,m_s=-\tfrac{1}{2}\rangle\,
\end{bmatrix}
=
\begin{bmatrix}
\boldsymbol{M}_{lsp} & \boldsymbol{0} \\
\boldsymbol{0} & \boldsymbol{M}_{lsp}
\end{bmatrix}
\begin{bmatrix}
|s\uparrow\rangle \\
|p_x\uparrow\rangle \\
|p_y\uparrow\rangle \\
|p_z\uparrow\rangle \\
|s\downarrow\rangle \\
|p_x\downarrow\rangle \\
|p_y\downarrow\rangle \\
|p_z\downarrow\rangle
\end{bmatrix}
\tag{3.17}
$$

where $\boldsymbol{M}_{lsp}$ is given by

$$
\boldsymbol{M}_{lsp}=
\begin{bmatrix}
1 & 0 & 0 & 0 \\
0 & -\frac{1}{\sqrt{2}} & -\frac{i}{\sqrt{2}} & 0 \\
0 & 0 & 0 & 1 \\
0 & \frac{1}{\sqrt{2}} & -\frac{i}{\sqrt{2}} & 0
\end{bmatrix}
\tag{3.18}
$$

and the empty matrix $\boldsymbol{0}$ is also $4\times4$.

Now we transform from the $|m_l,m_s\rangle$ basis to the $|\,j,m_j\rangle$ basis, while still keeping track of $l$. To this end, we simply expand the matrix in Eq. (2.32) and augment it to account for the $l=0$ states as well as the $m_j=-\tfrac{1}{2}$ and $m_j=-\tfrac{3}{2}$ states. With phases assigned in the conventional way, we obtain

$$
\begin{bmatrix}
\,|l=0,j=\tfrac{1}{2},m_j=\tfrac{1}{2}\rangle\, \\
\,|l=0,j=\tfrac{1}{2},m_j=-\tfrac{1}{2}\rangle\, \\
\,|l=1,j=\tfrac{3}{2},m_j=\tfrac{3}{2}\rangle\, \\
\,|l=1,j=\tfrac{3}{2},m_j=\tfrac{1}{2}\rangle\, \\
\,|l=1,j=\tfrac{3}{2},m_j=-\tfrac{1}{2}\rangle\, \\
\,|l=1,j=\tfrac{3}{2},m_j=-\tfrac{3}{2}\rangle\, \\
\,|l=1,j=\tfrac{1}{2},m_j=\tfrac{1}{2}\rangle\, \\
\,|l=1,j=\tfrac{1}{2},m_j=-\tfrac{1}{2}\rangle\,
\end{bmatrix}
$$

$$
=
\begin{bmatrix}
1 & 0 & 0 & 0 & 0 & 0 & 0 & 0 \\
0 & 0 & 0 & 0 & 1 & 0 & 0 & 0 \\
0 & 1 & 0 & 0 & 0 & 0 & 0 & 0 \\
0 & 0 & \sqrt{\tfrac{2}{3}} & 0 & 0 & \frac{1}{\sqrt{3}} & 0 & 0 \\
0 & 0 & 0 & \frac{1}{\sqrt{3}} & 0 & 0 & \sqrt{\tfrac{2}{3}} & 0 \\
0 & 0 & 0 & 0 & 0 & 0 & 0 & 1 \\
0 & 0 & -\frac{1}{\sqrt{3}} & 0 & 0 & \sqrt{\tfrac{2}{3}} & 0 & 0 \\
0 & 0 & 0 & -\sqrt{\tfrac{2}{3}} & 0 & 0 & \frac{1}{\sqrt{3}} & 0
\end{bmatrix}
\begin{bmatrix}
\,|\,l=0,m_l=0,m_s=\tfrac{1}{2}\rangle\, \\
\,|\,l=1,m_l=1,m_s=\tfrac{1}{2}\rangle\, \\
\,|\,l=1,m_l=0,m_s=\tfrac{1}{2}\rangle\, \\
\,|\,l=1,m_l=-1,m_s=\tfrac{1}{2}\rangle\, \\
\,|\,l=0,m_l=0,m_s=-\tfrac{1}{2}\rangle\, \\
\,|\,l=1,m_l=1,m_s=-\tfrac{1}{2}\rangle\, \\
\,|\,l=1,m_l=0,m_s=-\tfrac{1}{2}\rangle\, \\
\,|\,l=1,m_l=-1,m_s=-\tfrac{1}{2}\rangle\,
\end{bmatrix}
$$

$$
= M_{jls}
\begin{bmatrix}
| l = 0, m_l = 0, m_s = \tfrac{1}{2} \rangle \\
| l = 1, m_l = 1, m_s = \tfrac{1}{2} \rangle \\
| l = 1, m_l = 0, m_s = \tfrac{1}{2} \rangle \\
| l = 1, m_l = -1, m_s = \tfrac{1}{2} \rangle \\
| l = 0, m_l = 0, m_s = -\tfrac{1}{2} \rangle \\
| l = 1, m_l = 1, m_s = -\tfrac{1}{2} \rangle \\
| l = 1, m_l = 0, m_s = -\tfrac{1}{2} \rangle \\
| l = 1, m_l = -1, m_s = -\tfrac{1}{2} \rangle
\end{bmatrix}
\tag{3.19}
$$

To transform the $|s, p_x, p_y, p_z\rangle$ basis to the $|j, m_j\rangle$ basis, we multiply the two 8×8 matrices $M_{jls}$ and $\begin{bmatrix} M_{lsp} & 0 \\ 0 & M_{lsp} \end{bmatrix}$:

$$
\begin{bmatrix}
| l = 0, j = \tfrac{1}{2}, m_j = \tfrac{1}{2} \rangle \\
| l = 0, j = \tfrac{1}{2}, m_j = -\tfrac{1}{2} \rangle \\
| l = 1, j = \tfrac{3}{2}, m_j = \tfrac{3}{2} \rangle \\
| l = 1, j = \tfrac{3}{2}, m_j = \tfrac{1}{2} \rangle \\
| l = 1, j = \tfrac{3}{2}, m_j = -\tfrac{1}{2} \rangle \\
| l = 1, j = \tfrac{3}{2}, m_j = -\tfrac{3}{2} \rangle \\
| l = 1, j = \tfrac{1}{2}, m_j = \tfrac{1}{2} \rangle \\
| l = 1, j = \tfrac{1}{2}, m_j = -\tfrac{1}{2} \rangle
\end{bmatrix}
= M_{jls}
\begin{bmatrix} M_{lsp} & 0 \\ 0 & M_{lsp} \end{bmatrix}
\begin{bmatrix}
|s \uparrow\rangle \\
|p_x \uparrow\rangle \\
|p_y \uparrow\rangle \\
|p_z \uparrow\rangle \\
|s \downarrow\rangle \\
|p_x \downarrow\rangle \\
|p_y \downarrow\rangle \\
|p_z \downarrow\rangle
\end{bmatrix}
= M_{jsp}
\begin{bmatrix}
|s \uparrow\rangle \\
|p_x \uparrow\rangle \\
|p_y \uparrow\rangle \\
|p_z \uparrow\rangle \\
|s \downarrow\rangle \\
|p_x \downarrow\rangle \\
|p_y \downarrow\rangle \\
|p_z \downarrow\rangle
\end{bmatrix}
\tag{3.20}
$$

where the total transformation matrix $M_{jsp}$ is given by

$$
M_{jsp} =
\begin{bmatrix}
1 & 0 & 0 & 0 & 0 & 0 & 0 & 0 \\
0 & 0 & 0 & 0 & 1 & 0 & 0 & 0 \\
0 & -\tfrac{1}{\sqrt{2}} & -\tfrac{i}{\sqrt{3}} & 0 & 0 & 0 & 0 & 0 \\
0 & 0 & 0 & \sqrt{\tfrac{2}{3}} & 0 & -\tfrac{1}{\sqrt{6}} & -\tfrac{i}{\sqrt{6}} & 0 \\
0 & \tfrac{1}{\sqrt{6}} & -\tfrac{i}{\sqrt{6}} & 0 & 0 & 0 & 0 & \sqrt{\tfrac{2}{3}} \\
0 & 0 & 0 & 0 & 0 & \tfrac{1}{\sqrt{2}} & -\tfrac{i}{\sqrt{2}} & 0 \\
0 & 0 & 0 & -\tfrac{1}{\sqrt{3}} & 0 & -\tfrac{1}{\sqrt{3}} & -\tfrac{i}{\sqrt{3}} & 0 \\
0 & -\tfrac{1}{\sqrt{3}} & \tfrac{i}{\sqrt{3}} & 0 & 0 & 0 & 0 & \tfrac{1}{\sqrt{3}}
\end{bmatrix}
\tag{3.21}
$$

Starting with the spin–orbit Hamiltonian in Eq. (2.29), doubling it, and including the $l = 0$ states that are unaffected by spin–orbit coupling, the spin–orbit interaction for the $|j, m_j\rangle$ basis becomes

$$H(\mathbf{L\cdot S}) \quad \begin{bmatrix} |l=0,j=\tfrac{1}{2},m_j=\tfrac{1}{2}\rangle \\ |l=0,j=\tfrac{1}{2},m_j=-\tfrac{1}{2}\rangle \\ |l=1,j=\tfrac{3}{2},m_j=\tfrac{3}{2}\rangle \\ |l=1,j=\tfrac{3}{2},m_j=\tfrac{1}{2}\rangle \\ |l=1,j=\tfrac{3}{2},m_j=-\tfrac{1}{2}\rangle \\ |l=1,j=\tfrac{3}{2},m_j=-\tfrac{3}{2}\rangle \\ |l=1,j=\tfrac{1}{2},m_j=\tfrac{1}{2}\rangle \\ |l=1,j=\tfrac{1}{2},m_j=-\tfrac{1}{2}\rangle \end{bmatrix}$$

$$= \begin{bmatrix} 0 & 0 & 0 & 0 & 0 & 0 & 0 & 0 \\ 0 & 0 & 0 & 0 & 0 & 0 & 0 & 0 \\ 0 & 0 & \frac{\Delta}{3} & 0 & 0 & 0 & 0 & 0 \\ 0 & 0 & 0 & \frac{\Delta}{3} & 0 & 0 & 0 & 0 \\ 0 & 0 & 0 & 0 & \frac{\Delta}{3} & 0 & 0 & 0 \\ 0 & 0 & 0 & 0 & 0 & \frac{\Delta}{3} & 0 & 0 \\ 0 & 0 & 0 & 0 & 0 & 0 & -\frac{2\Delta}{3} & 0 \\ 0 & 0 & 0 & 0 & 0 & 0 & 0 & -\frac{2\Delta}{3} \end{bmatrix} \begin{bmatrix} |l=0,j=\tfrac{1}{2},m_j=\tfrac{1}{2}\rangle \\ |l=0,j=\tfrac{1}{2},m_j=-\tfrac{1}{2}\rangle \\ |l=1,j=\tfrac{3}{2},m_j=\tfrac{3}{2}\rangle \\ |l=1,j=\tfrac{3}{2},m_j=\tfrac{1}{2}\rangle \\ |l=1,j=\tfrac{3}{2},m_j=-\tfrac{1}{2}\rangle \\ |l=1,j=\tfrac{3}{2},m_j=-\tfrac{3}{2}\rangle \\ |l=1,j=\tfrac{1}{2},m_j=\tfrac{1}{2}\rangle \\ |l=1,j=\tfrac{1}{2},m_j=-\tfrac{1}{2}\rangle \end{bmatrix} \tag{3.22}$$

Now we can use the prescription in Eq. (2.35), with the transformation matrix given by Eq. (3.21), to convert the 8×8 matrix of Eq. (3.22) into the $|s, p_x, p_y, p_z\rangle$ basis:

$$H(\mathbf{L\cdot S}) \begin{bmatrix} |s\uparrow\rangle \\ |p_x\uparrow\rangle \\ |p_y\uparrow\rangle \\ |p_z\uparrow\rangle \\ |s\downarrow\rangle \\ |p_x\downarrow\rangle \\ |p_y\downarrow\rangle \\ |p_z\downarrow\rangle \end{bmatrix} = \begin{bmatrix} 0 & 0 & 0 & 0 & 0 & 0 & 0 & 0 \\ 0 & 0 & -i\frac{\Delta}{3} & 0 & 0 & 0 & 0 & \frac{\Delta}{3} \\ 0 & i\frac{\Delta}{3} & 0 & 0 & 0 & 0 & 0 & -i\frac{\Delta}{3} \\ 0 & 0 & 0 & 0 & 0 & -\frac{\Delta}{3} & i\frac{\Delta}{3} & 0 \\ 0 & 0 & 0 & 0 & 0 & 0 & 0 & 0 \\ 0 & 0 & 0 & -\frac{\Delta}{3} & 0 & 0 & i\frac{\Delta}{3} & 0 \\ 0 & 0 & 0 & -i\frac{\Delta}{3} & 0 & -i\frac{\Delta}{3} & 0 & 0 \\ 0 & \frac{\Delta}{3} & i\frac{\Delta}{3} & 0 & 0 & 0 & 0 & 0 \end{bmatrix} \begin{bmatrix} |s\uparrow\rangle \\ |p_x\uparrow\rangle \\ |p_y\uparrow\rangle \\ |p_z\uparrow\rangle \\ |s\downarrow\rangle \\ |p_x\downarrow\rangle \\ |p_y\downarrow\rangle \\ |p_z\downarrow\rangle \end{bmatrix}$$

$$\tag{3.23}$$

Now we are ready for the final step of simply adding the spin–orbit Hamiltonian in Eq. (3.23) to the spin-doubled $\mathbf{k\cdot p}$ Hamiltonian in Eq. (3.15):

$$
\begin{bmatrix} H_{4\times4} & 0 \\ 0 & H_{4\times4} \end{bmatrix} + H(\boldsymbol{L}\cdot\boldsymbol{S})
\begin{bmatrix} |s\uparrow\rangle \\ |p_x\uparrow\rangle \\ |p_y\uparrow\rangle \\ |p_z\uparrow\rangle \\ |s\downarrow\rangle \\ |p_x\downarrow\rangle \\ |p_y\downarrow\rangle \\ |p_z\downarrow\rangle \end{bmatrix}
= E_n
\begin{bmatrix} |s\uparrow\rangle \\ |p_x\uparrow\rangle \\ |p_y\uparrow\rangle \\ |p_z\uparrow\rangle \\ |s\downarrow\rangle \\ |p_x\downarrow\rangle \\ |p_y\downarrow\rangle \\ |p_z\downarrow\rangle \end{bmatrix}
\qquad (3.24)
$$

Equation (3.24) is a matrix equation that can be solved numerically for its eigenvalues $E_n$ and eigenvectors corresponding to different bands $n$. Figure 3.2 plots the conduction and valence-band dispersions for GaAs and InSb that have moderately wide and narrow energy gaps, respectively.

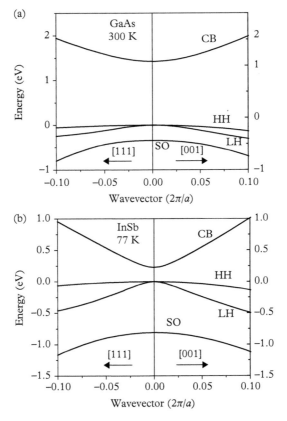

**Figure 3.2** *Numerical solutions to the Hamiltonian of Eq. (3.24) for the conduction- and valence-band dispersion relations of GaAs (at a temperature of 300 K) and InSb (at a temperature of 77 K) near $k = 0$. Here [111] stands for $k_x = k_y = k_z$ and [001] for $k_x = k_y = 0$.*

We could fit the numerically obtained dispersions in Fig.3.2 to measurable quantities such as the energy gap and effective masses. This is not wrong, but a little unsatisfying, because we remain in the dark about the general form of the dispersions near $k = 0$. We can understand the physics much better by solving Eq. (3.24) analytically, to parallel what we did with Kane's model. As always, we advise the reader to look at numerical solutions only once you have developed good analytical insight into the problem!

We could also transform the $\mathbf{k} \cdot \mathbf{p}$ Hamiltonian of Eq. (3.15) into the total angular momentum $(|j, m_j\rangle)$ basis instead of casting the spin–orbit Hamiltonian into the $|s, p_x, p_y, p_z\rangle$ basis. Of course, the spin–orbit interaction is diagonal in the $|j, m_j\rangle$ basis (Eq. (3.22)), and the HH, LH, and SO states at $k = 0$ are distinguishable by their values of $j$ and $m_j$. In Section 3.2, we will take advantage of this approach. Numerically, it is no easier to solve than Eq. (3.24) for $k \neq 0$, and we find its full form to be less memorable and more prone to coding errors. For this reason, we will work primarily with Eq. (3.24) and only use the total angular momentum basis to generate reduced forms of the Hamiltonian.

Another way to include the effects of remote bands is to include the antibonding $p$ states to the first order in the $\mathbf{k} \cdot \mathbf{p}$ theory by extending the basis to 14 bands. One disadvantage of this approach when applied to quantum structures in Chapter 9 is the potentially longer run time that goes up superlinearly with the number of bands. It is a matter of taste whether the second-order terms in the 8-band approach or additional bands in the 14-band method are simpler conceptually. Arguably, extra bands create an illusion of completeness that can scarcely be justified. The rest of Chapter 3 continues with the 8-band approach and its simplified variants.

## 3.2 Meaning of the Terms in the Second-Order Hamiltonian

At this point, we have the full Hamiltonian, but are not yet sure how much its solutions improve on Kane's model of Section 2.3. At a minimum, we would like a more realistic depiction of the HH band dispersion along the $z$-axis: $\mathbf{k} = (0, 0, k_z)$. Also, we will find that when the Hamiltonian includes second-order terms, the band structure is no longer independent of wavevector direction. In a cubic system such as the zinc-blende lattice, the $x$, $y$, and $z$ axes are all equivalent, but they differ from, e.g., [111] with $k_x = k_y = k_z$, or [110] with $k_x = k_y$ and $k_z = 0$. We already know that the anisotropy should be negligible for the spherically symmetric electrons, but how much does it affect the holes?

Our main interest is in the valence-band masses, i.e., the leading term in the $k^2$ series expansion. In this context, the conduction band is "remote" and adds second-order terms due to conduction–valence interactions to the reduced valence-band Hamiltonian. The machinery developed in Section 3.1 tells us that the spherically symmetric $s$-like states contribute the terms already present in the matrix elements of Eq. (3.15). This is because the extra terms are due to the $\langle i|p_i|s\rangle\,\langle s|p_i|i\rangle\,k_i^2$ interaction in the diagonal matrix elements and the $\langle i|p_i|s\rangle\,\langle s|p_j|j\rangle\,k_i k_j$ interaction in the off-diagonal ones, where $i$ and $j$ are $x, y, z$. The denominator in Eq. (3.10) reduces to the energy gap between the conduction and valence bands $E_g$. Putting these pieces together, we arrive at the approximate Hamiltonian without the spin–orbit interaction:

$$\boldsymbol{H}_{4x4} \approx \begin{bmatrix} E_g + Dk^2 & 0 & 0 & 0 \\ 0 & -Ak_x^2 - B\left(k_y^2 + k_z^2\right) & -Ck_x k_y & -Ck_x k_z \\ 0 & -Ck_x k_y & -Ak_y^2 - B\left(k_x^2 + k_z^2\right) & -Ck_y k_z \\ 0 & -Ck_x k_z & -Ck_y k_z & -Ak_z^2 - B\left(k_x^2 + k_y^2\right) \end{bmatrix}$$

$$(3.25)$$

where the four new coefficients contain the ratio of the Kane energy to the energy gap $\frac{E_P}{E_g} = \frac{2|\boldsymbol{p}_{cv}|^2}{mE_g}$:

$$A \equiv L + \frac{\hbar^2}{m^2}\frac{|\boldsymbol{p}_{cv}|^2}{E_g} = L + \frac{\hbar^2}{2m}\frac{E_P}{E_g} \qquad (3.26a)$$

$$B \equiv M \qquad (3.26b)$$

$$C \equiv N + \frac{\hbar^2}{m^2}\frac{|\boldsymbol{p}_{cv}|^2}{E_g} = N + \frac{\hbar^2}{2m}\frac{E_P}{E_g} \qquad (3.26c)$$

$$D \equiv \frac{\hbar^2}{2m}(1 + 2F) + \frac{\hbar^2}{m^2}\frac{|\boldsymbol{p}_{cv}|^2}{E_g} = \frac{\hbar^2}{2m}\left(1 + 2F + \frac{E_P}{E_g}\right) \qquad (3.26d)$$

The Hamiltonian in Eqs. (3.25) and (3.26) does not yet include spin–orbit terms. For this reason, $D$ in Eq. (3.26d) does not in general provide a good estimate of the inverse electron effective mass, as we already know from Chapter 2 (see Eq. (2.44)). For wavevectors along the $z$ axis, we have $k_x = k_y = 0$, and the Hamiltonian without the decoupled conduction-band states becomes

$$\boldsymbol{H}_{4\times4}\,[001] \begin{bmatrix} |p_x\rangle \\ |p_y\rangle \\ |p_z\rangle \end{bmatrix} \approx - \begin{bmatrix} Bk^2 & 0 & 0 \\ 0 & Bk^2 & 0 \\ 0 & 0 & Ak^2 \end{bmatrix} \begin{bmatrix} |p_x\rangle \\ |p_y\rangle \\ |p_z\rangle \end{bmatrix} \qquad (3.27)$$

To add the spin–orbit coupling, we switch to the $|\,m_l, m_s\rangle$ basis, using the results leading to Eq. (2.41). This procedure is not strictly correct, because knowing that some of the valence bands are degenerate, we should have added the spin–orbit terms before decoupling the CB. Nevertheless, the dispersions for the top of the valence band turn out to be unaffected by this sleight-of-hand. The reduced Hamiltonian, with its rows rearranged, is

$$
\mathbf{H}_{3\times3}
\begin{bmatrix}
|m_l = 1, m_s = -\tfrac{1}{2}\rangle \\
|m_l = 0, m_s = \tfrac{1}{2}\rangle \\
|m_l = 1, m_s = \tfrac{1}{2}\rangle
\end{bmatrix}
=
\begin{bmatrix}
-\tfrac{2\Delta}{3} - Bk^2 & \tfrac{\sqrt{2}\Delta}{3} & 0 \\
\tfrac{\sqrt{2}\Delta}{3} & -\tfrac{\Delta}{3} - Ak^2 & 0 \\
0 & 0 & -Bk^2
\end{bmatrix}
\begin{bmatrix}
|m_l = 1, m_s = -\tfrac{1}{2}\rangle \\
|m_l = 0, m_s = \tfrac{1}{2}\rangle \\
|m_l = 1, m_s = \tfrac{1}{2}\rangle
\end{bmatrix}
$$

(3.28)

The coefficients $A$ and $B$ are related to the inverse effective-mass parameter. The $A$ coefficient appears only for the matrix element multiplying the $m_l = 0$ state, which corresponds to the $p_z$ orbital. The other basis states are linear combinations of $p_x$ and $p_y$, as specified by Eqs. (2.30a) and (2.30b).

At $k = 0$, we recover the eigenenergies and eigenfunctions from Kane's model discussed in Section 2.3. A look at the dispersions shows that, as desired, the HH band is no longer free-electron-like. Equation (3.16b) shows that the HH dispersion comes entirely from its interaction with the remote $p$-like bands, dominated by the antibonding $p$-like bands above the conduction band. This is why the HH mass along the [001] direction is indeed heavy: $\frac{m_{HH}[001]}{m} = \frac{\hbar^2}{2mB}$.

What about the LH and SO masses? The upper $2\times2$ matrix in Eq. (3.28) is cleanly decoupled from the HH states with the diagonal element $-Bk^2$, so we can simply solve for the eigenvalue of this matrix:

$$
\left(\frac{2\Delta}{3} + Bk^2 + E\right)\left(\frac{\Delta}{3} + Ak^2 + E\right) = \frac{2\Delta^2}{9}
$$

(3.29)

We seek the eigenenergies associated with $E_{LH}(k) = -ak^2$ and $E_{SO}(k) = -\Delta - bk^2$. Substituting $E_{LH}$ and $E_{SO}$ for $E$ in Eq. (3.29) and neglecting any terms higher than quadratic in $k$, we obtain

$$
a = \frac{2A + B}{3} = \frac{\hbar^2}{2m}\frac{m}{m_{LH}[001]}
$$

(3.30a)

$$
b = \frac{A + 2B}{3} = \frac{\hbar^2}{2m}\frac{m}{m_{SO}}
$$

(3.30b)

The hole masses are implicitly negative here.

In hindsight, it is obvious that the HH mass should be proportional to $1/M'$, since that band contains only $x$- and $y$-oriented orbitals. We also know from Eqs. (2.38) and (2.39) that $p_z$ contributes twice the total amount of the $p_x$ and $p_y$ orbitals to the LH band. On the other hand, the SO band contains an equal mixture of all three. If we take these fractions of the diagonal matrix elements in Eq. (3.27), we obtain exactly the result of Eq. (3.30).

How do these results compare to Kane's model as given by Eqs. (2.47) and (2.49)? Both $A$ and $B$ in Eq. (3.26) depend on $\frac{E_P}{E_g}$, which is correct for light holes, but not for SO holes according to Eq. (2.49). This difficulty is due to the simplifying assumption we made to obtain Eq. (3.28). Fortunately, all we need to do to correct Eq. (3.30b) is to replace $A$ given by Eq. (3.26a) with $A'$ defined as

$$A' \equiv L + \frac{\hbar^2}{2m} E_P \left( \frac{1}{E_g + \Delta} - \frac{1}{E_g} \right) \tag{3.31a}$$

$$\frac{\hbar^2}{2m} \frac{m}{m_{SO}} = \frac{L + 2B}{3} - \frac{\hbar^2}{2m} \frac{E_P}{3E_g} \frac{\Delta}{E_g + \Delta} \tag{3.31b}$$

Clearly, as $\Delta \to 0$ we retrieve the form in Eq. (3.30b). If these manipulations are not very satisfying, the extra work needed for a proper derivation is not really justified because the SO hole mass is not well known for many III–V materials (see Chapter 6) and rarely affects the performance of semiconductor quantum devices.

At this point, we can better appreciate the physical significance of $A$ and $B$, and by implication, $L$ and $M$. The hole masses along [001] near $k = 0$ can all be expressed simply in terms of these quantities (as well as $\frac{E_P}{E_g}$ and $\frac{E_P}{E_g + \Delta}$).

We have not yet examined the anisotropy of the hole masses. Since the off-diagonal terms are proportional to $C$, we expect them to affect the dispersions along directions other than [001]. To discover exactly how, we focus on wavevectors along the [111] direction with $k_x = k_y = k_z = \frac{k}{\sqrt{3}}$. The reduced Hamiltonian from Eq. (3.25) is then

$$\boldsymbol{H}_{3 \times 3}[111] \begin{bmatrix} |p_x\rangle \\ |p_y\rangle \\ |p_z\rangle \end{bmatrix} \approx -\frac{k^2}{3} \begin{bmatrix} A + 2B & C & C \\ C & A + 2B & C \\ C & C & A + 2B \end{bmatrix} \begin{bmatrix} |p_x\rangle \\ |p_y\rangle \\ |p_z\rangle \end{bmatrix} \tag{3.32}$$

Since the SO band has equal contributions from all three orbitals, its mass should be nearly independent of direction in $k$ space. This line of reasoning is analogous to that for the CB states that are dominated by spherically symmetric $s$ orbitals. Therefore, the SO mass along [111] is also given by Eq. (3.31b).

Amazingly, we can obtain the correct HH and LH masses along [111] by simply dropping one row and one column from Eq. (3.32)! These bands are degenerate at $k = 0$, which means that the spin–orbit splitting $\Delta$ should not affect the HH and LH dispersions near $k = 0$. Using the $2 \times 2$ Hamiltonian matrix

$$\boldsymbol{H}_{2 \times 2}[111] \approx -\frac{k^2}{3} \begin{bmatrix} A + 2B & C \\ C & A + 2B \end{bmatrix} \tag{3.33}$$

we find the two (spin-degenerate) eigenvalues:

$$E_{HH}[111] = -\frac{A+2B-C}{3}k^2 = -\frac{\hbar^2}{2m}\frac{m}{m_{HH}[111]}k^2 \tag{3.34a}$$

$$E_{LH}[111] = -\frac{A+2B+C}{3}k^2 = -\frac{\hbar^2}{2m}\frac{m}{m_{LH}[111]}k^2 \tag{3.34b}$$

The result in Eq. (3.34) is merely a guess at this point, but we will show that it is correct in Section 3.3. Alternatively, we can substitute Eq. (3.34a) into the eigenvalue formulation of Eq. (3.32) to confirm the solution. We will defer until Section 3.3 the task of finding the dispersions for directions other than [001] and [111].

We have shown that the hole effective masses can be expressed solely in terms of $A$, $B$, and $C$ (with $E_P$, $E_g$, and $\Delta$). The literature often cites an alternative set of three constants that make the hole masses along various directions much easier to memorize. These parameters bear the name of Joaquin Luttinger, who performed pioneering work on this topic in collaboration with Walter Kohn in the 1950s. The Luttinger parameters are

$$\gamma_1 \equiv \frac{2m}{\hbar^2}\frac{A+2B}{3} \tag{3.35a}$$

$$\gamma_2 \equiv \frac{2m}{\hbar^2}\frac{A-B}{6} \tag{3.35b}$$

$$\gamma_3 \equiv \frac{2m}{\hbar^2}\frac{C}{6} \tag{3.35c}$$

The HH and LH masses along the [001] and [111] directions can be expressed in terms of the Luttinger parameters:

$$m_{HH}[001] = \frac{m}{\gamma_1 - 2\gamma_2} \tag{3.36a}$$

$$m_{LH}[001] = \frac{m}{\gamma_1 + 2\gamma_2} \tag{3.36b}$$

$$m_{HH}[111] = \frac{m}{\gamma_1 - 2\gamma_3} \tag{3.36c}$$

$$m_{LH}[111] = \frac{m}{\gamma_1 + 2\gamma_3} \tag{3.36d}$$

The HH mass is still large, since it is inversely proportional to the difference between two quantities with similar values ($\gamma_1$ and $2\gamma_2$ or $2\gamma_3$). As we already saw, this is because the HH dispersion is due entirely to the second-order interaction with remote bands rather than the much stronger conduction–valence band coupling. Since it arises from interactions with directed $p$-like bands, the HH mass is strongly anisotropic.

Just as in Kane's model, the LH mass is much smaller than the HH mass, and its dependence on the energy gap is analogous (but not identical!) to the electron mass. Because $\gamma_3 - \gamma_2 \ll \gamma_2$, the LH mass is much more isotropic in $k$ space than the HH mass. Figure 3.3 illustrates the anisotropy of the HH and LH effective masses for three III–V semiconductors, as calculated for an arbitrary angle from the complete Hamiltonian of Eq. (3.24). Typically, the heavy holes are at least twice as heavy along [111] than along [100]. The LH mass looks like a slightly heavier and very slightly anisotropic cousin of the electron mass.

We obtain the SO mass in terms of the Luttinger parameters by setting $L = A$ in Eq. (3.31b) and find that it depends only on $\gamma_1$:

$$\frac{m}{m_{SO}} \approx \gamma_1 - \frac{E_P}{3E_g}\frac{\Delta}{E_g + \Delta} \tag{3.37}$$

As we mentioned in Chapter 2, the SO mass is intermediate between the HH and LH masses. Experimental values for the hole masses of various III–V materials are discussed in Chapters 6 and 7.

Because the common way to specify the valence-band parameters of zinc-blende semiconductors is in terms of the Luttinger parameters of Eq. (3.35), we would like to express the $L, M, N$ parameters of Eq. (3.15) in terms of $\gamma_1, \gamma_2$, and

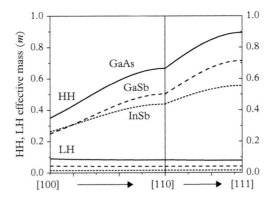

**Figure 3.3** *Heavy-hole and light-hole effective mass as a function of direction in k space for GaAs (solid curves), GaSb (dashed curves), and InSb (dotted curves). The directions are continuously varied from [100] to [110] and then to [111].*

$\gamma_3$ (as well as $E_P/E_g$). We can do this using Eqs. (3.26a), (3.26b), and (3.26c) as well as Eq. (3.35):

$$L = \frac{\hbar^2}{2m}\left(\gamma_1 + 4\gamma_2 - \frac{E_P}{E_g}\right) \tag{3.38a}$$

$$M = \frac{\hbar^2}{2m}(\gamma_1 - 2\gamma_2) \tag{3.38b}$$

$$N = \frac{\hbar^2}{2m}\left(6\gamma_3 - \frac{E_P}{E_g}\right) \tag{3.38c}$$

This completes the full 8-band model, although in many cases we can derive much physical insight from working with reduced forms of this model, as we will see in Section 3.3.

## 3.3   Simplified Forms of the Hamiltonian

The full 8-band Hamiltonian is given by Eq. (3.24). We have not written it in the total angular momentum basis $|j, m_j\rangle$ to avoid tedious transformations, but we should not give up on that basis just yet. For example, could we use it to eliminate putative "weak" interactions? In moderate- or wide-gap semiconductors, we might as well include the conduction–valence interaction to second order because the conduction band becomes quite "remote". This leaves a one-band CB problem and a 6×6 valence-band Hamiltonian that can be solved in either the $p_x, p_y, p_z$ basis or the $|j, m_j\rangle$ basis. To obtain that Hamiltonian, just take the lower right 3×3 block of Eq. (3.25) in place of Eq. (3.15). Unfortunately, the 6-band problem is still a bit tedious, so we look for a lower-order approximation.

We observe that the energies of interest are almost invariably much closer to the top of the valence band than to the split-off band. Then what if we disregard the SO states and write the matrix in terms of the $\psi_{HH}^+, \psi_{HH}^-, \psi_{LH}^+, \psi_{LH}^-$ basis from Eq. (2.38)? For convenience, the relevant part of that equation is

$$\psi_{HH}^+ = \left|\frac{3}{2},\frac{3}{2}\right\rangle = -\frac{1}{\sqrt{2}}(p_x + ip_y)\uparrow \tag{3.39a}$$

$$\psi_{HH}^- = \left|-\frac{3}{2},-\frac{3}{2}\right\rangle = \frac{1}{\sqrt{2}}(p_x - ip_y)\downarrow \tag{3.39b}$$

$$\psi_{LH}^+ = \left|\frac{3}{2},\frac{1}{2}\right\rangle = -\frac{1}{\sqrt{6}}(p_x + ip_y)\downarrow + \sqrt{\frac{2}{3}}\,p_z\uparrow \tag{3.39c}$$

$$\psi_{\overline{LH}} = \left|\frac{3}{2}, -\frac{1}{2}\right\rangle = \frac{1}{\sqrt{6}} \left(p_x - ip_y\right) \uparrow + \sqrt{\frac{2}{3}}\, p_z \downarrow \qquad (3.39\text{d})$$

To obtain the Hamiltonian in this basis, we transform the lower $3\times3$ block of Eq. (3.25) using Eq. (3.39) to transition from the $p_x, p_y, p_z$ basis. To do so, we denote the six distinct matrix elements in Eq. (3.25) as $H_{xx}, H_{yy}, H_{zz}, H_{xy} = H_{yx}, H_{xz} = H_{zx}, H_{yz} = H_{zy}$. Now we calculate the new matrix elements between the HH and LH states in terms of these six quantities by expanding the wavefunction products with the same spin using Eq. (3.39) (in fact, the matrix elements connecting different spins all vanish). To simplify the notation we use $|HH, \pm\rangle$ and $|LH, \pm\rangle$ rather than $\psi_{HH}^+, \psi_{HH}^-, \psi_{LH}^+, \psi_{LH}^-$, and obtain

$$\langle HH, +|\boldsymbol{H}|HH, +\rangle = \langle HH, -|\boldsymbol{H}|HH, -\rangle = \frac{1}{2}\left(H_{xx} + H_{yy}\right) \qquad (3.40\text{a})$$

$$\langle LH, +|\boldsymbol{H}|LH, +\rangle = \langle LH, -|\boldsymbol{H}|LH, -\rangle = \frac{1}{6}\left(H_{xx} + H_{yy}\right) + \frac{2}{3}H_{zz} \qquad (3.40\text{b})$$

$$\langle HH, +|\boldsymbol{H}|LH, +\rangle = -\langle HH - |\boldsymbol{H}|LH, -\rangle^* = \frac{1}{\sqrt{3}}\left(-H_{xz} + iH_{yz}\right) \qquad (3.40\text{c})$$

$$\langle HH, +|\boldsymbol{H}|LH, -\rangle = \langle LH, +|\boldsymbol{H}|HH, -\rangle = \frac{1}{2\sqrt{3}}\left(-H_{xx} + H_{yy}\right) + \frac{i}{\sqrt{3}}\,H_{xy} \qquad (3.40\text{d})$$

The rest follow from the Hermiticity requirement.

Now we can use Eq. (3.25) to write these matrix elements explicitly in terms of $A, B,$ and $C$:

$$\langle HH, +|\boldsymbol{H}|HH, +\rangle = \langle HH, -|\boldsymbol{H}|HH, -\rangle = -\frac{A}{2}\left(k_x^2 + k_y^2\right) - \frac{B}{2}\left(k_x^2 + k_y^2 + 2k_z^2\right) \qquad (3.41\text{a})$$

$$\langle LH, +|\boldsymbol{H}|LH, +\rangle = \langle LH, -|\boldsymbol{H}|LH, -\rangle$$
$$= -\frac{A}{6}\left(k_x^2 + k_y^2 + 4k_z^2\right) - \frac{B}{6}\left(5k_x^2 + 5k_y^2 + 2k_z^2\right) \qquad (3.41\text{b})$$

$$\langle HH, +|\boldsymbol{H}|LH, +\rangle = -\langle HH - |\boldsymbol{H}|LH, -\rangle^* = \frac{1}{\sqrt{3}}C\left(k_x - ik_y\right)k_z \qquad (3.41\text{c})$$

$$\langle HH, +|\boldsymbol{H}|LH, -\rangle = \langle LH, +|\boldsymbol{H}|HH, -\rangle = \frac{1}{2\sqrt{3}}(A - B)\left(k_x^2 - k_y^2\right) - \frac{i}{\sqrt{3}}Ck_xk_y \qquad (3.41\text{d})$$

Alternatively, we can use Eq. (3.35) to express Eq. (3.41) in terms of the Luttinger parameters. The full matrix equation becomes

$$
\boldsymbol{H}_{4\times4}
\begin{bmatrix}
|HH,+\rangle \\
|LH,+\rangle \\
|LH,-\rangle \\
|HH,-\rangle
\end{bmatrix}
= -
\begin{bmatrix}
P+Q & -S & R & 0 \\
-S^* & P-Q & 0 & R \\
R^* & 0 & P-Q & S \\
0 & R^* & S^* & P+Q
\end{bmatrix}
\begin{bmatrix}
|HH,+\rangle \\
|LH,+\rangle \\
|LH,-\rangle \\
|HH,-\rangle
\end{bmatrix}
\tag{3.42}
$$

where

$$
P = \frac{\hbar^2}{2m}\gamma_1\left(k_x^2 + k_y^2 + k_z^2\right) = \frac{\hbar^2}{2m}\gamma_1 k^2
\tag{3.43a}
$$

$$
Q = \frac{\hbar^2}{2m}\gamma_2\left(k_x^2 + k_y^2 - 2k_z^2\right)
\tag{3.43b}
$$

$$
R = \frac{\hbar^2}{2m}\left[-\sqrt{3}\gamma_2\left(k_x^2 - k_y^2\right) + i2\sqrt{3}\gamma_3 k_x k_y\right]
\tag{3.43c}
$$

$$
S = \frac{\hbar^2}{2m}2\sqrt{3}\gamma_3\left(k_x - ik_y\right)k_z
\tag{3.43d}
$$

To confirm that this Hamiltonian is consistent with what we already know, we calculate the HH and LH masses along the [001] direction and find that they indeed agree with Eq. (3.36). Equation (3.42) is the lowest-order form we have found yet, and is valid for computing the HH and LH states for any direction in $k$ space.

Is Eq. (3.42) really much simpler than Eq. (3.24)? After all, solving for the eigenvalues of a 4×4 matrix is not straightforward analytically. To reduce it to a simpler form, we use the fact that all the states are spin degenerate. Indeed, the characteristic polynomial of Eq. (3.42) factors into two identical quadratic equations. To streamline the calculation, we can subtract $P$ from the diagonal elements and add it back to the final form of $E_{HH,LH}$. The matrix has two blocks connected by the $R$ terms, where each block has the eigenvalues $\pm\sqrt{(Q^2 + |S|^2)}$. Assembling all the pieces leads to

$$
E_{HH,LH}(\boldsymbol{k}) = -P \pm \sqrt{Q^2 + |R|^2 + |S|^2},
\tag{3.44}
$$

where the plus sign is for heavy holes, and the minus sign for light holes. In terms of wavevector components and Luttinger parameters, we have

$$E_{HH,LH}(\mathbf{k}) = \frac{\hbar^2}{2m}\left[-\gamma_1 k^2 \pm 2\sqrt{\gamma_2^2\left(k_x^4 + k_y^4 + k_z^4\right) + \left(-\gamma_2^2 + 3\gamma_3^2\right)\left(k_x^2 k_y^2 + k_x^2 k_z^2 + k_y^2 k_z^2\right)}\right]$$

(3.45)

If we collect terms and use the relation $\gamma_3 - \gamma_2 \ll \gamma_2$ (see Chapter 6), we obtain

$$E_{HH,LH}(\mathbf{k}) = \frac{\hbar^2}{2m}\left[-\gamma_1 k^2 \pm 2\sqrt{\gamma_2^2 k^4 + 3\left(\gamma_3^2 - \gamma_2^2\right)\left(k_x^2 k_y^2 + k_x^2 k_z^2 + k_y^2 k_z^2\right)}\right]$$

$$\approx \frac{\hbar^2}{2m}\left[-\gamma_1 k^2 \pm 2\gamma_2\sqrt{k^4 + 6\frac{\gamma_3 - \gamma_2}{\gamma_2}\left(k_x^2 k_y^2 + k_x^2 k_z^2 + k_y^2 k_z^2\right)}\right]$$

$$\approx \frac{\hbar^2 k^2}{2m}\left[-\gamma_1 \pm 2\gamma_2 \pm 6\left(\gamma_3 - \gamma_2\right)\frac{k_x^2 k_y^2 + k_x^2 k_z^2 + k_y^2 k_z^2}{k^4}\right]$$

(3.46)

Using the first equality in Eq. (3.46) and the relation $k_x^2 = k_y^2 = k_z^2 = k^2/3$ for [111], we see that Eq. (3.36) for the HH and LH effective masses along [111] was indeed correct. In fact, Eq. (3.46) allows us to deduce the approximate masses for any arbitrary direction. For example, along [110] with $k_x^2 = k_y^2 = k^2/2$, and $k_z^2 = 0$, we obtain from the first equality in Eq. (3.46)

$$m_{HH}[110] \approx \frac{m}{\gamma_1 - \sqrt{\gamma_2^2 + 3\gamma_3^2}}$$

(3.47a)

$$m_{LH}[110] \approx \frac{m}{\gamma_1 + \sqrt{\gamma_2^2 + 3\gamma_3^2}}$$

(3.47b)

The resulting HH and LH masses are plotted in Fig. 3.3.

What if we want a single estimate for the HH mass that averages over the anisotropy, but does not break it down by direction in $k$ space? To find the "average" hole dispersion, we integrate over all possible directions, using polar angles $\theta$ and $\varphi$ ($k_x = k \sin\varphi \sin\theta$, $k_y = k \cos\varphi \sin\theta$, $k_z = k \cos\theta$), and assign equal weight to every direction. With $\gamma_3 - \gamma_2 \ll \gamma_2$ in Eq. (3.46), the polar integral yields

$$\overline{E}_{HH,LH} = \approx \frac{\hbar^2 k^2}{2m}$$

$$\times\left[-\gamma_1 \pm 2\gamma_2 \pm \frac{6\left(\gamma_3 - \gamma_2\right)}{4\pi}\int_0^{2\pi} d\varphi \int_{-1}^{1} d\left(\cos\theta\right) \sin^2\theta\left(\sin^2\theta\frac{1}{4}\sin^2 2\varphi + \cos^2\theta\right)\right]$$

$$= \frac{\hbar^2 k^2}{2m}\left[-\gamma_1 \pm \left(0.8\gamma_2 + 1.2\gamma_3\right)\right]$$

(3.48)

Because the Hamiltonian of Eq. (3.42) gives spin-degenerate eigenvalues for the HH and LH states, in the right basis it can be factored into two 2×2 matrices. Using the property that a linear combination of two spin-degenerate states is also an eigenstate, we rotate the matrix in Eq. (3.42) to take the form

$$
\mathbf{H}_{4\times4}
\begin{bmatrix} |HH,a\rangle \\ |LH,a\rangle \\ |LH,s\rangle \\ |HH,s\rangle \end{bmatrix}
= -
\begin{bmatrix}
P+Q & T & 0 & 0 \\
T^* & P-Q & 0 & 0 \\
0 & 0 & P-Q & T \\
0 & 0 & T^* & P+Q
\end{bmatrix}
\begin{bmatrix} |HH,a\rangle \\ |LH,a\rangle \\ |LH,s\rangle \\ |HH,s\rangle \end{bmatrix}
\tag{3.49}
$$

The new states are linear combinations of the old $| HH,\pm\rangle$ and $| LH,\pm\rangle$ states, with equal coefficients of magnitude $\frac{1}{\sqrt{2}}$ and direction-dependent phases. We require that the eigenvalues be the same as in Eq. (3.42). Since the determinant of the 2×2 blocks in Eq. (3.49) is $E_{HH,LH}(\mathbf{k}) = -P \pm \sqrt{Q^2 + |T|^2}$, the eigenvalue constraint translates into

$$
T = |R| - i|S| \tag{3.50}
$$

Of course, $T = |R| + i|S|$ works just as well because in both cases $|T|^2 = |R|^2 + |S|^2$. The savings in matrix size will become much more valuable once we need to solve for the eigenvalues in quantum structures.

We can make the expression for $T$ much simpler by using another approximation. If we rewrite the $R$ term in Eq. (3.43c) as

$$
R = -\frac{\hbar^2}{2m}\sqrt{3}\left[ \frac{\gamma_2+\gamma_3}{2}\left(k_x - ik_y\right)^2 + \frac{\gamma_2-\gamma_3}{2}\left(k_x + ik_y\right)^2 \right] \tag{3.51}
$$

and neglect the difference between $\gamma_2$ and $\gamma_3$ *in this expression only*, we can replace Eq. (3.50) with

$$
\tilde{T} = \frac{\hbar^2}{2m}\sqrt{3}\left[ \tilde{\gamma}k_t^2 - 2i\gamma_3 k_t k_z \right] \tag{3.52}
$$

where $\tilde{\gamma} \equiv \frac{\gamma_2+\gamma_3}{2}$ and $k_t = \sqrt{k_x^2 + k_y^2}$ is the in-plane wavevector perpendicular to the [001] axis. In this case, all the terms of this Hamiltonian except the first retain their dependences on $\gamma_2$ and $\gamma_3$ rather than $\tilde{\gamma}$. Since both $P$ and $Q$ in Eq. (3.43) depend only on $|k_t|$, the Hamiltonian of Eq. (3.49) becomes independent of direction in the $x$–$y$ plane. This *axial approximation* sometimes proves useful in the calculations of quantum structures. Figure 3.4 illustrates that in GaSb, the difference in valence-band structure between the axial approximation and the 8-band calculation is small.

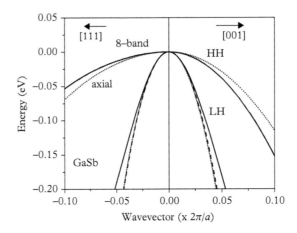

**Figure 3.4** *Heavy-hole and light-hole dispersions for GaSb, obtained using the full 8-band Hamiltonian (solid lines), the reduced 4-band Hamiltonian (dashed), and the axial approximation (dotted).*

## 3.4  Inclusion of Strain Effects

In real III–V structures, the underlying lattice can be perturbed by elastic deformations that affect the conduction- and valence-band structures. Most often, this happens when a thin epitaxial layer is grown on a substrate with a different lattice constant. Of course, the deformed lattice ceases to be invariant in all three dimensions, but in practice strain can be present even when the electron quantization is weak. Here we will describe how strain affects the bulk band structure and save the treatment of quantum effects for Part III.

Strain appears when the lattice is deformed by some outside force or *stress*, as shown in Fig. 3.5. Stress can be applied to any given face in three different directions, two in the plane of the face (*shear stress*) and one perpendicular to it (*axial stress*). Both the location of the force application and the direction of the force are vector quantities. This fact tells us to describe stress mathematically as a second-rank tensor with the nine components shown in Fig. 3.5. Fortunately, in most practical cases, only a few of these components are distinct and nonzero. In a (Cartesian) coordinate system, we write the stress tensor as a $3\times3$ matrix $\sigma_{ij}$, with $i,j = x,y,z$.

When a zinc-blende lattice is deformed, we define strain in terms of how the cube unit vectors are altered. A unit vector pointing along a given axis can be not only stretched or compressed along that axis, but also tilted toward the other axes. Therefore, a full description of the deformation requires that we specify three numbers for each unit vector, for a total of nine. This means that strain is also a second-rank tensor, written as the $3\times3$ matrix $\epsilon_{ij}$.

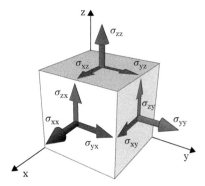

**Figure 3.5** *Schematic illustrating the application of a stress tensor $\sigma_{ij}$ to the crystal lattice.*

In fact, we can define strain more narrowly as the symmetric part of the general deformation. It turns out that the antisymmetric part vanishes because it represents a pure rotation. If the original unit vectors are $x_i$ (representing $x, y, z$) and the deformed unit vectors are $x_i'$, we can define the strain tensor as

$$\epsilon_{ij} = \frac{1}{2}\left(\frac{\partial x_i'}{\partial x_j} + \frac{\partial x_j'}{\partial x_i}\right) \tag{3.53}$$

Because the strain tensor is symmetric, $\epsilon_{ij} = \epsilon_{ji}$, there are only six independent components: $\epsilon_{xx}, \epsilon_{yy}, \epsilon_{zz}, \epsilon_{xy}, \epsilon_{xz}, \epsilon_{yz}$.

Since the stress and strain tensors are both second order, they are generally related by a fourth-order tensor:

$$\sigma_{kl} = \sum_{ij} c_{klij}\epsilon_{ij} \tag{3.54}$$

The $c_{klij}$ matrix simply tells us how stiff the material is (or how compliant, from the inverse of $c_{klij}$). The stiffer the material, with larger $c_{klij}$ components, the smaller the strain that results from a given stress. The components of $c_{klij}$ are the *elastic constants* of the material. In a cubic crystal, only three components of the $c_{klij}$ tensor are nonzero.

In practice, we are often interested in a layer of zinc-blende crystal grown epitaxially on top of a substrate with a different lattice constant, as illustrated in Fig. 3.6. If the strain is not to induce dislocations in the lattice, the layer must assume the substrate lattice constant in the plane of the structure. This happens only when the layer does not exceed some *critical thickness* that depends on the difference in lattice constants and other parameters.

What kind of deformation do we expect along the growth direction? We may guess that if the layer is forced to expand in the plane, it becomes compressed along

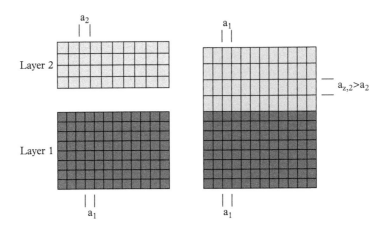

**Figure 3.6** *Schematic illustration of the strain that results when an epitaxial layer is grown on a substrate with a different lattice constant.*

the growth direction and vice versa. Certainly, this kind of deformation seems more likely to lead to a minimum of the elastic energy than if the deformation is the same along both axes. Most commonly, a zinc-blende crystal is grown along the [001] direction. Figure 3.6 shows that for this *biaxial* strain, the unit cube in the original lattice is transformed to a rectangular prism, with the right angles preserved. This means that there is no shear component to the stress and strain, and the off-diagonal elements of the $\epsilon_{ij}$ matrix vanish. We can write the typical equation for a non-vanishing stress component:

$$\sigma_{xx} = c_{11}\epsilon_{xx} + c_{12}\epsilon_{yy} + c_{12}\epsilon_{zz} \tag{3.55}$$

where $c_{11}$ and $c_{12}$ denote two of the surviving components of $c_{klij}$. Since any in-plane direction is as good as any other when a zinc-blende crystal is grown along the [001] axis, we know that $\epsilon_{xx} = \epsilon_{yy}$.

We now calculate the strain components in Eq. (3.55) using the fact that no stress is applied along the growth direction:

$$\sigma_{zz} = c_{12}\epsilon_{xx} + c_{12}\epsilon_{yy} + c_{11}\epsilon_{zz} = 2c_{12}\epsilon_{xx} + c_{11}\epsilon_{zz} = 0 \tag{3.56}$$

This gives us a relation between the in-plane and out-of-plane strain:

$$\epsilon_{zz} = -\frac{2c_{12}}{c_{11}}\epsilon_{xx} \tag{3.57}$$

The two strain components do in fact have opposite signs! Now we need to evaluate $\epsilon_{xx}$. If the original lattice constant of the epitaxial layer is $a$, the deformed lattice constant along an in-plane axis is $a(1 + \epsilon_{xx})$, which must equal the lattice constant $a_0$ of the substrate. Therefore, the in-plane strain is

$$\epsilon_{xx} = \epsilon_{yy} = \frac{a_0 - a}{a} \qquad (3.58)$$

By the same token, the volume of the deformed epitaxial lattice cell is the product of the deformed lattice constants:

$$V + \Delta V = a^3 (1 + \epsilon_{xx}) (1 + \epsilon_{yy}) (1 + \epsilon_{zz}) \qquad (3.59)$$

Since the deformation is small (or the crystal will not grow), the relative volume change is simply the sum of the strain components:

$$\frac{\Delta V}{V} = \epsilon_{xx} + \epsilon_{yy} + \epsilon_{zz} = 2\epsilon_{xx} + \epsilon_{zz} \qquad (3.60)$$

Wait, what happened to the idea of minimizing the elastic energy to find the equilibrium configuration? Actually, this is what we did implicitly in obtaining Eqs. (3.56) and (3.57). The mismatched substrate fixes the in-plane strain to the value in Eq. (3.58). Therefore, the only adjustable parameter is the out-plane strain $\epsilon_{zz}$. Using $\frac{\partial U}{\partial \epsilon_{zz}} = \sigma_{zz} = 0$ to minimize the elastic energy then leads exactly to Eq. (3.56). Throw in the little bit of insight needed to eliminate the shear components from the start, and the minimization produces Eq. (3.57) even more directly.

Another scenario of interest is a bulk crystal compressed (or stretched) along the [001] direction, which occurs when the crystal is compressed in an anvil cell. Clearly, the out-of-plane stress is then fixed externally, and the in-plane stress is zero. We still have $\epsilon_{xx} = \epsilon_{yy} \neq 0$, since the lattice is free to deform in that direction to accommodate the out-of-plane stress. Minimizing the elastic energy, this time by setting $\frac{\partial U}{\partial \epsilon_{xx}} = \sigma_{xx} = 0$, then leads to

$$\epsilon_{zz} = -\frac{c_{11} + c_{12}}{c_{11}} \epsilon_{xx} \qquad (3.61)$$

What if we grow on a substrate with a different orientation, say (111)? Our conclusions about the strain along certain directions still hold, but these directions no longer coincide with the crystal axes. We can remedy this by taking the simple strain matrix of Eqs. (3.57) and (3.58), and rotating it in real space by the polar angles $\theta = \cot^{-1}\left(\frac{1}{\sqrt{2}}\right)$ and $\varphi = \sin^{-1}\left(\frac{1}{\sqrt{2}}\right) = 45°$. The stiffness and stress matrices are rotated as well, by the same angles. For the (111) orientation, this leads to three identical diagonal strain components, which are now mixtures of the in-plane and out-of-plane strain, as well as identical, but non-zero, shear components. The third non-zero component of the stiffness matrix $c_{44}$ now enters the formulas.

The resulting strain matrix can be found in the reference materials at the end of the chapter.

We can now calculate the strain components $\epsilon_{ij}$ in a few practical situations. But we have not yet answered the question of how strain affects the 8-band $\mathbf{k} \cdot \mathbf{p}$ Hamiltonian (or any other Hamiltonian, for that matter). To begin with, a small deformation of the crystal should not have a large effect on the "kinetic-energy" wavevector-dependent terms, but is much more likely to modify the "potential-energy" term. For small strain, the deformation shifts the band-edge energies by an amount proportional to the strain components $\epsilon_{xx}$ and $\epsilon_{zz}$. These are dimensionless, so the proportionality constants, which are called *deformation potentials*, have the units of energy. Therefore, the deformation potentials are simply coefficients of the Taylor expansion of the band-edge energies as a function of strain. In practice, the deformation potentials are fitted to measurements of the band structure under strain.

How many deformation potentials do we need to account for the full crystal symmetry? You may be surprised to learn we have already solved that problem! The crystal symmetry is already built into the Hamiltonian (Eq. (3.25) in the *s–p* basis or Eqs. (3.42) and (3.43) in the total-angular-momentum basis). It enters via the second-order wavevector terms $k_i k_j \equiv \varepsilon_{ij}$. Therefore, if we substitute $\epsilon_{ij}$ for $\varepsilon_{ij}$: $k_x^2 \rightarrow \epsilon_{xx}$, $k_x k_y \rightarrow \epsilon_{xy}$, etc., the symmetry is preserved as long as the relative phases and magnitudes of the terms in the same matrix element are unaltered. For example, $k_x^2 + k_y^2 + k_z^2 \rightarrow \epsilon_{xx} + \epsilon_{yy} + \epsilon_{zz}$, $k_x^2 + k_y^2 - 2k_z^2 \rightarrow \epsilon_{xx} + \epsilon_{yy} - 2\epsilon_{zz}$, $k_x k_z + i k_y k_z \rightarrow \epsilon_{xz} + i\epsilon_{yz}$, etc.

As long as the states are still spin-degenerate, the form of the deformation-potential matrix is analogous to Eq. (3.25), apart from the $E_g$ term. Equation (3.25) requires four different parameters $A$, $B$, $C$, and $D$, which are here replaced by four deformation potentials. Therefore, the $4 \times 4$ blocks of the deformation-potential matrix have the form

$$
H_\epsilon \approx \begin{bmatrix} a_c\left(\epsilon_{xx} + \epsilon_{yy} + \epsilon_{zz}\right) & 0 & 0 & 0 \\ 0 & l_\epsilon \epsilon_{xx} + m_\epsilon\left(\epsilon_{yy} + \epsilon_{zz}\right) & n_\epsilon \epsilon_{xy} & n_\epsilon \epsilon_{xz} \\ 0 & n_\epsilon \epsilon_{xy} & l_\epsilon \epsilon_{yy} + m_\epsilon\left(\epsilon_{xx} + \epsilon_{zz}\right) & n_\epsilon \epsilon_{yz} \\ 0 & n_\epsilon \epsilon_{xz} & n_\epsilon \epsilon_{yz} & l_\epsilon \epsilon_{zz} + m_\epsilon\left(\epsilon_{xx} + \epsilon_{yy}\right) \end{bmatrix}
$$

$$(3.62)$$

where $a_c$, $l_\epsilon$, $m_\epsilon$, and $n_\epsilon$ are the deformation potentials.

Of course this procedure does not tell us the magnitudes of the deformation potentials, any more than the effective masses in a given material can be derived from the general form of the $\mathbf{k} \cdot \mathbf{p}$ Hamiltonian. However, we can infer which terms are responsible for certain effects. For example, the $a_c$ term shifts the position of the conduction band, which responds only to the overall volume deformation owing to its *s*-like symmetry near $k = 0$. We can describe the valence band using the 4-band form of Eq. (3.42), with $P$, $Q$, $R$, and $S$ replaced by

$$P_\epsilon = \frac{\hbar^2}{2m}\gamma_1\left(k_x^2 + k_y^2 + k_z^2\right) - a_v\left(\epsilon_{xx} + \epsilon_{yy} + \epsilon_{zz}\right) \tag{3.63a}$$

$$Q_\epsilon = \frac{\hbar^2}{2m}\gamma_2\left(k_x^2 + k_y^2 - 2k_z^2\right) - \frac{b}{2}\left(\epsilon_{xx} + \epsilon_{yy} - 2\epsilon_{zz}\right) \tag{3.63b}$$

$$R_\epsilon = \frac{\hbar^2}{2m}\left[-\sqrt{3}\gamma_2\left(k_x^2 - k_y^2\right) + i2\sqrt{3}\gamma_3 k_x k_y\right] + \frac{\sqrt{3}}{2}b\left(\epsilon_{xx} - \epsilon_{yy}\right) - id\epsilon_{xy} \tag{3.63c}$$

$$S_\epsilon = \frac{\hbar^2}{2m}2\sqrt{3}\gamma_3\left(k_x - ik_y\right)k_z - d\left(\epsilon_{xz} - i\epsilon_{yz}\right) \tag{3.63d}$$

As a result, the strain-related parts of $R_\epsilon$ and $S_\epsilon$ vanish for (100)-oriented substrates. The transformed deformation potentials are defined as

$$a_v = \frac{l_\epsilon + 2m_\epsilon}{3} \tag{3.64a}$$

$$b = \frac{l_\epsilon - m_\epsilon}{3} \tag{3.64b}$$

$$d = \frac{n_\epsilon}{\sqrt{3}} \tag{3.64c}$$

Even though it seems too easy to derive Eqs. (3.62) and (3.63) in this manner, the results are rigorously correct! The deformation potential $a_v$ determines the shift of the valence-band edge with a volume deformation, as does $a_c$ for the conduction band. Equation (3.63) shows that for a crystal grown along [001], the splitting between the HH and LH states at $k = 0$ is proportional to $b$, and the terms with $d$ vanish. That is, biaxial strain *lifts the degeneracy* between the heavy and light holes, because the growth and in-plane directions now look different. We can use Eq. (3.57) to calculate the splitting energy:

$$\Delta E_{HH,LH} = 2b\frac{c_{11} + 2c_{12}}{c_{11}}\epsilon_{xx} \tag{3.65}$$

Figure 3.7 illustrates the impact of compressive $(a > a_0, \epsilon_{xx} < 0)$ and tensile $(a < a_0, \epsilon_{xx} > 0)$ biaxial strain on the valence-band structure of GaSb. Since the deformation potentials are negative, compressive strain makes the energy gap wider and moves the HH band above the LH band. On the other hand, the gap becomes narrower, with LH above HH, in the presence of tensile strain. The so-called *hydrostatic* strain, which deforms the volume equally in all directions (i.e., $\epsilon_{xx} = \epsilon_{yy} = \epsilon_{zz}$), shifts the energy gap, but does not split the HH and LH states at $k = 0$. Typically, the electron and hole effective masses near the band edges change in the same direction as the energy gap.

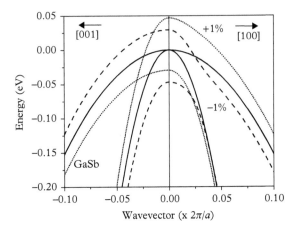

**Figure 3.7** *Effect of biaxial strain on the heavy- and light-hole dispersions for GaSb, using the full 8-band Hamiltonian. Results are shown for no strain (solid), 1% compressive strain (dashed), and 1% tensile strain (dotted). Note that strain lifts the degeneracy of the HH and LH bands at $k = 0$.*

The deformation due to biaxial strain breaks the cubic symmetry of a zinc-blende crystal. It follows that a non-cubic wurtzite crystal, which has one special axis (the $c$ axis), should not have degenerate valence bands at $k = 0$ (apart from spin degeneracy). This attribute will prove useful in Section 3.5.

## 3.5   Bulk Band Structure of Wurtzite Semiconductors

Several important wide-gap III–V semiconductors crystallize in the wurtzite crystal structure rather than zinc-blende. These include GaN, AlN, InN, and their alloys under most growth conditions, but not the very dilute nitride alloys composed mostly of other group V semiconductors. We will see that the wurtzite III–V materials can be described in a framework quite similar to that developed in Section 3.3 for the zinc-blende crystals. The crucial distinction is that one axis ($z$ in our coordinate system) is now different from the $x$ and $y$ in-plane axes. Figure 3.8 shows the zinc-blende and wurtzite crystals side by side. Because of the special nature of the $z$ axis, matrix elements involving the $|p_z\rangle$ orbital now differ from those connecting the $|p_x\rangle$ and $|p_y\rangle$ orbitals.

Most wurtzite materials (with the notable exception of InN) have wide energy gaps (see Chapter 6). Therefore, from the outset we will treat the conduction band as "remote" when describing the valence bands, as we did in arriving at Eq. (3.25). The conduction band can be described by the parabolic model for energies close to the band edge and, more generally, by an energy-dependent effective mass in Eq. (2.21). As in the other materials with moderate and wide energy gaps, nonparabolic effects are of concern only when the energy above the

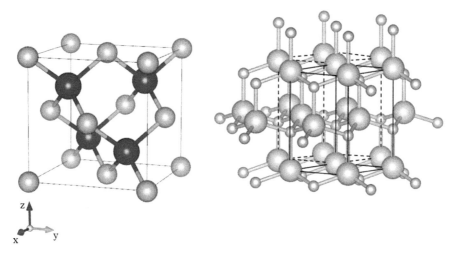

**Figure 3.8** *Atomic arrangement in the zinc-blende and wurtzite crystals.*

conduction-band edge is a large fraction of the energy gap. Therefore, we continue with the derivation of a 6-band Hamiltonian for the valence band, which cannot be reduced to 4 bands because of weak spin–orbit coupling in the nitrides. It is straightforward to extend the 6-band model below to 8 bands by adding the *s*-like band and the $\pm i\frac{\hbar}{m}k_{x,y,z}p_{cv}$ off-diagonal terms that couple it to the $|p_x\rangle$, $|p_y\rangle$, and $|p_z\rangle$ bands, by analogy with Eq. (3.15), but we will not show the explicit form of such a Hamiltonian here.

As before, we can separate the Hamiltonian into a sum of **k·p** and spin–orbit parts, with the **k·p** Hamiltonian identical for up and down spins, If we recognize the special nature of the $z$ axis, we "generalize" the lower $3\times3$ sub-block in Eq. (3.25) as

$$\boldsymbol{H}_{3\times3}\begin{bmatrix} |p_x\rangle \\ |p_y\rangle \\ |p_z\rangle \end{bmatrix}$$

$$= -\begin{bmatrix} L_1 k_x^2 + M_1 k_y^2 + M_2 k_z^2 & N_1 k_x k_y & N_2 k_x k_z \\ N_1 k_x k_y & M_1 k_x^2 + L_1 k_y^2 + M_2 k_z^2 & N_2 k_y k_z \\ N_2 k_x k_z & N_2 k_y k_z & M_3\left(k_x^2 + k_y^2\right) + L_2 k_z^2 \end{bmatrix}\begin{bmatrix} |p_x\rangle \\ |p_y\rangle \\ |p_z\rangle \end{bmatrix}$$

$$\text{(3.66)}$$

The lower symmetry certainly multiplies the number of parameters! Instead of three in a zinc-blende crystal, there are now seven. $L_1$ and $L_2$ (as well as $N_1$ and $N_2$) differ because one of them involves $|p_z\rangle$. $M_1$ and $M_2$ are different because the $k_z^2$ term includes coupling to $p_z$-like remote bands, as discussed in Section 3.2. To top it off, $M_3$ is distinct from both $M_1$ and $M_2$ because its initial and final states are $p_z$-like.

Can we use the symmetry of the wurtzite lattice to reduce the number of independent parameters in Eq. (3.66)? For example, if we rotate the hexagonal lattice by 60° in the $x-y$ plane it is unaltered, so the transformed and original Hamiltonians should be equal. Since the rotation modifies only the upper 2×2 sub-block, we expect that any dependent parameters should involve $L_1$, $M_1$, and $N_1$, with $M_2$ held fixed. Denoting the $|p_x\rangle$ and $|p_y\rangle$ orbitals as $|x\rangle$ and $|y\rangle$, the transformation rules for an ordinary rotation imply that

$$|x'\rangle = \cos 60°\,|x\rangle + \sin 60°\,|y\rangle = \frac{1}{2}\,|x\rangle \; + \frac{\sqrt{3}}{2}\,|y\rangle \qquad (3.67a)$$

$$|y'\rangle = -\sin 60°\,|x\rangle + \cos 60°\,|y\rangle = -\frac{\sqrt{3}}{2}\,|x\rangle + \frac{1}{2}\,|y\rangle \qquad (3.67b)$$

The momentum components $p_x$ and $p_y$ are transformed in exactly the same way:

$$p'_x = \cos 60°\,p_x + \sin 60°\,p_y = \frac{1}{2}p_x + \frac{\sqrt{3}}{2}\,p_y \qquad (3.67c)$$

$$p'_y = -\sin 60°\,p_x + \cos 60°\,p_y = -\frac{\sqrt{3}}{2}p_x + \frac{1}{2}\,p_y \qquad (3.67d)$$

The second-order perturbation formalism of Section 3.1 defines $L_1$, $M_1$, and $N_1$ as

$$L_1 = \langle x|p_x\,P\,p_x|x\rangle = \langle y|p_y\,P\,p_y|y\rangle \qquad (3.68a)$$

$$M_1 = \langle x|p_y\,P\,p_y|x\rangle = \langle y|p_x\,P\,p_x|y\rangle \qquad (3.68b)$$

$$N_1 = \langle x|p_x\,P\,p_y|y\rangle + \langle x|p_y\,P\,p_x|y\rangle \qquad (3.68c)$$

$$N_1^* = \langle y|p_y\,P\,p_x|x\rangle + \langle y|p_x\,P\,p_y|x\rangle = N_1 \qquad (3.68d)$$

For compactness, we denoted the interior sums over remote states as $\sum_{r=s,p,d} A_{r,ij}\,p_i$ $|r\rangle\langle r|p_j \equiv p_i\,P\,p_j$.

Rotation by 60° in the $x-y$ plane cannot affect the values of $L_1$, $M_1$, and $N_1$, so we can add primes to $x, y, p_x, p_y$ in Eq. (3.68) without changing anything else. The sums over the remote states are also unaltered. Now if we substitute Eq. (3.67) into the first equality of Eq. (3.68a):

$$L_1 = \langle x' | \, p_x' \, P \, p_x' | x' \rangle = \left\langle x' \left| \left( \frac{1}{2} p_x + \frac{\sqrt{3}}{2} \, p_y \right) P \left( \frac{1}{2} p_x + \frac{\sqrt{3}}{2} \, p_y \right) \right| x' \right\rangle$$

$$= \left( \frac{1}{2} \langle x | + \frac{\sqrt{3}}{2} \langle y | \right) \left( \frac{1}{2} p_x + \frac{\sqrt{3}}{2} \, p_y \right) P \left( \frac{1}{2} p_x + \frac{\sqrt{3}}{2} \, p_y \right) \left( \frac{1}{2} | x \rangle + \frac{\sqrt{3}}{2} \, | y \rangle \right)$$

$$(3.69)$$

Using the definitions in Eq. (3.68), equating the left and right sides of Eq. (3.69) and a similar transformation for $M_1$ yields

$$L_1 = \frac{5}{8} L_1 + \frac{3}{8} M_1 + \frac{3}{16} \left( N_1 + N_1^* \right) \tag{3.70a}$$

$$M_1 = \frac{5}{8} M_1 + \frac{3}{8} L_1 - \frac{3}{16} \left( N_1 + N_1^* \right) \tag{3.70b}$$

The final equality in Eq. (3.68d) tells us that $N_1$ is real. If we subtract Eq. (3.70a) from Eq. (3.70b), we have

$$L_1 - M_1 = \frac{1}{4} \left( L_1 - M_1 \right) + \frac{3}{4} N_1 \tag{3.71}$$

which simplifies to

$$L_1 - M_1 = N_1 \tag{3.72}$$

This leaves six independent parameters $(L_1 + M_1, \ L_1 - M_1 = N_1, \ L_2, \ M_2, \ M_3, \ N_2)$, which is a slight improvement. We cannot use the hexagonal symmetry to reduce the number further because the other parameters involve the $z$ component of the momentum or the $|p_z\rangle$ orbital.

Now we can include the spin–orbit interaction. If it were not for the special nature of the $z$ axis, we could have again used Eq. (2.36). But perhaps we can find a way to fix Eq. (2.36) so that it works for the wurtzite crystal. To see if this is possible, we transform the $|p_x, p_y, p_z\rangle$ basis into the $|m_l, m_s\rangle$ basis. This transformation is similar to Eq. (2.30):

$$\left| m_l = 1, m_s = \frac{1}{2} \right\rangle = -\frac{1}{\sqrt{2}} \left( |p_x\rangle + i \, |p_y\rangle \right) \uparrow \tag{3.73a}$$

$$\left| m_l = -1, m_s = -\frac{1}{2} \right\rangle = \frac{1}{\sqrt{2}} \left( |p_x\rangle - i \, |p_y\rangle \right) \downarrow \tag{3.73b}$$

$$\left| m_l = 0, m_s = \frac{1}{2} \right\rangle = |p_z\rangle \uparrow \qquad (3.73\text{c})$$

$$\left| m_l = -1, m_s = \frac{1}{2} \right\rangle = \frac{1}{\sqrt{2}} \left( |p_x\rangle - i|p_y\rangle \right) \uparrow \qquad (3.73\text{d})$$

$$\left| m_l = 1, m_s = -\frac{1}{2} \right\rangle = -\frac{1}{\sqrt{2}} \left( |p_x\rangle + i|p_y\rangle \right) \downarrow \qquad (3.73\text{e})$$

$$\left| m_l = 0, m_s = -\frac{1}{2} \right\rangle = |p_z\rangle \downarrow \qquad (3.73\text{f})$$

The special nature of the $z$ axis means that the valence-band edge associated with the $p_z$ orbital in Eq. (3.73c) can differ from those in Eqs. (3.73a) and (3.73b). So it seems that the only needed modification of Eq. (2.36) is to insert the potential energy splitting between the states composed of the $p_x, p_y$ orbitals and the $p_z$ orbitals into the diagonal matrix elements. This splitting, known as the *crystal-field potential*, is denoted $\Delta_c$. Following convention, we also set $\Delta_s \equiv \Delta/3$. Referring back to Eq. (2.36) and using these definitions, we obtain

$$\mathbf{H}_{3\times 3,l,s}\,(H_0 + \boldsymbol{L}\cdot\boldsymbol{S}) \begin{bmatrix} |1,-\frac{1}{2}\rangle \\ |0,\frac{1}{2}\rangle \\ |1,\frac{1}{2}\rangle \end{bmatrix} = \begin{bmatrix} \Delta_c - \Delta_s & \sqrt{2}\Delta_s & 0 \\ \sqrt{2}\Delta_s & 0 & 0 \\ 0 & 0 & \Delta_c + \Delta_s \end{bmatrix} \begin{bmatrix} |1,-\frac{1}{2}\rangle \\ |0,\frac{1}{2}\rangle \\ |1,\frac{1}{2}\rangle \end{bmatrix}$$
$$(3.74)$$

It appears that a wurtzite crystal has three non-degenerate valence-band states at $k = 0$. Conventionally, these are designated as the heavy holes (HH), light holes (LH), and crystal holes (CH), although these names do not mean the same thing as in zinc-blende materials. We add a reference energy $E_v$ to the diagonal elements and find the eigenvalues of Eq. (3.74):

$$E_{HH}\,(k = 0) = E_v + \Delta_c + \Delta_s \qquad (3.75\text{a})$$

$$E_{LH}\,(k = 0) = E_v + \frac{\Delta_c - \Delta_s}{2} + \sqrt{\frac{(\Delta_c - \Delta_s)^2}{4} + 2\Delta_s^2} \qquad (3.75\text{b})$$

$$E_{CH}\,(k = 0) = E_v + \frac{\Delta_c - \Delta_s}{2} - \sqrt{\frac{(\Delta_c - \Delta_s)^2}{4} + 2\Delta_s^2} \qquad (3.75\text{c})$$

If the HH band is uppermost, we could choose $E_v \equiv -\Delta_c - \Delta_s$, so that $E_{HH}\,(k = 0) = 0$. We can also find the eigenfunctions for the three bands at $k = 0$. The HH wavefunction is the same as for a zinc-blende material. On the other

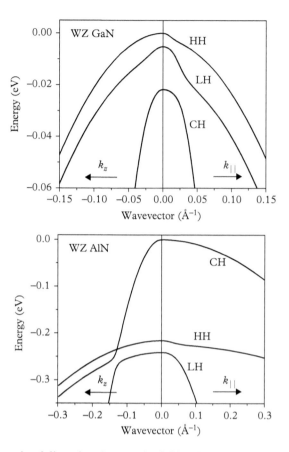

**Figure 3.9** *Valence-band dispersions for wurtzite GaN and AlN crystals.*

hand, both LH and CH are mixtures of the $|1, -\frac{1}{2}\rangle$ and $|0, \frac{1}{2}\rangle$ (or $|1, -\frac{1}{2}\rangle$ and $|0, \frac{1}{2}\rangle$) states.

The lower symmetry of the wurtzite crystal removes the double degeneracy of the HH and LH states that occurs in a zinc-blende material. If $\Delta_c \sim \Delta_s$ or $\Delta_c \gg \Delta_s > 0$, as in GaN and InN, HH is the uppermost valence band, followed by the LH band, and then the CH band. However, the CH band can rise to the top when $\Delta_c < 0$ and $|\Delta_c| \gg \Delta_s$ as in AlN. When $|\Delta_c| \gg \Delta_s$, the LH band has the $m_j = \pm\frac{1}{2}$ character ($|1, -\frac{1}{2}\rangle$ or $|-1, \frac{1}{2}\rangle$) at $k = 0$, while the CH band is primarily $p_z$-like. Of course, as in zinc-blende materials, the bands mix away from the zone center. Figure 3.9 shows the valence-band structures of GaN and AlN, with each band identified by its state at $k = 0$.

How do we actually derive the dispersions in Fig. 3.9? First, we need to transform Eq. (3.66) to the $|m_l, m_s\rangle$ basis. The diagonal terms connecting the states in Eqs. (3.73a) and (3.73b) are simply averages of the terms for the $|p_x\rangle$ and $|p_y\rangle$

orbitals in Eq. (3.66). Below, we will denote this term as $F_w$. The diagonal term between the $|p_z\rangle$ orbitals in Eq. (3.73c) is unchanged and is designated $G_w$. The off-diagonal terms, which we will call $K_w$ and $H_w$, have essentially the same form as $R$ and $S$ in Eqs. (3.43c) and (3.43d), apart from an unimportant phase factor:

$$
H
\begin{bmatrix}
|1,\frac{1}{2}\rangle \\
|-1,\frac{1}{2}\rangle \\
|0,\frac{1}{2}\rangle \\
|1,-\frac{1}{2}\rangle \\
|-1,-\frac{1}{2}\rangle \\
|0,-\frac{1}{2}\rangle
\end{bmatrix}
=
\begin{bmatrix}
F_w+\Delta_s & K_w & -H_w & 0 & 0 & 0 \\
K_w^* & F_w-\Delta_s & H_w^* & 0 & 0 & \sqrt{2}\Delta_s \\
-H_w^* & H_w & G_w & 0 & \sqrt{2}\Delta_s & 0 \\
0 & 0 & 0 & F_w+\Delta_s & K_w^* & H_w^* \\
0 & 0 & \sqrt{2}\Delta_s & K_w & F_w-\Delta_s & -H_w \\
0 & \sqrt{2}\Delta_s & 0 & H_w & -H_w^* & G_w
\end{bmatrix}
\begin{bmatrix}
|1,\frac{1}{2}\rangle \\
|-1,\frac{1}{2}\rangle \\
|0,\frac{1}{2}\rangle \\
|1,-\frac{1}{2}\rangle \\
|-1,-\frac{1}{2}\rangle \\
|0,-\frac{1}{2}\rangle
\end{bmatrix}
\tag{3.76}
$$

where

$$
F_w = \Delta_c + \left(\frac{L_1+M_1}{2}\right)\left(k_x^2+k_y^2\right) + M_2 k_z^2
\tag{3.77a}
$$

$$
G_w = M_3\left(k_x^2+k_y^2\right) + L_2 k_z^2
\tag{3.77b}
$$

$$
K_w = \frac{1}{2}\left[-\left(L_1-M_1\right)\left(k_x^2-k_y^2\right)+i2N_1 k_x k_y\right] = -\frac{1}{2}N_1\left(k_x-ik_y\right)^2
\tag{3.77c}
$$

$$
H_w = \frac{1}{\sqrt{2}}N_2\left(k_x-ik_y\right)k_z
\tag{3.77d}
$$

Notice that in contrast to Eq. (3.42), for example, there is no minus sign in front of Eq. (3.76).

The six independent parameters are often expressed as

$$
\frac{\hbar^2}{2m}A_1 = L_2
\tag{3.78a}
$$

$$
\frac{\hbar^2}{2m}A_2 = M_3
\tag{3.78b}
$$

$$
\frac{\hbar^2}{2m}A_3 = M_2 - L_2
\tag{3.78c}
$$

$$\frac{\hbar^2}{2m}A_4 = \frac{L_1 + M_1}{2} - M_3 \tag{3.78d}$$

$$\frac{\hbar^2}{2m}A_5 = \frac{N_1}{2} \tag{3.78e}$$

$$\frac{\hbar^2}{2m}A_6 = \frac{N_2}{\sqrt{2}} \tag{3.78f}$$

We can use the definitions in Eq. (3.78) to rewrite Eq. (3.77) in a better-known form:

$$F_w = \Delta_c + \frac{\hbar^2}{2m}\left(A_2 + A_4\right)\left(k_x^2 + k_y^2\right) + \frac{\hbar^2}{2m}\left(A_1 + A_3\right)k_z^2 \tag{3.79a}$$

$$G_w = \Delta_c + \frac{\hbar^2}{2m}A_2\left(k_x^2 + k_y^2\right) + \frac{\hbar^2}{2m}A_1 k_z^2 \tag{3.79b}$$

$$K_w = -\frac{\hbar^2}{2m}A_5\left(k_x - ik_y\right)^2 \tag{3.79c}$$

$$H_w = \frac{\hbar^2}{2m}A_6\left(k_x - ik_y\right)k_z \tag{3.79d}$$

By analogy with zinc-blende materials, we can use Eqs. (3.76) and (3.79) to recast four of these hole effective-mass parameters in terms of hole effective masses for different bands along different wavevector directions. The result, given in Appendix B, is used for interpolating the parameters of wurtzite alloys in Chapters 6 and 7.

How do we treat the effect of strain in a wurtzite crystal? Such crystals have five independent components of the stiffness tensor (elastic constants) rather than the three in a zinc-blende material, but only two of them are needed if growth is along the $c$ axis. The in-plane strain is then the same as in the zinc-blende case. As in the derivation in Section 3.4, we set the growth-direction strain to zero:

$$c_{13}\epsilon_{xx} + c_{13}\epsilon_{yy} + c_{33}\epsilon_{zz} = 2c_{13}\epsilon_{xx} + c_{33}\epsilon_{zz} = 0 \tag{3.80}$$

The out-of-plane strain then becomes:

$$\epsilon_{zz} = -\frac{2c_{13}}{c_{33}}\epsilon_{xx} \tag{3.81}$$

These are directly analogous to Eqs. (3.56) and (3.57).

Six independent valence-band parameters mean that six deformation potentials should be enough to describe the effects of strain on the valence band. The conduction band requires two ($a_1$ and $a_2$) rather than one for zinc-blende materials because of the reduced symmetry.

Once again, we mimic the symmetry of the $k$-dependent terms to obtain the modified forms of Eq. (3.79):

$$F_\epsilon = \Delta_c + \frac{\hbar^2}{2m}\left(A_2 + A_4\right)\left(k_x^2 + k_y^2\right) + \frac{\hbar^2}{2m}\left(A_1 + A_3\right)k_z^2$$

$$+ \left(D_2 + D_4\right)\left(\epsilon_{xx} + \epsilon_{yy}\right) + \left(D_1 + D_3\right)\epsilon_{zz} \tag{3.82a}$$

$$G_\epsilon = \Delta_c + \frac{\hbar^2}{2m}A_2\left(k_x^2 + k_y^2\right) + \frac{\hbar^2}{2m}A_1 k_z^2 + D_2\left(\epsilon_{xx} + \epsilon_{yy}\right) + D_1\epsilon_{zz} \tag{3.82b}$$

$$K_\epsilon = -\frac{\hbar^2}{2m}A_5\left(k_x - ik_y\right)^2 - D_5\left(\epsilon_{xx} - 2i\epsilon_{xy} - \epsilon_{yy}\right) \tag{3.82c}$$

$$H_\epsilon = \frac{\hbar^2}{2m}A_6\left(k_x - ik_y\right)k_z + D_6\left(\epsilon_{xz} - i\epsilon_{yz}\right) \tag{3.82d}$$

The symmetry of the $k$-dependent and strain terms makes them easier to derive!

## Suggested Reading

- For details of the degenerate perturbation theory needed to derive the 8-band **k·p** method, see J. P. Loehr, *Physics of Strained Quantum Well Lasers* (Kluwer, Dordrecht, 1997) and S. L. Chuang, *Physics of Photonic Devices*, 2nd edn (Wiley, New York, 2009). Chuang's book also contains an introduction to the 6-band **k·p** approach for wurtzite semiconductors.

- For everything you always wanted to know (and more) about the various **k·p** approximations, including a detailed derivation of the second-order terms and the "correct" operator ordering, try L. C. Lew Yan Voon and M. Willatzen, *The k·p Method* (Springer, Berlin, 2009).

# 4

# Absorption and Emission of Light in III–V Semiconductors

Previous chapters discussed the crystal structure and band structure of III–V semiconductors. We now shift to the book's second major topic: electronic interactions with light. We introduce the main ideas about how light waves propagate in semiconductor crystals and induce absorption, spontaneous emission, and stimulated emission in bulk semiconductors. We will consider the differences between the electronic interactions with light in zinc-blende and wurtzite crystals and what happens as the energy gap of the semiconductor is reduced to zero or when the crystal is two-dimensional.

## 4.1 Propagating Light and Interband Electronic Transitions

An electromagnetic wave propagates in a given direction in free space as a plane wave, with its electric and magnetic fields orthogonally polarized and periodically varying in space and time. Of course, the speed of its propagation is the speed of light in vacuum $c$. The spatial frequency is given by the wavevector of the light, which yields the temporal frequency $\omega$ when multiplied by the speed of light. As the wave propagates, the overall energy is continuously exchanged between the out-of-phase electric and magnetic fields. For example, if the electric field's time dependence is $\boldsymbol{E} \propto \cos(\omega t)$, the magnetic field is $\boldsymbol{H} \propto \sin(\omega t)$.

If the wave propagates in a medium characterized by constant permittivity $\varepsilon$ and permeability $\mu$, the average electric and magnetic energy densities are the same because the instantaneous electric/magnetic densities vanish when the respective fields cross zero, and energy cannot be stored anywhere else. Therefore, we have

$$E_m = \frac{1}{4}\mu|\boldsymbol{H}|^2 = E_e = \frac{1}{4}\varepsilon|\boldsymbol{E}|^2 \qquad (4.1)$$

*Bands and Photons in III–V Semiconductor Quantum Structures.* Vurgaftman, Lumb, and Meyer,
Oxford University Press (2021). © Vurgaftman, Lumb, and Meyer.
DOI: 10.1093/oso/9780198767275.003.0004

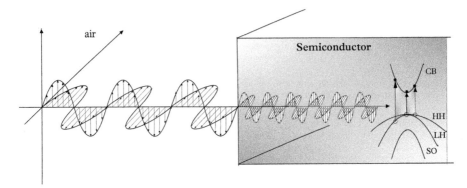

**Figure 4.1** *Schematic illustration of light propagation in a vacuum (left) and within a semiconductor with dispersion (right). Absorption of the propagating light can create electron–hole pairs via interband electronic transitions.*

This is equivalent to writing the electric-field amplitude as

$$|\boldsymbol{E}| = \sqrt{\frac{\mu}{\varepsilon}}|\boldsymbol{H}| = Z|\boldsymbol{H}| \qquad (4.2)$$

where $Z$ is the impedance of the medium, e.g., in vacuum $Z_0 \approx 377\ \Omega$.

An electromagnetic wave has a real impact only once it interacts with some object, in our case a semiconductor crystal, as shown in Fig. 4.1. Inside the semiconductor the transmitted component of the wave continues to propagate, even though the permittivity of the medium is higher than in free space: $\varepsilon(\omega) > \varepsilon_0$. The enhancement is due to back-and-forth "virtual" transitions between the conduction and valence bands, which do not move the carriers from band to band, but rather screen the electric field and slow down the light propagation. The effect becomes stronger at higher frequencies, before it eventually disappears when the photon energy $\hbar\omega \gg E_g$. In other words, the permittivity varies as a function of $\omega$, or it has dispersion. Dispersion is inevitably present if the medium interacts with the field, or in other words, can dissipate or temporarily store some energy.

How does the dispersion affect the electric-field energy in Eq. (4.1)? We expand $\partial\varepsilon/\partial t$, Fourier transformed as $-i\omega\varepsilon(\omega)$, in a Taylor series around $\omega$, and then retain only the first-order term. This shows that the electric energy density $E_e$ is now proportional to the group permittivity $\varepsilon_g \equiv \frac{\partial(\omega\varepsilon)}{\varepsilon_0\partial\omega}$, rather than to $\varepsilon$ itself. As long as the material is non-magnetic and described by the vacuum permeability, the magnetic energy density remains the same as in a medium with constant permittivity ($E_m \propto \varepsilon$). This result is also easily derived from Maxwell's equations for a plane wave. In general, the permittivity is now complex, with the imaginary part representing dissipation. The energy is proportional to the real part, so we can replace $\varepsilon/\varepsilon_0$ with $\varepsilon_r$. The total, time-averaged electromagnetic energy density of the light wave becomes

$$E_{em} = E_e + E_m = \frac{1}{4}\left(\varepsilon_g + \varepsilon_r\right)\varepsilon_0|\boldsymbol{E}|^2 \tag{4.3}$$

The first term in parentheses is the electric energy density, while the second is the magnetic energy density from Eq. (4.1). Equation (4.3) can also be written in terms of the *refractive index* $n_r \equiv \sqrt{\varepsilon_r}$:

$$E_{em} = \frac{n_r n_g}{2}\varepsilon_0|\boldsymbol{E}|^2 \tag{4.4}$$

where the group index is defined: $n_g \equiv n_r + \omega\frac{\partial n_r}{\partial \omega}$. Near the energy gap, the group index is only somewhat larger than $n_r$ (by 10–20 percent).

Equation (4.3) shows that in the presence of dispersion, the electric and magnetic energies for a propagating wave are no longer equal as they were in free space. If $\varepsilon_r < 0$, they can even have different signs (the magnetic energy is negative). Since the total electromagnetic energy cannot be negative, when $\varepsilon_r < 0$ we must have $\varepsilon_g > |\varepsilon_r|$, i.e., $\frac{\partial \varepsilon}{\partial \omega} > 0$. For example, negative permittivity can occur in the region below the plasmon frequency in a metal or heavily doped semiconductor, as well as between the transverse and longitudinal optical phonon frequencies in polar semiconductors and dielectrics. The permittivity in a semiconductor also becomes negative when $\hbar\omega \gg E_g$. The negative sign simply means that the material response is out of phase with the driving field, just as the magnetic field is out of phase with the electric field for a wave propagating in free space.

If the light incident on the semiconductor has a frequency such that $\hbar\omega > E_g$, real electronic transitions between the conduction and valence band exist alongside virtual transitions and cause absorption of the light. Real transitions must satisfy conservation of energy, and absorption due to these transitions is described by the imaginary part $\varepsilon_i$ of the permittivity. Here we deal mostly with energies not much higher than the gap, for which virtual transitions dominate, i.e., $\varepsilon_i \ll \varepsilon_r - 1$. This means we can neglect $\varepsilon_i$ when we calculate the group permittivity or group index.

Figure 4.2 shows the frequency-dependent permittivities associated with interband transitions in several III–V semiconductors. As a consequence of interband absorption, the imaginary part $\varepsilon_i$ increases dramatically at $\hbar\omega > E_g$. As already mentioned, at low frequencies, the absorption features due to optical phonons and the carrier plasma can become important. These phenomena are quite fascinating in their own right, but they fall outside of the scope of this book.

To calculate the light absorption due to interband transitions, we need some way to account mathematically for the quantity of incident photons. To this end, we define the *photon flux* $\Phi$ as the number of photons passing through a unit area within a unit time interval. In electromagnetic theory, the energy flux is given by the Poynting vector $\boldsymbol{E} \times \boldsymbol{H}$. If we include a factor of ½ for the time average of the sinusoidal fields, we obtain

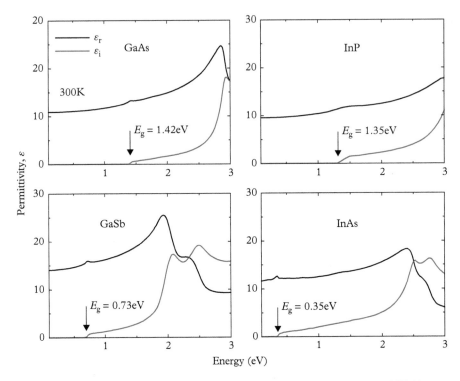

**Figure 4.2** *Real ($\varepsilon_r$) and imaginary ($\varepsilon_i$) components of the permittivity for several III–V semiconductor below and above the interband absorption edge, as measured by spectroscopic ellipsometry.*

$$\Phi = \frac{|E \times H|}{2\hbar\omega} = \frac{cn_r\varepsilon_0}{2\hbar\omega}|E|^2 = \frac{c}{n_g}\frac{E_{em}}{\hbar\omega} \qquad (4.5)$$

We see from the last equality that the photon flux is given by the electromagnetic energy density divided by the energy of each photon, and multiplied by the group velocity $v_g \equiv \frac{c}{n_g}$. We can loosely think of the incident wave as a stream of photons propagating in accordance with Maxwell's equations. The intensity of the incident wave is simply the photon flux multiplied by the photon energy $\hbar\omega$. In Section 4.2 we will find the rate at which photons are removed from this stream by absorption.

The absorption of a photon promotes an electron from an occupied state in the valence band to an unoccupied state in the conduction band, as illustrated in Fig. 4.1. In the process, it creates a "free" electron and leaves behind a hole. We can also think of the absorption as an interaction of the incoming wave's electric field with an interband dipole that extends throughout the semiconductor crystal. The large number of electrons in the crystal means that the interaction is really with a high density of such *transition dipoles*. In a bulk material, the spatial distribution of the dipoles is uniform, with their energy distribution determined by the joint density of states, as discussed in Section 1.4. Mathematically, the interaction Hamiltonian is given by the dot product of the electric-field vector and the transition dipole moment $r$:

$$H_{el} = e\boldsymbol{E} \cdot \boldsymbol{r} \tag{4.6}$$

The electric field $\boldsymbol{E}$ varies sinusoidally in both space and time.

How do we calculate the transition rate between the conduction and valence bands due to the Hamiltonian in Eq. (4.6)? We start by estimating the probability that an electron from one of the bands ends up in a different band. The probability is zero at first and increases linearly with time if the interaction is weak and time invariant. Since the time-dependent amplitude of the electron wavefunction is proportional to the interaction, the transition probability scales as $|H_{el}|^2$:

$$P_{el} \propto |H_{el}|^2 t \tag{4.7}$$

If the interaction stays on for a long time, the total energy must be conserved. That is, $E_i = E_f + \hbar\omega$, where $E_i$ and $E_f$ are the energies of the initial and final electron states. This is expressed by multiplying Eq. (4.7) by a delta function:

$$P_{el} \propto |H_{el}|^2 t\, \delta\left(E_i - E_f - \hbar\omega\right) \tag{4.8}$$

To make the probability dimensionless, we divide Eq. (4.8) by a constant with the units of energy × time, i.e., $\hbar$. The transition rate is then simply the derivative of the probability with respect to time. The rigorous derivation found in most books on quantum mechanics leads to a "circumference" factor of $2\pi$ in the transition rate:

$$W_{if} = \frac{2\pi}{\hbar}|H_{el}|^2\, \delta\left(E_i - E_f - \hbar\omega\right) \tag{4.9}$$

We will use this relation, known as the *Fermi golden rule*, to compute the rates for quantum transitions.

If there are multiple initial and final states, and the occupation probability of the initial (final) states is not necessary one (zero), we have an obvious extension:

$$W_{total} = \sum_{i,f} \frac{2\pi}{\hbar}|H_{el}|^2\, \delta\left(E_i - E_f - \hbar\omega\right) f_i \left(1 - f_f\right) \tag{4.10}$$

The Fermi distribution factor $1 - f_f$ enforces the Pauli exclusion rule (that the same state cannot be occupied by two electrons at the same time).

The wavevector of the light wave is inversely proportional to its wavelength, on the order of 1 μm for the visible and infrared. Therefore, it is three to four orders of magnitude smaller than the typical electronic wavevectors of 0.1–1 Å$^{-1}$ and can be ignored in most calculations. This means the electron wavevectors in the conduction and valence bands for the transitions described by Eq. (4.10) must match, i.e., $\mathbf{k}_c \approx \mathbf{k}_v$. If they do not, fast phase variation leads to cancellation in the spatial integrals. Notice that the electron wavevector in the valence band points in a direction opposite to the hole wavevector $\mathbf{k}_v = -\mathbf{k}_h$. This is because electrons and holes in the valence band move in opposite directions (recall that a hole is just a missing electron). Therefore, generally the sum of the electron and hole wavevectors involved in an interband transition is very close to zero. The same idea holds for the spin: the spins of the initial and final electronic states involved in an optical transition are the same for linearly polarized light, but we can think of a hole as an electron with the opposite spin. We will see how to deal with circularly polarized light in Section 4.3.

Before proceeding with the derivation, we digress briefly to entertain the following thought: what if we can confine the light to a length scale comparable to $1/k_c$ (or $1/k_v$)? In fact, it may be enough for just a small fraction of the Fourier components of the electric field of the light to be of the same order of magnitude as $k_c$ or $k_v$. In this case, to avoid a fast-varying phase, the matching condition becomes $\mathbf{k}_c = \mathbf{k}_v + \mathbf{k}_l$, where $\mathbf{k}_l$ is the wavevector of the light, and transitions can be "diagonal" rather than "vertical" in $k$ space. But how do we confine the light to a subwavelength scale, thereby beating the diffraction limit?

In fact, subwavelength light is very much possible in waveguides and cavities that deviate greatly from free space. The confinement is enabled by coupling the light to an excitation with a much smaller spatial extent, e.g., in plasmons (light and electrons), phonon polaritons (light and phonons), and other kinds of coupled light-matter particles. That said, by far the most common situation is when the wavevector is small, and the effect of matter on the propagating light wave is weak. This means that the light may slow down by a factor of $n_r > 1$ and/or become attenuated if $\alpha > 0$, but its propagation is not modified in a more fundamental way.

In some semiconductors, the bottom of the conduction band has a very different $\mathbf{k}_c$ than the top of the valence band. For example, the lowest electron states may reside near the edge rather than at the center of the Brillouin zone. In this case, optical transitions cannot occur without the involvement of another mechanism. Weak absorption can still take place if another excitation (such as phonons or impurities) take up the missing momentum. Diagonal transitions in $k$ space induced by subwavelength modes could fill that role.

With the assumption of equal electron and hole wavevectors, we are ready to calculate the transition dipole moment $\mathbf{r}$ between the initial and final states: $\langle f | \mathbf{r} | i \rangle = \int d^3 r \, \varphi_f^* \, \mathbf{r} \, \varphi_i$. But how do we do that for the electron and hole wavefunctions that extend throughout the macroscopic crystal? The clue is the dual relationship between $\mathbf{k}$ and $\mathbf{r}$ from Chapter 1. As long as the electrons reside in a bulk crystal, taking the spatial gradient $i\nabla$ is equivalent to multiplication by $\mathbf{k}$. Conversely, the gradient with respect to $\mathbf{k}$ ($i\nabla_{\mathbf{k}}$) is equivalent to multiplying by $\mathbf{r}$.

This means that the matrix element $\langle f|\mathbf{r}|i\rangle = -i\langle f|\nabla_{\mathbf{k}}|i\rangle$. Since the Hamiltonian matrix elements depend on the wavevector components, we seem to be on the right track! We just need to convert the $\mathbf{k}$-gradient of the eigenfunctions in $\langle f|\nabla_{\mathbf{k}}|i\rangle$ into a gradient of the Hamiltonian matrix elements to make the evaluation of the dipole straightforward.

To do so, we start from any Hamiltonian:

$$H_e|(n)\rangle = E_n|(n)\rangle \tag{4.11}$$

To avoid multiple components of the wavevector, we consider a 1D crystal. Now we multiply by $\langle m|$ and then differentiate with respect to $k$:

$$\frac{\partial}{\partial k}\langle m|E_n|n\rangle = \frac{\partial}{\partial k}\langle m|H_e|n\rangle \tag{4.12}$$

Equation (4.12) can be expanded by applying the derivative to each of the three factors on the left and right sides:

$$\frac{\partial E_n}{\partial k}\langle m|n\rangle + E_n\frac{\partial\langle m|}{\partial k}|n\rangle + E_n\langle m|\frac{\partial|n\rangle}{\partial k}$$
$$= \frac{\partial\langle m|}{\partial k}H_e|n\rangle + \langle m|H_e\frac{\partial|n\rangle}{\partial k} + \left\langle m|\frac{\partial H_e}{\partial k}|n\right\rangle \tag{4.13}$$

The first term on the left is zero because the eigenstates $|m\rangle$ and $|n\rangle$ have orthogonal wavefunctions. If we apply Eq. (4.11), the second term on the left cancels the first term on the right. In the second term on the right, we can apply the Hamiltonian operator to $\langle m|$ first, which gives $E_m\langle m|\frac{\partial|n\rangle}{\partial k}$. Combining all of these, Eq. (4.13) becomes

$$(E_n - E_m)\left\langle m|\frac{\partial}{\partial k}|n\right\rangle = \left\langle m|\frac{\partial H_e}{\partial k}|n\right\rangle \tag{4.14}$$

The duality of $\mathbf{k}$ and $\mathbf{r}$ means that the dipole for any arbitrary Hamiltonian describing the band structure of a solid-state material is given by

$$\langle m|\mathbf{r}|n\rangle = -\frac{i}{E_n - E_m}\langle m|\nabla_{\mathbf{k}}H_e|n\rangle \tag{4.15}$$

Equation (4.14) is in fact an off-diagonal form of the *Hellmann–Feynman theorem*. Richard Feynman derived one form of this theorem while he was an undergraduate at MIT in the late 1930s to directly calculate molecular forces. At the time he did not realize that a few other authors had beat him to the punch a few years earlier!

To recap, as long as we know the Hamiltonian and can solve for its eigenstates, the dipole follows immediately from the right-hand side of Eq. (4.15). If we pick a different pair of initial and final states, the dipole can change.

## 4.2 Simplified Treatment of the Bulk Absorption Coefficient

For practice, we apply Eq. (4.15) to a simple Hamiltonian, namely that of Eq. (2.7) with the free-electron terms omitted:

$$H_e = \begin{bmatrix} E_{c0} & \frac{\hbar}{m}\boldsymbol{k}\cdot\boldsymbol{p}_{cv} \\ \frac{\hbar}{m}\boldsymbol{k}\cdot\boldsymbol{p}_{cv}^* & E_{v0} \end{bmatrix} \tag{4.16}$$

Since $\nabla_{\boldsymbol{k}}H_e$ is purely off-diagonal in this case,

$$\nabla_{\boldsymbol{k}}H_e = \begin{bmatrix} 0 & \frac{\hbar}{m}\boldsymbol{p}_{cv} \\ \frac{\hbar}{m}\boldsymbol{p}_{cv}^* & 0 \end{bmatrix} \tag{4.17}$$

there is only one distinct dipole matrix element:

$$\langle c|\boldsymbol{r}|v\rangle = -\frac{i}{E_{n\boldsymbol{k}}-E_{m\boldsymbol{k}}}\frac{\hbar}{m}\langle c|\boldsymbol{p}_{cv}|v\rangle \tag{4.18}$$

Here we make it explicit that initial and final states have the same $\boldsymbol{k}$.

Using Eq. (4.15) we can show that the gradient of the Hamiltonian with respect to $\boldsymbol{k}$, $\nabla_{\boldsymbol{k}}H_e$, is equivalent to the momentum operator (up to some constants). In the literature, the interaction Hamiltonian $H_{el}$ is often written alternatively as proportional to $\boldsymbol{A}\cdot\boldsymbol{p}$, where $\boldsymbol{A}$ is the vector potential, rather than $\boldsymbol{E}\cdot\boldsymbol{r}$. Then the momentum matrix elements rather than the dipoles are sought from the very beginning. This approach gives the same results for the purposes of this book, but we find working with the $\boldsymbol{E}\cdot\boldsymbol{r}$ interaction more intuitive. Also, there is a greater potential for confusing the general momentum operator $\boldsymbol{p}\propto\nabla_{\boldsymbol{k}}H_e$ with the momentum matrix elements between the $s$-like and $p$-like states $\boldsymbol{p}_{cv}$. Equation (4.17) shows that whereas this identification is correct for a simple 2-band Hamiltonian, the general multiband Hamiltonian has higher order correction terms that cannot be neglected away from $k=0$.

We can use Eq. (4.10) to evaluate the transition rate from the valence band to the conduction band for the Hamiltonian of Eq. (4.16). For an incident electric field with amplitude $E_0 = |\boldsymbol{E}|$ and unit polarization vector $\hat{\boldsymbol{e}}$, the interaction matrix element is

$$H_{el} = \frac{eE_0}{2}\langle c|\hat{\boldsymbol{e}}\cdot\boldsymbol{r}|v\rangle = -\frac{ieE_0}{E_{c\boldsymbol{k}}-E_{v\boldsymbol{k}}}\frac{\hbar}{2m}\langle c|\hat{\boldsymbol{e}}\cdot\boldsymbol{p}_{cv}|v\rangle = -\frac{ieE_0}{2\left(E_{c\boldsymbol{k}}-E_{v\boldsymbol{k}}\right)}\langle c|\hat{\boldsymbol{e}}\cdot\nabla_{\boldsymbol{k}}H|v\rangle \tag{4.19}$$

The factor of $\frac{1}{2}$ arises from the sinusoidal time variation that has two Fourier frequency components. Only one of them contributes to Eq. (4.19) because the other has a rapidly varying phase that integrates to zero. To evaluate the momentum matrix element $\langle c|\boldsymbol{p}_{cv}|v\rangle$, we need the eigenfunctions of the Hamiltonian in Eq. (4.16). We assume again that the wavevector and $\boldsymbol{p}_{cv}$ are collinear. The eigenenergies for this case were already specified in Eq. (2.9):

$$E_{c,vk} = \frac{E_{c0} + E_{v0}}{2} \pm \sqrt{\frac{(E_{c0} - E_{v0})^2}{4} + \frac{\hbar^2 k^2}{m^2} p_{cv}^2} \tag{4.20}$$

We set $E_{v0} = 0$, $E_{c0} = E_g$, and $P \equiv \frac{\hbar p_{cv}}{m}$ to obtain

$$E_{c,vk} = \frac{E_g}{2} \pm \sqrt{\frac{E_g^2}{4} + k^2 P^2} \tag{4.21}$$

Figure 4.3 shows the resulting dispersions, which are identical to those in Fig. 2.1a, along with a few possible transitions due to interaction with the light.

What if absorption is relatively close to the band edge, i.e., $\hbar\omega$ only slightly larger than $E_g$ and $k$ relatively small? We approximate the energies in Eq. (4.21) as

$$E_{ck} \approx E_g + \frac{k^2 P^2}{E_g} \tag{4.22a}$$

$$E_{vk} \approx -\frac{k^2 P^2}{E_g} \tag{4.22b}$$

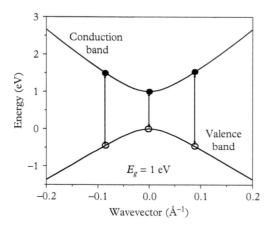

**Figure 4.3** *Conduction- and valence-band dispersions from Eq. (4.21), and interband transitions at different wavevectors k due to interaction with the light.*

The relations between the two components of the conduction- and valence-band eigenfunctions follow from substitution into Eq. (4.16):

$$\psi_{ck}^v \approx \frac{kP}{E_g}\, \psi_{ck}^c \tag{4.23a}$$

$$\psi_{vk}^c \approx -\frac{kP}{E_g}\, \psi_{vk}^v \tag{4.23b}$$

Here the subscript refers to the band, and the superscript to the eigenfunction component. For the conduction (valence)-band states we take $\psi_{ck}^c \approx 1$ ($\psi_{vk}^v \approx 1$). The momentum matrix element then becomes

$$\frac{\hbar}{m}\langle c|p_{cv}|v\rangle = \begin{bmatrix} \psi_{ck}^c & \psi_{ck}^v \end{bmatrix}\begin{bmatrix} 0 & P \\ P & 0 \end{bmatrix}\begin{bmatrix} \psi_{vk}^c \\ \psi_{vk}^v \end{bmatrix} = P\left(\psi_{ck}^{c\,*}\psi_{vk}^v + \psi_{ck}^{v\,*}\psi_{vk}^c\right) \approx P \tag{4.24}$$

In a 3D crystal we need to average over all spatial dimensions to account for the misalignment of $\boldsymbol{k}$ and $\boldsymbol{p}_{cv}$. We can still assume that the dispersions are isotropic, i.e., independent of wavevector direction. We defer the details until Section 4.3, but the upshot is that the square of the interaction matrix element is reduced by a factor of 1/3:

$$|H_{el}|^2 \approx \frac{1}{3}\left(\frac{eE_0 P}{2\,(E_{ck} - E_{vk})}\right)^2 = \frac{E_P}{12\left(E_g + \frac{\hbar^2 k^2}{m_e}\right)^2}\frac{\hbar^2}{2m}e^2 E_0^2 \tag{4.25}$$

Here we have used the definitions for $E_P$ and $m_e$ from Eq. (2.12).

The sum over initial and final states in Eq. (4.10) can be recast as an integration over $\boldsymbol{k}$, which remains the same for both bands. We assume that the crystal is cubic, with side length $L$, so that all the $\boldsymbol{k}$ components are integer multiples of $\frac{2\pi}{L}$. In this case, the sum is equal to the integral over $\boldsymbol{k}$ multiplied by $\left(\frac{L}{2\pi}\right)^3$. An extra factor of 2 is needed because both the initial and final states are spin degenerate, although only the states with the same spin can be connected by the transition. To begin with, we assume that the valence band is fully populated ($f_i = 1$) and the conduction band completely empty ($f_f = 0$). The impact of the actual carrier occupation will be included in Section 4.4. Accounting for all of these factors, the transition rate in Eq. (4.10) using the interaction in Eq. (4.25) becomes

$$W_{total} = \frac{2\pi}{\hbar}\frac{L^3}{(2\pi)^3}\int d^3k \frac{2E_P}{12(E_{ck} - E_{vk})^2}\frac{\hbar^2}{2m}e^2 E_0^2\,\delta\left(E_{ck} - E_{vk} - \hbar\omega\right) \tag{4.26}$$

Below we will calculate the transition rate per unit volume, for which we divide Eq. (4.26) by $L^3$ or simply assume that the side length is 1 cm.

By analogy with Eq. (1.45), we can convert this to an integral over energies, but now over the energy separation between conduction- and valence-band states at the same $\boldsymbol{k}$:

$$W_{total} = \frac{2\pi}{\hbar} \frac{1}{(2\pi)^3} \frac{\hbar^2}{2m}$$

$$\times e^2 E_0^2 \int d\,(E_{ck} - E_{vk}) \frac{E_P}{6(E_{ck} - E_{vk})^2} \left[ \frac{4\pi k^2}{d\,(E_{ck} - E_{vk})\,/dk} \right] \delta\,(E_{ck} - E_{vk} - \hbar\omega)$$

$$(4.27)$$

where the expression in brackets is known as the *joint density of states* (JDOS). For the eigenenergies in Eq. (4.22), the electron and hole masses $m_e$ and $m_h$ are the same, but more generally $m_e \neq m_h$, as we saw in Chapter 3. If $E_{ck} = E_g + \frac{\hbar^2 k^2}{2m_e}$ and $E_{vk} = \frac{\hbar^2 k^2}{2m_h}$, $E_{cv} \equiv E_{ck} - E_{vk} - E_g$ is still parabolic, but the *reduced mass* $m_r = \left( \frac{1}{m_e} + \frac{1}{m_h} \right)^{-1} = \frac{m_e m_h}{m_e + m_h}$ is smaller than either the electron mass or the hole mass. In this case, the expression for the JDOS is nearly identical to Eq. (1.46):

$$JDOS = \frac{1}{(2\pi)^3} \left[ \frac{4\pi k^2}{dE_{cv}/dk} \right] = \frac{1}{2\pi^2} \sqrt{2E_{cv} m_r^3 / \hbar^3} \qquad (4.28)$$

Figure 4.4 plots the JDOS from Eq. (4.28) for a few representative gaps and reduced masses.

Using Eq. (4.28) with $E_{ck} - E_{vk} = \hbar\omega$ and $E_{cv} = \hbar\omega - E_g$ in the integrand of Eq. (4.27), the integral over the delta function becomes trivial. The transition rate per unit volume is

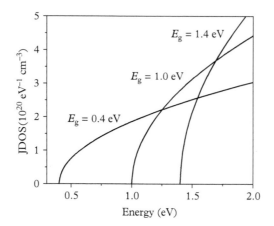

**Figure 4.4** *Joint densities of states vs energy for parabolic dispersions in semiconductors with energy gaps of 0.4, 1.0, and 1.4 eV, and reduced masses of 0.02m, 0.03m, and 0.04m, respectively.*

$$W_{total} = \frac{1}{12\pi\hbar^2} e^2 E_0^2 \frac{E_P}{(\hbar\omega)^2} \frac{\sqrt{2(\hbar\omega - E_g)m_r^3}}{m} \tag{4.29}$$

The transition rate is proportional to the pump intensity (or photon flux) via the optical electric field term $E_0^2$. We can use Eq. (4.5) to show this explicitly:

$$W_{total} = \frac{e^2}{12\pi\hbar^2} \frac{2\Phi}{cn_r\varepsilon_0} \frac{E_P}{\hbar\omega} \frac{\sqrt{2(\hbar\omega - E_g)m_r^3}}{m} \tag{4.30}$$

Equation (4.30) represents the rate at which the radiation excites electrons in the valence band to the conduction band. Energy conservation says that a single photon can excite a single electron, which removes the photon from the beam. Therefore, $W_{total}$ also represents the rate at which the optical beam intensity decreases as it propagates through the semiconductor along some axis $z$ (not necessarily a crystal axis). In other words, the photon flux diminishes with distance at the constant rate given by Eq. (4.30), which is the (negative) spatial derivative of the flux:

$$\frac{d\Phi}{dz} = -W_{total} = -\frac{e^2}{6\pi cn_r\varepsilon_0\hbar^2} \frac{E_P}{\hbar\omega} \frac{\sqrt{2(\hbar\omega - E_g)m_r^3}}{m} \Phi = -\alpha\Phi \tag{4.31}$$

Thus, the *absorption coefficient* $\alpha$ is

$$\alpha(\hbar\omega) = \frac{e^2}{6\pi cn_r\varepsilon_0\hbar^2} \frac{E_P}{\hbar\omega} \frac{\sqrt{2(\hbar\omega - E_g)m_r^3}}{m} \tag{4.32}$$

The solution of the simple differential equation in Eq. (4.31) is the exponential function $\Phi(z) = \Phi(z = 0)e^{-\alpha z}$, and the dependence is known as *Beer's law*.

Equation (4.32) has many constants, but we can group most of them into the *fine-structure constant* from quantum electrodynamics:

$$\alpha_f = \frac{e^2}{4\pi c\varepsilon_0\hbar} \approx \frac{1}{137.036} \tag{4.33}$$

The fine-structure constant determines the strength of the interaction between electrons and photons, so naturally the (linear) absorption in a semiconductor is proportional to $\alpha_f$. The value is so close to $\frac{1}{137}$ and of such importance in quantum electrodynamics that some physicists believed for a time that the denominator *had to be* an integer! Using Eq. (4.33) in Eq. (4.32), we obtain

$$\alpha(\hbar\omega) = \frac{2\alpha_f}{3n_r} \frac{E_P}{\hbar\omega} \frac{\sqrt{2(\hbar\omega - E_g)m_r^3}}{\hbar m} = \frac{8\pi^2\alpha_f}{3n_r} \frac{E_P}{\hbar\omega} \left[\frac{\hbar^2}{2m}\mathcal{J}DOS\right] \tag{4.34}$$

The quantity in brackets "normalizes" the joint density of states, so that it has the same units of $(distance)^{-1}$ as the absorption coefficient (in fact, it is usually the same order of magnitude). But remember that the square-root form in Eq. (4.34) is valid only for parabolic dispersion.

We can substitute typical parameters into Eq. (4.34), e.g., $E_P = 25$ eV, $m_r = 0.04m$, $n_r = 3.5$, and $E_g = 1$ eV, to estimate the absorption coefficient close to the bandgap:

$$\alpha\,(\hbar\omega) \approx 1.4 \times 10^4 \sqrt{\frac{\hbar\omega - E_g}{E_g}} \quad cm^{-1} \tag{4.35}$$

For $\frac{\hbar\omega - E_g}{E_g} = 0.1$ this yields $\alpha \sim 5000$ cm$^{-1}$ and an absorption depth of 2 μm (or about twice the free-space wavelength). This is a reasonable estimate, but somewhat lower than the absorption typically measured in a semiconductor with a similar gap (e.g., the InGaAs alloy lattice-matched to InP). We will see later that a better estimate should include separate transitions to the HH and LH bands, which can be mimicked by summing two terms with different $m_r$ in Eq. (4.34). The main value of Eq. (4.34) is in predicting trends, e.g., that the absorption becomes stronger in wider-gap materials with large masses and weaker in narrow-gap semiconductors. Figure 4.5 shows the result of using Eq. (4.34) to estimate the absorption coefficient versus photon energy for the same energy gaps and reduced masses as in Fig. 4.4.

The inverse dependence of the absorption coefficient in Eq. (4.34) on the refractive index $n_r$ is simply a consequence of how we define the absorption coefficient. For example, the imaginary part of the permittivity is given by

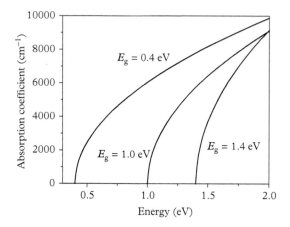

**Figure 4.5** *Absorption coefficients estimated from Eq. (4.34) for semiconductors with parabolic dispersions and the same energy gaps and reduced masses used to obtain Fig. 4.4.*

$$\varepsilon_i = \frac{\alpha \lambda n_r}{2\pi} \tag{4.36}$$

When the absorption is weak, the imaginary part is nearly independent of the real part $\varepsilon_r \approx n_r^2$.

The energy dispersions in a real semiconductor are non parabolic, and we saw in Chapters 2 and 3 that the effect becomes stronger as the gap narrows. We can test this by applying our method to a zero-gap semiconductor, described by the Hamiltonian of Eq. (4.16). No bulk III–V semiconductor actually has a zero (or near zero) energy gap (see Chapters 6 and 7), but this example is not completely irrelevant because near-zero gap can occur in other material systems, e.g., a particular composition of the II–VI HgCdTe alloy!

Equations (2.9) and (4.20) show that the dispersions in a zero-gap material are nearly linear. Even in a wider-gap material, the dispersions are asymptotically linear at energies $>> E_g$ away from the band edges. In either of these limits, the dispersions become approximately

$$E_{c,vk} \approx \pm kP \tag{4.37}$$

and the wavefunctions are

$$\psi_{ck}^v = \psi_{ck}^c = \frac{1}{\sqrt{2}} \tag{4.38a}$$

$$\psi_{vk}^c = -\psi_{vk}^v = \frac{1}{\sqrt{2}} \tag{4.38b}$$

As before, the subscripts designate the electronic states, while the superscripts refer to the components of the wavefunction. We know from Chapter 2 that "$c$" corresponds to $i|(s)$, and "$v$" is $|(p_x)$-like. Therefore, $\psi_{ck}^c{}^* \psi_{vk}^v \propto -ip_{cv} \propto -iP$ and $\psi_{ck}^v{}^* \psi_{vk}^c \propto ip_{cv} \propto iP$. The sign change cancels the minus sign in Eq. (4.38b):

$$\frac{\hbar}{m} \langle c | p_{cv} | v \rangle = \begin{bmatrix} \psi_{ck}^c & \psi_{ck}^v \end{bmatrix} \begin{bmatrix} 0 & P \\ P & 0 \end{bmatrix} \begin{bmatrix} \psi_{vk}^c \\ \psi_{vk}^v \end{bmatrix} = iP \left( \psi_{ck}^c{}^* \psi_{vk}^v + \psi_{ck}^v{}^* \psi_{vk}^c \right) = iP \tag{4.39}$$

We must be careful with the phases of the different components here! In fact, within the two-band Hamiltonian of Eq. (4.16) we always have

$$\psi_{ck}^v = -\psi_{vk}^c = a \tag{4.40a}$$

$$\psi_{ck}^c = \psi_{vk}^v = \sqrt{1 - a^2} \tag{4.40b}$$

where $a \approx 0$ for quadratic dispersion and $a = \frac{1}{\sqrt{2}}$ if the dispersion is linear. Using Eq. (4.40) in Eq. (4.39), we obtain the same result as in the zero-gap case, apart from an unimportant phase factor. This is a trivial example of the more general *sum rule* concept.

For the dispersions in Eq. (4.37) with $E_{cv} \approx 2kP$, the joint density of states becomes

$$\mathcal{J}DOS = \frac{1}{(2\pi)^3}\left[\frac{4\pi k^2}{dE_{cv}/dk}\right] = \frac{1}{2\pi^2}\frac{E_{cv}^2}{8P^3} \tag{4.41}$$

from which the absorption coefficient is

$$\alpha(\hbar\omega) = \frac{8\pi^2\alpha_f E_P}{3n_r}\frac{E_P}{\hbar\omega}\left[\frac{\hbar^2}{2m}\mathcal{J}DOS\right] = \frac{8\pi^2\alpha_f E_P}{3n_r}\frac{\hbar^2}{\hbar\omega}\frac{\left(\hbar\omega - E_g\right)^2}{2m}\frac{}{8P^3} \tag{4.42}$$

We see that the increase of the absorption coefficient with energy is faster than for a parabolic dispersion! In fact, it is not difficult to calculate the JDOS corresponding to the dispersions in Eq. (4.21):

$$\mathcal{J}DOS = \frac{1}{2\pi^2}\frac{k\left(E_{cv} + E_g\right)}{4P^2} = \frac{1}{2\pi^2}\frac{\left(E_{cv} + E_g\right)\sqrt{\left(E_{cv} + E_g\right)^2 - E_g^2}}{8P^3} \tag{4.43}$$

Equation (4.43) reduces to Eq. (4.41) when $E_g = 0$, but varies more slowly with energy than Eq. (4.41) when $E_g \neq 0$ and $E_{cv} \ll E_g$.

Before we improve the accuracy by using the full Hamiltonian in Section 4.3, we close with the instructive example of a two-dimensional semiconductor. The full framework for modeling quantum wells will be developed in Chapter 11, but for now we simply use the 2D JDOS:

$$\mathcal{J}DOS = \frac{1}{(2\pi)^2}\left[\frac{2\pi k}{d\left(E_{ck} - E_{vk}\right)/dk}\right] \tag{4.44}$$

To move from 3D to 2D, we simply removed a factor $1/(2\pi)$ and performed the integration in $k$ space over a circle rather than a sphere. Using Eq. (4.21) in Eq. (4.44), we have

$$\frac{d\left(E_{ck} - E_{vk}\right)}{dk} = \frac{4kP^2}{E_{ck} - E_{vk}} \tag{4.45}$$

Since the numerators of Eqs. (4.44) and (4.45) both contain $k$, the joint density of states is independent of $k$:

$$\frac{\hbar^2}{2m}\mathcal{J}DOS = \frac{\hbar^2}{2m}\frac{E_{ck} - E_{vk}}{8\pi P^2} = \frac{1}{8\pi}\frac{E_{ck} - E_{vk}}{E_P} \tag{4.46}$$

where this JDOS is normalized as in Eq. (4.34).

Now we substitute Eq. (4.46) into the second part of Eq. (4.34). The factor of 3 in the denominator is replaced with a factor of 2 to reflect the number of dimensions. We do not include the refractive index because a 2D crystal has no optical thickness to speak of. In the surrounding free space, we obtain

$$\alpha\left(\hbar\omega\right) = \frac{8\pi^2 \alpha_f}{2} \frac{E_P}{\hbar\omega} \left[\frac{\hbar^2}{2m} \mathcal{J}DOS\right] H\left(\hbar\omega - E_g\right) = \frac{\pi\alpha_f}{2} H\left(\hbar\omega - E_g\right) \quad (4.47)$$

where $H\left(\hbar\omega - E_g\right)$ is the Heaviside step function.

What exactly does Eq. (4.47) tell us? First, we should measure how much is absorbed in traveling vertically through a 2D crystal *not* per unit length, but *per pass*. Whether the 2D crystal is one or more atoms thick, the absorption does not care! 2D crystals such as graphene have two equivalent valleys contributing to absorption near the Dirac point, so in that case we multiply Eq. (4.47) by 2. The result is the constant absorption per pass of $\pi\alpha_f \approx 2.3$ percent, *independent of both the photon energy and the crystal's energy gap*! This is the well-known THz-infrared absorption strength of graphene, which does feature zero bandgap. This bears repeating: as long as the band structure is described by the Hamiltonian of Eq. (4.16), the absorption is independent of energy regardless of the value of $E_g$. The "flat-lining" of the absorption follows from $\frac{d(E_{ck} - E_{vk})}{dk}$ being linear in $k$. The result applies to some extent to thin III–V quantum wells discussed in Chapter 11, with the caveat that the finite "flat-absorption" region extends to much higher energies in graphene.

## 4.3   Detailed Evaluation of the Absorption Coefficient

Rather than remaining content with the simple expressions of Section 4.2, we will now outline a full calculation, paying particular attention to the matrix elements and including the transitions that result from spin–orbit coupling. Before turning to the full Hamiltonian of Eq. (3.24), we first use the Kane Hamiltonian of Eq. (3.1) as a springboard.

Recall that for both Hamiltonians, the eigenstates at $k = 0$ are given by Eq. (2.38) and (2.39), which are reproduced here for convenience:

$$\psi_{HH}^+ = \left|\frac{3}{2}, \frac{3}{2}\right\rangle = \left|1, \frac{1}{2}\right\rangle = -\frac{1}{\sqrt{2}}\left(p_x + ip_y\right) \uparrow \quad (4.48a)$$

$$\psi_{HH}^- = \left|-\frac{3}{2}, -\frac{3}{2}\right\rangle = \left|-1, -\frac{1}{2}\right\rangle = \frac{1}{\sqrt{2}}\left(p_x - ip_y\right) \downarrow \quad (4.48b)$$

$$\psi_{LH}^+ = \left|\frac{3}{2}, \frac{1}{2}\right\rangle = \frac{1}{\sqrt{3}}\left|1, -\frac{1}{2}\right\rangle + \sqrt{\frac{2}{3}}\left|0, \frac{1}{2}\right\rangle = -\frac{1}{\sqrt{6}}\left(p_x + ip_y\right) \downarrow + \sqrt{\frac{2}{3}} \, p_z \uparrow$$

$$(4.48c)$$

$$\psi_{\overline{LH}} = \left|\frac{3}{2}, -\frac{1}{2}\right\rangle = \frac{1}{\sqrt{3}}\left|-1, \frac{1}{2}\right\rangle + \sqrt{\frac{2}{3}}\left|0, -\frac{1}{2}\right\rangle = \frac{1}{\sqrt{6}}\left(p_x - ip_y\right)\uparrow + \sqrt{\frac{2}{3}}\,p_z\downarrow$$

$$(4.48d)$$

$$\psi_{SO}^+ = \left|\frac{1}{2}, \frac{1}{2}\right\rangle = -\sqrt{\frac{2}{3}}\left|1, -\frac{1}{2}\right\rangle + \frac{1}{\sqrt{3}}\left|0, \frac{1}{2}\right\rangle = \frac{1}{\sqrt{3}}\left(p_x + ip_y\right)\downarrow + \frac{1}{\sqrt{3}}\,p_z\uparrow$$

$$(4.48e)$$

$$\psi_{\overline{SO}} = \left|\frac{1}{2}, -\frac{1}{2}\right\rangle = \sqrt{\frac{2}{3}}\left|-1, \frac{1}{2}\right\rangle - \frac{1}{\sqrt{3}}\left|0, -\frac{1}{2}\right\rangle = \frac{1}{\sqrt{3}}\left(p_x - ip_y\right)\downarrow - \frac{1}{\sqrt{3}}\,p_z\uparrow$$

$$(4.48f)$$

The interaction term in Eq. (4.6) and the dipole in Eq. (4.18) for Kane's model tell us that light polarized along the $x$ axis couples only the $|s\rangle$ and $|p_x\rangle$ states (with the same spin), etc. In other words, the $p$ orbital's orientation must match the polarization of the optical electric field. This means that light polarized in the $x - y$ plane interacts strongly with HH states, whereas light polarized along the $z$ axis does not induce CB–HH transitions.

How can this be, considering that we chose the $z$ axis arbitrarily? Even though the choice was indeed arbitrary, once made, all the expressions must be consistent with it. For Eq. (4.48) to remain valid, both the wavevector and spin for any state must point along the $z$ axis. This may seem to be a strange assertion because Eq. (4.48) holds for $k = 0$. However, had the wavevector pointed along a different axis, there would have been an unphysical jump between the eigenfunctions at $k = 0$ and $k > 0$. The true meaning of the *selection rule* embodied in the form of the dipole matrix element is that CB–HH transitions are induced only by light orthogonal to the wavevector of those states. On the other hand, CB–LH transitions couple more strongly when the polarization aligns with $\boldsymbol{k}$.

We can express these ideas mathematically. For a unit polarization vector $\hat{\boldsymbol{e}}$ and unit wavevector $\hat{\boldsymbol{k}} \equiv \boldsymbol{k}/k$, $k_z = \hat{\boldsymbol{e}} \cdot \hat{\boldsymbol{k}}$ if we keep the same convention as before. We can rewrite the square of the $\frac{\hbar}{m}\langle c|p_{cv}|v\rangle \propto P$ matrix element in terms of $\hat{\boldsymbol{e}} \cdot \hat{\boldsymbol{k}}$ using the unit vector property $e_x^2 + e_y^2 + e_z^2 = 1$ and defining $\hat{\boldsymbol{e}} \cdot \hat{\boldsymbol{k}} \equiv e_z$:

$$|M_{HH}|^2 = \frac{1}{2}\left(e_x^2 + e_y^2\right)P^2 = \left(\frac{1}{2} - \frac{|\hat{\boldsymbol{e}} \cdot \hat{\boldsymbol{k}}|^2}{2}\right)P^2 \qquad (4.49a)$$

$$|M_{LH}|^2 = \left[\frac{1}{6}\left(e_x^2 + e_y^2\right) + \frac{2}{3}e_z^2\right]P^2 = \left(\frac{1}{6} + \frac{|\hat{\boldsymbol{e}} \cdot \hat{\boldsymbol{k}}|^2}{2}\right)P^2 \qquad (4.49b)$$

The first equality in Eq. (4.49) follows immediately from Eq. (4.48). The second is valid for an arbitrary direction of $\boldsymbol{k}$. Since the wavefunctions for the SO band have the same amplitudes along all axes, the matrix element is

$$|M_{SO}|^2 = \frac{1}{3}P^2 \tag{4.50}$$

Of course, the incident light actually interacts with states having all possible $k$ directions at the same time. To evaluate the net absorption coefficient, we average the squared matrix elements for the HH and LH bands over all directions in $k$ space. Since $\left|\hat{e}\cdot\hat{k}\right|^2$ is a positive scalar and there are 3 spatial dimensions, a reasonable guess is that the average is $\frac{1}{3}$. We can check this by performing an angular average, with $\theta$ being the angle between $\hat{e}$ and $\hat{k}$, and the integrand being independent of azimuthal angle $\varphi$:

$$\left|\hat{e}\cdot\hat{k}\right|^2_{ave} = \frac{1}{4\pi}\int d\Omega\, \cos^2\theta = \frac{1}{2}\int_{-1}^{1} d(\cos\theta)\, \cos^2\theta = \frac{1}{3} \tag{4.51}$$

When Eq. (4.51) is substituted into Eq. (4.49), the averages of $|M_{HH}|^2$ and $|M_{LH}|^2$ both reduce to $\frac{1}{3}P^2$, which is also the value of $|M_{SO}|^2$ in any direction. We conclude that the absorption coefficient of the bulk cubic crystal remains the same regardless of the polarization of the incident light. This is a direct consequence of the cubic symmetry.

But what about a 2D crystal with a small, but non-zero, thickness $t$ along the $z$ axis? In this case, it helps to distinguish two light polarizations: (1) TE (s-polarized, in the plane of the crystal) and (2) TM (p-polarized, with a component out of the crystal plane). Apart from the in-plane wavevector, we must not forget about $k_z \sim \pi/t$, which is large if the well is narrow.

For TE polarization, $\hat{e}\cdot\hat{k} \approx 0$, and the squared matrix elements become

$$|M_{HH}|^2 = \frac{1}{2}P^2 \quad (TE) \tag{4.52a}$$

$$|M_{LH}|^2 = \frac{1}{6}P^2 \quad (TE) \tag{4.52b}$$

Their sum is still $\frac{2}{3}P^2$ (or $P^2$ if the CB–SO transitions are included). This is a better example of the sum rule, in which the overall transition strength is fixed, but can be redistributed between different transitions.

For TM polarization, $\left|\hat{e}\cdot\hat{k}\right|^2 = \sin^2\theta_i$, where $\theta_i$ is the angle of incidence. This is because the TM polarization has an out-of-plane component. When $\theta_i = 0$ there is no difference between the TE and TM polarizations, and the squared matrix elements are given by Eq. (4.52). More generally, we use Eq. (4.49) to obtain

$$|M_{HH}|^2 = \frac{1}{2}\cos^2\theta_i P^2 \quad (TM) \tag{4.53a}$$

$$|M_{LH}|^2 = \left(\frac{1}{6} + \frac{1}{2}\sin^2\theta_i\right) P^2 \quad (TM) \tag{4.53b}$$

We will return to Eqs. (4.52) and (4.53) when quantum wells are considered in Chapter 11.

What if the light is not linearly polarized? For example, it may be circularly polarized instead (for simplicity, in the $x - y$ plane). One way to deal with this case is to go back to Eq. (4.48) and determine the correct selection rules. We may save ourselves the bother by recalling that a circularly polarized photon carries $m_l = \pm 1$. Therefore, it connects the states that differ by $m_j = \pm 1$, e.g., $\left|\frac{3}{2},\frac{3}{2}\right\rangle$ (HH) with $\left|\frac{1}{2},\frac{1}{2}\right\rangle$ (CB), $\left|\frac{3}{2},-\frac{3}{2}\right\rangle$ (HH) with $\left|\frac{1}{2},-\frac{1}{2}\right\rangle$ (CB), $\left|\frac{3}{2},\frac{1}{2}\right\rangle$ (LH) with $\left|\frac{1}{2},-\frac{1}{2}\right\rangle$ (CB), etc. We can deduce the matrix elements by following the same steps as for linear polarization earlier.

Now we may substitute the average matrix elements in Eqs. (4.52) and (4.53) back into Eq. (4.34). The only problem is that the joint density of states for a real III–V semiconductor is generally a complicated function of energy. In fact, the apparent simplicity of Eq. (4.34) is often illusory, since all the hard work has been shifted to the evaluation of the JDOS. This being the case, we might as well skip the intermediate step of computing the JDOS and derive the general expression for the absorption coefficient. To do so, we backtrack to Eqs. (4.10) and (4.26) and insert the matrix element from Eqs. (4.15) and (4.19):

$$W_{total} = \frac{2\pi}{\hbar} \frac{L^3}{(2\pi)^3} \int d^3k \frac{|\langle c|\nabla_{\boldsymbol{k}}H_e|v\rangle|^2}{4(E_{ck} - E_{vk})^2} e^2 E_0^2 \, \delta\left(E_{ck} - E_{vk} - \hbar\omega\right) \tag{4.54}$$

Using the results from Section 4.2 and Eq. (4.5) to express the photon flux $\Phi$ in terms of $E_0^2 = |\boldsymbol{E}|^2$, the absorption coefficient becomes

$$\alpha\,(\hbar\omega) = \frac{W_{total}}{\Phi L^3} = \frac{2\pi}{\hbar} \frac{1}{(2\pi)^3} \frac{2\hbar\omega e^2}{cn_r\varepsilon_0} \int d^3k \frac{|\langle c|\nabla_{\boldsymbol{k}}H_e|v\rangle|^2}{4(E_{ck} - E_{vk})^2} \, \delta\left(E_{ck} - E_{vk} - \hbar\omega\right) \tag{4.55}$$

How do we compute the matrix element $\langle c|\nabla_{\boldsymbol{k}}H_e|v\rangle$? We follow much the same route that led to Eq. (4.24), except that now we calculate $\nabla_{\boldsymbol{k}}H_e$ for the full Hamiltonian (Kane or 8-band). The off-diagonal terms proportional to $P$ dominate even in the 8-band Hamiltonian, but there are higher order corrections due to the interactions with remote bands.

The states involved in the interband transition are not in fact completely static. The electrons continually scatter to and from these states via phonons, impurities, other disorder, etc. If the scattering time is $\tau_c$ (typically, several hundred femtoseconds), we can neatly account for the finite lifetime in a given state by replacing the delta function in Eq. (4.55) with a broadening line-shape of width $\approx \hbar/\tau_c$. What this function should be remains a matter of some debate, but the choice is usually not very consequential. Some of the possibilities include Lorentzian and Gaussian profiles as well as their convolution. Generally, we expect the tails of the broadening function to decay exponentially far away from the transition energy, which helps to reduce spurious features below the energy gap. A simple choice with exponential tails is the hyperbolic secant function $\text{sech}\left(\frac{E}{\Delta_l}\right) \equiv \frac{1}{\cosh\left(\frac{E}{\Delta_l}\right)}$, where $\Delta_l$ is the linewidth parameter. This function integrates to $\pi\Delta_l$, just like the Lorentzian, which gives the normalized form

$$\sigma(E) = \frac{1}{\pi\Delta_l}\frac{1}{\cosh\left(\frac{E}{\Delta_l}\right)} = \frac{\text{sech}\left(\frac{E}{\Delta_l}\right)}{\pi\Delta_l} \qquad (4.56)$$

This broadening function, along with the simplified version of Eq. (4.55) by means of Eq. (4.33), leads to a general expression for the absorption coefficient:

$$\alpha(\hbar\omega) = \frac{\alpha_f}{2\pi n_r}\int d^3k\,|\langle c|\nabla_{\boldsymbol{k}}H_e|v\rangle|^2\,\frac{\hbar\omega}{(E_{ck}-E_{vk})^2}\,\sigma(E_{ck}-E_{vk}-\hbar\omega) \qquad (4.57)$$

We do not include an extra factor of 2 to account for spin degeneracy, since it is already present in the full Hamiltonian. Of course, it is still a matter of taste whether to use Eq. (4.57) or to obtain the absorption coefficient from the JDOS convolved with the broadening function of Eq. (4.56). In practice, the broadening function governs how abruptly the absorption sets in near the bandgap and has little effect elsewhere (unless other abrupt transitions occur in the spectrum).

One case for which an analytical JDOS can be derived is Kane's model of Eq. (2.51) without the free-electron terms and with a flat HH band at $E_{hk}=0$. This is a good example of including the contributions from several bands to obtain a more accurate result. We will treat only the HH and LH bands, so the validity is restricted to $E_g < \hbar\omega < E_g+\Delta$. The electron and light-hole energies are given by

$$E_{ck} = \frac{E_g}{2} + \sqrt{\frac{E_g^2}{4} + \frac{\hbar^2 k^2 E_P}{3m}} \qquad (4.58a)$$

$$E_{lk} = \frac{E_g}{2} - \sqrt{\frac{E_g^2}{4} + \frac{\hbar^2 k^2 E_P}{3m}} \qquad (4.58b)$$

From Eq. (4.58) we can obtain the CB–HH and CB–LH transition energies:

$$E_{ch}(k) = \frac{E_g}{2} + \sqrt{\frac{E_g^2}{4} + \frac{\hbar^2 k^2 E_P}{3m}} \qquad (4.59a)$$

$$E_{cl}(k) = \sqrt{E_g{}^2 + \frac{4\hbar^2 k^2 E_P}{3m}} \qquad (4.59b)$$

The two contributions to the JDOS are

$$\frac{\hbar^2}{2m}\mathcal{JDOS}_{ch} = \frac{\hbar^2}{2m}\frac{1}{(2\pi)^3}\left[\frac{4\pi k^2}{\frac{dE_{ch}}{dk}}\right] = \frac{3}{4\pi^2}\frac{k\left(E_{ch}-\frac{E_g}{2}\right)}{E_P}$$

$$= \frac{3\sqrt{3}}{4\pi^2}\sqrt{\frac{m}{E_P^3\hbar^2}}\left(E_{ch}-\frac{E_g}{2}\right)\sqrt{\left(E_{ch}-\frac{E_g}{2}\right)^2 - \frac{E_g^2}{4}} \qquad (4.60a)$$

$$\frac{\hbar^2}{2m}\mathcal{JDOS}_{cl} = \frac{\hbar^2}{2m}\frac{1}{(2\pi)^3}\left[\frac{4\pi k^2}{dE_{cl}/dk}\right] = \frac{3}{16\pi^2}\frac{kE_{cl}}{E_P} = \frac{3\sqrt{3}}{32\pi^2}\sqrt{\frac{m}{E_P^3\hbar^2}}E_{cl}\sqrt{E_{cl}^2 - E_g^2}$$

$$\qquad (4.60b)$$

We now insert this JDOS into the last equality of Eq. (4.34) to obtain the absorption coefficient. Is this procedure more accurate than the parabolic form of Eq. (4.34)? Even though Eq. (4.60) takes the HH mass to be infinite, and the LH mass is not quite right because spin–orbit splitting is neglected, the comparison in Fig. 4.6 shows that it is much closer to the full 8-band calculation than the parabolic model.

Even this detailed description of the absorption in a bulk semiconductor based on Eq. (4.55) turns out to be incomplete. One reason has to do with the Coulomb interactions between electrons and holes in the semiconductor that we have not

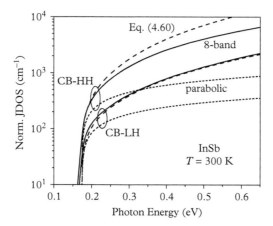

**Figure 4.6** *CB–HH and CB–LH joint densities of states for InSb at 300 K, as obtained from the parabolic approximation (dotted), Eq. (4.60) (dashed), and the full 8-band calculation (solid).*

included in our theory. These interactions become more important as the energy gap widens and have the strongest effect for energies near the gap. We will discuss some effects of the interactions in Section 4.5.

Thus far, we have made no effort to fit the observed absorption spectra in III–V semiconductors. Instead, we fitted the measured gap and masses, i.e., the curvature of the bands. Historically, these were known to a much higher precision that the absorption coefficient in many semiconductors. The result is that we do not expect the absorption spectra to match the experimental values perfectly. We also know that the semiconductor states near the bottom of the conduction and the top of the valence bands are not true *s* and *p* atomic orbitals, so different fitting procedures could yield slightly different results. As the semiconductor gap becomes narrower, the 8-band basis becomes a more self-contained model, and the discrepancy should disappear. The Coulomb effects are also much weaker in narrow-gap materials.

Figure 4.7 compares the 8-band calculations and ellipsometry measurements of the absorption spectra (obtained in part from Fig. 4.1) for several representative semiconductors. For wider-gap and moderate-gap materials, it appears that the experimental absorption is stronger than calculated by $\approx 40$ percent. Some of the extra absorption and its apparent below-gap onset are due to the Coulomb effects discussed in Section 4.5. As mentioned earlier, a separate material-dependent discrepancy is also present. The calculations match the experiment much better for narrow-gap materials, which are a particular focus of this book. Again, this is likely because the 8-band model is sufficient to describe *both the energies and the wavefunctions* of the electronic states of these materials. The optical properties are discussed in more detail in Chapter 7.

The absorption coefficient in a wurtzite semiconductor such as GaN can be calculated by analogy with the zinc-blende case. The HH wavefunction is the same in both types of crystals, which means that the band-edge absorption in GaN and GaN-rich alloys (AlGaN and InGaN) is due to the matrix element in Eq. (4.49a). Because $\Delta_s$ and $\Delta_c$ are quite small, the LH and CH bands begin to contribute at only somewhat higher energies. In principle, there are two distinct values of $P$: $P_x$ and $P_z$ because of the special nature of the $c$ axis, but the difference between them is not large in most materials of interest (see also Chapter 6). Practical applications focus on thin quantum wells of the GaN-rich alloys, as we will see in Chapters 10 and 11.

Sometimes we are interested in the absorption of light propagating in an optical waveguide within the semiconductor crystal rather than incident from the outside. We will return to the problem of absorption in a waveguide in Chapter 11.

## 4.4  Optical Gain and Radiative Recombination

We have been assuming up to now that the valence band is full and the conduction band is empty, which has allowed us to drop the carrier occupation factors from

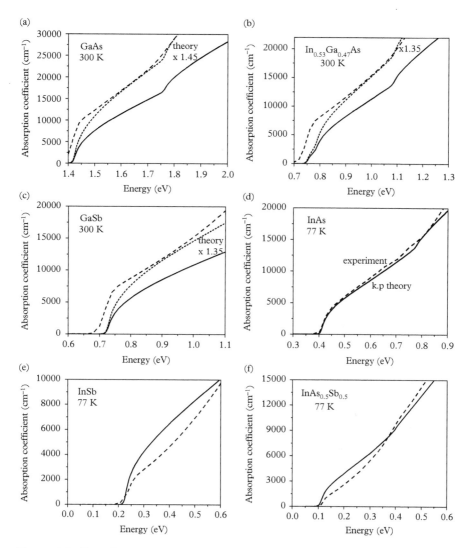

**Figure 4.7** *Absorption coefficients calculated (solid lines) [with indicated scaling (dotted lines)] for GaAs, In$_{0.53}$Ga$_{0.47}$As, GaSb, InAs, InSb, and InAs$_{0.5}$Sb$_{0.5}$ compared to the results of ellipsometric measurements (dashed lines). For further discussion of the parameters, see Chapter 7.*

Eq. (4.10). This approximation certainly makes sense if the carrier densities are low. But what if they become relatively high, e.g., due to doping or non-equilibrium injection of carriers, either electrical or optical? Or what if the carrier occupations matter because the energy gap is not much larger than $k_B T$, and the electrons populate the conduction band of an intrinsic material even in equilibrium?

The obvious fix is to bring back the occupation factors $f_i (1 - f_f)$ in Eq. (4.10). But recall that these factors apply to a process that promotes electrons from the valence band to the conduction band, at the same time removing a photon from the incident beam. In other words, Eq. (4.10) technically gives the upward transition rate rather than the absorption coefficient. The missing piece is the inverse process that transfers an electron from the conduction band down to the valence band, *adding* a photon to the beam (photon *emission*). The net absorption is then the difference between the upward and downward electron transition rates or, equivalently, the difference between the rates of photon absorption and emission:

$$\alpha (\hbar\omega) = \alpha_0 (\hbar\omega) [f_v (1 - f_c) - f_c (1 - f_v)] = \alpha_0 (\hbar\omega) (f_v - f_c) \tag{4.61}$$

Here $\alpha_0 (\hbar\omega)$ is the absorption coefficient for a fully occupied valence band and empty conduction band. We have already computed this quantity in Sections 4.2 and 4.3.

When the semiconductor is in thermal equilibrium, the electrons in the conduction band and holes in the valence band share the same distribution function. Figure 4.8 shows the absorption coefficient in InAs at room temperature for several values of the equilibrium electron density due to $n$-type doping. We see that when electrons are degenerate, the *Burstein–Moss shift* (named after Elias Burstein and Trevor Moss) bleaches the absorption near the band edge by blocking the transitions into occupied states (with $f_c \approx 1$). The Burstein–Moss shift can also occur for degenerate holes, but this requires a higher density because the holes are heavier.

We can also connect a semiconductor crystal with oppositely doped sides (see Chapters 12 and 14) to an external circuit so that some voltage drops over it. Often in such cases, an electron in the conduction band interacts rapidly enough with the other electrons (via equilibrium-restoring scattering processes) to maintain *quasi-equilibrium* within the conduction band. This can occur even though the electron population is much higher than in thermal equilibrium. The same can be true of the holes in the valence band, but not necessarily of the electrons and holes taken as a whole; i.e., the two species may no longer follow the same distribution function. In this scenario, the electron and hole statistics have the Fermi–Dirac form of Eq. (1.41), but with different quasi-Fermi levels $E_{Fc}$ and $E_{Fv}$:

$$f_v = \frac{1}{1 + \exp [(E_v - E_{Fv}) / k_B T]} \equiv \frac{1}{1 + \exp \Delta_v} \tag{4.62a}$$

$$f_c = \frac{1}{1 + \exp [(E_c - E_{Fc}) / k_B T]} \equiv \frac{1}{1 + \exp \Delta_c} \tag{4.62b}$$

where $E_v < 0$ and $E_c > 0$ are the hole and electron energies. Again, the equilibrium situation in Fig. 4.8 is different in that there is a single Fermi level for both electrons

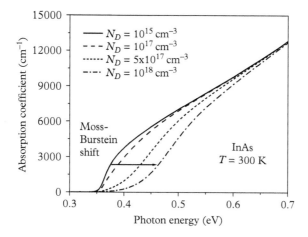

**Figure 4.8** *Absorption spectrum of bulk InAs at T = 300 K for several electron densities in thermal equilibrium. Non-equilibrium electron and hole densities are assumed to be negligible.*

and holes, $E_{Fc} = E_{Fv}$. We assume as before that the corresponding states have the same $\boldsymbol{k}$.

A hole is nothing more than a missing electron, so the hole distribution function can be written

$$f_h = 1 - f_v = \frac{1}{1 + \exp\left[(E_{Fv} - E_v)/k_B T\right]} = \frac{1}{1 + \exp\left(-\Delta_v\right)} \quad (4.63)$$

We can think of holes as electrons with negative energies in the valence band! This is a useful shorthand and fully consistent with holes in semiconductors being akin to positrons in vacuum.

Substituting Eq. (4.62) into Eq. (4.61), we obtain

$$\alpha\left(\hbar\omega\right) = \alpha_0\left(\hbar\omega\right)\left(f_v - f_c\right) = \alpha_0\left(\hbar\omega\right)\left(\frac{1}{1 + \exp\Delta_v} - \frac{1}{1 + \exp\Delta_c}\right)$$

$$= \alpha_0\left(\hbar\omega\right)\frac{\exp\Delta_c - \exp\Delta_v}{\left(1 + \exp\Delta_c\right)\left(1 + \exp\Delta_v\right)} \quad (4.64)$$

We found in Sections 4.2 and 4.3 that the electron and hole states in this expression are separated by an energy equal (or very close if broadening is included) to the photon energy $\hbar\omega$. The absorption coefficient decreases as the electron quasi-Fermi level rises and the hole quasi-Fermi level falls relative to the band edges. Figure 4.9 illustrates the behavior of the quasi-Fermi-level dependent factor in Eq. (4.64). The same approach is then applied to Fig. 4.10, which shows absorption spectra for bulk InAs at several equal electron and hole densities in quasi-equilibrium.

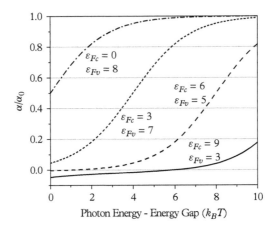

**Figure 4.9** *Quasi-Fermi-level dependent factor in Eq. (4.64). The energies are measured in units of $k_B T$, $\left|\frac{E_c}{E_v}\right| = 3$, and $\varepsilon_{Fc} = (E_{Fc} - E_{c,0})/k_B T$, $\varepsilon_{Fv} = (E_{Fv} - E_{v,0})/k_B T$. The curves are shown for degenerate electrons and non-degenerate holes with $\varepsilon_{Fc} > 0$ and $\varepsilon_{Fv} > 0$. Negative absorption coefficients correspond to optical gain.*

We see from Figs. 4.9 and 4.10 that for high carrier densities the absorption coefficient just above the bandgap becomes negative! The change in sign can happen only when $\Delta_v > \Delta_c$ or $E_v - E_{Fv} > E_c - E_{Fc}$ in Eq. (4.64). Clearly, this condition is equivalent to

$$E_{Fc} - E_{Fv} > E_c - E_v = \hbar\omega \tag{4.65}$$

That is, whenever the splitting between the electron and hole quasi-Fermi levels exceeds the photon energy, downward transitions become more likely than upward transitions. When Eq. (4.65) is satisfied, the optical flux at photon energy $\hbar\omega$ *increases* (rather than decreases as in Beer's law) exponentially with distance as it propagates through the sample. This case of *optical gain*, with coefficient $g(\hbar\omega) \equiv -\alpha(\hbar\omega)$, requires that either or both of the electron and hole densities be high enough to achieve a sufficient separation between the electron and hole quasi-Fermi levels. Clearly, there is no gain in equilibrium, for which $E_{Fc} = E_{Fv}$, no matter how high the electron or hole density is. Gain and absorption are not mutually exclusive and in fact coexist for transitions at different energies in the same non-equilibrium material, as illustrated in Fig. 4.11.

Equation (4.65) implies that there is optical gain at *some* photon energy as long as

$$E_{Fc} - E_{Fv} > E_g \tag{4.66}$$

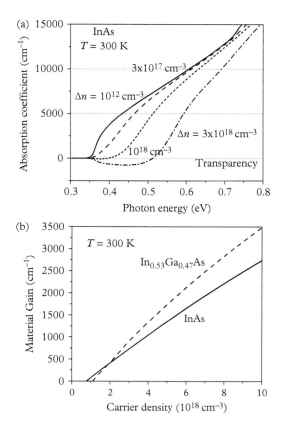

**Figure 4.10** *Absorption spectra for bulk InAs at 300 K, for several non-equilibrium carrier densities (a). The electron and hole densities are assumed to be equal, and no doping is present. A negative absorption coefficient represents gain rather than loss of the optical signal as it propagates through the material. The maximum material gain as a function of electron/hole density is also shown for InAs and $In_{0.53}Ga_{0.47}As$ lattice matched to an InP substrate (b).*

When the two sides of Eq. (4.66) are equal, the material becomes transparent at $\hbar\omega = E_g$ (it is still absorbing at higher energies). Figure 4.10b shows that transparency occurs at a carrier density of $7.5 \times 10^{17}$ cm$^{-3}$ for InAs and $1.1 \times 10^{18}$ cm$^{-3}$ for $In_{0.53}Ga_{0.47}As$ at room temperature (assuming equal electron and hole densities). The gain is nearly linear for carrier densities above the transparency carrier density, with sublinear behavior setting in at high densities.

If the electron and hole quasi-Fermi levels are separated by more than the energy gap, the optical gain peaks at some photon energy $E_{Fc} - E_{Fv} > \hbar\omega > E_g$. However, for $\hbar\omega \gg E_{Fc} - E_{Fv}$, the curve in Fig. 4.10 always converges to the equilibrium absorption coefficient $\alpha_0$ in the absence of carriers.

We can describe interband absorption and gain without quantizing the field, i.e., without assuming that the light comes in discrete packets (quanta), each

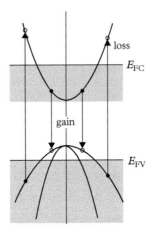

**Figure 4.11** *Schematic of gain and absorption coexisting at different photon energies with the electrons and holes present in the semiconductor.*

containing energy $\hbar\omega$. This picture does not resolve all the questions because emission happens even in the absence of any photons, as first postulated by Albert Einstein in 1916. How is that possible if no beam travels through the sample? In the modern view, this *spontaneous* emission is caused by the vacuum itself (assuming the vacuum admits the presence of light). Following Einstein, we can draw a distinction between *stimulated* emission of the light into an optical beam that excites a given mode in the sample and spontaneous emission into all possible modes of the system, most of which contain no or very few photons. Even at the time, Einstein's assumption was hardly arbitrary or fanciful. In fact, he wanted to make the results consistent with the measured dependence of blackbody emission on wavelength and temperature, which Planck had explained by introducing light quanta.

Can the vacuum field induce upward transitions as well as downward transitions of the electrons? No, because there are no photons to remove from vacuum so that energy is conserved (you can add to nothing, but there is nothing to take away)! This fundamental, thermodynamic difference between the emission and absorption processes was Einstein's key insight, even though the concept of the vacuum field had not been developed at that time.

How do we calculate the spontaneous emission rate? Fortunately, this calculation does not require the full machinery of quantum field theory. Spontaneous downward transitions from conduction to valence-band states can occur only when both electrons and holes are present. For a given optical mode, the transition rate per unit time follows from Eq. (4.57) multiplied by the group velocity of the propagating light $c/n_g$, where $n_g$ is the group index defined after Eq. (4.4):

$$r_{sp}(\hbar\omega) = \frac{\alpha_f}{2\pi n_r} \frac{c}{n_g} \int d^3k \, |\langle c|\nabla_{\boldsymbol{k}} H_e|v\rangle|^2 \frac{\hbar\omega}{(E_{ck} - E_{vk})^2}$$
$$\times \sigma\,(E_{ck} - E_{vk} - \hbar\omega) f_c\,(E_{ck}) \, [1 - f_v\,(E_{vk})] \tag{4.67}$$

The occupation factors account for the requirement that the electron and hole participating in a given transition must have the same wavevector.

Equation (4.67) represents the rate for spontaneous emission into a given optical mode with photon energy $\hbar\omega$. Of course, vacuum fields induce the spontaneous emission of photons into *all possible* modes of the field spanning *all* photon energies. This means we must determine how many modes are present per unit energy, and then integrate over energy. For a large crystal, the modes are indeed quasi-continuous. To obtain the net emission rate at $\hbar\omega$ per unit volume, we multiply the right side of Eq. (4.67) by the density of modes (DOM) per unit energy, which is the optical equivalent of the electronic DOS in Eq. (1.46). The difference is that the photon dispersion is approximately linear: $\omega = cq/n_r\,(\omega)$, where $q$ is the photon wavevector:

$$DOM = \frac{2}{(2\pi)^3} \left[ \frac{4\pi q^2}{d\,(\hbar\omega)/dq} \right] = \frac{1}{\pi^2} \frac{(\hbar\omega)^2}{\hbar^3 c^3} n_r^2 n_g \tag{4.68}$$

The extra factor of 2 accounts for the two orthogonal polarizations of a plane wave.

The density of modes in Eq. (4.68) works well for uniform or nearly uniform media, but not necessarily when the refractive index changes rapidly! For cavities with sizes not much larger than the wavelength, the DOM can be strongly modified by a few cavity modes (or even one mode) and becomes much larger at the resonant wavelength(s) and weaker elsewhere. This is the famous *Purcell effect* named after Edward Purcell, who originally reported the result in a 24-line abstract published in the *Physical Review* in 1946.

To estimate the magnitude of the Purcell effect, we write the DOM for a single mode at energy $\hbar\omega_c$ with Lorentzian broadening of the line:

$$DOM = \frac{1}{\pi} \frac{\hbar\Delta\omega/2}{(\hbar\omega - \hbar\omega_c)^2 + (\hbar\Delta\omega_c/2)^2} \frac{1}{V_{eff}} = \frac{1}{\pi} \frac{\hbar\omega_c/2Q}{(\hbar\omega - \hbar\omega_c)^2 + (\hbar\omega_c/2Q)^2} \frac{1}{V_{eff}} \tag{4.69}$$

where the full width at half-maximum of the cavity mode $\hbar\Delta\omega_c$ can also be described in terms of cavity $Q = \omega_c/\Delta\omega_c$. The *effective volume* of the cavity $V_{eff} = \int d^3r \, E_{em}\,(\boldsymbol{r})/E_{em}\,(\boldsymbol{r}_{max})$, where $E_{em}$ is the electromagnetic energy distribution of the cavity mode, which has a form similar to Eq. (4.3). However, in subwavelength cavities, the magnetic field becomes very small, and only the first term in Eq. (4.3) applies. $E_{em}\,(\boldsymbol{r}_{max})$ is its maximum value at some point in the cavity. For example, for a cosine-like energy profile, $\boldsymbol{r}_{max}$ corresponds to one of the peaks that occur

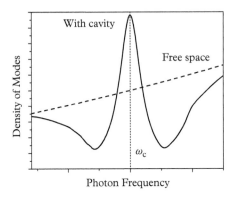

**Figure 4.12** *Illustration of the density of modes in free space and in a cavity. The differences near the resonant frequency drive the Purcell effect as discussed in the text.*

along the cavity axis. The difference between the density of modes in and out of the cavity is illustrated in Fig. 4.12.

The emission rate clearly depends on the position of the dipole with respect to the electric-field distribution of the mode. For example, if the dipole happens to be near the node of the cavity field, the emission into the cavity mode is strongly suppressed. The same happens if the dipole orientation is orthogonal to the cavity field polarization. These arguments lead to the concept of the *local density of modes* given by $LDOM = DOM\left(\frac{E(r_d)\cdot r}{|E(r_{max})||r|}\right)^2$, where $E(r_d)$ is the electric field at the dipole position and $r$ is the transition dipole moment. LDOM replaces DOM in the expression for the emission rate of a localized dipole. Below we assume that $E(r_d)\cdot r = |E(r_{max})||r|$ and $LDOM = DOM$.

The canonical expression for the Purcell enhancement is obtained by assuming that the emission occurs at a single-photon energy $\hbar\omega = \hbar\omega_c$, where $DOM = \frac{2Q}{\pi\hbar\omega_c}\frac{1}{V_{eff}}$, and dividing Eq. (4.69) by Eq. (4.68). If we average the relative directions of the dipole and the electric field in free space, the "free-space" rate is reduced by a factor of 3, but the ingredients other than the DOM remain the same regardless of whether we have a "macrocavity" or a "microcavity". As a result, we obtain

$$F_P = \frac{6\pi Q\hbar^3 c^3}{n_r^2 n_g (\hbar\omega_c)^3}\frac{1}{V_{eff}} = \frac{3}{4\pi^2}\frac{Q}{V_{eff}}\frac{\lambda_c^3}{n_r^2 n_g} \tag{4.70}$$

For large $Q$ and small $V_{eff}/\lambda_c^3$, the Purcell factor can be large, but that requires that the width of the emission line remain below the width of the cavity mode $\Delta\omega_c = \omega_c/Q$. If the emitter is much wider than the cavity, the cavity $Q$ in Eq. (4.70) should be replaced by the "emitter $Q$", given by $\omega_c/\Delta\omega_e$, where $\hbar\Delta\omega_e$

is the emitter bandwidth. Even more generally, we replace $Q$ in Eq. (4.70) with $Q_{eff} \equiv (\Delta\omega_e/\omega_c + \Delta\omega_c/\omega_c)^{-1}$, provided the cavity mode is tuned to the peak emission energy. This means we should aim for a cavity $Q \geq \omega_c/\Delta\omega_e$.

Even though we consider only discrete modes of optical cavities here, similar considerations apply to emission enhancement into a quasi-continuum of waveguide modes. The crucial difference is that the waveguide effect is broadband (i.e., present for any wavelength that supports the waveguide mode) and occurs only in strongly confined geometries.

Equation (4.70) and its generalizations show that the strength of spontaneous emission is not a fixed property of the semiconductor medium, but an emergent characteristic of both the medium and its environment. Even if the spontaneous emission from some material is weak in free space, it may become very strong when the DOM at the emission wavelength has a peak dictated by the surroundings. Here we will continue with the free-space DOM relevant for semiconductor cavities that are much larger than the emission wavelength, but keep $F_P$ as a reminder of the potential cavity effects. The device applications reviewed in Part IV are limited to cavities that are many wavelengths in size.

Multiplying Eqs. (4.67) and (4.68), we obtain for the total rate of spontaneous emission at a given photon energy:

$$R_{sp}(\hbar\omega) = \frac{\alpha_f n_r}{2\pi^3} \frac{F_P}{\hbar^3 c^2} \int d^3 k \, |\langle c|\nabla_{\boldsymbol{k}} H_e|v\rangle|^2 \frac{(\hbar\omega)^3}{(E_{ck} - E_{vk})^2}$$
$$\times \sigma \left(E_{ck} - E_{vk} - \hbar\omega\right) f_c(E_{ck})\left[1 - f_v(E_{vk})\right] \qquad (4.71)$$

We can simplify by assuming an isotropic band-structure, delta-function-like broadening, a constant matrix element $|\langle c|\nabla_{\boldsymbol{k}} H_e|v\rangle|^2 \approx \frac{\hbar^2}{6m} E_P$ as in Eq. (4.25), and two identical sets of transitions with different spins. Alternatively, we could have multiplied the product of Eqs. (4.34) and (4.68) by the group velocity and occupation factors. Either approach leads to

$$R_{sp}(\hbar\omega) = \frac{8\alpha_f n_r}{3} \frac{E_P \hbar\omega}{\hbar^3 c^2} F_P \left[\frac{\hbar^2}{2m} \mathcal{J}DOS\right] f_c(1 - f_v) \qquad (4.72)$$

The quantity $R_{sp}(\hbar\omega)$ represents the total spontaneous emission rate per unit energy. If this rate is then integrated over photon energy, the inverse is the net radiative carrier lifetime. The lower bound is set to $E_g$, since photons are not absorbed below the energy gap:

$$R_{sp} = \frac{8\alpha_f n_r}{3} \frac{E_P}{\hbar^3 c^2} F_P \int_{E_g}^{\infty} \hbar\omega \, d(\hbar\omega) \left[\frac{\hbar^2}{2m} \mathcal{J}DOS\right] f_c(1 - f_v) \qquad (4.73)$$

The units for the rate per unit time per unit volume are easier to follow in the slightly modified form

$$R_{sp} = \frac{8 \alpha_f n_r}{3} \frac{E_P}{\hbar} F_P \int_{E_g/\hbar c}^{\infty} \frac{\omega}{c} \, d\left(\frac{\omega}{c}\right) \left[\frac{\hbar^2}{2m} \mathcal{J}DOS\right] f_c(1 - f_v) \tag{4.74}$$

where $\frac{\omega}{c} = \frac{2\pi}{\lambda}$. Recall that the normalized JDOS (in brackets) has units of inverse length, and that its form for parabolic bands with reduced mass $m_r$ is given by Eq. (4.34):

$$\frac{\hbar^2}{2m} \mathcal{J}DOS = \frac{\sqrt{2\left(\hbar\omega - E_g\right)m_r{}^3}}{4\pi^2 \hbar m} \tag{4.75}$$

If the quasi-Fermi levels of the electrons and holes do not extend into the bands, i.e., they are nondegenerate, the carrier densities are given by Eq. (1.50) with the zero energies shifted to the band edges $E_{c0}$ and $E_{v0}$ (with $E_{c0} - E_{v0} = E_g$):

$$n = N_c \exp\left(\frac{E_{Fc} - E_{c0}}{k_B T}\right), \quad p = N_v \exp\left(-\frac{E_{Fv} - E_{v0}}{k_B T}\right) \tag{4.76}$$

where

$$N_c \equiv \frac{\sqrt{(k_B T m_e)^3 / (2\pi^3)}}{\hbar^3}, \quad N_v \equiv \frac{\sqrt{(k_B T m_h)^3 / (2\pi^3)}}{\hbar^3} \tag{4.77}$$

Equations (4.76) and (4.77) are strictly valid for parabolic dispersions, but can also be applied if the nonparabolicity is small by fitting to the densities of states. Here holes are described as electrons with negative energy, as implied by Eq. (4.63).

We can now rewrite the occupation factors in Eq. (4.74) in terms of the carrier densities $n$ and $p$:

$$\begin{aligned} f_c(1 - f_v) = f_c f_h &\approx \exp\left(-\frac{E_{ck} - E_{Fc}}{k_B T}\right) \exp\left(\frac{E_{vk} - E_{Fv}}{k_B T}\right) \\ &= \exp\left(-\frac{\hbar\omega - E_g}{k_B T}\right) \exp\left(\frac{E_{Fc} - E_{c0}}{k_B T}\right) \exp\left(-\frac{E_{Fv} - E_{v0}}{k_B T}\right) \\ &= \exp\left(-\frac{\hbar\omega - E_g}{k_B T}\right) \frac{n}{N_c} \frac{p}{N_v} \end{aligned} \tag{4.78}$$

As long as the carriers are nondegenerate, the exponential dependence of carrier density on quasi-Fermi level holds even when the dispersions are nonparabolic.

Equations (4.73) and (4.78) tell us that the radiative recombination rate for non-degenerate electrons and holes is proportional to the product of their densities:

$$R_{sp} = B(E_g, T)\, np \qquad (4.79)$$

with the density-independent radiative coefficient given by

$$B(E_g, T) = \frac{8\alpha_f n_r}{3} \frac{E_P F_P}{N_c N_v \hbar^3 c^2} \int_{E_g}^{\infty} \hbar\omega\, d(\hbar\omega) \left[ \frac{\hbar^2}{2m} \mathcal{J}DOS \right] \exp\left( -\frac{\hbar\omega - E_g}{k_B T} \right) \qquad (4.80)$$

Equation (4.79) makes sense because each radiative recombination process requires the presence of both an electron and a hole! However, we will see later that this dependence breaks down when at least one type of carrier is very likely to be present.

From Eq. (4.79), we see that spontaneous emission occurs even in thermal equilibrium, i.e., when $E_{Fc} = E_{Fv}$ and $np = n_0 p_0 = n_i^2$. Here $n_0$ and $p_0$ are the equilibrium electron and hole concentrations, and $n_i$ is the *intrinsic carrier density* that is obtained when $E_{Fc} = E_{Fv}$:

$$\begin{aligned}
n_i &= \left[ N_c \exp\left( \frac{E_{Fc} - E_{c0}}{k_B T} \right) N_v \exp\left( -\frac{E_{Fv} - E_{v0}}{k_B T} \right) \right]^{\frac{1}{2}} \\
&= (N_c N_v)^{1/2} \exp\left( -\frac{E_{c0} - E_{v0}}{2k_B T} \right) \\
&= \left( \frac{k_B T}{2^{1/3} \pi \hbar^2} \right)^{3/2} (m_e m_h)^{3/4} \exp\left( -\frac{E_g}{2k_B T} \right) \qquad (4.81)
\end{aligned}$$

Equation (4.81) is strictly correct only for parabolic bands. Even if the bands are non parabolic, we may still use it for fitting, as illustrated in Fig. 4.13. The relation $np = n_0 p_0 = n_i^2$ is valid in most situations. One exception is when either the electron or hole density is high enough to push the Fermi levels past one of the band edges. This condition of *strong degeneracy* reduces the equilibrium value so that $n_0 p_0 < n_i^2$. Finally, for some III–V semiconductors, the condition $E_g \gg k_B T$ is not fulfilled. A case in point is InAs$_{0.52}$Sb$_{0.48}$ at room temperature, shown in Fig. 14.13b, where the ratio of the electron and hole DOS is so small that the Fermi level corresponding to the intrinsic carrier density of $n_i \approx 10^{17}$ cm$^{-3}$ crosses over into the conduction band. Chapter 7 shows how narrow the gap of the InAsSb alloy can become at various temperatures.

How can spontaneous emission occur if we are not pumping any extra carriers into the system? In fact, some pumping does occur and is due to the *blackbody radiation* emitted by any object with a non-zero temperature. As long as the semi-conductor is in thermal equilibrium with its environment, every emitted photon

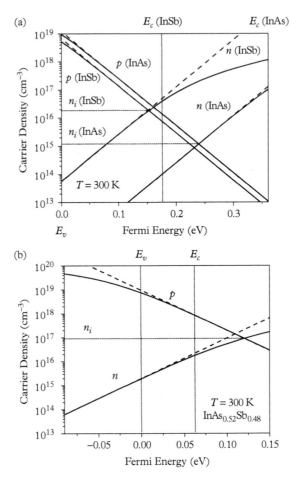

**Figure 4.13** *Calculated electron and hole densities vs the Fermi level position in InAs and InSb (a) as well as InAs$_{0.52}$Sb$_{0.48}$ (b) at room temperature. Exponential fits to the carrier densities using Eq. (4.76) are shown as dashed lines. The intrinsic carrier density is obtained as the crossing point of the electron and hole density curves. The deviation from the exponential fits is due to degeneracy near and beyond the band edges.*

is balanced by a photon absorbed from the outside. The absorption generates an electron–hole pair to replenish the carrier pool depleted by the emission. Both processes occur at the same rate $Bn_i^2$, so the *net* emission rate vanishes. When the environment is at a higher or lower temperature, the semiconductor experiences a net influx (outflow) of blackbody photons from the environment. As a result, the semiconductor heats up or cools down to match the temperature of the surroundings (assuming it is small in comparison with the outside world). Thermal conduction and convection also act eventually to restore equilibrium.

If we apply external pumping to create carrier densities $n > n_0$ and $p > p_0$, the emission and absorption processes no longer balance, and there is a net spontaneous emission rate:

$$R_{sp,net} = B(E_g, T)\left(np - n_i^2\right) \tag{4.82}$$

In this case, thermal equilibrium is reached only when the pump is turned off, which allows the $np$ product to fall to its equilibrium value $n_i^2$. In some semiconductor structures, the carrier densities in a certain region can be reduced below their equilibrium values (e.g., by applying an electric field that sweeps out the carriers) so that $np < n_i^2$, and there is a net influx of radiation, just as expected if that region were cooler than its surroundings ("negative" luminescence!).

What is the typical value of $B(E_g, T)$ in Eq. (4.80)? To make things simpler, we assume parabolic bands and use the analytical JDOS in Eq. (4.75). We also convert to a dimensionless integration variable: $x \equiv (\hbar\omega - E_g)/k_B T$, and split the integral into two parts, with the first proportional to $E_g$ and the second to $\hbar\omega - E_g$. With these changes, and using the definitions of $N_c$ and $N_v$ in Eq. (4.77) as well as $m_r = m_e m_h/(m_e + m_h)$, Eq. (4.80) yields

$$B = \frac{4\sqrt{2\pi}\alpha_f n_r}{3} \frac{E_P \hbar^2 F_P}{c^2(m_e + m_h)^{3/2} m(k_B T)^{3/2}} \left[ E_g \int_0^\infty x^{1/2} e^{-x} dx + k_B T \int_0^\infty x^{3/2} e^{-x} dx \right] \tag{4.83}$$

The two terms in brackets integrate to $E_g \Gamma(3/2) + k_B T \Gamma(5/2) = \frac{\sqrt{\pi}}{2}(E_g + \frac{3}{2}k_B T)$. The first dominates unless the semiconductor has a very narrow gap relative to $k_B T$. Equation (4.83) becomes

$$B = \frac{2\sqrt{2\pi}^{3/2}\alpha_f n_r}{3} \frac{E_P}{mc^2} F_P \left( \frac{\hbar^2}{(m_e + m_h)k_B T} \right)^{3/2} \frac{1}{\hbar} \left( E_g + \frac{3}{2}k_B T \right) \tag{4.84}$$

The first two factors in Eq. (4.84) are dimensionless, and the third has the units of volume. The combined value is on the order of 1 Å$^3$, i.e., atomic dimensions. The rest has a value slightly larger than the energy gap in frequency units, i.e., on the order of $10^{15}$ s$^{-1}$.

Substituting the band parameters for GaAs at $T = 300$ K (see Chapter 6), $E_P = 25.75$ eV, $E_g = 1.42$ eV, and $m_e + m_h \approx 0.45$, we calculate $B = 2.5 \times 10^{-10}$ cm$^3$/s from Eq. (4.84) for $F_P = 1$. A numerical integration of Eq. (4.71) over all energies using the **k·p** states yields $B \approx 1.8 \times 10^{-10}$ cm$^3$/s, which we can correct to $B \approx 2.6 \times 10^{-10}$ cm$^3$/s with the empirical 45 percent enhancement from the absorption spectra in Fig. 4.7. The experimental values in the literature generally fall in the range $2-4 \times 10^{-10}$ cm$^3$/s, although the reabsorption of emitted photons in the semiconductor requires an estimated correction factor. For In$_{0.53}$Ga$_{0.47}$As lattice matched to InP at room temperature, we calculate $B = 1.4 \times 10^{-10}$ cm$^3$/s using Eq. (4.84) and $B \approx 1.2 \times 10^{-10}$ cm$^3$/s numerically (with the correction factor of 1.35 included). In carrying out these calculations, we do not include the spectral dependence of the refractive index, which works well only for low carrier densities.

Equation (4.84) also shows that $B \propto T^{-3/2}$ in semiconductors with gaps much wider than $k_B T$, i.e., almost all the III–V compounds and alloys at room temperature. The recombination is more rapid at lower temperatures because a carrier in a state near $k = 0$ is much more likely to find its opposite-charge partner at the same $k$. At higher temperatures, the distribution becomes more spread out, and the mating odds are lower.

The $np$ dependence in Eq. (4.79) breaks down and the temperature dependence is suppressed when either the electrons or holes become degenerate. In this limit, the exponential tails are no longer sufficient to describe the carrier occupations and densities. Can we use another simple model then? In fact, the Fermi function at high densities approaches a step function at the quasi-Fermi level. This means that nearly all of the transitions with $\hbar\omega < E_{Fc} - E_{Fv}$ can take place, but little emission will occur at higher photon energies. For equal electron and hole densities $(n = p)$, the recombination rate then increases as $\left(E_{Fc} - E_{Fv} - E_g\right)^{3/2}$, which is approximately proportional to $(E_{Fc} - E_{c0})^{3/2}$ or $(E_{v0} - E_{Fv})^{3/2}$, depending on which carrier has the lower density of states. The recombination rate in this limit asymptotically approaches a linear variation with carrier density: $R_{sp} \propto n$. In other words, the radiative lifetime $\tau_R \equiv n/R_{sp}$ is independent of carrier density. In this scenario, an injected electron or hole can usually find a carrier of the opposite type to recombine with, so adding carriers does not improve the odds appreciably. Of course, the carriers do not need to become degenerate for the radiative lifetime to be nearly constant. If the equilibrium carrier density $n_0$ (due to doping) is much higher than the injected density $\Delta n$ (but not necessarily degenerate!), the radiative lifetime pins at $\tau_R \approx 1/(Bn_0)$ for much the same reason.

Figure 4.14 shows the results of numerically integrating Eq. (4.71) over photon energy for the examples of GaAs (dotted), InGaAs lattice-matched to InP (dashed), and InAs (solid) at room temperature, assuming equal electron and hole densities. Since $R_{sp} \propto n^2$ at low densities, the emission rates are normalized by $n^2$. For materials with moderate and wide gaps, the behavior is nearly quadratic up to high carrier densities. For InAs, the room-temperature intrinsic carrier density $n_i \approx 1.2 \times 10^{15}$ cm$^{-3}$, which means that we must take it into account in deducing the radiative coefficient.

We can connect the low-density and high-density regimes with the approximation $R_{sp} \approx Bn^2/\left(1 + \frac{n}{n_s}\right)$, which will also prove useful in Parts III and IV. This form straddles the quadratic (non-degenerate) and linear (degenerate) limits as long as the electron and hole densities are equal. For GaAs, In$_{0.53}$Ga$_{0.47}$As, and InAs, the "saturation" carrier density at room temperature is estimated to be $n_s \approx 6.3, 4.5, 4.7 \times 10^{18}$ cm$^{-3}$, respectively. The fits are shown as dashed curves in Fig. 4.14.

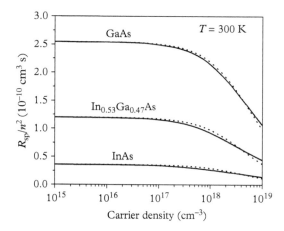

**Figure 4.14** *Numerically calculated radiative recombination rates normalized to the carrier density squared vs carrier density for bulk GaAs, InGaAs lattice matched to InP, and InAs at 300 K. The dashed curves represent fits for the three bulk materials using the functional form that spans the degenerate and non-degenerate limits, with $n_s \approx 6.3, 4.5, 4.7 \times 10^{18}$ cm$^{-3}$ for GaAs, InGaAs, and InAs, respectively.*

What is the spectral width of the spontaneous emission? Assuming nonde-generate statistics and taking the JDOS from Eq. (4.75), we find that $R_{sp}(\hbar\omega)$ has a peak at $\hbar\omega = E_g + k_B T/2$. This position is determined by a competition between the increase of the JDOS with photon energy and the decrease of the Boltzmann factor $\exp\left(-\frac{\hbar\omega - E_g}{k_B T}\right)$ in Eq. (4.78). More generally, for a JDOS varying as $(\hbar\omega - E_g)^a$ in the absence of broadening and degeneracy, the peak emission occurs at $\hbar\omega = E_g + ak_B T$ when $a \geq 0$, and at $\hbar\omega = E_g$ when $a \leq 0$. From the form of Eq. (4.80), the full width at half-maximum (FWHM) should be on the order of $\hbar\Delta\omega_e \sim k_B T$, with numerical estimates ranging from $\approx 1.8 k_B T$ for $a = \frac{1}{2}$ in bulk to $\approx 2.5 k_B T$ for $a = 1$ and down to $\ln 2 k_B T$ for $a = 0$. These results will be useful when we consider quantum structures. Generally, the emission spectrum is narrower when the JDOS has a more gradual onset above the band edge. Figure 4.15 shows several numerically calculated radiative emission spectra for bulk InAs at $T = 300$ K for several values of the injected carrier density.

What, exactly, is the connection between the spontaneous emission rate and the absorption coefficient? Recall that we multiplied the rate of emission into a particular mode Eq. (4.67) by the density of modes Eq. (4.68) to obtain the radiative rate for photons of some energy $\hbar\omega$. Therefore, the rates for emission and absorption into a given mode are related by a simple ratio of the occupation factors:

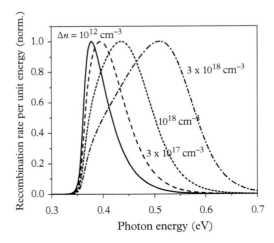

**Figure 4.15** *Radiative recombination spectrum for a series of non-equilibrium carrier densities in bulk InAs at T = 300 K. The maximum rates per unit energy are normalized to unity for all the curves.*

$$\frac{r'_{sp}}{c\alpha/n_g} = \frac{f_c(1-f_v)}{f_v-f_c} = \frac{\frac{1}{1+\exp\Delta_c}\frac{\exp\Delta_v}{1+\exp\Delta_v}}{\frac{\exp\Delta_c-\exp\Delta_v}{(1+\exp\Delta_c)(1+\exp\Delta_v)}} = \frac{1}{\exp(\Delta_c-\Delta_v)-1}$$

$$= \frac{1}{\exp\left[(E_c - E_{Fc} - E_v + E_{Fv})/k_B T\right] - 1} \tag{4.85}$$

If we electrically inject both electrons and holes into an appropriately doped semiconductor structure, the quasi-Fermi levels are split by the voltage applied across the structure: $E_{Fc} - E_{Fv} = eV$. Using $E_c - E_v = \hbar\omega$, we rewrite Eq. (4.85) as

$$\frac{r'_{sp}}{c\alpha/n_g} = \frac{1}{\exp\left[(\hbar\omega - eV)/k_B T\right] - 1} \tag{4.86}$$

Equation (4.86) should remind us of the Boltzmann occupation factor from Eq. (1.40). For example, the probability of finding $n$ photons in a particular optical mode is proportional to

$$P_n(\hbar\omega) \propto \exp\left[n(\mu - \hbar\omega)/k_B T\right] \tag{4.87}$$

where $\mu$ is the chemical potential. In Chapter 1 we derived the Fermi–Dirac distribution using the property that electrons, being fermions, cannot occupy the same state at the same time. On the other hand, there may be as many bosons (such as photons) in the same mode as you please! Therefore, we normalize Eq. (4.87) by summing $\sum_{n=0}^{\infty} P_n(\hbar\omega)$ and setting the sum to one. We obtain

$$P_n(\hbar\omega) = \{1 - \exp[(\mu - \hbar\omega)/k_B T]\} \exp[n(\mu - \hbar\omega)/k_B T] \qquad (4.88)$$

Setting $x \equiv (\mu - \hbar\omega)/k_B T$, we find that the average photon number in the mode $\bar{n}$ is

$$\bar{n} = \sum_n n P_n = (1 - e^x)\sum_n n \exp(nx) = (1 - e^x)\sum_n \frac{\partial}{\partial x}\exp(nx) \qquad (4.89)$$

Interchanging the order of summation and differentiation leads to

$$\bar{n} = (1 - e^x)\frac{\partial}{\partial x}\sum_n \exp(nx) = (1 - e^x)\frac{\partial}{\partial x}\left(\frac{1}{1 - e^x}\right) = \frac{e^x}{1 - e^x} = \frac{1}{e^{-x} - 1} \qquad (4.90)$$

Here the chemical potential is nothing but the applied bias (multiplied by the electron charge) $eV$, so the average photon number becomes

$$\bar{n} = \frac{1}{\exp[(\hbar\omega - eV)/k_B T] - 1} \qquad (4.91)$$

It may not be very clear why photons are described with a chemical potential if they can be created or destroyed at will; i.e., the photon number is generally not conserved. This is because the photons here are in quasi-equilibrium with the carrier distribution that *is* described by an (electro)-chemical potential. In this case, the photons acquire some of the electron properties such as the chemical potential that they lack in true thermal equilibrium.

If we compare Eq. (4.91) and Eq. (4.86), we see that the spontaneous emission rate is equal to the absorption rate multiplied by the average photon number:

$$r'_{sp} = \bar{n} c\alpha/n_g \qquad (4.92a)$$

$$\alpha = \frac{n_g}{c}\{\exp[(\hbar\omega - eV)/k_B T] - 1\} r'_{sp} \qquad (4.92b)$$

In other words, we can derive the spontaneous emission rate from the absorption coefficient and vice versa. Of course, to obtain the emission into all modes, we must multiply by the density of modes in Eq. (4.68). The tight connection only makes sense if spontaneous and stimulated processes are in fact due to *the same phenomenon*! Why then do we draw a distinction between different aspects of this phenomenon? Because a typical optical cavity supports so many modes that spontaneous emission into just one is of little importance (even though the total rate certainly matters!). On the other hand, stimulated emission needs to involve only a single mode for lasing to be observed, so we cannot make do with the total rate in this case. When $\bar{n} = 1$, the net upward transition rate due to excitation by a photon (absorption) is the same as the downward transition rate due to the vacuum field, again assuming a single optical mode.

What happens when $V > \hbar\omega/e$ in Eq. (4.91)? We already know that absorption then turns into gain, but it seems that the average photon number also becomes negative! Since this is impossible, we know that the quasi-equilibrium approach has broken down. The photon number does not follow the thermal distribution of Eq. (4.87), so Eq. (4.92) does not hold. In fact, the average photon number keeps rising as the pump intensity increases above the lasing threshold. Semiconductor lasers working on this principle are discussed in detail in Chapter 12.

On the other hand, if $V < \hbar\omega/e$ and other non-idealities are neglected, the emitted photon removes an amount of energy greater than the applied potential. If that photon is then absorbed in a different material, the emitting structure can cool below its initial temperature. This effect is difficult (but not impossible!) to observe in practice because extra energy is often dissipated in the injection process. Nevertheless, it follows directly from the non-equilibrium nature of spontaneous emission in a semiconductor structure.

When reverse bias is applied ($V < 0$), the photon number becomes lower than in equilibrium, which means that emission is suppressed (but absorption still present). Conversely, in some situations emission can be much *stronger* than absorption (e.g., if a semiconductor is at a much higher temperature than the surroundings into which it emits). In that case, a negative voltage can develop over the terminals of a semiconductor diode structure (discussed in Part IV) to counteract the effect. These examples illustrate the power of introducing chemical potential into Eq. (4.91).

Experimentally, we observe the radiation that emerges from the surface of the sample. If the semiconductor slab is thick, radiation generated near the top (i.e., within the absorption depth of a few µm) can escape before it is re-absorbed within the material, but the rest may go through one or more absorption and re-emission steps. To calculate the emitted flux, we multiply the radiative emission rate from Eq. (4.92) by the mode density in Eq. (4.68). Then we integrate over the solid angle accessible from the outside (i.e., within the total-internal-reflection cone) and the photon energy. To obtain a rough estimate, we may neglect the photon-energy dependence of the absorption coefficient:

$$
R_{sp,app} = \int_{E_g}^{\infty} d(\hbar\omega) \, \frac{\Omega_a}{4\pi} \frac{c\alpha}{n_g} \bar{n} \, DOM \approx \frac{\Omega_a}{4\pi} \frac{n_r^2\alpha}{\pi^2\hbar^3 c^2} \int_{E_g}^{\infty} d(\hbar\omega) \, (\hbar\omega)^2 \exp\left[\frac{eV - \hbar\omega}{k_B T}\right]
$$

$$
= \frac{\Omega_a}{4\pi} \frac{n_r^2\alpha}{\pi^2} \frac{k_B T}{\hbar^3 c^2} \left(E_g^2 + 2E_g k_B T + 2k_B T\right) \exp\left[\frac{eV - E_g}{k_B T}\right] \tag{4.93}
$$

What is the typical value of the solid angle $\Omega_a$? If we assume a perfect anti-reflection coating on top and a perfectly reflecting mirror on the bottom, $\Omega_a \approx 2\pi$, emission is into the top half-space. The emission increases exponentially in forward bias, and essentially disappears at sufficiently strong reverse bias $|V| > 5k_B T/e$. At zero bias, we subtract a blackbody emission term that leads to vanishing net recombination in equilibrium ($V = 0$). The radiative recombination rate as a function of voltage is then

$$R_{sp}(V) \propto \exp\left(\frac{eV}{k_B T}\right) - 1 \qquad (4.94)$$

The expression in Eq. (4.93) is not exactly equivalent to what we deduce from Eq. (4.84) because we have taken a few shortcuts, particularly assuming flat absorption. Nonetheless, we can now state quite generally that the observed radiative emission must be smaller than the radiation within the material because at least some of the light is lost to re-absorption.

## 4.5   Excitonic Optical Effects

Of course, the electrons in a semiconductor do not just whiz by each other with no interaction. In reality, each electron (or hole) experiences not only the potentials of the ionic cores, but also the Coulomb potential of all the other electrons and holes. Does that mean we must start afresh and construct a more comprehensive theory?

Fortunately, it turns out that we often *can* regard the electronic states in semiconductors as single-particle entities. The full reasons were worked out by Lev Landau and others starting in the 1950s. We will not go into the details here, but simply note that most of the electrons in a semiconductor are well below the Fermi level and cannot be easily excited or scattered at room temperature. Despite the relative accuracy of the single-particle theory, it clearly does miss some important phenomena that we will briefly discuss in this section.

First, the presence of other electrons in the core states screens the potential of the ionic cores, but these results are already incorporated into the band parameters. Secondly, a high carrier density shifts the band gap downward, but this shift is large only in wider-gap material and can be accounted for by tinkering with the gap energy (*gap renormalization*). The one important and non-trivial many-body effect on the optical properties is the Coulomb interaction between electrons and holes near the band edges. The two carriers can be treated on a similar footing, but the description is much easier if we assume that the hole mass is much larger than the electron mass so that the hole is nearly stationary by comparison with the electron. This assumption makes the problem look very much like that of the hydrogen atom from textbook quantum mechanics, and the details of the valence band are in any event more complicated than a parabolic hole with a somewhat smaller mass, so we should not lose much sleep over rigor in this motivating discussion.

We would like to find the ground state of the electron–hole pair in the presence of the Coulomb interaction $e^2 / (4\pi\varepsilon_s r)$, where $r$ is the distance between the electron and the hole. It is plausible that such a state is spherically symmetric; i.e., we only need to consider the radial term of the kinetic energy with the origin at the massive hole. Of course, this corresponds to the $s$ orbital in the quantum theory of the hydrogen atom shown in Fig. 3.1. Also, as we saw in Section 4.1, in almost all cases only states with opposite electron and hole wavevectors $\mathbf{k}_c \approx -\mathbf{k}_h$ interact with the optical field. The combined wavevector of the optically active electron–hole pair is then $\mathbf{k}_c + \mathbf{k}_h \approx 0$, and we do not need to include it into the description. The Schrödinger equation becomes

$$\left[ -\frac{\hbar^2}{2m_r} \frac{1}{r^2} \frac{\partial}{\partial r}\left(r^2 \frac{\partial}{\partial r}\right) - \frac{e^2}{4\pi\varepsilon_s r} \right]\psi(r) = (E_{ex} - E_g)\,\psi(r) \tag{4.95}$$

where we have substituted the reduced mass $m_r$ from Eq. (4.28) for the electron mass to correct for our neglect of the hole mass, and the energy $E_{ex}$ of excitons is measured with respect to the energy gap of the material $E_g$. This is because an exciton is created when an electron is removed from the valence band and transferred to the conduction band, a process that involves an extra energy of $E_g$ before the Coulomb attraction is taken into account. Also, remember that the permittivity $\varepsilon_s$ in the Coulomb potential is the static (zero-frequency) permittivity rather than the square of the refractive index at optical frequencies.

If we expand Eq. (4.95), we obtain a second-derivative term and a first-derivative term that are compensated by a constant term and the $1/r$ term, respectively:

$$\left[\frac{\hbar^2}{2m_r}\frac{\partial^2}{\partial r^2} + (E_{ex} - E_g)\right]\psi(r) + \frac{1}{r}\left[\frac{\hbar^2}{m_r}\frac{\partial}{\partial r} + \frac{e^2}{4\pi\varepsilon_s}\right]\psi(r) = 0 \tag{4.96}$$

Because of the form of this expression, a natural trial solution is an exponential that decreases radially from the origin $\psi \propto e^{-r/a_0}$. It is easy to check that it indeed satisfies Eq. (4.96) when

$$a_0 = \frac{4\pi\varepsilon_s\hbar^2}{m_r e^2} = \frac{\hbar(\varepsilon_s/\varepsilon_0)}{\alpha_f m_r c} \tag{4.97a}$$

$$E_{ex} = E_g - \frac{\hbar^2}{2m_r a_0^2} = E_g - \frac{m_r e^4}{2\hbar^2(4\pi\varepsilon_s)^2} = E_g - \frac{m_r c^2 \alpha_f^2}{2(\varepsilon_s/\varepsilon_0)^2} \tag{4.97b}$$

where Eq. (4.97a) follows from the second term in the brackets, and Eq. (4.97b) from the first. The distance in Eq. (4.97a) is known as the modified *Bohr radius*. The lowest exciton state occurs *below* the energy gap, and the difference $E_{ex} - E_g$ is referred to as the *binding energy*. Excited states of the exciton also exist, as well as the full exciton dispersion with respect to the "hole" and combined electron–hole wavevectors. As we mentioned earlier, only the excitons with the total wavevector $\mathbf{k}_c + \mathbf{k}_h \approx 0$ contribute optically (in almost all circumstances). Because the reduced mass is smaller for the exciton that combines an electron and a light hole than an electron and a heavy hole, the LH exciton states are much less prominent than HH states.

What are the typical exciton binding energies and Bohr radii for bulk III–V semiconductors? Clearly, the binding energy goes up with reduced mass, so that it is larger for wider-gap semiconductors. With typical parameters $m_r = 0.04m$ and $\varepsilon_s = 10\varepsilon_0$, we obtain $E_{ex} - E_g = 5$ meV and $a_0 = 130$ Å, close to the experimental values in GaAs. Since the modified Bohr radius is much larger than the lattice spacing, we are a posteriori justified in the assumption of starting with free-carrier-like band states. However, the binding energy is no larger than the typical broadening at higher temperatures. This means that excitons are clearly observable at room temperature only in wide-gap materials such as GaN.

A Bohr radius of 100 Å does *not* mean that the exciton is localized to a physical region of that size! The holes may be much heavier than the electrons, but their states still extend throughout the crystal. It follows that the excitons are no more localized than the holes. The Bohr radius simply tells us that electrons and holes are more likely to be found near each other, but the probability of finding *both* is uniform over the crystal. The extended nature of excitons makes them very different from truly localized states such as those bound to point defects including vacancies, interstitials, and impurities.

What is the absorption due to the exciton transitions? To estimate its strength, we return to Eq. (4.57) and instead of integrating in $k$ space, we divide by the effective volume of the exciton state involved in the transition (not forgetting the factors of $1/(2\pi)$ that go with each $k$ integration). We also assume the emission is from an HH exciton and into a TE-polarized mode, so we can use Eq. (4.52a). Each transition involves two spin-degenerate exciton states and is sufficiently narrow that we can approximate $\frac{\hbar\omega}{E_{ex}^2} \approx \frac{1}{\hbar\omega}$:

$$\alpha(\hbar\omega) = \frac{2\pi^2 \hbar a_f E_P}{m\omega n_r V_{ex}} \sigma(E_{ex} - \hbar\omega) \tag{4.98}$$

What is the effective volume of the extended exciton state? It is just the spatial extent of the exciton probability density assuming the hole is at any point in the crystal, which should be proportional to $a_0^3$. The exact relation turns out to be $V_{ex} = 1/|\psi(r=0)|^2$. To complete the calculation, we normalize $|\psi|^2 \propto e^{-2r/a_0}$:

$$4\pi \int |\psi|^2 \, r^2 \, dr = 1 \tag{4.99}$$

from which we obtain

$$\psi = \sqrt{\frac{1}{\pi a_0^3}} e^{-r/a_0} \tag{4.100}$$

Hence the effective exciton volume is $V_{ex} = \pi a_0^3$, and the absorption coefficient in Eq. (4.98) becomes

$$\alpha\,(\hbar\omega) = \frac{2\pi\alpha_f}{n_r}\frac{E_P}{\hbar\omega}\frac{\hbar^2}{ma_0^3}\sigma\,(E_{ex} - \hbar\omega) \qquad (4.101)$$

Substituting the parameters $E_P = 25$ eV, $\hbar\omega = 1.5$ eV, $a_0 = 130$ Å, $n_r = 3.5$, and the lineshape function from Eq. (4.56) with $\Delta_l = 2$ meV, we calculate a peak absorption of $\approx 10^4$ cm$^{-1}$ at $\hbar\omega = E_{ex}$. Because wider-gap semiconductors have smaller Bohr radii, the excitons in those materials absorb more strongly, with the approximate scaling $\alpha\,(\hbar\omega = E_{ex}) \propto E_g^2$.

Absorption peaks due to excited states are also present and eventually merge into a quasi-continuous distribution close to the energy gap even in wide-gap semiconductors. This is one reason the absorption coefficient does not vanish at the energy gap for practical semiconductors. Another is that even the best semiconductor materials are imperfect and often have defect states extending into the energy gap. The combined exciton and single-particle absorption spectra in a parabolic material with parameters similar to those in GaAs at low temperatures are illustrated in Fig. 4.16. Exciton resonances are much more pronounced at low temperatures, where the broadening is reduced. Transitions from excited states of the excitons typically cannot be resolved for non-nitride III–V materials.

Exciton features explain some of the differences in Fig. 4.7 between the single-particle and experimental absorption spectra near the energy gap for bulk semiconductors. In these plots, the measured data at room temperature were extrapolated to low temperatures and unusual alloy compositions. Therefore, we do not expect to resolve the exciton peaks in the "experimental" spectra of

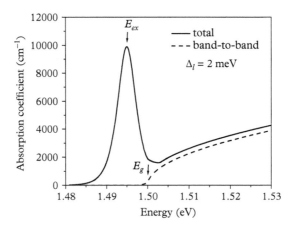

**Figure 4.16** *Combined exciton and single-particle absorption spectra in a parabolic material.*

Fig. 4.7. Real semiconductors have a series of higher exciton lines that merge with the "single-particle" spectrum.

As the carrier density in the semiconductor increases, the exciton peaks become less distinct. This is because the rest of the carriers screen the Coulomb potential of the electron–hole pair. In general, excitons disappear when the carrier density approaches the *Mott density* $n_{max} \sim 1/V_{ex}$ or, equivalently, when the screening length is close to the modified Bohr radius. The Mott density in materials such as GaAs is on the order $10^{17}$ cm$^{-3}$ (higher in wider-gap materials, lower in narrow-gap materials). The single-particle absorption picture described in this chapter is valid when the excitons are quenched. Let us briefly summarize how excitons fit into the overall picture: at low densities of electron–hole pairs, the Coulomb interaction between them causes electrons and holes to be correlated, and the characteristic exciton peak to form below the single-particle energy gap. As the carrier density increases, the Coulomb interaction is screened and the excitons dissociate. At this point, we should not speak of excitons proper, but only of electron–hole pairs that can recombine radiatively. Since optical gain in most semiconductors is reached for carrier densities well above the Mott value (see Fig. 4.9), we do not need to revise our description in Section 4.4 in any fundamental way.

We can treat a two-dimensional exciton by analogy with the bulk problem. For the ground state, we assume circular symmetry with radial coordinate $r$. Equation (4.95) is replaced with

$$\left[ -\frac{\hbar^2}{2m_r} \frac{1}{r} \frac{\partial}{\partial r} \left( r \frac{\partial}{\partial r} \right) - \frac{e^2}{4\pi \varepsilon r} \right] \psi(r) = \left( E_{ex} - E_g \right) \psi(r) \tag{4.102}$$

and Eq. (4.96) becomes

$$\left[ \frac{\hbar^2}{2m_r} \frac{\partial^2}{\partial r^2} + \left( E_{ex} - E_g \right) \right] \psi(r) + \frac{1}{r} \left[ \frac{\hbar^2}{2m_r} \frac{\partial}{\partial r} + \frac{e^2}{4\pi \varepsilon_s} \right] \psi(r) = 0 \tag{4.103}$$

These are solved by direct analogy with the bulk problem, with $a_{0,2D} = a_0/2$, and $E_{ex,2D} = E_g - \frac{\hbar^2}{2m_r a_{0,2D}^2}$. Therefore, the modified Bohr radius is half of that in 3D, and the binding energy is four times the bulk value. We will learn in Chapter 11 how these elementary results fit into the physics of real quantum-well and thin-layer systems.

There is much more to say about the story of the emission and absorption of light in bulk semiconductors! Here are two examples: (1) The Coulomb interaction between electrons and holes has a modest effect on the magnitude of the absorption *above* the energy gap in moderate- and wide-gap semiconductors (due to the series of exciton lines mentioned earlier). (2) High carrier densities modify the energy gap and the structure of optical excitations well beyond the quenching of the excitons. *Ars longa, liber brevis.*

## Suggested Reading

- Many books cover the optical properties of bulk semiconductors, but some of the clearest expositions are in S. L. Chuang, *Physics of Photonic Devices*, 2nd edn (Wiley, New York, 2009), and L. A. Coldren, S. W. Corzine, and M. L. Mašanović, *Diode Lasers and Photonic Integrated Circuits*, 2nd edn (Wiley, New York, 2012).

- The derivation of the optical matrix element for the general Hamiltonian is discussed in the Ph.D. dissertation of L. C. Lew Yan Voon, "Electronic and Optical Properties of Semiconductors: A Study Based on the Empirical Tight-Binding Method" (Worcester Polytechnic Institute, 1993).

- For those interested in taking a deeper plunge into the theory of excitonic effects, another useful reference is J. Singh, *Electronic and Optoelectronic Properties of Semiconductor Structures* (Cambridge University Press, New York, 2007), as well as Chuang's book cited above.

- If you feel like pursuing the trail of the Fermi liquid theory originated by Landau, start with the classic: N. W. Ashcroft and N. D. Mermin, *Solid State Physics* (Holt, Rinehard and Winston, New York, 1976).

# 5

# Other Techniques for Calculating Semiconductor Band Structure

This chapter will expand our toolbox to include a few other techniques for calculating the band structure of a III–V semiconductor: the empirical tight-binding method, its effective bond-orbital form for zinc-blende materials, and the pseudopotential method. These methods allow the band structure over the full Brillouin zone to be described quantitatively using a limited set of parameters. We will also answer the question of how easy it is to relate the parameters for each of the computational approaches to measurable quantities such as the gaps and effective masses.

## 5.1 Basics of the Tight-Binding Method

In Chapter 1, we showed that the Bloch wavefunctions can be expanded in a basis set of atomic orbitals. If we replace the scalars with vectors in Eq. (1.29), we obtain

$$\Psi_{\boldsymbol{k}}(\boldsymbol{r}) = \sum_{l,\boldsymbol{R}_j} b_{l,\boldsymbol{k}} e^{i\boldsymbol{k}\cdot\boldsymbol{R}_j} \phi_l(\boldsymbol{r} - \boldsymbol{R}_j) \tag{5.1}$$

The lattice sites $\boldsymbol{R}_j$ are now in 3D rather along a line. We can separate Eq. (5.1) into two separate sums over the orbitals and the lattice sites:

$$\Psi_{\boldsymbol{k}}(\boldsymbol{r}) = \sum_{l} b'_{l,\boldsymbol{k}} \chi_{l,\boldsymbol{k}}(\boldsymbol{r}) \tag{5.2a}$$

$$\chi_{l,\boldsymbol{k}}(\boldsymbol{r}) = \frac{1}{\sqrt{N}} \sum_{l,\boldsymbol{R}_j} e^{i\boldsymbol{k}\cdot\boldsymbol{R}_j} \phi_l(\boldsymbol{r} - \boldsymbol{R}_j) \tag{5.2b}$$

where we have pulled the (very large) number of lattice sites in the crystal $N$ out of the expansion coefficient $b_{l,\boldsymbol{k}}$ to make the normalization explicit. At first blush, an approach of this kind seems very different from the $\mathbf{k}\cdot\mathbf{p}$ method developed in Chapter 3.

*Bands and Photons in III–V Semiconductor Quantum Structures.* Vurgaftman, Lumb, and Meyer,
Oxford University Press (2021). © Vurgaftman, Lumb, and Meyer.
DOI: 10.1093/oso/9780198767275.003.0005

We already know from Section 1.3 that the expansion of Eqs. (5.1) and (5.2) does not need to use actual atomic orbitals. Of course, the $\phi_l$ functions should still be orthogonal so that the expansion is unique, and enough of them should be included that we are not shortchanging the problem we are trying to solve. We retain the minimal set of Chapter 3, namely, states that have the symmetries of $s$, $p_x$, $p_y$, and $p_z$ orbitals. The practical consequence is that the angular dependence of the wavefunctions follows that of the atomic orbitals: the $\phi_s$ functions are spherically symmetric, while $\phi_{px,y,z}$ point along a specific axis and are antisymmetric; i.e., they undergo a sign change about the origin along that axis. We will not include $d$-like states, since in III–V semiconductors they act solely as higher-order corrections. In Section 3.1, we already used these matrix elements to obtain a better fit to the valence-band dispersions.

We can write the matrix elements of any Hamiltonian with the periodicity of the crystal potential as

$$\langle \chi_{l,k}|H|\chi_{m,k}\rangle = \frac{1}{N} \sum_{R_j,R_n} e^{ik\cdot(R_n - R_j)} \int dr\, \phi_l^* \left(r - R_j\right) H\left(r\right)\, \phi_m \left(r - R_n\right) \quad (5.3)$$

Since $H(r) = H\left(r - R_j\right)$ for the lattice sites $R_j$, Eq. (5.3) can be simplified by a redefinition of the variables:

$$S \equiv R_n - R_j \qquad\qquad (5.4a)$$

$$s \equiv r - R_j \qquad\qquad (5.4b)$$

Then we can use one of the sums over $N$ sites in Eq. (5.3) to cancel the $\frac{1}{N}$ factor in front:

$$\langle \chi_{l,k}|H|\chi_{m,k}\rangle = \sum_{S} e^{ik\cdot S} \int ds\, \phi_l^* \left(s\right) H\left(s\right)\, \phi_m \left(s - S\right) \quad (5.5)$$

The sum over a phase factor and a matrix element does not look easy to evaluate. Of course, we learned in Chapter 3 that, generally, integrals such as that of Eq. (5.5) do not actually need to be *computed* since in the end they are fitted to the measured band structure, i.e. to the energy gap and effective masses. The 4×4 Hamiltonian in the $s$-$p$ basis takes the form

$$\boldsymbol{H}_{4\times4,TB}\begin{bmatrix}|\chi_s\rangle \\ |\chi_{px}\rangle \\ |\chi_{py}\rangle \\ |\chi_{pz}\rangle\end{bmatrix} = \begin{bmatrix} \langle\chi_s|H|\chi_s\rangle & \langle\chi_s|H|\chi_{px}\rangle & \langle\chi_s|H|\chi_{py}\rangle & \langle\chi_s|H|\chi_{pz}\rangle \\ \langle\chi_{px}|H|\chi_s\rangle & \langle\chi_{px}|H|\chi_{px}\rangle & \langle\chi_{px}|H|\chi_{py}\rangle & \langle\chi_{px}|H|\chi_{pz}\rangle \\ \langle\chi_{py}|H|\chi_s\rangle & \langle\chi_{py}|H|\chi_{px}\rangle & \langle\chi_{py}|H|\chi_{py}\rangle & \langle\chi_{py}|H|\chi_{pz}\rangle \\ \langle\chi_{pz}|H|\chi_s\rangle & \langle\chi_{pz}|H|\chi_{px}\rangle & \langle\chi_{pz}|H|\chi_{py}\rangle & \langle\chi_{pz}|H|\chi_{pz}\rangle \end{bmatrix}$$

$$\times \begin{bmatrix} |\chi_s\rangle \\ |\chi_{px}\rangle \\ |\chi_{py}\rangle \\ |\chi_{pz}\rangle \end{bmatrix} \tag{5.6}$$

This Hamiltonian can be spin doubled to produce an $8\times8$ matrix and augmented with the spin–orbit interaction from Eq. (3.23), exactly as in Eq. (3.24). If there are two atoms per basis (as in the diamond and zinc-blende structures), we double the number of states in the Hamiltonian one more time to account for the orbitals residing on both sublattices. Thus, the dimensionality of the tight-binding Hamiltonian is eight (sixteen) bands without (with) spin. We would like to determine the matrix elements in Eq. (5.5) in terms of measurable properties of the semiconductor band structure.

We computed similar matrix elements for the single-band 1D case in Eqs. (1.31) and (1.32). There we observed that the 1D band structure periodically repeats the values in the range $-\frac{\pi}{a} < k < \frac{\pi}{a}$, and that the dispersion is symmetric about the origin. The 2D and 3D cases are analogous, but we need to work a little harder to find the irreducible space of the $\boldsymbol{k}$ vectors.

What $\boldsymbol{k}$ vectors exactly reproduce the state at $k=0$? In 1D, these are clearly $k_n = \frac{2\pi n}{a}$, where $n$ is an integer, because the plane-wave component of the wavefunction must be periodic (the $u_k$ component is always periodic): $\exp(ik_n a) = 1$, i.e., $k_n a = \pi n$. In 2D or 3D, the condition becomes $\exp(i\boldsymbol{b}_i \cdot \boldsymbol{a}_i) = 1$, where $\boldsymbol{b}_i$ is the smallest such $\boldsymbol{k}$ vector (corresponding to $n=1$), and $\boldsymbol{a}_i$ are the lattice vectors defined in Sections 1.1 and 1.2. For a 2D square lattice or 3D simple cubic lattice, the extension is obvious: all the lattice vectors are orthogonal, and each component of $\boldsymbol{k}$ ($k_x, k_y, k_z$) ranges between $-\frac{\pi}{a}$ and $\frac{\pi}{a}$. It seems that we need some kind of systematic procedure for more complicated cases such as the 2D hexagonal lattice or the 3D fcc lattice.

The condition $\boldsymbol{b}_i \cdot \boldsymbol{a}_i = 2\pi$ does not determine $\boldsymbol{b}_i$ uniquely because these vectors have multiple components; i.e., this is a single equation with multiple unknowns. There will be as many $\boldsymbol{b}_i$ vectors as $\boldsymbol{a}_i$, $d$ in total, where $d$ is the number of dimensions. Therefore, we can supply enough equations to find $\boldsymbol{b}_i$ if we introduce $d-1$ additional constraints $\boldsymbol{b}_i \cdot \boldsymbol{a}_j = 0$ for $i \neq j$. The literature refers to the $\boldsymbol{b}_i$ obtained in this manner as *reciprocal lattice vectors*.

As an example, we calculate the reciprocal lattice vectors for the 2D hexagonal lattice with the lattice vectors of Eq. (1.3), which is shown in Fig. 5.1:

$$\boldsymbol{a}_1 = a(1,0) \qquad \boldsymbol{a}_2 = a\left(\frac{1}{2}, \frac{\sqrt{3}}{2}\right) \tag{5.7}$$

We see immediately from the form of $\boldsymbol{a}_1$ that $\boldsymbol{b}_2$ can only have a $y$ component, which must be $b_{2y} = \frac{2\pi}{a_{2y}} = \frac{4\pi}{a\sqrt{3}}$. Also, the $x$ component must be $b_{1x} = \frac{2\pi}{a}$ to fulfill

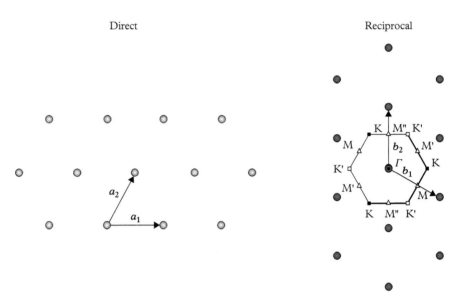

**Figure 5.1** *Real and reciprocal 2D hexagonal lattices with the elementary lattice vectors.*

the $b_1 \cdot a_1 = 2\pi$ condition. As for $b_{1y}$, it follows from $b_1 \cdot a_2 = 0$ that $\pi + \frac{b_{1y}a\sqrt{3}}{2} = 0$ and $b_{1y} = -\frac{2\pi}{\sqrt{3}a}$. Putting all the pieces together, we have

$$b_1 = \frac{2\pi}{a}\left(1, -\frac{1}{\sqrt{3}}\right) \qquad b_2 = \frac{2\pi}{a}\left(0, \frac{2}{\sqrt{3}}\right) \tag{5.8}$$

We see from Fig. 5.1 that the reciprocal of the 2D hexagonal lattice is also hexagonal, but rotated by 90° (or 30°). This should not be surprising, since the reciprocal of the 2D square lattice is also a square lattice, and there are not so many distinct lattices to choose from in two dimensions.

In 3D we can construct a more general expression for $b_i$, based on the property that the cross product $a_j \times a_k$ is orthogonal to the original lattice vectors $a_j$ and $a_k$. If $b_i \propto a_j \times a_k$, the orthogonality condition is satisfied automatically, and we only need to normalize $b_i \cdot a_i$:

$$b_1 = 2\pi \frac{a_2 \times a_3}{a_1 \cdot (a_2 \times a_3)} \tag{5.9a}$$

$$b_2 = 2\pi \frac{a_3 \times a_1}{a_1 \cdot (a_2 \times a_3)} \tag{5.9b}$$

$$b_3 = 2\pi \frac{a_1 \times a_2}{a_1 \cdot (a_2 \times a_3)} \tag{5.9c}$$

We used the property $a_1 \cdot (a_2 \times a_3) = a_2 \cdot (a_3 \times a_1) = a_3 \cdot (a_1 \times a_2)$ here, so that the denominator is the same and equal to the unit cell volume for the original crystal. We could also apply Eq. (5.9) to a 2D lattice, by making $a_3$ a unit vector along the $z$ axis and ignoring $b_3$.

We can use Eq. (5.9) to calculate the reciprocal lattice vectors for the fcc lattice shown in Fig. 5.2a from the direct lattice vectors from Eq. (1.8):

$$a_1 = \frac{a}{2}(0,1,1) \quad a_2 = \frac{a}{2}(1,0,1) \quad a_3 = \frac{a}{2}(1,1,0) \tag{5.10}$$

Keeping in mind that, e.g., $(0, 1, 1)$ is a sum of $\hat{y}$ and $\hat{z}$ unit vectors, we obtain a unit-cell volume $a_1 \cdot (a_2 \times a_3) = \frac{a^3}{4}$. The reciprocal lattice vectors are

$$b_1 = \frac{2\pi}{a}(-1,1,1) \quad b_2 = \frac{2\pi}{a}(1,-1,1) \quad b_3 = \frac{2\pi}{a}(1,1,-1) \tag{5.11}$$

These are the same as the direct lattice vectors of the bcc lattice in Eq. (1.7), provided we take the "lattice constant" of that reciprocal bcc lattice to be $\frac{4\pi}{a}$! Since the bcc lattice is reciprocal to the fcc lattice, we can guess that the converse is also true. This is easily verified using Eq. (5.9). The choice of lattice vectors is not unique: their linear combinations are also lattice vectors. For example, an alternative set for the reciprocal fcc lattice is $b_1' = b_2 + b_3$, $b_2' = b_1 + b_3$, $b_3' = b_1 + b_2 + b_3$ with the explicit result

$$b_1' = \frac{2\pi}{a}(2,0,0) \quad b_2' = \frac{2\pi}{a}(0,2,0) \quad b_3' = \frac{2\pi}{a}(1,1,1) \tag{5.12}$$

Armed with these definitions, we can now deduce the repeating region of the $k$ space. Imagine planes perpendicular to the reciprocal lattice vectors and passing through their midpoints, as shown in Fig. 5.2b. The region we want is completely enclosed by the planes and their intersections, with its interior closer to the origin than to any other lattice point in the reciprocal lattice. This is known as the *(first) Brillouin zone*. In fact, only a small fraction of the Brillouin zone is distinct, and the rest can be transformed into that irreducible part by reflections or rotations. For example, we only need 1/48th of a cube to reproduce all the unique $k$ states in a simple cubic lattice.

In 2D, planes become lines and volume becomes area. For the hexagonal lattice, Eq. (5.8) tells us that the reciprocal unit cell is the hexagon illustrated in Fig. 5.1a, which repeats when rotated by 60°. This means we only need to consider 1/12th of the hexagon contained between points $M$ and $K$ in Fig. 5.1. In fact, there are three pairs of $M$-like and two pairs of $K$-like points that should be distinguished ($M$, $M'$, $M''$ and $K$, $K'$). This is because points connected by a reciprocal lattice vector are in fact physically the same (equivalent) in $k$ space, but points such as $M$ and $M''$ are geometrically similar, but represent different positions in $k$ space.

In spite of that distinction, the physical band structure is exactly the same at all similar ($M$-like and $K$-like) points, e.g., $K$ and $K'$. Since $M$ is rotated clockwise by $30°$ from the $x$ axis (and counterclockwise by $60°$ from $M''$) with the same distance from the origin, its coordinates are $\frac{\pi}{\sqrt{3}a}\left(\sqrt{3},-1\right)$. The $K$ point lies at the intersection of the two lines separating neighboring points of the reciprocal lattice with the coordinates $\frac{4\pi}{3a}(1,0)$. Figure 5.1 shows that in $\boldsymbol{k}$ space the $K'$ point has the coordinates $-\frac{4\pi}{3a}(1,0)$.

The same ideas can be applied to the 3D bcc lattice reciprocal to the fcc lattice. The plane intersections follow more readily from the lattice-vector set in Eq. (5.12), as illustrated in Fig. 5.2b. There is a plane corresponding to the midpoint of $\boldsymbol{b_3}'$, and seven other planes that differ only in the signs of the vector components $\frac{\pi}{a}(\pm1,\pm1,\pm1)$. Along the $x$, $y$, and $z$ axes, the intersections of these planes are cut off by the six planes perpendicular to the corresponding axes at the midpoints of $\pm\boldsymbol{b_1}'$, $\pm\boldsymbol{b_2}'$, and $\pm(\boldsymbol{b_1}+\boldsymbol{b_2})$. The important high-symmetry points are the $L$ point at $\frac{\pi}{a}(1,1,1)$ and similar coordinates and the $X$ point at $\frac{2\pi}{a}(1,0,0)$, etc. Figure 5.2b shows that the Brillouin zone is bounded by 8 hexagons and 6 squares. The irreducible region is only 1/48th of the Brillouin zone.

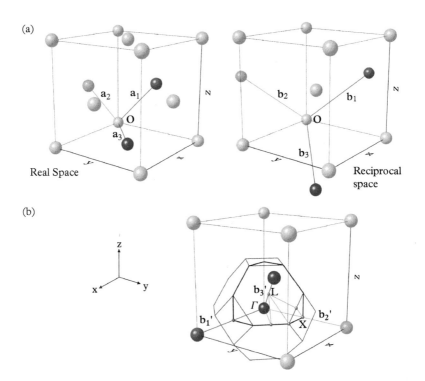

**Figure 5.2** *(a) Real space lattice vectors and reciprocal lattice vectors for the fcc lattice. (b) The Brillouin zone and its irreducible region.*

Why is this terminology useful? To see it applied in practice, we consider a 2D honeycomb lattice illustrated in Fig. 1.3 with two atoms in the basis, i.e., sublattices $A$ and $B$ in graphene. The tight-binding method is easy to apply if we include only the $p_z$ orbitals for the two sublattices and neglect the spin–orbit interaction with the Hamiltonian:

$$\boldsymbol{H}_{graphene,TB} \begin{bmatrix} |\chi_A\rangle \\ |\chi_B\rangle \end{bmatrix} = \begin{bmatrix} \langle\chi_A|H|\chi_A\rangle & \langle\chi_A|H|\chi_B\rangle \\ \langle\chi_B|H|\chi_A\rangle & \langle\chi_B|H|\chi_B\rangle \end{bmatrix} \begin{bmatrix} |\chi_A\rangle \\ |\chi_B\rangle \end{bmatrix} \tag{5.13}$$

The simplest variant of the tight-binding method neglects any integrals in Eq. (5.5) that do not involve the nearest neighbors of a given lattice point. We can also take the $\phi_l(\boldsymbol{r}-\boldsymbol{R}_j)$ functions centered on different lattice sites $\boldsymbol{R}_j$ to be orthogonal, so that we do not need to renormalize the results. The same assumption will be used in the 3D tight-binding models discussed in Section 5.2.

To perform the summation in Eq. (5.5), we need the three nearest-neighbor positions from Eq. (1.4):

$$\boldsymbol{\delta}_1 = \frac{a}{2}\left(1,\frac{1}{\sqrt{3}}\right) \quad \boldsymbol{\delta}_2 = \frac{a}{2}\left(-1,\frac{1}{\sqrt{3}}\right) \quad \boldsymbol{\delta}_3 = a\left(0,-\frac{1}{\sqrt{3}}\right), \tag{5.14}$$

which can be substituted one by one for the $\boldsymbol{S}$ vector in Eq. (5.5). Since the $p_z$ orbital is orthogonal to the crystal plane, the overlap integral is the same regardless of the origin:

$$t_{AB} = \int d^2r\,\phi_A^*(\boldsymbol{r})\,H(\boldsymbol{r})\,\phi_B(\boldsymbol{r}-\boldsymbol{\delta}_i) = \int d^2r\,\phi_B^*(\boldsymbol{r})\,H(\boldsymbol{r})\,\phi_A(\boldsymbol{r}-\boldsymbol{\delta}_i) \tag{5.15}$$

If the $\phi_{A,B}(\boldsymbol{r}-\boldsymbol{R}_j)$ are atomic orbitals, this integral accounts for perturbations introduced by the other sublattice.

With these substitutions, the nearest-neighbor contribution to Eq. (5.5) becomes

$$\langle\chi_B|H|\chi_A\rangle = \left(e^{i\boldsymbol{k}\cdot\boldsymbol{\delta}_1} + e^{i\boldsymbol{k}\cdot\boldsymbol{\delta}_2} + e^{i\boldsymbol{k}\cdot\boldsymbol{\delta}_3}\right)t_{AB} \tag{5.16}$$

Using Eqs. (5.7) and (5.14) we can factor out the common phase factor that does not affect the band structure:

$$\langle\chi_B|H|\chi_A\rangle = e^{i\boldsymbol{k}\cdot\boldsymbol{\delta}_3}\left(1 + e^{i\boldsymbol{k}\cdot\boldsymbol{a}_2} + e^{i\boldsymbol{k}\cdot(\boldsymbol{a}_2-\boldsymbol{a}_1)}\right)t_{AB} \equiv f(\boldsymbol{k})\,t_{AB} \tag{5.17}$$

where

$$\boldsymbol{a}_2 = \frac{a}{2}\left(1,\sqrt{3}\right) \qquad \boldsymbol{a}_2 - \boldsymbol{a}_1 = \frac{a}{2}\left(-1,\sqrt{3}\right) \tag{5.18}$$

We reference zero energy to the $p_z$ orbitals in isolated atoms so they do not contribute to the diagonal matrix elements. Our eigenproblem then becomes

$$\begin{vmatrix} -E(\mathbf{k}) & t_{AB}\, f^*(\mathbf{k}) \\ t_{AB}\, f(\mathbf{k}) & -E(\mathbf{k}) \end{vmatrix} = 0 \tag{5.19}$$

It has a simple solution that encompasses both the electron and hole branches:

$$E_{\pm}(\mathbf{k}) = \pm |f(\mathbf{k})|\, t_{AB} \tag{5.20}$$

Setting the common phase factor in Eq. (5.17) to unity, we write the real and imaginary parts of $f(\mathbf{k})$ as

$$\mathrm{Re}(f) = 1 + \cos\left[\frac{a}{2}\left(k_x + \sqrt{3}k_y\right)\right] + \cos\left[\frac{a}{2}\left(-k_x + \sqrt{3}k_y\right)\right] \tag{5.21a}$$

$$\mathrm{Im}(f) = \sin\left[\frac{a}{2}\left(k_x + \sqrt{3}k_y\right)\right] + \sin\left[\frac{a}{2}\left(-k_x + \sqrt{3}k_y\right)\right] \tag{5.21b}$$

Do the electron and hole dispersions of Eq. (5.20) cross anywhere in the Brillouin zone? Equation (5.21b) tells us that $\mathrm{Im}(f) = 0$ when $k_y = 0$, with the real part being

$$\mathrm{Re}(f) = 1 + 2\cos\left(\frac{a}{2}k_x\right) \tag{5.22}$$

Setting Eq. (5.22) to zero, we obtain $k_x = \pm\frac{4\pi}{3a}$. These are exactly the $K$ and $K'$ points of the reciprocal lattice!

Figure 5.3 plots the full band structure according to Eqs. (5.20) and (5.21), moving from the origin of the reciprocal lattice, i.e., the $\Gamma$ point to the $K$ point. We make no claim that the dispersion is accurate for graphene over the entire Brillouin zone, but only that it works well in reproducing the band crossing at the $K$ point. The dispersions of both bands become linear there. We verify this by defining the wavevector as $\mathbf{k} = \mathbf{K} + \mathbf{q}$, with $\mathbf{K} = \frac{4\pi}{3a}(1,0)$, and in the spirit of $\mathbf{k} \cdot \mathbf{p}$ theory assume that $\mathbf{q}$ is relatively small compared to $\mathbf{K}$. Expanding $f(\mathbf{k})$ to first order in $\mathbf{q}$

$$f(\mathbf{K} + \mathbf{q}) = 1 + e^{i\mathbf{K}\cdot\mathbf{a}_2 + i\mathbf{q}\cdot\mathbf{a}_2} + e^{i\mathbf{K}\cdot(\mathbf{a}_2 - \mathbf{a}_1) + i\mathbf{q}\cdot(\mathbf{a}_2 - \mathbf{a}_1)}$$

$$\approx i\mathbf{q}\cdot\mathbf{a}_2\, e^{\frac{2\pi i}{3}} + i\mathbf{q}\cdot(\mathbf{a}_2 - \mathbf{a}_1)\, e^{-\frac{2\pi i}{3}} \tag{5.23}$$

and then substituting the lattice vectors, we obtain

$$f(\mathbf{K} + \mathbf{q}) \approx i\frac{a}{2}\left[\left(q_x + \sqrt{3}q_y\right)e^{\frac{2\pi i}{3}} + \left(-q_x + \sqrt{3}q_y\right)e^{-\frac{2\pi i}{3}}\right]$$

$$= a\left[-q_x\sin\left(\frac{2\pi}{3}\right) + \sqrt{3}iq_y\cos\left(\frac{2\pi}{3}\right)\right] = -\frac{\sqrt{3}a}{2}(q_x + iq_y) \tag{5.24}$$

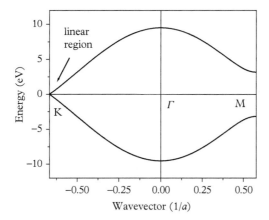

**Figure 5.3** *Band structure of graphene near the crossover at the K point, from Eqs. (5.20) and (5.21).*

According to Eqs. (5.19) and (5.24), the Hamiltonian near the $K$ point is

$$H_{graphene,K}\begin{bmatrix} |\chi_A\rangle \\ |\chi_B\rangle \end{bmatrix} = -\frac{\sqrt{3}a}{2}t_{AB}\begin{bmatrix} 0 & q_x - iq_y \\ q_x + iq_y & 0 \end{bmatrix}\begin{bmatrix} |\chi_A\rangle \\ |\chi_B\rangle \end{bmatrix},$$ (5.25)

which has the eigenvalues

$$E_{\pm}(\boldsymbol{q}) = \pm\left(-\frac{\sqrt{3}a}{2}t_{AB}\right)\sqrt{q_x^2 + q_y^2} \equiv \pm\hbar v_F q$$ (5.26)

Here $v_F = -\frac{\sqrt{3}a}{2\hbar}t_{AB}$ is the Fermi velocity whose experimental value of $\approx 10^8$ cm/s approaches the upper limit predicted in Section 2.1. This Fermi velocity implies that $t_{AB} \approx -3.2$ eV.

Equation (5.25) expresses the "chiral" form of the graphene Hamiltonian, which leads to linear-in-$k$ dispersions for both electrons and holes. Of course, the linear dispersion by itself is hardly a surprise, since Eq. (2.17) already predicted this form near the crossing point of a zero-gap material. In Chapter 10 we will find out about quasi-2D III–V analogs of graphene. For now, it is enough that a single orbital model with nearest neighbors reproduces the much-touted graphene band structure!

## 5.2 Effective Bond-Orbital Method

We know from Chapter 3 that four $s$ and $p$ orbitals are needed to describe the band structure of a bulk III–V semiconductor near the $\Gamma$ point. With spin included, this gives the eight bands used in the **k·p** Hamiltonian. Of course, like graphene, the zinc-blende lattice has two atoms in its basis, i.e., two sublattices. Does this mean that a minimum of sixteen bands are required to apply a tight-binding method to the zinc-blende lattice? Not quite, because the **k·p** approximation reduced the zinc-blende crystal to the fcc lattice; i.e., the two sublattices were merged without losing any critically important features of the band structure.

Here we develop a tight-binding analog of the 8-band **k·p** method, including remote-band contributions, as constructed in Chapter 3. It may help to think back to Section 5.1, in which we found a tight-binding Hamiltonian that generated "**k·p**"-like linear dispersions near the $K$ point of graphene. In that case, we guessed from the outset that a zero-gap material could be modeled with only two bands, and then showed that the two-band nearest-neighbor model is indeed sufficient. In fact, this was possible because a two-band eigenvalue problem has one off-diagonal and two diagonal parameters. Not counting the diagonal parameter that sets the absolute energy scale, we have one parameter that determines the gap (zero for graphene) and another that quantifies the strength of the nearest-neighbor-only interaction.

How many parameters do we need for the more complicated case of a bulk III–V semiconductor? Two define the relative positions of the conduction and valence bands at the $\Gamma$ point, and a third determines the magnitude of the spin–orbit coupling. A fourth one accounts for the electron effective mass, and three serve as stand-ins for the Luttinger parameters, i.e., the valence-band masses. This gives a total of seven parameters when spin–orbit coupling is included (and six if it is not). A tight-binding calculation constructed with these parameters should converge to the **k·p** result near the $\Gamma$ point. To go beyond the **k·p** method, we can fix the band energies away from the center of the Brillouin zone, e.g., the energies at the $X$ point. This is important because quantum wells are typically grown so that quantum confinement is along the $\Gamma$ direction.

By examining the full band structures of common III–V materials such as GaAs, GaSb, or InSb, as shown in Fig. 5.4, we can spot the three energies at the $X$ point that we might like to fit, namely, $E_{X1c}$, $E_{X3v}$, and $E_{X5v}$. This adds to ten (nine) free parameters with (without) the spin–orbit interaction. Can we construct a tight-binding method with that many parameters, as well as eight (four) bands? Does this require anything beyond the nearest-neighbor interactions?

Moving from words to calculations, we know from Chapter 1 that an atom at the origin of the fcc lattice has twelve nearest neighbors: $\frac{a}{2}(\pm1, \pm1, 0)$, $\frac{a}{2}(\pm1, 0, \pm1)$, and $\frac{a}{2}(0, \pm1, \pm1)$. We can sum over them and the atom at the origin using the exponential factors in Eq. (5.5). We identify two parameters due to the $s$ band alone. The first corresponds to $\boldsymbol{S} = 0$ and represents the energy of $s$-like states in an individual atom, which we denote $E_{ss}(0,0,0)$. The second corresponds to the nearest neighbor with $\boldsymbol{S} = \frac{a}{2}(1,1,0)$ and is referred to as $E_{ss}(1,1,0)$. The integrals involving orbitals at the other eleven nearest neighbors are identical, because the spherically symmetric $s$ orbital looks the same in all directions. As a result, we obtain the matrix element

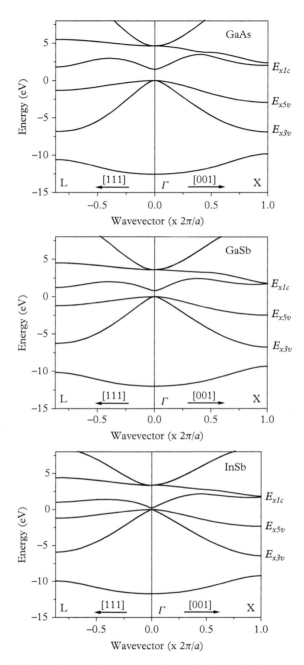

**Figure 5.4** *Full tight-binding band structures of GaAs, GaSb, and InSb, as obtained using the methods described in Section 5.3.*

$$\langle \chi_s|H|\chi_s \rangle = e^{i\mathbf{k}\cdot 0} E_{ss}(0,0,0) + \left( e^{i\frac{a}{2}\mathbf{k}\cdot(1,1,0)} + e^{-i\frac{a}{2}\mathbf{k}\cdot(1,1,0)} + e^{i\frac{a}{2}\mathbf{k}\cdot(1,-1,0)} \right.$$

$$+ e^{-i\frac{a}{2}\mathbf{k}\cdot(1,-1,0)} + e^{i\frac{a}{2}\mathbf{k}\cdot(1,0,1)} + e^{-i\frac{a}{2}\mathbf{k}\cdot(1,0,1)} + e^{i\frac{a}{2}\mathbf{k}\cdot(1,0,-1)} + e^{-i\frac{a}{2}\mathbf{k}\cdot(1,0,-1)}$$

$$+ e^{i\frac{a}{2}\mathbf{k}\cdot(0,1,1)} + e^{-i\frac{a}{2}\mathbf{k}\cdot(0,1,1)} + e^{i\frac{a}{2}\mathbf{k}\cdot(0,1,-1)} + e^{-i\frac{a}{2}\mathbf{k}\cdot(0,1,-1)} \left. \right) E_{ss}(1,1,0) \quad (5.27)$$

We can compress the notation by identifying the adjacent pairs of phase factors as twice the cosine function and defining $c_x \equiv \cos(k_x a/2)$, $c_y \equiv \cos(k_y a/2)$, and $c_z \equiv \cos(k_z a/2)$. Simplifying with trigonometric identities such as $\cos\left((k_x + k_y)a/2\right) + \cos\left((k_x - k_y)a/2\right) = 2c_x c_y$, we finally obtain

$$\langle \chi_s|H|\chi_s \rangle = E_{ss}(0,0,0) + 4E_{ss}(1,1,0)\left(c_x c_y + c_z c_x + c_y c_z\right) \quad (5.28)$$

The matrix elements between $p$ orbitals are similar if the same orbital direction is involved. The schematic in Fig. 5.5a shows that the overlap integrals for a $p_x$ orbital displaced along the $x$ axis are not necessarily the same for the same orbital translated in the $y - z$ plane. Therefore, we distinguish between $E_{xx}(1,1,0)$ and $E_{xx}(0,1,1)$, but $E_{xx}(1,1,0) = E_{xx}(1,0,1)$. Then by analogy with Eqs. (5.27) and (5.28), we obtain

$$\langle \chi_x|H|\chi_x \rangle = E_{xx}(0,0,0) + 4E_{xx}(1,1,0)\left(c_x c_y + c_z c_x\right) + 4E_{xx}(0,1,1) c_y c_z \quad (5.29)$$

Equations (5.28) and (5.29) add three parameters to the mix. For matrix elements between the $p_y$ and $p_z$ orbitals, we substitute $y$ (or $z$) for $x$ and vice versa in the cosines. For example, $\langle \chi_y|H|\chi_y \rangle = E_{xx}(0,0,0) + 4E_{xx}(1,1,0)\left(c_y c_x + c_z c_y\right) + 4E_{xx}(0,1,1) c_x c_z$ etc.

We now tackle the matrix elements between different orbitals. One example is $\langle \chi_s|H|\chi_x \rangle$, an off-diagonal element that has no zero-displacement term that is independent of $k$. Figure 5.5b shows that these matrix elements flip their signs when displaced along the axis of the orbital with respect to the origin (here, the $x$ axis). This is because the spherically symmetric $s$ orbital overlaps with opposite lobes of the $p$ orbital. On the other hand, for translations in the $y$–$z$ plane the contributions from positive and negative lobes cancel out, so the corresponding matrix elements can be assumed to vanish from the outset. The overall matrix element then has the form

$$\langle \chi_s|H|\chi_x \rangle = \left( e^{i\frac{a}{2}\mathbf{k}\cdot(1,1,0)} - e^{-i\frac{a}{2}\mathbf{k}\cdot(1,1,0)} + e^{i\frac{a}{2}\mathbf{k}\cdot(1,-1,0)} - e^{-i\frac{a}{2}\mathbf{k}\cdot(1,-1,0)} + e^{i\frac{a}{2}\mathbf{k}\cdot(1,0,1)} \right.$$

$$\left. - e^{-i\frac{a}{2}\mathbf{k}\cdot(1,0,1)} + e^{i\frac{a}{2}\mathbf{k}\cdot(1,0,-1)} - e^{-i\frac{a}{2}\mathbf{k}\cdot(1,0,-1)} \right) E_{sx}(1,1,0) \quad (5.30)$$

Simplifying by converting to terms with sines and cosines with $\sin\left((k_x + k_y)a/2\right) + \sin\left((k_x - k_y)a/2\right) = 2s_x c_y$ and similar identities, where $s_x \equiv \sin\left(\frac{k_x a}{2}\right)$ etc., we obtain

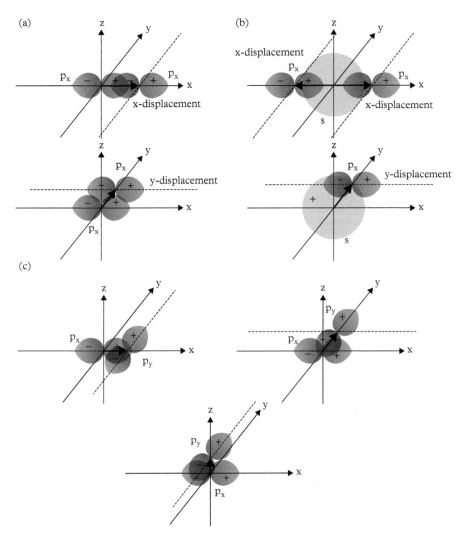

**Figure 5.5** *Schematic illustrations of the overlap integrals used in the tight-binding method for (a) two $p_x$ orbitals in two possible configurations; (b) s and $p_x$ orbitals in two possible configurations; and (c) $p_x$ and $p_y$ orbitals in three possible configurations.*

$$\langle \chi_s | H | \chi_x \rangle = 4iE_{sx}(1,1,0)\left(s_x c_y + c_z s_x\right) \tag{5.31}$$

The *s–y* and *s–z* matrix elements feature the same parameter $E_{sx}(1,1,0)$, and have the same functional form when $y$ and $z$ are substituted for $x$, e.g., $\langle \chi_s | H | \chi_y \rangle = 4iE_{sx}(1,1,0)\left(s_y c_x + c_z s_y\right)$. To account for the possibility that the last four phase terms do not cancel, we could generalize Eq. (5.31) to read

$$\langle \chi_s | H | \chi_x \rangle = 4iE_{sx} (1,1,0) \left( s_x c_y + c_z s_x \right) - 4E_{sx} (0,1,1) \, s_y s_z \tag{5.32}$$

Of course, in this fcc-based model we take $E_{sx} (0,1,1) = 0$. This is not the case in the zinc-blende model to be considered below, in which the asymmetric potential from the two dissimilar atoms in the basis prevents cancellation.

This leaves the $\langle \chi_x | H | \chi_y \rangle$ matrix element to be determined. Now opposite-sense translations along both the $x$ and $y$ axes flip the signs of the integrals. A displacement along the $z$ axis produces equal and opposite terms from the two lobes, as illustrated in Fig. 5.5c, so the terms proportional to $E_{xy} (0,1,1) = E_{xy} (1,0,1)$ vanish:

$$\langle \chi_x | H | \chi_y \rangle = \left( e^{i\frac{a}{2}\mathbf{k}\cdot(1,1,0)} + e^{-i\frac{a}{2}\mathbf{k}\cdot(1,1,0)} - e^{i\frac{a}{2}\mathbf{k}\cdot(1,-1,0)} - e^{-i\frac{a}{2}\mathbf{k}\cdot(1,-1,0)} \right) E_{xy} (1,1,0) \tag{5.33}$$

The second term is positive because of two sign flips. Equation (5.33) is equivalent to

$$\langle \chi_x | H | \chi_y \rangle = -4E_{xy} (1,1,0) \, s_x s_y \tag{5.34}$$

From this discussion, it should be clear how to write matrix elements involving the other $p$ orbitals ($x$–$z$ and $y$–$z$) as well.

We now have seven free parameters. If we do not try to fit the $X$-point energies this is one more than we need for a four-band model without the spin–orbit interaction, but not quite enough to move significantly beyond the **k·p** description. To fix this problem without adding bands, we resort to *next-nearest-neighbor* interactions, with six next-nearest neighbors at $a(\pm1,0,0)$, $a(0,\pm1,0)$, $a(0,0,\pm1)$. Denoting $cc_x \equiv \cos(k_x a)$, $ss_x \equiv \sin(k_x a)$, etc., we can easily add the terms to the $s$–$s$ and $x$–$x$ matrix elements:

$$\langle \chi_s | H | \chi_s \rangle = E_{ss} (0,0,0) + 4E_{ss} (1,1,0) \left( c_x c_y + c_z c_x + c_y c_z \right)$$

$$+ 2E_{ss} (2,0,0) \left( cc_x + cc_y + cc_z \right) \tag{5.35}$$

$$\langle \chi_x | H | \chi_x \rangle = E_{xx} (0,0,0) + 4E_{xx} (1,1,0) \left( c_x c_y + c_z c_x \right)$$

$$+ 4E_{xx} (0,1,1) \, c_y c_z + 2E_{xx} (2,0,0) \, cc_x + 2E_{xx} (0,0,2) \left( cc_y + cc_z \right) \tag{5.36}$$

In Eq. (5.36), two next-nearest-neighbor parameters appear because the integrals over $p_x$ orbitals depend on whether there is translation along the $x$ axis.

In the $s$–$x$ matrix element, we must account for the sign flip of the integral and the cancellation when there is no displacement along the $x$ axis:

$$\langle \chi_s | H | \chi_x \rangle = 4iE_{sx}(1,1,0)\, s_x\left(c_y + c_z\right) + 2iE_{sx}(2,0,0)\, ss_x \qquad (5.37)$$

Finally, Eq. (5.34) still holds for the $x$–$y$ matrix element. All the integrals vanish because the next-nearest neighbors are positioned one lattice constant away along a given axis, but never along two different axes at the same time.

Overall, we have added three parameters to produce a total of ten (no spin–orbit coupling so far). Recall that we needed only two more to fix the $X$-point energies. It is easier to reduce the number of parameters than to increase it, e.g., by expressing $E_{xx}(1,1,0)$, $E_{xx}(0,1,1)$, and $E_{xy}(1,1,0)$ in terms of two quantities (see Appendix C for details).

We have just described the *effective bond-orbital method* (EBOM), using the parameters shown in Table 5.1. This approach is practically useful because all free parameters can be written as closed-form expressions involving the experimental quantities, as we will see later. This is a rare feat among the existing tight-binding variants.

To see the connection between the EBOM and **k·p** approaches, we expand the sines and cosines in Eqs. (5.34)–(5.37) in a Taylor series around the $\Gamma$ point:

**Table 5.1** *Parameters and their reduction to a smaller set for the second-nearest-neighbor EBOM described in the text and Appendix C.*

| Parameter | Approximation |
|---|---|
| $E_{ss}(0,0,0)$ | $V_{ss0}$ |
| $E_{ss}(1,1,0)$ | $V_{ss1}$ |
| $E_{ss}(2,0,0)$ | $V_{ss2}$ |
| $E_{xx}(0,0,0)$ | $V_{pp\sigma 0}$ |
| $E_{xx}(1,1,0)$ | $\left(V_{pp\sigma 1} + V_{pp\pi 1}\right)/2$ |
| $E_{xx}(0,1,1)$ | $V_{pp\pi 1}$ |
| $E_{xx}(2,0,0)$ | $V_{pp\sigma 2}$ |
| $E_{xx}(0,0,2)$ | $V_{pp\pi 2}$ |
| $E_{sx}(1,1,0)$ | $V_{sp\sigma 1}/\sqrt{2}$ |
| $E_{sx}(2,0,0)$ | $0$ |
| $E_{xy}(1,1,0)$ | $\left(V_{pp\sigma 1} - V_{pp\pi 1}\right)/2$ |
| $\Delta$ | $\Delta$ |

$$s_x \approx \frac{a}{2} k_x \tag{5.38a}$$

$$ss_x \approx a k_x \tag{5.38b}$$

$$c_x \approx 1 - \frac{a^2}{8} k_x^2 \tag{5.38c}$$

$$cc_x \approx 1 - \frac{a^2}{2} k_x^2 \tag{5.38d}$$

We neglect any higher-order terms in $k$ in the spirit of the **k·p** method. For example, the products of sines and cosines will retain only the first terms in Eqs. (5.38c) and (5.38d); i.e., the cosines are approximated as unity. The $s$–$x$ matrix element in Eq. (5.37) becomes

$$\langle \chi_s | H | \chi_x \rangle \approx 2ai\, [2E_{sx}(1,1,0) + E_{sx}(2,0,0)]\, k_x \tag{5.39}$$

$E_{sx}(1,1,0)$ and $E_{sx}(2,0,0)$ occur only in this sum, so no information is lost if we set one of them to zero, e.g., $E_{sx}(2,0,0) = 0$.

The $x$–$y$ matrix element in Eq. (5.34) takes the form

$$\langle \chi_x | H | \chi_y \rangle = -a E_{xy}(1,1,0)\, k_x k_y \tag{5.40}$$

and the diagonal matrix elements in Eqs. (5.35) and (5.36) can be simplified as

$$\langle \chi_s | H | \chi_s \rangle = E_{ss}(0,0,0) + 12 E_{ss}(1,1,0) + 6 E_{ss}(2,0,0)$$
$$- a^2\, [E_{ss}(1,1,0) + E_{ss}(2,0,0)]\left( k_x^2 + k_y^2 + k_z^2 \right) \tag{5.41}$$

$$\langle \chi_x | H | \chi_x \rangle = E_{xx}(0,0,0) + 8 E_{xx}(1,1,0) + 4 E_{xx}(0,1,1) + 2 E_{xx}(2,0,0)$$
$$+ 4 E_{xx}(0,0,2) - a^2\, [E_{xx}(1,1,0) + E_{xx}(2,0,0)]\, k_x^2$$
$$- \frac{a^2}{2}\, [E_{xx}(1,1,0) + E_{xx}(0,1,1) + 2 E_{xx}(0,0,2)]\left( k_y^2 + k_z^2 \right) \tag{5.42}$$

We can use Eq. (3.15) to compare how the matrix elements in Eqs. (5.39)–(5.42) vary with $k_x$, $k_y$, and $k_z$. Of course, there are no linear-in-$k$ terms because the fcc lattice has inversion symmetry. As discussed in Section 2.2, the $\boldsymbol{k}$ and $-\boldsymbol{k}$ states are degenerate for both spins. We can readily identify the tight-binding parameters with $E_g$, $p_{cv}$, $L$, $M$, and $N$ in Eq. (3.15). The last four can be re-expressed in terms of the electron effective mass and Luttinger parameters, once we have used the technique discussed in Chapter 3 to include the spin–orbit interaction. This implicitly assumes that the spin–orbit interaction is local, i.e., it exists only on a given lattice site, and that the spin–orbit terms connecting different lattice sites are negligible. The details of this approach (see Appendix C) were first presented by John Loehr. The EBOM is also amenable to application of the deformation-potential theory from Section 3.4, as long as we only care about the effects of strain close to the $\Gamma$ point.

Figure 5.6 compares the band structures of GaAs, GaSb, and InSb as obtained using the 8-band EBOM and $\mathbf{k \cdot p}$ approaches. In these plots, as in Fig. 5.4, zero energy is referenced to the valence band maximum in each material. As expected, the $\mathbf{k \cdot p}$ approximation performs well over the initial $\approx 20\%$ of the distance from $\Gamma$ to $X$, and the two techniques coincide in that limit because of how we fitted the EBOM parameters. Of course, the EBOM and $\mathbf{k \cdot p}$ approaches are fundamentally related via the symmetry of the fcc lattice and the $sp^3$ basis set. We see by comparing with Fig. 5.4 that the EBOM dispersions are more realistic along the entire distance, since the energies at the $X$ point were fitted. In fact, the EBOM functions as a useful interpolation scheme, since we do not attempt to reproduce the nature of the CB states at the $X$ point, but simply aim for a better depiction of the states partway through the zone. It will become clear in Chapter 9 why we might want to extend the $\mathbf{k \cdot p}$ description to larger wavevectors.

We did not have enough parameters to fit the energies at the $L$ point as well, nor would we want to attempt such a tedious fit. Therefore, it is no surprise that the band structures in Fig. 5.6 do not represent these energies accurately. The most glaring error is for GaSb, which is shown as an indirect-gap material with $L$ valleys $\approx 280$ meV *below* the $\Gamma$ valley. We will see in Chapter 6 that in reality the energy gap in GaSb is weakly direct, with the $L$ valleys only 63 meV *above* the $\Gamma$ valley. Other materials such as InAs may actually appear *semimetallic*, because the $L$ valleys are incorrectly positioned *below* the valence band maximum. These are clear limitations of the EBOM.

## 5.3 Second-Nearest-Neighbor Tight-Binding Method

Can we do even better than EBOM and $\mathbf{k \cdot p}$ in describing the full-zone dispersions? For example, we have not yet taken into account the full zinc-blende symmetry of the crystal in place of the reduced fcc symmetry. The catch is that in return for potentially more accurate results, we lose the ability to write analytical expressions for all tight-binding parameters and must fit the experimental energies and effective masses the hard way. This kind of multi-parameter fitting is hardly illuminating and prone to missing the global minimum. In spite of this drawback, the full tight-binding method can be a powerful tool and well worth describing here.

The zinc-blende tight-binding method can be cast in many forms that differ in what neighboring atoms and orbitals are included. Here we discuss a sixteen-band $sp^3$ basis that includes both nearest and next-nearest neighbors, i.e. an extension of the EBOM presented in Section 5.2, as well as a purely nearest-neighbor technique that achieves sufficient accuracy by adding a dummy band to the $sp^3$ basis.

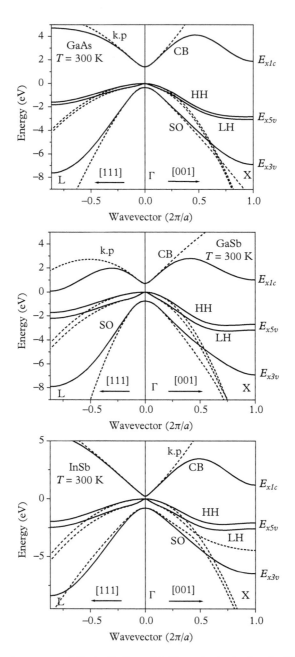

**Figure 5.6** *Band structures of GaAs, GaSb, and InSb, as obtained using the 8-band second-nearest-neighbor effective band-orbital method (EBOM) described in the text (solid curves) and the 8-band **k·p** calculation described in Chapter 3 (dashed curves).*

The nearest neighbors in the zinc-blende lattice are at $\frac{a}{4}(1,1,1)$, $\frac{a}{4}(1,-1,-1)$, $\frac{a}{4}(-1,1,-1)$, and $\frac{a}{4}(-1,-1,1)$. The fcc nearest neighbors, at $\frac{a}{2}(\pm 1, \pm 1, 0)$, $\frac{a}{2}(\pm 1, 0, \pm 1)$, and $\frac{a}{2}(0, \pm 1, \pm 1)$, become second nearest neighbors in the zinc-blende lattice. Fortunately, we already found their matrix elements in Section 5.2. Any terms involving the atoms at $a(\pm 1, 0, 0)$, $a(0, \pm 1, 0)$, and $a(0, 0, \pm 1)$ are ignored, since they are no longer second nearest neighbors.

Since there are two distinct sublattices, we must distinguish between the cation (group III elements) and anion (group V elements) orbitals. Equations (5.28), (5.29), (5.32), and (5.34) will be used for both cations and anions, with necessary substitutions for the other orbitals. Now the more general form of Eq. (5.32) will prove useful (the $E_{sx}(0, 1, 1)$ term in Eq. (5.32) is sometimes shown with a different sign). We have 8 parameters for cations and 8 for anions, a total of 16.

The matrix elements for the nearest neighbors in the zinc-blende lattice are not present in the EBOM, so we need to derive them from scratch. Setting the origin at a cation position, the $s$–$s$ matrix element becomes

$$\langle \chi_{s,a} | H | \chi_{s,c} \rangle$$

$$= \left( e^{i\frac{a}{4}\boldsymbol{k}\cdot(1,1,1)} + e^{i\frac{a}{4}\boldsymbol{k}\cdot(1,-1,-1)} + e^{i\frac{a}{4}\boldsymbol{k}\cdot(-1,1,-1)} + e^{i\frac{a}{4}\boldsymbol{k}\cdot(-1,-1,1)} \right) E_{ss}\left(\frac{1}{2},\frac{1}{2},\frac{1}{2}\right)$$

$$(5.43)$$

Denoting $C_x \equiv \cos(k_x a/4)$, $S_x \equiv \sin(k_x a/4)$, etc., Eq. (5.43) can be rewritten in a somewhat more pleasing form:

$$\langle \chi_{s,a} | H | \chi_{s,c} \rangle = 4E_{ss}\left(\frac{1}{2},\frac{1}{2},\frac{1}{2}\right)(C_x C_y C_z - iS_x S_y S_z) \equiv 4E_{ss}\left(\frac{1}{2},\frac{1}{2},\frac{1}{2}\right)g_0 \quad (5.44)$$

Equations (5.43) and (5.44) are equivalent because the terms with even numbers of positive components in the exponents cancel.

The cation and anion orbitals are the same when the phase factors have the same form, since the shift from cation to anion, or vice versa, involves equal displacements of all three coordinates. For example:

$$\langle \chi_{x,a} | H | \chi_{x,c} \rangle = 4E_{xx}\left(\frac{1}{2},\frac{1}{2},\frac{1}{2}\right)g_0 \quad (5.45)$$

On the other hand, for $s$–$x$ matrix elements we flip the signs in front of the phase factors with a negative $x$ displacement:

$$\langle \chi_{s,a} | H | \chi_{x,c} \rangle$$

$$= \left( e^{i\frac{a}{4}\boldsymbol{k}\cdot(1,1,1)} + e^{i\frac{a}{4}\boldsymbol{k}\cdot(1,-1,-1)} - e^{i\frac{a}{4}\boldsymbol{k}\cdot(-1,1,-1)} - e^{i\frac{a}{4}\boldsymbol{k}\cdot(-1,-1,1)} \right) E_{sx}^{ac}\left(\frac{1}{2},\frac{1}{2},\frac{1}{2}\right)$$

$$(5.46)$$

This transforms to

$$\langle \chi_{s,a}|H|\chi_{x,c}\rangle = 4E_{sx}^{ac}\left(\frac{1}{2},\frac{1}{2},\frac{1}{2}\right)\left(-C_x S_y S_z + i S_x C_y C_z\right) \equiv 4E_{sx}^{ac}\left(\frac{1}{2},\frac{1}{2},\frac{1}{2}\right)g_1$$

(5.47)

From the viewpoint of an anion, the nearest neighbors are at $\frac{a}{4}(-1,1,1)$, $\frac{a}{4}(1,-1,1)$, $\frac{a}{4}(1,1,-1)$, and $\frac{a}{4}(-1,-1,-1))$. The form corresponding to Eq. (5.46) is then

$$\langle \chi_{s,c}|H|\chi_{x,a}\rangle$$

$$= \left(e^{i\frac{a}{4}\boldsymbol{k}\cdot(1,-1,1)} + e^{i\frac{a}{4}\boldsymbol{k}\cdot(1,1,-1)} - e^{i\frac{a}{4}\boldsymbol{k}\cdot(-1,1,1)} - e^{i\frac{a}{4}\boldsymbol{k}\cdot(-1,-1,-1)}\right) E_{sx}^{ca}\left(\frac{1}{2},\frac{1}{2},\frac{1}{2}\right)$$

(5.48)

Expanding the complex exponentials in terms of sines and cosines, this expression can be rewritten as

$$\langle \chi_{s,c}|H|\chi_{x,a}\rangle = 4E_{sx}^{ca}\left(\frac{1}{2},\frac{1}{2},\frac{1}{2}\right)\left(C_x S_y S_z + i S_x C_y C_z\right) = -4E_{sx}^{ca}\left(\frac{1}{2},\frac{1}{2},\frac{1}{2}\right)g_1^*$$

(5.49)

To obtain the $s$–$y$ and $s$–$z$ matrix elements for either $a$–$c$ or $c$–$a$ coupling, we substitute $y$ and $z$ for $x$ in the sines and cosines. For example,

$$\langle \chi_{s,a}|H|\chi_{y,c}\rangle = 4E_{sx}^{ac}\left(\frac{1}{2},\frac{1}{2},\frac{1}{2}\right)\left(-S_x C_y S_z + i C_x S_y C_z\right) \equiv 4E_{sx}^{ac}\left(\frac{1}{2},\frac{1}{2},\frac{1}{2}\right)g_2$$

(5.50a)

$$\langle \chi_{s,a}|H|\chi_{z,c}\rangle = 4E_{sx}^{ac}\left(\frac{1}{2},\frac{1}{2},\frac{1}{2}\right)\left(-S_x S_y C_z + i C_x C_y S_z\right) \equiv 4E_{sx}^{ac}\left(\frac{1}{2},\frac{1}{2},\frac{1}{2}\right)g_3$$

(5.50b)

Obviously, there is no difference between $E_{sx}^{ac}$ and $E_{sy}^{ac}$ (or $E_{sz}^{ac}$), since the nearest-neighbor distance is the same along any of the axes. For a diamond lattice, we could also ignore any differences between, say, $E_{sx}^{ac}$ and $E_{sx}^{ca}$. This does not strictly apply to a zinc-blende crystal, but the relative anion and cation positions are unimportant whenever the two orbitals are both $s$-like or $p$-like. They matter only for the mixed ($s$–$p$) interaction. The anion–cation $s$–$y$ and $s$–$z$ matrix elements follow from the analogy with Eq. (5.49).

The final set of matrix elements is $x$–$y$, $z$–$x$, and $y$–$z$. We assume that $E_{xy}^{ac}\left(\frac{1}{2},\frac{1}{2},\frac{1}{2}\right) \approx E_{xy}^{ca}\left(\frac{1}{2},\frac{1}{2},\frac{1}{2}\right)$. From the discussion preceding Eqs. (5.33) and (5.34), as well as Fig. 5.5c, when the displacements along the first and second axes have opposite signs, the sign is flipped compared to displacements with the same sign. This leads to

$$\langle \chi_{x,a}|H|\chi_{y,c}\rangle = \left(e^{i\frac{a}{4}\boldsymbol{k}\cdot(1,1,1)} - e^{i\frac{a}{4}\boldsymbol{k}\cdot(1,-1,-1)} - e^{i\frac{a}{4}\boldsymbol{k}\cdot(-1,1,-1)} + e^{i\frac{a}{4}\boldsymbol{k}\cdot(-1,-1,1)}\right)$$

$$E_{xy}\left(\frac{1}{2},\frac{1}{2},\frac{1}{2}\right) \qquad (5.51)$$

The phase term in parentheses is identical to that for the $s$–$z$ matrix element in Eq. (5.50b), i.e. $g_3$. The other matrix elements are similar:

$$\langle \chi_{x,a}|H|\chi_{y,c}\rangle = 4E_{xy}\left(\frac{1}{2},\frac{1}{2},\frac{1}{2}\right)g_3 \qquad (5.52a)$$

$$\langle \chi_{x,a}|H|\chi_{z,c}\rangle = 4E_{xy}\left(\frac{1}{2},\frac{1}{2},\frac{1}{2}\right)g_2 \qquad (5.52b)$$

$$\langle \chi_{y,a}|H|\chi_{z,c}\rangle = 4E_{xy}\left(\frac{1}{2},\frac{1}{2},\frac{1}{2}\right)g_1 \qquad (5.52c)$$

Finally, if we interchange the axes but not anions and cations, the phase terms do not change:

$$\langle \chi_{y,a}|H|\chi_{x,c}\rangle = 4E_{xy}\left(\frac{1}{2},\frac{1}{2},\frac{1}{2}\right)g_3 \qquad (5.53a)$$

$$\langle \chi_{z,a}|H|\chi_{x,c}\rangle = 4E_{xy}\left(\frac{1}{2},\frac{1}{2},\frac{1}{2}\right)g_2 \qquad (5.53b)$$

$$\langle \chi_{z,a}|H|\chi_{y,c}\rangle = 4E_{xy}\left(\frac{1}{2},\frac{1}{2},\frac{1}{2}\right)g_1 \qquad (5.53c)$$

We can stop here because the remaining terms can be derived from the Hermiticity condition.

What is the form of the Hamiltonian and the number of free parameters? Apart from the sixteen cation–cation and anion–anion parameters that resemble those from the EBOM (with the extra $s$–$x$ parameter), we require five additional parameters to characterize the cation–anion coupling, as summarized in Table 5.2. We can organize the matrix elements to be inserted into the Hamiltonian in several different ways. For example, we can group the cation or anion terms in the same block:

$$\boldsymbol{H}_{sp^3} = \begin{bmatrix} h_{aa} & h_{ac} \\ h_{ca} & h_{cc} \end{bmatrix} \qquad (5.54)$$

**Table 5.2** *Parameters for the zinc-blende tight-binding model described in the text.*

| Parameter | Notes |
|---|---|
| $E_{ss}(0,0,0)$a, c | one for anions and one for cations |
| $E_{ss}(1,1,0)$a, c | one for anions and one for cations |
| $E_{xx}(0,0,0)$a, c | one for anions and one for cations |
| $E_{xx}(1,1,0)$a, c | one for anions and one for cations |
| $E_{xx}(0,1,1)$a, c | one for anions and one for cations |
| $E_{sx}(1,1,0)$a, c | one for anions and one for cations |
| $E_{sx}(0,1,1)$a, c | one for anions and one for cations |
| $E_{xy}(1,1,0)$a, c | one for anions and one for cations |
| $E_{ss}\left(\frac{1}{2},\frac{1}{2},\frac{1}{2}\right)$ | |
| $E_{xx}\left(\frac{1}{2},\frac{1}{2},\frac{1}{2}\right)$ | |
| $E_{sx}^{ac}\left(\frac{1}{2},\frac{1}{2},\frac{1}{2}\right)$ | |
| $E_{sx}^{ca}\left(\frac{1}{2},\frac{1}{2},\frac{1}{2}\right)$ | |
| $E_{xy}\left(\frac{1}{2},\frac{1}{2},\frac{1}{2}\right)$ | |

where each $h_{ij}$ sub-block is a 4×4 (8×8 with spin) matrix with $h_{aa}$ and $h_{cc}$ given by Eq. (5.6). The off-diagonal sub-blocks are

$$
h_{ac} = \begin{bmatrix}
\langle\chi_{s,a}|H|\chi_{s,c}\rangle & \langle\chi_{s,a}|H|\chi_{x,c}\rangle & \langle\chi_{s,a}|H|\chi_{y,c}\rangle & \langle\chi_{s,a}|H|\chi_{z,c}\rangle \\
\langle\chi_{x,a}|H|\chi_{s,c}\rangle & \langle\chi_{x,a}|H|\chi_{x,c}\rangle & \langle\chi_{x,a}|H|\chi_{y,c}\rangle & \langle\chi_{x,a}|H|\chi_{z,c}\rangle \\
\langle\chi_{y,a}|H|\chi_{s,c}\rangle & \langle\chi_{y,a}|H|\chi_{x,c}\rangle & \langle\chi_{y,a}|H|\chi_{y,c}\rangle & \langle\chi_{y,a}|H|\chi_{z,c}\rangle \\
\langle\chi_{z,a}|H|\chi_{s,c}\rangle & \langle\chi_{z,a}|H|\chi_{x,c}\rangle & \langle\chi_{z,a}|H|\chi_{y,c}\rangle & \langle\chi_{z,a}|H|\chi_{z,c}\rangle
\end{bmatrix}, \qquad (5.55)
$$

and the matrix for $h_{ca}$ is similar.

Another route to more accurate results with minimal complexity is to keep just the nearest-neighbor terms proportional to $g_0$, $g_1$, $g_2$, $g_3$ (i.e. neglecting the next-nearest-neighbor terms that are present in the EBOM) and introduce a dummy band to correct the conduction-band dispersion near the upper valleys. The dummy orbital can have the $s$ symmetry because we do not need very many matrix elements for the correction. It is referred to as $s^*$ to distinguish it from the original $s$-like states. The approach is then called the $sp^3s^*$ tight-binding method. Typically, we need only the off-diagonal $E_{s^*x}^{ac}\left(\frac{1}{2},\frac{1}{2},\frac{1}{2}\right)$ and $E_{s^*x}^{ca}\left(\frac{1}{2},\frac{1}{2},\frac{1}{2}\right)$ elements as well as the cation and anion diagonal elements for the $s^*$ orbital. The band parameters for this implementation of the tight-binding method are given in Chapter 6.

In contrast with the EBOM, some tight-binding implementations ignore the spin–orbit interaction to focus on the "big picture" of dispersion far from the $\Gamma$ point. However, for greater accuracy, the spin–orbit terms should be included. As in the EBOM, the key assumption is again that the local interactions only affect terms coupling the same lattice site. When treated in this manner, the spin–orbit terms enter the cation–cation and anion–anion sub-blocks separately and do not influence the cation–anion terms. The full Hamiltonian has the arrangement

$$
H_{sp^3,so} =
\begin{bmatrix}
h_{aa} + h_{aa}^{so}(\uparrow\uparrow) & h_{ac} & h_{aa}^{so}(\uparrow\downarrow) & 0 \\
h_{ca} & h_{cc} + h_{cc}^{so}(\uparrow\uparrow) & 0 & h_{cc}^{so}(\uparrow\downarrow) \\
h_{aa}^{so}(\downarrow\uparrow) & 0 & h_{aa} + h_{aa}^{so}(\downarrow\downarrow) & h_{ac} \\
0 & h_{cc}^{so}(\downarrow\uparrow) & h_{ca} & h_{cc} + h_{cc}^{so}(\downarrow\downarrow)
\end{bmatrix}
\tag{5.56}
$$

Chapter 6 will tabulate tight-binding parameters for the III–V semiconductors. Their band structures obtained using the $sp^3$ next-nearest-neighbor model without spin–orbit splitting are shown in Fig. 5.4. The calculations use parameters from the literature that rely on explicit formulas for the critical-point energies.

An obvious advantage of the full tight-binding model is that it can accurately fit the $L$-valley and $X$-valley energies, even in a material such as GaSb that has multiple conduction-band valleys at nearly the same energy. We also see that the peculiar change in curvature of the $s$-like conduction band along the $\Gamma$-$X$ direction is due to the interaction with a higher $p$-like band. This is quite a contrast with the EBOM, which imposed the same change via the boundary condition at the $X$ point! Only the lowest $s$-like band in Fig. 5.4 is immaterial to the characteristics of III–V semiconductors.

## 5.4 The Empirical Pseudopotential Method

At this point, our toolbox contains multiple techniques for calculating the band structure of III–V semiconductors. We have seen that they require a deep insight into the symmetries of the basis functions and, in some cases, involve drastic approximations. Is all this work really necessary? Could we not attack the problem more directly, e.g. by expanding the potential and wavefunctions as Fourier series, as we did for a single band in 1D using Eqs. (1.16) and (1.17)? There is indeed a catch in that the rapid potential variations near the atoms make it difficult to treat core electrons. But if what happens near the atoms stays near the atoms, such a reciprocal-space approach might just work.

To that end, we might try to fit the lowest-order expansion coefficients of the potential to the band-structure region of interest. Since this potential is only realistic far away from the atoms, it is known as a *pseudopotential,* and the corresponding solution of the Schrödinger equation in $k$ space is called the *empirical pseudopotential method* (EPM). Of course, we want to include the full zinc-blende symmetry of the crystal. We will see that while the EPM is in fact feasible, it has a few disadvantages: (1) like most tight-binding variations, it requires fitting parameters that do not relate easily to measured quantities such as energy gap and effective masses; and (2) the spin–orbit interaction is more awkward in reciprocal space than in an orbital-like basis. These drawbacks are balanced by a pedagogical advantage in that the EPM overcomes the limitations of the **k·p** model differently from the tight-binding method.

By analogy with Eq. (1.20), the (pseudo-)wavefunction for a wavevector $\boldsymbol{k}$ can be expanded as a Fourier series in the reciprocal lattice vectors:

$$\Psi_{\boldsymbol{k}}(\boldsymbol{r}) = \sum_{\boldsymbol{G}} a_{\boldsymbol{k},\boldsymbol{G}} e^{i\boldsymbol{k}\cdot\boldsymbol{r}} e^{i\boldsymbol{G}\cdot\boldsymbol{r}} \tag{5.57}$$

The pseudopotential is also expanded as a Fourier series:

$$V(\boldsymbol{r}) = \sum_{\boldsymbol{G}'} V_{\boldsymbol{G}'} \, e^{i\boldsymbol{G}'\cdot\boldsymbol{r}} \tag{5.58}$$

The potential has a constant (dc) term $V_0$ that fixes the reference point for the overall energy scale.

We now repeat the steps that led to Eqs. (1.19) and (1.22), but in 3D; i.e., we multiply by $\exp(-i(\boldsymbol{k}+\boldsymbol{G})\cdot\boldsymbol{r})$ and integrate over the crystal cube with length $L$. Only the terms with vanishing phases in the exponentials survive, and the Schrödinger equation becomes

$$\left(\frac{\hbar^2}{2m}|\boldsymbol{k}+\boldsymbol{G}|^2 + V_0\right) a_{\boldsymbol{k},\boldsymbol{G}} + \sum_{\boldsymbol{G}'} V_{\boldsymbol{G}-\boldsymbol{G}'} a_{\boldsymbol{k},\boldsymbol{G}'} = E_{\boldsymbol{k}} a_{\boldsymbol{k},\boldsymbol{G}} \tag{5.59}$$

A useful mnemonic is that all the terms should have the same sum of the subscripts.

Equation (5.59) must be solved separately for each $\boldsymbol{k}$. The number of coupled equations is equal to the number of retained $\boldsymbol{G}'$s, and we hope to use just a few vectors to simplify the solution. In practice, we start from a set of reliable $V_{\boldsymbol{G}}$ that were determined previously for a similar material and then make small adjustments to improve the fit. The Fourier-transformed pseudopotential $V_{\boldsymbol{G}}$ falls off as $G^{-2} \propto a^2$; i.e., any correlation on the scale of many lattice constants is suppressed even for the outer-shell electrons. In fact, this is why the method works in the first place: the $V_{\boldsymbol{G}}$ decrease rapidly with $\boldsymbol{G}$!

The $\boldsymbol{G}'$s with the smallest magnitudes are just multiples of the fcc reciprocal lattice vectors, $\boldsymbol{b}_1$, $\boldsymbol{b}_2$, and $\boldsymbol{b}_3$ in Eq. (5.11). Ranked and subscripted by their magnitude in units of $\frac{2\pi}{a}$, they are

$$\boldsymbol{G}_0 = \frac{2\pi}{a}(0,0,0) \tag{5.60a}$$

$$\boldsymbol{G}_3 = \frac{2\pi}{a}[(1,1,1),(1,1,-1),(1,-1,1),\ (-1,1,1)\cdots(-1,-1,-1)] \tag{5.60b}$$

$$\boldsymbol{G}_4 = \frac{2\pi}{a}[(2,0,0),(0,2,0),(0,0,2),(-2,0,0)\cdots(0,0,-2)] \tag{5.60c}$$

$$\boldsymbol{G}_8 = \frac{2\pi}{a}[(2,2,0),(2,0,2),(0,2,2),(2,-2,0)\cdots(0,-2,-2)] \tag{5.60d}$$

When all permutations of the sign and coordinate are applied, we obtain a total of eight $\boldsymbol{G}_3$, six $\boldsymbol{G}_4$, six $\boldsymbol{G}_8$, and twenty-four $\boldsymbol{G}_{11}$. No other $\boldsymbol{G}'$s with $|\boldsymbol{G}| \le 11\frac{2\pi}{a}$ are possible. In fact, only the $\boldsymbol{G}_0$, $\boldsymbol{G}_3$, and $\boldsymbol{G}_4$ plane waves will be used in the EPM variety discussed here. This means that 15 $a_{\boldsymbol{k},\boldsymbol{G}}$ coefficients are solved at each value of $\boldsymbol{k}$ (or 30 accounting for spin, as discussed later).

How many different $V_{\boldsymbol{G}}$ parameters do we need? We position the origin of the zinc-blende midway between the anion and the cation, which then reside at $\frac{a}{8}(1,1,1) \equiv \boldsymbol{d}$ and $\frac{a}{8}(-1,-1,-1) \equiv -\boldsymbol{d}$. The total pseudopotential is

$$V(\boldsymbol{r}) = V_a(\boldsymbol{r}+\boldsymbol{d}) + V_c(\boldsymbol{r}-\boldsymbol{d}), \tag{5.61}$$

and its Fourier coefficients are

$$
\begin{aligned}
V_{\boldsymbol{G}} &= \frac{1}{L^3}\int d^3r\,[V_a(\boldsymbol{r}+\boldsymbol{d}) + V_c(\boldsymbol{r}-\boldsymbol{d})]\,e^{-i\boldsymbol{G}\cdot\boldsymbol{r}} \\
&= \frac{1}{L^3}\int d^3r\left[V_a(\boldsymbol{r})\,e^{-i\boldsymbol{G}\cdot\boldsymbol{d}} + V_c(\boldsymbol{r})\,e^{i\boldsymbol{G}\cdot\boldsymbol{d}}\right]e^{-i\boldsymbol{G}\cdot\boldsymbol{r}} \\
&= \frac{1}{L^3}\int d^3r\{[V_a(\boldsymbol{r}) + V_c(\boldsymbol{r})]\cos(\boldsymbol{G}\cdot\boldsymbol{d}) - i\,[V_a(\boldsymbol{r}) - V_c(\boldsymbol{r})]\sin(\boldsymbol{G}\cdot\boldsymbol{d})\}\,e^{-i\boldsymbol{G}\cdot\boldsymbol{r}}
\end{aligned}
\tag{5.62}
$$

To see how many non-zero parameters need to be fitted, we take the symmetric and antisymmetric parts of the cation and anion pseudopotentials as the fundamental parameters:

$$V_{\boldsymbol{G}}^s \equiv \frac{1}{L^3}\int d^3r\,[V_a(\boldsymbol{r}) + V_c(\boldsymbol{r})]\,e^{-i\boldsymbol{G}\cdot\boldsymbol{r}} \tag{5.63a}$$

$$V_{\boldsymbol{G}}^a \equiv \frac{1}{L^3}\int d^3r\,[V_a(\boldsymbol{r}) - V_c(\boldsymbol{r})]\,e^{-i\boldsymbol{G}\cdot\boldsymbol{r}}, \tag{5.63b}$$

so that the total Fourier component of the pseudopotential is

$$V_{\boldsymbol{G}} = V_{\boldsymbol{G}}^s \cos(\boldsymbol{G}\cdot\boldsymbol{d}) - iV_{\boldsymbol{G}}^a \sin(\boldsymbol{G}\cdot\boldsymbol{d}) \tag{5.64}$$

The pseudopotentials $V_G^s$ and $V_G^a$ are independent of direction because of the cubic symmetry; i.e., they depend only on the magnitude of $\boldsymbol{G}$. The antisymmetric pseudopotentials vanish in a diamond crystal with two identical atoms and inversion symmetry. In a zinc-blende crystal, the "odd" pseudopotentials have both symmetric and antisymmetric parts, but the symmetric contribution dominates since the anions and cations in a III–V material are not *all that* different. For "even" pseudopotentials, only one contribution survives as dictated by the lattice symmetry. For example, we may set $V_4^s = 0$ because $\cos(\boldsymbol{G} \cdot \boldsymbol{d}) = 0$ when $\boldsymbol{G} = \frac{2\pi}{a}(2,0,0)$. Similarly, $V_8^a = 0$ because $\sin(\boldsymbol{G} \cdot \boldsymbol{d}) = 0$ when $\boldsymbol{G} = \frac{2\pi}{a}(2,2,0)$. Reasoning in this manner, we can deduce the six non-zero pseudopotential parameters listed in Table 5.3 for the 15-plane-wave expansion. The valence-band offset on an absolute scale (see Chapter 6) is set by the dc component $V_0$.

Since the wavefunction is expanded in plane waves, the $\boldsymbol{L} \cdot \boldsymbol{S}$ spin–orbit interaction cannot be expressed compactly in the EPM. We saw in Section 2.2 that this term results from coupling between the radially varying potential and the electron velocity (or equivalently, momentum). The momentum expectation value is a matrix element involving the derivative of the wavefunction in Eq. (5.57); i.e., it is proportional to $\boldsymbol{k} + \boldsymbol{G}$. The electric field in Eq. (2.26) is proportional to the gradient of the potential, and assuming we can substitute the gradient of the pseudopotential, we have another factor of the pseudopotential expansion index $\boldsymbol{G}$. Combining, the spin–orbit coupling term is proportional to $\boldsymbol{L} \cdot \boldsymbol{S} = \boldsymbol{r} \times \boldsymbol{p} \cdot \boldsymbol{S} \propto \boldsymbol{S} \cdot \nabla V \times \boldsymbol{p}$.

We would like to write the matrix element of the spin–orbit interaction between two states that may have different spins, $s$ and $s'$. We expect this will take the form of a spin–orbit "correction" to the Fourier term $V_{\boldsymbol{G}-\boldsymbol{G'}}$ in Eq. (5.59). The extra

Table 5.3 *Parameters for the 15-plane-wave empirical pseudopotential model (EPM) described in the text.*

| Parameter | Notes |
|---|---|
| $V_3^s$ | symmetric, large |
| $V_3^a$ | antisymmetric, small |
| $V_4^a$ | antisymmetric, small |
| $V_8^s$ | high-order, small |
| $V_{11}^s$ | high-order, small |
| $V_{11}^a$ | high-order, small |
| $C'$ | order of 1,000–10,000 |
| $m_{EPM}$ | $0.9m$ |

term is proportional to the product of the gradient of the pseudopotential and the
momentum:

$$V_{\boldsymbol{G}-\boldsymbol{G'}} \to V_{\boldsymbol{G}-\boldsymbol{G'}} + e^{-i\boldsymbol{G}\cdot\boldsymbol{r}} C \left[ V_{\boldsymbol{G''}}\, e^{i\boldsymbol{G''}\cdot\boldsymbol{r}} \boldsymbol{G''} \times (\boldsymbol{k}+\boldsymbol{G'})\, e^{i\boldsymbol{G'}\cdot\boldsymbol{r}} \right] \cdot \langle s'|\boldsymbol{S}|s\rangle \quad (5.65)$$

The motivation for the "fudge factor" $C$ will be given below. As before, the phases
in the complex exponentials must vanish, i.e. $\boldsymbol{G''} = \boldsymbol{G} - \boldsymbol{G'}$. Therefore, we replace
the pseudopotential Fourier component in Eq. (5.59) with

$$V_{\boldsymbol{G}-\boldsymbol{G'}} \to V_{\boldsymbol{G}-\boldsymbol{G'}} + C \left[ V_{\boldsymbol{G}-\boldsymbol{G'}}\, (\boldsymbol{G}-\boldsymbol{G'}) \times (\boldsymbol{k}+\boldsymbol{G'}) \right] \cdot \langle s'|\boldsymbol{S}|s\rangle \quad (5.66)$$

Things are progressing, but recall that we want to include only a few lowest-order
expansion terms into the formalism. Because $V_{\boldsymbol{G}}$ in the spin–orbit terms falls off
as $G^{-2}$, the last term in Eq. (5.66) does not converge so easily. What are we to do?

Time to apply the "fudge factor" $C$! This amounts to multiplying the spin–
orbit term by a large factor and fitting the band structure with the 15 Fourier
components in Eqs. (5.60a)–(5.60c). We define a dimensionless quantity $C'$ and
assume that the spin operator is normalized to $\frac{\hbar}{2}$:

$$C \equiv \frac{i\hbar^2}{4m^2c^2} C' = \frac{iC'}{2mc^2}\frac{\hbar^2}{2m} \quad (5.67)$$

The ratio of $\frac{\hbar^2}{2m}\left(\frac{2\pi}{a}\right)^2$ to the vacuum gap $2mc^2$ is $\approx 1/24{,}000$. Empirically, $C'$
values of several thousand lead to sufficiently fast convergence of the spin–orbit
term. This approach may not be esthetically appealing, but the results speak for
themselves. In partial defense of this species of EPM, we can say that any practical
band-structure method involves approximations, some of them drastic, and others
less so. Here the fitting works primarily because the lattice symmetry remains
embedded in the problem, via the choice of the reciprocal lattice vectors and
Fourier components.

Now we only need to specify the matrix elements of the (normalized) electron
spin operator. Spin adopts discrete values of $+1$ (up) or $-1$ (down) when
measured along the $z$ axis (or $\frac{1}{2}$ and $-\frac{1}{2}$, when normalized to $\hbar$ as in Section 2.2).
When spin is prepared in either an up or down state and then measured along an
orthogonal axis ($x$ or $y$), the result is equally likely to be positive or negative. These
properties are described mathematically by the Pauli spin matrices $\sigma_x$, $\sigma_y$, and $\sigma_z$,
which act on vectors composed of up and down states $\begin{bmatrix} \uparrow \\ \downarrow \end{bmatrix}$ and express the spin
operator:

$$S_x = \sigma_x \equiv \begin{bmatrix} 0 & 1 \\ 1 & 0 \end{bmatrix} \quad (5.68a)$$

$$S_y = \sigma_y \equiv \begin{bmatrix} 0 & -i \\ i & 0 \end{bmatrix} \tag{5.68b}$$

$$S_z = \sigma_z \equiv \begin{bmatrix} 1 & 0 \\ 0 & -1 \end{bmatrix} \tag{5.68c}$$

The Pauli matrices must be complex to satisfy in 3D the requirements mentioned earlier. Like the Hamiltonian itself, the Pauli matrices are Hermitian. They are substituted for the spatial components of the spin vector $S$, with $s$ and $s'$ either $\uparrow$ or $\downarrow$.

Counting $C'$, we have seven parameters available to fit the band structure. Empirically, better fits are obtained when the electron mass is taken to be somewhat smaller than the free-electron mass (typically, $m_{EPM} = 0.9m$). Figure 5.7

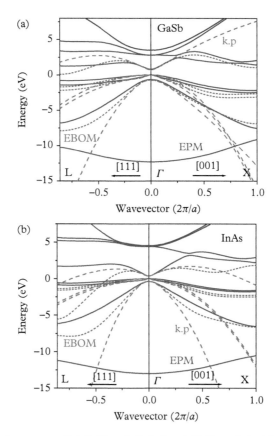

**Figure 5.7** *Band structures of (a) GaSb and (b) InAs at 300 K, as derived from the empirical pseudopotential method (EPM) described in the text, as compared to the 8-band EBOM and* $k \cdot p$ *method.*

plots the results for GaSb and InAs with a mass of $0.9m$ and the pseudopotential Fourier parameters tabulated in Chapter 6. As with the tight-binding method, there is no analytical mapping onto the energy gap, electron mass, and Luttinger parameters. The result does resemble the tight-binding model (Fig. 5.4) in that the low-lying "$s$-like" band has a nearly-free-electron dispersion with the dominant $G_0$ contribution. At the edges of the Brillouin zone, this band interacts with the SO band, which is its folded continuation. A few other points of note are: (1) the EPM bands do not match precisely with the EBOM at the $X$ point due to the limitations of the EPM fitting parameters; (2) the interaction with the higher-lying bands partway through the Brillouin zone is present in the tight-binding and EPM band structures, but not for a reduced-basis method such as EBOM or $\mathbf{k} \cdot \mathbf{p}$; (3) the $L$ valley is represented much more faithfully by the EPM than by the EBOM and $\mathbf{k} \cdot \mathbf{p}$; and (4) the complete failure of the $\mathbf{k} \cdot \mathbf{p}$ dispersion for InAs when $k \sim 0.3 - 0.4 \times (2\pi/a)$ in Fig. 5.7b!

The pseudopotential Fourier components can also be derived wholly or in part from a first-principles formalism such as the density-functional theory. We will see in Chapter 9 that modeling a quantum-confined structure requires many multiples of the number of bulk bands. These computational demands make the 30-band EPM much less practical than the 8-band $\mathbf{k} \cdot \mathbf{p}$ and EBOM for routine optical and electronic device modeling.

## Suggested Reading

- The effective bond-orbital method with next-nearest-neighbor interactions is derived and described in J. P. Loehr , *Physics of Strained Quantum Well Lasers* (Kluwer, Dordrecht, 1997).

- The best references on the tight-binding method for semiconductors are the Ph.D. dissertation of L. C. Lew Yan Voon, "Electronic and Optical Properties of Semiconductors: A Study Based on the Empirical Tight-Binding Method" (Worcester Polytechnic Institute, 1993) and an article by J. P. Loehr and D. N. Talwar, "Exact parameter relations and effective masses within $sp^3$ zinc-blende tight-binding models," *Physical Review B***55**, 4353 (1997).

- The $sp^3 s^*$ tight-binding method is described in P. Vogl, H. P. Hjalmarson, and J. D. Dow, "Semi-empirical tight-binding theory of the electronic structure of semiconductors," *Journal of the Physics and Chemistry of Solids* **44**, 365 (1983).

- The formulation of the empirical pseudopotential method found here is described by G. C. Dente and M. L. Tilton, "Pseudopotential methods for superlattices: Application to mid-infrared semiconductor lasers," *Journal of Applied Physics* **96**, 420 (1999).

- The tight-binding description of the graphene band structure is presented in more detail in J.-N. Fuchs and M. O. Goerbig, "Introduction to the physical properties of graphene", which can be retrieved at http://web.physics.ucsb.edu/~phys123B/w2015/pdf_CoursGraphene2008.pdf.

# Part II

# Band Parameters

# 6

# Binary Compound Semiconductors

If we are to apply the band-structure models discussed in Part I (extended to structures with quantum confinement in Part III and applied to photonic devices in Part IV), we need a reliable set of input parameters. Here we overview the literature related to these parameters that has appeared since our published reviews. We also recommend specific values for all of them, including the dependence on temperature and alloy composition. If reliable experimental reports are available, they are used preferentially in the recommendations. Otherwise, we fall back to extrapolations from the existing data and theoretical estimates to fill in the gaps. Chapter 6 will start by reviewing and tabulating band parameters for the III–V compound semiconductors GaAs, AlAs, InAs, GaSb, AlSb, InSb, GaP, AlP, InP, GaN, AlN, and InN. The parameters include energy gaps, electron and hole mass parameters, deformation potentials, elastic constants, band offsets, and their temperature dependences (if any). Both zinc-blende and wurtzite forms of the nitrides are reviewed.

## 6.1 Introduction

The palette of available III–V materials is limited by nature. The simplest building block is a "binary" material like GaSb, which is formed by the ionic bonding of Ga and Sb atoms. The next step up in complexity is the ternary alloy, a physical mixture of two binary compounds. For example, $In_xGa_{1-x}Sb$ is a mixture of InSb and GaSb, with the InSb (or In) mole fraction given by $x$. The result of mixing more than two binary semiconductors, e.g., $A_xB_yC_{1-x-y}D$ and $A_xB_{1-x}C_yD_{1-y}$, is a quaternary whose composition must be specified with two different mole fractions. We can keep going, but quinternary (also known as penternary) alloys are rare in practice.

Why are we interested in semiconductor alloys? As mentioned earlier, many of the quantum structures used in devices, as described in Parts III and IV, cannot be constructed using only the available binary III–V materials. We may

*Bands and Photons in III–V Semiconductor Quantum Structures.* Vurgaftman, Lumb, and Meyer,
Oxford University Press (2021). © Vurgaftman, Lumb, and Meyer.
DOI: 10.1093/oso/9780198767275.003.0006

wish to select not only a particular energy gap, but also a particular position of the conduction or valence band with respect to a reference energy (or another material), a higher or lower density of states, specific optical properties, etc. Modern epitaxial techniques such as molecular beam epitaxy (MBE) and metal organic chemical vapor deposition (MOCVD) make it possible not only to grow high-quality materials, but also to mix and match them with considerable freedom. Another benefit of these epitaxial techniques is that the band parameters can be measured with much higher precision than at the dawn of the semiconductor age.

However, our freedom to combine different III–V materials to form a desired epitaxial structure is not unlimited. Most important is that approximately the same (average) lattice constant must be maintained throughout all the grown layers. Failure to meet this stringent condition causes strain to build up rapidly, which induces dislocations as discussed in Section 3.4. This is usually fatal for devices that rely on low nonradiative recombination rates (see Chapter 12 for details). But we will see in Chapter 10 that the flexibility needed to realize a given device property can often be gained by growing very thin layers of each material, and then taking advantage of quantum confinement.

Any reliable design of a multilayer semiconductor structure must naturally begin with accurate inputs, i.e., the band parameters for the different III–V semiconductors and their alloys. It also helps to understand how these parameters vary from one material to the next, and what physical mechanisms account for the variations. Here we provide a comprehensive and mutually consistent set of parameters that are critical to the band-structure computations, including: (1) direct and indirect energy gaps and their temperature dependences; (2) spin–orbit splitting; (3) electron effective mass; (4) Luttinger parameters and hole masses; (5) the interband matrix element $E_P$ and associated $F$ parameter described in Chapters 2 and 3; (6) the conduction- and valence-band deformation potentials described in Section 3.4; and (7) band offsets on an absolute scale, which specify the band alignments between any combination of materials.

The earlier sets of these parameters, reported in 2001, 2003, and 2007 [1–3], form the basis for these tabulations. However, the new information from literature reports published since then makes it necessary to thoroughly update these parameters. Instead of repeating the arguments for specific values from the earlier reviews, we highlight the updated recommendations with appropriate citations, and present revised versions of the complete band-parameter tables. We also expand the consideration to dilute bismides, and tabulate tight-binding and pseudopotential parameters for some of the approaches discussed in Chapter 5.

We discussed relationships between the band parameters and measured quantities in Chapters 2, 3, and 5. It remains to point out which parameters are taken to depend on temperature. Most temperature variations ultimately derive from expansion of the semiconductor crystal. Hence the energy gap, as well as the electron, light-hole, and split-off masses, also vary due to their tight connection to the gap as discussed in Section 2.3. The temperature dependence of the energy gap is usually fit to the empirical Varshni form:

$$E_g(T) = E_g(0) - \frac{\alpha T^2}{T + \beta} \tag{6.1}$$

where $T$ is temperature in Kelvin, $\alpha$ and $\beta$ are empirical constants, and $E_g(0)$ is the gap at absolute zero.

There are many ways to account for the temperature dependence of the effective masses. The variation of the effective mass follows from Eq. (3.2) and Eq. (6.1). On the other hand, we do not expect any lowest-order dependence of the heavy-hole mass on temperature, because the effects of remote bands should remain much the same over the temperatures of interest. We assume the same for $E_P$ and $F$, even though Hübner *et al.* [4] have proposed that a better fit to temperature-dependent effective-mass data for GaAs can be achieved when assuming temperature-dependent interband matrix elements. However, these matrix elements cannot be measured directly and instead are derived from other measurements using the physical relationships given in Chapters 2 and 3. Furthermore, there is often a spread in literature values for effective mass reported at any given temperature. These facts combined make it difficult to state authoritatively what the temperature dependence of the matrix elements should be (with the possible important exception of GaAs). In light of this, we recommend adopting a constant value for $E_P$ and $F$ for each binary compound. To account for the variation of the light-hole mass, we adopt the following simple approach: we fix the ratio of the light-hole and electron masses to the value at $T = 0$. Then the temperature dependence of the light-hole mass follows that of the electron mass. While this approach may not be rigorous, it has the virtue of being easy to implement and displays the correct trend. We also neglect any temperature dependences of the deformation potentials and elastic constants. Because the shift in valence-band position is much smaller than that for the conduction band, we also take the valence-band offset to be temperature independent. This is justifiable in view of the significant uncertainties in band alignments that remain for some of the III–V semiconductors.

## 6.2   Zinc-Blende Non-nitride Compounds

The most common, and by far the most technologically important, binary III–V materials have one of three group III elements Ga, In, or Al and one of three group V elements Sb, As, or P. Namely, they are GaSb, GaAs, GaP, InSb, InAs, InP, AlSb, AlAs, and AlP. Many of the band parameters tabulated in the following are derived directly from experimental data, sometimes with small, well-defined corrections. Other parameters can be inferred from the data by fits to established models. In principle, the band parameters can also be deduced from first-principles calculations, although the precision of that approach currently falls well short of the experimental methods, even when they are calibrated against other materials. Therefore, we resort to computations only when few or no experimental data are available. We are most interested in the band parameters entering the eight-band **k·p** approach of Chapter 3.

Although precise values of the relevant parameters are not always easy to come by, there are obvious trends in how they vary with lattice constant and/or energy gap. For example, Fig. 6.1a plots electron, light-hole, and split-off hole effective masses for the nine As-, P-, and Sb-containing binary compounds against Å-valley energy gap. As expected from Eqs. (2.45), (2.47), and (2.49), the masses scale roughly with gap, with that dependence being more pronounced when the gap is narrow. This is because $E_P$ for these III–V compounds falls within a relatively narrow range, as illustrated in Fig. 6.1b.

It is perhaps more surprising that the valence-band offset (VBO) increases almost linearly with lattice constant, as shown in Fig. 6.1c. This stems from the

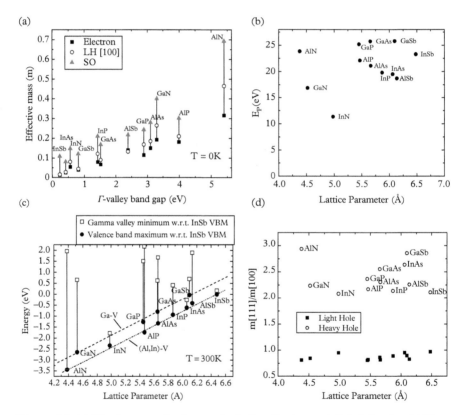

**Figure 6.1** *(a) Effective masses for electrons in the Γ valley (squares), light holes along [100] (circles), and split-off holes (triangles) for the zinc-blende III–V binary compounds, plotted against the Γ-valley energy gap. (b) Interband momentum matrix element $E_P$ for the III–V binary semiconductors. (c) Valence-band maximum relative to InSb (circles), which is seen to scale roughly linearly with binary lattice parameter. This is the Γ-valley conduction-band minima (open squares). (d) Ratio of [111] and [100] direction hole effective masses for III–V compounds with the zinc-blende crystal lattice.*

approximate inverse-square dependence of the off-diagonal matrix elements in the tight-binding Hamiltonian on bond length, which is directly related to the lattice constant [5]. However, this general trend conceals a great deal of complexity. In particular, we see from Fig. 6.1c that the trend line for the valence-band maxima of the Ga–V compounds is displaced in energy with respect to the other III–V materials. While such differences may seem minor in the broad-brush context of the semiconductor band structure, they can have important consequences in practical applications. For example, the type-I alignment of GaAs and AlAs (with the conduction band of GaAs below the conduction band of AlAs, and the valence band above) enables many of the quantum and heterostructure confinement effects discussed in Parts III and IV, while the broken-gap alignment between GaSb and InAs proves at least equally important in midwave- and longwave-infrared photonic device applications.

Figure 6.1c also shows that the conduction-band minima are less predictable, apart from the usual weak correlation between energy gap and lattice constant. Figure 6.1d confirms that the heavy holes are strongly anisotropic, with masses differing by more than a factor of 2 along different directions in the Brillouin zone. On the other hand, the light holes in all the non-nitride zinc-blende compounds are relatively isotropic (also see Fig. 3.3).

While the energy gap correlates only weakly with the lattice constant, there is a much tighter connection in its variation with the product of lattice constant and total atomic number (for both group III and group V elements), $a_L Z_{tot}$. Figure 6.2 shows that a linear dependence emerges for almost all zinc-blende III–V semiconductors, including the semimetals GaBi and InBi that will be discussed in Chapter 7. The simple rule embodied in Fig. 6.2 states that the gaps for two materials with similar lattice constants, e.g., GaAs and AlAs or GaSb and InAs,

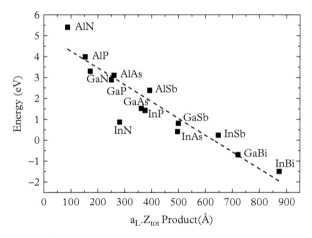

**Figure 6.2** *Room temperature Γ-valley energy gap versus the product of the lattice parameter and atomic number for zinc-blende binary compounds.*

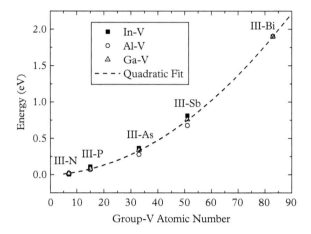

**Figure 6.3** *Spin–orbit splitting energy $\Delta$ for III–V binary compounds with In, Ga, and Al versus group V atomic number.*

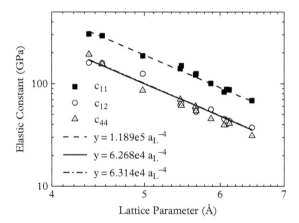

**Figure 6.4** *Log–log plot of the $c_{11}$, $c_{12}$, and $c_{44}$ elastic constants versus lattice parameter for the zinc-blende binaries.*

are inversely proportional to the total atomic number. There appear to be only two significant exceptions to this rule: InN and AlN.

Figure 6.3 illustrates another useful rule of thumb: the spin–orbit splitting $\Delta$ is similar for materials with the same group V atom. For completeness, on the far right we include theoretical estimates for GaBi and InBi [6]. The trend in Fig. 6.3 arises because the core electrons in a material with a heavier group V element (see Section 2.2) provide better screening, along with the group V element's dominance in determining the ionic electrostatic potential of the valence band.

The elastic properties of III–V zinc-blende crystals relate straightforwardly to their lattice constants, as illustrated in Fig. 6.4 for $c_{11}$, $c_{12}$, and $c_{44}$. Dimensional analysis predicts scaling of the form $b^{-4}$, where $b$ is the nearest-neighbor separation [7]. This suggests it is also straightforward to estimate the crystal deformations due to lattice mismatch, which proves useful, e.g., in analyzing X-ray diffraction data. Unfortunately, there appear to be no clear trends in the deformation potentials, beyond the fact that most of the strain-induced energy shift takes place in the conduction band. This book takes the sign of the valence-band deformation potential to be opposite that in [1], as is common in the literature. This better reflects the opposite shift of the conduction- and valence-band edges in response to strain. With this convention, the energy-gap deformation potential becomes $a_c - a_v$.

We already know from Fig. 6.1b that the Kane energy $E_P$ shows little variation across all the III–V materials. Is that also true of the remote-band correction $F$ from Eq. (3.2)? Characterizing $F$ requires that even if we measure the electron mass, we still need an independent determination of $E_P$. One possibility is to fit the absorption spectra as discussed in Chapter 4. However, this was not done traditionally because absorption measurements were considered of lower quality than those for gap and mass. Instead, it was customary to fit $E_P$ from simultaneous measurements of the electron mass and effective g-factor [8]. Assuming dominance of the nearest anti-bonding $p$ band over remote-band effects (see Fig. 2.3), that approach often yielded large $F$ values, some of which were adopted in [1]. Unfortunately, the fits are difficult to reconcile with pseudopotential calculations [9] and other experimental studies [10] that imply much smaller values of $F$. In fact, it is difficult to fully reconcile the measured mass and g-factor using second-order perturbation theory [11]. Large $F$ values can also produce artifacts when quantum structures are simulated by numerical methods as discussed in Chapter 9. While the main motivation for defining separate $E_P$ and $F$ values in the first place was to allow the absorption coefficient to be calculated more accurately using a band structure that depends mostly on the electron mass alone, we saw in Chapter 4 that discrepancies with experiment persist even when the parameter choices are optimal. To summarize, at this writing we see no compelling rationale for enforcing $|F| > 1$. Instead, we take $F = -1$ for GaAs, InAs, InP, and GaSb, and $F = -0.5$ for the indirect-gap materials with poorly known $\Gamma$-valley electron masses (GaP, AlP, AlSb, and AlAs). We then adjust $E_P$ for consistency with the chosen value of $F$.

The uncertainty in $F$ translates into a range of possible values for $E_P$. For the example of GaAs, feasible values of $E_P$ range from 28.8 to 25.6 eV, i.e., on the order of 10 percent. The uncertainties for other materials are similar, and may be higher when few or no measurements have been reported, as in the cases of AlP, AlSb, and AlAs.

While the trends summarized earlier inform our understanding, ultimately we would like to use the best available value for each III–V band parameter. These are tabulated in the following sections and take into account the relevant works published in the past two decades. As mentioned earlier, we do not reproduce the arguments from [1] unless they need to be revised in light of more recent data.

## 6.2.1 GaAs, AlAs, and InAs

Because the band parameters of GaAs have been known to high precision for some time, we recommend only three updates to [1]. Apart from a reduced value of $F$ and the corresponding adjustment of $E_P$, we also recommend a slight increase to the low-temperature electron effective mass after the work by Hübner *et al.* [4] which is based on experimental data from Hopkins *et al.* [12] and Hazama *et al.* [13]. For AlAs, only the room-temperature lattice constant of 5.66139 Å [14] has been updated. Besides the adjusted values of $F$ and $E_P$ for InAs, we introduce an improved low-temperature $X$-valley gap of 1.87 eV [15]. We also adopt revised estimates for the direct-gap Varshni parameters and spin–orbit splitting energies, which were obtained by Lin *et al.* [16] from fits to temperature-dependent photoreflectance data. Minor tweaks to the Luttinger parameters and VBO (based on data for the type-II superlattices discussed in Chapter 10) are also recommended, along with updated values from Saidi *et al.* [15] for the longitudinal and transverse electron masses in the $X$ and $L$ valleys. We also include the bulk spin-splitting parameter, $B_{sp}$ (see Chapter 8), for each binary in units of eV·Å$^3$. For consistency, we recommend the values provided by Jancu *et al.* [17] for the nine common III–V binaries. These are close to the available experimental results. Table 6.1 summarizes the recommended band parameters for GaAs, AlAs, and InAs.

**Table 6.1** *Recommended band parameters for the III–As binary compounds.*

| Parameters | GaAs | AlAs | InAs |
|---|---|---|---|
| $a_L$ (Å) | $5.65325+3.88\times10^{-5}$ $(T-300)$ | $5.66139+2.90\times10^{-5}$ $(T-300)$ | $6.0583+2.7\times10^{-5}$ $(T-300)$ |
| $E_g(\Gamma)$ (eV) | 1.519 | 3.099 | 0.417 |
| $\alpha(\Gamma)$ (eV/K) | $5.41\times10^{-4}$ | $8.85\times10^{-4}$ | $3.07\times10^{-4}$ |
| $\beta$ $(\Gamma)$ (K) | 204 | 530 | 191 |
| $E_g$ $(X)$ (eV) | 1.981 | 2.24 | 1.87 |
| $\alpha$ $(X)$ (eV/K) | $4.60\times10^{-4}$ | $7.00\times10^{-4}$ | $2.76\times10^{-4}$ |
| $\beta$ $(X)$ (K) | 204 | 530 | 93 |
| $E_g$ $(L)$ (eV) | 1.815 | 2.46 | 1.133 |
| $\alpha$ $(L)$ (eV/K) | $6.05\times10^{-4}$ | $6.05\times10^{-4}$ | $2.76\times10^{-4}$ |
| $\beta$ $(L)$ (K) | 204 | 204 | 93 |
| $\Delta$ (eV) | 0.341 | 0.28 | 0.367 |
| $m_e/m$ $(\Gamma)$ | 0.0675 | 0.15 | 0.026 |
| $m_l/m$ $(L)$ | 1.9 | 1.32 | 0.64 |
| $m_t/m$ $(L)$ | 0.0754 | 0.15 | 0.05 |

| | | | |
|---|---|---|---|
| $m_{DOS}/m$ (L) | 0.557 | 0.78 | 0.29 |
| $m_l/m$ (X) | 1.3 | 0.97 | 1.13 |
| $m_t/m$ (X) | 0.23 | 0.22 | 0.16 |
| $m_{DOS}/m$ (X) | 0.852 | 0.75 | 0.64 |
| $\gamma_1$ | 6.98 | 3.76 | 19.7 |
| $\gamma_2$ | 2.06 | 0.82 | 8.4 |
| $\gamma_3$ | 2.93 | 1.42 | 9.3 |
| $m_{SO}/m$ | 0.172 | 0.28 | 0.09 |
| $E_P$ (eV) | 25.6 | 21.1 | 19.5 |
| $F$ | −1.0 | −0.48 | −1.0 |
| VBO (eV) | −0.8 | −1.33 | −0.62 |
| $a_c$ (eV) | −7.17 | −5.64 | −5.08 |
| $a_v$ (eV) | 1.16 | 2.47 | 1 |
| $b$ (eV) | −2.0 | −2.3 | −1.8 |
| $d$ (eV) | −4.8 | −3.4 | −3.6 |
| $c_{11}$ (GPa) | 122.1 | 125 | 83.29 |
| $c_{12}$ (GPa) | 56.6 | 53.4 | 45.26 |
| $c_{44}$ (GPa) | 60.0 | 54.2 | 39.59 |
| $B_{sp}$ (eVÅ$^3$) | 23.7 | 11.4 | 40.5 |

## 6.2.2   GaP, AlP, and InP

We recommend a few minor changes for GaP, including an improved estimate of the $X$-valley bandgap [18, 19] as well as revised values of $F$ and $E_P$. Several refinements are also suggested for AlP, including a revised lattice constant of 5.4635 Å, as determined from an X-ray diffraction fit of AlP grown on GaP [14]. We also adopt Adachi's [20] recommendation of 3.91 eV for the $\Gamma$-valley energy gap at room temperature, based on interpolations from experimental studies of AlGaP. Using the Varshni coefficients from [1], the low-temperature gap becomes 3.99 eV. We also adopt the updated set of elastic coefficients provided by Adachi [20]. The low-temperature $X$- and $L$-valley gaps in InP are modified slightly to 2.32 and 2.12 eV, respectively, for better agreement with Adachi [20]. We also adopt longitudinal and transverse masses for these valleys from the 40-band **k·p** approach of Saidi *et al.* [15]. Table 6.2 summarizes the recommended band parameters for GaP, AlP, and InP.

**Table 6.2** *Recommended band parameters for the III–P binary compounds.*

| Parameters | GaP | AlP | InP |
|---|---|---|---|
| $a_L$ (Å) | $5.4505+2.92\times10^{-5}$ $(T-300)$ | $5.4635+2.92\times10^{-5}$ $(T-300)$ | $5.8687+2.79\times10^{-5}$ $(T-300)$ |
| $E_g$ ($\Gamma$) (eV) | $2.886+0.1081$ $[1\text{-}\coth(164/T)]$ | 3.99 | 1.4236 |
| $\alpha$ ($\Gamma$) (eV/K) | — | $5.77\times10^{-4}$ | $3.63\times10^{-4}$ |
| $\beta$ ($\Gamma$) (K) | — | 372 | 162 |
| $E_g$ ($X$) (eV) | 2.34 | 2.52 | $2.32\text{-}3.70\times10^{-4}T$ |
| $\alpha$ ($X$) (eV/K) | $5.77\times10^{-4}$ | $3.18\times10^{-4}$ | — |
| $\beta$ ($X$) (K) | 372 | 588 | — |
| $E_g$ ($L$) (eV) | 2.72 | 3.57 | 2.12 |
| $\alpha$ ($L$) (eV/K) | $5.77\times10^{-4}$ | $3.18\times10^{-4}$ | $3.63\times10^{-4}$ |
| $\beta$ ($L$) (K) | 372 | 588 | 162 |
| $\Delta$ (eV) | 0.08 | 0.07 | 0.108 |
| $m_e/m$ ($\Gamma$) | 0.13 | 0.22 | 0.0795 |
| $m_l/m$ ($L$) | 2.0 | — | 1.82 |
| $m_t/m$ ($L$) | 0.253 | — | 0.132 |
| $m_{DOS}/m$ ($L$) | 1.27 | — | 0.47 |
| $m_l/m$ ($X$) | 1.2 | 2.68 | 1.35 |
| $m_t/m$ ($X$) | 0.25 | 0.155 | 0.275 |
| $m_{DOS}/m$ ($X$) | 0.624 | 1.01 | 0.88 |
| $\gamma_1$ | 4.05 | 3.35 | 5.08 |
| $\gamma_2$ | 0.49 | 0.71 | 1.6 |
| $\gamma_3$ | 1.25 | 1.23 | 2.1 |
| $m_{SO}/m$ | 0.25 | 0.28 | 0.14 |
| $E_P$ (eV) | 25.2 | 22.1 | 19.8 |
| $F$ | $-0.5$ | $-0.5$ | $-1.0$ |
| VBO (eV) | $-1.27$ | $-1.74$ | $-0.94$ |
| $a_c$ (eV) | $-8.2$ | $-5.7$ | $-6$ |
| $a_v$ (eV) | 1.7 | 3 | 0.6 |

| | | | |
|---|---|---|---|
| $b$ (eV) | −1.6 | −1.5 | −2 |
| $d$(eV) | −4.6 | −4.6 | −5 |
| $c_{11}$ (GPa) | 140.5 | 150 | 101.1 |
| $c_{12}$ (GPa) | 62.03 | 64.2 | 56.1 |
| $c_{44}$ (GPa) | 70.33 | 61.1 | 45.6 |
| $B_{sp}$ (eV·Å$^3$) | −2.4 | 2.1 | −10.1 |

## 6.2.3　GaSb, AlSb, and InSb

The parameters for AlSb are largely unchanged. The relatively few updates for GaSb include revised values of $F$ and $E_P$. Based on Madelung's review [14], we recommend a low-temperature $X$-valley gap of 1.79 eV for InSb. Transverse and longitudinal masses for the $X$ and $L$ valleys are taken from Saidi *et al.* [15].

**Table 6.3** *Recommended band parameters for the III–Sb binary compounds.*

| Parameters | GaSb | AlSb | InSb |
|---|---|---|---|
| $a_L$ (Å) | 6.0959+4.72×10$^{-5}$ | 6.1355+2.60×10$^{-5}$ | 6.4794+3.48×10$^{-5}$ |
| | $(T-300)$ | $(T-300)$ | $(T-300)$ |
| $E_g(\Gamma)$ (eV) | 0.812 | 2.386 | 0.235 |
| $\alpha(\Gamma)$ (meV/K) | 4.17×10$^{-4}$ | 4.20×10$^{-4}$ | 3.20×10$^{-4}$ |
| $\beta$ ($\Gamma$) (meV/K) | 140 | 140 | 170 |
| $E_g$ ($X$) (eV) | 1.141 | 1.696 | 1.79 |
| $\alpha$ ($X$) (meV/K) | 4.75×10$^{-4}$ | 3.90×10$^{-4}$ | — |
| $\beta$ ($X$) (meV/K) | 94 | 140 | — |
| $E_g$ ($L$) (eV) | 0.875 | 2.329 | 0.93 |
| $\alpha$ ($L$) (meV/K) | 5.97×10$^{-4}$ | 5.80×10$^{-4}$ | — |
| $\beta$ ($L$) (meV/K) | 140 | 140 | — |
| $\Delta$ (eV) | 0.76 | 0.676 | 0.81 |
| $m_e/m$ ($\Gamma$) | 0.039 | 0.14 | 0.0135 |
| $m_l/m$ ($L$) | 1.3 | 1.64 | 0.96 |
| $m_t/m$ ($L$) | 0.10 | 0.23 | 0.088 |

*(continued)*

**Table 6.3** *Continued*

| Parameters | GaSb | AlSb | InSb |
|---|---|---|---|
| $m_{DOS}/m$ (L) | 0.592 | 1.116 | 0.492 |
| $m_l/m$ (X) | 1.51 | 1.357 | 1.27 |
| $m_t/m$ (X) | 0.22 | 0.123 | 0.224 |
| $m_{DOS}/m$ (X) | 0.87 | 0.569 | 0.831 |
| $\gamma_1$ | 13.4 | 5.18 | 34.8 |
| $\gamma_2$ | 4.7 | 1.19 | 15.5 |
| $\gamma_3$ | 6 | 1.97 | 16.5 |
| $m_{so}/m$ | 0.12 | 0.22 | 0.11 |
| $E_P$ (eV) | 25.8 | 18.7 | 23.3 |
| $F$ | −1.0 | −0.56 | −0.23 |
| VBO (eV) | −0.03 | −0.41 | 0 |
| $a_c$ (eV) | −7.5 | −4.5 | −6.94 |
| $a_v$ (eV) | 0.8 | 1.4 | 0.36 |
| $b$ (eV) | −2 | −1.35 | −2 |
| $d$ (eV) | −4.7 | −4.3 | −4.7 |
| $c_{11}$ (GPa) | 88.42 | 87.69 | 68.47 |
| $c_{12}$ (GPa) | 40.26 | 43.41 | 37.35 |
| $c_{44}$ (GPa) | 43.22 | 40.76 | 31.11 |
| $B_{sp}$ (eV·Å³) | 167 | 40.9 | 326 |

Table 6.3 summarizes the recommended band parameters for GaSb, AlSb, and InSb.

Table 6.4 shows the $L, M, N$ and $W_s, W_p, W_d$ parameter sets that can be used in place of the Luttinger parameters, computed at room temperature. The different valence-band mass parameters are connected via Eqs. (3.16) and (3.38). Table 6.4 shows that while it is usually true that $| W_d | \ll | W_p |$, in many cases we cannot

**Table 6.4** *Calculated values for the $L, M, N$ parameters and associated valence-band coupling coefficients calculated at $T = 300$ K. The values are given in units of $\hbar^2/2m$.*

| Parameter | GaAs | AlAs | InAs | GaP | AlP | InP | GaSb | AlSb | InSb |
|---|---|---|---|---|---|---|---|---|---|
| $L$ | −1.7443 | 0.2642 | 7.0771 | −1.0480 | 0.6823 | −2.4710 | 0.4303 | 2.2131 | −5.6385 |
| $M$ | 2.8600 | 2.1200 | 2.9000 | 2.4600 | 1.9300 | 1.8800 | 4.0000 | 2.8000 | 3.8000 |
| $N$ | 0.6157 | 1.7442 | 9.5771 | 0.7520 | 1.8723 | −1.3510 | 4.2303 | 4.0931 | −3.4385 |
| $W_s$ | 2.0776 | 0.1625 | −6.8104 | 1.4880 | −0.1889 | 2.9777 | −0.2970 | −1.5998 | 6.7051 |
| $W_p$ | −2.8600 | −2.1200 | −2.9000 | −2.4600 | −1.9300 | −1.8800 | −4.0000 | −2.8000 | −3.8000 |
| $W_d$ | −0.1667 | −0.2133 | −0.1333 | −0.2200 | −0.2467 | −0.2533 | −0.0667 | −0.3067 | −0.5333 |

assume that $|W_s| < |W_p|$. We do not attempt to adjust the parameters to satisfy these rules.

While the quantum structures discussed in Parts III and IV generally do not require any parameters beyond associated with bulk materials, an important exception is the delta-function-like interface terms discussed in Section 8.3. The interface parameters for the InAs/GaSb superlattice are specified in that section. The parameters for other interfaces remain mostly unknown.

## 6.3 Tight-Binding and Pseudopotential Parameters for the III–V Binary Compounds

As we saw in Section 5.2, the EBOM is attractive because it can use the same band parameters (i.e., those in Tables 6.1–6.3) as the 8-band $\mathbf{k \cdot p}$ method. While other tight-binding approaches are more general, they require a new, non-intuitive parameter set for each material. For example, the second-nearest-neighbor tight-binding method of Section 5.3 extracts 21 different parameters by fitting a full-zone calculation to measured (or first-principles) data at various symmetry points. However, we can reduce the parameter space using analytical formulas to relate some of the diagonal matrix elements to critical-point energies [21].

Here we provide parameters for the popular $sp^3s^*$ form of the tight-binding method [5] (see Section 5.3). While the basis set of this approach is extended to 10 bands, the fitting procedure is simplified when only nearest neighbors are used. We need to specify additional diagonal matrix elements for the $s^*$ states of each anion or cation atom, as well as non-zero off-diagonal matrix elements for the interactions between nearest-neighbor $p$ states. For consistency, we adopt the parameters given by Klimeck *et al.* [22], which are listed in Table 6.5.

It is known that the $sp^3s^*$ model has trouble predicting the transverse electron mass at the $X$-valley minimum [23]. Also, the electron mass predictions for both $X$ and $L$ valleys are somewhat hit or miss. More accurate results can be obtained by including second-nearest neighbors [24] or five $d$ orbitals [25]. However, we see from Tables 6.6–6.8 and Fig. 6.5 that the simple $sp^3s^*$ approach describes well many important features of the band structure. This remains the case even if we use parameters that were derived almost twenty years ago [22], partly because many of the underlying band parameters have not changed since then, as discussed in Section 6.1.

Table 6.9 lists recommended parameters for the empirical pseudopotential method described in Section 5.4. Dente *et al.* [26] and Chelikowsky *et al.* [27] fit parameters for some of the III–V materials at a few temperatures. While incomplete, these should provide a useful starting point for fitting other materials and temperatures. Alternative values for InAs and InSb were also proposed by Varea de Alvarez *et al.* [28].

Table 6.5 *Fitted nearest-neighbor sp³s* tight-binding parameters (in eV) from Klimeck et al. [22] for the nine common III–V compounds.*

| Vogl notation | GaAs | AlAs | InAs | GaP | AlP | InP | GaSb | AlSb | InSb |
|---|---|---|---|---|---|---|---|---|---|
| $E(s,a)$ | -3.53284 | -3.21537 | -9.57566 | -8.63163 | -8.93519 | -7.91404 | -7.16208 | -4.5572 | -7.80905 |
| $E(p,a)$ | 0.27772 | -0.09711 | 0.02402 | 0.77214 | 1.13009 | 0.08442 | -0.17071 | 0.01635 | -0.14734 |
| $E(s,c)$ | -8.11499 | -9.52462 | -2.21525 | -1.778 | 0.06175 | -2.76662 | -4.77036 | -4.118 | -2.83599 |
| $E(p,c)$ | 4.57341 | 4.97139 | 4.64241 | 4.17259 | 4.55816 | 4.75968 | 4.06643 | 4.87411 | 3.91522 |
| $E(s^*,a)$ | 12.3393 | 12.0555 | 7.44461 | 11.9005 | 12.8247 | 9.88869 | 7.3219 | 9.84286 | 7.43195 |
| $E(s^*,c)$ | 4.31241 | 3.99445 | 4.12648 | 7.9967 | 9.41477 | 7.66966 | 3.1233 | 7.43245 | 3.5454 |
| $V(s,s)$ | -6.87653 | -8.84261 | -5.06858 | -7.21087 | -6.68397 | -6.16976 | -6.60955 | -6.63365 | -4.89637 |
| $V(x,x)$ | 1.33572 | -0.01434 | 0.84908 | 1.83129 | 2.2863 | 0.75617 | 0.58073 | 1.10706 | 0.7526 |
| $V(x,y)$ | 5.07596 | 4.25949 | 4.68538 | 4.87432 | 5.12891 | 4.2337 | 4.7652 | 4.8996 | 4.4803 |
| $V(sa,pc)$ | 2.85929 | 2.42476 | 2.51793 | 6.12826 | 9.44286 | 3.62283 | 3.00325 | 4.58724 | 3.33714 |
| $V(sc,pa)$ | 11.09774 | 13.20317 | 6.18038 | 6.10944 | 5.93164 | 6.9039 | 7.78033 | 8.53398 | 5.60426 |
| $V(s^*a,pc)$ | 6.31619 | 5.83246 | 3.79662 | 6.69771 | 10.08057 | 4.61375 | 4.69778 | 7.38446 | 4.59953 |
| $V(pa,s^*c)$ | 5.02335 | 4.60075 | 2.45537 | 6.33303 | 4.80831 | 6.18932 | 4.09285 | 6.29608 | -2.53756 |
| $\Delta_a$ | 0.32703 | 0.29145 | 0.38159 | 0.05379 | 0.046 | 0.094 | 0.75773 | 0.70373 | 0.85794 |
| $\Delta_c$ | 0.12 | 0.03152 | 0.37518 | 0.21636 | 0.01608 | 0.54 | 0.15778 | 0.03062 | 0.51 |

Table 6.6 *Computed gaps and masses for GaAs, AlAs, and InAs, as generated using the* $rsp^3s^*$ *nearest-neighbor tight-binding model with recommended parameters from Table 6.5 compared to the values recommended in Table 6.1.*

| Parameter | GaAs | | AlAs | | InAs | |
|---|---|---|---|---|---|---|
| | TB | Recomm. | TB | Recomm. | TB | Recomm. |
| $E_g$ ($\Gamma$) (eV) | 1.424 | 1.4225 | 3.018 | 3.003 | 0.368 | 0.3607 |
| $\Delta$ (eV) | 0.312 | 0.341 | 0.291 | 0.28 | 0.381 | 0.367 |
| $m_e/m$ | 0.068 | 0.0631 | 0.154 | 0.1456 | 0.024 | 0.0227 |
| $m_{lh}/m$ [100] | 0.08 | 0.0837 | 0.151 | 0.1796 | 0.028 | 0.024 |
| $m_{lh}/m$ [110] | 0.073 | 0.0806 | 0.131 | 0.1587 | 0.027 | 0.0264 |
| $m_{lh}/m$ [111] | 0.072 | 0.0779 | 0.127 | 0.1515 | 0.027 | 0.0261 |
| $m_{hh}/m$ [100] | 0.389 | 0.3497 | 0.52 | 0.4717 | 0.364 | 0.345 |
| $m_{hh}/m$ [110] | 0.663 | 0.6431 | 1.1 | 0.8197 | 0.657 | 0.645 |
| $m_{hh}/m$ [111] | 0.838 | 0.8929 | 1.578 | 1.087 | 0.883 | 0.909 |
| $m_{so}/m$ | 0.159 | 0.172 | 0.262 | 0.28 | 0.098 | 0.094 |
| $E_g$ ($X$) (eV) | 1.932 | 1.8989 | 2.285 | 2.164 | 2.345 | 1.807 |
| $m_l/m$ ($X$) | 1.301 | 1.3 | 1.006 | 0.97 | 1.103 | 1.13 |
| $m_t/m$ ($X$) | 3.99 | 0.23 | 2.009 | 0.22 | $\infty$ | 0.16 |
| $E_g$ ($L$) (eV) | 1.708 | 1.707 | 2.351 | 2.352 | 1.46 | 1.07 |
| $m_l/m$ ($L$) | 1.775 | 1.9 | 2.609 | 1.32 | 1.852 | 0.64 |
| $m_t/m$ ($L$) | 0.713 | 0.0754 | 0.86 | 0.15 | 0.304 | 0.05 |

## 6.4 III–N Binary Compounds

### 6.4.1 Wurtzite

III–N compounds with the wurtzite crystal structure, for example wurtzite GaN, are often referred to as hexagonal, or $\alpha$-GaN. We learned in Section 3.5 that the lower wurtzite symmetry leads to additional band parameters such as electron masses along two different directions, spin–orbit and crystal-field splitting $\Delta_s$ and $\Delta_c$, and six different valence-band effective-mass parameters $A_1$–$A_6$. In assigning these parameters, we start from the values published over ten years ago in [2] and [3]. The following updates are recommended for GaN: (1) better estimates of the electron mass from Adachi [20], (2) an improved set of $A$ parameters from Rinke et al. [29], and (3) deformation potentials from Yan et al. [30]. For completeness,

**Table 6.7** *Computed gaps and masses for GaP, AlP, and InP, generated using the $sp^3s^*$ nearest-neighbor tight-binding model and the recommended parameters from Table 6.5, compared to the values recommended in Table 6.2.*

| Parameter | GaP | | AlP | | InP | |
|---|---|---|---|---|---|---|
| | **TB** | **Database** | **TB** | **Database** | **TB** | **Database** |
| $E_g$ (Γ) (eV) | 0.2779 | 2.777 | 3.62 | 3.9127 | 1.345 | 1.3529 |
| $\Delta$ (eV) | 0.08 | 0.08 | 0.04 | 0.07 | 0.105 | 0.108 |
| $m_e/m$ | 0.219 | 0.1231 | 0.249 | 0.2383 | 0.078 | 0.0752 |
| $m_{lh}/m$ [100] | 0.16 | 0.1885 | 0.184 | 0.2053 | 0.082 | 0.1142 |
| $m_{lh}/m$ [110] | 0.142 | 0.1621 | 0.164 | 0.1802 | 0.076 | 0.1107 |
| $m_{lh}/m$ [111] | 0.138 | 0.1527 | 0.159 | 0.1721 | 0.075 | 0.1078 |
| $m_{hh}/m$ [100] | 0.494 | 0.3257 | 0.552 | 0.5181 | 0.48 | 0.5319 |
| $m_{hh}/m$ [110] | 0.809 | 0.5181 | 0.864 | 0.8696 | 0.886 | 0.885 |
| $m_{hh}/m$ [111] | 0.982 | 0.6452 | 1.026 | 1.1236 | 1.187 | 1.1364 |
| $m_{so}/m$ | 0.248 | 0.25 | 0.279 | 0.28 | 0.15 | 0.14 |
| $E_g$ (X) (eV) | 2.432 | 2.2727 | 2.505 | 2.4878 | 2.314 | 2.495 |
| $m_l/m$ (X) | 0.91 | 1.2 | 3.052 | 2.68 | 5.985 | 1.35 |
| $m_t/m$ (X) | 0.254 | 0.25 | ∞ | 0.155 | 11.452 | 0.275 |
| $E_g$ (L) (eV) | 2.635 | 2.6427 | 3.54 | 3.5378 | 1.958 | 1.9433 |
| $m_l/m$ (L) | 24.323 | 2 | >100 | — | 2.588 | 1.82 |
| $m_t/m$ (L) | 2.306 | 0.253 | <0 | — | 0.636 | 0.132 |

as in the earlier reviews we also list the $A_7$ parameter for GaN. This parameter controls the spin splitting away from the zone center.

For AlN and InN, we also recommend the deformation potentials of Yan *et al.* [30]. Additional revisions are made for InN. First, the in-plane and out-of-plane lattice constants are adjusted to 3.55 and 5.76 Å, respectively. A revised energy gap and its temperature dependence are based on the works of Wu *et al.* [31] and Yu *et al.* [32]. We adopt a low-temperature gap $E_g = 0.73$ eV, which leads to the currently accepted value of 0.7 eV at room temperature. Furthermore, we adopt the longitudinal and transverse electron effective masses provided by Adachi [20]. With several sets of elastic coefficients for InN provided in the literature, and no strong reason to favor one over another, we recommend the values in [2].

**Table 6.8** *Computed gaps and masses for GaSb, AlSb, and InSb, generated using the $sp^3s^*$ nearest-neighbor tight-binding model and the recommended parameters from Table 6.5, compared to the values recommended in Table 6.3.*

| Parameter | GaSb | | AlSb | | InSb | |
|---|---|---|---|---|---|---|
| | TB | Database | TB | Database | TB | Database |
| $E_g$ ($\Gamma$) (eV) | 0.751 | 0.7267 | 2.3 | 2.3 | 0.169 | 0.1737 |
| $\Delta$ (eV) | 0.747 | 0.76 | 0.675 | 0.676 | 0.847 | 0.81 |
| $m_e/m$ | 0.042 | 0.035 | 0.121 | 0.1352 | 0.014 | 0.0102 |
| $m_{lh}/m$ [100] | 0.043 | 0.0394 | 0.099 | 0.1277 | 0.014 | 0.0115 |
| $m_{lh}/m$ [110] | 0.04 | 0.0404 | 0.089 | 0.1145 | 0.014 | 0.0149 |
| $m_{lh}/m$ [111] | 0.04 | 0.0394 | 0.086 | 0.1096 | 0.014 | 0.0147 |
| $m_{hh}/m$ [100] | 0.3 | 0.25 | 0.363 | 0.3571 | 0.287 | 0.2632 |
| $m_{hh}/m$ [110] | 0.559 | 0.4878 | 0.632 | 0.6135 | 0.531 | 0.4348 |
| $m_{hh}/m$ [111] | 0.759 | 0.7143 | 0.8 | 0.8065 | 0.732 | 0.5556 |
| $m_{so}/m$ | 0.134 | 0.12 | 0.196 | 0.22 | 0.132 | 0.11 |
| $E_g$ ($X$) (eV) | 1.21 | 1.0325 | 1.632 | 1.6162 | 1.524 | 0.93 |
| $m_l/m$ ($X$) | 1.424 | 1.51 | 1.576 | 0.123 | 1.181 | 1.79 |
| $m_t/m$ ($X$) | 3.183 | 0.22 | 2.734 | 0.569 | $\infty$ | 0.224 |
| $E_g$ ($L$) (eV) | 0.833 | 0.7529 | 2.211 | 2.2104 | 0.93 | 0.93 |
| $m_l/m$ ($L$) | 1.421 | 1.3 | 24.866 | 1.64 | 1.838 | 0.96 |
| $m_t/m$ ($L$) | 0.405 | 0.1 | 1.125 | 0.23 | 0.312 | 0.088 |

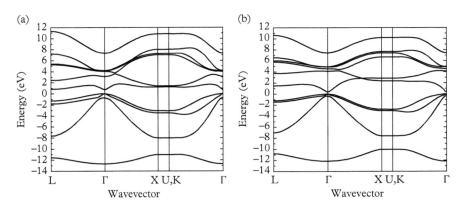

**Figure 6.5** *Band structures for (a) GaSb and (b) InAs, as computed by the $sp^3s^*$ nearest-neighbor tight-binding model.*

**Table 6.9** Parameters employed in the empirical pseudopotential method described in Section 5.4. According to the convention used in the field, all energies are expressed in Rydberg units ($\approx 13.6$ eV).

| | InAs (77 K) [26] | GaSb (77 K) [26] | AlSb (77 K) [26] | Al$_{0.32}$Ga$_{0.68}$As (77 K) [26] | GaAs (77 K) [26] | GaAs (300 K) [26] | GaP (300 K) [27] | InSb (300 K) [27] |
|---|---|---|---|---|---|---|---|---|
| $V_3^s$ | $-0.266$ | $-0.249$ | $-0.23$ | $-0.247$ | $-0.235$ | $-0.248$ | $-0.249$ | $-0.25$ |
| $V_8^s$ | 0.018 | 0.05 | 0.013 | 0.011 | 0 | 0 | 0.017 | 0.01 |
| $V_{11}^s$ | 0.047 | 0.032 | 0.094 | 0.088 | 0.09 | 0.0907 | 0.083 | 0.044 |
| $V_3^a$ | 0.07 | 0.038 | 0.071 | 0.088 | 0.089 | 0.09 | 0.081 | 0.049 |
| $V_4^a$ | 0.038 | 0.004 | 0 | 0.0337 | 0 | 0 | 0.055 | 0.038 |
| $V_{11}^a$ | 0.01 | 0.035 | 0 | 0.00982 | 0.0118 | 0.014 | 0.003 | 0.01 |
| $C'$ | 4100 | 13,800 | 12,000 | 3079 | 4100 | 3900 | | |
| $m_{EPM}/m$ | 0.9 | 0.9 | 0.9 | 0.9 | 0.9 | 0.9 | 0.930 | 0.839 |

**Table 6.10** *Recommended band parameters for III–N binary compounds with the wurtzite structure.*

| Parameters | GaN | AlN | InN |
|---|---|---|---|
| $a_L$ (Å) | 3.189 | 3.112 | 3.55 |
| $c_L$ (Å) | 5.185 | 4.982 | 5.76 |
| $E_g$ ($\Gamma$) (eV) | 3.510 | 6.1 | 0.73 |
| $\alpha$ ($\Gamma$) (meV/K) | 0.914 | 2.63 | 0.31 |
| $\beta$ ($\Gamma$) (K) | 825 | 2082 | 580 |
| $\Delta_c$ (eV) | 0.010 | −0.227 | 0.024 |
| $\Delta_s$ (eV) | 0.017 | 0.036 | 0.005 |
| $m_e^{\parallel}/m$ | 0.22 | 0.32 | 0.039 |
| $m_e^{\perp}/m$ | 0.21 | 0.30 | 0.047 |
| $E_P$ (eV) | 19.8 | 13.6 | 11.4 |
| $F$ | −0.82 | 0 | −1.63 |
| $A_1$ | −6.75 | −3.86 | −8.21 |
| $A_2$ | −0.592 | −0.25 | −0.68 |
| $A_3$ | 6.2 | 3.58 | 7.57 |
| $A_4$ | −2.83 | −1.32 | −5.23 |
| $A_5$ | −2.9 | −1.47 | −5.11 |
| $A_6$ | −3.8 | −1.64 | −5.96 |
| $A_7$ | 0.0937 | 0 | 0 |
| $a_1$ (eV) | −6.1 | −4.4 | −3.6 |
| $a_2$ (eV) | −8.9 | −12.4 | −4.6 |
| $D_1$ (eV) | −3.6 | −2.9 | −1.5 |
| $D_2$ (eV) | 1.7 | 4.9 | 1.2 |
| $D_3$ (eV) | 5.4 | 9.2 | 2.7 |
| $D_4$ (eV) | −2.7 | −3.7 | −1.8 |
| $D_5$ (eV) | −2.6 | −2.9 | −2.1 |
| $D_6$ (eV) | −3.9 | −4.6 | −3 |
| $c_{11}$ (GPa) | 390 | 396 | 223 |
| $c_{12}$ (GPa) | 145 | 137 | 115 |

| | | | |
|---|---|---|---|
| $c_{13}$ (GPa) | 106 | 108 | 92 |
| $c_{33}$ (GPa) | 398 | 373 | 224 |
| $c_{44}$ (GPa) | 105 | 116 | 48 |
| $d_{13}$ (pm/V) | −1.0 | −2.1 | −3.5 |
| $d_{33}$ (pm/V) | 1.9 | 5.4 | 7.6 |
| $d_{15}$ (pm/V) | 3.1 | 3.6 | 5.5 |
| $P_{sp}$ (C/m$^2$) | −0.034 | −0.090 | −0.042 |
| VBO (eV) | 0 | −0.7 | 1.05 |

The valence-band offsets reported for wurtzite nitrides cover a range of values, partly because polarization effects at the interface can complicate the measurements. These lead to apparent non-transitivity [33], and differences for the polar and non-polar directions [34]. The set provided by Martin *et al.* [33], which is corrected for strain-induced piezoelectric fields, is relatively close to most in the literature. Of course, spontaneous and piezoelectric polarization effects should be included separately, as will be discussed in Section 13.5 (see also [35–37]). Depending on the chosen simulation approach, it may be useful to consult the other measured offsets for different III–N heterointerfaces in polar and non-polar directions [34, 38–40]. Table 6.10 summarizes the recommended band parameters for wurtzite GaN, AlN, and InN. Note that the valence-band offset is referenced to that in GaN rather than InSb, since the nitride family remains largely self-contained.

## 6.4.2   Zinc-Blende

While the wurtzite phase of GaN, AlN, and InN has the lowest energy, the III-nitrides also occur in a zinc-blende phase that is often referred to as the cubic, or $\beta$-phase. Cubic III–N compounds are generally of lower quality, however, and they contain high defect densities or even phase mixtures. The practical importance of the zinc-blende III–N compounds is also limited by their narrower direct gaps compared to the wurtzite counterparts, although this does not apply to all potential applications [41].

looseness-1We recommend a few changes to the parameters listed in [2] for the zinc-blende III–N materials. For GaN, we recommend a slightly different lattice constant [42] and an updated spin–orbit splitting based on the studies of Feneberg *et al.* [43]. We also adopt the set of effective-mass and Kane parameters provided by Rinke *et al.* [29]. For AlN, we take the energy gap from As *et al.* [44], the effective-mass and Kane parameters from Rinke *et al.* [29], and the estimated $\Gamma$- and $X$-valley effective masses compiled by Adachi [20]. Finally, we recommend

**Table 6.11** *Recommended band parameters for III–N binary compounds with the zinc-blende structure.*

| Parameters | GaN | AlN | InN |
|---|---|---|---|
| $a_{lc}$ (A) | 4.52 | 4.38 | 4.98 |
| $E_g$ ($\Gamma$) (eV) | 3.299 | 5.99 | 0.56 |
| $\alpha$ ($\Gamma$) (meV/K) | 0.593 | 0.593 | 0.245 |
| $\beta$ ($\Gamma$) (meV/K) | 600 | 600 | 624 |
| $E_g$ (X) (eV) | 4.52 | 4.9 | 2.51 |
| $\alpha$ (X) (meV/K) | 0.593 | 0.593 | 0.245 |
| $\beta$ (X) (meV/K) | 600 | 600 | 624 |
| $E_g$ (L) (eV) | 5.59 | 9.3 | 5.82 |
| $\alpha$ (L) (meV/K) | 0.593 | 0.593 | 0.245 |
| $\beta$ (L) (meV/K) | 600 | 600 | 624 |
| $\Delta$ (eV) | 0.015 | 0.019 | 0.005 |
| $m_e/m$ ($\Gamma$) | 0.193 | 0.316 | 0.054 |
| $m_l/m$ (X) | 0.58 | 0.52 | 0.48 |
| $m_t/m$ (X) | 0.3 | 0.32 | 0.27 |
| $\gamma_1$ | 2.506 | 1.450 | 6.817 |
| $\gamma_2$ | 0.636 | 0.349 | 2.810 |
| $\gamma_3$ | 0.977 | 0.597 | 3.121 |
| $m_{so}/m$ | 0.40 | 0.69 | 0.15 |
| $E_P$ (eV) | 16.86 | 23.84 | 11.37 |
| $F$ | −0.46 | −1.125 | −1.30 |
| VBO (eV) | −2.64 | −3.14 | −2.34 |
| $a_c$ (eV) | −6.71 | −4.5 | −2.65 |
| $a_v$ (eV) | 0.69 | 4.9 | 0.7 |
| $b$ (eV) | −2.0 | −1.7 | −1.2 |
| $d$ (eV) | −3.7 | −5.5 | −9.3 |
| $c_{11}$ (GPa) | 293 | 304 | 187 |
| $c_{12}$ (GPa) | 159 | 160 | 125 |
| $c_{44}$ (GPa) | 155 | 193 | 86 |

the VBO value for AlN (relative to GaN) reported by Mietze *et al.* [45]. For InN, we revise the zero-temperature energy gap to 0.56 eV based on Schormann *et al.* [46]. Table 6.10 summarizes the recommended band parameters for these zinc-blende materials.

# References

[1] I. Vurgaftman, *et al.*, "Band parameters for III-V compound semiconductors and their alloys," *Journal of Applied Physics* **89**, 5815–75 (2001).

[2] I. Vurgaftman and J. R. Meyer, "Band parameters for nitrogen-containing semiconductors," *Journal of Applied Physics* **94**, 3675–96 (2003).

[3] I. Vurgaftman and J. Meyer, "Band parameters for wurtzite nitride semiconductors," in J. Piprek (Ed.), *Nitride Semiconductor Devices: Principles and Simulation*, vol. 590 (Wiley Online Library, New York, 2007).

[4] J. Hübner, *et al.*, "Temperature-dependent electron Landé g factor and the interband matrix element of GaAs," *Physical Review B* **79**, 193–307 (2009).

[5] P. Vogl, *et al.*, "A Semi-empirical tight-binding theory of the electronic structure of semiconductors," *Journal of Physics and Chemistry of Solids* **44**, 365–78 (1983).

[6] H. Li and Z. M. Wang (Eds.), *Bismuth Containing Compounds* (Springer Series in Material Science (Springer, 2013).

[7] R. W. Keyes, "Elastic properties of diamond-type semiconductors," *Journal of Applied Physics* **33**, 3371–2 (1962).

[8] C. Hermann and C. Weisbuch, "k.p perturbation theory in III–V compounds and alloys: a reexamination," *Physical Review B* **15**, 823–33 (1977).

[9] C. M. O. Bastos, *et al.*, "A comprehensive study of g-factors, elastic, structural and electronic properties of III-V semiconductors using hybrid-density functional theory," *Journal of Applied Physics* **123**, 065702 (2018).

[10] A. Adams, *et al.*, "Pressure induced changes in the effective mass of electrons in GaAs, InP and (GaIn)(AsP)/InP," *Physica B+ C* **139**, 401–3 (1986).

[11] M.-H. Serre, *et al.*, "Inconsistency of standard k·p band parameters," *Nanophotonics* 61951B (2006).

[12] M. Hopkins, *et al.*, "A study of the conduction band non-parabolicity, anisotropy and spin splitting in GaAs and InP," *Semiconductor Science and Technology* **2**, 568 (1987).

[13] H. Hazama, *et al.*, "Temperature dependence of the effective masses in III–V semiconductors," *Journal of the Physical Society of Japan* **55**, 1282–93 (1986).

[14] O. Madelung, *Semiconductors: Data Handbook*, 3rd edn. (Berlin: Springer, 2012).

[15] I. Saïdi, *et al.*, "Band parameters of GaAs, InAs, InP, and InSb in the 40-band k.p model," *Journal of Applied Physics* **107**, 043701 (2010).

[16] C. H. Lin, *et al.*, "Infrared photoreflectance of InAs," *Semiconductor Science and Technology* **12**, 1619–24 (1997).

[17] J. M. Jancu, *et al.*, "Atomistic spin-orbit coupling and k.p parameters in III–V semiconductors," *Physical Review B* **72**, 193201 (2005).

[18] H. Sonomura, *et al.*, "Composition dependences of the energy gap and the green-band emission peak for the $Al_xGa_{1-x}P$ ternary system," *Applied Physics Letters* **24**, 77–8 (1974).

[19] R. Nelson and N. Holonyak Jr, "Exciton absorption, photoluminescence and band structure of N-free and N-DOPED $In_{1-x}Ga_xP$," *Journal of Physics and Chemistry of Solids* **37**, 629–37 (1976).

[20] S. Adachi, *Properties of Semiconductor Alloys* (Chichester, UK: Wiley, 2009).

[21] J. Loehr and D. Talwar, "Exact parameter relations and effective masses within sp3s zinc-blende tight-binding models," *Physical Review B* **55**, 4353 (1997).

[22] G. Klimeck, *et al.*, "sp3s★Tight-binding parameters for transport simulations in compound semiconductors," *Superlattices and Microstructures* **27**, 519–24 (2000).

[23] T. B. Boykin, *et al.*, "Effective-mass reproducibility of the nearest-neighbor sp3s★ models: Analytic results," *Physical Review B - Condensed Matter and Materials Physics* **56**, 4102–7 (1997).

[24] T. B. Boykin, "Improved fits of the effective masses at G in the spin-orbit, second-nearest-neighbor sp3s★ model: Results from analytic expressions," *Physical Review B* **56**, 9613–18 (1997).

[25] J.-M. Jancu, *et al.*, "Empirical spds★ tight-binding calculation for cubic semiconductors: General method and material parameters," *Physical Review B* **57**, 6493–507 (1998).

[26] G. C. Dente and M. L. Tilton, "Pseudopotential methods for superlattices: Applications to mid-infrared semiconductor lasers," *Journal of Applied Physics* **86**, 1420–9 (1999).

[27] J. Chelikowsky, *et al.*, "Calculated valence-band densities of states and photoemission spectra of diamond and zinc-blende semiconductors," *Physical Review B* **8**, 2786–94 (1973).

[28] C. Varea De Alvarez, "Calculated band structures, optical constants and electronic charge densities for InAs and InSb," *J. Chem. Phys. Solids* **34**, 337–45 (1973).

[29] P. Rinke, *et al.*, "Consistent set of band parameters for the group-III nitrides AlN, GaN, and InN," *Physical Review B* **77**, 075202 (2008).

[30] Q. Yan, *et al.*, "Effects of strain on the band structure of group-III nitride," *Physical Review B* **90**, 125118 (2014).

[31] J. Wu, *et al.*, "Temperature dependence of the fundamental band gap of InN," *Journal of Applied Physics* **94**, 4457–60 (2003).

[32] K. M. Yu, *et al.*, "On the crystalline structure, stoichiometry and band gap of InN thin films," *Applied Physics Letters* **86**, 071910 (2005).

[33] G. Martin, *et al.*, "Valence-band discontinuities of wurtzite GaN, AlN, and InN heterojunctions measured by x-ray photoemission spectroscopy," *Applied Physics Letters* **68**, 2541–3 (1996).

[34] H. Li, *et al.*, "Determination of polar C-plane and nonpolar A-plane AlN/GaN heterojunction band offsets by X-ray photoelectron spectroscopy," *physica status solidi (b)* **251**, 788–91 (2014).

[35] L. Sang, *et al.*, "Band offsets of non-polar A-plane GaN/AlN and AlN/GaN heterostructures measured by X-ray photoemission spectroscopy," *Nanoscale Research Letters* **9**, 470 (2014).

[36] N. Binggeli, *et al.*, "Band-offset trends in nitride heterojunctions," *Physical Review B* **63**, 245306 (2001).

[37] J.-J. Du, *et al.*, "Determination of N-polar AlN/GaN heterojunction valance-band offsets by X-ray photoelectron spectroscopy," *Materials Letters* **230**, 135–8 (2018).

[38] J. Waldrop and R. Grant, "Measurement of AlN/GaN (0001) heterojunction band offsets by x-ray photoemission spectroscopy," *Applied Physics Letters* **68**, 2879–81 (1996).

[39] P. King, *et al.*, "InN/GaN valence band offset: High-resolution x-ray photoemission spectroscopy measurements," *Physical Review B* **78**, 033308 (2008).

[40] C.-L. Wu, *et al.*, "Valence band offset of wurtzite InN/AlN heterojunction determined by photoelectron spectroscopy," *Applied Physics Letters* **88**, 032105 (2006).

[41] C. J. Stark, *et al.*, "Green cubic GaInN/GaN light-emitting diode on microstructured silicon (100)," *Applied Physics Letters* **103**, 232107 (2013).

[42] V. Bougrov, *et al.*, *Gallium nitride (GaN)*. New York: Wiley, 2001.

[43] M. Feneberg, *et al.*, "Optical properties of cubic GaN from 1 to 20 eV," *Physical Review B* **85**, 155207 (2012).

[44] D. As, "Recent developments on non-polar cubic group III nitrides for optoelectronic applications," *Quantum Sensing and Nanophotonic Devices VII*, 76080G (2010).

[45] C. Mietze, *et al.*, "Band offsets in cubic GaN/AlN superlattices," *Physical Review B* **83**, 195301 (2011).

[46] J. Schörmann, *et al.*, "Molecular beam epitaxy of phase pure cubic InN," *Applied Physics Letters* **89**, 261903 (2006).

# 7

# Alloys and Exotic Materials

We next describe how the band parameters of ternary and quaternary alloys can be interpolated over the entire range of compositions, and tabulate the non-vanishing bowing parameters for most of the common alloys with both zinc-blende and wurtzite lattice structure. We also describe ordering in some of the ternary alloys, and how ordering affects the energy gap. The band parameters of dilute nitrides, dilute bismides, and hexagonal boron nitride are also examined. We finally present schemes for interpolating the optical parameters of III–V alloys, i.e., the real and imaginary parts of the permittivity or dielectric function.

## 7.1 Interpolation Approach for Ternaries

The simplest III–V alloy is a ternary $A_x B_{1-x}$, i.e., a mixture of two binary compounds A and B that share either a group III or group V element. Examples are $Ga_{1-x}In_x Sb$ and $InAs_{1-x}Sb_x$. If the mixture is completely random and no other interactions come into play, a simple average of the binary band parameters over composition should be sufficient. This *virtual crystal approximation* is sufficiently accurate for certain alloy properties such as the lattice constant and elastic constants. However, it does not reliably describe other band parameters, in particular the energy gap and electron mass, which may exhibit a nonlinear dependence on the lattice constant. Fortunately, the band parameter $P_{AB}$ in an alloy often requires only a single correction term:

$$P_{AB}(x) = x P_A + (1-x) P_B - x(1-x) C_{AB}. \qquad (7.1)$$

Since Thompson and Woolley first used this expression to describe the nonlinear variation of energy gaps in ternary III–V alloys, it is sometimes referred to as the Thompson–Woolley formula [1]. $C_{AB}$ in this expression is known as the *bowing parameter,* which quantifies the deviation from the virtual crystal approximation or Vegard's law. For the energy gap we usually have $C_{AB} > 0$; i.e., the gap in the alloy is narrower than would be expected from simple averaging. However, more generally the bowing parameters can be either positive or negative.

*Bands and Photons in III–V Semiconductor Quantum Structures.* Vurgaftman, Lumb, and Meyer,
Oxford University Press (2021). © Vurgaftman, Lumb, and Meyer.
DOI: 10.1093/oso/9780198767275.003.0007

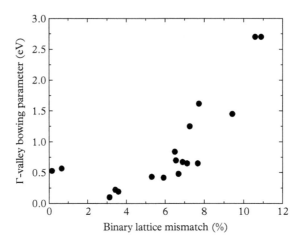

**Figure 7.1** *Energy-gap bowing parameters of the zinc-blende III–V ternary alloys against lattice constant mismatch of the constituent binary compounds. The latter quantity is defined here as $(a_L - a_S)/a_L$, where $a_L$ and $a_S$ are the larger and smaller of the two lattice constants, respectively. If the bowing is not constant, the value at 50% composition is used.*

It is usually difficult to determine accurate bowing parameters from first principles, even though some of the main trends have been explained [2]. In particular, Fig. 7.1 illustrates that the bowing tends to be more pronounced when the bond length difference between the two constituent binaries, in other words the lattice constant mismatch, is large. This is intuitively reasonable if the energy gap's deviation from linearity results mostly from its nonlinear dependence on lattice parameter. Nevertheless, the bowing of the gap does not disappear even when the two endpoint binaries have nearly the same lattice constant, e.g., as in AlGaAs.

Figure 7.2 plots the recommended bandgaps for each ternary in the III–As/P/Sb family as a function of lattice constant. The conduction band minimum in a given alloy may occur in the $\Gamma$ (solid curves), $X$ (dashed), or $L$ (dotted) valley(s). Since high dislocation densities inevitably occur when strain builds up in a thick, non-lattice-matched layer (see Section 6.1), practical limitations usually force the device designer to work near the lattice constant of a common substrate such as GaAs, GaSb, InP, InAs, or InSb. Figure 7.3 plots the valence and $\Gamma$-valley conduction band alignments of the alloys that are lattice-matched to each of those substrates.

Not all III–V alloys are random. Some resist alloying by segregating into distinct phases, in which one or the other binary strongly predominates. Others exhibit spontaneous ordering of the elements within the crystal. A common example of spontaneous order is $In_{0.49}Ga_{0.51}P$ grown on GaAs, in which monolayers of InP and GaP alternate along the [111] direction (CuPt-B ordering). This

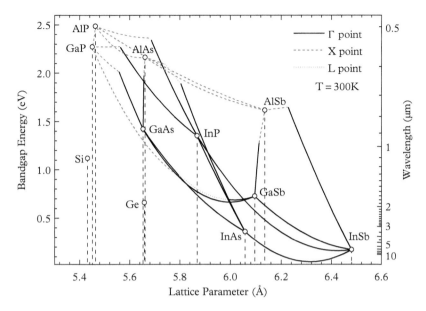

**Figure 7.2** *Lowest direct or indirect energy gap (Γ, X, or L valley) versus lattice parameter for the non-nitride zinc-blende III–V binaries and ternaries.*

most prevalent and well-studied type of ordering is related to specific MBE and MOCVD growth characteristics.

Chapter 10 will explore how superlattices modify the band structure. For now it is enough to note that a random alloy usually produces the widest gap achievable for a particular mixture of binaries. Any periodic arrangement creates additional potential wells that confine carriers below the random-alloy band edges. The degree of ordering is quantified via the $\eta$ parameter (where $\eta = 0$ represents completely disordered and $\eta = 1$ completely ordered material). The most important ordering corrections are to the energy gap and spin–orbit splitting, which are quadratic in the ordering parameter. For example, the shift in energy gap for $0 \leq \eta \leq 1$ is given by $\eta^2 \left( E_g \left( \eta = 1 \right) - E_g \left( \eta = 0 \right) \right) \equiv \Delta E_g \left( \eta \right)$, and the shift in spin–orbit splitting by $\eta^2 \left( \Delta \left( \eta = 1 \right) - \Delta \left( \eta = 0 \right) \right)$, where $E_g \left( \eta = 0 \right)$ and $\Delta \left( \eta = 0 \right)$ are the values tabulated below which correspond to the absence of ordering. Since no alloy is ever perfectly ordered, $E_g \left( \eta = 1 \right)$ and $\Delta \left( \eta = 1 \right)$ are usually calculated theoretically [3]. Table 7.1 summarizes shifts of the low-temperature gaps in several fully ordered alloys, based on [3] and also extensive experimental work on $In_{0.5}Ga_{0.5}P$ [4]. With limited available data, we recommend the same shift at all temperatures.

Most experimental data for the Γ-valley electron mass are given in terms of the mass bowing parameter (see Eq. (7.1)). However, it is important to understand theoretically how this bowing arises from more fundamental quantities such as $E_P$ and $F$. Given the uncertainty in $F$ discussed in Chapter 6, we adopt a linear

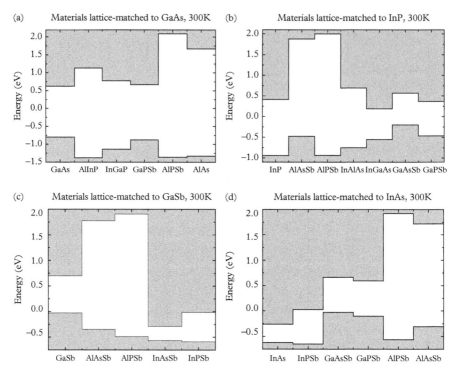

**Figure 7.3** *Alignments of the valence band maximum and Γ-valley conduction band minimum at 300 K of materials (nearly) lattice matched to (a) GaAs, (b) InP, (c) GaSb, and (d) InAs.*

**Table 7.1** *Low-temperature Γ-valley energy gap and gap shift for the best-known fully ordered alloys.*

| Material | $E_g\,(\eta = 1)\,(\mathrm{eV})$ | $\Delta E_g\,(\eta = 1)\,(\mathrm{eV})$ |
|---|---|---|
| $In_{0.5}Ga_{0.5}P$ | 1.7 | 0.49 |
| $InAs_{0.5}Sb_{0.5}$ | −0.28 | 0.40 |
| $In_{0.5}Ga_{0.5}Sb$ | 0.09 | 0.51 |
| $In_{0.5}Ga_{0.5}As$ | 0.55 | 0.30 |
| $GaAs_{0.5}Sb_{0.5}$ | 0.36 | 0.45 |
| $Al_{0.5}Ga_{0.5}As$ | 1.97 | 0.21 |

interpolation for all the III–V alloys. In the tables to follow, the bowing parameter of $E_P$ is adjusted for consistency with Eq. (3.2). This works well for most of the III–V alloys, with the notable exception of InAsSb. In that case, Eq. (7.1) fails to give consistent results at all temperatures. If accurate values for InAsSb are needed over a wide range of temperatures, we recommend using Eq. (3.2).

## 7.2   Recommended Bowing Parameters for the III–V Ternary Alloys

### 7.2.1   AlGaAs

The recommended bowing parameters for the technologically important and well-studied alloy $Al_xGa_{1-x}As$ are unchanged from [5]. Table 7.2 lists the non-zero values. Note that the bowing specified for the $\Gamma$-valley energy gap depends on composition, since a cubic dependence is known to provide a better fit to the experimental data. Here and below, we recommend a zero bowing parameter for any quantity that does not appear in the tables.

**Table 7.2** *Non-zero recommended bowing parameters for AlGaAs.*

| Parameter | Recommended Value |
|---|---|
| $E_g$ ($\Gamma$) (eV) | $-0.127+1.310x$ |
| $E_g$ ($X$) (eV) | 0.055 |
| $E_P$ (eV) | 3.2 |

### 7.2.2   InGaAs

$In_xGa_{1-x}As$ with $x < 0.2 - 0.25$ is often used in light-emitting devices on GaAs substrates. The strain with respect to GaAs remains low enough that quantum wells with narrower gaps can be grown without dislocations. Furthermore, $In_{0.53}Ga_{0.47}As$ is lattice-matched to InP, so compositions around that value are particularly important for devices operating in the telecommunications range. Recommended revisions of the parameter set in [5] include the bowing parameter for $E_P$, as described above, a slight increase in the electron mass for $In_{0.53}Ga_{0.47}As$ to better fit the accepted value at room temperature, and the valence band offset, which has been deduced from a fit of values reported for different binary compounds and ternary alloys [6–11]. The recommended non-zero bowing parameters are given in Table 7.3.

### 7.2.3   InAlAs

$In_xAl_{1-x}As$ is lattice-matched to InP when $x = 0.52$, and it has a direct gap at that composition. It is in fact the widest direct-gap ternary that can be grown on InP. InAlAs is often used as a barrier in devices based on InP substrates, as discussed in Chapter 10. In addition to the updated values of $E_P$ and $F$ discussed in Section 7.1, we also recommend a revised VBO bowing parameter, based on fitting to experimental data from Sacilotti *et al.* [12] and Kawamura *et al.* [13]. The recommended non-zero bowing parameters are listed in Table 7.4.

Table **7.3** *Recommended non-zero bowing parameters for InGaAs.*

| Parameter | Recommended Value |
|---|---|
| $E_g$ ($\Gamma$) (eV) | 0.477 |
| $E_g$ (X) (eV) | 1.4 |
| $E_g$ (L) (eV) | 0.33 |
| VBO (eV) | −0.61 |
| $m_{HH}(100)$ | −0.145 |
| $m_{LH}(100)$ | 0.0202 |
| $m_e$ ($\Gamma$) | 0.005 |
| $a_c$ (eV) | 2.61 |
| $E_P$ (eV) | −4.2 |
| $\Delta$ (eV) | 0.15 |

Table **7.4** *Recommended non-zero bowing parameters for InAlAs.*

| Parameter | Recommended Value |
|---|---|
| $E_g$ ($\Gamma$) (eV) | 0.7 |
| VBO (eV) | −0.824 |
| $m_e$ ($\Gamma$) | 0.049 |
| $a_c$ (eV) | −1.4 |
| $E_P$ (eV) | −10.2 |
| $\Delta$ (eV) | 0.15 |

## 7.2.4 InGaP

$In_xGa_{1-x}P$ maintains a wide direct energy gap when grown lattice-matched to GaAs with $x = 0.49$. For this reason, it has found applications in red light-emitting devices and solar cells (see Chapters 12 and 15). Section 7.1 mentioned the common observation of ordering in InGaP. At room temperature, the $\Gamma$-valley energy gap of disordered $In_{0.49}Ga_{0.51}P$ is 1.91 eV, although for material grown by MOCVD it can be as low as 1.86 eV. The degree of ordering is controlled by growth temperature and substrate, with surfactants such as Sb and Bi commonly used to promote the growth of disordered InGaP. We recommend revising the bowing parameter of the $L$-valley bandgap downward, to 0.34 eV [14], and also adopt Adachi's value for the electron mass bowing [6]. The VBO bowing parameter has been revised to fit the available band offset data [15, 16]. The recommended non-zero bowing parameters are given in Table 7.5.

**Table 7.5** *Recommended non-zero bowing parameters for InGaP.*

| Parameter | Recommended Value |
|---|---|
| $E_g$ (Γ) (eV) | 0.65 |
| $E_g$ (X) (eV) | 0.2 |
| $E_g$ (L) (eV) | 0.34 |
| VBO (eV) | 0.127 |
| $m_e$ (Γ) | 0.022 |
| $E_P$ (eV) | −2.0 |

## 7.2.5 AlInP

Like InGaP, the most common composition of $Al_{1-x}In_xP$ is lattice-matched to GaAs. At this composition ($x = 0.47$), AlInP has an indirect gap and is mainly used as a (transparent) barrier in photonic devices. It becomes direct for somewhat lower Al compositions of 36–39%. Kim *et al.* [17] fit spectroscopic ellipsometry data to extract the compositional dependence of the main critical points in AlInP, although the direct gap was estimated from model predictions rather than a direct fit to the data. Nonetheless, given the generally good agreement with higher critical-point data, we adopt their bowing parameters for the direct bandgap and spin–orbit splitting. In view of the band alignment data from Gudovskikh *et al.* [18] and Kuo *et al.* [15], we recommend a VBO bowing of 104 meV. The recommendations for non-zero bowing parameters are summarized in Table 7.6.

**Table 7.6** *Recommended non-zero bowing parameters for AlInP.*

| Parameter | Recommended Value |
|---|---|
| $E_g$ (Γ) (eV) | 0.672 |
| $E_g$ (X) (eV) | 0.38 |
| VBO (eV) | 0.104 |
| Δ (eV) | −0.043 |

## 7.2.6  AlGaP

Because $Al_{1-x}Ga_xP$ exhibits an indirect gap over its entire compositional range, it has largely been ignored. Our only recommended non-zero bowing parameter is for the $X$-valley gap, which is the lowest point in the conduction band. While uncertainty remains about whether it differs meaningfully from zero [19, 20], we continue to recommend the bowing parameter from [5]. Assuming linear variations of $E_P$ and $F$, we deduce a small downward bowing of the electron mass. Table 7.7 lists these recommendations.

**Table 7.7**  *Recommended non-zero bowing parameters for AlGaP.*

| Parameter | Recommended Value |
|---|---|
| $E_g$ ($X$) (eV) | 0.13 |
| $m_e$ ($\Gamma$) | 0.009 |

## 7.2.7  InGaSb

Like InGaAs, $In_xGa_{1-x}Sb$ retains a direct gap over its entire composition range. While its lattice constant matches a commonly available substrate only at the endpoints, the $In_xGa_{1-x}Sb$ composition in infrared light emitters and detectors is often varied to compensate strain in other layers. As described in Section 7.1, we recommend revision of the $E_P$ bowing parameter for consistency with the slight adjustment of the mass bowing. Recommended non-zero bowing parameters are summarized in Table 7.8.

**Table 7.8**  *Recommended non-zero bowing parameters for InGaSb.*

| Parameter | Recommended Value |
|---|---|
| $E_g$ ($\Gamma$) (eV) | 0.415 |
| $E_g$ ($X$) (eV) | 0.33 |
| $E_g$ ($L$) (eV) | 0.4 |
| $m_{LH}(100)$ | 0.011 |
| $m_e$ ($\Gamma$) | 0.01 |
| $E_P$ (eV) | 7.8 |
| $\Delta$ (eV) | 0.1 |

## 7.2.8  AlInSb

As in the case of InGaSb, no intermediate composition of $Al_{1-x}In_xSb$ is lattice-matched to a readily available substrate. It is used primarily as a barrier layer in strain-compensated designs such as those considered in Chapter 10. Our revisions of the parameters are limited to a new electron mass bowing that assumes linear variations of both $E_P$ and $F$. The recommendations are summarized in Table 7.9.

**Table 7.9** *Recommended non-zero bowing parameters for AlInSb.*

| Parameter | Recommended Value |
| --- | --- |
| $E_g$ ($\Gamma$) (eV) | 0.43 |
| $m_e$ ($\Gamma$) | 0.05 |
| $\Delta$ (eV) | 0.25 |

## 7.2.9  AlGaSb

$Al_xGa_{1-x}Sb$ nearly matches the lattice constant of GaSb, and its gap remains direct below $x \approx 0.5$. As in the case of AlGaAs, the best fit for the $\Gamma$-valley gap is cubic rather than quadratic; i.e., the most accurate bowing parameter is a function of composition. It is used mainly as a barrier material in GaSb-based devices. Revisions are limited to the electron-mass bowing that now follows from linear interpolations of $E_P$ and $F$. Table 7.10 summarizes the recommendations.

**Table 7.10** *Recommended non-zero bowing parameters for AlGaSb.*

| Parameter | Recommended Value |
| --- | --- |
| $E_g$ ($\Gamma$) (eV) | $-0.044+1.22x$ |
| $m_e$ ($\Gamma$) | 0.083 |
| $\Delta$ (eV) | 0.3 |

## 7.2.10  InAsSb

$InAs_{1-x}Sb_x$ boasts the narrowest bandgap of any III–V semiconductor (not counting the dilute nitrides and dilute bismides), which can be as narrow as 100 meV at room temperature. $InAs_{0.91}Sb_{0.09}$ is lattice-matched to GaSb, and has the same broken-gap alignment with GaSb as InAs. For these reasons InAsSb is widely used in long-wave IR detectors (see Chapter 14). We have slightly adjusted the energy gap bowing based on the work of Svensson *et al.* [21], who grew InAsSb on graded buffer layers to produce unstrained and unrelaxed films. Section 7.1 mentioned that effective-mass bowing must be used with care, and that Eq. (3.2) always provides more reliable values. The recommended non-zero bowing parameters for InAsSb are listed in Table 7.11.

**Table 7.11** *Recommended non-zero bowing parameters for InAsSb.*

| Parameter | Recommended Value |
|---|---|
| $E_g$ ($\Gamma$) (eV) | 0.84 |
| $E_g$ ($X$) (eV) | 0.6 |
| $E_g$ ($L$) (eV) | 0.6 |
| VBO (eV) | −0.02 |
| $m_e$ ($\Gamma$) | 0.05 |
| $\Delta$ (eV) | 1.2 |

## 7.2.11  GaAsSb

Different compositions of $GaAs_{1-x}Sb_x$ can be lattice-matched to either InP or InAs. The gap remains direct over all compositions, and exhibits strong bowing. However, it is likely that some literature reports of the bowing have been affected by atomic ordering. Kawamura *et al.* [22] investigated the extent of ordering by comparing partially ordered $GaAs_{0.5}Sb_{0.5}$, grown on (001) InP, with "fully disordered" $GaAs_{0.5}Sb_{0.5}$ grown on (111)B InP. The gap of 0.785 eV for the latter implies $C = 1.16$ eV, although there was a slight mismatch to the substrate. Since this agrees well with $C = 1.25$ eV that Adachi [6] obtained by fitting numerous data sets spanning a wide range of compositions, we continue to recommend Adachi's value. We have revised the VBO bowing for consistency with several values reported for materials matched to InP [7, 9, 23]. As in the case of other alloys for which no mass bowing has been reported, we derive a new recommended value by assuming linear variations of $E_P$ and $F$. Table 7.12 lists the recommended non-zero bowing parameters for GaAsSb.

## 7.2.12  AlAsSb

$AlAs_{1-x}Sb_x$ can be grown lattice-matched to an InP, GaSb, or InAs substrate, and like most other Al-containing materials it is used mainly as a barrier layer. Only recently has a detailed understanding of the energy levels in $AlAs_{1-x}Sb_x$ become available. From in situ ellipsometry measurements over a range of compositions,

**Table 7.12** *Recommended non-zero bowing parameters for GaAsSb.*

| Parameter | Recommended Value |
|---|---|
| $E_g$ ($\Gamma$) (eV) | 1.25 |
| $E_g$ ($X$) (eV) | 1.2 |
| $E_g$ ($L$)(eV) | 1.2 |
| VBO (eV) | −0.895 |
| $m_e$ ($\Gamma$) | 0.058 |
| $\Delta$ (eV) | 0.6 |

Kim *et al.* [24] observed strong downward bowing of the $\Gamma$-valley and $X$-valley bandgaps, and we adopt their values. Based on fits to data from several sources [10, 11, 13, 23], we have revised the VBO bowing parameter to be composition dependent and result in strong upward bowing for compositions near the InP lattice constant and weak bowing for compositions near the GaSb lattice constant. We estimate the electron-mass bowing by assuming linear variations of $E_P$ and $F$. Table 7.13 summarizes all the non-zero bowing parameters for AlAsSb.

**Table 7.13** *Recommended non-zero bowing parameters for AlAsSb.*

| Parameter | Recommended Value |
|---|---|
| $E_g$ ($\Gamma$) (eV) | 1.55 |
| $E_g$ ($X$) (eV) | 0.85 |
| $E_g$ ($L$) (eV) | 0.28 |
| VBO (eV) | −3.41+3.59$x$ |
| $m_e$ ($\Gamma$) | 0.081 |
| $\Delta$ (eV) | 0.15 |

## 7.2.13  GaAsP

GaAs$_{1-x}$P$_x$ was important in early light-emitting devices, and is still used as a strain-compensating barrier for growths on GaAs. We recommend a small electron-mass bowing based on linear variations of $E_P$ and $F$. The recommendations are listed in Table 7.14.

**Table 7.14** *Recommended non-zero bowing parameters for GaAsP.*

| Parameter | Recommended Value |
| --- | --- |
| $E_g$ ($\Gamma$) (eV) | 0.19 |
| $E_g$ ($X$) (eV) | 0.24 |
| $E_g$ ($L$) (eV) | 0.16 |
| $m_e$ ($\Gamma$) | 0.0006 |

## 7.2.14  InAsP

The gap for $InAs_{1-x}P_x$ is direct over its entire range, with values spanning 0.35 to 1.35 eV. However, only its endpoints match the lattice constants of common substrates. We recommend a small electron-mass bowing based on linear variations of $E_P$ and $F$. Table 7.15 summarizes the recommended non-zero bowing parameters.

**Table 7.15** *Recommended non-zero bowing parameters for InAsP.*

| Parameter | Recommended Value |
| --- | --- |
| $E_g$ ($\Gamma$) (eV) | 0.1 |
| $E_g$ ($X$) (eV) | 0.27 |
| $E_g$ ($L$) (eV) | 0.27 |
| $m_e$ ($\Gamma$) | 0.0071 |
| $\Delta$ (eV) | 0.16 |

## 7.2.15  AlAsP

Very little is known about $AlAs_{1-x}P_x$. It is used sparingly, most often for strain compensation. The energy gap is indirect, with conduction-band minimum at the $X$ point for all compositions. We have revised the $X$- and $L$-valley gap bowing parameters based on band-structure calculations [2]. While uncertainty is high, we retain the $\Gamma$-valley gap bowing from [5] because of this alloy's similarity to InAsP and GaAsP. We recommend a small electron-mass bowing based on linear variations of $E_P$ and $F$. Table 7.16 summarizes the recommended non-zero bowing parameters.

Table **7.16** *Recommended non-zero bowing parameters for AlAsP.*

| Parameter | Recommended Value |
|---|---|
| $E_g$ ($\Gamma$) (eV) | 0.22 |
| $E_g$ ($X$) (eV) | 0.40 |
| $E_g$ ($L$) (eV) | 0.38 |
| $m_e$ ($\Gamma$) | 0.0089 |

## 7.2.16 InPSb

The interesting $InP_{1-x}Sb_x$ alloy spans a wide range of gaps, exhibits strong bowing, and is lattice-matched to both InAs and GaSb at intermediate compositions. However, a miscibility gap is predicted, and few machines allow the simultaneous epitaxial growth of antimonides and phosphides. Studies by Reihlen *et al.* [25] and Duncan *et al.* [26] suggest a slightly smaller bowing parameter for the $\Gamma$-valley bandgap than the low-temperature value from Jou *et al.* [27]. Adachi's least-squares fit to these data implies $C = 1.45$ eV, which we adopt for the $X$ and $L$ valleys as well. Table 7.17 summarizes the recommendations.

Table **7.17** *Recommended non-zero bowing parameters for InPSb.*

| Parameter | Recommended Value |
|---|---|
| $E_g$ ($\Gamma$) (eV) | 1.45 |
| $E_g$ ($X$) (eV) | 1.45 |
| $E_g$ ($L$) (eV) | 1.45 |
| $m_e$ ($\Gamma$) | 0.088 |
| $\Delta$ (eV) | 0.75 |

## 7.2.17 GaPSb

$GaP_{1-x}Sb_x$ has an indirect gap over much of its composition range. Also because phosphide antimonides are difficult to grow, it has received little attention. The bowing parameters recommended in [5] were deduced from PL data that were published separately by Loualiche *et al.* [28] and Shimomura [29]. Although those values are much greater than the theoretical predictions of 0.7–0.8 eV [2], we continue to recommend them. Table 7.18 summarizes the non-zero bowing parameters.

Table 7.18 *Recommended non-zero bowing parameters for GaPSb.*

| Parameter | Recommended Value |
|---|---|
| $E_g$ ($\Gamma$) (eV) | 2.7 |
| $E_g$ ($X$) (eV) | 2.7 |
| $E_g$ ($L$) (eV) | 2.7 |
| $m_e$ ($\Gamma$) | 0.102 |

## 7.2.18  AlPSb

Hardly any information is available about this indirect-gap alloy. Therefore, Ref. [5] suggested adoption of the bowing parameters for GaPSb. Although we continue to recommend that approach for the $\Gamma$-valley gap, based on the trend in Fig. 7.1 we employ calculated values from Chen and Sher [2] for the indirect-valley bowing. Table 7.19 summarizes the recommended non-zero bowing parameters.

Table 7.19 *Recommended non-zero bowing parameters for AlPSb.*

| Parameter | Recommended Value |
|---|---|
| $E_g$ ($\Gamma$) (eV) | 2.7 |
| $E_g$ ($X$) (eV) | 0.28 |
| $E_g$ ($L$) (eV) | 0.76 |
| $m_e$ ($\Gamma$) | 0.125 |

# 7.3  Bowing Parameters for the III–N Ternary Alloys

## 7.3.1  Wurtzite

### 7.3.1.1  *AlGaN*

$Al_xGa_{1-x}N$ spans a range of gaps from the visible to the deep ultraviolet (UV). Besides functioning as a barrier, like most other Al-containing materials, it is also an important light-emitting material for the UV. Only the $\Gamma$-valley gap has been studied in great detail, with conflicting reports finding both upward and downward bowing. Experimental luminescence spectra are often difficult to interpret because the fundamental gap can be masked by defect or impurity-related transitions.

We recommend the bowing parameter in [30], which agrees with the data set of Yun *et al.* [31]. The bowing parameters for spontaneous polarization from [30] are recommended for this and the other ternaries. Table 7.20 summarizes all the recommended non-zero bowing parameters for both wurtzite and zinc-blende AlGaN, InGaN, and AlInN.

### 7.3.1.2 *InGaN*

Because of its central role in blue and green nitride-based light emitters, $In_xGa_{1-x}N$ is one of the most technologically significant materials to emerge in recent decades. However, phase segregation and high dislocation densities in the epitaxial films of this alloy make it difficult to characterize its bulk properties. Several recent studies have provided improved data for estimating the bowing of the Γ-valley gap. We adopt the value 1.64 eV obtained by Adachi [6] by fitting to a range of data sets, which agrees quite well with results from a recent spectroscopic ellipsometry study by Sakalauskas *et al.* [32].

### 7.3.1.3 *AlInN*

Although $Al_xIn_{1-x}N$ is the least studied of the III–N ternaries, it is important because it can be lattice-matched to GaN. The composition dependence implied by early characterizations of the energy gap in AlInN, when combined with uncertainties in the gaps of the endpoint binaries, appeared to fit poorly to a quadratic bowing model. While [33] recommended a cubic expression for the bandgap, Adachi [6] found that a constant bowing of 3.7 eV gave reasonable agreement with more recent studies. This also agrees with a recent study by Cramer *et al.* [34]. Some experimental studies of the VBO between AlInN and GaN [35–38], in conjunction with endpoint values from Martin *et al.* [39], suggest a strong upward bowing of the VBO. We recommend $C = -2.23$ eV, with the additional note that any reliable simulation should include spontaneous and piezoelectric polarization effects.

## 7.3.2  Cubic

We mentioned earlier that the zinc-blende III–N alloys are significantly less studied than their wurtzite counterparts, because of the greater difficulty of growing homogeneous, single-phase crystalline films. It follows that the uncertainty in their band parameters is substantial. Figure 7.4 plots the recommended dependences of energy gap on lattice constant for all the wurtzite and zinc-blende nitride ternaries.

### 7.3.2.1 *AlGaN*

Reports of the energy gap for AlGaN range from modest downward bowing [40] to a linear trend [41]. Because a wider composition range was studied in the former case, we recommend $C = 1.0$ eV.

Table 7.20 Recommended bowing parameters for wurtzite and zinc-blende III–N ternary alloys.

| Parameters | InGaN | | AlGaN | | AlInN | |
|---|---|---|---|---|---|---|
| | Wurtzite | Zinc-blende | Wurtzite | Zinc-blende | Wurtzite | Zinc-blende |
| $E_g$ (Γ) (eV) | 1.64 | 0 | 0.8 | 1.0 | 3.7 | 0 |
| $E_g$ (X) (eV) | 0.69 | 0 | 0.61 | 0 | 0.61 | 0 |
| $E_g$ (L) (eV) | 1.84 | 0 | 0.8 | 0 | 0.8 | 0 |
| $\alpha$ (meV/K) | 0 | 0 | 2.15 | 0 | 0 | 0 |
| $\beta$ (K) | 0 | 0 | 1561 | 0 | 0 | 0 |
| VBO (eV) | 0 | 0 | 0 | 0 | −2.23 | 0 |
| $P_{sp}$ (C/m$^2$) | −0.037 | — | −0.021 | — | −0.070 | — |

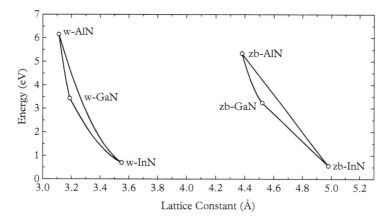

**Figure 7.4** *Room temperature Γ-valley bandgap versus lattice constant for all the III–N compounds and ternary alloys with both wurtzite and zinc-blende lattice arrangements.*

### 7.3.2.2 *InGaN*

Reflectivity and transmission data, along with spectroscopic ellipsometry studies of cubic InGaN by Mullhauser *et al.* [42] and Goldhahn *et al.* [43], suggest a linear variation in the fundamental energy gap. In the absence of better information, we recommend assuming all other parameters do so as well.

### 7.3.2.3 *AlInN*

The lack of experimental reports for cubic AlInN leads us to recommend the linear interpolation of all parameters between the endpoints.

## 7.4 Quaternary Alloys

Quaternary alloys may have two group III and two group V elements, three group III and one group V element, or one group III and three group V elements. Of these, only the first two cases are commonly encountered. In principle, a quaternary with a given lattice constant can more flexibly offer a desired energy gap and band offset. But this often comes at the expense of greater uncertainty, particularly when the quaternary mixes group V elements. Many quaternaries also feature wide miscibility gaps, that is, composition ranges in which thermal equilibrium favors a binary or ternary mixture rather than the desired quaternary. While the miscibility gap does not fundamentally limit a non-equilibrium growth technique such as MBE, it nonetheless becomes more challenging to obtain a high-quality random alloy with the composition in question.

Among the various techniques for calculating the band parameters of quaternary alloys with mixed group III and V elements from the constituent binaries, one preferred approach is that of Donati *et al.* [44]. This method employs a bi-quadratic interpolation algorithm with coefficient $D$ that describes the composition-dependent bowing between different ternaries. Assuming the binary parameters $B_1, B_2, B_3$, and $B_4$, and that the ternary bowing parameters in Eq. (7.1) are $C_{12}, C_{14}, C_{23}$, and $C_{34}$, the quaternary band parameters are

$$
\begin{aligned}
Q(x,y) = {} & y(1-x)B_1 + xyB_2 + x(1-y)B_3 + (1-x)(1-y)B_4 \\
& - x(1-x)(1-y)C_{34} - xy(1-x)C_{12} - y(1-x)(1-y)C_{14} \\
& - xy(1-y)C_{23} - xy(1-x)(1-y)D
\end{aligned}
\tag{7.2}
$$

For the example of $In_xGa_{1-x}As_ySb_{1-y}$, $x$ and $y$ denote the mole fractions of the cationic and anionic species. We can also use Eq. (7.2) for alloys with three group III or three group V elements, e.g., $Al_x[Ga_yIn_{1-y}]_{1-x}As$, by making two of the binary parameters equal and setting the corresponding bowing parameter to zero, e.g., $B_2 = B_3, C_{23} = 0$.

Most technologically important quaternaries are grown to match the lattice constant of a common III–V substrate. We can take advantage of this, e.g., by writing the quaternary $In_xGa_{1-x}As_{1-y}Sb_y$ lattice-matched to InAs as $(InAs)_z(GaAs_{0.08}Sb_{0.92})_{1-z}$. It is then easier to compute the alloy's band parameters by interpolating between the binary or ternary endpoints with the same lattice constant, rather than from Eq. (7.2). For binary and ternary endpoints $\rho$ and $\sigma$ and bowing parameter $C_{\rho\sigma}$ (if one or both of the endpoints is a ternary, $C_{\rho\sigma}$ is not listed in the preceding tables), the quaternary band parameters are given by a variant of Eq. (7.1):

$$
Q_{LM}(z) = zP_\rho + (1-z)P_\sigma - z(1-z)C_{\rho\sigma}
\tag{7.3}
$$

This expression is not strictly equivalent to Eq. (7.2) for several reasons. First, the quaternary's lattice constant is not the same for all $z$. Figure 7.5 shows that the contours for fixed lattice constant are slightly curved rather than straight lines. Second, Eq. (7.3) neglects the higher-order (cubic and fourth-order) terms present in Eq. (7.2). Nevertheless, considering that quaternary band parameters tend to be uncertain anyway, Eq. (7.3) is reasonably accurate and simplifies the thinking about these alloys. Whenever reliable bowing parameters are available for Eq. (7.3), $D$ can be chosen to minimize the quantitative difference between the two methods.

Table 7.21 lists the recommended non-zero bowing parameters for the lattice-matched quaternaries, as computed from Eq. (7.3). Revisions of the bowings for $(In_{0.49}Ga_{0.51}P)_{1-z}(GaAs)_z$ and AlGaAsSb on InP are based on recent experimental data [45]. The electron-mass bowings assume linear variations of $E_P$ and $F$, except for AlGaInAs where $E_P$ has been fit to match the experimentally determined mass.

Table 7.21 *Recommended bowing parameters for quaternary alloys matched to various substrates.*

| Alloy | $C_\Gamma$ (eV) | $C_X$ (eV) | $C_L$ (eV) | $C_{me}$ (m) | $C_\Delta$ (eV) | $C_{EP}$ (eV) |
|---|---|---|---|---|---|---|
| $(Al_{0.52}In_{0.48}P)_z(In_{0.49}Ga_{0.51}P)_{1-z}$ | 0.18 | 0 | 0 | 0.0135 | 0 | 0 |
| $(In_{0.49}Ga_{0.51}P)_z(GaAs)_{1-z}$ | 0.18 | 0.53 | 0 | 0.0094 | 0 | 0 |
| $(Al_{0.48}In_{0.52}As)_z(In_{0.53}Ga_{0.47}As)_{1-z}$ | 0.22 | 0 | 0 | −0.016 | 0 | 9.24 |
| $(InP)_z(In_{0.53}Ga_{0.47}As)_{1-z}$ | 0.13 | 0 | 0 | 0.013 | −0.06 | 0 |
| $(In_{0.53}Ga_{0.47}As)_z(GaAs_{0.5}Sb_{0.5})_{1-z}$ | 0.22 | 0 | 0 | 0.01 | 0 | 0 |
| $(GaAs_{0.5}Sb_{0.5})_{1-z}(AlAs_{0.56}Sb_{0.44})_z$ | 1.14 | 0 | 0 | 0.067 | 0 | 0 |
| $(InAs)_{1-z}(GaAs_{0.08}Sb_{0.92})_z$ | 0.6 | 0.15 | 0.6 | 0.027 | 0 | 0 |
| $(GaAs_{0.08}Sb_{0.92})_{1-z}(AlAs_{0.16}Sb_{0.84})_z$ | 0.48 | 0.807 | 1.454 | 0.04 | 0 | 0 |
| $(InAs)_{1-z}(InSb_{0.31}P_{0.69})_z$ | 0 | 0 | 0 | −0.009 | −0.75 | 0 |
| $(GaSb)_{1-z}(InAs_{0.91}Sb_{0.09})_z$ | 0.75 | 0.43 | 0.85 | 0 | 0.25 | 0 |
| $(GaSb)_{1-z}(AlAs_{0.08}Sb_{0.92})_z$ | 0.48 | 0.807 | 1.454 | 0.043 | 0 | 0 |

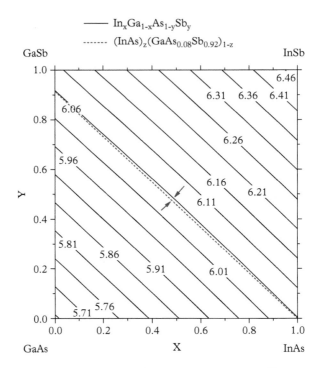

**Figure 7.5** *Contours of the $In_xGa_{1-x}As_{1-y}Sb_y$ lattice constant at 300 K, compared for: (1) bilinear interpolation; and (2) the contour for $(InAs)_z(GaAs_{0.08}Sb_{0.92})_{1-z}$ assuming the value 6.0583 Å over the entire range of compositions.*

Table 7.22 lists recommended values of the $D$ parameter in Eq. (7.2) for the energy gaps, electron mass, and valence-band offset (the other $D$ parameters are assumed to vanish) for selected quaternaries. Because these values are derived from the limited experimental data available for lattice-matched alloys, some results may not be accurate over the entire range of compositions. For example, the band parameters for AlInGaP on GaAs are most reliable near $(Al_{0.53}In_{0.47}P)_z(In_{0.49}Ga_{0.51}P)_{1-z}$. Figure 7.6 plots the $\Gamma$-valley gaps at room temperature for seven of the quaternary alloys lattice-matched to different substrates, as computed using both the Thompson-Woolley (Eq. (7.3)) and Donati (Eq. (7.2)) models. Note that the D parameters for InGaAsP apply to that quaternary alloy when lattice-matched to either GaAs or InP.

## 7.5  Dilute Nitrides

A III–V alloy with As, P, or Sb mixed with a small amount of N represents a special case that is not well described by the simple bowing model of Eq. (7.1).

**Table 7.22** *Donati D parameters for the Γ-, X-, and L-valley gaps, deduced from a least-squares fit to the energy gap values along fixed lattice constant lines.*

| Alloy | $D_\Gamma$ (eV) | $D_X$ (eV) | $D_L$ (eV) | $D_{me}$ (m) | $D_{VBO}$ (eV) |
|---|---|---|---|---|---|
| AlInGaP | 0.35 | −0.4 | −0.33 | 0.04 | −0.13 |
| InGaAsP | 0.27 | −1.57 | −1.67 | 0 | −0.08 |
| AlInGaAs | 0.605 | −1.45 | −0.32 | −0.11 | 0.62 |
| InGaAsSb | −0.694 | 0 | 0 | −0.1 | 2.11 |
| AlGaAsSb | −2.1 | 1.76 | 2.6 | 0.04 | −0.02 |
| AlGaAsP | 5.6 | 0 | 0 | 0 | 0 |

The energy gap typically decreases very fast, even when the N fraction $x$ is very low. Because such alloys can only be grown stably to N concentrations $< 10\%$, they are known as *dilute nitrides*. They are of technological interest because the rapid bandgap narrowing allows GaAs-based devices to operate in the telecommunications wavelength range, or to function as the narrow-gap regions of multi-junction solar cells (see Chapters 12 and 15). Similarly, InP-based materials can be extended to longer wavelengths, and InAsSbN alloy lattice-matched to GaSb can absorb photons out to a wavelength of 10 μm.

Fortunately, another simple model successfully describes the effect of a dilute nitride concentration on the energy gap: the band anti-crossing (BAC) model. This model neglects any effect of N on the valence band in the host III–V compound. It focuses instead on interactions with the conduction band, assuming an isoelectronic impurity level with fixed energy $E_N$ relative to the conduction-band states $E_C(k)$. The impurity's interaction strength with the conduction band is $V_N\sqrt{x}$, where the fitting parameter $V_N$ has a different value for each dilute nitride. While clearly the BAC model is approximate, its accuracy and simplicity make it the default for describing the band parameters. The BAC model can also be incorporated into the **k·p** method, by including two extra bands corresponding to one N level for each spin, plus off-diagonal (and spin-conserving) $V_N\sqrt{x}$ terms that interact with the conduction band only.

For wide- and moderate-gap materials in which the conduction band near $k = 0$ is well described by a single band, we can use Eq. (1.25) to calculate the energies of the two modified states:

$$E_\pm(k) = \frac{1}{2}\left\{[E_C(k) + E_N] \pm \sqrt{[E_C(k) - E_N]^2 + 4(V_N)^2 x}\right\} \qquad (7.4)$$

Figure 7.7 illustrates this dependence for the important example of GaAsN.

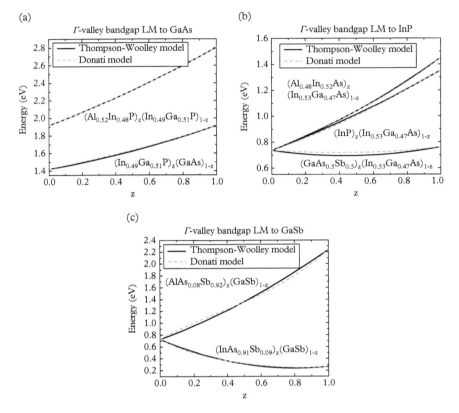

**Figure 7.6** *Thompson–Woolley and Donati models for the Γ-valley gaps at room temperature versus composition in lattice-matched (LM) quaternary alloys on (a) GaAs, (b) InP, and (c) GaSb substrates.*

The BAC model takes any temperature dependence to arise from the conduction-band dispersion $E_C(k)$ of the host material, while $E_N$ is independent of temperature. Therefore, the temperature variation of the energy gap in a dilute nitride is slower than in the host material. Also, strain effects must be applied to $E_C(k)$ via deformation potentials, rather than directly to $E_\pm(k)$, because the impurity level $E_N$ does not shift much with applied pressure. Accurate results are usually obtained when the **k·p** method described in Chapter 3 is augmented by the nitrogen levels as described earlier. We recommend adopting the elastic constants of the host material for a dilute nitride, even though there are no data confirming the validity of this procedure. Table 7.23 lists recommended $E_N$ and $V_N$ parameters for the dilute nitride alloys formed by incorporating N into the nine common III–V semiconductors. We adopt the values from [33], but include a more recent study of GaSbN by Jefferson *et al.* [46] The unstrained energy gap of

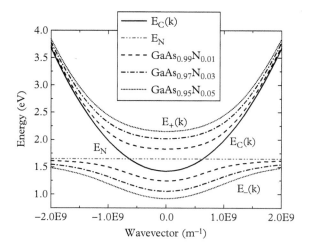

**Figure 7.7** *The conduction band dispersion for GaAs with dilute concentrations of nitrogen, as calculated using Eq. (7.4). The solid curve is the unperturbed conduction band of GaAs $E_C(k)$, relative to the nitrogen energy $E_N$.*

each alloy is shown as a function of nitrogen content (0–10%) and lattice constant in Figs. 7.8a and 7.8b, respectively.

Several dilute nitride quaternaries have been explored. By far the most important is $In_{1-y}Ga_yAs_{1-x}N_x$, which can be grown lattice-matched to GaAs. The material holds promise for telecommunication lasers and other light-emitting devices, since conventional InGaAs quantum wells have too much compressive strain and produce too many dislocations when pushed to such narrow gaps. This material has also been used in state-of-the-art multi-junction photovoltaics as a narrow-gap bulk semiconductor absorber grown on GaAs, and at one time held the record for solar cell efficiency [47]. We employ the values in [33] for InGaAsN and the less important InGaPN quaternary. GaAsSbN has been studied as an alternative to InGaAsN for narrowing the bandgap in a GaAs-based material. Recent models for the band structure of this alloy favor a double-band anti-crossing approach, where the nitrogen and antimony incorporations lead to separate band anti-crossing phenomena in the conduction and valence bands, respectively. Due to the somewhat special nature of this alloy, we refer the reader to [48, 49] rather than providing recommended values here.

Broderick *et al.* [50] recently provided a compilation of dilute nitride parameters, following a review by O'Reilly *et al.* [51], by fitting an atomistic tight-binding model. However, that analysis includes the composition variations of the host conduction-band edge and nitrogen defect state energy level, which are not included in our simple approach. Therefore, the results are inaccurate if values from [51] are substituted without accounting for composition-dependent effects.

**Table 7.23** *Recommended nitrogen energy level with respect to the valence-band maximum in the host semiconductor, and the interaction potential, for dilute nitrides formed by incorporating N into the Ga- and In-containing binary III–V semiconductors.*

| Compound | $E_N$ (eV) | $V_N$ (eV) |
|---|---|---|
| GaAsN | 1.65 | 2.7 |
| InAsN | 1.44 | 2.0 |
| GaPN | 2.18 | 3.05 |
| InPN | 1.79 | 3.0 |
| GaSbN | 0.78 | 2.6 |
| InSbN | 0.65 | 3.0 |
| $Ga_{1-x}In_xAsN$ | $1.65(1-x)+1.44x-0.38x(1-x)$ | $2.7(1-x)+2.0x-3.5x(1-x)$ |
| $Ga_{1-x}In_xPN$ | $2.18(1-x)+1.79x$ | $3.05(1-x)+3.0x-3.3x(1-x)$ |

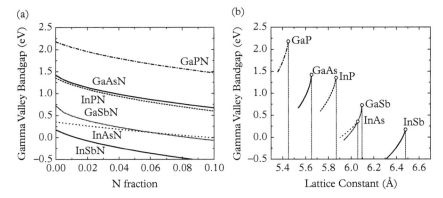

**Figure 7.8** *(a) Energy gap versus N fraction and (b) versus lattice constant for Ga and In-containing dilute nitride ternary alloys with N fractions of 0–10%.*

The reader who desires the best available calibration is referred to [50]. The suggestions in Table 7.23 provide a simpler and relatively accurate alternative.

## 7.6    Dilute Bismides

Another interesting class of zinc-blende III–V alloys incorporate a small fraction of Bi, i.e., *dilute bismides*. As with the dilute nitrides, they are of interest primarily because they can extend the wavelength range of GaAs-, InP-, and GaSb-based materials. The most studied to date are GaAsBi, InAsBi, InSbBi, and the quaternary InAsSbBi. By analogy with dilute nitrides, the introduction of limited

Bi content results in significant narrowing of the gap. But now it is primarily the valence band of the host semiconductor that is shifted by the alloying. Bismuth incorporation shifts the valence band maximum upward, and also induces a large spin–orbit splitting by moving the split-off band downward. We will learn in Chapter 12 that this can suppress nonradiative Auger recombination in some long-wavelength materials. Another intriguing, albeit exotic, possibility is to incorporate N and Bi into the same host semiconductor, to allow independent control of both conduction- and valence-band edges. As with the dilute nitrides, it is difficult to incorporate the Bi atoms in a random manner. The resulting material often features clustering and high defect densities.

Since neither GaBi nor InBi has been grown successfully in bulk, as with the dilute nitrides the dilute bismide band structure is specified in terms of deviation from that of the III–V host material, typically an arsenide or antimonide. From theoretical modeling and extrapolation, GaBi should have a lattice constant of 6.324 Å, and should be semi-metallic with an estimated band overlap in the range $-0.7$ to $-1.6$ eV. While lattice constants ranging from 6.500 to 7.292 Å have been derived for InBi, a recent analysis by Shalindar *et al.* [52] gave the recommended value of 6.6107 Å. InBi should also be semi-metallic, with estimated overlap in the range $-1.5$ to $-1.89$ eV.

To apply the BAC model to the valence band of a dilute bismide, we must specify the position of the Bi impurity with respect to the top of the valence band, $E_{Bi}$, which generally is not well known. Alberi *et al.* [53] took the experimental value $E_{Bi} \approx 0.1$ eV above the GaP valence-band maximum, and then deduced its position in the As- and Sb-containing alloys by assuming it remains fixed with respect to vacuum. That simple approach neglects the chemical and size differences between Bi and the other group V atoms. For example, Usman *et al.* [54] obtained $E_{Bi} \approx -180$ meV for GaAsBi from a tight-binding model, which compares to $-400$ mV from Alberi's approach. That said, in the absence of better information we recommend Alberi's method, which was also followed in a comprehensive review by Polak *et al.* [55]. Table 7.24 lists the recommended $E_{Bi}$ and coupling parameter $V_{Bi}$ for the most common dilute bismides, as well as the gap reduction, band-edge shifts, and increase in spin–orbit splitting [55]. The table also lists experimental values for the gap reduction. The agreement is generally good for the InGa–V–Bi alloys, with the exception of GaPBi where substitutional Bi creates localized impurity states that lie within the GaP gap. The detailed simulations and experimental characterizations performed on this alloy by Bushell *et al.* [56] found that an optically active band of Bi-hybridized states leads to strong bowings of $E_g$ and $\Delta$, particularly at the lowest compositions $x < 1\%$. Agreement for the lesser studied In–V–Bi alloys is somewhat mixed, with Svensson *et al.* [57] pointing out that the uncertain lattice parameter of InBi creates uncertainty in the Bi fraction.

A theoretical analysis of dilute bismide ternary alloys by Samajdar and Dhar [65] agrees well with measured energy gap reductions, and provides an alternative to the band anti-crossing model. Table 7.25 compiles their quadratic fits to the energy gaps.

The quaternary alloy InAsSbBi has been investigated over a range of compositions as a candidate material for high-performance mid- and long-wavelength

**Table 7.24** *Recommended band parameters for the most commonly studied dilute bismide ternary alloys, based on the simulations of Polak [55].*

| | CB shift (meV/%Bi) | VB shift (meV/%Bi) | $E_g$ reduction (meV/%Bi) | Experimental (meV/%Bi) | $\Delta$ increase (meV/%Bi) | $E_{Bi}$ (eV) | $V_{Bi}$ (eV) |
|---|---|---|---|---|---|---|---|
| $GaP_{1-x}Bi_x$ | 16 | 82 | 66 | 200[56] | 99 | −0.1 | N/A |
| $GaAs_{1-x}Bi_x$ | −29 | 62 | 91 | 84[58]–88[59] | 66 | 0.37 | 1.9 |
| $GaSb_{1-x}Bi_x$ | −16 | 16 | 32 | 35[60] | 18 | 1.14 | 0.9 |
| $InP_{1-x}Bi_x$ | −26 | 78 | 104 | 50[61]–83[62] | 86 | 0.23 | 1.9 |
| $InAs_{1-x}Bi_x$ | −15 | 48 | 63 | 38[57]–55[63] | 54 | 0.58 | 1.7 |
| $InSb_{1-x}Bi_x$ | −10 | 16 | 26 | 35[64] | 2 | 1.17 | 0.7 |

**Table 7.25** *Recommended quadratic expressions for the dilute bismide gaps, as derived by Samajdar and Dhar [65].*

| | $E_g$ (eV) |
|---|---|
| $GaP_{1-x}Bi_x$ | $2.26-9.92x+5.31x^2$ |
| $GaAs_{1-x}Bi_x$ | $1.43-6.12x+2.34x^2$ |
| $GaSb_{1-x}Bi_x$ | $0.81-3.71x+0.55x^2$ |
| $InP_{1-x}Bi_x$ | $1.42-8.95x+5.91x^2$ |
| $InAs_{1-x}Bi_x$ | $0.35-4.72x+2.75x^2$ |
| $InSb_{1-x}Bi_x$ | $0.17-2.21x+0.42x^2$ |

infrared detection and emission, with largely independent control over the energy gap and strain. Schaefer *et al.* [66] fit temperature-dependent optical data for InAsSbBi quaternary alloys grown by MBE. They determined an empirical temperature dependence of the gap from results for the constituent ternary alloys, including the composition-dependent bowing parameter for InAsBi, as shown in Eq. (7.5). Table 7.26 lists the fitting parameters. The low-temperature gaps of InAs and InSb, and the bowing parameter of InAsSb in that work, are close to those given in Tables 6.6 and 7.11 and the authors assumed a low-temperature energy gap of $-1.63$ eV for InBi.

$$E_{g,InAsSbBi}(x,y,T) = (1-x-y)\,E_{g,InAs}(0) + yE_{g,InSb}(0) + xE_{g,InBi}(0)$$
$$- y(1-y)\,C_{InAsSb} - x(1-x)\,C_{InAsBi,0}\,e^{\frac{x}{x_a}}$$
$$- \left(S_{0,InAs} - \frac{y}{y_0} - \frac{x}{x_0}\right) kT_E / \left(e^{T_E/T} - 1\right) \qquad (7.5)$$

## 7.7 Boron Nitride

This fascinating material appears in nature in one of three principal crystallographic forms:

- Hexagonal BN or h-BN, which is sometimes referred to as graphitic boron nitride or white graphene because of its similarity to graphite and graphene in its single-monolayer form. This is the most stable crystalline form of BN.
- Cubic BN, which has the zinc-blende structure.

**Table 7.26.** *Parameter values for the temperature- and composition-dependent bandgap model for InAsSbBi.*

| Parameter | Value |
| --- | --- |
| $C_{InAsBi,0}$ (eV) | 3.629 |
| $x_a$ | 0.229 |
| $S_{0,InAs}$ | 3.231 |
| $x_0$ | 0.254 |
| $y_0$ | 0.183 |
| $T_E$ (K) | 200 |

- Wurtzite BN, the least commonly encountered natural form of crystalline BN.

Cubic BN and wurtzite BN are notable for their hardnesses comparable to diamond [67]. Also remarkable is their chemical and thermal stability in air to temperatures in excess of 1000°C, as well as their resistance to acid. BN also occurs with a rhombohedral lattice arrangement, although that uncommon form is rarely studied.

The main forms of BN are all indirect, wide-gap semiconductors, although it has been proposed that h-BN has a direct transition at one of the symmetry points. The recent study by Cassebois *et al.* [68] showed the gap of 5.955 eV to be indirect, but also noted the existence of phonon-assisted optical transitions and a large exciton binding energy of 130 meV.

An improved estimate by Evans *et al.* [69] for the indirect energy gap of c-BN, with conduction-band minimum in the $X$ valley, is in reasonable agreement with first-principles calculations by Gao [70]. The gap for wurtzite BN is not well known, although the consensus has it to be indirect with a $K$-valley conduction-band minimum and $\Gamma$-point valence-band maximum. We adopt the gap and electron mass values from a recent theoretical study [71]. All other band parameters are taken from the reviews of Madelung [72] and Levinshtein, Rumyantsev, and Shur [73]. For c-BN, we derive Luttinger parameters from the light- and heavy-hole mass estimates given by Madelung [72]. There is a high degree of uncertainty, however, since these do not agree well with more recent Luttinger parameters [74]. Because the electron mass in the $\Gamma$ valley is unknown, we derive it by assuming $E_P = 25$ eV. Also unknown is the spin–orbit splitting, so we adopt a value similar to that in the III–V binaries (0.02 eV). Note that when $F = -0.5$ and the gap is much larger than $\Delta$, Eqs. (3.2) and (3.37) reduce to $m_e/m \approx E_g/E_P$, and $m_{so}/m \approx (\gamma_1)^{-1}$.

**Table 7.27** *Some recommended basic band parameters for the three common crystallographic forms of boron nitride.*

| Parameters | Wurtzite | Zinc-blende | Hexagonal |
|---|---|---|---|
| $a_{lc}$(Å) | 2.55 | 3.616 | 2.5072 |
| $c_{lc}$(Å) | 4.17 | | 6.87 |
| $E_g$ (Γ) (eV) | 10.2 | 8.7 | 9 |
| $E_g$ (indirect) (eV) | 6.8 | 6.36 | 5.955 |
| Δ (eV) | | 0.02 | |
| $m_e/m$ (Γ) | | 0.195 | |
| $m_l/m$ | 0.52 | 1.2 | |
| $m_t/m$ | 0.33 | 0.26 | |
| $\gamma_1$ | | 5.13 | |
| $\gamma_2$ | | 1.23 | |
| $\gamma_3$ | | 2.04 | |
| $m_{SO}/m$ | | 0.195 | |
| $E_P$ (eV) | | 11.58 | |
| $F$ | | −0.5 | |
| VBO (eV) | | −4.84 | |
| $a_c$ (eV) | | −36.5 | |
| $a_v$(eV) | | −7.3 | |
| $b$ (eV) | | −3.41 | |
| $d$ (eV) | | −3.75 | |
| $c_{11}$ (GPa) | 982 | 820 | 750 |
| $c_{12}$(GPa) | 134 | 190 | 150 |
| $c_{13}$ (GPa) | 74 | | |
| $c_{33}$ (GPa) | 1077 | | 32 |
| $c_{44}$ (GPa) | 388 | 480 | 3 |

We recommend the set of deformation potentials provided by Adachi [75]. For the VBO of this material with respect to InSb, we extrapolate the fit to (Al,In)-V binaries in Fig. 6.1c. Table 7.27 summarizes the recommended band parameters for the three common forms of BN.

## 7.8  Optical Properties of Alloys

The complex dielectric function or permittivity of a semiconductor material, $\varepsilon$, can be measured using reflectivity, transmission, or variable angle spectroscopic ellipsometry (VASE). It is directly related to the band structure via optical transitions, as discussed in Chapter 4. We also saw that optical beam propagation in a semiconductor may be conveniently described in terms of the complex refractive index with a real part $n_r$ and imaginary part $k_r \equiv \frac{\lambda \alpha}{4\pi}$, where $\lambda$ is the wavelength and $\alpha$ is the absorption coefficient. Continuing the development in Section 4.1, we can relate the dielectric function and complex index:

$$\varepsilon = \varepsilon_r + i\varepsilon_i \equiv (n_r + ik_r)^2 \tag{7.6}$$

from which we find

$$\varepsilon_r = n_r^2 - k_r^2 \tag{7.7}$$

$$\varepsilon_i = 2n_r k_r \tag{7.8}$$

$$n_r = \sqrt{\frac{\sqrt{\varepsilon_r^2 + \varepsilon_i^2} + \varepsilon_r}{2}} \tag{7.9a}$$

$$k_r = \sqrt{\frac{\sqrt{\varepsilon_r^2 + \varepsilon_i^2} - \varepsilon_r}{2}} \tag{7.9b}$$

How do we estimate the dielectric function of a semiconductor alloy, given those of the endpoints? One approach is the so-called model dielectric function developed by Adachi [76, 77]. Its starting point is an empirical description of the direct transitions at critical points, as shown for AlSb in Fig. 7.9. These can be studied using modulated spectroscopic measurements, by looking for a large change in the joint density of states near a critical point [77], or in line-shape fits to the second derivatives of the dielectric function.

In the usual critical-point nomenclature, structure associated with the direct bandgap is denoted the $E_0$ transition, while a transition from the split-off valence band to the conduction band at the $\Gamma$ point is $E_0 + \Delta$. The transition to the next conduction band at the $\Gamma$ point is $E_0'$. Transitions at the $L$ point are denoted $E_1$ and $E_1 + \Delta_1$, while the $X$-point transition is $E_2$. Tables 7.28 and 7.29 show recommended higher-order critical-point energies for the binaries, and their recommended bowing parameters for the III–V ternaries.

The Adachi model uses the critical-point energies and semi-empirical line-shape functions to assemble the dielectric function. Unfortunately, it requires a

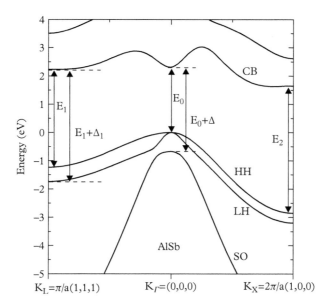

**Figure 7.9** *Some important transitions between valence-band and conduction-band states for the example of AlSb, which are reflected in the crystal's dielectric function.*

**Table** 7.28 *Recommended critical points for nine III–V binary compounds at room temperature.*

|  | $E_1$ (eV) | $E_1 + \Delta_1$ (eV) | $E_0'$ (eV) | $E_2$ (eV) |
|---|---|---|---|---|
| GaAs | 2.89 | 3.13 | 4.52 | 5.13 |
| AlAs | 3.8 | 4 | 4.62 | 4.87 |
| InAs | 2.5 | 2.78 | 4.44 | 4.7 |
| GaP | 3.71 | | 4.74 | 5.28 |
| AlP | 4.3 | | 5.14 | 4.63 |
| InP | 3.17 | 3.29 | 4.7 | 5.1 |
| GaSb | 2.05 | 2.5 | 3.56 | 4.14 |
| AlSb | 2.84 | 3.23 | 3.76 | 4.23 |
| InSb | 1.8 | 2.3 | 3.26 | 3.9 |

large number of parameters, and particularly near the absorption edge its precision is limited. Other interpolation methods similarly require a lot of parameters. We can alternatively use empirical polynomial line shapes that flexibly resemble the various interband transitions [78] in Fig. 7.9 to extract the optical constants

**Table 7.29** *Recommended bowing parameters for critical points in the III–V ternary alloys.*

|         | $E_1$ (eV) | $E_1 + \Delta_1$ (eV) | $E_0'$ (eV) | $E_2$ (eV) |
|---------|-----------|----------------------|------------|-----------|
| AlGaAs  | 0.392     | 0.391                | 0          | 0         |
| InGaAs  | 0.55      | 0.32                 | 0          | 0         |
| InAlAs  | 0.97      | 0.64                 | 0          | 0.285     |
| InGaP   | 0.86      | 0.86                 | 0          | 0         |
| AlInP   | 0         | 0                    | 0          | 0         |
| AlGaP   | 0         | 0                    | 0          | 0         |
| InGaSb  | 0.38      | 0.28                 | 0          | 0.24      |
| AlInSb  | 0.34      | 0.21                 | 0          | 0         |
| AlGaSb  | 0.306     | 0.291                | 0          | 0.339     |
| InAsSb  | 0         | 0                    | 0          | 0         |
| GaAsSb  | 0.55      | 0.59                 | 0.32       | 0.19      |
| AlAsSb  | 0.84      | 0.47                 | 0          | 0         |
| GaAsP   | 0.29      | 0.29                 | 0.04       | 0         |
| InAsP   | 0.17      | 0.11                 | 0.03       | 0.11      |
| AlAsP   | 0         | 0                    | 0          | 0         |
| InPSb   | 0.8       | 0.8                  | 0.4        | 0.4       |
| GaPSb   | 1.35      | 1.35                 | 0.68       | 0.68      |
| AlPSb   | 0         | 0                    | 0          | 0         |

from experiments such as VASE. Although this approach does not provide any deep insight into the band structure, it can generate reliable optical constants. For example, Fig. 7.10 shows a polynomial model for the imaginary part of the dielectric function of InP that is built into the Woollam WVASE32 ellipsometer software (see also Fig. 4.2 for the room-temperature dielectric function of other III–V materials). Appendix D tabulates experimental results for the nine common III–V binaries that were obtained using this and other approaches.

Lumb *et al.* [79] outlined a simple, interpolative approach that moves away from model fitting, following a similar technique presented in Snyder *et al.* [80]. This method uses only empirical functions to describe the composition dependence of a critical-point energy. It also exploits the approximately linear variation between two composition points, $X_0$ and $X_1$, of the real and imaginary components of the

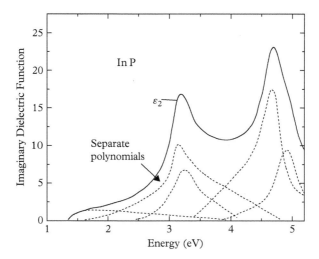

**Figure 7.10** *Fitted polynomial oscillator functions for the imaginary part of the InP dielectric function at room temperature*

dielectric function, when read off at a given critical-point energy. The interpolation method uses two dielectric function data sets with an energy basis to generate the results at an intermediate composition $X$, where $X_0 < X < X_1$.

The algorithm's first requirement is to map the energy data points $E_i(X_0)$ to their corresponding energies at a different composition $X$. For data points in the $X_0$ data set lying between two critical-point energies, $CP_i(X_0)$ and $CP_{i+1}(X_0)$, it is assumed that the shifted energy at composition $X$ is in the same ratio to $CP_i(X)$ and $CP_{i+1}(X)$ for all $X$. Therefore, the shifted energy for an energy between $CP_i$ and $CP_{i+1}$ is

$$E_j(X) = \frac{E_j(X_0) - CP_i(X_0)}{CP_{i+1}(X_0) - CP_i(X_0)} [CP_{i+1}(X) - CP_i(X)] + CP_i(X). \quad (7.10)$$

The next step is to read off the real and imaginary parts of the $X_1$ dielectric function at each of the transformed energy values, $E_i(X_1)$. Linear interpolation between $\varepsilon(E_j(X_0))$ and $\varepsilon(E_j(X_1))$ is then used to predict the dielectric function at the desired alloy composition:

$$\varepsilon(E_j(X)) = \frac{X - X_0}{X_1 - X_0} [\varepsilon(E_j(X_1)) - \varepsilon(E_j(X_0))] + \varepsilon(E_j(X_0)) \quad (7.11)$$

Figure 7.11 shows that the model (curves) accurately reproduces experimental data for InGaSb (points). Furthermore, the approach's accuracy improves considerably if data are available at compositions intermediate between the two binaries. Figure 7.11c shows the evolution with composition of InGaSb's dielectric function

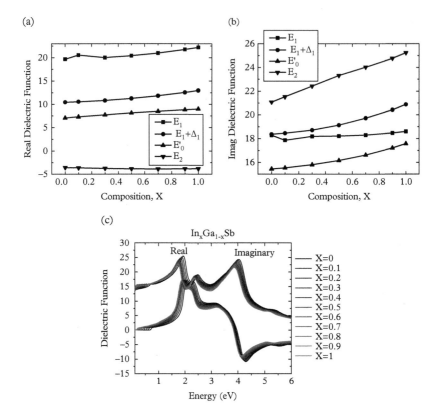

**Figure 7.11** *(a) Real and (b) imaginary parts of the dielectric function for InGaSb, evaluated at different critical points. (c) Interpolated dielectric functions for InGaSb at different compositions.*

computed using this method. Even though the resulting dielectric function does not fully satisfy the Kramers–Kronig relation derived from fundamental causality arguments, we still recommend this approach for ternaries and lattice-matched quaternaries (with one or both endpoints being ternaries) because of its simplicity and ease of fitting to measured data. While this book cannot reproduce the optical constants of all possible alloys, numerous published compilations [81, 82] can serve as intermediate points in the routine.

Currently, there is no validated method for determining the optical constants of quaternary alloys with arbitrary strain. Provisionally, we recommend following the same procedure as for lattice-matched quaternaries, but interpolating between ternaries with the same mismatched lattice constant. For example, we might expect the optical constants for InGaAsSb compositions along the lattice constant contour $a_L = 5.91$ Å to be approximated reasonably well by those for the endpoint ternaries $(GaAs_{0.41}Sb_{0.59})_{1-z}(In_{0.64}Ga_{0.36}As)_z$, as shown graphically in Fig. 7.12. Finally, the same method can predict temperature-dependent optical constants, where the composition terms in Eqs. (7.10) and (7.11) are replaced by

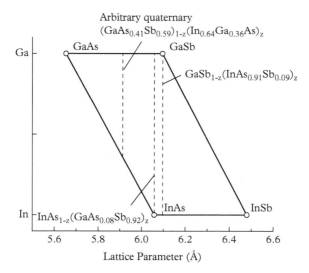

**Figure 7.12** *Suggested optical constant interpolation scheme for quaternaries with arbitrary lattice constant: Endpoints should be taken along contours with the same lattice constant, as with lattice-matched quaternaries. The vertical dashed curves indicate the contours* $(GaAs_{0.41}Sb_{0.59})_{1-z}(In_{0.64}Ga_{0.36}As)_z$, $InAs_{1-z}(GaAs_{0.08}Sb_{0.92})_z$, *and* $GaSb_{1-z}(InAs_{0.91}Sb_{0.09})_z$.

temperature terms, using the same assumption of linear dependence. Unfortunately, the experimental data at temperatures other than 300 K are quite sparse.

Generally, the methods of Chapter 11 provide reasonably good estimates of the band-edge absorption coefficients in superlattices and quantum wells. At energies well above the band discontinuities, the material can be treated as an alloy with composition obtained from the fractional layer thicknesses. Because the refractive index is determined mostly by transitions at energies well above the energy gap, the "effective alloy" approach can be used to determine a very rough first cut. Unfortunately, no completely consistent method for treating the dielectric constants of arbitrary quantum structures over a wide range of energies has been developed at the time of publication.

# References

[1] A. G. Thompson and J. C. Woolley, "Electroreflectance measurements on $Ga_xIn_{1-x}As$ alloys," *Canadian Journal of Physics* **45**, 2597–607 (1967).
[2] A. Chen and A. Sher, *Semiconductor Alloys: Physics and Materials EngineerinSemiconductor Alloys: Physics and Materials Engineeringg*. New York: Springer, 1995.
[3] S.-H. Wei and A. Zunger, "Optical properties of zinc-blende semiconductor alloys: Effects of epitaxial strain and atomic ordering," *Physical Review B* **49**, 14337–51 (1994).

[4] P. Ernst, *et al.*, "Band-gap reduction and valence-band splitting of ordered GaInP$_2$," *Applied Physics Letters* **67**, 2347–9 (1995).

[5] I. Vurgaftman, *et al.*, "Band parameters for III-V compound semiconductors and their alloys," *Journal of Applied Physics* **89**, 5815–75 (2001).

[6] S. Adachi, *Properties of Semiconductor AlloyProperties of Semiconductor Alloyss*. Chichester, UK: Wiley, 2009.

[7] J. Klem, *et al.*, "Growth and properties of GaAsSb/InGaAs superlattices on InP," *Journal of Crystal Growth* **111**, 628–32 (1991).

[8] Y. Kawamura, *et al.*, "In$_{0.53}$Ga$_{0.47}$As/GaAs$_{0.5}$Sb$_{0.5}$/In$_{0.52}$Al$_{0.48}$As asymmetric type II quantum well structures lattice-matched to InP grown by molecular beam epitaxy," *Journal of Crystal Growth* **237**, 1499–503 (2002).

[9] J. Hu, *et al.*, "Type II photoluminescence and conduction band offsets of GaAsSb/InGaAs and GaAsSb/InP heterostructures grown by metalorganic vapor phase epitaxy," *Applied Physics Letters* **73**, 2799–801 (1998).

[10] N. Georgiev and T. Mozume, "Photoluminescence study of InGaAs/AlAsSb heterostructure," *Journal of Applied Physics* **89**, 1064–9 (2001).

[11] Y. Nakata, *et al.*, "InGaAs/AlAsSb heterostructures lattice-matched to InP grown by molecular beam epitaxy," *MRS Online Proceedings Library Archive* **198** (1990).

[12] M. Sacilotti, *et al.*, "Structural and optical properties of AlInAs/InP and GaPSb/InP type II interfaces," *Canadian Journal of Physics* **74**, 202–208 (1996).

[13] Y. Kawamura, *et al.*, "InAlAs/AlAsSb type II multiple quantum well layers lattice-matched to InP grown by molecular beam epitaxy," *Japanese Journal of Applied Physics* **36**, L757 (1997).

[14] R. Hill, "Energy-gap variations in semiconductor alloys," *Journal of Physics C: Solid State Physics* **7**, 521 (1974).

[15] H. Kuo, *et al.*, "Determination of the band offset of GaInP–GaAs and AlInP–GaAs quantum wells by optical spectroscopDetermination of the band offset of GaInP–GaAs and AlInP–GaAs quantum wells by optical spectroscopyy," *Journal of Electronic MaterialJournal of Electronic Materialss* **26**, 944–8 (1997).

[16] J. Chen, *et al.*, "Band offset of GaAs/In$_{0.48}$Ga$_{0.52}$P measured under hydrostatic pressure," *Applied Physics Letters* **58**, 744–6 (1991).

[17] T. J. Kim, *et al.*, "Dielectric functions and interband transitions of In$_x$Al$_{1-x}$P alloys," *Current Applied Physics* **14**, 1273–6 (2014).

[18] A. Gudovskikh, *et al.*, "III-phosphides heterojunction solar cell interface properties from admittance spectroscopy," *Journal of Physics D: Applied Physics* **42**, 165307 (2009).

[19] L. K. Vodop'yanov, *et al.*, "Cathodoluminescence and Raman scattering in Ga$_{1-x}$Al$_x$P epitaxial films," *Semiconductors* **34**, 405–9 (2000).

[20] H. Sonomura, *et al.*, "Composition dependences of the energy gap and the green-band emission peak for the Al$_x$Ga$_{1-x}$P ternary system," *Applied Physics Letters* **24**, 77–8 (1974).

[21] S. P. Svensson, *et al.*, "Band gap of InAs$_{1-x}$Sb$_x$ with native lattice constant," *Physical Review B* **86**, 245205 (2012).

[22] Y. Kawamura, *et al.*, "Photoreflectance characterization of ordered/disordered GaAs$_{0.5}$Sb$_{0.5}$ layers grown on InP substrates by molecular beam epitaxy," *Japanese Journal of Applied Physics* **42**, 4288 (2003).

[23] O. Ostinelli, *et al.*, "Photoluminescence and band offset of type-II AlGaAsSb/InP heterostructures," *Semiconductor Science and Technology* **21**, 681 (2006).

[24] T. J. Kim, *et al.*, "Parametrization of the optical constants of AlAs$_x$Sb$_{1-x}$ alloys in the range 0.74-6.0 eV," *Journal of the Optical Society of Korea* **18**, 359–64 (2014).

[25] E. H. Reihlen, *et al.*, "Optical absorption and emission of InP$_{1-x}$Sb$_x$ alloys," *Journal of Applied Physics* **68**, 4604–9 (1990).

[26] W. J. Duncan, *et al.*, "Metalorganic vapour phase epitaxy growth of InPAsSb alloys lattice matched to InAs," *Journal of Crystal Growth* **143**, 155–61 (1994).

[27] M. J. Jou, *et al.*, "OMVPE growth of the new semiconductor alloys GaP$_{1-x}$Sb$_x$ and InP$_{1-x}$Sb$_x$," *Journal of Crystal Growth* **93**, 62–9 (1988).

[28] S. Loualiche, *et al.*, "GaPSb: A new ternary material for Schottky diode fabrication on InP," *Applied Physics Letters* **59**, 423–4 (1991).

[29] H. Shimomura, *et al.*, "Growth of AlPSb and GaPSb on InP by gas-source molecular beam epitaxy," *Journal of Crystal Growth* **162**, 121–5 (1996).

[30] I. Vurgaftman and J. Meyer, "Band parameters for wurtzite nitride semiconductors," in, J. Piprek (Ed.), *Nitride Semiconductor Devices: Principles and Simulation*, vol. 590. New York: Wiley Online Library, 2007.

[31] F. Yun, *et al.*, "Energy band bowing parameter in Al$_x$Ga$_{1-x}$N alloys," *Journal of Applied Physics* **92**, 4837–9 (2002).

[32] E. Sakalauskas, *et al.*, "Dielectric function and bowing parameters of InGaN alloys," *physica status solidi (b)* **249**, 485–8 (2012).

[33] I. Vurgaftman and J. R. Meyer, "Band parameters for nitrogen-containing semiconductors," *Journal of Applied Physics* **94**, 3675–96 (2003).

[34] R. C. Cramer, *et al.*, "Band gap bowing for high In content InAlN films," *Journal of Applied Physics* **121266**, 035703 (2019).

[35] M. Akazawa, *et al.*, "Measurement of valence-band offsets of InAlN/GaN heterostructures grown by metal-organic vapor phase epitaxy," *Journal of Applied Physics* **109**, 013703 (2011).

[36] M. Akazawa, *et al.*, "Small valence-band offset of In$_{0.17}$Al$_{0.83}$N/GaN heterostructure grown by metal-organic vapor phase epitaxy," *Applied Physics Letters* **96**, 132104 (2010).

[37] J. Wang, *et al.*, "Evidence of type-II band alignment in III-nitride semiconductors: Experimental and theoretical investigation for In$_{0.17}$Al$_{0.83}$N/GaN heterostructures," *Scientific Reports* **4**, 6521 (2014).

[38] W. Jiao, *et al.*, "Characterization of MBE-grown InAlN/GaN heterostructure valence band offsets with varying In composition," *AIP Advances* **6**, 035211 (2016).

[39] G. Martin, *et al.*, "Valence-band discontinuities of wurtzite GaN, AlN, and InN heterojunctions measured by x-ray photoemission spectroscopy," *Applied Physics Letters*, **68**, 2541–3 (1996).

[40] T. Suzuki, *et al.*, "Optical constants of cubic GaN, AlN, and AlGaN alloys," *Japanese Journal of Applied Physics* **39**, L497 (2000).

[41] A. Kasic, *et al.*, "Optical phonon modes and interband transitions in cubic Al$_x$Ga$_{1-x}$N films," *Physical Review B* **65**, 184302 (2002).

[42] J. R. Müllhäuser, et al., "Green photoluminescence from cubic In$_{0.4}$Ga$_{0.6}$N grown by radio frequency plasma-assisted molecular beam epitaxy," *Applied Physics Letters* **73**, 1230–2 (1998).

[43] R. Goldhahn, *et al.*, "Refractive index and gap energy of cubic InxGa1−xN," *Applied Physics Letters* **76**, 291–3 (2000).

[44] G. P. Donati, *et al.*, "Interpolating semiconductor alloy parameters: Application to quaternary III-V band gaps," *Journal of Applied Physics* **94**, 5814–19 (2003).

[45] S. Shirakata, *et al.*, "Electroreflectance and photoluminescence studies of $In_{1-x}Ga_xP_{1-y}As_y$ lattice-matched to GaAs," *Japanese Journal of Applied Physics* **25**, 435 (1986).

[46] P. H. Jefferson *et al.*, "Band anticrossing in $GaN_xSb_{1-x}$," *Applied Physics Letters* **89**, 111921 (2006).

[47] D. Derkacs, *et al.*, "Lattice-matched multijunction solar cells employing a 1 eV GaInNAsSb bottom cell," *Journal of Photonics for Energy* **2**, 021805–1 (2012).

[48] K.-I. Lin *et al.*, "Double-band anticrossing in GaAsSbN induced by nitrogen and antimony incorporation," *Applied Physics Express* **6**, 121202 (2013).

[49] Y.-T. Lin, *et al.*, "Energy gap reduction in dilute nitride GaAsSbN," *Applied Physics Letters* **93**, 171914 (2008).

[50] C. Broderick, *et al.*, "Dilute Nitride Alloys," in J. Piprek (Ed.), *Handbook of Optoelectronic Device Modeling and Simulation*, 1st edn. Boca Raton, FL: CRC Press, 2017.

[51] E. O'Reilly, *et al.*, "Trends in the electronic structure of dilute nitride alloys," *Semiconductor Science and Technology* **24**, 033001 (2009).

[52] A. J. Shalindar, *et al.*, "Measurement of InAsBi mole fraction and InBi lattice constant using Rutherford backscattering spectrometry and X-ray diffraction," *Journal of Applied Physics* **120**, 145704 (2016).

[53] K. Alberi, *et al.*, "Valence-band anticrossing in mismatched III-V semiconductor alloys," *Physical Review B* **75**, 045203 (2007).

[54] M. Usman, *et al.*, "Tight-binding analysis of the electronic structure of dilute bismide alloys of GaP and GaAs," *Physical Review B* **84**, 245202 (2011).

[55] M. P. Polak, *et al.*, "First-principles calculations of bismuth induced changes in the band structure of dilute Ga–V–Bi and In–V–Bi alloys: chemical trends versus experimental data," *Semiconductor Science and Technology* **30**, 094001 (2015).

[56] Z. Bushell *et al.*, "Valence band-anticrossing in GaPBi dilute bismide alloys: giant bowing of the band gap and spin-orbit splitting energy" (2017).

[57] S. P. Svensson, *et al.*, "Molecular beam epitaxy control and photoluminescence properties of InAsBi," *Journal of Vacuum Science & Technology B* **30**, 02B109 (2012).

[58] S. Tixier, *et al.*, "Molecular beam epitaxy growth of $GaAs_{1-x}Bi_x$," *Applied Physics Letters* **82**, 2245–7 (2003).

[59] S. Francoeur, *et al.*, "Band gap of $GaAs_{1-x}Bi_x$, $0<x<3.6\%$," *Applied Physics Letters* **82**, 3874–6 (2003).

[60] M. K. Rajpalke, *et al.*, "High Bi content GaSbBi alloys," *Journal of Applied Physics* **116**, 043511 (2014).

[61] T. Das, "The effect of Bi composition on the properties of $InP_{1-x}Bi_x$ grown by liquid phase epitaxy," *Journal of Applied Physics* **115**, 173107 (2014).

[62] J. Kopaczek, *et al.*, "Contactless electroreflectance and theoretical studies of band gap and spin-orbit splitting in $InP_{1-x}Bi_x$ dilute bismide with $x \leq 0.034$," *Applied Physics Letters* **105**, 222104 (2014).

[63] K. Ma, *et al.*, "Organometallic vapor phase epitaxial growth and characterization of InAsBi and InAsSbBi," *Applied Physics Letters* **55**, 2420–2 (1989).

[64] M. Rajpalke, *et al.*, "Bi-induced band gap reduction in epitaxial InSbBi alloys," *Applied Physics Letters* **105**, 212101 (2014).

[65] D. P. Samajdar and S. Dhar, "Estimation of Bi-induced changes in the direct E0 band gap of III–V-Bi alloys and comparison with experimental data," *Physica B: Condensed Matter* **484**, 27–30 (2016).

[66] S. T. Schaefer, *et al.*, "Molecular beam epitaxy growth and optical properties of InAsSbBi," *Journal of Applied Physics* **126**, 083101 (2019).

[67] Z. Pan, *et al.*, "Harder than diamond: Superior indentation strength of wurtzite BN and Lonsdaleite," *Physical Review Letters* **102**, 055503 (2009).

[68] G. Cassabois, *et al.*, "Hexagonal boron nitride is an indirect bandgap semiconductor," *Nature Photonics* **10**, 262 (2016).

[69] D. A. Evans, *et al.*, "Determination of the optical band-gap energy of cubic and hexagonal boron nitride using luminescence excitation spectroscopy," *Journal of Physics: Condensed Matter* **20**, 075233 (2008).

[70] S.-P. Gao, "Cubic, wurtzite, and 4H-BN band structures calculated using GW methods and maximally localized Wannier functions interpolation," *Computational Materials Science* **61**, 266–9 (2012).

[71] M. Zhang and X. Li, "Structural and electronic properties of wurtzite $B_xAl_{1-x}N$ from first-principles calculations," *physica status solidi (b)* **254**, 1600749 (2017).

[72] O. Madelung, *Semiconductors: Data Handbook*, 3rd edn. Berlin: Springer, 2012.

[73] M. E. Levinshtein, *et al.*, *Properties of Advanced Semiconductor Materials: GaN, AlN, InN, BN, SiC, SiGe*. New York: Wiley, 2001.

[74] R. de Paiva and S. Azevedo, "Cubic $(BN)_xC_{2(1-x)}$ ordered alloys: a first-principles study of the structural, electronic, and effective mass properties," *Journal of Physics: Condensed Matter* **18**, 3509 (2006).

[75] S. Adachi, "Hexagonal boron nitride (h-BN)," in *Optical Constants of Crystalline and Amorphous Semiconductors: Numerical Data and Graphical Information*, pp. 127–36. Boston, MA: Springer US, 1999.

[76] S. Adachi, "Model dielectric constants of GaP, GaAs, GaSb, InP, InAs, and InSb," *Physical Review B* **35**, 7454–63 (1987).

[77] M. Cardona, *Modulation Spectroscopy*. San Diego, CA: Academic Press, 1969.

[78] B. Johs, *et al.*, "Development of a parametric optical constant model for $Hg_{1-x}Cd_xTe$ for control of composition by spectroscopic ellipsometry during MBE growth," *Thin Solid Films* **313–14**, 137–42 (1998).

[79] M. P. Lumb, *et al.*, "Simulation of novel InAlAsSb solar cells," in *Proceedings of SPIE, Physics and Simulation of Optoelectronic Devices*, San Francisco, 2012, p. 82560S.

[80] P. G. Snyder, *et al.*, "Modeling $Al_xGa_{1-x}As$ optical constants as functions of composition," *Journal of Applied Physics* **68**, 5925–6 (1990).

[81] S. Adachi, *The Handbook on Optical Constants of Semiconductors*. Singapore: World Scientific Publishing, 2012.

[82] E. D. Palik, *Handbook of Optical Constants of Solids*, vol. 3. San Diego, CA: Academic Press, 1998.

# Part III

# Band Structure and Optical Transitions in Quantum Structures

# 8

# Basics of Envelope-Function Theory

In this chapter, we show how the bulk theory described in Part I can be generalized within the envelope-function framework to model the band structure of
layered materials with quantum confinement of carriers such as quantum wells
or superlattices. In practice, the approach amounts to substituting derivatives for
wavevector components in suitably chosen Hamiltonians as well as augmenting
them with interface terms. We also discuss the spin splitting of the states of the
quantum structures that arises from structural and intrinsic asymmetries.

## 8.1 From Bulk to Quantum-Layer Hamiltonian

In Part I, we described various methods for modeling the band structure of a bulk
material that extends to infinity in all dimensions. The methods of Part I work well
when the spatial extent of the semiconductor region $L$ is sufficiently large, i.e., the
structure is in the *bulk-like* limit. How large is large enough? There is no universal
rule, but the best criterion may be the spacing between the electronic states in
units of the thermal energy $k_B T$. For example, if the confinement wavevector is
a tiny fraction of the Brillouin zone extent, $\frac{\pi}{L} \ll 0.1 \frac{2\pi}{a}$, the confinement energies
should be much less than $k_B T$, and the occupations of the closely spaced states
are then nearly the same. In this case, it serves no useful purpose to distinguish
them, and we may as well resort to the bulk description of Part I. In practical terms,
this approach is quite accurate when confinement lengths are at least a hundred
atomic layers, or $t \sim 300 - 1000 \overset{\circ}{A}$ in all directions. Because the confinement
energy depends on the mass, and the mass is smaller in narrow-gap materials (see
Chapters 2, 3, and 6), confinement effects appear at larger spatial scales in these
materials.

In the 1980s, layer-by-layer epitaxial semiconductor growth became more
mature, and heterostructures with layer thicknesses $\leq 100$ Å could be grown
routinely. This development permitted electrons and holes to be confined by band
offsets in adjacent materials to very thin layers, as illustrated in Fig. 8.1. Now

*Bands and Photons in III–V Semiconductor Quantum Structures.* Vurgaftman, Lumb, and Meyer,
Oxford University Press (2021). © Vurgaftman, Lumb, and Meyer.
DOI: 10.1093/oso/9780198767275.003.0008

if we think of electrons and holes in the material as lighter versions of the free electron, the plane-wave part of the wavefunctions transforms into a standing wave along the direction of confinement. Mathematically, $\psi \propto \cos(\kappa z)$ in the well $\left(|z| < \frac{t}{2}\right)$. The wavefunctions become evanescent, with $\psi \propto \exp\left(\mp\gamma\left(z - \frac{t}{2}\right)\right)$ in the barriers $\left(|z| > \frac{t}{2}\right)$. We assume that the cell-periodic component of the wavefunction is unchanged and refer to the confined component as the *envelope function*. At this point, we can use elementary quantum mechanics to calculate the electronic states in quantum wells.

Broadly speaking, this approach is correct, but lacking in many crucial details. For example, Chapter 2 showed that the band structure is approximately parabolic only in the conduction band of a wide-gap material, so that in general we need a multiband model to treat both bands accurately. This means we cannot settle for the solution of a one-dimensional version of the Schrödinger equation. Instead, we must seek a more general method. It may come as a slight surprise that a fully rigorous formalism is not easy to derive. This is due to fundamental tension between the continuous description of the crystal via the envelope function and the discrete nature of the individual atoms that comprise that crystal. In an important example discussed in Section 8.3, we expect "new" atomic bonds to appear at the interfaces between dissimilar materials (e.g., InAs and GaSb), and it is not obvious that a continuous envelope function can represent the resulting electronic states.

What exactly is the problem? Can we simply join two piecewise continuous envelope functions at an interface between two materials? For one thing, an abrupt transition in the band parameters generates high-spatial-frequency (large $k$) Fourier components, and we know from Part I that these should be avoided in any useful **k·p** picture of the band structure. Of course, in reality it takes at least one atomic thickness to make the transition, but the envelope-function formalism has a great deal of trouble accounting for this fact rigorously! Nevertheless, quasi-continuous models are very powerful, and they match the physical reality quite closely in most cases.

It seems plausible that despite the abrupt change in the band parameters, the envelope functions are well behaved, as shown in Fig. 8.1. If we assume they vary relatively slowly, we recover the standard approach to treating quantum-confined structures. It turns out that to correct for the absence of high frequencies, we will need to insert extra delta-function terms at interfaces. These *interface terms* are *the* major non-trivial result of the envelope-function formalism. Their form is also implicit in the tight-binding description of Chapter 5.

Another well-known corollary of the theory is correct ordering of the Hamiltonian's coordinate-dependent terms and derivatives. We will see that it is in fact a natural by-product of the careful **k·p** development of Chapter 3. We will also find that if we use the most complete 8-band model of the band structure, the choice of the ordering has little impact on the subband dispersions.

This is then the roadmap for our journey into the effects of quantum confinement. To fill in the details, we must first decide how to deal with the cell-periodic

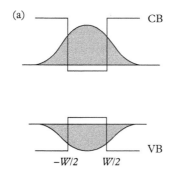

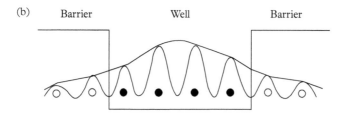

**Figure 8.1** *(a) A quantum well created by band offsets in a semiconductor heterostructure and schematic envelope functions of the confined electron and hole states in the well. (b) The full wavefunction exhibits fast variation near the atoms, but the envelope function is often smooth on the scale of many atomic periods. In real crystals, both anions and cations are present, so the exact atomic-scale spatial dependence is more complicated than shown.*

component of the wavefunction. We could assume that it: (1) coincides with the bulk $v_{nk}$ in one of the layers, (2) varies from layer to layer in accordance with the bulk $v_{nk}$, or (3) remains invariant over the entire structure, but is not given by any of the bulk solutions. The third option leads to the formal envelope-function theory pioneered by Michael Burt in the late 1980s. When the entire structure uses a single set of periodic functions, the envelope function becomes a formal representation of the true wavefunction, rather than an approximation in need of justification.

We do not develop Burt's formal theory in the intuitive description that follows. Instead, we proceed with layer-specific cell-periodic functions and make a few key approximations. Working in one dimension and bearing in mind that the cell-periodic component $v_{nk}$ is a linear combination of the $u_{j0}$ functions for various bands $j$ at $k = 0$, as in Eq. (2.1), we can write the *bulk* wavefunction of band $n$ at arbitrary $k$:

$$\varphi_{nk}(z) = e^{ikz} v_{nk}(z) = e^{ikz} \sum_{j} a_j(z) \, u_{j0}(z) \tag{8.1}$$

The $u_{j0}$ functions could have the symmetry of $s$ and $p$ atomic orbitals. Their linear combination $v_{nk}$ represents the actual cell-periodic function at some $k$ in the bulk crystal. For example, the light-hole state at $k = 0$ is a linear combination of $p_x$-like, $p_y$-like, and $p_z$-like cell-periodic functions in the bulk crystal. We avoid explicit normalization unless necessary and assume that the spatial integrals are normalized by their total extents. For example, in Eq. (8.1) we can think of the entire crystal structure as having a length of one unit, and measure all the distances in these units.

In a structure containing different materials separated by interfaces, we expect the wavefunction in Eq. (8.1) to combine Fourier components corresponding to different $k$ as well as several distinct bulk bands. Thus, the general form of a given state's wavefunction in a quantum-confined structure is

$$\psi(z) = \sum_k \sum_j \phi_j(k) \varphi_{jk}(z) = \sum_k \sum_j \phi_j(k) \, e^{ikz} \, v_{jk}(z) \tag{8.2}$$

where the $\phi_j(k)$ are Fourier coefficients. In a 3D material with quantum confinement along only one direction, we should multiply by a plane-wave factor $\exp(i\mathbf{k}_\parallel \cdot \mathbf{r}_\parallel)$ to account for the in-plane dimensions. If the confinement is multidimensional, we replace the scalars $k$ and $z$ with vectors. The "envelope function" we would like to derive is a slowly varying function superimposed on the cell-periodic $v_{jk}(z)$ component, as illustrated in Fig. 8.1b. We can always expand $v_{jk}(z)$ in terms of $u_{j0}(z)$, as in Eq. (8.1), but leave this step implicit here.

How many Fourier components $k$ are needed in Eq. (8.2) to faithfully represent the wavefunction? Not too many, we hope, because we would like to make use of the **k·p** theory that becomes unreliable at large $k$. Of course, if the envelope variation is sufficiently slow, Fourier coefficients become negligibly small well before the boundary of the first Brillouin zone.

We insert the wavefunction in Eq. (8.2) into the Schrödinger equation:

$$(H_0 + V)\,\psi(z) = E\psi(z) \tag{8.3}$$

where we have separated the potential term $V$ associated with spatially varying conduction and valence-band edges (or an applied electric field) from the "kinetic" part $H_0$ that includes the periodic potential of a bulk crystal and contains spatial derivatives. To isolate a single Fourier component on the right, we multiply by $\exp(-ik'z)\,v_{j'k'}^*$ and integrate over the entire structure, by analogy with the steps leading from Eq. (1.18) to Eq. (1.19). Then we exchange the primed and non-primed $j$ and $k$. The result is equivalent to the original Eq. (8.3):

$$\sum_{k'} \sum_{j'} \int dz\, e^{-i(k-k')z} v_{jk}^*(z)\,(H_0 + V)\,v_{j'k'}(z)\,\phi_{j'}(k') = E\phi_j(k) \tag{8.4}$$

To make further progress, we consider matrix elements of the potential energy $V(z)$ with respect to $k$ and bulk band index $j$:

$$\langle jk|V|j'k'\rangle = \int dz\, e^{-i(k-k')z}\, v_{jk}^*(z)\, V(z)\, v_{j'k'}(z) \tag{8.5}$$

Expanding the product of the cell-periodic functions in terms of reciprocal lattice vectors $G$, we obtain

$$v_{jk}^*(z)\, v_{j'k'}(z) = \sum_G C_G^{jk,j'k'}\, e^{iGz} \tag{8.6}$$

This expansion is always possible in a crystal with periodic $v_{jk}(z)$. We can also invert Eq. (8.6) to write an expression for the Fourier coefficients:

$$C_G^{jk,j'k'} = \int dz\, e^{-iGz}\, v_{jk}^*(z)\, v_{j'k'}(z) \tag{8.7}$$

On the other hand, we can write the Fourier transform of the potential $V$ as

$$\tilde{V}(q) = \int dz\, e^{-iqz}\, V(z) \tag{8.8}$$

Now we substitute Eq. (8.6) into Eq. (8.5) and make use of Eq. (8.8) to simplify

$$\langle jk|V|j'k'\rangle = \sum_G C_G^{jk,j'k'} \int dz\, e^{-i(k-k'-G)z}\, V(z) = \sum_G C_G^{jk,j'k'}\, \tilde{V}(k-k'-G) \tag{8.9}$$

So far we have merely rewritten Eq. (8.5), but now we simplify drastically and retain only the $G=0$ term in the expansion of the potential $V(z)$; i.e., we take $\tilde{V}(k-k'-G) \approx \tilde{V}(k-k')$. Of course, an abrupt heterointerface in reality generates many $G$ components, but those with $G = \frac{2\pi m}{a} > 0$ oscillate so fast at distances $\gg a/2$ from an interface that their net effect there is negligible. Even though this reasoning does not apply very close to interfaces, we will develop an interface-term approach to fix that problem later. Keeping only the $G=0$ term in Eq. (8.9) also makes subsequent calculations much easier. Equation (8.9) becomes

$$\langle jk|V|j'k'\rangle = \sum_G C_G^{jk,j'k'}\, \tilde{V}(k-k') \tag{8.10}$$

For $G=0$, the coefficients $C_{G=0}^{jk,j'k'}$ can be calculated using the expansion in Eq. (8.1) and Eq. (8.7). Since the $u_{j0}$ functions are orthogonal, so are (approximately) $v_{jk}$ for small $k$. This means that the potential components with small $k$ do not couple different bands:

$$\langle jk|V|j'k'\rangle \approx \tilde{V}\left(k-k'\right)\delta_{jj'} \tag{8.11}$$

We now repeat the procedure for $H_0$. Following the same logic, but now including $H_0$ into the expansion similar to Eq. (8.6), we arrive at

$$\langle jk|H_0|j'k'\rangle = \sum_G A_G^{jk,j'k'} \int dz\, e^{-i(k-k'-G)z} \tag{8.12}$$

where $A_G^{jk,j'k'} = \int dz\, e^{-iGz}\, v_{jk}^*(z)\, H_0(z)\, v_{j'k'}(z)$. Because $k$ and $k'$ are restricted to the first Brillouin zone, the components with $G > 0$ can again be neglected away from the interfaces. Also, rapid phase oscillations cause the integral in Eq. (8.12) to vanish unless $k = k'$. As a result, the matrix elements have the form

$$\langle jk|H_0|j'k'\rangle = \delta_{kk'} \int dz\, v_{jk}^*(z) H_0 v_{j'k}(z) \tag{8.13}$$

Within $\mathbf{k}\cdot\mathbf{p}$ theory, Eq. (2.2) defines $H_0$ as

$$H_0 = \frac{\boldsymbol{p}^2}{2m} + \frac{\hbar}{m}\boldsymbol{k}\cdot\boldsymbol{p} + \frac{\hbar^2 k^2}{2m} + V_{per}(\boldsymbol{r}) = \frac{(\boldsymbol{p}+\hbar\boldsymbol{k})^2}{2m} + V_{per}(\boldsymbol{r}) \tag{8.14}$$

where $\boldsymbol{p} \equiv -i\hbar\nabla$, the subscript in $V_{per}(\boldsymbol{r})$ reminds us to include the periodic potential of the crystal (assumed to be almost the same in all the layers for these purposes), and $\boldsymbol{k}$ is the wavevector. In 1D, the first term of Eq. (8.14) is simply $\frac{(\boldsymbol{p}+\hbar k)^2}{2m}$. The matrix elements in Eq. (8.13) should be very familiar to us, since we spent a good part of Chapter 3 deriving them. In the envelope-function theory, these matrix elements are a function of position along the quantum-confined direction.

Even though the $G > 0$ components of $H_0$ are left out of this treatment, their approximate effect can be reintroduced via the *interface terms* discussed in more detail in Section 8.3. To put this model on a rigorous footing, we could turn, e.g., to Burt's formalism. The detailed form of interface terms follows from the reduction of the full zinc-blende symmetry to that of the fcc lattice in the underlying $\mathbf{k}\cdot\mathbf{p}$ theory. Finally, we will see in Section 8.2 that discontinuities in the material parameters can lead to delta-function-like contributions at the interfaces in a numerical solution, but these corrections are usually minor.

With the matrix elements in Eqs. (8.11) and (8.13), Eq. (8.4) can be rewritten

$$\sum_{k'}\sum_{j'}\left[\langle jk|H_0|j'k'\rangle\delta_{kk'} + \tilde{V}\left(k-k'\right)\delta_{jj'}\right]\phi_{j'}\left(k'\right) = E\phi_j(k) \tag{8.15}$$

We now multiply by $\exp(ikz)$, sum over $k$, and substitute Eq. (8.8) with $z'$ as the integration variable for $\tilde{V}\left(k-k'\right)$. This transforms Eq. (8.15) into

$$\sum_{j'} \left[ \sum_k \langle jk|H_0|j'k' \rangle e^{ikz} \delta_{kk'} + \delta_{jj'} \sum_k \sum_{k'} \int dz' e^{ik(z-z')+ik'z'} V(z') \right] \phi_{j'}(k')$$

$$= E \sum_k e^{ikz} \phi_j(k) \qquad (8.16)$$

The spatially varying term on the right-hand side is defined as the envelope function $F_j(z)$:

$$F_j(z) = \sum_k e^{ikz} \phi_j(k) \qquad (8.17)$$

The second term on the left in Eq. (8.16) clearly contains the envelope function above. To simplify it, we recall that the sum of the phase factor $\exp[ik(z-z')]$ over *all* $k$ yields a delta function $\delta(z-z')$. If $k$ is limited to the first Brillouin zone, the sum is still peaked sharply at $z = z'$, and we are justified in approximating it as a delta function when it is integrated over slowly varying functions. Renaming the index $k'$ as $k$ in the second term of Eq. (8.16), the entire equation becomes

$$\sum_{j'} \sum_k \langle jk|H_0|j'k \rangle e^{ikz} \phi_{j'}(k) + V(z) F_j(z) = E F_j(z) \qquad (8.18)$$

The first term on the left is also the envelope function multiplied by some $k$-dependent coefficients. Since $F_j(z)$ is expanded in plane waves $\exp(ikz)$, taking its derivative is equivalent to multiplying by $ik$ (with a sum over $k$). In fact, for any polynomial function of $k$, we can replace $k$ by $-i\partial/\partial z$ (or $ik$ by $\partial/\partial z$) without affecting the result. Writing $\langle jk|H_0|j'k \rangle$ as $H_{0,jj'}(k \to -i\partial/\partial z)$, we transform Eq. (8.18) into a gratifyingly simple final form:

$$\sum_{j'} H_{0,jj'}(k \to -i\partial/\partial z) F_{j'}(z) + V(z) F_j(z) = E F_j(z) \qquad (8.19)$$

It is worth restating what we have found: to model the quantum-confined structure, we substitute $-i\partial/\partial z$ for $k$ (more precisely, for $k_z$ when there is 1D confinement in a 3D structure) into the bulk Hamiltonian in each layer, add the potential associated with the electric field and different conduction-/valence-band positions in the constituent layers, and solve the resulting ordinary differential equations. This procedure is called *derivative substitution*. Since the order of the terms matters, $A(z)\hat{k}_z F_j(z)$ is transformed into $-iA(z)\frac{\partial F_j(z)}{\partial z}$, while $\hat{k}_z A(z) F_j(z)$ becomes $-i\frac{\partial [A(z) F_j(z)]}{\partial z}$, where $A(z)$ is the coefficient of a linear-in-$k$ term in $H_0$. We can now see why the operators in the **k·p** matrix elements derived in Chapter 3 were carefully ordered. Keeping in mind the approximations we have made, the derivative substitution seems a remarkably straightforward and appealing concept.

Does this recipe maintain Hermiticity? The Fourier components of $A(z)$ are in general imaginary. We saw in Section 3.1 that for every $H_{ij}$ term in the bulk Hamiltonian with $A(z)\hat{k}_z$, there is a corresponding $H_{ji}$ term with $\hat{k}_z A^*(z)$. With derivative substitution, $H_{ji}$ is obtained by multiplying $-i\frac{\partial[A^*(z)F_j(z)]}{\partial z}$ by $F_i^*(z)$ on the left and integrating by parts over $z$ with bounds at $z_1$ and $z_2$:

$$H_{ji} = -i\int_{z_1}^{z_2} dz F_i^*(z) \frac{\partial\left[A^*(z)F_j(z)\right]}{\partial z}$$

$$= -i\left[F_i^*(z)A^*(z)F_j(z)\right]_{z_1}^{z_2} + i\int_{z_1}^{z_2} dz F_j(z)A^*(z)\frac{\partial F_i^*(z)}{\partial z} \qquad (8.20)$$

The first term vanishes, e.g., in a quantum well, if both $z_1$ and $z_2$ are far enough from the well that the wavefunction at those points is negligible. The second term is clearly equal to the complex conjugate of $H_{ij}$, since we have

$$H_{ij} = -i\int dz F_j^*(z)\, A(z)\frac{\partial F_i(z)}{\partial z} = H_{ji}^* \qquad (8.21)$$

It is trivial to prove that symmetric terms of the form $\hat{k}_z B(z)\hat{k}_z$ are Hermitian as well, so everything appears in order. Obviously, replacing $k_z$ with an operator maintains Hermiticity so long as we are careful to keep the original ordering, or if we have only symmetric terms.

## 8.2   Examples of Hamiltonians in Quantum Structures

Before deriving the interface terms in Section 8.3, we consider a few examples of the derivative substitution in Eq. (8.19). The simplest imaginable Hamiltonian represents a parabolic band near $k = 0$ in a semiconductor with $E_g > 0$, as in Eq. (2.11):

$$E_k(k \to 0) \approx E_0 + \frac{\hbar^2 k^2}{2m_e} \qquad (8.22)$$

The origin of energy is set to the band minimum in one of the materials: $E_{0,ref} = 0$, typically the lowest one in energy. The potential energy $V(z) = E_0(z) - E_{0,ref}$ describes the variation of the band minimum from material to material.

What is the correct form of the "kinetic" energy in a one-band model with spatially dependent effective mass $m_e(z)$? From Section 8.1, we know that $\hat{k}_z\left[1/m_e(z)\right]\hat{k}_z$ is an acceptable model (where we omit the constant $\hbar^2/2$). In fact, this is the simplest possible form because $m_e(z)$ appears only once, as a power of $-1$. Logically, nothing prevents us from splitting off additional powers of $m_e(z)$ to symmetrically flank the outer sides of the $\hat{k}_z$s, with the total exponent still summing to $-1$ (e.g., $m_e^{-(1+a)/2}(z)\hat{k}_z m_e^{a}(z)\hat{k}_z m_e^{-(1+a)/2}$), but we lack strong grounds for doing so. Therefore, the Hamiltonian for the quantum-confined system becomes

$$\frac{\hbar^2}{2}\left[-\frac{\partial}{\partial z}\frac{1}{m_e(z)}\frac{\partial}{\partial z} + \frac{k_\parallel^2}{m_e(z)} + V(z)\right]F(z) = E\left(k_\parallel\right)F(z) \tag{8.23}$$

where $k_\parallel = \sqrt{k_x^2 + k_y^2}$ is the *in-plane wavevector*. Apart from the spatially varying mass and the in-plane wavevector term, this Hamiltonian is identical to the standard Schrödinger equation for a free particle, which explains its evergreen popularity in expository texts. In practice, this model is not very useful, apart from being an approximate description of the low-energy states in the conduction band of a wide-gap semiconductor.

Since the band structure is non-parabolic, we could also "correct" the one-band model by making the mass energy dependent, as in Eq. (2.20) or (2.21). The solutions might then be approximated iteratively, e.g., by first solving a Hermitian problem with a fixed mass and then modifying the mass based on the energy of the desired eigenstate and solving the adjusted problem again. However, we prefer a more general approach that explicitly introduces multiple bands.

For example, Section 2.1 shows that including a single valence band can make the conduction subband dispersions more accurate without changing the nature of the eigenproblem. Disregarding the less important free-electron terms and using the phase convention of Eq. (3.1), we have the Hamiltonian

$$H\begin{bmatrix}F_c\\F_v\end{bmatrix} = \begin{bmatrix}E_{c0} & i\frac{\hbar}{m}\boldsymbol{k}\cdot\boldsymbol{p}_{cv}\\ -i\frac{\hbar}{m}\boldsymbol{k}\cdot\boldsymbol{p}_{cv} & E_{v0}\end{bmatrix}\begin{bmatrix}F_c\\F_v\end{bmatrix} \tag{8.24}$$

where $\boldsymbol{p}_{cv}$ is real.

For consistency with the original **k·p** derivation that leads to Eq. (2.2), we can write Eq. (8.24) in a symmetric form for states with $k_\parallel = 0$:

$$H\begin{bmatrix}F_c\\F_v\end{bmatrix} = \begin{bmatrix}E_{c0} & i\frac{\hbar}{2m}\left(p_{cv}\hat{k}_z + \hat{k}_z p_{cv}\right)\\ -i\frac{\hbar}{2m}\left(p_{cv}\hat{k}_z + \hat{k}_z p_{cv}\right) & E_{v0}\end{bmatrix}\begin{bmatrix}F_c\\F_v\end{bmatrix} \tag{8.25}$$

Following the derivative-substitution rule in Eq. (8.19), we obtain

$$H = \begin{bmatrix}E_{c0}(z) & \frac{\hbar}{2m}\left[p_{cv}(z)\frac{\partial}{\partial z} + \frac{\partial}{\partial z}p_{cv}(z)\right]\\ -\frac{\hbar}{2m}\left[p_{cv}(z)\frac{\partial}{\partial z} + \frac{\partial}{\partial z}p_{cv}(z)\right] & E_{v0}(z)\end{bmatrix} \tag{8.26}$$

If the momentum matrix element $p_{cv}$ varies with position, Eq. (8.26) is certainly different from the other obvious possibilities:

$$H = \begin{bmatrix} E_{c0}(z) & \frac{\hbar}{m}\frac{\partial}{\partial z}p_{cv}(z) \\ -\frac{\hbar}{m}p_{cv}(z)\frac{\partial}{\partial z} & E_{v0}(z) \end{bmatrix} \tag{8.27}$$

and

$$H = \begin{bmatrix} E_{c0}(z) & \frac{\hbar}{m}p_{cv}(z)\frac{\partial}{\partial z} \\ -\frac{\hbar}{m}\frac{\partial}{\partial z}p_{cv}(z) & E_{v0}(z) \end{bmatrix} \tag{8.28}$$

In fact, all three versions of the Hamiltonian (Eqs. (8.26), (8.27), and (8.28)) are Hermitian, even though Eq. (8.26) is more symmetric. For compactness, we can combine all the Hermitian possibilities into a single expression:

$$H = \begin{bmatrix} E_{c0}(z) & \frac{\hbar}{m}\left[fp_{cv}(z)\frac{\partial}{\partial z} + (1-f)\frac{\partial}{\partial z}p_{cv}(z)\right] \\ -\frac{\hbar}{m}\left[(1-f)p_{cv}(z)\frac{\partial}{\partial z} + f\frac{\partial}{\partial z}p_{cv}(z)\right] & E_{v0}(z) \end{bmatrix} \tag{8.29}$$

where $0 \leq f \leq 1$. We see from Eq. (8.29) that the actual value of $f$ matters only when $p_{cv}$ changes on opposite sides of a heterointerface; i.e., it quantifies the magnitude of the term $f\Delta p_{cv}\delta(z - z_{IF})$ at an abrupt interface $z = z_{IF}$. This is *not* one of the promised interface terms, which will make their appearance in Section 8.3, but a close analog that arises from the off-diagonal terms in Eq. (8.24). Equation (8.29) is clearly symmetric around $f = 1/2$; e.g., the band structure for $f = 0$ and $f = 1$ is the same with only the phase of the envelope function potentially affected.

The functions $E_{c0}(z)$ and $E_{v0}(z)$ represent the position-dependent conduction- and valence-band edges (e.g., if the band offsets vary piecewise from layer to layer) plus any electrostatic potential from an external or internal electric field. With weak confinement, the ground state of Eq. (8.29) is close to that of Eq. (8.23), but in narrow and deep wells, nonparabolicity (as captured e.g. in Eq. (2.9)) plays a major role.

What is the effect of the off-diagonal delta-function terms arising from Eq. (8.29) on the band structure? First, they affect only the magnitude of the mass at the interface. If the envelope function probes a region much larger than the atomic thickness, a modest change in the mass near the interface can hardly amount to a large impact on the subband positions. This intuition is borne out by the calculation of the energy gap of a 50-Å-thick quantum well (between the bottom conduction subband and the top valence subband) for different $f$ in Fig. 8.2 using the methods described in Chapter 9. Although dependent on the step in $p_{cv}$ (expressed in energy units $E_P$), the differences are tiny, $\sim 0.1$ meV. We conclude that the choice of $f$ generally has a negligible effect on the band structure. For example, the uncertainty in the $E_P$ values and other parameters discussed in Chapter 6 has a larger effect on the determination of the gap in quantum wells and superlattices.

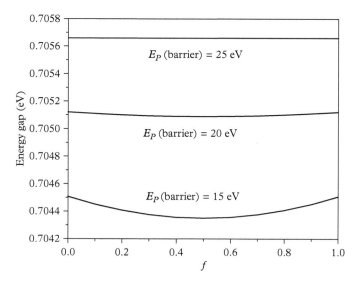

**Figure 8.2** *Dependence on interface-term parameter f of the energy gap calculated for a quantum well of 50 Å thickness formed by sandwiching a hypothetical material with 0.5-eV energy gap between two thick barrier layers with 1.25-eV energy gap and 0.25-eV relative valence-band offset. Curves are shown for three values of the momentum matrix element $E_P$ in the barrier, with a fixed value of $E_P = 25$ eV in the well.*

The next level of accuracy takes us to the quantum-confined version of the Kane Hamiltonian of Eq. (2.41) or (2.51) that includes spin–orbit coupling and accurately represents the CB, LH, and SO (but not HH!) states at $k_\parallel = 0$. Starting with Eq. (2.41) and proceeding by analogy with Eq. (8.25), the Hamiltonian along the [001] direction becomes

$$
\mathbf{H}_K(z)
\begin{bmatrix}
F_s(z) \\
F_{x-iy}(z) \\
F_z(z) \\
F_{x+iy}(z)
\end{bmatrix}
=
$$

$$
\begin{bmatrix}
E_c(z) - \frac{\hbar^2}{2m}\frac{\partial^2}{\partial z^2} & 0 & \frac{\hbar}{m}\left[ f p_{cv}\frac{\partial}{\partial z} + (1-f)\frac{\partial}{\partial z} p_{cv} \right] & 0 \\
0 & E_v(z) - \frac{2\Delta(z)}{3} - \frac{\hbar^2}{2m}\frac{\partial^2}{\partial z^2} & \frac{\sqrt{2}\Delta(z)}{3} & 0 \\
-\frac{\hbar}{m}\left[ (1-f)p_{cv}\frac{\partial}{\partial z} + f\frac{\partial}{\partial z} p_{cv} \right] & \frac{\sqrt{2}\Delta(z)}{3} & E_v(z) - \frac{\Delta(z)}{3} - \frac{\hbar^2}{2m}\frac{\partial^2}{\partial z^2} & 0 \\
0 & 0 & 0 & E_v(z) - \frac{\hbar^2}{2m}\frac{\partial^2}{\partial z^2}
\end{bmatrix}
$$

$$
\times
\begin{bmatrix}
F_s(z) \\
F_{x-iy}(z) \\
F_z(z) \\
F_{x+iy}(z)
\end{bmatrix}
\quad (8.30)
$$

where $F_s, F_{x \pm iy}$, and $F_z$ are the envelope-function components corresponding to the $|0,0,\frac{1}{2}\rangle, |1,1,\pm\frac{1}{2}\rangle$, and $|1,0,\frac{1}{2}\rangle$ states in bulk. The free-electron terms can be omitted if we redefine $p_{cv}$ to recover the correct effective mass in the low-energy limit.

Equation (8.30) consists of four coupled differential equations, including the standalone free-particle equation that can be modified to represent HH states along the [001] growth direction. To do so, we would use the HH mass along [001] from Eq. (3.36a) in place of the free-electron mass in the last row of Eq. (8.30). Since the states are spin degenerate at the zone center, we do not need to solve the equations for the other spin. We will learn how to solve differential equations for arbitrary quantum structures numerically in Chapter 9.

To describe $k_\parallel > 0$ states, we start with the Hamiltonians of Eqs. (3.15) and (3.24) and replace $\hat{k}_z$ with $-i\partial/\partial z$. To keep the original ordering, we must also return to Eqs. (3.13) and (3.14) and cast them in terms of $W_s, W_p$, and $W_d$ rather than $M, L$, and $N$. On the other hand, if we did not know how Eq. (3.15) was derived, we might proceed to naively "symmetrize" all the terms (for an analogy, consider the difference between Eqs. (8.26) and (8.29)). Since the spin–orbit terms do not include any derivatives, the quantum confinement changes only the spin–orbit-free part:

$$\boldsymbol{H}_{SOF} = \begin{bmatrix} H_{ss} & H_{sx} & H_{sy} & H_{sz} \\ H_{xs} & H_{xx} & H_{xy} & H_{xz} \\ H_{ys} & H_{yx} & H_{yy} & H_{yz} \\ H_{zs} & H_{zx} & H_{zy} & H_{zz} \end{bmatrix} \tag{8.31}$$

The matrix elements $H_{sx}, H_{sy}, H_{xs}, H_{ys}, H_{xy}$, and $H_{yx}$ are the same as in Eq. (3.15) because they do not contain $\hat{k}_z$. Some of the other matrix elements are similar to the Hamiltonians of Eqs. (8.29) and (8.30):

$$H_{ss} = E_c(z) - \frac{\hbar^2}{2m}\frac{\partial^2}{\partial z^2} \tag{8.32a}$$

$$H_{sz} = \frac{\hbar}{m}\left[ f p_{cv}(z)\frac{\partial}{\partial z} + (1-f)\frac{\partial}{\partial z}p_{cv}(z) \right] \tag{8.32b}$$

$$H_{zs} = -\frac{\hbar}{m}\left[ (1-f) p_{cv}(z)\frac{\partial}{\partial z} + f\frac{\partial}{\partial z}p_{cv}(z) \right] \tag{8.32c}$$

The remaining matrix elements are obtained from Eqs. (3.13) and (3.14) using derivative substitution:

$$H_{xx} = E_v(z) - \frac{\hbar^2}{2m}\frac{\partial^2}{\partial z^2} + [W_s(z) + 2W_d(z)]\, k_x^2 + W_p(z)k_y^2 - \frac{\partial}{\partial z}W_p(z)\frac{\partial}{\partial z} \tag{8.33a}$$

$$H_{yy} = E_v(z) - \frac{\hbar^2}{2m}\frac{\partial^2}{\partial z^2} + W_p(z)k_x^2 + [W_s(z) + 2W_d(z)]\,k_y^2 - \frac{\partial}{\partial z}W_p(z)\frac{\partial}{\partial z}$$

$$\text{(8.33b)}$$

$$H_{zz} = E_v(z) - \frac{\hbar^2}{2m}\frac{\partial^2}{\partial z^2} + W_p(z)\left(k_x^2 + k_y^2\right) - \frac{\partial}{\partial z}[W_s(z) + 2W_d(z)]\frac{\partial}{\partial z} \quad \text{(8.33c)}$$

$$H_{xz} = -ik_x\,[W_s(z) - W_d(z)]\,\frac{\partial}{\partial z} - ik_x\frac{\partial}{\partial z}W_p(z) \quad \text{(8.33d)}$$

$$H_{zx} = -ik_x\frac{\partial}{\partial z}[W_s(z) - W_d(z)] - ik_x W_p(z)\frac{\partial}{\partial z} \quad \text{(8.33e)}$$

$$H_{yz} = -ik_y\,[W_s(z) - W_d(z)]\,\frac{\partial}{\partial z} - ik_y\frac{\partial}{\partial z}W_p(z) \quad \text{(8.33f)}$$

$$H_{zy} = -ik_y\frac{\partial}{\partial z}[W_s(z) - W_d(z)] - ik_y W_p(z)\frac{\partial}{\partial z} \quad \text{(8.33g)}$$

The difference between Eqs. (8.33a) and (8.33b) (as well as between Eqs. (8.33d) and (8.33f) and between Eqs. (8.33e) and (8.33g)) is that we have swapped $k_y$ for $k_x$ and vice versa.

Equation (8.33) clearly differs from the symmetrized form obtained directly from Eq. (3.15):

$$H_{xx} = E_v(z) - \frac{\hbar^2}{2m}\frac{\partial^2}{\partial z^2} - L(z)k_x^2 - M(z)k_y^2 + \frac{\partial}{\partial z}M(z)\frac{\partial}{\partial z} \quad \text{(8.34a)}$$

$$H_{yy} = E_v(z) - \frac{\hbar^2}{2m}\frac{\partial^2}{\partial z^2} - L(z)k_x^2 - M(z)k_y^2 + \frac{\partial}{\partial z}M(z)\frac{\partial}{\partial z} \quad \text{(8.34b)}$$

$$H_{zz} = E_v(z) - \frac{\hbar^2}{2m}\frac{\partial^2}{\partial z^2} - M(z)\left(k_x^2 + k_y^2\right) + \frac{\partial}{\partial z}L(z)\frac{\partial}{\partial z} \quad \text{(8.34c)}$$

$$H_{xz} = -H_{zx}^* = \frac{ik_x}{2}\left[N(z)\frac{\partial}{\partial z} + \frac{\partial}{\partial z}N(z)\right] \quad \text{(8.34d)}$$

$$H_{yz} = -H_{zy}^* = \frac{ik_y}{2}\left[N(z)\frac{\partial}{\partial z} + \frac{\partial}{\partial z}N(z)\right] \quad \text{(8.34e)}$$

where $M = -W_p$, $L = -(W_s + 2W_d)$, and $N = -(W_s + W_p - W_d)$. To see the difference, we recast Eq. (8.33) in terms of $M$, $L$, and $N$:

$$H_{xx} = E_v(z) - \frac{\hbar^2}{2m}\frac{\partial^2}{\partial z^2} - L(z)k_x^2 - M(z)k_y^2 + \frac{\partial}{\partial z}M(z)\frac{\partial}{\partial z} \quad \text{(8.35a)}$$

$$H_{yy} = E_v(z) - \frac{\hbar^2}{2m}\frac{\partial^2}{\partial z^2} - L(z)k_x^2 - M(z)k_y^2 + \frac{\partial}{\partial z}M(z)\frac{\partial}{\partial z} \tag{8.35b}$$

$$H_{zz} = E_v(z) - \frac{\hbar^2}{2m}\frac{\partial^2}{\partial z^2} - M(z)\left(k_x^2 + k_y^2\right) + \frac{\partial}{\partial z}L(z)\frac{\partial}{\partial z} \tag{8.35c}$$

$$H_{xz} = ik_x\left[N(z) - M(z)\right]\frac{\partial}{\partial z} + ik_x\frac{\partial}{\partial z}M(z) \tag{8.35d}$$

$$H_{zx} = ik_x\frac{\partial}{\partial z}\left[N(z) - M(z)\right] + ik_xM(z)\frac{\partial}{\partial z} \tag{8.35e}$$

$$H_{yz} = ik_y\left[N(z) - M(z)\right]\frac{\partial}{\partial z} + ik_y\frac{\partial}{\partial z}M(z) \tag{8.35f}$$

$$H_{zy} = ik_y\frac{\partial}{\partial z}\left[N(z) - M(z)\right] + ik_yM(z)\frac{\partial}{\partial z} \tag{8.35g}$$

The diagonal elements in Eqs. (8.34) and (8.35) are identical, but the off-diagonal terms differ even if we take $W_s = 0$. Therefore, the interface delta-function terms resulting from the two approaches are also different. The expressions for $L$, $M$, and $N$ in terms of the more widely used Luttinger parameters are given in Eq. (3.38).

Because the value $f$ in Eq. (8.29) did not matter, we may wonder whether the distinction between Eqs. (8.34) and (8.35) is more important. It turns out that the differences are still negligible as long as we solve the full 8-band Hamiltonian. As an example, Fig. 8.3a shows the valence-band dispersions obtained from the Hamiltonians based on Eqs. (8.34) and (8.35) with $f = 1/2$ for an InSb/AlInSb quantum well along the [11] in-plane direction. The energy gap does not change because the two approximations are equivalent at $k_\parallel = 0$. Even for $k_\parallel > 0$ the differences are too small to be visible on the scale of the plot.

Does this mean that we should learn to stop worrying and love the 8-band Hamiltonian? When it comes to narrow-gap semiconductors, this is indeed good advice. But reduced models of the valence-band structure do have their niches. To see that, we start with the $4 \times 4$ Hamiltonian of Eq. (3.42). Three of the expressions in Eq. (3.43) are replaced with

$$P = \frac{\hbar^2}{2m}\left[\gamma_1(z)\left(k_x^2 + k_y^2\right) - \frac{\partial}{\partial z}\gamma_1(z)\frac{\partial}{\partial z}\right] \tag{8.36a}$$

$$Q = \frac{\hbar^2}{2m}\left[\gamma_2(z)\left(k_x^2 + k_y^2\right) + 2\frac{\partial}{\partial z}\gamma_1(z)\frac{\partial}{\partial z}\right] \tag{8.36b}$$

$$S = -i\frac{\hbar^2}{2m}\sqrt{3}\left(k_x - ik_y\right)\left[\gamma_3(z)\frac{\partial}{\partial z} + \frac{\partial}{\partial z}\gamma_3(z)\right] \tag{8.36c}$$

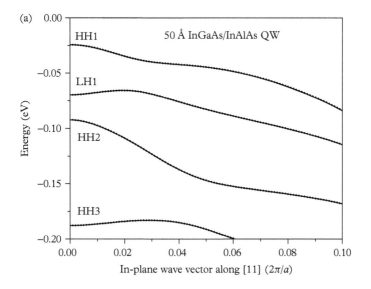

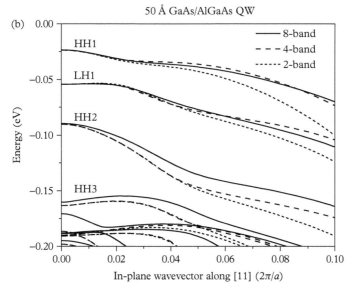

**Figure 8.3** *Valence-band structures along the [11] in-plane direction for: (a) 50 Å*
*In$_{0.53}$Ga$_{0.47}$As/100 Å Al$_{0.48}$In$_{0.52}$As and (b) 50 Å GaAs/100 Å Al$_{0.3}$Ga$_{0.7}$As quantum wells,*
*calculated with f = 1/2. The 8-band calculations were performed for the InGaAs quantum well*
*using both the symmetrized (points) and "correct" (lines) ordering schemes, but the differences*
*between the two of < 0.1 meV are too small to be visible on the scale of the plot. For the GaAs*
*quantum well, the 8-band theory (solid curves) is compared to the 4-band Hamiltonian (dashed)*
*and the 2-band Hamiltonian (dotted) using symmetrized operators. In the 2-band case, the*
*dispersions are always isotropic in the plane of the quantum well.*

but $R$ remains unchanged:

$$R = \frac{\hbar^2}{2m}\left[-\sqrt{3}\gamma_2(z)\left(k_x^2 - k_y^2\right) + i2\sqrt{3}\gamma_3(z)k_x k_y\right] \tag{8.36d}$$

As for any symmetrized Hamiltonian, the same equations are used for $H_{HH+,LH+}$ and $H_{LH+,HH+}$ etc.

The quantum-confined symmetrized version of Eq. (3.52) for the 2×2 Hamiltonian of Eq. (3.49) follows from the "symmetrized" transformation rule of Eq. (8.26):

$$T = \frac{\hbar^2}{2m}\sqrt{3}\left[\tilde{\gamma}k_{\parallel}^2 + k_{\parallel}\left[\gamma_3(z)\frac{\partial}{\partial z} + \frac{\partial}{\partial z}\gamma_3(z)\right]\right] \tag{8.37a}$$

$$T^{\dagger} = \frac{\hbar^2}{2m}\sqrt{3}\left[\tilde{\gamma}k_{\parallel}^2 - k_{\parallel}\left[\gamma_3(z)\frac{\partial}{\partial z} + \frac{\partial}{\partial z}\gamma_3(z)\right]\right] \tag{8.37b}$$

Figure 8.3b plots the valence-band structure along the [11] in-plane direction for a GaAs/AlGaAs quantum well calculated with the reduced 4-band Hamiltonian of Eq. (3.42) and the symmetrized operators of Eq. (8.36). For comparison, we also show the results for the 8-band Hamiltonian and the 2-band Hamiltonian contained in Eq. (3.49) with the elements given by Eqs. (8.36a), (8.36b), and (8.37). The 2-band and 4-band reduced models deviate from the full solution for $k_{\parallel} > 0$ and usually predict slightly lower hole masses and densities of states. Nevertheless, if what we are after is the valence-band structure near $k_{\parallel} = 0$ for moderate- and wide-gap semiconductors, either approach is often good enough and can save some valuable simulation time.

So far we have assumed quantum confinement along the $z$ axis, e.g., a quantum well grown along the [001] direction. Can we derive such a Hamiltonian for a different substrate orientation, say [111]? Naïve derivative substitution leaves us with three partial spatial derivatives, which does not seem promising. Instead, we should rotate the coordinate system so that the direction of quantum confinement is aligned with the $z$ axis to make the substitution easy. For example, for [111] growth, this direction is $\hat{k}'_z = \frac{\hat{k}_x}{\sqrt{3}} = \frac{\hat{k}_y}{\sqrt{3}} = \frac{\hat{k}_z}{\sqrt{3}}$. In general, the coordinate system is rotated using the matrix

$$U = \begin{bmatrix} \cos\theta\cos\varphi & \cos\theta\sin\varphi & -\sin\theta \\ -\sin\varphi & \cos\varphi & 0 \\ \sin\theta\cos\varphi & \sin\theta\sin\varphi & \cos\theta \end{bmatrix} \tag{8.38}$$

where $\theta$ and $\varphi$ are the polar (with respect to the $z$ axis, as rotated toward the $x - y$ plane) and azimuthal (with respect to the $x$ axis, as rotated toward the $y$ axis in the $x - y$ plane) angles illustrated in Fig. 8.4. For [111] growth, $\cos\theta = 1/\sqrt{3}$ and $\cos\varphi = \sin\varphi = \sin\left(45°\right) = 1/\sqrt{2}$, and the rotation matrix becomes

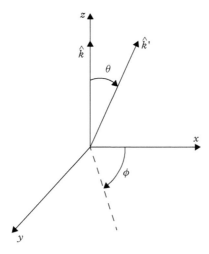

**Figure 8.4** *Illustration of the transformation needed to apply derivative substitution to growth directions other than [001].*

$$U_{[111]} = \begin{bmatrix} \frac{1}{\sqrt{6}} & \frac{1}{\sqrt{6}} & -\sqrt{\frac{2}{3}} \\ -\frac{1}{\sqrt{2}} & \frac{1}{\sqrt{2}} & 0 \\ \frac{1}{\sqrt{3}} & \frac{1}{\sqrt{3}} & \frac{1}{\sqrt{3}} \end{bmatrix} \tag{8.39}$$

This matrix transforms the wavevector and the Hamiltonian to

$$k_i' = \sum_j U_{ij} k_j \tag{8.40a}$$

$$H_{ik}' = \sum_{j,l} U_{ij} U_{kl} H_{jl} \tag{8.40b}$$

It should be clear from Eq. (8.39) that multiplying $U_{[111]}$ by the original wavevec-

tor $\begin{bmatrix} \frac{k}{\sqrt{3}} \\ \frac{k}{\sqrt{3}} \\ \frac{k}{\sqrt{3}} \end{bmatrix}$ , as specified by Eq. (8.40a), leads to a wavevector directed along the

$z$ axis. Since the $s$-like states are spherically symmetric, $H_{ss}$ in the Hamiltonian of Eq. (8.31) remains unchanged. $H_{sx}$, $H_{sy}$, etc. transform as the wavevector in Eq. (8.40a), while the rest of the terms transform according to Eq. (8.40b).

To take another example, for growth along [101] with $\hat{k}_z' = \frac{\hat{k}_x}{\sqrt{2}} = \frac{\hat{k}_z}{\sqrt{2}}, \hat{k}_y = 0$, the angles are $\cos\theta = \sin\theta = \sin\left(45°\right) = 1/\sqrt{2}$ and $\varphi = 0°$, and the rotation matrix becomes

$$U_{[101]} = \begin{bmatrix} \frac{1}{\sqrt{2}} & 0 & -\frac{1}{\sqrt{2}} \\ 0 & 1 & 0 \\ \frac{1}{\sqrt{2}} & 0 & \frac{1}{\sqrt{2}} \end{bmatrix} \qquad (8.41)$$

Once the Hamiltonian is transformed, the derivative-substitution method ($\hat{k}_z \rightarrow -i\partial/\partial z$) works exactly as before, even if the algebra is thorny. This method also works in solving for the band structure of wurtzite quantum wells, e.g., GaN/AlGaN, using symmetrized Hamiltonians. Finally, strain terms do not contain any wavevector components, so all we have to do is to include the $z$ dependence of the strain tensor and deformation potential into the calculation.

## 8.3  Additional Interface Terms

As mentioned in Section 8.1, the primary result of the envelope-function theory is the appearance of new terms at heterointerfaces not present in the bulk **k·p** Hamiltonian. For example, at an interface $z_i$ the Hamiltonian acquires a sharply peaked element, modeled by a delta function, that connects bands $a$ and $b$:

$$\langle a|\delta V|b \rangle = \Omega_{ab}\delta(z - z_i) \qquad (8.42)$$

with the perturbation $\delta V$, due to the broken crystal symmetry at the interface. We already saw in Section 8.2 that discontinuities in the **k·p** matrix elements at an abrupt interface between two dissimilar materials give rise to delta-function terms in the Hamiltonian for the quantum structure. But the terms in Eq. (8.42) do not even appear in the original Hamiltonian! Of course, this does not mean they are completely unexpected. For example, bulk terms corresponding to $H_{ab}$ in Eq. (8.42) make an appearance in microscopic models such as the tight-binding formalism of Section 5.3.

Why should that be the case? Recall that the tight-binding Hamiltonian builds on the zinc blende lattice rather than fcc. Symmetry breaking at an interface means that these terms can be of the same order of magnitude in the envelope-function theory as the plain-vanilla **k·p** matrix elements. The magnitude of the new interface terms depends on the properties of a particular interface, rather than those of any bulk material.

The interface terms are relatively small when the materials on the two sides have a common atom, e.g., GaAs/AlGaAs. By far the most interesting case is when they have neither a common cation (group III element) nor a common anion (group V element), as in the narrow-gap heterojunctions InAs/GaSb and InAs/AlSb. These interfaces can have either In–Sb or Ga(Al)–Sb bond types. That is to say, for the In–Sb bond, InAs growth ends with a plane of In atoms and the GaSb growth begins with a plane of Sb atoms, as illustrated in Fig. 8.5. Conversely, the Ga–As bond occurs when the InAs growth ends with As and the GaSb growth begin with Ga. In real materials, the bond types may be mixed to a certain extent

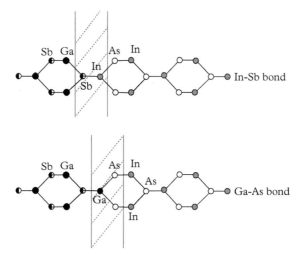

**Figure 8.5** *Two different interface bond types for interfaces without a common anion or cation.*

over the entire growth plane, but one type often dominates. We now list the resulting interface terms and outline their magnitudes and impacts on the band structure for the case of [001] growth.

We look for the extra interface terms in the elements of the 16-band tight-binding Hamiltonian of Section 5.3 that are neglected in the 8-band formulation, namely, the "cation–anion" and "anion–cation" terms in Eqs. (5.43)–(5.53). Only the elements do not vanish when $k_\parallel = 0$ are significant. This eliminates all the terms proportional to $\sin(k_x a/4)$ or $\sin(k_y a/4)$ ($S_x$ and $S_y$ in our notation). We see immediately from the definitions of $g_0$, $g_1$, $g_2$, and $g_3$ in Eqs. (5.44), (5.47), and (5.50) that only $g_0$ and $g_3$ survive. The $s$–$s$, $x$–$x$, $y$–$y$, and $z$–$z$ matrix elements are proportional to $g_0$, and the $s$–$z$ and $x$–$y$ matrix elements are proportional to $g_3$. For an interface with a common atom, the interface profile is an odd function along the $z$ axis. Therefore, the two complex exponentials that form the $C_z$ term on both sides of the interface cancel, and the $s$–$s$, $x$–$x$, $y$–$y$, and $z$–$z$ elements disappear. This does not happen for an interface with no common atoms. Here the even components dominate, and the $s$–$s$, $x$–$x$, $y$–$y$, and $z$–$z$ elements become larger than the $s$–$z$ and $x$–$y$ terms. Since there is no physical difference between the $x$ and $y$ axes in a cubic lattice grown along [001], the $x$–$x$ and $y$–$y$ elements are equal. Based on the discussion in Section 5.3, the $x$–$y$ and $y$–$x$, as well as the $s$–$z$ and $z$–$s$ elements, are also equal. In the general case, the contribution to the Hamiltonian from the interfaces is written as

$$\boldsymbol{H}_{IF} = \sum_i \delta\left(z - z_i\right) \begin{bmatrix} \Omega_{ss} & 0 & 0 & \pi_i \Omega_{sz} \\ 0 & \Omega_{xx} & \pi_i \Omega_{xy} & 0 \\ 0 & \pi_i \Omega_{xy} & \Omega_{xx} & 0 \\ \pi_i \Omega_{sz} & 0 & 0 & \Omega_{zz} \end{bmatrix} \tag{8.43}$$

For the odd terms located off the main diagonal, the order in which the materials are grown matters. We can take one particular sequence (say, GaAs on AlAs or InAs on GaSb) to have $\pi_i = 1$, and the inverted sequence (AlAs on GaAs or GaSb on InAs) to correspond to $\pi_i = -1$, or vice versa.

We have already pointed out that $\Omega_{ss} \approx \Omega_{xx} \approx \Omega_{zz} \approx 0$ for a common-atom interface. But the tables are turned when the interface has no common anion or cation, so that $\Omega_{ss}, \Omega_{xx}, \Omega_{zz} \gg \Omega_{xy}, \Omega_{sz}$. The actual magnitudes of the parameters depend on which interface bond type dominates, which has the regrettable side effect of requiring a new set of parameters for each particular growth technique. Finally, interface terms of higher orders in the potential perturbation do exist, but Eq. (8.43) is generally more than sufficient for a practical description of narrow-gap superlattices and quantum wells. In fact, much of the literature does not consistently include even the five interface terms shown in Eq. (8.43).

The effects of the interface terms on the energy gap of the InAs/GaSb superlattice with variable InAs thickness are illustrated in Fig. 8.6. We use the parameters specified by Livneh *et al.* (see "Suggested Reading"): $\Omega_{ss} = 3$ eV, $\Omega_{xx} = 1.3$ eV, $\Omega_{zz} = 1.1$ eV, $\Omega_{xy} = \Omega_{sz} = 0.2$ eV at 77 K. The interface terms speed up the variation of the gap with InAs thickness and push it a little wider when the InAs is thin. These trends have been found in better agreement with experimental data. The largest differences arise for thin InAs and GaSb layers, but even then they hardly exceed the typical shifts from uncertainties in the layer thicknesses and

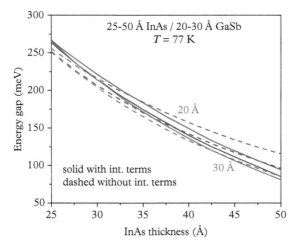

**Figure 8.6** *Energy gap of the InAs/GaSb superlattice with variable InAs thickness and three different GaSb thicknesses (20, 25, and 30 Å). The dashed curves are for the Hamiltonian without any interface terms, and the solid curves are for the Hamiltonian that includes the terms of Eq. (8.43). The curves are for 20 (red), 25 (blue), 30 Å (green) GaSb layers in both cases. All gaps are calculated at a temperature of 77 K. The convergence issues associated with the calculations in this plot are discussed further in Chapter 10.*

composition grading near the interfaces discussed in Chapter 10. Finally, it may be tempting to insert a half-monolayer of InSb or GaAs at the InAs/GaSb interface as a shortcut to approximating the impact of the interfaces. But the interface bond between two atoms (In and Sb or Ga and As) is nothing like bulk InSb or GaAs, which means that this procedure is not expected to be very accurate.

We already mentioned that the largest interface terms vanish for structures with a common atom (this is usually a group V atom). Therefore, the effect of interface terms in Fig. 8.6 is at the extreme end of the scale. In fact, interface terms remain mostly undetermined for all superlattices other than InAs/Ga(In)Sb. Similarly, it is not well known how fast they change with In composition for GaInSb hole wells.

## 8.4  Spin Splitting

In structures with quantum confinement, the two spin states at a given in-plane wavevector are in general not degenerate: $E_+(\mathbf{k}) \neq E_-(\mathbf{k})$. Following Section 2.2, this happens whenever inversion symmetry is broken. By time-reversal symmetry we still have $E_+(\mathbf{k}) = E_-(-\mathbf{k})$. In other words, any state with a given $\mathbf{k}$ must be accompanied by a state with the same energy and opposite spin at $-\mathbf{k}$ (but not necessarily at $\mathbf{k}$ itself).

How exactly does the lack of inversion symmetry give rise to spin splitting? We take the cue from the interface terms of Section 8.3 and again look for tight-binding terms that are missing from the bulk 8-band $\mathbf{k \cdot p}$ model. Focusing on the matrix elements on the same atom, we find that in Eq. (5.32) we set $E_{sx}(0,1,1) = 0$ in the $4E_{sx}(0,1,1)\,s_y s_z$ term, even though we do not expect this identity to hold in the zinc-blende crystal. Expanding in a Taylor series within the framework of the $\mathbf{k \cdot p}$ model and retaining the lowest-order term, we obtain $B_{sx}k_y k_z$, where $B_{sx} \equiv 4E_{sx}(0,1,1)$. Not to be outdone, the $s$–$y$ and $s$–$z$ matrix elements also acquire their respective $B_{sx}k_z k_x$ and $B_{sx}k_x k_y$ terms.

These extra terms are sufficient to derive the spin splitting from the full Hamiltonian that includes spin–orbit coupling, but we will not take that route. Instead, we use physical arguments to arrive at the possible form of the splitting in the conduction band. For simplicity, we confine our discussion to subbands with parabolic dispersion, i.e., quantum wells with relatively wide energy gaps. However, spin splitting in either band is always due to interactions between the $s$-like and $p$-like states.

We reason as follows:

1. The spin splitting should be of higher order in $k$ than the quadratic leading term of the expansion, so it can be treated as a perturbation for small $k$. Therefore, the lowest allowed term of the spin splitting is cubic in $k$.

2. The spin splitting cannot single out a particular crystal axis. If one component is known, the other two follow by cyclic permutation, i.e., shifting the

$x, y, z$ indices by one to the left or right and mentally connecting the tail end of the expression with the front.

3. Since the general form of the spin–orbit interaction is $\boldsymbol{L} \cdot \boldsymbol{S} \propto \boldsymbol{\sigma} \cdot \boldsymbol{p} \times \boldsymbol{r}$, where $\boldsymbol{\sigma}$ are the Pauli spin matrices in Eq. (5.68), the reduced Hamiltonian should have the form $\boldsymbol{\sigma} \cdot \boldsymbol{\kappa}$. Here $\boldsymbol{\kappa}$ is orthogonal to $\boldsymbol{k}$, i.e., $\boldsymbol{k} \cdot \boldsymbol{\kappa} = 0$, because it is derived from $\boldsymbol{p} \times \boldsymbol{r}$.

Combining these conditions suggests that $|\boldsymbol{\kappa}| = \boldsymbol{k} \cdot \boldsymbol{\xi}$, where $\boldsymbol{\xi}$ is quadratic in $k$. Since $\boldsymbol{k} \cdot \boldsymbol{\kappa}$ is quartic in wavevector components, the most natural choice to satisfy $\boldsymbol{k} \cdot \boldsymbol{\kappa} = 0$ is for $\kappa_x$ to be linear in $k_x$, etc. Therefore, $\boldsymbol{k} \cdot \boldsymbol{\kappa}$ contains a sum of three terms with corresponding factors $k_x^2, k_y^2, k_z^2$. If $\kappa_x \propto \mathcal{O}\left(k_x k_y^2, k_x k_z^2\right)$, the wavevector components can be cancelled by including both positive and negative terms, i.e., $\kappa_x = k_x\left(k_y^2 - k_z^2\right)$, etc.

The sign of the leading spin-splitting term is arbitrary, and the magnitude must be determined experimentally or from first principles. Its overall form in bulk is

$$B_{sp}\boldsymbol{\sigma} \cdot \boldsymbol{\kappa} = B_{sp}\left[\sigma_x k_x\left(k_y^2 - k_z^2\right) + \sigma_y k_y\left(k_z^2 - k_x^2\right) + \sigma_z k_z\left(k_x^2 - k_y^2\right)\right] \qquad (8.44)$$

where $B_{sp} \propto B_{sx}$. It is easy to check that Eq. (8.44) satisfies the third condition $\boldsymbol{k} \cdot \boldsymbol{\kappa} = 0$. In many bulk semiconductors, spin splitting is weak enough that all higher orders beyond Eq. (8.44) can be neglected. The leading-order splitting vanishes along high-symmetry directions such as [001] and [111].

Curiously enough, the form of Eq. (8.44) is similar to the magnetic-field interaction $-\boldsymbol{\sigma} \cdot \boldsymbol{B}$, with the energy lower for electron spin aligned with the magnetic field! Therefore, we can think of $\boldsymbol{\kappa}$ as a "quasi-magnetic field" induced by the spin–orbit interaction in a crystal without inversion symmetry. It goes without saying that no actual magnetic field is present, and even the quasi-field depends on the direction in $k$ space. If no magnetic field is present, there is no preferred axis, and, strictly speaking, we cannot associate the two spin states with up and down spins, although this is a common informal practice.

The spin splitting in the valence band also exists, and its dependence on the wavevector components is more complex because the cell-periodic part of the wavefunction is not spherically symmetric. We will not give the details here, but accurate results can be obtained by solving the full Hamiltonian with the $B_{sx}$ term.

How does the conduction-band spin splitting change in a structure with quantum confinement? Because the symmetry is reduced, we expect the spin splitting to become more prominent. To estimate its magnitude, consider the electron wavevector along the growth direction in a quantum well with infinite barriers. It is quantized in units of $\pi/W$, where $W$ is the width of the well, because the wavefunction must vanish at the boundaries of the well. In other words, restricting the accessible volume induces a discrete spectrum in its dual,

Fourier-pair quantity. It turns out that Eq. (8.44) still applies if we replace $k_z$ and $k_z^2$ by their expectation values $\langle k_z \rangle$ and $\langle k_z^2 \rangle$. The expectation value of the growth-direction wavevector $\langle k_z \rangle = 0$ for any state of the well because the sine-like wavefunction has equal contributions from $\pi/W$ and $-\pi/W$. On the other hand, the square of the wavevector has a non-zero expectation value, $\langle k_z^2 \rangle = \frac{n^2 \pi^2}{W^2}$, where $n = 1, 2, 3, \ldots$. For a realistic quantum well, $\langle k_z^2 \rangle$ is a little smaller, but the same principle applies. For the [001] growth direction, Eq. (8.44) then reduces to

$$B_{sp}\boldsymbol{\sigma} \cdot \boldsymbol{\kappa}_{QW[001]} = B_{sp}\left[-\sigma_x k_x \langle k_z^2 \rangle + \sigma_y k_y \langle k_z^2 \rangle\right] \qquad (8.45)$$

We see that the spin splitting in a quantum well is linear rather than cubic in $k$. The spin splitting in Eqs. (8.44) and (8.45) is known as the Dresselhaus splitting, after Gene Dresselhaus who was the first to derive Eq. (8.44) in 1955.

The typical magnitude of $B_{sp}$ is a few tens to a few hundred eV Å$^3$ in most semiconductors, with higher values usually observed in narrower-gap materials (see Chapter 6). The order of magnitude follows from a product of three momentum matrix elements including $P = \frac{\hbar}{m} \langle c|p_{cv}|v \rangle$ from Section 4.3, which is a few hundred (eV Å)$^3$, divided by the square of the energy separations that tend to exceed the energy gap and hence are in the range 1–10 eV$^2$. Therefore, if $\langle k_z^2 \rangle \sim 10^{-3}$Å$^{-2}$, this corresponds to spin splitting that scales as 10–100 $(k/\text{Å}^{-1})$ meV and amounts to a few meV for typical occupied states near the bottom of the conduction band in common materials such as GaAs and InGaAs.

A naïve inspection of Eq. (8.44) appears to show that the spin splitting is zero for [111] growth. This cannot be correct because the spin splitting along any given direction vanishes only for wavevectors perpendicular to that direction, which follows from the form of the spin–orbit interaction $\boldsymbol{L} \cdot \boldsymbol{S} \sim \boldsymbol{\sigma} \cdot \boldsymbol{k} \times \boldsymbol{r}$. Because [111] is not perpendicular to any crystal axis and forms the same angle with all of them, the spin splitting is also the same for all crystal-axis directions. Requiring that the spin splitting be orthogonal to the wavevector, we can deduce the functional form without transforming the coordinates:

$$B_{sp}\boldsymbol{\sigma} \cdot \boldsymbol{\kappa}_{QW[111]} \approx B_{sp}\langle k_n^2 \rangle \, \boldsymbol{\sigma} \times \boldsymbol{k} \qquad (8.46)$$

A complete calculation leads to an extra factor of $2/\sqrt{3}$ on the right.

In fact, Eqs. (8.45) and Eq. (8.46) are not necessarily the only contribution to spin splitting in the conduction band of a quantum well. These spin-related terms appear in the Hamiltonian because the crystal symmetry is reduced by singling out the quantum-well axis. In fact, any structural variation along a single axis is sufficient, whether or not its profile is invariant under reflection. The quantum structure may also have built-in asymmetry of the types illustrated in Fig. 8.7. For

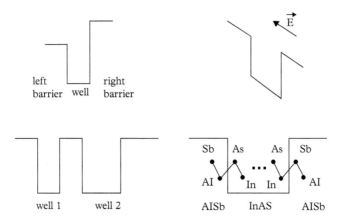

**Figure 8.7** *Various types of asymmetry that may introduce Rashba spin splitting.*

example, the two barriers may be different, or else two wells of different width may be coupled via a thin barrier. When the interfaces have no common atom, the interface bond types on opposite sides of the well may be different, e.g., InSb and GaAs for the InAs/GaSb quantum well. The most visually obvious asymmetry is from an electric field applied across the quantum well. Are these possibilities qualitatively different from single symmetric quantum wells when it comes to observable spin splitting?

In fact, when any of the asymmetries shown in Fig. 8.7 are present, there is an additional spin–orbit term of the form $\boldsymbol{L} \cdot \boldsymbol{S} \sim \boldsymbol{\sigma} \cdot \boldsymbol{k} \times \boldsymbol{r} = \boldsymbol{\sigma} \times \boldsymbol{k} \cdot \boldsymbol{r}$, where $\boldsymbol{r}$ is directed along the growth direction ($\hat{\boldsymbol{z}}$ in the case of [001] growth):

$$H_R = \alpha_R \left( \boldsymbol{\sigma} \times \boldsymbol{k} \right) \cdot \hat{\boldsymbol{z}} \tag{8.47}$$

This term leads to *Rashba* spin splitting, named after Emmanuel Rashba, who introduced it to treat bulk uniaxial solids (such as wurtzite crystals or cubic crystals with an applied electric field) in the late 1950s. The Rashba coefficient $\alpha_R$ is determined by *both* the amount of asymmetry and the strength of the spin–orbit interaction. Even though the spin–orbit coupling mostly affects the valence band, Rashba showed that its effect on the conduction-band states via the term in Eq. (8.47) is not negligible. At the time this came as a surprise to the semiconductor physics community. In fact, some confusion about its origin persisted well after the initial discovery. For diamond crystals and other materials with a weak Dresselhaus effect, the Rashba term is the leading one in the $k$ expansion and strongly modifies the electron dispersions, as we will see below. Structural asymmetry also affects the spin splitting in the valence band, but the details are more complicated, as for bulk asymmetry. The general conclusion is that that the linear spin terms are small in the valence band, and the spin splitting tends to vary with the cube of the wavevector for the HH states.

In general, the quantum-well Hamiltonian in the parabolic subband model contains both Rashba and Dresselhaus terms for the $s$-like electrons:

$$H_{SS} = \beta_D \left( \sigma_x k_x - \sigma_y k_y \right) + \alpha_R \left( \sigma_x k_y - \sigma_y k_x \right) \tag{8.48}$$

where $\beta_D = -B_{sp} \langle k_z^2 \rangle$. Figure 8.8 illustrates the spin-splitting orientations due to Rashba and Dresselhaus terms, both isolated and combined at different ratios. Pure Rashba splitting in Fig. 8.8a locks the spin to point along an in-plane direction perpendicular to the wavevector. On the other hand, pure Dresselhaus splitting aligns (or anti-aligns) the spin with respect to $\boldsymbol{k}$.

For [001] and most other growth orientations, spin splitting is always present regardless of the relative magnitudes of $\beta_D$ and $\alpha_R$. However, when the growth direction is [111], Eqs. (8.46) and (8.47) show that complete cancellation is possible for $\beta_D = -\alpha_R$. Even though the sign of $\beta_D$ is determined by the bulk band structure, the Rashba coefficient $\alpha_R$ depends on the orientation of the asymmetry, e.g., on the direction of the applied electric field, and is generally tunable. The magnitudes of $\alpha_R$ (and even $\beta_D$) are determined empirically to a limited precision, which means that in practice it takes a good deal of tinkering with the structural parameters to minimize the spin splitting.

To visualize the conduction-band energy spectrum with spin splitting given by Eq. (8.46) or (8.47), we start from a parabolic unperturbed dispersion. Augmented with the Rashba term in Eq. (8.47), the Hamiltonian has two eigenvalues, corresponding to $\alpha_R k_{\parallel}$ and $-\alpha_R k_{\parallel}$. The two dispersions are separated by $2\alpha_R k_{\parallel}$ at all wavevectors, and the energy minima are displaced from $k_{\parallel} = 0$, as shown in Fig. 8.9 (compare with Fig. 2.5). The same result holds for the Dresselhaus splitting, albeit with different eigenfunction phases. For [111] growth, the splitting is proportional to the sum of $\alpha_R$ and $\beta_D$. For the [001] substrate, it depends on the wavevector orientation rather than just its magnitude, as illustrated in Fig. 8.8. In narrow-gap materials, higher-order terms are more prominent, and Eq. (8.48) is no longer an accurate approximation. An improved model might involve the 8-band Hamiltonian with the Dresselhaus terms such as $B_{sx} k_y k_z$, structural asymmetry, and a full complement of calibrated interface terms from Eq. (8.43). We will return to the discussion of realistic spin splitting in narrow-gap quantum wells in Chapter 10. The broad field of exploring the spin degree of freedom in semiconductors and other materials is called spintronics.

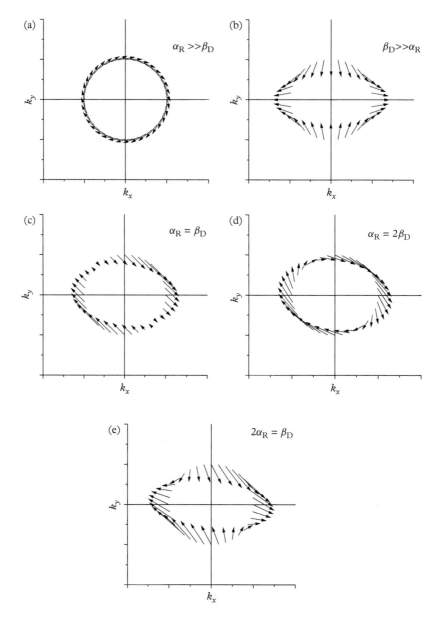

**Figure 8.8** *Spin orientations in the growth plane for Rashba, Dresselhaus, and combined* $(\beta_D =\alpha_R, \beta_D =2\alpha_R, \text{ and } 2\beta_D =\alpha_R)$ *contributions to the spin splitting. Not shown is the second spin subband with the opposite orientation (i.e., counter-clockwise rather than clockwise in the case of the Rashba effect, etc.).*

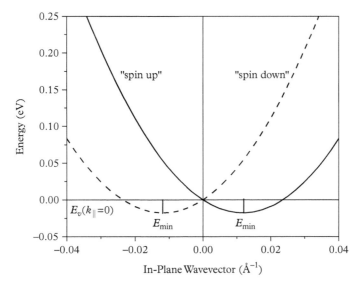

**Figure 8.9** *Spin-split dispersions of the Rashba Hamiltonian of Eq. (8.48) as a function of in-plane wavevector.*

## 8.5 Boundary Conditions for Quantum Structures

In Eq. (8.17), the envelope function $F_j(z)$ was expanded in terms of plane waves. Can $k$ in that equation be a discrete rather than continuous quantity? If so, can we put a limit on the number of $k's$? We will see in Chapter 9 that if we can answer both questions affirmatively, several powerful computational techniques become available to us.

But are the affirmative responses justified? Depending on the boundary conditions, discretization in $k$ space can be equivalent either to isolating the wavefunction in a finite spatial region (Fig. 8.1) or to periodic dependence with a period equal to the extent of the region (Fig. 8.10a). This is not too restrictive, since we can indeed divide all quantum structures into two broad classes: (1) isolated quantum systems such as quantum wells, wires, and dots; and (2) repeated layer configurations or superlattices. One binding constraint for using the envelope-function description is that the period should be much larger than the atomic spacing. If it is not satisfied, we advise exploring the quasi-atomistic approaches of Section 9.5.

Of course, real superlattices do not extend to infinity. The superlattice description makes sense when the structures have enough periods to neglect most edge effects. If the structure is not repeating, but the region of interest is not too large, we can isolate it from the surroundings using very high artificial barriers. This case can also be treated as a superlattice if we assume that the layers repeat themselves beyond the high barriers, as shown in Fig. 8.10b. As a shorthand, we refer to

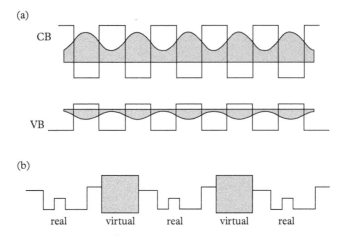

**Figure 8.10** *(a) Conduction- and valence-band profiles, along with superimposed electron and hole wave functions, for a superlattice; (b) Conduction-band profile for two coupled quantum wells, configured as a superlattice with artificial high barriers inserted between the periods. The purpose is to model it as a periodic superlattice structure.*

the case illustrated in Fig. 8.10a as a superlattice, while Fig. 8.10b shows the conduction band for a quantum-well structure treated as a virtual superlattice.

To understand the effect of periodic boundary conditions in Fig. 8.10a, we focus on $k_\parallel = 0$ states. By analogy with a 1D semiconductor crystal in Section 1.3, the spatial dependence in a superlattice is factored into two components: (1) slow variation along the growth axis described by $\exp(ik_z z)$, where $k_z$ ranges from $-\pi/d$ to $\pi/d$, and $d$ is the superlattice period; (2) fast variation within each period (repeated in other periods) that is Fourier-expanded in multiples of $G = 2\pi/d$. Now these two components are present in the envelope function of Eq. (8.17) rather than the full wavefunction of Eq. (8.2) (of course, the variation on the scale of the atomic spacing in the cell-periodic function is still present). The envelope-function expansion for the $j$th bulk band component of any given state becomes

$$F_j(z) = \sum_l e^{i2\pi(k'+l)z/d}\phi_{jl} \qquad (8.49)$$

where $k' = k_z d/(2\pi)$ ranges continuously from $-1/2$ to $1/2$, and $l$ is an integer that counts the multiples of $G = 2\pi/d$. When dealing with an isolated well structure, there is no interaction between periods, i.e., no slow variation on the multi-period scale, and we immediately set $k_z = k' = 0$.

What is the appropriate range of $l$ values? For example, to span a fraction $f_B/2$ of the Brillouin zone along the growth axis, $l$ should vary (in integer steps) from $-\left\lceil \frac{f_B d}{a} \right\rceil$ to $\left\lceil \frac{f_B d}{a} \right\rceil$, where the brackets indicate rounding up to the nearest integer. If we use the **k·p** method, we might choose $f_B \sim 0.2 - 0.3$. The full wavefunction is a sum of the products of the envelope and cell-periodic functions:

$$\psi(z) = \sum_j F_j(z) u_{j0}(z) \qquad (8.50)$$

with $F_j(z)$ given by Eq. (8.49). In the rest of the book, we will sometimes use the term "wavefunction" to refer to the envelope function when it is clear that the cell-periodic component is of no import. Even though most problems can be solved in the superlattice geometry, a viable alternative to modeling quantum wells is to solve the coupled differential equations for the envelope-function components directly, with $F_j(z) = 0$ at the edges of the simulation region.

In general, we also need to specify how to connect the envelope functions at the interfaces between different layers. For example, if the equations involve second-order derivatives, we need two interface boundary conditions for each envelope-function component.

The first question is whether all the components of the envelope function are continuous across the interface, just as they are within any given layer:

$$F_a\Big|_{z_0^+} = F_a\Big|_{z_0^-} \qquad (8.51)$$

Of course, if the wavefunction were discontinuous in space, there would be a jump in the probability of finding an electron an infinitesimal distance away, which is unphysical. This does not necessarily apply when we talk about the envelope part of the wavefunction only. In fact, if the cell-periodic parts on the two sides of the interface are different, Eq. (8.51) generally does not hold. As we know from Section 8.1, the exact envelope-function theory *does* in fact maintain continuity across interfaces, but the various approximate forms of that theory used in calculations may not. Fortunately, this turns out to be of little consequence in almost all cases for which the envelope-function theory gives accurate answers, so we may consider Eq. (8.51) to be at least approximately true regardless of the exact implementation.

The second condition must involve the derivative of the wavefunction. Starting with the one-band Hamiltonian of Eq. (8.23), we integrate the left- and right-hand sides over an infinitesimal distance $2\varepsilon$ around the interface at $z = z_0$:

$$-\frac{\hbar^2}{2} \int_{z_0-\varepsilon}^{z_0+\varepsilon} \frac{\partial}{\partial z} \frac{1}{m_e(z)} \frac{\partial F(z)}{\partial z} \, dz = \int_{z_0-\varepsilon}^{z_0+\varepsilon} \left[ E(k_\parallel) - V(z) - \frac{k_\parallel^2}{m_e(z)} \right] F(z) \, dz \qquad (8.52)$$

The integration on the left cancels out the first differentiation. Even if the potential and effective mass are discontinuous at the interface, the right-hand side vanishes when $\varepsilon \to 0$ as long as the integrand is finite. As a result, we obtain

$$\left[\frac{1}{m_e(z)}\frac{\partial F(z)}{\partial z}\right]\Big|_{z_0-\varepsilon}^{z_0+\varepsilon} = 0 \tag{8.53}$$

which can be rewritten as the desired boundary condition:

$$\left[\frac{1}{m_e(z)}\frac{\partial F(z)}{\partial z}\right]\Big|_{z_0^+} = \left[\frac{1}{m_e(z)}\frac{\partial F(z)}{\partial z}\right]\Big|_{z_0^-} \tag{8.54}$$

When the same approach is applied to the Hamiltonian of Eq. (8.29), we obtain the boundary conditions

$$[p_{cv}(z)F_v(z)]\,|_{z_0^+} = [p_{cv}(z)F_v(z)]\,|_{z_0^-} \tag{8.55a}$$

$$[p_{cv}(z)F_c(z)]\,|_{z_0^+} = [p_{cv}(z)F_c(z)]\,|_{z_0^-} \tag{8.55b}$$

where only terms containing the derivative of $p_{cv}(z)F_{c,v}(z)$ survive. The prescriptions in Eqs. (8.54) and (8.55) are all we need for any Hamiltonian with only first and second derivatives. For example, including a free-electron term in Eq. (8.30) leads to

$$\left[\frac{\hbar}{2}\frac{\partial F_c(z)}{\partial z} + (1-f)p_{cv}(z)F_v(z)\right]\Big|_{z_0^+} = \left[\frac{\hbar}{2}\frac{\partial F_c(z)}{\partial z} + (1-f)p_{cv}(z)F_v(z)\right]\Big|_{z_0^-} \tag{8.56}$$

A similar expression obtained by interchanging the $v$ and $c$ subscripts and substituting $f$ for $(1-f)$ replaces Eq. (8.55b).

We can now write the second boundary condition for a general Hamiltonian represented by the differential equation

$$\sum_b\left[\frac{\partial}{\partial z}A_{ab}(z)\frac{\partial}{\partial z} + \frac{\partial}{\partial z}B_{ab}(z) + C_{ab}(z)\frac{\partial}{\partial z} + D_{ab}\right]F_b(z) = EF_a(z) \tag{8.57}$$

where the various coefficients comprise matrices $A, B, C,$ and $D$. We already know from Section 8.2 that with 1D confinement, the matrix $A$ is diagonal, and $B$ and $C$ are related via $C = -B^\dagger$ (hence $B = -C^\dagger$). Symmetrization further demands that $B = C$, and hence the two matrices are anti-Hermitian ($B = -B^\dagger = C = -C^\dagger$), but this is not a necessary condition from a physical point of view, as we discovered in Chapter 3 and Section 8.2.

Why are the matrices $B$ and $C$ anti-Hermitian rather than Hermitian in the symmetrized approach? Also, why do we have $B = -C^\dagger$ rather than $B = C^\dagger$? This is simply because they represent the coefficients of the antisymmetric first derivative, which is odd for an even envelope-function component and vice versa. In other words, the first derivative leads to a parity flip. Alternatively, we could have chosen to pull out an imaginary factor $i$ from each matrix to make the remaining matrix Hermitian in the symmetrized case. Regardless of which choice we make, Chapter 9 shows that the Hamiltonian obtained by expanding the envelope functions in some convenient basis is always Hermitian provided the derivation is fully consistent.

The boundary condition that includes the potential interface terms from Eqs. (8.42) and (8.43) is

$$\sum_b \left( \delta_{ab} A_{bb} \frac{\partial}{\partial z} + B_{ab} + \Omega_{ab} \right) F_b \bigg|_{z_0^+} = \sum_b \left( \delta_{ab} A_{bb} \frac{\partial}{\partial z} + B_{ab} \right) F_b \bigg|_{z_0^-} \qquad (8.58)$$

Equation (8.58) applies when the derivative of the wavefunction is finite; i.e., the wavefunction is continuous as specified by Eq. (8.51). From Eq. (8.43), we observe that $\Omega_{ba} = \Omega_{ab}$; i.e., the matrix of the interface terms is symmetric. Therefore, Eq. (8.58) is equivalent to

$$\sum_a \left( \delta_{ba} A_{aa} \frac{\partial}{\partial z} - C_{ab}^* + \Omega_{ab} \right) F_a \bigg|_{z_0^+} = \sum_a \left( \delta_{ba} A_{aa} \frac{\partial}{\partial z} - C_{ab}^* \right) F_a \bigg|_{z_0^-} \qquad (8.59)$$

Remember that there is no specific meaning attached to the summation indexes $a$ and $b$.

It turns out that imposing the correct boundary conditions in Eqs. (8.58) and (8.59) has generally a minor effect on the subband energies. In many situations, we can simply require that the derivative of the envelope function be continuous, i.e.,

$$\frac{\partial}{\partial z} F_a \bigg|_{z_0^+} = \frac{\partial}{\partial z} F_a \bigg|_{z_0^-} \qquad (8.60)$$

without the error becoming excessively large. In fact, the derivatives of the envelope-function components are continuous in the rigorous theory developed by Burt and Foreman. As we will see in Chapter 9, the interface boundary conditions often remain implicit when we introduce a particular computational technique for approximating the solutions. Nevertheless, it is important to understand how to apply them correctly.

## Suggested Reading

- The envelope-function theory for quantum structures is covered in a number of books and review articles. The presentation in Section 8.1 hews closely to J. P. Loehr, *Physics of Strained Quantum Well Lasers* (Kluwer, Dordrecht, 1997), Chapter 3.

- For more rigor and important caveats, see the book by L. C. Lew Yan Voon and M. Willatzen, *The k·p Method*, Chapter 12, the review article by Burt, "Fundamentals of envelope function theory for electronic states and photonic modes in nanostructures," *J. Phys.: Condens. Matter* **11**, R53 (1999), and B. A. Foreman, "Connection rules versus differential equations for envelope functions in abrupt heterostructures," *Phys. Rev. Lett.* **80**, 3823 (1998).

- The most complete presentation of the interface terms is set out in the works of Philip Klipstein and collaborators, e.g., Y. Livneh, "k·p model for the energy dispersions and absorption spectra of InAs/GaSb superlattices," *Physical Review B* **86**, 235,311 (2012).

- For more on spin effects in quantum structures, see R. Winkler, *Spin–Orbit Coupling Effects in Two-Dimensional Electron and Hole Systems* (Springer, Berlin, 2003).

# 9

# Methods for Computing the States of Quantum Structures

This chapter presents a detailed development of several numerical methods for calculating the band structure of semiconductor quantum wells and superlattices. These include the transfer-matrix method, the finite-difference method, and the reciprocal-space approach. The relative merits and drawbacks of each approach are briefly considered. We point out that real-space methods often introduce spurious states for the most common forms of the Hamiltonian. We also discuss how the tight-binding and pseudopotential methods can be applied to model quantum structures.

## 9.1 Transfer-Matrix Method

As in Part I, we will compute the band structure by solving a matrix equation for eigenvalues and eigenvectors. This means that we will need to reduce the set of differential equations developed in Chapter 8 to an algebraic eigenproblem.

What is the dimension of the matrix in that eigenproblem? We would like to include a certain number of bulk bands, e.g., $M = 8$ in the trusty 8-band $\mathbf{k \cdot p}$ approach of Chapter 3. We would also like to represent the spatial variation within the quantum structure in a faithful manner. To do so, the spatial dependence of every bulk band component is rendered by the values at certain points. Even though the points remain the same, the values at those points are in general different for each quantum state. Therefore, the dimension of the matrix is always larger (and typically much larger, particularly if we need to represent a function with many oscillations) than the number of bands. For example, in the 8-band $\mathbf{k \cdot p}$ treatment of a quantum structure, the matrix dimension can reach a few hundred. To solve such a large eigenproblem, we must resort to numerical techniques. Fortunately, there are many software packages such as MATLAB, as well as public-domain libraries within programming languages such as Python and Fortran, that can perform this task quickly and efficiently.

One instructive approach that leads to an eigenproblem is to solve the differential equations analytically in a given layer, and then use the boundary conditions

*Bands and Photons in III–V Semiconductor Quantum Structures.* Vurgaftman, Lumb, and Meyer,
Oxford University Press (2021). © Vurgaftman, Lumb, and Meyer.
DOI: 10.1093/oso/9780198767275.003.0009

from Section 8.5 to connect the wavefunctions in different layers. The resulting *transfer-matrix method* is close kin to textbook examples of solving the Schrödinger equation.

In any given layer, the coefficients in Eq. (8.57) are constants:

$$\sum_b \left[ A_{ab} \frac{\partial^2}{\partial z^2} + (B_{ab} - B_{ba}^*) \frac{\partial}{\partial z} + D_{ab} \right] F_b(z) = E F_a(z) \qquad (9.1)$$

In Chapter 8 we showed that $C_{ab} = -B_{ba}^*$, which explains the form of Eq. (9.1). However, in general $\boldsymbol{B} \neq -\boldsymbol{B}^\dagger$; i.e., the Hamiltonian is not symmetrized. To proceed, we define a vector $\boldsymbol{\Phi}(z)$ that has twice as many components as there are *bulk bands*, i.e., $2M$. This vector includes the envelope-function components for all the bulk bands and their first derivatives:

$$\boldsymbol{\Phi}(z) = \begin{bmatrix} \cdots \\ F_{a-1}(z) \\ F_a(z) \\ F_{a+1}(z) \\ \cdots \\ \frac{\partial}{\partial z} F_{a-1}(z) \\ \frac{\partial}{\partial z} F_a(z) \\ \frac{\partial}{\partial z} F_{a+1}(z) \\ \cdots \end{bmatrix} \qquad (9.2)$$

To compress the notation, we rewrite Eq. (9.1) in the matrix form

$$\left[ A \frac{\partial^2}{\partial z^2} + \left( B - B^\dagger \right) \frac{\partial}{\partial z} + D \right] F(z) = E F(z) \qquad (9.3)$$

or, equivalently,

$$\left[ \frac{\partial^2}{\partial z^2} + A^{-1} \left( B - B^\dagger \right) \frac{\partial}{\partial z} + A^{-1} D \right] F(z) = A^{-1} I E F(z) \qquad (9.4)$$

Substituting directly into Eq. (9.1), we see that the vector in Eq. (9.2) satisfies the differential equation

$$\boldsymbol{\Phi}'(z) = \begin{bmatrix} 0 & I \\ A^{-1}(IE - D) & -A^{-1}(B - B^\dagger) \end{bmatrix} \boldsymbol{\Phi}(z) = \Lambda \boldsymbol{\Phi}(z) \qquad (9.5)$$

The dimension of the matrices $A$, $B$, $D$, and the identity matrix $I$ equals the number of bands $M$. The top row of Eq. (9.5) is just the definition in Eq. (9.2). The bottom row follows from Eq. (9.4) if we observe from Eq. (9.2) that the bottom half of $\boldsymbol{\Phi}'$ contains the second derivative of the envelope function. The $2M \times 2M$ matrix $\boldsymbol{\Lambda}$ defined in Eq. (9.5) is the same for any given material, but it varies from layer to layer because the coefficients in the matrices $\boldsymbol{A}, \boldsymbol{B}$, and $\boldsymbol{D}$ are material dependent.

If we ignore the fact that $\boldsymbol{\Lambda}$ is a matrix, we can write the solution of Eq. (9.5) in terms of the envelope function at some arbitrary point to which we assign the coordinate $z = 0$:

$$\boldsymbol{\Phi}(z) = \exp\left(\boldsymbol{\Lambda} z\right) \boldsymbol{\Phi}(0) \equiv \boldsymbol{T}(z) \boldsymbol{\Phi}(0) \tag{9.6}$$

Here $\boldsymbol{T}(z)$ is the *transfer matrix*, since it connects different slices of space within the same material. Equation (9.6) looks simple enough, but conceals a good deal of complexity behind the definition of the transfer matrix. In fact, it is not clear what the exponential of a matrix in Eq. (9.6) even means!

It seems plausible that we can express the solution of Eq. (9.5) as a linear combination of $\exp\left(ik^{(i)} z\right)$, with various $k^{(i)}$. The matrix $\boldsymbol{\Lambda}$ is not necessarily diagonal, but we can bring it to a diagonal form using a *similarity transformation*:

$$\boldsymbol{P}^{-1} \boldsymbol{\Lambda} \boldsymbol{P} = \boldsymbol{Q} \tag{9.7}$$

where $\boldsymbol{P}$ is constructed from the eigenvectors of $\boldsymbol{\Lambda}$ arranged in columns:

$$\boldsymbol{P} = \begin{bmatrix} \xi_1^{(1)} & \cdots & \xi_1^{(2M)} \\ \vdots & \ddots & \vdots \\ \xi_m^{(1)} & \cdots & \xi_m^{(2M)} \end{bmatrix} \tag{9.8}$$

and $\xi_i^{(j)}$ is the $i$th component of the $j$th eigenvector of $\boldsymbol{\Lambda}$. Equation (9.7) is just a rearrangement of the eigenvalue problem with all the eigenvectors present at once: $\boldsymbol{\Lambda} \boldsymbol{P} = \boldsymbol{P} \boldsymbol{Q}$. Now $\boldsymbol{Q}$ is a diagonal matrix that contains the eigenvalues of $\boldsymbol{\Lambda}$:

$$\boldsymbol{Q} = \begin{bmatrix} ik^{(1)} & 0 & \cdots & 0 \\ 0 & ik^{(2)} & \cdots & 0 \\ \vdots & \vdots & \ddots & \vdots \\ 0 & 0 & \cdots & ik^{(m)} \end{bmatrix} \tag{9.9}$$

The matrix $\boldsymbol{P}^{-1} \boldsymbol{T} \boldsymbol{P}$ is also diagonal, since Eq. (9.6) shows that it represents a simple exponentiation of $\boldsymbol{Q}$:

$$\boldsymbol{P}^{-1} \boldsymbol{T} \boldsymbol{P} = \exp\left(\boldsymbol{Q}\right) = \begin{bmatrix} \exp\left(ik^{(1)} z\right) & 0 & \cdots & 0 \\ 0 & \exp\left(ik^{(2)} z\right) & \cdots & 0 \\ \vdots & \vdots & \ddots & \vdots \\ 0 & 0 & \cdots & \exp\left(ik^{(m)} z\right) \end{bmatrix} \tag{9.10}$$

This is the real meaning of the matrix exponential in Eq. (9.6). We can substitute the diagonal matrix $P^{-1}TP$ for our transfer matrix to obtain the differential equation $X' = P^{-1}\Lambda PX$ that has the same eigenvalues as Eq. (9.5). The eigenvectors must be post-processed with those in $P$ to obtain $\Phi$ via the relation $\Phi = PX$.

We now have a straightforward recipe for solving Eq. (9.1) within a layer having a given set of band parameters: (1) calculate the eigenvalues and eigenvectors of $\Lambda$ in Eq. (9.5); (2) solve the differential equations in $X' = P^{-1}\Lambda PX$, where $P^{-1}\Lambda P$ is a diagonal matrix and $P$ is given by Eq. (9.8); and (3) calculate the wavefunctions using $\Phi = PX$. The next order of business is to connect the solutions in different layers. At an interface between two layers (see Fig. 9.1), we apply the boundary conditions of Eqs. (8.51) and (8.58). In fact, the continuity of the wavefunction components in Eq. (8.51) is built into the first row of Eq. (9.5). The boundary condition corresponding to Eq. (8.58) can be written as the matrix equation

$$\begin{bmatrix} I & 0 \\ B+\Omega & A \end{bmatrix} \Phi(z)\Big|_{z_0^+} = \begin{bmatrix} I & 0 \\ B & A \end{bmatrix} \Phi(z)\Big|_{z_0^-} \tag{9.11}$$

Multiplying by the inverse of the first matrix on both sides, we obtain

$$\Phi(z)\Big|_{z_0^+} = \begin{bmatrix} I & 0 \\ B+\Omega & A \end{bmatrix}^{-1} \begin{bmatrix} I & 0 \\ B & A \end{bmatrix} \Phi(z)\Big|_{z_0^-} = \Xi_{z_0} \Phi(z)\Big|_{z_0^-} \tag{9.12}$$

This expression allows us to propagate the wavefunction from layer to layer. The propagation within any given layer requires solving Eq. (9.5), as described above.

How do we enforce the periodic boundary condition for the superlattice geometry? As a simple illustration, we take a two-layer superlattice with layers 1 and 2 and interfaces 21 and 12. If layer 1 starts at $z = 0$, the wavefunction at $z = 0^+$ is connected with that at $z = d$, where $d$ is the superlattice period

$$\Phi(d^+) = \Xi_{12} T_2 \Xi_{21} T_1 \Phi(0^+) \tag{9.13}$$

As before, we indicate the right side of the interface with $0^+$ and $d^+$.

Since the structure is a superlattice, we also know that the wavefunctions at the same position in neighboring *periods* are connected by the Bloch factor $e^{ik_z d}$:

$$\Phi(d^+) = e^{ik_z d} \Phi(0^+) \tag{9.14}$$

Therefore, the superlattice band structure follows from the eigenvalue problem

$$\Xi_{12} T_2 \Xi_{21} T_1 \Phi(0^+) = e^{ik_z d} \Phi(0^+) \tag{9.15}$$

where the transfer and interface matrices on the left are known. We can always normalize the envelope function after the fact, so it can be unity at the (arbitrary) starting point.

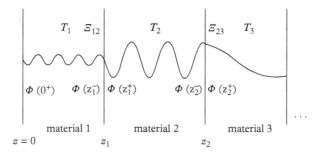

**Figure 9.1** *Schematic of how to form the transfer matrix that solves the Hamiltonian of Eqs. (9.1) and (9.3).*

The eigenvalue problem is more easily solved in terms of $X(z)$ rather than $\Phi(z)$:

$$P^{-1}\Xi_{12}P\exp\left(Q_{2}\right)P^{-1}\Xi_{21}P\exp\left(Q_{1}\right)X\left(0^{+}\right)=e^{ik_{z}z}X\left(0^{+}\right) \qquad (9.16)$$

after which the $X$ eigenfunctions are transformed into $\Phi$ using $\Phi = PX$.

This essentially completes the prescription for the transfer-matrix method applied to quantum structures for any Hamiltonian with second derivatives. If the Hamiltonian contains even higher derivatives, we can triple, quadruple, etc., the number of bands to solve it, but we do not encounter such Hamiltonians within common multiband models. Another extension practiced by many is to search for solutions that decay exponentially in the outer barriers, so as to treat the quantum-well problem more directly. We will see in Section 9.4 that the main disadvantage of solving the problem in real space is that some Hamiltonians can produce physically unreasonable ("spurious") states. Of course, the transfer-matrix method is suitable only for problems with quantum confinement in one dimension only, so it cannot be used for quantum wires and dots.

## 9.2 Finite-Difference Method

The finite-difference method takes a very different route to solving the differential equations in Eq. (8.57) than the transfer-matrix method of Section 9.1. Instead of developing accurate relations for the wavefunction at various points in the structure, the finite-difference method attempts to approximate the solutions at discrete slices within the structure, as shown in Fig. 9.2. The wavefunction at a single mesh point represents its average value over the slice. The derivatives of the wavefunction are approximated by differences between the values in neighboring slices. In principle, as we make the thickness of the slices smaller and smaller, the solution of the finite-difference problem becomes closer and closer to the solution of the differential equations.

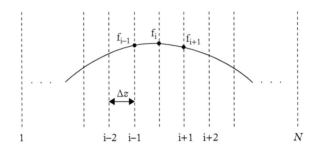

**Figure 9.2** *Approximating a function with finite differences in the discretizing space.*

Since we can imagine making the slices arbitrarily thin, the finite-difference method is in some sense the most intuitive, if not the most efficient, approximation to the continuous problem. The popularity of the finite-difference method has waned somewhat recently, but it is still an instructive technique.

To understand why derivatives can be approximated as finite differences, we start with the Taylor-series expansion at any point:

$$f(z+\Delta z) = f(z) + \Delta z\, f'(z) + \frac{(\Delta z)^2}{2} f''(z) + O\big((\Delta z)^3\big) \qquad (9.17)$$

It seems that to order $O\big((\Delta z)^2\big)$ we can write

$$f'(z) \approx \frac{f(z+\Delta z) - f(z)}{\Delta z} \qquad (9.18)$$

Better yet, we can approximate the first derivative to order $O\big((\Delta z)^3\big)$ if we include both $f(z+\Delta z)$ from Eq. (9.17) and

$$f(z-\Delta z) = f(z) - \Delta z\, f'(z) + \frac{(\Delta z)^2}{2} f''(z) + O\big((\Delta z)^3\big) \qquad (9.19)$$

The second-order approximation is

$$f'(z) \approx \frac{f(z+\Delta z) - f(z-\Delta z)}{2\Delta z} \qquad (9.20)$$

More generally, *central differences* such as in Eq. (9.20) have a higher-order convergence, which usually (but not always!) leads to more accurate solutions for a given mesh pitch $\Delta z$.

From Eqs. (9.17) and (9.19) we see that the second derivative can be approximated to the same order $O\big((\Delta z)^3\big)$ using

$$f''(z) \approx \frac{f(z+\Delta z) - 2f(z) + f(z-\Delta z)}{(\Delta z)^2} \qquad (9.21)$$

To solve Eq. (8.57), we also need a recipe for approximating derivatives with spatially varying coefficients: $[A(z)f'(z)]'$, i.e., $\partial [A(z)\partial f/\partial z]/\partial z$. To this end, we imagine that before we take the second derivative, we have a "diminished" central difference defined on the points halfway between the original mesh points (on which $A(z)$ itself is defined):

$$A(z)\frac{\partial f}{\partial z} \approx A(z)\frac{f(z+\Delta z/2) - f(z - \Delta z/2)}{\Delta z} \tag{9.22}$$

Taking another diminished difference with constant $A$ reproduces Eq. (9.21).

To avoid writing too many symbols, we denote $f_i \equiv f(z), f_{i+1} \equiv f(z+\Delta z), f_{i-1} \equiv f(z-\Delta z), A_i \equiv A(z)$, etc. We would like to take another diminished central difference, but we do not know the values of $A(z)$ at "intermediate" points. These points may fall at interfaces between different materials. It seems that the best we can do is to approximate such values by averaging the function at adjacent true mesh points:

$$\left\{ \frac{\partial}{\partial z}\left[ A(z)\frac{\partial f}{\partial z}\right]\right\}\Bigg|_{z_i} \approx A_{i+\frac{1}{2}}\frac{f_{i+1}-f_i}{(\Delta z)^2} - A_{i-\frac{1}{2}}\frac{f_i - f_{i-1}}{(\Delta z)^2}$$
$$\approx \frac{A_{i+1}+A_i}{2}\frac{f_{i+1}-f_i}{(\Delta z)^2} - \frac{A_i + A_{i-1}}{2}\frac{f_i - f_{i-1}}{(\Delta z)^2} \tag{9.23}$$

To approximate $\partial [B(z)f(z)]/\partial z$, we apply the central difference in Eq. (9.20) to the product $Bf$:

$$\left\{ \frac{\partial}{\partial z}[B(z)f(z)]\right\}\Bigg|_{z_i} \approx \frac{B_{i+1}f_{i+1} - B_{i-1}f_{i-1}}{2\Delta z} \tag{9.24}$$

The same idea applies to $B^{\dagger}(z)\partial f/\partial z$, except that $B^{\dagger}$ is evaluated at the mesh point $i$ where the value of the derivative is approximated:

$$\left[ B^{\dagger}(z)\frac{\partial f(z)}{\partial z}\right]\Bigg|_{z_i} \approx B_i^{\dagger}\frac{f_{i+1}-f_{i-1}}{2\Delta z} \tag{9.25}$$

Obviously, $D(z)f(z) \to D_i f_i$, and we have a full prescription for transforming Eq. (8.57) into a finite-difference equation:

$$\frac{\partial}{\partial z}\left[ A(z)\frac{\partial f}{\partial z}\right] + \frac{\partial}{\partial z}[B(z)f(z)] - B^{\dagger}(z)\frac{\partial f(z)}{\partial z} + D(z)f(z) \approx \sum_i \frac{A_{i+1}+A_i}{2}\frac{f_{i+1}-f_i}{(\Delta z)^2}$$
$$- \frac{A_i + A_{i-1}}{2}\frac{f_i - f_{i-1}}{(\Delta z)^2} + \frac{B_{i+1}f_{i+1} - B_{i-1}f_{i-1}}{2\Delta z} - B_i^{\dagger}\frac{f_{i+1}-f_{i-1}}{2\Delta z} + D_i f_i = E f_i \tag{9.26}$$

Here $A(z), B(z)$, and $D(z)$ are matrices with the same dimensionality as the Hamiltonian and spatially varying elements, and $f(z)$ is a vector, whose components are $F_j(z)$ with different $j$. The finite-difference equation in Eq. (9.26) is just the type of eigenvalue problem we are looking for in this chapter. When we solve it, we obtain not just the energies, but also the eigenvectors $F_j^n(z)$ that are distinct for any given subband.

The approach that leads to Eq. (9.26) incorporates interface boundary conditions, with the exception of the delta-function interface terms. One way to put them back in is to position a mesh point at the interface and insert $\Omega_{ab}/\Delta z$ at that *point only* to represent $\Omega_{ab}\delta (z - z_i)$. Of course, we no longer have a delta function in that case, but a finite approximation to that function extended over one mesh spacing. However, the approximation becomes better for smaller mesh pitches, just as for all the other terms in the Hamiltonian.

How precisely should we form the final Hamiltonian matrix? One way is to group the band components into blocks, with each block corresponding to position $z$. The blocks are then linked by the non-local portions of the finite differences, i.e. the factors in Eq. (9.26) that multiply $f_{i+1}$ and $f_{i-1}$. To wit, the non-local parts are

$$H_{i,i+1} = \frac{A_{i+1} + A_i}{2(\Delta z)^2} + \frac{B_{i+1} - B_i^\dagger}{2\Delta z} \tag{9.27a}$$

$$H_{i,i-1} = \frac{A_i + A_{i-1}}{2(\Delta z)^2} - \frac{B_{i-1} - B_i^\dagger}{2\Delta z} \tag{9.27b}$$

and the "local" part is

$$H_{i,i} = -\frac{A_{i+1} + 2A_i + A_{i-1}}{2(\Delta z)^2} + D_i \tag{9.28}$$

So far we have worked under the assumption of a constant mesh spacing $\Delta z$. But what if we choose a mesh with a spatially varying pitch? Certainly, the wavefunction varies faster in some regions than in others, and it may be more efficient to sample such regions more densely. To this end, we can generalize the previous formulas as follows: (1) $2\Delta z$ in the denominators of Eqs. (9.24) and (9.25) becomes $\Delta z_{i-1/2} + \Delta z_{i+1/2}$ to represent the first central difference, and (2) $(\Delta z)^2$ is replaced by $\Delta z_{i+1/2}(\Delta z_{i-1/2} + \Delta z_{i+1/2})/2$ in the first term and by $\Delta z_{i-1/2}(\Delta z_{i-1/2} + \Delta z_{i+1/2})/2$ in the second term of Eq. (9.23).

As a further illustration, we write schematically the complete finite-difference Hamiltonian for the simple case of one conduction and one valence band:

$$\boldsymbol{H}_{FDM} = \begin{bmatrix} \cdots & \cdots & \cdots & \cdots & \cdots & \cdots & \cdots & \cdots \\ \cdots & H_{i-1,i-1}^{cc} & H_{i-1,i-1}^{cv} & H_{i-1,i}^{cc} & H_{i-1,i}^{cv} & 0 & 0 & \cdots \\ \cdots & H_{i-1,i-1}^{vc} & H_{i-1,i-1}^{vv} & H_{i-1,i}^{vc} & H_{i-1,i}^{vv} & 0 & 0 & \cdots \\ \cdots & H_{i,i-1}^{cc} & H_{i,i-1}^{cv} & H_{i,i}^{cc} & H_{i,i}^{cv} & H_{i,i+1}^{cc} & H_{i,i+1}^{cv} & \cdots \\ \cdots & H_{i,i-1}^{vc} & H_{i,i-1}^{vv} & H_{i,i}^{vc} & H_{i,i}^{vv} & H_{i,i+1}^{vc} & H_{i,i+1}^{vv} & \cdots \\ \cdots & 0 & 0 & H_{i+1,i}^{cc} & H_{i+1,i}^{cv} & H_{i+1,i+1}^{cc} & H_{i+1,i+1}^{cv} & \cdots \\ \cdots & 0 & 0 & H_{i+1,i}^{vc} & H_{i+1,i}^{vv} & H_{i+1,i+1}^{vc} & H_{i+1,i+1}^{vv} & \cdots \\ \cdots & \cdots & \cdots & \cdots & \cdots & \cdots & \cdots & \cdots \end{bmatrix} \tag{9.29}$$

Does the form of Eq. (9.29) guarantee Hermiticity? We can confirm this explicitly by taking, e.g., $H_{i,i-1}^{cv}$ from Eq. (9.27b) and $H_{i-1,i}^{cv}$ from a renumbering of Eq. (9.27a):

$$H_{i-1,i}^{cv} = \frac{A_i^{cv} + A_{i-1}^{cv}}{2(\Delta z)^2} + \frac{B_i^{cv} - \left(B_{i-1}^{vc}\right)^*}{2\Delta z} = \frac{B_i^{cv} - B_{i-1}^{cv}}{2\Delta z} \tag{9.30a}$$

$$H_{i,i-1}^{vc} = \frac{A_i^{vc} + A_{i-1}^{vc}}{2(\Delta z)^2} - \frac{B_{i-1}^{vc} - \left(B_i^{cv}\right)^*}{2\Delta z} = \frac{B_i^{vc} - B_{i-1}^{vc}}{2\Delta z} \tag{9.30b}$$

The first term in Eq. (9.30) vanishes because the **A** matrix is diagonal. Therefore, if the original Hamiltonian is Hermitian, $B^{cv} = (B^{vc})^*$, so is its finite-difference version. In fact, Hermiticity is guaranteed by

$$\mathbf{C} = -\mathbf{B}^\dagger \tag{9.31}$$

in agreement with the derivation in Chapter 8.

We define the (uniform) mesh by partitioning the simulation region into $N$ slices with thicknesses $\Delta z = d/N$ and indices $i = 1, 2, \cdots, N$. It remains to specify the boundary conditions at the edges of the simulation region, $z = 0$ and $z = d$. For a quantum well, we simply set the wavefunction to zero for $z < 0$ and $z > d$ to model negligible wavefunction penetration far from the well. For the Hamiltonian matrix in Eq. (9.29), this amounts to neglecting the non-local terms $H_{0,1}$ and $H_{N,N+1}$ that fall outside of the square matrix.

The boundary conditions for a superlattice can be enforced by setting the wavefunction at $z = d$ to $f(d) = \exp(ik_z d) f(0)$. For the 2-band Hamiltonian matrix similar to Eq. (9.29), we have

$$\mathbf{H}_{FDM} = \begin{bmatrix}
H_{1,1}^{cc} & H_{1,1}^{cv} & H_{1,2}^{cc} & H_{1,2}^{cv} & \cdots & 0 & 0 & H_{1,N}^{cc} & H_{1,N}^{cv} \\
H_{1,1}^{vc} & H_{1,1}^{vv} & H_{1,2}^{vc} & H_{1,2}^{vv} & \cdots & 0 & 0 & H_{1,N}^{vc} & H_{1,N}^{vv} \\
H_{2,1}^{cc} & H_{2,1}^{cv} & H_{2,2}^{cc} & H_{2,2}^{cv} & \cdots & 0 & 0 & 0 & 0 \\
H_{2,1}^{vc} & H_{2,1}^{vv} & H_{2,2}^{vc} & H_{2,2}^{vv} & \cdots & 0 & 0 & 0 & 0 \\
\cdots & \cdots & \cdots & \cdots & \cdots & \cdots & \cdots & \cdots & \cdots \\
0 & 0 & 0 & 0 & \cdots & H_{N-1,N-1}^{cc} & H_{N-1,N-1}^{cv} & H_{N-1,N}^{cc} & H_{N-1,N}^{cv} \\
0 & 0 & 0 & 0 & \cdots & H_{N-1,N-1}^{vc} & H_{N-1,N-1}^{vv} & H_{N-1,N}^{vc} & H_{N-1,N}^{vv} \\
H_{N,1}^{cc} & H_{N,1}^{cv} & 0 & 0 & \cdots & H_{N,N-1}^{cc} & H_{N,N-1}^{cv} & H_{N,N}^{cc} & H_{N,N}^{cv} \\
H_{N,1}^{vc} & H_{N,1}^{vv} & 0 & 0 & \cdots & H_{N,N-1}^{vc} & H_{N,N-1}^{vv} & H_{N,N}^{vc} & H_{N,N}^{vv}
\end{bmatrix} \tag{9.32}$$

where $H_{1,N}^{ab} = \exp(-ik_z d) H_{1,0}^{ab}$ and $H_{N,1}^{ab} = \exp(ik_z d) H_{N,N+1}^{ab}$. This is because the coordinate $z = Z$, which corresponds to "mesh point 1," is exactly one period away from "mesh point $N+1$" with $z = Z + d$. The 8-band Hamiltonian has

8×8 blocks for each mesh point rather than 2×2, but the idea remains the same. The overall dimension of the matrix is $N \times M$, where $M$ is the total number of bands included in the bulk Hamiltonian. For example, if we choose to divide the superlattice into $N = 20$ slices, we must solve for the eigenvalues of a 160×160 matrix.

The maximum spatial frequency allowed by the discretization of space along the quantum direction is the Nyquist frequency $f_N = \pi/\Delta z$. For a superlattice with a uniform mesh and $N$ mesh points so that $\Delta z = d/N$, $f_N = N\pi/d$.

Equation (9.32) shows that most of the matrix elements in the finite-difference solution are equal to zero. This fact allows us to take advantage of sparse-matrix techniques to determine its eigenvalues and eigenvectors, an important advantage of this approach. On the downside, it is not always easy to see how to generate the optimum mesh. Of course, to avoid sampling too much of $k$ space, the number of mesh points should not exceed $N \sim 2f_B d/a$, where $f_B \sim 0.2 - 0.3$, as defined in Section 8.5. Also, the number of mesh points should be much larger than the highest subband number of interest so that the spatial oscillations characteristic of those states are rendered faithfully. For example, with 20 mesh points, we may hope to produce reasonably good approximations for the energies and envelope functions of the first 4 or 5 subbands.

Like the transfer-matrix method, the finite-difference approximation can generate spurious states that will be discussed further in Section 9.4. Because of this defect, we *do not* recommend using the finite-difference method to solve the 8-band Hamiltonian. However, we *do* find it useful for the reduced Hamiltonians in the valence band that do not feature the conduction–valence coupling (see Section 9.4 for details). To give but one example, consider the valence-band structure of the $In_{0.53}Ga_{0.47}As/In_{0.52}Al_{0.48}As$ quantum well that follows from the 4-band Hamiltonian solved with different pitches. Figure 9.3a shows the results that indicate convergence for $\Delta z$ on the order of the lattice constant. The dependence of the valence subband energies at $k_{\parallel} = 0$ on the number of points in the well (inversely proportional to mesh pitch) is displayed in Fig. 9.3b.

Unlike the transfer-matrix method, the finite-difference approach can be applied to multidimensional confinement, at the expense of increasing the final size of the matrix. For example, if we choose $N$ mesh points to model a quantum wire, we must find the eigenvalues and eigenvectors of an $N \times N \times M$ square matrix. Such problems can easily get out of hand (particularly, for 3D confinement) if we do not pick a judicious distribution of mesh points. Mixed partial derivatives are approximated by following the line of reasoning that leads to Eq. (9.23):

$$\left\{ \frac{\partial}{\partial y} \left[ A(y,z) \frac{\partial f}{\partial z} \right] \right\} \Bigg|_{y_i, z_i} \approx \frac{1}{4\Delta y \, \Delta z} \Big[ \left( A_{i+1,j+1} + A_{i,j+1} \right) \left( f_{i+1,j+1} - f_{i,j+1} \right)$$
$$- \left( A_{i,j+1} + A_{i-1,j+1} \right) \left( f_{i,j+1} - f_{i-1,j+1} \right) - \left( A_{i+1,j-1} + A_{i,j-1} \right) \left( f_{i+1,j-1} - f_{i,j-1} \right)$$
$$+ \left( A_{i,j-1} + A_{i-1,j-1} \right) \left( f_{i,j-1} - f_{i-1,j-1} \right) \Big] \tag{9.33}$$

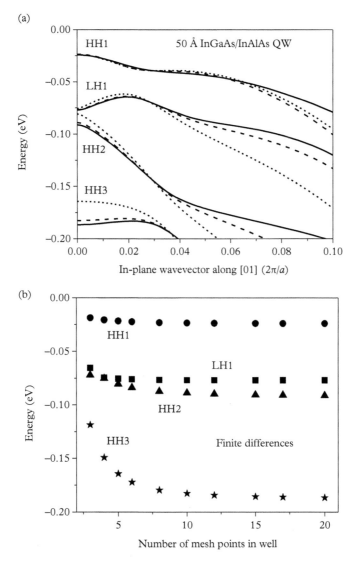

**Figure 9.3** *(a) Valence-band dispersions for the 50-Å $In_{0.53}Ga_{0.47}As/In_{0.52}Al_{0.48}As$ quantum well obtained using the 4-band Hamiltonian solved by the finite-difference method with 5 (dotted), 10 (dashed), and 20 (solid) points in the InGaAs well that correspond to $\Delta z = 10, 5,$ and 2.5 Å, respectively. (b) Valence subband energies at $k_\parallel = 0$ obtained from the 4-band Hamiltonian by the finite-difference method for the same well as a function of the number of points in the InGaAs well.*

In all other respects, the finite-difference method works for any number of dimensions. In 2D apart from matrix elements like $H_{1,2}^{cv}$ in Eq. (9.32) connecting neighboring mesh points, we also have $H_{1,N+1}^{cv}$, and instead of $H_{1,N}^{cv}$, we have $H_{1,N^2}^{cv}$, etc. Generally, the matrix elements belonging to the same mesh point are adjacent, while the mesh points corresponding to the same position along one of the confinement axes form the next grouping level.

Even though we limit ourselves to confinement along Cartesian coordinates in this book, we can extend these calculations to polar or spherical coordinates without much trouble. Unfortunately, such geometries are difficult to realize using epitaxial growth of III–V materials, but they can arise as a result of self-assembly in other material systems.

For quantum wires and dots that include strained materials, we must also extend the approach described in Section 3.4 to multiple confinement dimensions. Since the deformation potential theory is still applicable, our main task is to calculate the strain tensor for a general confined geometry. This is not an easy task, but we can always find the solution by minimizing the elastic energy, whether by including the atomic bonds explicitly or via approximate models validated by microscopic physics.

## 9.3  Reciprocal-Space Method

In Eq. (8.2), we expanded the wavefunction in a Fourier series multiplied by a cell-periodic component. By analogy with the pseudopotential approach to the bulk band structure described in Section 5.4, could we solve for the states of a superlattice by directly applying this expansion? In fact, the transition to reciprocal space proves convenient for several reasons. First, the **k·p** theory is fundamentally a low-spatial-frequency approximation to the true band structure. If we eliminate high frequencies from the start, they would not be able to contaminate our numerical solutions. In one fell swoop, we would also solve the issue of optimal mesh spacing raised in Section 9.2. Also not to be undervalued is that the reciprocal-space method can be concisely stated and straightforwardly implemented.

For a superlattice with periodic boundary conditions, we can always expand the envelope function in multiples of $2\pi/d$ so that the component $j$ associated with subband $n$ becomes

$$F_j^n(z) = \sum_m \phi_{m,j}^n e^{i(\mathbf{k}+\mathbf{G}_m)\cdot\mathbf{r}} \tag{9.34}$$

where $\mathbf{k} = k_x\hat{x} + k_y\hat{y} + k_z\hat{z}$ is a 3D vector and $\mathbf{G}_m = \hat{z}\, 2\pi m/d$, $m = 0, \pm1, \pm2, \ldots$

We can insert the envelope function given by Eq. (9.34) into Eq. (8.57), which we rewrite here with state $n$ explicitly labeled:

$$\sum_b \left[ \frac{\partial}{\partial z} A_{ab}(z) \frac{\partial}{\partial z} + \frac{\partial}{\partial z} B_{ab}(z) - B_{ba}^*(z) \frac{\partial}{\partial z} + D_{ab}(z) \right] F_b^n(z) = E_n F_a^n(z) \quad (9.35)$$

We can also substitute the Fourier-series expansions for the spatially varying coefficients:

$$c_{ab}(z) = \sum_m c_m^{ab} e^{i(k_z + G_m)z} \quad (9.36)$$

where $c_{ab}$ stand for $A_{ab}$, $B_{ab}$, or $D_{ab}$, and the Fourier coefficients $c_m^{ab}$ are averaged over the superlattice period

$$c_m^{ab} = \frac{1}{d} \int_0^d dz\, c_{ab}(z)\, e^{-i(k_z + G_m)z} \quad (9.37)$$

Finally, we multiply Eq. (9.35) by $e^{-i(k_z + G_l)z}$, with the coefficients expanded using Eq. (9.36), and integrate over the superlattice period to eliminate non-commensurate phase terms. This integration step allows us to express all the spatially varying band parameters in terms of the Fourier coefficients from Eq. (9.37):

$$\sum_{m,b} \left[ -(k_z + G_l) A_{l-m}^{ab} (k_z + G_m) + i(k_z + G_l) B_{l-m}^{ab} - i B_{l-m}^{*ba} (k_z + G_m) + D_{l-m}^{ab} \right] \phi_{m,b}^n$$

$$= E_n \phi_{l,a}^n \quad (9.38)$$

If we start from a model with $M$ bands and limit $m \le N$ (so that the maximum spatial frequency is $G_N = \pm 2\pi N/d$, or $2f_N$ in the notation of Section 9.2), the matrix dimension in Eq. (9.38) is $(2N+1) \times M$. For superlattices and quantum wells grown along [001], we set $N = f_B d/a$, with $f_B = 0.2 - 0.3$, as in Section 9.2.

For a superlattice or quantum well with abrupt interfaces, the Fourier coefficients in Eq. (9.37) are straightforward to calculate

$$c_{l-m}^{ab} = \sum_j e^{-i(G_l - G_m)s_j} \frac{2 \sin\left( \frac{(G_l - G_m)t_j}{2} \right)}{(G_l - G_m) d} c_{ab,j} = \sum_j e^{-i(G_l - G_m)s_j} \frac{\sin\left( \frac{(G_l - G_m)t_j}{2} \right)}{\pi (l - m)} c_{ab,j}$$

$$(9.39)$$

where the sum is over all the layers in the superlattice, $c_{ab,j}$ is the Hamiltonian matrix element in layer $j$, and $s_j$ and $t_j$ are the center coordinate and thickness, respectively, of layer $j$. Therefore, for abrupt interfaces the coefficients decay as the sinc function. If interface broadening is present, e.g., due to interdiffusion, the actual interface profile enters Eq. (9.37). For example, the convolution with Gaussian interface broadening multiplies Eq. (9.39) by a Gaussian, and the

Fourier coefficients drop off faster with $l - m$. The $l = m$ "dc" coefficient in Eq. (9.39) is simply the parameter's average over all the layers:

$$c_0^{ab} = \sum_j \frac{t_j}{d} c_{ab,j} \qquad (9.40)$$

The delta-function interface term has an even simpler representation in reciprocal space with a constant magnitude for all $G_m$. The sum of the interface terms over multiple interfaces at points $z_j$, $\sum_j \Omega_{ab}\delta(z - z_j)$, yields the Fourier coefficients

$$c_{l-m}^{ab}(interface) = \sum_j e^{-i(G_l - G_m)z_j} \frac{\Omega_{ab}}{d} \qquad (9.41)$$

For a quantum well with an external field modeled as a superlattice or, even more pertinently, for a polar nitride superlattice with internal fields discussed more fully in Chapters 10 and 12, the potential energy contains a term that varies linearly with $z$: $V_e(z) = E_{zj}(z - s_j)$. We can find the corresponding Fourier coefficients using integration by parts:

$$V_{l-m}^e = \sum_j i E_{zj} t_j\, e^{-i(G_l - G_m)s_j} \left[ \frac{\cos\left(\frac{(G_l - G_m)t_j}{2}\right)}{(G_l - G_m)\,d} - \frac{2\sin\left(\frac{(G_l - G_m)t_j}{2}\right)}{(G_l - G_m)^2\, dt_j} \right] \qquad (9.42)$$

Another case with internal fields in a quantum structure occurs when carriers and ions are spatially separated. In that case, the fields are no longer constant in each layer, but we can still obtain the solutions by adding an electrostatic energy term $eV(z)$ to the diagonal elements of the Hamiltonian. The corresponding Fourier coefficients may need to be calculated numerically in general. To compute $V(z)$, we solve the 1D Poisson's equation with the spatial distributions of the charge densities:

$$\frac{\partial^2 V}{\partial z^2} = -e\frac{n(z) - p(z) - N_D(z) + N_A(z)}{\varepsilon_s(z)} \qquad (9.43)$$

Here $\varepsilon_s(z)$ is the position-dependent static permittivity, and $N_D(z)$ and $N_A(z)$ are the position-dependent charge densities due to ionized donors and acceptors. To see how the carrier densities vary in space, we average the square of the wavefunction over all occupied states:

$$n(z) = \sum_{kn,j} f_{kn} \left| F_j^{kn}(z) \right|^2 \qquad (9.44)$$

where $f_{kn} = \{1 + \exp[(E_{kn} - E_F)/k_B T]\}^{-1}$ is the Fermi–Dirac factor for subband $n$ with wavevector $k$ from Eq. (1.41). We can find the envelope function in Eq. (9.44) using the coefficients $\phi^n_{m,j}$ from the solution of Eq. (9.38) and the Fourier-series representation of Eq. (9.34). If holes are also present, they are included with the opposite sign as the $p(z)$ term. Since the wavefunction depends on $V(z)$ from Eq. (9.43), the problem must be solved iteratively and "self-consistently", i.e., using successive solutions of the Hamiltonian and Eq. (9.43) until the changes in the potential become negligible. If this does not happen after many iterations, we can try adding only a fraction of $V(z)$ computed from Eq. (9.43) to the diagonal terms in each iteration, a procedure known as *under-relaxation*. There are also other techniques to achieve and speed up the convergence.

We can arrange the Hamiltonian in the reciprocal-space method by analogy with Eq. (9.33), but the matrix is usually *not* sparse because the Fourier coefficients decrease gradually or not at all with spatial frequency $G_l - G_m = (l - m) 2\pi/d$, as shown by Eqs. (9.39), (9.41), and (9.42). Because we cannot take advantage of sparse-matrix methods, the run time using ordinary computing resources can become uncomfortably long even if all we care about are a few eigenstates near the energy gap.

Does the reciprocal-space method respect the boundary conditions in Section 8.5? The wavefunctions are certainly continuous, but discontinuities in the first derivative do not allow us to place a hard limit on the number of spatial frequencies. Then it becomes a matter of whether the truncated basis is "good enough" by comparing the results to experimental data, other solution methods, and atomistic approaches such as the pseudopotential method.

As with the finite-difference method, we can extend the reciprocal-space approach to quantum confinement in multiple dimensions, i.e., to quantum wires and dots. To this end, two sets of reciprocal vectors $G$ are used, and the size of the matrix increases from $(2N + 1) \times M$ to $(2N + 1) \times (2N + 1) \times M$, assuming the same period along the two confinement axes (of course, this constraint can be relaxed in most practical cases). To outline the extension to a 2D superlattice, we rewrite Eq. (9.34) as

$$F^n_j(z) = \sum_m \phi^n_{k,m,j} e^{i(k + G_m + G_k)\cdot r} \tag{9.45}$$

and the envelope-function equations as

$$\sum_b \left[ \frac{\partial}{\partial z} A_{ab}(y,z) \frac{\partial}{\partial z} + \frac{\partial}{\partial z} B'_{ab}(y,z) - B^{*\prime}_{ba}(y,z) \frac{\partial}{\partial z} + \frac{\partial}{\partial y} Z_{ab}(y,z) \frac{\partial}{\partial y} + \frac{\partial}{\partial y} Y_{ab}(y,z) \right.$$

$$\left. - Y^*_{ba}(y,z) \frac{\partial}{\partial y} + \frac{\partial}{\partial y} X_{ab}(y,z) \frac{\partial}{\partial z} + \frac{\partial}{\partial z} X^*_{ba}(y,z) \frac{\partial}{\partial y} + D'_{ab}(y,z) \right] F^n_b(y,z) = E_n F^n_a(y,z) \tag{9.46}$$

where the matrix elements that contain $k_z^2$ ($k_y^2$) in the bulk Hamiltonian are denoted $A_{ab}$ ($Z_{ab}$), and the elements that contain $k_z$ ($k_y$), but not $k_y$ ($k_z$), are represented as $B'_{ab}$ ($Y_{ab}$), the elements that contain both $k_y$ and $k_z$ are denoted $X_{ab}$, and the ordering in the bulk Hamiltonian is maintained. Note that $B'_{ab}$ and $D'_{ab}$ are not necessarily the same as $B_{ab}$ and $D_{ab}$ because the terms that contain both $k_y$ and $k_z$ are excluded from the primed elements. Of course, we could have chosen the axes of quantum confinement differently (e.g., either as $x$ and $y$ or as $x$ and $z$). As long as the wavefunctions and optical matrix elements are calculated using consistent definitions of the axes, any of these choices leads to the same physically observable results, e.g., for the eigenenergies, the absorption of polarized light, etc.

The Fourier-series expansion for a 2D superlattice becomes

$$c_{ab}(z) = \sum_{m,k} c_{k,m}^{ab} e^{i(k_y + G_k)y + i(k_z + G_m)z} \tag{9.47}$$

with the Fourier coefficients $c_{k,m}^{ab}$ averaged over the superlattice unit cell (assumed to be of square or rectangular shape):

$$c_{k,m}^{ab} = \frac{1}{d_y d_z} \int_0^{d_z} dz \int_0^{d_y} dy \, c_{ab}(y,z) \, e^{-i(k_y + G_k)y - i(k_z + G_m)z} \tag{9.48}$$

In place of Eq. (9.38), we have

$$\sum_{k,m,b} \Big[ -(k_z + G_l) A_{p-k,l-m}^{ab} (k_z + G_m) + i(k_z + G_l) B'^{ab}_{p-k,l-m} - iB'^{ab*}_{p-k,l-m} (k_z + G_m)$$
$$+ D_{p-k,l-m}^{ab} - (k_y + G_p) Z_{p-k,l-m}^{ab} (k_y + G_k) + i(k_y + G_p) Y_{p-k,l-m}^{ab}$$
$$- iY_{p-k,l-m}^{ab*} (k_y + G_k) - (k_y + G_p) X_{p-k,l-m}^{ab} (k_z + G_k)$$
$$- (k_z + G_p) X_{p-k,l-m}^{ab*} (k_y + G_k) \Big] \phi_{k,m,b}^n = E_n \phi_{p,l,a}^n \tag{9.49}$$

Let us calculate the Fourier coefficients for a superlattice of quantum wires (easily converted to the case of an isolated wire by making the barriers thick) with a rectangular cross section of widths $t_y$ and $t_z$ centered at the origin, as illustrated in Fig. 9.4. The Fourier coefficients for $p \neq k$, $l \neq m$ are

$$c_{p-k,l-m}^{ab} = \frac{2\sin\left(\frac{(G_p - G_k)t_y}{2}\right)}{(G_p - G_k) d_y} \frac{2\sin\left(\frac{(G_l - G_m)t_z}{2}\right)}{(G_l - G_m) d_z} (c_{ab,w} - c_{ab,b}) \tag{9.50}$$

while the dc coefficients along at least one axis are

$$c_{0,0}^{ab} = c_{ab,b} + \frac{t_y t_z}{d_y d_z} (c_{ab,w} - c_{ab,b}) \tag{9.51}$$

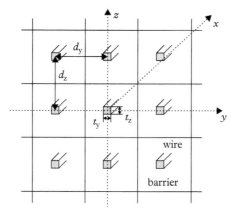

**Figure 9.4** *Illustration of a square or rectangular array of quantum wires with a square or rectangular cross section. In this case, only one well and one barrier material are present.*

and

$$c_{p-k,0}^{ab} = c_{ab,b} + \frac{t_z}{d_z} \frac{2\sin\left(\frac{(G_p-G_k)t_y}{2}\right)}{(G_p - G_k)\, d_y} \left(c_{ab,w} - c_{ab,b}\right) \tag{9.52}$$

$$c_{0,l-m}^{ab} = c_{ab,b} + \frac{t_y}{d_y} \frac{2\sin\left(\frac{(G_l-G_m)t_z}{2}\right)}{(G_l - G_m)\, d_z} \left(c_{ab,w} - c_{ab,b}\right) \tag{9.53}$$

We can easily extend this approach to more than two materials in a supercell. The various matrix elements in the Hamiltonian are grouped as discussed in Section 9.2.

Figure 9.5a shows the 8-band reciprocal-space results for the InGaAs wells that we solve as a superlattice with a 100-Å InAlAs barrier as a function of the number of reciprocal-lattice vectors. We see reasonably good convergence (to < 1 meV) for $N \geq 5$. As with finite differences, the energies converge faster for states close to the valence-band maximum. The results for two different well widths in Fig. 9.5b illustrate that the convergence of $E_g$ is mostly driven by CB1.

In the rest of the simulations in this book, we will use the reciprocal-space method unless otherwise indicated. Special difficulties arise for narrow-gap super-lattices that will be discussed in detail in Chapter 10.

We will not attempt to cover all the techniques developed to compute the band structures of quantum wells and superlattices. For structures with 2D and 3D quantum confinement, i.e., wires and dots, the finite-element method can work well. In this approach, the envelope functions are expanded in polynomials in each section of the structure. The unknown coefficients are then determined by minimizing the action integral derived from the Hamiltonian. Because of the

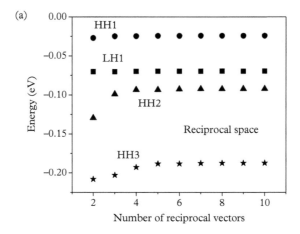

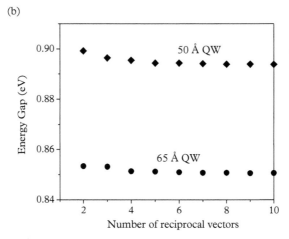

**Figure 9.5** *(a) Valence subband energies at $k_\parallel = 0$ obtained from the 8-band **k·p** Hamiltonian by the reciprocal-space method for the 50 Å $In_{0.53}Ga_{0.47}As/In_{0.52}Al_{0.48}As$ well as a function of the number of reciprocal-lattice vectors. (b) Energy gap as a function of the number of reciprocal-lattice vectors for the $In_{0.53}Ga_{0.47}As/In_{0.52}Al_{0.48}As$ quantum well with widths of 50 and 65 Å, obtained using the 8-band reciprocal-space method.*

additional freedom in the choice of finite elements, this can be more efficient than the finite-difference and reciprocal-space methods for complicated structures in multiple dimensions (see "Suggested Reading" for details).

## 9.4   Problem of Spurious States

We have mentioned that some methods for solving the Hamiltonians of quantum structures can generate unphysical (or spurious) states. These states can appear through either the choice of the discretization scheme or the particular form of the bulk Hamiltonian. To see how they arise, we take a simple two-band Hamiltonian with parabolic diagonal and linear off-diagonal terms:

$$H_{2b} = \begin{bmatrix} E_g - A_c k^2 & iPk \\ -iPk & -A_v k^2 \end{bmatrix} \tag{9.54}$$

The effective mass near the bottom of the conduction band is given by

$$m_e = \frac{\hbar^2}{2} \left( \frac{P^2}{E_g} - A_c \right)^{-1} \tag{9.55}$$

Generally, $A_c \geq 0$ (e.g., when $\Delta = 0$ in Eq. (3.2), $A_c = -2F - 1$ with $F$ tabulated in Chapter 6) and $A_v \geq 0$ because remote interactions tend to flatten the conduction band and to steepen the valence bands. The Hamiltonian of (Eq. (9.54) is not a bad model for the semiconductor band structure when the spin–orbit coupling is weak.

The eigenenergies $E$ of this Hamiltonian satisfy the condition

$$c_4 k^4 + c_2 k^2 + c_0 = 0 \tag{9.56}$$

where

$$c_4 = A_c A_v \tag{9.57a}$$

$$c_2 = E A_c + A_v (E - E_g) - P^2 \tag{9.57b}$$

$$c_0 = E (E - E_g) \tag{9.57c}$$

Reversing the usual procedure, we can solve Eq. (9.56) for the wavevector $k$ at any energy $E$:

$$k^2 = \frac{-c_2 \pm \sqrt{c_2^2 - 4 c_4 c_0}}{2 c_4} \tag{9.58}$$

Not surprisingly, the energy is the same for the states at $k$ and $-k$.

What are the solutions at the bottom of the conduction band ($E = E_g$, i.e., $c_0 = 0$)? With the positive sign in Eq. (9.58), we obtain $k = 0$; i.e., as expected, the bottom of the conduction band is at the zone center. But what if we choose the negative sign? That leads to a very different result:

$$k^2 = \frac{P^2/E_g - A_c}{A_c A_v/E_g} = \frac{\hbar^2}{2 m_e} \frac{E_g}{A_c A_v} \tag{9.59}$$

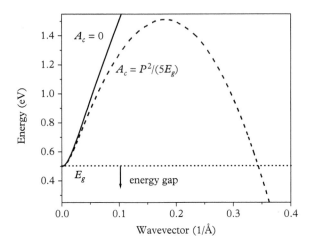

**Figure 9.6** *Conduction-band dispersions obtained from the Hamiltonian of Eq. (9.54), for $E_g = 0.5\ eV, m_e = 0.025m$: $E_P = 25\ eV, A_c = A_v = \frac{P^2}{5E_g}$ (solid), and for $E_P = 20\ eV, A_c = 0$, $A_v = \frac{P^2}{5E_g}$ (dashed).*

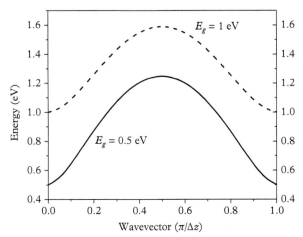

**Figure 9.7** *Dispersion of the conduction-band states in Eq. (9.63) for $E_g = 0.5\ eV$ (solid) and $E_g = 1.0\ eV$ (dashed), assuming $A_v = \frac{P^2}{5E_g}$ and $E_P = 25\ eV$.*

Unexpectedly, the conduction-band dispersion dips back down at higher $k$, as illustrated in Fig. 9.6. For many parameter values, Eq. (9.59) even produces states within the gap ($0 < E < E_g$), which is unphysical and reminiscent of the behavior of the bands for bulk InAs in the 8-band $\mathbf{k} \cdot \mathbf{p}$ model (see Fig. 5.7).

This example illustrates what we already know: the $\mathbf{k} \cdot \mathbf{p}$ method becomes questionable at large wavevectors. Unfortunately, for quantum structures with abrupt interfaces, some computational methods fold the spurious branch in Fig. 9.6 *back* into the zone-center solution near the conduction- or valence-band edges. This folding is due to spatial discretization and can contaminate the true states of the quantum structure. It seems important to find a way to eliminate the dip or, at the very least, to move it as far away in $k$ as possible.

Within the two-band model of Eq. (9.54) we could eliminate spurious states by setting either $A_c$ or $A_v$ to zero, and then adjusting $P$(or $E_P$) to fit the experimental conduction- and valence-band effective masses. In doing so, we would choose to neglect the effect of remote bands on that particular dispersion branch and may underestimate the interband matrix element, as discussed in Chapter 6. Fortunately, we can simply scale the final optical transition rates to correct the error. The correction does not necessarily guarantee high accuracy because we saw in Chapter 6 that there are still considerable uncertainties in the interband matrix elements.

However, even setting $A_c = 0$ may not be enough to see us through safely in some computational methods. To see why, we make that assignment in Eq. (9.54):

$$\boldsymbol{H}_{2b} = \begin{bmatrix} E_g & iPk \\ -iPk & -A_v k^2 \end{bmatrix}$$

(9.60)

Using the prescription of Section 9.2 for the finite-difference method, we substitute derivatives and then approximate them with central finite differences. The off-diagonal terms become

$$iPkF \rightarrow P\frac{\partial F}{\partial z} \rightarrow P\frac{F_{j+1} - F_{j-1}}{2\Delta z}$$

(9.61)

The envelope function for a conduction-band state $E_c(k)$ is characterized by the dependence $\exp(ikz) = \exp(ik\Delta zj)$, for which the central difference becomes

$$P\frac{F_{j+1} - F_{j-1}}{2\Delta z} \rightarrow P\exp(ik\Delta zj)\frac{\exp(ik\Delta z) - \exp(-ik\Delta z)}{2\Delta z} \rightarrow iP\frac{\sin(k\Delta z)}{\Delta z}F_j$$

(9.62)

Substituting the same phase dependence into the diagonal term $-A_v k^2$, the finite-difference Hamiltonian is

$$\boldsymbol{H}_{2b,FDM} = \begin{bmatrix} E_g & iP\frac{\sin(k\Delta z)}{\Delta z} \\ -iP\frac{\sin(k\Delta z)}{\Delta z} & -2A_v\frac{1-\cos(k\Delta z)}{(\Delta z)^2} \end{bmatrix}$$

(9.63a)

Figure 9.7 plots the conduction-band dispersion that solves the Hamiltonian in Eq. (9.63) for representative values of $E_g$, $P$, and $A_v$. As $k$ approaches $\pi/\Delta z$, the wavefunction flips its phase from one data point to the next (i.e., oscillates with period $2\Delta z$), and the energy is barely higher than at the zone center $k = 0$ if $A_v$ is small. The "false valley" at $k = \pi/\Delta z$ is problematic because it can lead to spurious band-edge states in quantum structures that have oscillating phase, even if no states are generated within the gap.

An interesting approach to fixing this problem is to use first-order rather than second-order differences in the diagonal elements of Eq. (9.60):

$$iPkF \rightarrow P\frac{\partial F}{\partial z} \rightarrow P\frac{F_{j+1} - F_j}{\Delta z} \tag{9.63b}$$

$$-iPkF \rightarrow -P\frac{\partial F}{\partial z} \rightarrow -P\frac{F_j - F_{j-1}}{\Delta z} \tag{9.63c}$$

This assignment preserves Hermiticity because for $z = j\Delta z$, $H^{cv}_{j,j+1} = \frac{P}{\Delta z}$, while for $z = (j+1)\Delta z$, $H^{vc}_{j+1,j} = \frac{P}{\Delta z} = H^{cv}_{j,j+1}$. Even though the forward and backward differences are a lower-order approximation, we may not need to change the mesh much to reach the same accuracy as for the central differences.

If we use Eqs. (9.63b) and (9.63c) in Eq. (9.63a), we obtain

$$H_{2b,FDM} = \begin{bmatrix} E_g & P\frac{e^{ik\Delta z}-1}{\Delta z} \\ P\frac{e^{-ik\Delta z}-1}{\Delta z} & -2A_v\frac{1-\cos(k\Delta z)}{(\Delta z)^2} \end{bmatrix}$$

$$= \begin{bmatrix} E_g & Pe^{\frac{ik\Delta z}{2}}\frac{\sin\left(\frac{k\Delta z}{2}\right)}{\Delta z} \\ -Pe^{-\frac{ik\Delta z}{2}}\frac{\sin\left(\frac{k\Delta z}{2}\right)}{\Delta z} & -2A_v\frac{1-\cos(k\Delta z)}{(\Delta z)^2} \end{bmatrix}$$

$$= \begin{bmatrix} E_g & Pe^{ik\Delta z/2}\frac{\sqrt{1-\cos(k\Delta z)}}{\sqrt{2}\Delta z} \\ -Pe^{-ik\Delta z/2}\frac{\sqrt{1-\cos(k\Delta z)}}{\sqrt{2}\Delta z} & -2A_v\frac{1-\cos(k\Delta z)}{(\Delta z)^2} \end{bmatrix} \tag{9.63d}$$

We notice from Eq. (9.63d) that the conduction-band dispersion does not head down as quickly.

The full 8-band Hamiltonian is prone to the exact same problem, and we can address it in the same manner: by adjusting the remote-band parameters to minimize the effect of spurious states and substituting forward and backward differences for central differences. The choice of the mesh pitch looms large here: if the mesh is too crude, the rounding error can be large, but too fine a mesh can push spurious states closer to the band edges. The flaw is rarely fatal because any spurious states can be identified and removed after the fact, but only at the expense of convenience and ease of use. One telling indicator of spuriousness is a rapid change in the energy of the state as the mesh pitch is increased. Another is when the envelope function contains rapid unphysical oscillations characteristic of large wavevector components.

Spurious states do not appear in the reciprocal-space method if the wavevector cut-off is low enough to suppress the unphysical behavior at large $k$. Still all may not be well because the solutions could deviate from their "true" values if the cut-off is too low. Chapter 10 will show how we can steer between the two extremes for narrow-gap structures. Finally, we can combat spurious states in the finite-element method by choosing the right basis functions, and the quasi-atomistic approaches of Section 9.5 are unconditionally free of spurious states.

## 9.5   Tight-Binding and Pseudopotential Methods Applied to Quantum Structures

In the envelope-function theory of Chapter 8, we build the lattice symmetry into the basis functions and matrix elements from the start and then disregard the discrete nature of the atoms while plying the numerical techniques of Sections 9.1–9.3 (at least, until we run into the issues of proper cut-off and spurious states!). Unfortunately, difficulties with convergence arise for thin, deep wells and narrow-gap materials, as shown in Chapter 10. These problems can make us wonder whether we have made a devil's bargain and should return to the fold of the full-zone, quasi-atomistic techniques from Chapter 5.

Our tentative conclusion in this book is to decline such a belated conversion, at least for routine work with quantum structures. This is primarily because the $\mathbf{k} \cdot \mathbf{p}$-based envelope theory is so much more convenient to use. The parameters for the quasi-atomistic methods (apart from the next-nearest-neighbor EBOM discussed in Section 5.2) cannot be deduced analytically from measured quantities, i.e., gaps and masses. Instead, we must laboriously fit the parameters for each new material, and change them when these materials are grown with a different lattice constant and when the alloy composition changes even by a small amount. Not to mention these methods take much longer to run than the 8-band solution. Therefore, in our view, the benefits of using them on a routine basis justify the costs only in a few select cases. Nevertheless, it is well worth learning how to construct the tight-binding and pseudopotential models for quantum structures, in part as a benchmark for other computational techniques.

How do we generalize the EBOM for structures with quantum confinement? By analogy with Eq. (5.2), we can write the wavefunction as two nested sums:

$$\Psi_{\mathbf{k}_\parallel}(\mathbf{r}) = \sum_{l,\mathbf{Z}} c_{l,\mathbf{k}_\parallel}(\mathbf{Z}) \, \chi_{l,\mathbf{k}_\parallel}(\mathbf{r}, \mathbf{Z}) \tag{9.64a}$$

$$\chi_{l,\mathbf{k}_\parallel}(\mathbf{r}, \mathbf{Z}) = \frac{1}{\sqrt{N}} \sum_{l,\mathbf{R}_\parallel} e^{i(\mathbf{k}_\parallel \cdot \mathbf{R}_\parallel + k_z Z)} \phi_l(\mathbf{r} - \mathbf{R}_\parallel - \mathbf{Z}) \tag{9.64b}$$

where the plane-wave expansion is retained in the plane of a structure with "atoms" placed at the fcc sites $\mathbf{R}_\parallel$. The out-of-plane variation is also present

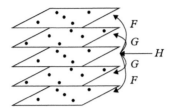

**Figure 9.8** *Schematic of the EBOM calculation for a quantum structure with confinement along the* [001] *direction.*

in $c_{l,k_\parallel}(\boldsymbol{Z})$, where $\boldsymbol{Z}$ denote "atomic" positions along the growth direction. To summarize the task ahead, terms that include the in-plane wavevector are unchanged from the bulk formulation, but we will need to deduce those involving $k_z$.

To keep the calculations simple, we consider only the most common case of growth along the [001] direction. We replace the bulk Hamiltonian of Eq. (5.6) with a matrix that has four (eight if spin is included) times as many rows and columns as the number of atomic layers in one period of the superlattice (or a quantum well in the periodic scheme of Fig. 8.10b). Each entry in the Hamiltonian is the matrix element between two atomic orbitals *and* between the same or two different monolayers of the superlattice, e.g., $\langle \chi_{s,z_i} | H | \chi_{px,z_{i+1}} \rangle$. Since the EBOM includes interactions between the nearest and next-nearest neighbors, each $4 \times 4$ (or $8 \times 8$) block couples to four other blocks. For example, the block corresponding to $i$ couples to $i-2$, $i-1$, $i+1$, and $i+2$, as shown in Fig. 9.8. The full Hamiltonian has the form

$H_{SL,EBOM}$

$$
=
\begin{bmatrix}
H_1 & G_{12} & F_{13} & 0 & 0 & 0 & \cdots & 0 & 0 & 0 & F_{N-1,1} & G_{N,1} \\
G_{21} & H_2 & G_{23} & F_{24} & 0 & 0 & \cdots & 0 & 0 & 0 & 0 & F_{N,2} \\
F_{31} & G_{32} & H_3 & G_{34} & F_{35} & 0 & \cdots & 0 & 0 & 0 & 0 & 0 \\
0 & F_{42} & G_{43} & H_4 & G_{45} & F_{46} & \cdots & 0 & 0 & 0 & 0 & 0 \\
\cdots & \cdots & \cdots & \cdots & \cdots & \cdots & \cdots & \cdots & \cdots & \cdots & \cdots & \cdots \\
0 & 0 & 0 & 0 & 0 & 0 & \cdots & G_{N-3,N-4} & H_{N-3} & G_{N-3,N-2} & F_{N-3,N-1} & 0 \\
0 & 0 & 0 & 0 & 0 & 0 & \cdots & F_{N-2,N-4} & G_{N-2,N-3} & H_{N-2} & G_{N-2,N-1} & F_{N-2,N} \\
F_{1,N-1} & 0 & 0 & 0 & 0 & 0 & \cdots & 0 & F_{N-1,N-3} & G_{N-1,N-2} & H_{N-1} & G_{N-1,N} \\
G_{1,N} & F_{2,N} & 0 & 0 & 0 & 0 & \cdots & 0 & 0 & F_{N,N-2} & G_{N,N-1} & H_N
\end{bmatrix}
$$
$$(9.65)$$

where $F_{ij} = F_{ji}^\dagger$, $G_{ij} = G_{ji}^\dagger$, and the individual blocks are given by

$$
\boldsymbol{H}_i =
\begin{bmatrix}
H_{ss} & H_{sx} & H_{sy} & 0 \\
H_{xs} & H_{xx} & H_{xy} & 0 \\
H_{ys} & H_{yx} & H_{yy} & 0 \\
0 & 0 & 0 & H_{zz}
\end{bmatrix}
\qquad (9.66a)
$$

$$G_{i,i+1} = \frac{h_g}{2} \begin{bmatrix} G_{ss} & G_{sx} & G_{sy} & G_{sz} \\ G_{xs} & G_{xx} & 0 & G_{xz} \\ G_{ys} & 0 & G_{yy} & G_{yz} \\ G_{zs} & G_{zx} & G_{zy} & G_{zz} \end{bmatrix} \tag{9.66b}$$

$$F_{i,i+2} = \frac{h_f}{2} \begin{bmatrix} F_{ss} & 0 & 0 & 0 \\ 0 & F_{xx} & 0 & 0 \\ 0 & 0 & F_{yy} & 0 \\ 0 & 0 & 0 & F_{zz} \end{bmatrix} \tag{9.66c}$$

Here $h_g$ and $h_f$ are phase factors that will be derived later. $G_{1,N}$ and $F_{1,N-1}, F_{2,N}$ are given by Eqs. (9.66b) and (9.66c) with $h_{gc}$ and $h_{fc}$ replacing $h_g$ and $h_f$, while $G_{N,1}$ and $F_{N-1,1}, F_{N,2}$ are the conjugate transposes (Hermitian adjoints) of the matrices in Eq. (9.66). The corner elements in Eq. (9.65) should be consistent with the definitions in Eqs. (9.66b) and (9.66c) and their adjoints.

The Hamiltonian in Eq. (9.65) operates on a vector that consists of the $c_l(\mathbf{Z})$ coefficients in Eq. (9.64a). The matrix elements correspond to different orbitals (as well as electron spins), and each block is written for a particular point $\mathbf{Z}$. The coefficients in general vary with in-plane wavevector $\mathbf{k}_\parallel$.

To derive the form of the elements in Eq. (9.66), we return to the phase terms in Section 5.2. The elements vanish when: (i) all of the terms in Eq. (9.66a) are $z$-dependent, i.e., there are no local interactions; (ii) none of the terms in Eq. (9.66b) are $z$-dependent, i.e., all of the interactions are local; or (iii) there are no next-nearest-monolayer interactions along the $z$ axis in Eq. (9.66c). The local (no $z$ dependence) terms in Eq. (9.66a) follow from Eqs. (5.34)–(5.37):

$$H_{ss} = E_{ss}(0,0,0) + 4E_{ss}(1,1,0)c_xc_y + 2E_{ss}(2,0,0)(cc_x + cc_y) \tag{9.67}$$

$$H_{xx} = E_{xx}(0,0,0) + 4E_{xx}(1,1,0)c_xc_y + 2E_{xx}(2,0,0)cc_x + 2E_{xx}(0,0,2)cc_y \tag{9.68}$$

$$H_{zz} = E_{xx}(0,0,0) + 4E_{xx}(0,1,1)c_xc_y + 2E_{xx}(0,0,2)(cc_x + cc_y) \tag{9.69}$$

$$H_{sx} = 4iE_{sx}(1,1,0)s_xc_y \tag{9.70}$$

$$H_{xy} = -4E_{xy}(1,1,0)s_xs_y \tag{9.71}$$

The remaining matrix elements in Eq. (9.66a) can be generated by substituting $y$ for $x$ etc., and taking Hermitian conjugates of the off-diagonal elements $H_{sx}$, $H_{sy}$, and $H_{xy}$. Since we set $E_{sx}(2,0,0) = 0$ in Section 5.2, we omit it from Eq. (9.70).

The Hamiltonian in Eq. (9.65) omits the spin–orbit terms. To include them, we simply double the number of orbitals and add the matrix in Eq. (3.23) to $\mathbf{H}_i$ from Eq. (9.66a). The same procedure works for any implementation of the tight-binding method, which is a distinct advantage of formulating the problem in terms of the $s$- and $p$-like states.

We obtain the non-local terms by multiplying the Bloch factors $\exp(ik_z a/2)$ and $\exp(ik_z a)$ by the $c_{l,\mathbf{k}_\parallel}(\mathbf{Z})$ expansion coefficients from Eq. (9.64a). Put another way, we return to the formulation of Eq. (5.27) and isolate the phase factors affected by translation along the growth axis. The simple recipe is to replace $c_z$ with $\exp\left(\pm ik_z a/2\right)/2$ and $s_z$ with $-i\exp\left(\pm ik_z a/2\right)/2$, for the $i+1$ and $i-1$ monolayers, as well as $cc_z$ with $\exp\left(\pm ik_z a\right)/2$ and $ss_z$ with $-i\exp\left(\pm ik_z a\right)/2$ for the $i+2$ and $i-2$ monolayers. From this recipe, it follows that the phase terms in Eqs. (9.66b) and (9.66c) are $h_g = e^{ik_z a/2}$ and $h_f = e^{ik_z a}$.

With these substitutions, the matrix elements that couple monolayers $i$ and $i+1$ can be deduced from Eqs. (5.34), (5.35), (5.36), and (5.37):

$$G_{ss} = 4E_{ss}\,(1,1,0)\left(c_x + c_y\right) \tag{9.72}$$

$$G_{xx} = 4E_{xx}\,(1,1,0)\,c_x + 4E_{xx}\,(0,1,1)\,c_y \tag{9.73}$$

$$G_{yy} = 4E_{xx}\,(1,1,0)\,c_y + 4E_{xx}\,(0,1,1)\,c_x \tag{9.74}$$

$$G_{zz} = 4E_{xx}\,(1,1,0)\left(c_x + c_y\right) \tag{9.75}$$

$$G_{sx} = 4iE_{sx}\,(1,1,0)\,s_x \tag{9.76}$$

$$G_{sy} = 4iE_{sx}\,(1,1,0)\,s_y \tag{9.77}$$

$$G_{sz} = 4E_{sx}\,(1,1,0)\left(c_x + c_y\right) \tag{9.78}$$

$$G_{xs} = -4iE_{sx}\,(1,1,0)\,s_x \tag{9.79}$$

$$G_{ys} = -4iE_{sx}\,(1,1,0)\,s_y \tag{9.80}$$

$$G_{zs} = -4E_{sx}\,(1,1,0)\left(c_x + c_y\right) \tag{9.81}$$

$$G_{xz} = 4iE_{xy}\,(1,1,0)\,s_x \tag{9.82}$$

$$G_{yz} = 4iE_{xy}\,(1,1,0)\,s_y \tag{9.83}$$

$$G_{zx} = 4iE_{xy}\,(1,1,0)\,s_x \tag{9.84}$$

$$G_{zy} = 4iE_{xy}\,(1,1,0)\,s_y \tag{9.85}$$

For the matrix elements between monolayers $i$ and $i-1$, the signs of $G_{sz}$ and $G_{zs}$, $G_{xz}$ and $G_{zx}$, $G_{yz}$ and $G_{zy}$ are flipped to account for the fact that the bulk elements in Eqs. (5.31) and (5.34) contain $s_z$.

For the next-nearest monolayers, we obtain

$$F_{ss} = 2E_{ss}\,(2,0,0) \tag{9.86}$$

$$F_{xx} = F_{yy} = 2E_{xx}\,(0,0,2) \tag{9.87}$$

$$F_{zz} = 2E_{xx}\,(2,0,0) \tag{9.88}$$

Finally, $F_{sz} = F_{zs} = 0$ in Eq. (9.66c) because we set $E_{sx}(2,0,0) = 0$ in Section 5.2.

The parameters in Eqs. (9.67)–(9.88) are distinct in different layers of the structure. For a matrix element that connects two monolayers straddling the boundary between different materials, we recommend a simple average.

The same applies to the upper-right and lower-left corners of the matrix in Eq. (9.65), but with Hermitian conjugates to account for the transposition of the elements. Because the $F$ matrix is diagonal, only the $G$ matrix is affected by the conjugation. The phase factors here are augmented by the period, e.g., $h_{gc} = e^{ik_z(\frac{a}{2}+d)}$ for $G_{1,N}$ and $h_{fc} = e^{ik_z(a+d)}$ for $F_{2,N}$ and $F_{1,N-1}$. These matrices can be set to zero for the confined states of a quantum well with thick barriers, along with any $k_z$ dependence.

The Hermiticity of $H_i$ in Eq. (9.65) follows from the bulk Hamiltonian, but what about $G$ and $F$? The $F$ blocks are Hermitian because the conjugation of the $h_f$ phase factor is the only difference between $F_{i,i+2}$ and $F_{i+2,i}$. The same applies to the diagonal terms of $G$, namely, $G_{ss}$, $G_{xx}$, $G_{yy}$, and $G_{zz}$. The off-diagonal elements of $G$ also obey Hermiticity because the signs of the bulk elements that contain $s_z$, i.e., $G_{sz}$, $G_{xz}$, and $G_{yz}$, flip between $G_{i,i+1}$ and $G_{i+1,i}$, as noted earlier. The elements in blocks like $G_{1,N}$ are Hermitian for the same reason, with the phase factor $h_{gc} = e^{ik_z(\frac{a}{2}+d)}$ in $G_{1,N}$ transformed into $h_{gc}^* = e^{-ik_z(\frac{a}{2}+d)}$ in $G_{N,1}$.

The EBOM intrinsically operates in units of monolayers. Of course, in real structures the thickness of any given layer is not so constrained because it represents an average over microscopic in-plane regions of varying thickness. To approximate such structures within the EBOM, we recommend solving for the band structure with the layers in question rounded down and up to the nearest number of monolayers and then interpolating between the results. To compare an envelope-function approach to the EBOM, it certainly helps to choose structures with an integer number of monolayers in each layer.

A superlattice may be so thin that it is unclear how to apply Eq. (9.65) directly. These cases are hardly irrelevant because envelope-function techniques are of dubious validity here. We start with a superlattice that contains just two monolayers in each period. To make progress, we need to realize that the $N$th monolayer in one period is equivalent to the second monolayer in the previous period. To account for this, we modify the Hamiltonian assuming that the two monolayers belong to different materials (otherwise, we are back in the bulk limit):

$$H_{2ML-SL,EBOM} = \begin{bmatrix} H_1 + F_{11} + F_{11}^\dagger & G_{12} + G_{12}^\dagger \\ G_{12}^\dagger + G_{12} & H_2 + F_{22} + F_{22}^\dagger \end{bmatrix} \quad (9.89)$$

The $F$ interaction with the next nearest neighbors is with the same monolayer in the neighboring periods! Of course, a matrix like $F_{11} + F_{11}^\dagger$ is real, so the Hermiticity is not affected. By the way, we have formulated the superlattice problem so that the Hamiltonian of a superlattice with just one monolayer per period (bulk) $H + G + G^\dagger + F + F^\dagger$ is exactly equal to the bulk form derived in Section 5.2.

Proceeding along similar lines for a 3-monolayer superlattice, the Hamiltonian becomes

$$
H_{3ML-SL,EBOM} =
\begin{bmatrix}
H_1 & G_{12} & G_{13}^\dagger + F_{13} + F_{13}^\dagger \\
G_{12}^\dagger & H_2 + F_{22} + F_{22}^\dagger & G_{23} \\
G_{13} + F_{13}^\dagger + F_{13} & G_{23}^\dagger & H_3
\end{bmatrix}
$$

(9.90)

If you wish to relate Eqs. (9.89) and (9.90) to Eq. (9.65), it is important to remember that $F_{31} = F_{13}^\dagger$, etc. The extension to a superlattice with a period of 4 monolayers should be obvious now. Starting with a period of 5 monolayers, we can apply the general Hamiltonian of Eq. (9.65) directly.

We can use the same approach to make the atomistic tight-binding method suitable for quantum wells and superlattices, except that the anion–cation matrix elements now take on the interlayer interactions of Eqs. (9.72)–(9.88). Because the details quickly become tedious, we recommend that the reader consult the references listed under "Suggested Reading." We could also extend these varieties of the tight-binding method to cases of multidimensional quantum confinement. Much like in the finite-difference method based on the envelope-function theory, we would introduce matrix elements that correspond to coupling along both axes. Whether such a complex computational scheme is really justified is another matter. In our opinion, atomistic treatments of quantum wires and dots should be reserved for those few rare cases that have a particularly compelling rationale, e.g., a very deep or a very small III–V quantum dot. At this writing, they are not very important technologically, so we will not elaborate further.

Quantum structures are also amenable to the empirical pseudopotential method of Section 5.4. One way to do so is to form the pseudopotentials of a given structure by superposition of the pseudopotentials for all the atoms in that structure. The unit cell is replaced by one period of the superlattice along the confinement direction (a *supercell*). The plane-wave expansion is in 3D rather than 1D. A useful shortcut is to fuse the bulk-like pseudopotentials in each layer with a variant of the reciprocal-space method of Section 9.3, which makes it straightforward to implement. The total pseudopotential is then the sum of those in all the layers multiplied by the square wave that delimits the layer boundaries. This approach leads to an abrupt transition between the pseudopotentials between layers rather than a more gradual variation in the atomistic EPM (produced by the influence of the nearby atoms in a different layer). The pseudopotential components of the simplified method resemble the Fourier-transformed parameters in Eq. (9.39):

$$
V(\mathbf{r}) = \sum_{\mathbf{G}} \sum_{m} V_{\mathbf{G},m}\, e^{i(\mathbf{G}_\parallel \cdot \mathbf{r}_\parallel + G_m z)}
$$

(9.91)

$$V_{\mathbf{G},m} = \sum_i e^{-iG_m z_i} \frac{2\sin\left(\frac{G_m l_i}{2}\right)}{G_m d} V_{\mathbf{G},i} \qquad V_{\mathbf{G},i} = \sum_i e^{-iG_m z_i} \frac{\sin\left(\frac{G_m l_i}{2}\right)}{\pi m} V_{\mathbf{G},i} \qquad (9.92)$$

where $G_m = G_z + 2\pi m/d$, $V_{\mathbf{G},i}$ is the bulk pseudopotential in layer $i$, and the sum in Eq. (9.92) is over all layers in the superlattice. The $G = 0$ pseudopotential is a weighted average of the valence band offsets in the different layers, which must be made consistent from layer to layer using the recommended values of Chapter 6. We can also set the absolute energy scale, e.g., by defining the origin at the top of the valence band in InSb, which is higher than for any other III–V semiconductor according to Chapter 6, or in any other arbitrary manner.

We expand the wavefunction in a Fourier series that incorporates both the bulk variation and the functional form in any given layer:

$$\Psi_{\mathbf{k}}(\mathbf{r}) = \sum_{\mathbf{G}} \sum_m a_{\mathbf{k},\mathbf{G},m} e^{i(\mathbf{G}_\parallel + \mathbf{k}_\parallel)\cdot\mathbf{r}_\parallel + i(G_m + k_z)z} \qquad (9.93)$$

As in Section 5.4, we substitute Eqs. (9.91)–(9.93) into the Schrödinger equation and integrate over the total in-plane area and one period of the superlattice:

$$\left[\frac{\hbar^2}{2m}\left(|\mathbf{k}_\parallel + \mathbf{G}_\parallel|^2 + (k_z + G_m)^2\right) + V_0\right] a_{\mathbf{k}_\parallel, k_z, \mathbf{G}_\parallel, m}$$
$$+ \sum_{\mathbf{G}'_\parallel, m'} V_{\mathbf{G}_\parallel - \mathbf{G}'_\parallel, m - m'} a_{\mathbf{k}_\parallel, k_z \mathbf{G}'_\parallel, m} = E_{\mathbf{k}_\parallel, k_z} a_{\mathbf{k}_\parallel, k_z, \mathbf{G}_\parallel, m} \qquad (9.94)$$

Now we can add the electron spin and spin–orbit coupling, just as we did in Section 5.3. Of course, the treatment of the spin–orbit interaction is still somewhat ad hoc, as for bulk III–V semiconductors. This is the big difference with any version of the tight-binding method (and envelope-function theory).

Strain elements may be taken as local within the 8-band envelope-function theory and the EBOM. They are included by augmenting the Hamiltonian, as was done in Section 3.4. It is more difficult to build strain accurately into the tight-binding and pseudopotential methods. In principle, a change in the lattice constants produces a non-linear shift of all the parameters and requires a separate fit. Fortunately, shortcuts using approximately linear shifts parameterized with deformation potentials also exist. Complexity escalates rapidly and bolsters our view that the 8-band **k·p** method should not be abandoned lightly. Whenever the **k·p** method is not accurate enough, we should consider first whether it can be fixed rather than completely replaced. An example is the introduction of interface terms in Section 8.3.

## Suggested Reading

- The computational techniques for solving the band structures of quantum wells and superlattices are covered rather sparsely by the existing references. This may be due to the unjustified assumption that once the Hamiltonian is defined, the solutions should be easy to compute accurately. For a treatment of the transfer-matrix theory, see B. Chen, M. Lazzouni, and L. R. Ram-Mohan, "Diagonal representation for the transfer-matrix method for obtaining electronic energy levels in layered semiconductor heterostructures," *Physical Review B* **45**, 1204 (1992).

- Transfer matrices are used to find solutions of the one-band Schrödinger equations in many quantum-mechanics textbooks and in the context of semiconductor physics in S. L. Chuang, *Physics of Photonic Devices*, 2nd edn. (Wiley, New York, 2009).

- A few articles that solve band structure problems using the finite-difference method are D. Z-Y. Ting, and T. C. McGill, "Efficient, numerically stable multiband $k \cdot p$ treatment of quantum transport in semiconductor heterostructures," *Physical Review B* **54**, 5675 (1996); and S. L. Chuang and C. S. Chang, "A band-structure model of strained quantum-well wurtzite semiconductors," *Semiconductor Science and Technology* **12**, 252 (1997).

- A good reference for applying the reciprocal-space method to practical problems is B. Vinter, "Auger recombination in narrow-gap semiconductor superlattices," *Physical Review B* **66**, 045324 (2002), although it is limited to first-order terms only.

- To learn more about solving quantum-mechanical problems using the finite-element method, a comprehensive reference is L. R. Ram-Mohan, *Finite Element and Boundary Element Applications in Quantum Mechanics* (Oxford University Press, Oxford, 2002).

- The origin of spurious states in the finite-difference method is discussed in D. Z.-Y. Ting, and T. C. McGill, "Numerical spurious solutions in the effective-mass approximation," *Journal of Applied Physics* **93**, 3974 (2003).

- A more general discussion that inspired the first part of Section 9.4 can be found in B. A. Foreman, "Elimination of spurious solutions from eight-band k·p theory," *Physical Review B* **56**, R12748 (1997).

- A well-rounded description of tight-binding approaches to the superlattice band structure is presented in the Ph.D. dissertation of L. C. Lew Yan Voon, "Electronic and Optical Properties of Semiconductors: A Study Based on the Empirical Tight-Binding Method," (Worcester Polytechnic Institute, 1993). Although it may be difficult to obtain, this work outlines how to extend the tight-binding method to superlattices grown along directions other than [001].

- A useful variant of the empirical pseudopotential method for superlattices is described in G. C. Dente and M. L. Tilton, "Pseudopotential methods for superlattices: Application to mid-infrared semiconductor lasers," *Journal of Applied Physics* **96**, 420 (1999); and G. C. Dente and M. L. Tilton, "Comparing pseudopotential predictions for InAs/GaSb superlattices," *Physical Review B* **66**, 165,307 (2002).

# 10

# Superlattice and Quantum-Well Band Structure

This chapter presents typical band structures for superlattices and quantum wells computed using the methods described in Chapter 9. We identify important features of the conduction and valence subbands and minibands, their dispersions, optical matrix elements, and characteristic dependences on the materials, thicknesses, and compositions. The changes that occur when the energy gap becomes very small are also discussed. To complete the picture, we consider how the band structure of wurtzite materials differs from their zinc-blende counterparts, as well as the band structure of quantum wires and dots that feature multidimensional confinement.

## 10.1 Multiband Description of Quantum Wells

In a single-band description with parabolic dispersion, the eigenstates and eigenfunctions of a quantum well are the same as for a particle with the appropriate mass that is confined in a one-dimensional box with finite-potential walls. Of course, if the barriers are infinite, the eigenstates are standing waves (cosine-like), with quantized energies: $E_{\nu 0} \equiv \frac{\hbar^2}{2m} \frac{\nu^2 \pi^2}{t^2}$, where $\nu$ is the state index and $t$ is the well width. The confinement energy is lower when the wall potentials are finite, but the wavefunctions still are cosine-like in the well and decay exponentially into the barriers. The details are available in many textbooks on quantum mechanics.

This is all rather unexceptionable, but we know from Chapters 2 and 3 that to describe the bands in semiconductors realistically, we need a multiband formalism. We could also introduce an energy-dependent effective mass, taking our cue from Eq. (2.21), but this approach is less general and accurate. Even if the states of only one band are of interest, the calculations are often much easier in the multiband framework. To develop a sense of how things are different with multiple bands, we return to our old standby, the two-band Hamiltonian at the zone center ($k_\parallel = 0$) with one $s$-like and one $p_z$-like components, similar to Eqs. (2.7) and (8.24):

*Bands and Photons in III–V Semiconductor Quantum Structures.* Vurgaftman, Lumb, and Meyer,
Oxford University Press (2021). © Vurgaftman, Lumb, and Meyer.
DOI: 10.1093/oso/9780198767275.003.0010

$$H_{2-band}(z)\begin{bmatrix} F_s(z) \\ F_z(z) \end{bmatrix} = \begin{bmatrix} E_{c0}(z) - \frac{\hbar^2}{2m}\frac{\partial^2}{\partial z^2} & \frac{\hbar}{m}\frac{\partial}{\partial z}p_{cv}(z) \\ -\frac{\hbar}{m}p_{cv}(z)\frac{\partial}{\partial z} & E_{v0}(z) - \frac{\hbar^2}{2m}\frac{\partial^2}{\partial z^2} \end{bmatrix}\begin{bmatrix} F_s(z) \\ F_z(z) \end{bmatrix} \tag{10.1}$$

The second derivative in the diagonal terms is the free-electron contribution. For clarity, we make a few simplifying assumptions that are easily removed in a numerical solution: (1) only the conduction-band edge described by $V(z)$ is discontinuous across the quantum-well interfaces at $z = 0, t$ (i.e., the valence-band offset is zero); and (2) the momentum matrix elements (and electron masses) are the same in both material constituents. Defining $P \equiv \frac{\hbar}{m}p_{cv}$, we obtain

$$H_{2-band}(z) = \begin{bmatrix} V(z) + E_g - \frac{\hbar^2}{2m}\frac{\partial^2}{\partial z^2} & P\frac{\partial}{\partial z} \\ -P\frac{\partial}{\partial z} & -\frac{\hbar^2}{2m}\frac{\partial^2}{\partial z^2} \end{bmatrix} \tag{10.2}$$

The function $V(z)$ is step-like, with $V(z) = 0$ in the wells and $V(z) = V_0$ in the barriers.

For states near the conduction-band edge (with $E_v \ll V_0$), we expect the $s$ component of the wavefunction to have a standing-wave form, i.e., $F_s \sim \sin\left(\frac{v\pi z}{t}\right)$ with $v = 1, 2, \cdots$. Hence, the parity around the center of the well flips for every consecutive state (from odd to even, and from even to odd), even when $E_v$ is not much smaller than $V_0$. The $p_z$-like component of the wavefunction is given by the second equation of the eigenvalue problem obtained from Eq. (10.2):

$$\left(-\frac{\hbar^2}{2m}\frac{\partial^2}{\partial z^2} - E_v\right)F_z = P\frac{\partial}{\partial z}F_s = P\frac{v\pi}{t}\cos\left(\frac{v\pi z}{t}\right) \tag{10.3}$$

from which we obtain

$$F_z = -\frac{Pv\pi}{t}\frac{1}{E_v - \frac{\hbar^2}{2m}\frac{v^2\pi^2}{t^2}}\cos\left(\frac{v\pi z}{t}\right) = -\frac{Pv\pi}{t}\frac{1}{E_v - E_{v0}}\cos\left(\frac{v\pi z}{t}\right) \tag{10.4}$$

These envelope-function components are *not* normalized, but the relative magnitudes are correct. The initial guess for $F_s$ also appears to be on target, since $F_s \propto \frac{\partial}{\partial z}F_z$ and $F_z \propto \frac{\partial}{\partial z}F_s$ from Eq. (10.2).

Figure 10.1 illustrates that the $p_z$-like component of the conduction-band wavefunction from Eq. (10.4) peaks at the interfaces. Since the $p_z$-like component is proportional to the derivative of $F_s$, as in Eq. (10.3), for a confined state it decays to zero far away from the quantum well. Then because of Eq. (10.3) and the exponential decay in the barriers, $F_z$ must peak at the interfaces, with the same value at both interfaces if the well is symmetric.

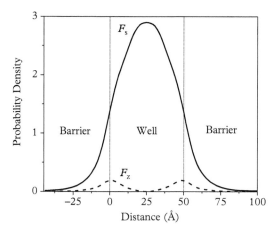

**Figure 10.1** *The s-like and $p_z$-like components of the conduction-band wavefunction for the lowest subband of a 50-Å-wide quantum well with $V_0 = 0.5\ eV$ and $E_P = 25\ eV$ within a 2-band model.*

We can solve for $E_\nu$ using the first equation contained in Eq. (10.2):

$$\left(E_\nu - E_{\nu 0} - E_g\right) = \frac{P^2 \nu^2 \pi^2}{t^2}\frac{1}{E_\nu - E_{\nu 0}} = \frac{2mP^2}{\hbar^2}\frac{E_{\nu 0}}{E_\nu - E_{\nu 0}} = \frac{E_P E_{\nu 0}}{E_\nu - E_{\nu 0}} \qquad (10.5)$$

where the last equality follows from the definition of $E_P$ in Eq. (2.12). We can solve Eq. (10.5) approximately by taking $E_\nu \approx E_g + (1 + x)E_{\nu 0}$ and solving for $x$. If $xE_{\nu 0} \ll E_g$, we have $E_\nu - E_{\nu 0} \approx E_g$ in the denominator of Eq. (10.5). This immediately tells us that $x \approx \frac{E_p}{E_g}$ and that

$$E_\nu \approx E_g + E_{\nu 0}\left(1 + \frac{E_p}{E_g}\right) \qquad (10.6)$$

This is identical to the particle-in-a-box prediction when the inverse mass is $\frac{1}{m_e} \approx \frac{E_p}{E_g}$. This is expected because the two-band model becomes parabolic in the limit of small energies (see Eq. (2.11)). Differences arise when $E_\nu - E_g \sim E_g$ and when $E_\nu - E_g$ approaches $V_0$. In this case, the particles are only weakly confined by the well. As always, finite barriers imply that the energies are somewhat smaller than in Eq. (10.6), and that they do not increase as fast with well width. If the subband energies are high, the nonparabolic bulk dispersions affect the quantum-confined states, and reduce their energies by comparison with Eq. (10.6). Despite that, the nature of the conduction–valence band coupling does not change.

Using the definitions leading to Eq. (10.6), we rewrite Eq. (10.4) as

$$F_z \approx \frac{\sqrt{E_{v0}E_p}}{E_g}\cos\left(\frac{v\pi z}{t}\right) = \frac{\sqrt{E_{v0}E_p}}{E_g}\cot\left(\frac{v\pi z}{t}\right)F_s \tag{10.7}$$

With increasing confinement and narrower energy gaps, $F_z$ becomes larger relative to $F_s$, but the trend does not hold when $E_g \approx 0$.

Since Eq. (10.2) is only valid at $k_\parallel = 0$, the energies in Eq. (10.6) do *not* correspond to discrete states like those of isolated atoms or molecules. The (regrettably frequent) references in the literature to $E_v\,(k_\parallel = 0)$ as "electron levels" make little sense in view of the dispersions in the other two dimensions! For quantum wells these are actually the minima of a series of electron *subbands* with quasi-continuous dispersions $E_v\,(k_\parallel)$, as illustrated in Fig. 10.2. For example, for a parabolic dependence on the in-plane wavevector, we have

$$E_v\,(k_\parallel) = E_v\,(k_\parallel = 0) + \frac{\hbar^2}{2m_e}k_\parallel^2 \tag{10.8}$$

where $m_e$ is again given by Eq. (2.12). More generally, the dispersions are not parabolic (see, e.g., Eq. (2.9)]:

$$E_v\,(k_\parallel) = E_v\,(k_\parallel = 0) + f\,(k_\parallel) \tag{10.9}$$

The function $f\,(k_\parallel)$ increases (decreases) monotonically from zero at $k_\parallel = 0$ for the CB1 (HH1) subbands. Other electron and hole subbands do not in general vary monotonically with $k_\parallel$, as illustrated in Fig. 10.2.

The subband minimum $E_v\,(k_\parallel = 0)$ is not necessarily given by Eq. (10.6). When the quantum well is not symmetric, the minimum shifts away from $k_\parallel = 0$, and the two spin-degenerate subbands split, as discussed in Section 8.4. Recall that in that case, the minima of $E_v\,(k_\parallel)$ occurs when $\frac{\partial f}{\partial k_\parallel} = 0$. For example for the linear-in-$k$ Rashba effect, the two minima are at $k_\parallel = \pm\frac{m_e\alpha_R}{\hbar^2}$, where $\alpha_R$ is the Rashba coefficient.

We can use the same trial function to calculate the subbands when $E_g \approx 0$. The Hamiltonian in Eq. (10.2) becomes

$$H = \begin{bmatrix} V(z) - \frac{\hbar^2}{2m}\frac{\partial^2}{\partial z^2} & P\frac{\partial}{\partial z} \\ -P\frac{\partial}{\partial z} & -\frac{\hbar^2}{2m}\frac{\partial^2}{\partial z^2} \end{bmatrix} \tag{10.10}$$

and Eq. (10.5) now reads

$$(E_v - E_{v0})^2 = E_P E_{v0} \tag{10.11}$$

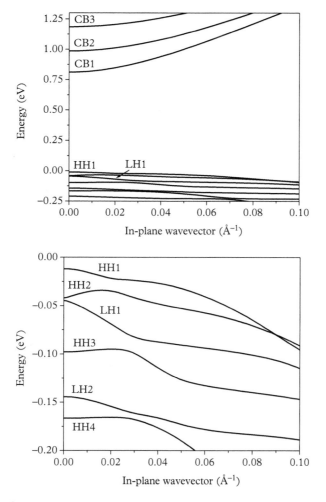

**Figure 10.2** *Calculated dispersions along the [01] in-plane direction using the 8-band* **k·p** *method for the 80-Å In$_{0.53}$Ga$_{0.47}$As/100-Å Al$_{0.48}$In$_{0.52}$As quantum well modeled as a superlattice. The bottom panel shows a magnified view of the valence-band structure near the top of the valence band.*

The energies of the subband minima follow from Eq. (10.11):

$$E_v = E_{v0} + \sqrt{E_P E_{v0}} \approx \sqrt{E_P E_{v0}} \tag{10.12}$$

As in Eq. (10.6), the apparent "correction" to $E_{v0}$ is much larger than $E_{v0}$ because $E_P \gg E_{v0}$. Equation (10.12) is a direct quantum-confinement analog of the Kane dispersion with $E_g = 0$ in Eq. (2.9).

When the gap is not exactly zero, but narrow enough that the confinement energy $E_v \sim E_g$, the subband minima do not follow the $E_v \propto v^2$ dependence expected for "nearly free" electrons. Instead, we see from Eq. (10.12) that their spacing becomes almost linear in the subband index $E_v \propto v$. If the confinement potential is shallow, the spacing is also reduced because the higher subbands spread out of the well. For them the "effective" well is wider, i.e., $t_{eff} > t$, with $t_{eff}$ replacing $t$ in the expression for $E_{v0}$.

Confined states in the well are not the only solutions of the quantum-well problem. Some states have energies higher than $V_0$ and extend far beyond the confined region. The well potential can affect these states if their energy is sufficiently enough to $V_0$, with the strongest impact observed in thin superlattices.

To describe the valence-band states accurately, we solve the 8-band Hamiltonian. This is equivalent to the Kane Hamiltonian in Eq. (2.51) at $k_\parallel = 0$, with the HH mass replaced by its value along [001] from Eq. (3.36a). The derivative substitution procedure lets us write

$$
\mathbf{H}_{Kane}(z)
\begin{bmatrix}
\left| l=0, j=\tfrac{1}{2}, m_j=\tfrac{1}{2} \right\rangle(z) \\[4pt]
\left| l=1, j=\tfrac{1}{2}, m_j=\tfrac{1}{2} \right\rangle(z) \\[4pt]
\left| l=1, j=\tfrac{3}{2}, m_j=\tfrac{1}{2} \right\rangle(z) \\[4pt]
\left| l=1, j=\tfrac{3}{2}, m_j=\tfrac{3}{2} \right\rangle(z)
\end{bmatrix}
$$

$$
=
\begin{bmatrix}
E_g + E_c(z) - \dfrac{\hbar^2}{2m}\dfrac{\partial^2}{\partial z^2} & -i\sqrt{\tfrac{2}{3}}P\dfrac{\partial}{\partial z} & i\sqrt{\tfrac{1}{3}}P\dfrac{\partial}{\partial z} & 0 \\[10pt]
-i\sqrt{\tfrac{2}{3}}P\dfrac{\partial}{\partial z} & E_v(z) - \dfrac{\hbar^2}{2m}\dfrac{\partial^2}{\partial z^2} & 0 & 0 \\[10pt]
i\sqrt{\tfrac{1}{3}}P\dfrac{\partial}{\partial z} & 0 & E_v(z) - \Delta - \dfrac{\hbar^2}{2m}\dfrac{\partial^2}{\partial z^2} & 0 \\[10pt]
0 & 0 & 0 & E_v(z) - \dfrac{\hbar^2}{2m_{HH}}\dfrac{\partial^2}{\partial z^2}
\end{bmatrix}
$$

$$
\times
\begin{bmatrix}
\left| l=0, j=\tfrac{1}{2}, m_j=\tfrac{1}{2} \right\rangle(z) \\[4pt]
\left| l=1, j=\tfrac{1}{2}, m_j=\tfrac{1}{2} \right\rangle(z) \\[4pt]
\left| l=1, j=\tfrac{3}{2}, m_j=\tfrac{1}{2} \right\rangle(z) \\[4pt]
\left| l=1, j=\tfrac{3}{2}, m_j=\tfrac{3}{2} \right\rangle(z)
\end{bmatrix}
\tag{10.13}
$$

Here we assume that $P$, $\Delta$, and $m_{HH}$ are the same on both sides of the interface, but it is not difficult to lift this assumption. We calculate the solutions for a single quantum well with $E_c = E_v = 0$ in the well with energy gap $E_g$, while $E_c > 0$ and $E_v < 0$ in the barrier material that has a bulk energy gap $E_g + E_c + E_v$.

The heavy-hole states $\left| j=\tfrac{3}{2}, m_j=\tfrac{3}{2} \right\rangle$ are of the particle-in-a-box type and do not couple to any other states. The HH wavefunction has one component $\left| j=\tfrac{3}{2}, m_j=\tfrac{3}{2} \right\rangle$ at $k_\parallel = 0$. The electron, light-hole, and split-off states are connected by the off-diagonal matrix elements of the matrix in Eq. (10.13). Figure 10.2 shows all the calculated confined states at the zone center.

We can also generalize the 2-band solution for the electron states well below the barrier by using the same trial form $F_s \sim \sin\left(\frac{v\pi z}{t}\right)$. From the second and third equations in Eq. (10.13), the LH and SO components of the *conduction-band* wavefunction are

$$F_{LH} \approx -i\sqrt{\frac{2}{3}}\frac{Pv\pi}{t}\frac{1}{E_{v,CB}-E_{v0}}\cos\left(\frac{v\pi z}{t}\right) \approx -i\sqrt{\frac{2}{3}}\frac{\sqrt{E_{v0}E_p}}{E_g}\cos\left(\frac{v\pi z}{t}\right)$$

(10.14a)

$$F_{SO} \approx i\sqrt{\frac{1}{3}}\frac{Pv\pi}{t}\frac{1}{E_{v,CB}+\Delta-E_{v0}}\cos\left(\frac{v\pi z}{t}\right) \approx i\sqrt{\frac{1}{3}}\frac{\sqrt{E_{v0}E_p}}{E_g+\Delta}\cos\left(\frac{v\pi z}{t}\right)$$

(10.14b)

Figure 10.3 shows the squares of the various components of the 8-band electron wavefunction for the CB1 and CB2 subbands for a 50-Å InAs/50-Å AlSb superlattice. Because the electron envelope function hardly penetrates to the other side of the AlSb barrier, the structure looks like an InAs quantum well for electrons. Curiously, from the point of view of the holes, it is the AlSb layer that acts as a well! We will say more about such structures in Section 10.2. The CB1 envelope-function profiles resemble the approximate forms in Eq. (10.14), but the LH and SO components penetrate into the finite barriers. Nor is it surprising they are similar to that of $F_z$ in the 2-band Hamiltonian of Eq. (10.2) and Fig. 10.1, with $F_{LH}$ and $F_{SO}$ peaking at the interfaces and close to zero at the center of the well ($z = t/2$) for the lowest electron subband. The LH component is larger than the SO component, especially in a material with spin–orbit splitting $\Delta \sim E_g$ (such as InAs). The LH and SO components for CB2 have two nodes in the well, as expected from Eq. (10.14). The nodes are at $z < t/4$ and $z > 3t/4$ because the wavefunction penetrates into the barriers, which makes the "effective" well a little wider than the physical one.

The electron mass now depends on the spin–orbit splitting, and $E_{v,CB}$ follows the form of Eq. (2.45):

$$E_{v,CB} \approx E_g + E_{v0}\left(1 + \frac{E_P\left(E_g + \frac{2\Delta}{3}\right)}{E_g\left(E_g+\Delta\right)}\right) \approx E_g + E_{v0}\frac{m}{m_e}$$

(10.15)

As in Eq. (10.6), if $E_{v0}\frac{m}{m_e} \ll E_g$ the energies of the conduction band states are approximately the same as for the 1D Schrödinger equation with mass $m_e$.

What about the confined LH states shown in Fig. 10.2? It is reasonable to suppose that the LH wavefunction component dominates and is given by $F_{LH} \sim \sin\left(\frac{v\pi z}{t}\right)$. Equation (10.13) shows that the LH states are coupled only to CB at $k_\parallel = 0$, so the $s$-like component becomes

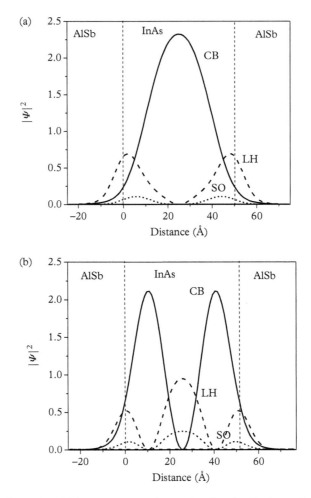

**Figure 10.3** *CB, LH, and SO components of the envelope function for the two lowest CB subbands (a) and (b) in the 8-band **k·p** model for a 50-Å InAs/50-Å AlSb superlattice at $k_\parallel = 0$. At the zone center, this model is equivalent to the quantum solution of the Kane Hamiltonian in Eqs. (2.51) and (10.13). The HH component for the CB subbands is identically zero at the zone center because the HH band decouples in the Kane Hamiltonian.*

$$F_s \approx i\sqrt{\frac{2}{3}}\frac{Pv\pi}{t}\frac{1}{E_g - E_{v,LH} + E_{v0}}\cos\left(\frac{v\pi z}{t}\right) \approx i\sqrt{\frac{2}{3}}\frac{\sqrt{E_{v0}E_p}}{E_g}\cos\left(\frac{v\pi z}{t}\right) \quad (10.16)$$

In this case, it is $F_s$ that peaks at the interfaces and vanishes at the center of the well! The LH energies in the limit of a deep well follow Eq. (10.15), with $m_{LH}$ from Eq. (2.47) replacing $m_e$. Figure 10.4 displays the various envelope-function components of a typical LH state for an InGaAs/InAlAs well. The SO states look

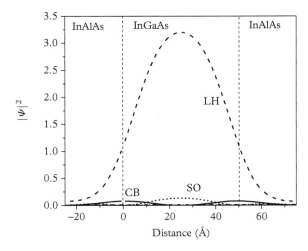

**Figure 10.4** *CB, LH, and SO components of the envelope function for the highest LH subband in the 8-band **k·p** model for a 50-Å In$_{0.53}$Ga$_{0.47}$As/100-Å In$_{0.52}$Al$_{0.48}$As quantum well at k$_{\parallel}$ = 0. As in Fig. 10.3, the HH component is identically zero at the zone center.*

much like the LH states, but are less confined because they are far below the top of the valence band.

If the well is not strained, the heavier HH mass puts the heavy holes at the summit of the valence band at $k_{\parallel} = 0$, as shown in Fig. 10.2. The only way for the LH states to come out on top is to reside in a well under tensile strain, which must be high enough to overcome the mass difference (see Section 3.4). The HH–LH splitting is then negative and described by Eq. (3.65), reproduced here:

$$\Delta E_{HH,LH} = 2b\frac{c_{11} + 2c_{12}}{c_{11}}\epsilon_{xx} \tag{10.17}$$

Recall that the biaxial deformation potential $b < 0$, and that $\epsilon_{xx}$ is negative for compressive strain and positive for tensile strain. Equation (10.17) implies that the splitting between the $\nu$th HH subband and the $\mu$th LH subband in a quantum well is approximately

$$\Delta E_{\nu,HH;\mu,LH} \approx \left(E_{\nu,HH,0} - E_{\mu,LH,0}\right) + 2b\frac{c_{11} + 2c_{12}}{c_{11}}\epsilon_{xx} \tag{10.18}$$

where $E_{\nu,HH,0}$ and $E_{\mu,LH,0}$ are the subband energies in the absence of strain.

Figure 10.5a illustrates the dependences of the HH and LH subband energies at $k_{\parallel} = 0$ on the sign and magnitude of the strain, for InGaAs quantum wells with InAlAs barriers on an InP substrate. If the In composition exceeds the lattice-matched (LM) level of 53%, the well is under compressive strain; otherwise, it is

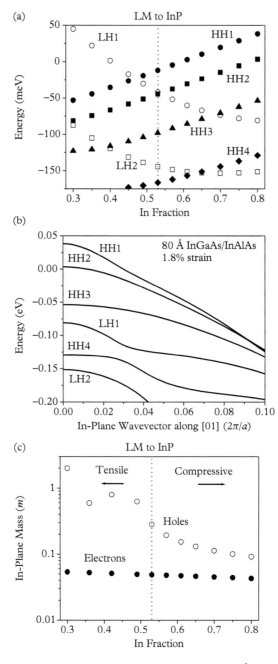

**Figure 10.5** *(a) HH and LH subband energies at $k_\parallel = 0$ for the 80-Å $In_xGa_{1-x}As/In_{0.52}Al_{0.48}As$ wells with various well compositions corresponding to different strain levels on the InP substrate. The energy is referenced to the unstrained valence-band maximum in the bulk InGaAs layer with the same composition. (b) In-plane dispersions of the valence subbands along the [01] direction for the maximum compressive strain in (a), corresponding to 80% In and 1.8% strain. (c) Average in-plane electron and hole masses near the band extrema weighted by the Boltzmann factor as a function of well composition (strain).*

under tensile strain. Clearly, the HH and LH subband edges in Fig. 10.5a move in opposite directions (at a rate that saturates as they approach the valence-band offset at the bottom of the figure). Another interesting observation is that strong compressive strain reduces the hole mass near the band edge so that it is much closer to the electron mass than in bulk. The resulting band structure is shown in Fig. 10.5b. The in-plane electron and hole masses averaged with the weighting by the Boltzmann factor are given as a function of strain in Fig. 10.5c. The electron mass shows little variation and is somewhat larger than the band-edge mass in bulk InGaAs. For structures with tensile strain, the hole mass is sensitive to details of the curvatures of the lower (LH1 and HH2 subbands), but is generally large with values similar to bulk III–V materials. As compressive strain becomes stronger, the hole mass drops gradually. The reduction is due to the repulsive interaction between the strain-split HH and LH subbands near $k_\parallel = 0$. The strain splitting changes for growth directions other than [001], as discussed in Chapter 3.

The strain reduces the hole mass near the top of the valence band in many other compressively strained quantum wells. Another important example is the wider-gap InGaAsP well grown on InP. Strained InGaAs with $\approx 20\%$ In is widely used to produce even wider-gap wells on GaAs substrates. For narrower gaps, InGaAsSb wells on GaSb substrates are common, in which the In fraction is set to obtain $E_g = 0.4$–$0.6$ eV and the As composition is dialed down for the right amount of compressive strain (most commonly, 1.5–2%). All of these wells are quite important for the semiconductor lasers that will be discussed in Chapter 12.

The rest of the book often mentions "strain-compensated" quantum structures. What this means is that the net strain weighted by the thickness of each layer is made to approximately vanish. For example, in a well-barrier pair with the same thickness of the two layers, this is equivalent to equal and opposite strains in the well and the barrier (if the well is compressively strained, the barrier has tensile strain and vice versa). To be sure, there are more sophisticated methods for the analysis of the strain balance, using stiffness parameters, etc. But their predictions are usually not much better than the simple approach outlined above. Epitaxial growers often fine tune empirically, simply because thicknesses and compositions are not known precisely. Hence they hardly ever clamor for highly accurate theoretical estimates.

## 10.2   Coupled Quantum Wells and Superlattices

Many important quantum structures are periodic superlattices rather than isolated quantum wells. If multiple wells are present, the transition between the two classes is gradual and depends on the thickness of the barriers between the wells. Superlattices with barriers just thick enough to allow the wavefunction to penetrate to the next well are often referred to as *coupled quantum wells*. Coupled wells may have the same or different thicknesses and compositions.

Two coupled wells with some wavefunction overlap resemble a diatomic molecule. At $k_\parallel = 0$, the isolated versions of the same wells may have two eigenstates $E_a$ and $E_b$, each associated with its own well:

$$H\varphi_a = E_a\varphi_a \tag{10.19a}$$

$$H\varphi_b = E_b\varphi_b \tag{10.19b}$$

with the wavefunctions $\varphi_a$ and $\varphi_b$ (we can think of them as the dominant components of the multiband wavefunction). If the overlap between the wavefunctions is not too large, we can use the overlap integral

$$V_{ab} = \int \varphi_b^* H\varphi_a \; dz \tag{10.20}$$

to measure of the strength of the interaction between the wells. Expanding the full Hamiltonian in a basis consisting of the two multiband wavefunctions leads to

$$\begin{bmatrix} E_a & V_{ab} \\ V_{ab}^* & E_b \end{bmatrix} \begin{bmatrix} \varphi_a \\ \varphi_b \end{bmatrix} = E \begin{bmatrix} \varphi_a \\ \varphi_b \end{bmatrix} \tag{10.21}$$

Mathematically the problem is identical to Eq. (1.24), and the energies of the two eigenstates of the coupled system are

$$E_\pm = \frac{E_a + E_b}{2} \pm \sqrt{\frac{(E_a - E_b)^2}{4} + |V_{ab}|^2} \tag{10.22}$$

If the initial energies $E_a$ and $E_b$ are the same, and the overlap integral in Eq. (10.20) is real, the two states (these are in fact the subband minima) are split by $2|V_{ab}|$, and the eigenfunctions are given by

$$\begin{bmatrix} \varphi_a \\ \varphi_b \end{bmatrix}_\pm = \frac{1}{\sqrt{2}} \begin{bmatrix} 1 \\ \mp 1 \end{bmatrix} \tag{10.23}$$

The wavefunction for the state with lower (higher) energy is symmetric (antisymmetric), as shown in Fig. 10.6.

If the energy in one well is much larger than in the other, $E_a - E_b \gg |V_{ab}|$, the energies and the wavefunctions are close to those of the isolated wells:

$$E_\pm = E_{a,b} \pm \frac{|V_{ab}|^2}{E_a - E_b} \tag{10.24a}$$

$$\begin{bmatrix} \varphi_a \\ \varphi_b \end{bmatrix}_+ \approx \begin{bmatrix} 1 \\ \frac{-V_{ab}^*}{E_a - E_b} \end{bmatrix} \tag{10.24b}$$

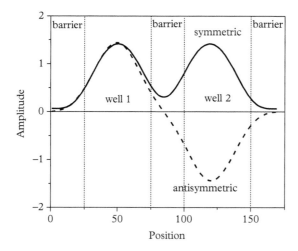

**Figure 10.6** *Wavefunctions for the symmetric and antisymmetric states of a coupled-well system.*

$$\begin{bmatrix} \varphi_a \\ \varphi_b \end{bmatrix}_- \approx \begin{bmatrix} \frac{V_{ab}}{E_a - E_b} \\ 1 \end{bmatrix} \tag{10.24c}$$

As we bring the two isolated wells closer to each other, it seems that the states in the individual wells *hybridize* so that the states of the combined system are superpositions of the original states. If the wells are the same, the superpositions are symmetric and antisymmetric, as shown in Fig. 10.6.

A similar idea applies to three wells, with coupling between the neighboring wells. To simplify, we assume that all the wells are identical, and the unperturbed ground-state energy is zero:

$$H_{3QW} = \begin{bmatrix} 0 & V & 0 \\ V^* & 0 & V \\ 0 & V^* & 0 \end{bmatrix} \tag{10.25}$$

The eigenproblem of Eq. (10.25) leads to $E^3 - 2E|V|^2 = 0$, with the three eigenenergies given by $-\sqrt{2}|V|, 0, \sqrt{2}|V|$.

Now we can guess the effect of additional wells. In fact, the problem is very similar to that illustrating the origin of bands in Section 1.3. The total number of states must equal the number of wells, and the states should be equidistant from one another because all the wells are the same. That is, for $M$ wells the total bandwidth is $U \equiv 2\sqrt{M}|V|$, and the spacing between the consecutive states becomes progressively smaller: $\frac{U}{M} = 2|V|/\sqrt{M}$. When there are many wells, we can use the general solution from Eq. (1.34) in a slightly different form:

$$E(k_z) = E_0 + \frac{U}{2}[1 - \cos(k_z d)] \tag{10.26}$$

where $d$ is the spacing between the wells (period of the resulting superlattice). Again, as we bring $M$ wells closer to each other, with all the wells and barriers being the same, the isolated states split into $M$ hybridized states, and the separation between those states scales with the interaction between the wells, i.e., with the barrier height and thickness.

The wavevector $k_z$ accounts for the phase variation between neighboring wells, since the envelope function is similar to Eq. (1.29):

$$\Psi_{k_z}(z) = \sum_{l,n} b_l e^{ik_z nd} \varphi_l(z - nd) \tag{10.27}$$

Here $l$ is the band index and $n$ is the well number in the chain. The wavefunction of the lowest-energy state in the *miniband* defined by Eq. (10.26) has the same phase in the neighboring wells ($k_z = 0$), as shown in Fig. 10.7. It is in fact the symmetric state of the coupled-well problem! At the other extreme, the top of the miniband corresponds to $k_z = \pi/d$, with the wavefunction alternating in sign between the neighboring wells. This state resembles the antisymmetric solution in Fig. 10.6.

The curvature of the miniband at $k_z = 0$ follows from Eq. (10.26) and is the reciprocal of the growth-direction electron effective mass:

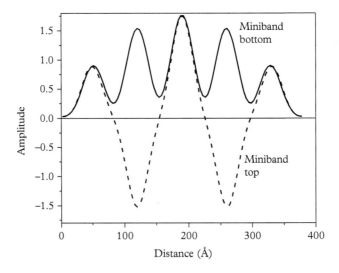

**Figure 10.7** *Wavefunctions for the states at the bottom ($k_z = 0$) and top ($k_z = \pi/d$) of the miniband for a superlattice structure with five identical wells.*

$$m_z \, (k_z \to 0) \equiv \hbar^2 \left( \frac{\partial^2 E}{\partial k_z{}^2} \right)^{-1} = \frac{2\hbar^2}{d^2 U} \tag{10.28}$$

However, this is not the whole story because the miniband dispersion is symmetric about $k_z = \pi/(2d)$. This means the positive mass near the bottom is exactly compensated by the negative mass near its top. In fact, the electrons in a full miniband do not contribute to transport along the growth axis (nor do the electrons in a full bulk band, e.g., the valence band at zero temperature and doping). The signs in Eqs. (10.26) and (10.28) are reversed for holes.

A full range of $k_\parallel$ miniband states exists in addition to $k_\parallel = 0$, just as for quantum wells. The complete band structure includes variation along $k_\parallel$ as well as $k_z$, as shown in Fig. 10.8 for an InAs/GaSb superlattice, as well as multiple minibands at different energies. In-plane dispersion fills the apparent gaps between subbands and minibands (i.e., these gaps are *not* devoid of electronic states). Nevertheless, when only the states near $k_\parallel = 0$ are important, it is useful to talk about *minigaps* (between minibands).

When the dispersion along the growth direction is very small, i.e., $U \to 0$, we have $m_z \, (k_z \to 0) \to \infty$. This happens in a series of nearly isolated (weakly coupled) quantum wells separated by thick barriers, with very little overlap between the wavefunctions in the neighboring wells. At the other end of the spectrum, the growth-direction dispersion is simply the folded version of the bulk dispersion along the [001] (growth) direction when the barriers are very thin. Out-of-plane dispersions that span the two extremes are illustrated in Fig. 10.9. Lighter particles, namely, electrons and light holes, are more likely to penetrate barriers than heavy holes. Hence, the topmost HH states are easily localized by even shallow potential wells.

The average band curvature for occupied states does not necessarily predict how smoothly carrier transport along the growth direction can proceed. Instead, we must resort to the transport theory, which falls outside of the scope of this book. Briefly stated, it turns out that the reciprocal of the *conductivity* electron mass is given by $\frac{1}{\hbar^2} \left( \frac{\partial E}{\partial k_z} \right)^2 / (k_B T)$ rather than the "curvature" mass $\frac{1}{\hbar^2} \frac{\partial^2 E}{\partial k_z{}^2}$. Since $\frac{1}{\hbar} \frac{\partial E}{\partial k_z}$ is the "instantaneous" electron velocity along the growth direction, we can think of the conductivity mass as derived from setting the average thermal energy equal to the kinetic energy. When the mass is very heavy near the band extrema, the conductivity mass can be finite because the states with faster growth-direction dispersion at $k_\parallel > 0$ contribute to the overall transport.

The InAs/GaSb superlattice in Fig. 10.8 is referred to as *type II* because the principal components of the electron and hole envelope functions peak in different layers (see Fig. 10.8b). The InAs layer acts as a well for electrons and a barrier for holes, whereas the opposite is true of the GaSb layer. The overlap between the electron and hole wavefunctions in a type II structure can be arbitrarily small. By

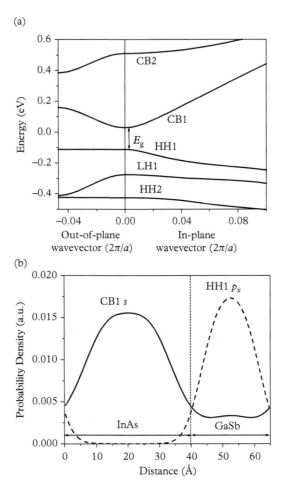

**Figure 10.8** *(a) In-plane and out-of-plane dispersions for the CB1, CB2, HH1, LH1, and HH2 subbands for a 40-Å InAs/25-Å GaSb superlattice on the GaSb substrate at a temperature of 77 K; (b) electron (s-like) and hole (p$_x$-like) envelope functions squared for CB1 and HH1 at $k_\parallel = 0$ for the same superlattice.*

contrast, when the wavefunctions peak in the same layer (type I alignment), e.g., in a GaAs/AlGaAs quantum wells or superlattices, the overlap is close to unity unless the wells are so narrow that the electron wavefunction spills out of them.

   The type II superlattice of Fig. 10.8 with InAs electron wells slows down the convergence of the reciprocal-space method (and of other computational approaches discussed in Chapter 9). The culpability lies with the abnormal curvature of the InAs conduction band at large wavevectors, needed to faithfully represent the electronic states of short-period superlattices, as discussed in Section

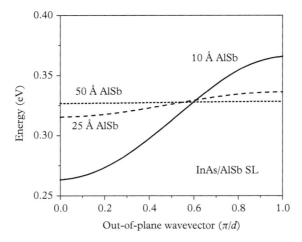

**Figure 10.9** *Out-of-plane electron (CB1) dispersions for the 50-Å InAs/50-Å AlSb, 50-Å InAs/25-Å AlSb, and 50-Å InAs/10-Å AlSb quantum wells on the GaSb substrate at room temperature.*

9.4. As we increase the cut-off to ensure the convergence, we begin to probe the spurious region of the bulk band structure.

To illustrate this behavior, we show the absolute deviation from the energy gap at the wavevector cut-off for three different structures with InAs and InAs-rich electron wells in Fig. 10.10a. These are: (1) the type I 80-Å $In_{0.8}Ga_{0.2}As$/100-Å $In_{0.52}Al_{0.48}As$ quantum well of Fig. 10.5; (2) the type II 40-Å InAs/25-Å GaSb superlattice of Fig. 10.8; and (3) a type II 103-Å InAs/23-Å $InAs_{0.5}Sb_{0.5}$ superlattice at $T = 77$ K. The quantum well has a gap of 634 meV at $T = 300$ K, while the two superlattices have gaps of 140 and 103 meV at $T = 77$ K, for the GaSb and InAsSb hole wells, respectively. Even though we plot the absolute deviation in Fig. 10.10a, in almost all cases the gaps for lower cut-offs are larger. The convergence is relatively fast for the wide-gap well (1), with $f_B = 0.05 - 0.1$ sufficient for a negligible error in the energy gap ($< 1\%$). The superlattices are a study in contrast: the gap at first drops rapidly, then decreases more slowly (but still noticeably) in the range $f_B = 0.1 - 0.3$. Is the decrease at large $f_B$ due to improved spatial resolution or is an unphysical large-$k$ bulk dispersion to blame?

In general, it is difficult to answer this question definitively. One clue comes from the spatial profiles of the envelope functions. As shown in Fig. 10.10b for CB1 for $k_\parallel = 0$, the envelope function develops additional oscillations far away from the boundaries when the wavevector cut-off is too large. This indicates some degree of coupling to unphysical bulk states and hence that the slight energy gap reduction at large cut-offs may be less rather than more accurate. Because a strict convergence criterion does not appear to exist, we suggest selecting $f_B = 0.2 - 0.3$ in routine simulations of type II superlattices. The accuracy of this cut-off should

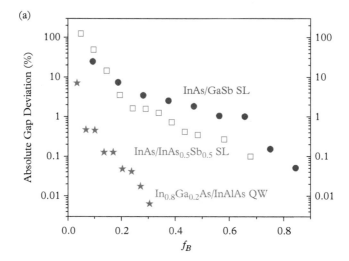

(a)

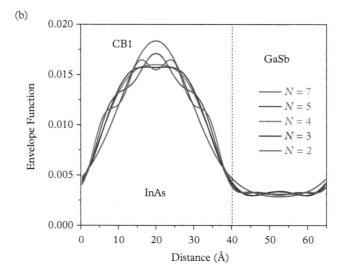

(b)

**Figure 10.10** *(a) Energy gap computed with the reciprocal-space 8-band envelope-function theory for three different structures: (1) the 80-Å $In_{0.8}Ga_{0.2}As$/100-Å $In_{0.52}Al_{0.48}As$ quantum well (modeled as a superlattice) of Fig. 10.5 at room temperature; (2) the 40-Å InAs/25-Å GaSb superlattice of Fig.10.8 at $T = 77$ K; and (3) a 103-Å InAs/23-Å $InAs_{0.5}Sb_{0.5}$ superlattice at $T = 77$ K as a function of wavevector cut-off expressed in units of $f_B$. (b) Envelope function of the lowest conduction subband CB1 at the zone center for several different wavevector cut-offs for the 40-Å InAs/25-Å GaSb superlattice of Fig.10.8 at $T = 77$ K. For comparison, $N = 2$ corresponds to $f_B \approx 0.19$, while $N = 7$ corresponds to $f_B \approx 0.66$.*

be checked in each case of interest. We will address the question of whether quasi-atomistic approaches can offer a way out of this dilemma in Section 10.5.

So far we have considered only structures with abrupt transitions between layers. This is the standard model for structures grown epitaxially, but in practice some diffusion and intermixing (interface grading) is observed. What are the consequences of this intermixing for the band structure, *assuming that the average composition of the layers remains unchanged*? It turns out that such intermixing usually shifts the conduction-band states to higher energies and the valence-band states to lower energies (in other words, the energy gap becomes wider). A simple way to understand this behavior is to notice that the states of the well are nearly always lower in the energy than the average of the bulk (conduction) band edges in a superlattice. The amount of the shift depends on the structure at hand and the extent of the intermixing, but the shift is relatively small for layers grown by modern epitaxial techniques. Even in narrow-gap superlattices, it rarely exceeds 10% of the original energy gap. Because the shift is limited, the abrupt-interface model can do the trick in many practical situations.

## 10.3  Quantum Structures in Electric Fields

What happens when a static electric field with amplitude $F \equiv E_z$ is applied to an isolated quantum well? Assuming that we started out with the lowest solution in the form

$$H_0\varphi_{1,0} = E_{1,0}\varphi_{1,0} \tag{10.29}$$

now we have to solve the problem

$$(H_0 + eFz)\,\varphi_{1,e} = E_{1,e}\varphi_{1,e} \tag{10.30}$$

As shown in Fig. 10.11, the electric field breaks the symmetry of the original well, so that the envelope function of the bottom subband no longer peaks at the center. If the field is not too strong, we can imagine that some of the second subband wavefunction is mixed in, with the shifted energy given by a modified version of Eq. (10.24a):

$$E_{1,e} \approx E_{1,0} - \frac{C(eFt)^2}{E_{2,0} - E_{1,0}} \tag{10.31}$$

where $C$ is a constant of order unity. However, in many cases, Eq. (10.31) does not provide accurate estimates of the conduction-band minimum in real wells. For one thing, we would need to introduce the coupling to all the other eigenstates of the original Hamiltonian $H_0$.

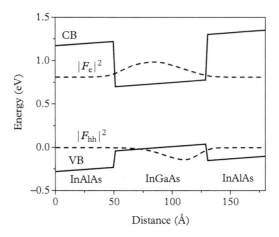

**Figure 10.11** *Band profiles and wavefunctions for the CB1 and HH1 subbands of an 80-Å* $In_{0.53}Ga_{0.47}As/In_{0.52}Al_{0.48}As$ *quantum well in an electric field of 100 kV/cm.*

However, Eq. (10.31) works reasonably well if we need to estimate the relative shift between two CB subbands of a quantum well under an applied electric field. Since the denominator of Eq. (10.31) changes sign when we determine $E_{2,e}$:

$$E_{2,e} - E_{1,e} \approx \frac{2C(eFt)^2}{E_{2,0} - E_{1,0}} \tag{10.32}$$

Fig. 10.11 shows that the electron and hole wavefunctions are shifted to the opposite sides of the well. This is because we can imagine the hole states to be mirror images of the electron states. Each finds a potential minimum that is lower when an electric field is applied, which has the effect of reducing the energy gap of the quantum well. The field also decreases the electron–hole wavefunction overlap, but this effect is minor for narrow type I wells. The shift of the electron and hole states in a quantum well in an electric field is called the *quantum-confined Stark effect*, after Johannes Stark who studied field phenomena in atoms and molecules in the early twentieth century.

The type II structure is similar in that the field shifts electrons and holes to the opposite sides of *their* respective wells. If the potential seen by the electrons and holes reaches its maximum at the edge of the opposite-carrier well, the energy gap drops, as for a type I well. But if the field direction is reversed, the gap increases instead. This situation is illustrated in Fig. 10.12 for a type II quantum well with $E_g \approx 150$ meV. For this structure, the gap shifts with electric field at a rate of 0.32 meV/(kV/cm), The rate increases when the wavefunction peaks become more widely separated (for wider electron and/or hole wells). Because the quantum confinement here is much stronger here than in a type I well, the asymmetry of the wavefunctions is less noticeable.

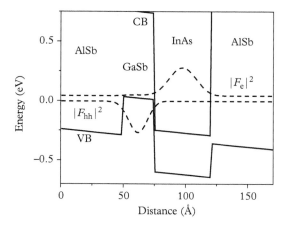

**Figure 10.12** *Band profiles and wavefunctions for the CB1 and HH1 subbands of a 45-Å InAs/25-Å GaSb type II quantum well with thick AlSb barriers in an electric field of 100 kV/cm.*

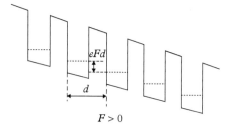

**Figure 10.13** *Schematic illustration of a Wannier–Stark ladder that results when an electric field is applied to a superlattice. Instead of forming a miniband, the subbands are discrete and equidistant in energy.*

What effect does the field have on a superlattice in which the subband minima in different wells are coupled to form a miniband? For narrow wells, the shift of the states in each well given by Eq. (10.31) is not large. If the field is weak, modified versions of the minibands are still present. However, if $eFd > U$, the subbands in the neighboring wells separated by the superlattice period $d$ are displaced with respect to each other by an energy close to $eFd$, as shown in Fig. 10.13. The subband minima now form a *Wannier–Stark ladder* with steps of magnitude $eFd$. The crossover between the two regimes may strongly affect the electron transport through the superlattice. For small fields and wide minibands, the main transport features are broadly the same as in a bulk band. As the field increases, the transport proceeds mostly via sequential scattering or scattering-assisted tunneling from one well to the next.

The electric field can also play the opposite role, by helping to *create* a miniband. Consider the structure of Fig. 10.14, in which the well width varies from left to

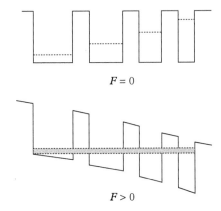

$F = 0$

$F > 0$

**Figure 10.14** *Schematic illustration of a chirped superlattice forming a ladder at zero applied electric field and a miniband at a higher value of the electric field.*

right so that the energy changes from well to well. In the absence of the field, the subband minima form a ladder similar to that in Fig. 10.13. However, when a field of the right magnitude is applied, all the states at $k_\parallel = 0$ can come into approximate alignment. This structure is sometimes called a *chirped superlattice*. We will explore the practical application of this arrangement in Chapters 12 and 13. The transport in a chirped superlattice becomes much faster as the applied field is increased to its target value.

Other asymmetric quantum structures that do not resemble a superlattice include two coupled wells of different thicknesses and a step quantum well with different compositions on the different sides of the well. Just as in the chirped superlattice, the two lowest subbands can be associated with different wells or with different layers of a step quantum well. In such cases, the relative shift between the two subbands is again linear in the electric field and given by $eF(z_2 - z_1)$, where $z_1$ and $z_2$ are the centers-of-mass of the envelope functions for the two subbands. The sign of the energy shift depends on the direction of the electric field.

As a practical matter, the calculations for quantum structures in an electric field can be treated using both quantum-well and virtual-superlattice (see Fig. 8.10b) geometries. In the latter case, the virtual barrier must be high enough to prevent wavefunction penetration from period to period for the strongest electric field of interest. Otherwise, the structure has no real physical meaning!

## 10.4 Coupled Conduction and Valence Bands in Quantum Wells

The conduction-band in-plane dispersion in a quantum well is similar to a bulk material. The CB subbands are parabolic near the zone center if the bulk gap is not

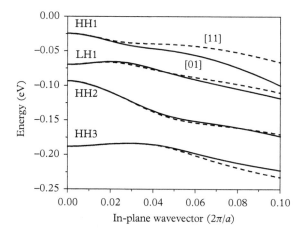

**Figure 10.15** *Valence-band dispersions along different in-plane directions for a 50-Å* $In_{0.53}Ga_{0.47}As/100$-$Å$ $Al_{0.48}In_{0.52}As$ *quantum well.*

too small, and gradually shift to linear as the in-plane wavevector increases. The different in-plane directions, e.g., [01] and [11] have nearly the same dispersion.

Figures 8.3 and 10.2 show that these statements do not apply in general to the valence subbands. The various HH and LH subbands are alternately repelled and attracted as $k_\parallel$ varies. This behavior makes more sense if we recall that the HH and LH bulk masses are very different, and that away from the zone center, the two bands become mixed by the terms in the Hamiltonian that are proportional to $k_x$ and $k_y$. Because they depend on both in-plane components rather than just on the magnitude of $k_\parallel$, the valence dispersions vary noticeably with in-plane direction, as illustrated in Fig. 10.15. We can think of this as another consequence of the direction-dependent HH mass in bulk III–V materials in Eq. (3.36).

Given that the valence band is much more complicated than the conduction band, what happens as the energy gap of a quantum well approaches zero? Figure 10.16 depicts the band structures of the InAs/GaInSb/InAs/AlSb type II quantum wells with energy gaps ranging from near zero to the mid-infrared (0.3 eV). Because the conduction-band edge has twin minima corresponding to the InAs layers, one of the authors of this book (J.M.) gave it the distinctive name "*W*" *structure* in the 1990s. Owing to the compressive strain in the GaInSb layer that acts as the hole well, the HH1 subband is well separated from the LH subbands. The results are for a symmetric structure with two identical InAs wells on both sides of the GaInSb layer. Because there are two InAs wells, the structure has two closely spaced CB subbands: symmetric (CB1) and anti-symmetric (CB2), by analogy with Fig. 10.6.

We can capture the essential features of the CB–HH interaction by a simple model that includes only the electron $i|C\rangle$ and heavy-hole $|p_x + ip_y\rangle$ states:

$$\begin{bmatrix} \frac{E_g}{2} + \frac{\hbar^2 k^2}{2m_e} & \hbar v_F \left( k_x - i k_y \right) \\ \hbar v_F \left( k_x + i k_y \right) & -\frac{E_g}{2} - \frac{\hbar^2 k^2}{2m_h} \end{bmatrix} \begin{bmatrix} F_c \\ F_v \end{bmatrix} = E \begin{bmatrix} F_c \\ F_v \end{bmatrix} \tag{10.33}$$

Here $E_g$ is the energy gap, $m_e$ and $m_h$ are the effective-mass parameters due to bands not included in the 2-band interaction, $v_F$ is the Fermi velocity that describes the slope of the linear dispersions at zero gap, and $i\,|\,C\rangle$ is the conduction-band wavefunction $F_c$ that is mostly $s$-like, but partially $p_z$-like (see Eq. (10.7)). As mentioned earlier, the valence-band wavefunction $F_v$ is made up of the heavy-hole $|p_x + i p_y\rangle$ linear combination of the $p$-like orbitals. The off-diagonal elements are analogous to the $\mathbf{k \cdot p}$ terms in the conventional theory. But, importantly, Eq. (10.33) describes the final band structure of the quantum well. It is *not* a Hamiltonian to be solved by derivative substitution and numerical methods.

When $m_e \to \infty$ and $m_h \to \infty$, the eigenenergies of Eq. (10.33) assume a simple form:

$$E_{\pm} (\mathbf{k}) = \pm \sqrt{\frac{E_g^2}{4} + \hbar^2 v_F^2 \left( k_x^2 + k_y^2 \right)} \tag{10.34}$$

where the plus and minus signs correspond to the conduction and valence bands, respectively. As in a zero-gap bulk semiconductor, when $E_g = 0$ the dispersions are linear in $k$ (there is hardly any parabolic region) and touch at a single energy ($E = 0$) at the zone center ($k_{\parallel} = 0$). In fact, they are the same as the graphene dispersions given by Eq. (5.26). To complete the analogy, we can write the wavefunctions

$$\begin{bmatrix} F_c \\ F_v \end{bmatrix}_+ = \frac{1}{\sqrt{2}} \begin{bmatrix} e^{i\theta_k/2} \\ -e^{-i\theta_k/2} \end{bmatrix} e^{i\mathbf{k \cdot r}} \tag{10.35a}$$

$$\begin{bmatrix} F_c \\ F_v \end{bmatrix}_- = \frac{1}{\sqrt{2}} \begin{bmatrix} e^{i\theta_k/2} \\ e^{-i\theta_k/2} \end{bmatrix} e^{i\mathbf{k \cdot r}} \tag{10.35b}$$

where $\theta_k = \tan^{-1} \left( k_y/k_x \right)$ is the in-plane angle. Even though the band structure of a quantum well with zero gap is the same as in graphene, it has many (rather than just one) atoms along the growth direction, and the region of validity of Eq. (10.35) is not as wide.

Can the gap in a type II quantum well be *negative*? If that simply means that CB1 starts out *below* HH1 at the zone center, then the answer is yes. Materials with such band arrangements are known as *semimetallic*. Of course, the actual gap is never negative. In Fig. 10.17, we can see that a very narrow (positive!) gap opens up at $k_{\parallel} > 0$ due to the interaction between the electron and hole subbands. In fact, within the 8-band $\mathbf{k \cdot p}$ model they may interact only away from the zone center because that is where the HH subbands acquire some LH character, and the CB–LH coupling follows from the Kane Hamiltonian of Eq. (2.51). Because

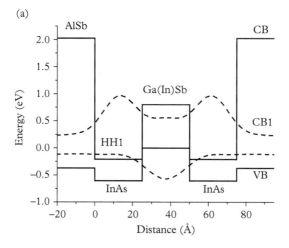

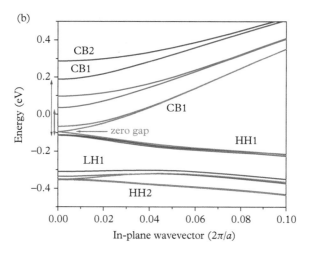

**Figure 10.16** *(a) Band profile and the dominant components of the envelope functions for the CB1 and HH1 subbands of a InAs/GaInSb/InAs/AlSb "W" structure. (b) Band structure of the "W" structures with gaps of 300 meV (25 Å InAs/25 Å GaSb/25 Å InAs, blue lines), 150 meV (41 Å InAs/25 Å GaSb/41 Å InAs, red lines), and near zero (85 Å InAs/25 Å Ga$_{0.87}$In$_{0.13}$Sb/85 Å InAs, green lines) at T = 77 K. Simulations of narrow-gap and zero-gap type II quantum wells present convergence challenges similar to those discussed in connection with Fig. 10.10.*

the InAs/GaInSb/AlSb quantum well in Fig. 10.17 is asymmetric (in contrast to the symmetric structures of Fig. 10.16 with two InAs wells), the dispersions show spin splitting for $k_\parallel > 0$, as discussed in Section 8.4. When the in-plane wavevector is small, and HH1 is higher than CB1, the spin splitting is mostly in CB1, but interestingly, the spin splitting remains in the lower subband once the two switch

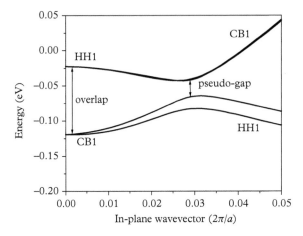

**Figure 10.17**  *Band structure of 85-Å InAs/30-Å Ga$_{0.7}$In$_{0.3}$Sb/100-Å AlSb wells with a "negative" (semimetallic) gap at T = 300 K.*

places. The gap observed for $k_\parallel \approx 0.03 \, \frac{2\pi}{a}$ is clearly smaller than it would have been in the absence of spin splitting.

No III–V materials can form a quantum well with both zero gap *and* a type I band alignment. To see this behavior in practice, we will take a detour to II–VI materials, in particular, quantum wells based on Hg(Cd)Te. What makes that alignment possible considering that quantum confinement pushes electron and hole subbands in the opposite directions? The key is the *inverted* nature of the conduction and valence bands in HgTe and HgTe-rich HgCdTe (as opposed to the *normal* arrangement in III–V semiconductors, i.e. with the *p*-like antibonding states below the *s*-like bonding states, see Chapter 2). The inverted bands resemble the type II quantum well of Fig. 10.16 that starts out with a semimetallic band lineup when there is no quantum confinement. As the HgTe layer becomes thinner, the subbands move toward each other and can balance on the thin edge between the inverted (semimetallic) and normal (semiconducting) regimes. Figure 10.18 shows the energy gap for HgTe/CdTe wells as a function of HgTe thickness at temperatures of 77 and 300 K. At lower temperatures, the semimetallic overlap in HgTe is larger, so the semimetal–semiconductor transition occurs for thinner layers that have additional quantum confinement to compensate the overlap.

How does the computed band structure of narrow-gap type II superlattices differ between atomistic and 8-band envelope-function approaches? There is no simple answer because, as we saw in Chapter 5, most atomistic approaches do not use exactly the same parameters as the envelope-function model. The sole exception is the EBOM, which is not truly atomistic. However, it does provide more realistic dispersions away from the zone center. The EBOM also avoids the convergence issues that plague the envelope-function modeling of superlattices with small periods, as described in Section 10.2.

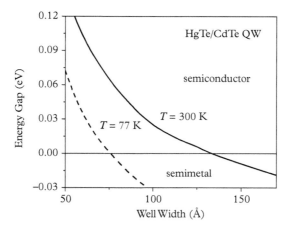

**Figure 10.18** *Energy gap for a HgTe well with CdTe barriers as a function of well width at temperatures of 77 and 300 K.*

Figure 10.19 compares the CB1, HH1, and LH1 energies at $k_\parallel = 0$ for the InAs/GaSb superlattice in which the two layers have the same thickness, as computed using the 8-band envelope-function theory solved by the reciprocal-space method and EBOM. We used our best guess for the optimal number of reciprocal lattice vectors, as discussed in Section 10.2, which has a slight effect on the energies computed for superlattices with the smallest periods in Fig. 10.19. We did not include the interface terms into either approach even though nothing prevents us from generalizing the EBOM by analogy with Section 8.3. As a reminder, the EBOM is based on fcc symmetry and ignores these terms just as the 8-band envelope-function theory does.

As expected, the results converge as the layer thickness increases, with the largest differences for the shortest superlattices. The EBOM predicts a somewhat lower confinement energy for electrons, but a slightly *larger* confinement energy for heavy holes in that limit. Therefore, the deviation of the bulk dispersions for the two methods in Fig. 5.7 seems to play at best a minor role here. Considering the convergence issues and the dubious validity of the envelope functions for short superlattices, the EBOM may represent the energy gap dependence on period more faithfully in the $d \rightarrow 0$ limit. However, the omission of the interface terms included in Fig. 8.6 probably means that the real gap is somewhere between the envelope-function and EBOM estimates.

We would also caution the readers against making conclusions about the accuracy of some computational method for type II superlattices by comparing the results to the experimental data for a single structure or even a limited number of them. This is because neither the input parameters nor the details of the grown structure are known precisely. A better way to judge whether a particular approach works well is to compare the *trends* from varying one or more of the parameters

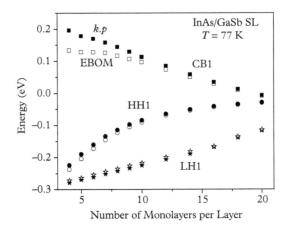

**Figure 10.19** *CB1, HH1, and LH1 energies at the zone center for two-layer InAs/GaSb superlattices, in which the two layers have the same thickness, as a function of layer thickness. The results computed using the 8-band envelope-function theory without interface terms (solid points) and EBOM (open points) at 77 K are compared.*

over a large enough range. Another matter to keep in mind is the great flexibility and near-universality of the envelope-function theory that often outweighs a slight reduction in accuracy. In that spirit, we would conclude that the results of Fig. 10.19 do not demonstrate that the EBOM performs much better than envelope-function theory for type II superlattices.

## 10.5   Wurtzite Quantum Wells

There are many possible nitride quantum wells with the wurtzite crystal structure. A practical wrinkle is that very few of the nitride materials can be truly lattice matched (the rare exceptions are GaN and $Al_{0.83}In_{0.27}N$, as well as some quaternaries) or even strain compensated. This is because other group V atoms do not mix well with nitrogen, as mentioned in Chapter 7. In fact, much of the initial development of nitrides was based on "foreign" substrates with a hexagonal crystal structure such as sapphire ($\alpha$-$Al_2O_3$) and SiC. Only later did relatively small GaN substrates become available. These practical difficulties might have spelled doom for the nitride technology had non-nitride III–V semiconductors offered a viable alternative for materials with wide direct energy gaps. Fortunately for the nitrides, their main competition from wide-gap II–VI compounds suffered from even greater practical constraints. Most of the work on optically active structures took advantage of compositions close to GaN, i.e., InGaN and AlGaN with large Ga fractions.

We already know from Parts I and II that wurtzite materials differ from zinc-blende crystals in that they have a preferred axis (the $c$ axis). The electronic and ionic charges in the crystal are not fully compensated along that axis because the geometric centers of the negative and positive charge distributions associated with the same unit cell do not coincide. This means that if we choose the $c$ axis (or another axis that is *not* perpendicular to it) as the growth direction, the layers exhibit strong built-in fields due to *spontaneous polarization* $P_{sp}$. The sign of the spontaneous polarization depends on the growth orientation, but that orientation usually remains fixed for a particular epitaxial technique. Apart from these fields, the well and barrier in nitride wells usually have different lattice constants, and the resulting strain leads to a *piezoelectric effect* that also contributes to polarization. The sign of both polarization components flips when the growth orientation is reversed.

Why should we be concerned about the polarization? We will see later in this section that it results in very strong ($\sim 1$ MV/cm) built-in fields. At this level, the energy shifts due to the fields are comparable to band offsets and quantum confinement energies. Because the field imposes a linear slope on the band profiles, the field-induced energy shift scales with well width. Therefore, the strong fields and heavy masses lead naturally to narrow (20–40 Å) quantum wells, in which subbands are more widely separated, and the shifts do not exceed band offsets.

To estimate the internal field for such a quantum well, we calculate the strain components and the resulting piezoelectric polarization and add the spontaneous part. The lattice mismatch between GaN and AlN is much smaller than between GaN and InN, which means that the fields in GaN/AlGaN quantum wells are dominated by spontaneous polarization, while the piezoelectric contribution is larger in InGaN/GaN wells. These are by far the most common well/barrier pairs used in practical applications, even though InGaN/AlGaN wells (with a small Al fraction in the barrier) are also quite important.

The piezoelectric effect may be cast in terms of either the piezoelectric constants $e_{ij}$ or the piezoelectric moduli $d_{ij}$ tabulated in Chapter 6. They are related via

$$e_{ij} = \sum_k d_{ik} c_{kj} \tag{10.36}$$

Here we mostly use $d_{ij}$, which has three distinct components in a wurtzite crystal. Assuming biaxial strain with $\epsilon_{xx}$ and $\epsilon_{zz}$ given by Eqs. (3.58) and (3.81) with

$$\epsilon_{zz} = -\frac{2c_{13}}{c_{33}} \epsilon_{xx} \tag{10.37}$$

the piezoelectric polarization is directed along the $c$ axis and given by the dot product of the piezoelectric constants and the strain:

$$P_{pz} = e_{33}\epsilon_{zz} + e_{31}\left(\epsilon_{xx} + \epsilon_{yy}\right) = e_{33}\epsilon_{zz} + 2e_{31}\epsilon_{xx} \tag{10.38}$$

Adding the piezoelectric and spontaneous polarizations, we obtain

$$P_{tot} = P_{pz} + P_{sp} = 2d_{31} \left[ \epsilon_{xx} (c_{11} + c_{12}) + \epsilon_{zz} c_{13} \right] + P_{sp}$$

$$= 2d_{31} \epsilon_{xx} \left( c_{11} + c_{12} - 2\frac{c_{13}^2}{c_{33}} \right) + P_{sp} \tag{10.39}$$

How do we include the polarization into the band-structure model? The only way to do so is to calculate its effect on the "macroscopic" electric field $E_{mc}$ in each layer. Once that is done, we can simply insert the field into the diagonal potential-energy elements, by analogy with Section 10.3. This electric field must terminate on electric charges, which are induced at heterointerfaces. This means that only *differences* between the polarizations in neighboring layers have physical consequences. Also, because $P \sim \varepsilon_s E_{mc}$, the electric field should be inversely proportional to the static permittivity $\varepsilon_s$, which in general varies (by a small amount) from layer to layer and is expressed in units of the vacuum permittivity $\varepsilon_0$. We start with the case of an isolated quantum well surrounded by very thick left and right barriers. If the two barriers are different, we need two distinct polarizations $P_L$ and $P_R$ to represent their effect.

The field in the well is given by

$$E_W = \left[ \frac{1}{2} (P_L + P_R) - P_W \right] / \varepsilon_W \tag{10.40}$$

where $\varepsilon_W$ is the *static* permittivity of the well. If the barriers are much thicker than the well, the polarization in the well $P_W$ does not affect the field in the barrier. Only the difference between the polarizations in the barriers is important:

$$E_{L,R} \approx \frac{1}{2} (P_{R,L} - P_{L,R}) / \varepsilon_{L,R} \tag{10.41}$$

We often want to find the fields for a two-constituent superlattice with layers $A$ and $B$, in which the layer thicknesses are comparable. The electric fields in the two layers are related by exchanging the subscripts:

$$E_A = \frac{t_B (P_B - P_A)}{t_A \varepsilon_B + t_B \varepsilon_A} \tag{10.42a}$$

$$E_B = \frac{t_A (P_A - P_B)}{t_A \varepsilon_B + t_B \varepsilon_A} \tag{10.42b}$$

If layer $B$ is much thinner than layer $A$ ($t_B \ll t_A$), we have $E_A \propto \frac{t_B}{t_A}$ and $E_B \approx \frac{(P_B - P_A)}{\varepsilon_B} \gg E_A$, which is consistent with Eq. (10.40). The ideas expressed in Eq. (10.42) also apply to superlattices with more than two layers as well as multilayer wells. For an arbitrary number of layers with arbitrary thicknesses, the polarizations should be weighted by each thickness and by the reciprocal of the static permittivity (with the polarization in that layer subtracted to find the difference):

$$E_i = \frac{\sum_j \frac{t_j P_j}{\varepsilon_j} - P_i \sum_j \frac{t_j}{\varepsilon_j}}{\varepsilon_i \sum_j \frac{t_j}{\varepsilon_j}} \qquad (10.43)$$

The polarization disappears when a nitride well is grown along a non-polar direction orthogonal to the $c$ axis. Some possible non-polar growth planes (with the growth direction normal to them) are illustrated in Fig. 10.20. For a non-polar growth direction, there is no internal field, just as in a zinc-blende well, and the wavefunction overlap is close to unity when the wells are thick. Nevertheless, the most common and technologically mature option to date is still to grow along the $c$ axis, a case covered in the remainder of this section.

Apart from the piezoelectric effect, biaxial strain affects the band edges via the deformation potentials $D_1, D_2, \ldots D_6$ discussed in Chapter 3. Figure 10.21 illustrates the effect of strain on the bulk GaN valence-band edges (HH, LH, and CH). Compressive strain separates the CH (crystal hole) band more widely from the HH and LH bands, but does not change the HH–LH splitting very much. For that reason, the hole density of states near the top of the valence band does not decrease significantly in compression. This is very different from what happens in zinc-blende wells, as illustrated in Fig. 10.5c! Tensile strain raises the LH band above the HH band, just as in zinc-blende materials, but has little impact on the HH and CH bands. Even though LH is now the top valence band, the density of states is much the same as in an unstrained material. We conclude that for wurtzite crystals grown along the $c$ axis, biaxial strain is a much less effective means of reducing the hole effective mass.

The behavior of the conduction band is much simpler and remains the same in all crystals considered in this book: compressive strain raises the CB (and widens the energy gap), while tensile strain has the opposite effect. The direction of the shift is the same as for hydrostatic strain and corresponds to the crystal lattice

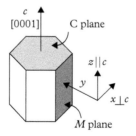

**Figure 10.20** *Illustration of nitride quantum wells grown along polar and nonpolar directions.*

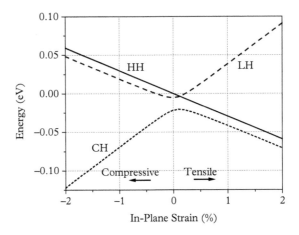

**Figure 10.21** *Shift of the HH, LH, and CH valence-band edges in GaN with in-plane strain component $\epsilon_{xx}$.*

becoming smaller (compressive strain) or larger (tensile strain) without changing its chemical character.

Figure 10.22a shows the band diagram and the envelope functions for the CB1 and HH1 subbands of a 30-Å-wide GaN/Al$_{0.3}$Ga$_{0.7}$N quantum well grown on a substrate or buffer with the GaN lattice constant. The electric field over the well is $\approx 1.6$ MV/cm from Eq. (10.42) and scales with the Al fraction. The field separates the electrons and holes, by analogy with the quantum-well structure in an external field depicted in Fig. 10.11. The result is a kind of a type II structure that depends on the well thickness and well/barrier compositions (as well as on the direction of the crystal growth, as we saw earlier). Another variable is the carrier density in the well. High carrier densities create an internal field that opposes the polarization-induced fields and increase the wavefunction overlap. To estimate this effect, we can take advantage of the "self-consistent" approach to electrostatic fields based on Eqs. (9.43) and (9.44).

Multiple electron and hole subbands are confined even in thin GaN wells because the electron and hole masses are large and the barriers are formed not just by the band offsets, but also by the slope changes of the internal fields at the heterointerfaces. The top valence subband in this case is HH1, followed by LH1 and CH1. The LH and CH subbands interact strongly away from the zone center.

With the addition of compressive strain, the CH1 subbands is shifted well away from the top of the valence band. If the compressive strain is achieved by incorporating some In into the well, the energy gap shrinks and, in wavelength units, moves from the ultraviolet (more precisely, UVA) range to the blue-violet part of the spectrum. In practical InGaN wells, the In fraction is not necessarily uniform in the well plane. Some regions may have more In than other regions, which potentially creates a quantum-dot-like texture. The electrons and holes

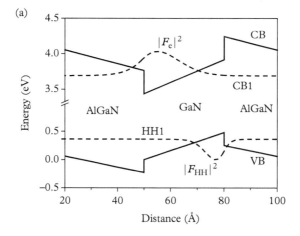

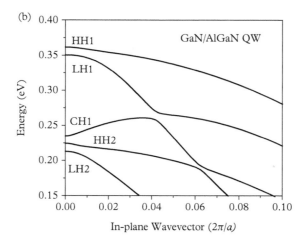

**Figure 10.22** *Band diagram and dominant envelope-function components for the lowest CB and HH subbands of the 30-Å GaN/100-Å Al$_{0.3}$Ga$_{0.7}$N (a). Band structure near the top of the valence band for the same well (b).*

tend to fall into the In-rich regions and participate in the band-to-band optical transitions there. We will discuss the optical characteristics of nitride wells in Chapter 11.

To move to shorter rather than longer wavelengths, an AlGaN well surrounded by AlGaN barriers that contain even more Al is commonly used. As the Al fraction in the well increases, the crystal-field splitting becomes negative so that eventually CH1 rises to the top of the valence band. Structures like that are useful because they can reach the deep-UV wavelengths, as we will see in Chapter 12.

## 10.6   Quantum Wires and Quantum Dots

The growth of thin layers using modern epitaxial growth techniques has become routine. It is much harder in practice to impose quantum confinement in multiple dimensions. Even though confinement in two (quantum wires) or three (quantum dots) directions, illustrated in Fig. 10.23, is less common, the envelope-function theory developed in Chapter 8 should work no matter how many confined dimensions there are. We already know from Chapter 9 that to generalize the theory we should perform multiple derivative substitutions, i.e., $k_x \to -i\frac{\partial}{\partial x}$, $k_y \to -i\frac{\partial}{\partial y}$ etc. Note that to be consistent with much of the literature, we switch the definitions in Chapter 9 and now take the $z$ axis to point along the wire axis.

We also mentioned in Chapter 9 that to represent the discrete spatial variation in two or three dimensions, we need additional quantities such as mesh point values along another axis or another set of reciprocal lattice vectors. If we assume the number of mesh points to be the same for each axis and given by $N$, we must find the eigenstates of an $N^2 \times M$ (wires) or $N^3 \times M$ (dots) matrix, where $M$ is the number of bulk bands. Unfortunately, the effort required to solve an eigenproblem scales superlinearly with the number of equations, so the computation time can vastly exceed what is needed to solve the $N \times M$ matrix of a quantum well or a superlattice. It helps somewhat that only one matrix diagonalization is required for an isolated quantum dot, which has no continuous degrees of freedom and hence no free wavevector components.

Because the run times are relatively long, it helps to know what the answer would look like. To this end, we start with a one-band model for the conduction band of a quantum wire. As before, we assume perfect confinement of the states within the wire (i.e., infinite potential barriers). This allows us to convert spatial derivatives into discrete wavevectors, e.g., $\frac{\nu\pi}{t_x}$, where $\nu$ is an integer. There will be a continuum

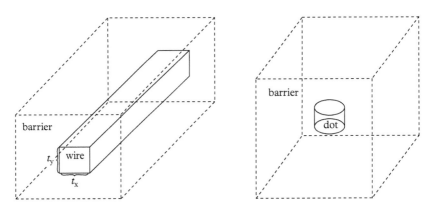

**Figure 10.23** *Schematics of quantum wires (with 2D quantum confinement) and quantum dots (with 3D quantum confinement).*

of electronic states characterized by the dispersion along one axis (here, $\hat{z}$) rather than two for a quantum well. Assuming parabolic dispersion, the subband energies in a wire with rectangular cross section ($t_x \times t_y$) are

$$E_{v,\mu}(k) = \frac{\hbar^2}{2m_e} \left[ \left(\frac{v\pi}{t_x}\right)^2 + \left(\frac{\mu\pi}{t_y}\right)^2 + k^2 \right] \qquad (10.44)$$

Each subband is now characterized by two integers, $v$ and $\mu$, that represent the number of antinodes along the $x$ and $y$ directions, respectively. The ground subband has indices $[v,\mu] = [1,1]$, and the second subband is doubly degenerate in a wire with square cross section ($t_x = t_y$): [1,2] or [2,1]. Of course, when the aspect ratio is very large (e.g., $t_y \gg t_x$), the subbands with different $\mu$ are spaced so closely that they can be well described by a continuous distribution, with $k_y \equiv \pi\mu/t_y$.

The conduction-band structure in a real quantum wire is similar to the idealized form of Eq. (10.44), with subbands that are now mixtures of three different bulk components (here, $s$-like, $p_x$-like, and $p_y$-like). In a wire with wide or moderate energy gap, the $s$-like component dominates. When the gap is narrow, nonparabolicity shifts the position of the subbands and changes the form of the continuous dispersion. For example, with the more realistic dispersion of Eq. (2.9), neglecting the free-electron term and setting $E_{v0} = 0$ for the material inside the quantum wire, we obtain for perfect confinement in the conduction band:

$$E_{v,\mu}(k) = \frac{E_g}{2} + \sqrt{\frac{E_g^2}{4} + \frac{\hbar^2}{2m}E_p \left[ \left(\frac{v\pi}{t_x}\right)^2 + \left(\frac{\mu\pi}{t_y}\right)^2 + k^2 \right]} \qquad (10.45)$$

We must work a little harder to describe the valence-band states in a quantum wire. The heavy holes and light holes can now mix even at the zone center ($k = 0$). The simplest geometry we can work with to understand the valence band is a quantum wire with cylindrical cross section. The valence subbands in this case are eigenstates of the total angular momentum $F = J + N$, where $J = L + S$ as in Chapter 2, and $N$ is the angular momentum along the wire axis (the $z$ axis). As before, we identify the states by their components along the wire axis, i.e., $m_f$, $m_n$, $m_j$ etc.

Here is an outline of the full calculation. If we reduce the basis to just the HH and LH bulk states, i.e., $j = \frac{3}{2}$, the lowest-order states in the quantum wire correspond to the wire-axis orbital-angular-momentum quantum numbers $n = 0$ and $n = 1$. In terms of the eigenstates of total angular momentum $F$, we have states with $m_f = \pm 1/2$ and $m_f = \pm 3/2$. Because the quantum confinement is two-dimensional, each value of $m_f$ corresponds to a separate series of subbands.

The topmost valence subband now has $m_f = \pm 1/2$, rather than $m_f = \pm 3/2$ that one might naively expect from the (incorrect) analogy with quantum wells.

Keeping in mind that there is no confinement along $\hat{z}$, the wavefunction at $k = 0$ is reminiscent of the light-hole states in Eqs. (2.38c) and (2.38d):

$$\psi_{1/2}^+ = \left( |f = \frac{3}{2}, m_F = \frac{1}{2} \right) = -\frac{1}{\sqrt{6}} (p_x + ip_y) \downarrow + \sqrt{\frac{2}{3}} \, p_z \uparrow \qquad (10.46a)$$

$$\psi_{1/2}^- = \left( |f = \frac{3}{2}, m_F = -\frac{1}{2} \right) = \frac{1}{\sqrt{6}} (p_x - ip_y) \uparrow + \sqrt{\frac{2}{3}} \, p_z \downarrow \qquad (10.46b)$$

The wavefunctions for the next series, with $m_f = \pm 3/2$, are just like the heavy-hole states in Eqs. (2.38a) and (2.38b):

$$\psi_{3/2}^+ = \left( |f = \frac{3}{2}, m_F = \frac{3}{2} \right) = -\frac{1}{\sqrt{2}} (p_x + ip_y) \uparrow \qquad (10.47a)$$

$$\psi_{3/2}^- = \left( |f = \frac{3}{2}, m_F = -\frac{3}{2} \right) = \frac{1}{\sqrt{2}} (p_x - ip_y) \downarrow \qquad (10.47b)$$

Here no $p_z$-like component is present.

We will use these results to evaluate the optical matrix elements in Chapter 11. In the meantime, Fig. 10.24 shows the conduction and valence subband dispersions calculated using the reciprocal-space method of Section 9.3 for a GaAs/AlGaAs quantum wire with square cross section ($t_x = t_y = 5$ nm). Curiously, some of the valence subbands (in particular, the $m_f = \pm 3/2$ subband that is second from the top) have electron-like dispersion that bends up away from the zone center for small wavevectors. This behavior affects the valence-band density of states, but has little impact on the optical properties of the wires, as we will see in Chapter 11.

In contrast to quantum wires and wells, quantum dots have electronic states that are truly discrete at energies below the band discontinuities. Assuming perfect confinement in a rectangular prism (the quantum box illustrated in Fig. 10.23), we can write by analogy with Eq. (10.44):

$$E_{\nu,\mu,\lambda} = \frac{\hbar^2}{2m_e} \left[ \left( \frac{\nu\pi}{t_x} \right)^2 + \left( \frac{\mu\pi}{t_y} \right)^2 + \left( \frac{\lambda\pi}{t_z} \right)^2 \right] \qquad (10.48)$$

The shapes of quantum dots can range from spheres to flat disks. One popular way to produce III–V quantum dots is by self-assembly that takes place during molecular beam epitaxy when strain exceeds certain bounds. The confinement in the self-assembled dots is much stronger along one axis (the growth direction $\hat{z}$) than in the plane, so we can think of the dot as a quantum well with some additional in-plane confinement. The topmost valence state is still mostly HH-like, but mixed with a $p_z$ component that scales with the ratio of in-plane to growth-direction confinement. Because higher-order dot states have larger

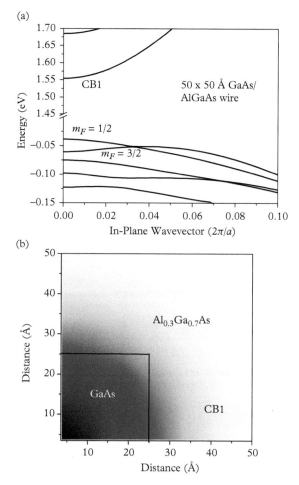

**Figure 10.24** *(a) Conduction and valence subband dispersions in a square-cross-section 50 Å ×50 Å GaAs quantum wire with Al$_{0.3}$Ga$_{0.7}$As barriers (modeled as a 2D superlattice with 100-Å-wide barriers along both axes). (b) One quadrant of the wire with the envelope function of the CB1 subband (blue corresponds to higher values).*

admixtures of components not present in a quantum well at $k_\| = 0$, they are analogous to the quantum-well states with larger $k_\|$. When the lateral dimensions become large enough, we can make a smooth transition to the quantum-well-like eigenenergies:

$$E_{v,\mu,\lambda} \approx E_{QW,\lambda} \left( \sqrt{\left(\frac{v\pi}{t_x}\right)^2 + \left(\frac{\mu\pi}{t_y}\right)^2} \right) \qquad (10.49)$$

where $E_{QW,\lambda}(k_\parallel)$ is the dispersion of the quantum-well subband with quantum number $\lambda$ (CB$\lambda$, HH$\lambda$, LH$\lambda$, SO$\lambda$).

Real dots and wires are often strained with respect to the underlying substrate. The modified elastic strain profile is not without consequences for the quantum-confined states. We can find the full profile, e.g., by minimizing the elastic energy as discussed in Section 3.4.

At low temperatures, a few drastic simplifications can streamline our thinking about quantum-dot states. By analogy with atomic states that are also confined in all three dimensions, we call the lowest electron and holes states the $s$ shell, the next lowest the $p$ shell, etc. Unless the quantum dot is deep, only two or three lowest shells lie below the potential barrier. Just beyond it are the 1D-confined states of the quantum well into which the dot is often embedded (or the unconfined states of the surrounding bulk-like matrix). Of course, each quantum dot state illustrated in Fig. 10.25a is not an atomic shell. Instead, the $s$ shell is the ground state with zero nodes in the dot, the $p$ shell is the first excited state with a single node near the center of the dot, etc. We will see how to use this convenient nomenclature in Chapter 11.

If a single quantum dot bears some resemblance to an atom (albeit with a much stronger confinement along one axis), the two side-by-side dots shown in Fig. 10.25b certainly look like a diatomic molecule. For the dots to interact, the barrier between them must be very thin (no more than a few tens of Å). This arrangement is difficult to carry off for dots in the growth plane, but modern epitaxial techniques allow routinely growing such coupled dots along the vertical axis. When the quantum dot molecule is oriented vertically, we recognize it as an analog of a coupled quantum-well system that has extra confinement in the growth plane. If the dots are of the same dimensions, the electronic $s$ shells hybridize, as in the quantum-well case. The interaction between the dots is generally small enough to be treated as a perturbation.

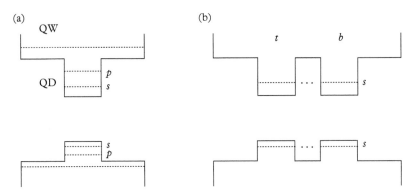

**Figure 10.25** *Illustration of the most relevant states in shallow quantum dots such as self-assembled InAs/InGaAs: (a) an isolated quantum dot; (b) two vertically coupled dots. The confinement extends in all three dimensions for the quantum dot inside the quantum well.*

## Suggested Reading

- Multiband approaches to the band structure of quantum wells, wires, and dots are routinely used for the valence band, but less common for the coupled conduction–valence problem. In the process, it is easy to forget about admixing the $p_z$-like component to the electron wavefunctions and the real origin of nonparabolicity in the conduction band. On the other hand, many works cover the superlattice band structure using the Kronig–Penney model, e.g., S. L. Chuang, *Physics of Photonic Devices*, 2nd edn. (Wiley, New York, 2009). This model works reasonably well for determining the zone-center HH energies, but is only qualitatively useful for other situations and is not covered here.

- The CB–HH interaction in zero-gap wells and wires is covered in J. B. Khurgin and I. Vurgaftman, "Electronic states, pseudo-spin, and transport in the zinc-blende quantum wells and wires with vanishing band gap," *Applied Physics Letters* **104**, 132,107 (2014).

- Analytical multiband models for cylindrical quantum wires and spherical quantum dots are introduced in P. C. Sercel and K. J. Vahala, "Analytical formalism for determining quantum-wire and quantum-dot band structure in the multiband envelope-function approximation," *Physical Review B* **42**, 3690 (1990).

# 11

# Absorption and Emission of Light in Quantum Structures

In this chapter, we learn how to calculate the absorption coefficient, optical gain, and radiative recombination rates in quantum wells and superlattices. We present a detailed treatment of both interband and intersubband transitions and consider in detail their differences and similarities. The optical properties of wurtzite quantum wells and zinc-blende quantum wires and dots are also discussed. Finally, we consider the interaction of excitonic transitions with incident light in quantum wells as a model for other two-dimensional materials.

## 11.1 Interband Transitions in Quantum Wells

We saw in Chapter 4 that the interband transitions in a bulk zinc-blende semiconductor crystal arise from matrix elements of the type $\frac{\hbar}{m}\langle c|p_{cv}|v\rangle = \frac{\hbar}{m}\langle s|p_{cv}|p_x,p_y,p_z\rangle = P$. The light's polarization selects the orientation of the orbital, i.e., whether it is $p_x, p_y, p_z$, but the magnitude of the matrix element is independent of polarization. In bulk crystals, transitions are possible only between initial (valence-band) and final (conduction-band) states with (nearly) identical $\boldsymbol{k}$ vectors because the incoming photon does not carry much momentum, at least by comparison with the massive electrons. Therefore, even though the conduction band acquires some $p$-orbital character at larger $k$, first-order transitions between two conduction-band states in bulk could never conserve both energy and momentum. With an assist from phonons or crystal defects, intraband optical transitions become possible, but generally remain much weaker than the "first-order" transitions we discuss in the rest of this chapter.

With translation symmetry broken along at least one direction, we need to rethink some of these conclusions. In a superlattice with periodicity along the confinement direction, $k_z$ is a conserved quantity, but now it describes variation on a scale larger than the superlattice period. And if periodicity in the growth direction is lacking (i.e., in a quantum well), $k_z$ is no longer meaningful (you may think of the period as the entire Universe!). Instead, optical transitions must satisfy

*Bands and Photons in III–V Semiconductor Quantum Structures.* Vurgaftman, Lumb, and Meyer,
Oxford University Press (2021). © Vurgaftman, Lumb, and Meyer.
DOI: 10.1093/oso/9780198767275.003.0011

the requirement for a non-zero wavefunction overlap, which is also the concealed mechanism behind wavevector conservation in bulk crystals.

How can we modify Eq. (4.57) to account for the differences between interband transitions in bulk semiconductors and quantum structures? First, we separate the integrations over in-plane and growth-direction wavevectors:

$$
\alpha\,(\hbar\omega) = \frac{\alpha_f}{2\pi\,n_r} \sum_{\mu,\nu} \int d^2 k_\| \int dk_z \, |\langle c, \mu | \nabla_{\boldsymbol{k}} H_e | \nu, v \rangle|^2
$$
$$
\frac{\hbar\omega}{\left(E_{ck,\mu} - E_{vk,v}\right)^2} \, \sigma \left(E_{ck,\mu} - E_{vk,v} - \hbar\omega\right) \tag{11.1}
$$

in which the sums over all conduction and valence subbands (or minibands) $\mu$ and $\nu$ are written explicitly. For a quantum well, we eliminate the integration over $k_z$ along with the factor of $\frac{1}{2\pi}$ that goes with it. This case is not much different from a 2D crystal, which we suppose makes it "quasi-2D". Instead of expressing the strength in terms of the absorption coefficient per unit length, we do so in terms of *the absorption per pass* (through the QW in the vertical direction or along the $z$ axis):

$$
\alpha_{QW}\,(\hbar\omega) = \frac{\alpha_f}{n_r} \sum_{\mu,\nu} \int d^2 k_\| \, |\langle c, \mu | \nabla_{\boldsymbol{k}} H_e | \nu, v \rangle|^2
$$
$$
\frac{\hbar\omega}{\left(E_{ck,\mu} - E_{vk,v}\right)^2} \, \sigma \left(E_{ck,\mu} - E_{vk,v} - \hbar\omega\right) \tag{11.2a}
$$

How do we evaluate the matrix element $\langle c, \mu | \nabla_{\boldsymbol{k}} H_e | \nu, v \rangle$ in practice? Each conduction and valence subband has a wavefunction with 8 different components in the standard 8-band Hamiltonian, as well as the full envelope-function variation along the confined direction (here, $z$). Therefore, the matrix element in Eq. (11.2a) becomes

$$
\boldsymbol{e} \cdot \langle c, \mu | \nabla_{\boldsymbol{k}} H_e | \nu, v \rangle = \boldsymbol{e} \cdot \left[ \sum_{i,j} \int dz \big(F_i^\mu\big)^* \nabla_{\boldsymbol{k}} H_{e,ij} \, F_j^\nu \right] \tag{11.2b}
$$

To evaluate the absorption of light polarized along a certain axis, we differentiate the Hamiltonian with respect to the wavevector *along that axis*, characterized by the unit vector $\boldsymbol{e}$. In Eq. (11.2b), we have a double sum over all the wavefunction components of the initial and final states, e.g., $F_j^\nu$, as well as the integration over the confined dimension.

If we start with the reduced Hamiltonians for the conduction and valence bands, we need to evaluate $\nabla_{\boldsymbol{k}} H_{e,ij}$ based on the original, full Hamiltonian. In either case, the matrix element is dominated by the interband term $\frac{\hbar}{m} \langle s | p_{cv} | p_i \rangle = P$. For quantum wells grown along the [001] direction, heavy-hole subbands couple to light polarized in the plane of the quantum well. This is because their wavefunctions include only $p_x$ and $p_y$ orbitals, and light-hole subbands interact most strongly with $z$-polarized light. These results follow from Eq. (4.49) that applies to any coordinate system as long as $\boldsymbol{k}$ is along the direction of quantum confinement. Of course, as we learned in Chapter 10, the HH and LH character is increasingly mixed as we travel away from the zone center, so these rules apply strictly only at $k_\parallel = 0$.

How strong is interband absorption in a quantum well? The simplest scenario is to consider a single electron subband (CB1) separated from the HH1 subband by an energy gap $E_g$ (and by much higher energies from any other subbands). To make things as simple as possible, we also assume parabolic subbands that are fully confined in the well; i.e., the normalized envelope functions along the growth direction are $F_s, F_{HH} = \frac{\sqrt{2}}{\sqrt{t}} \sin\left(\frac{\pi z}{t}\right)$, with the CB1 wavefunction being $s$-like (vanishingly small $p_z$-like component) and the HH1 wavefunction being $\frac{1}{\sqrt{2}}|p_x + ip_y\rangle$. The optical matrix element follows from Eq. (4.19):

$$\langle c, 1|\nabla_{\boldsymbol{k}} H_e|v, h1\rangle = \frac{\hbar}{m}\langle c, 1|\boldsymbol{p}_{cv}|v, h1\rangle \tag{11.3}$$

We are interested in the absorption of an optical plane wave at normal incidence, linearly polarized in the plane of the well. The polarization direction makes no difference for now because we assume that the subbands are isotropic in the plane. The square of the matrix element becomes

$$|\langle c, 1|\nabla_{\boldsymbol{k}} H_e|v, h1\rangle|^2 = \frac{\hbar^2 E_P}{4m} \frac{2}{t} \int_{-t/2}^{t/2} dz \sin^2 \frac{\pi z}{t} = \frac{\hbar^2 E_P}{4m} \tag{11.4}$$

Equation (11.4) can be generalized to include electrons and holes that are not perfectly confined to the same well by including a factor $|O_{eh}|^2 = \left|\int dz\, F_s^{CB*}(z) F_{x,y}^{HH}(z)\right|^2$. This is the square of the envelope-function overlap along the growth direction, which is nearly unity for a type I well, but can be made arbitrarily small for a type II structure by increasing the separation between the electron and hole envelope functions. We also assume that the matrix element does not vary with $k_\parallel$ and ignore the line-shape broadening. Spin conservation in the absorption process leads to two distinct transition series for every pair of subbands in a symmetric well.

We learned in Chapter 6 that $E_P$ has similar values near 25 eV in all non-nitride III–V semiconductors. Therefore, $\langle c, 1|\nabla_{\boldsymbol{k}} H_e|v, h1\rangle \sim 7$ eV Å for a type I well. Although we do not consider it good practice, this value is sometimes multiplied by $\sqrt{2}$ ($\sim 10$ eV Å) to include both degenerate transitions with different spins. If we divide the matrix element by the energy gap to recast it as a dipole (Eq. (4.18)), we obtain values between 5 and 20 Å, depending on the material system and well width.

The absorption per pass for this simple two-subband system follows from the following integration with an additional factor of 2 to account for both spins:

$$
\alpha_{QW}(\hbar\omega) = \frac{\alpha_f}{n_r\hbar\omega}\frac{\hbar^2 E_P}{4m} 2 \int d^2k_{\parallel}\,|O_{eh}|^2\delta\left(E_{ck,1}-E_{vk,1}-\hbar\omega\right)
$$

$$
= \frac{\alpha_f}{n_r\hbar\omega}\frac{\hbar^2 E_P}{2m} 2\pi\,|O_{eh}|^2 \int k_{\parallel}dk_{\parallel}\,\delta\left(E_{ck,1}-E_{vk,1}-\hbar\omega\right)
$$

$$
= \frac{\pi\alpha_f}{n_r\hbar\omega}\frac{\hbar^2 E_P}{2m}|O_{eh}|^2 \int d\left(E_{ck,1}-E_{vk,1}\right)\frac{1}{d\left(E_{ck}-E_{vk}\right)/d\left(k_{\parallel}^2\right)}\delta\left(E_{ck,1}-E_{vk,1}-\hbar\omega\right)
$$

$$(11.5)$$

Assuming parabolic dispersions with the effective masses $m_e$ and $m_h$, the reciprocal of the derivative is

$$
\frac{1}{d\left(E_{ck}-E_{vk}\right)/d\left(k_{\parallel}^2\right)} = \frac{2m_r}{\hbar^2} \tag{11.6}
$$

where the reduced mass is $m_r = \left(m_e^{-1}+m_h^{-1}\right)^{-1}$, just as it was defined in Section 4.2. As we saw in Fig. 10.5c, the electron mass tends to be smaller than the hole mass near the band edge, but the difference is not large for a well with heavy compressive strain. Substituting Eq. (11.6) into Eq. (11.5) yields

$$
\alpha_{QW}(\hbar\omega) = \frac{\pi\alpha_f}{n_r}\frac{E_P}{\hbar\omega}\frac{m_r}{m}|O_{eh}|^2 H\left(\hbar\omega-E_g\right) \tag{11.7}
$$

where $H\left(\hbar\omega-E_g\right)$ is the Heaviside step function needed to model the onset of absorption at $\hbar\omega = E_g$. Therefore, the absorption spectrum between two parabolic subbands in a quantum well is flat near the energy gap. Quite a contrast with the $\left(\hbar\omega-E_g\right)^{1/2}$ dependence in bulk!

How does Eq. (11.7) compare to the absorption of a 2D crystal? To compare apples to apples, we reinsert the refractive index into the denominator of Eq. (4.47). Also, we omit the factor of 2 that arises from the two equivalent $K$ valleys in graphene to make the result comparable to a single $\Gamma$ valley in III–V materials:

$$
\alpha_{2D}(\hbar\omega) = \frac{\pi\alpha_f}{2n_r}H\left(\hbar\omega-E_g\right) \tag{11.8}
$$

For symmetric electron and hole subbands, the reduced mass is approximately $m_r \approx m\frac{E_g}{2E_P}$. For a type I well with $|O_{eh}|^2 \approx 1$, at a photon energy close to the energy gap, we recover the correspondence $\alpha_{QW} \approx \alpha_{2D}$. This means that we expect the interband absorption in a quantum well to be of order $\sim |O_{eh}|^2 \times 0.5\%$ per pass.

Of course, in a real quantum well, we encounter many annoying complications: the $p_z$-like component is introduced into the CB subband by quantum confinement, the HH subband is mixed with LH away from the zone center, etc. Unfortunately, we cannot neglect them outright if we expect to match experimental results. Section 11.3 will develop rigorous methods to treat these phenomena.

How does the absorption in a quantum well depend on photon energy? Of course, there is the explicit inverse dependence in Eq. (11.7), but in practice, it is of little consequence if the gap is not too narrow, and we are only interested in energies near the gap. What matters more is the onset of higher-subband transitions that leads to a staircase-like pattern. The absorption spectra for GaAs- and InP-based InGaAs quantum wells are shown in Fig. 11.1. For wells with wide bulk gaps, the first steps of the staircase tend to be quite flat, as expected from Eq. (11.7). Quasi-3D states contribute beyond the barrier offsets, and the dependence converges to the quasi-bulk form at high energies. We observe another instance of convergence to the square-root bulk form for wide lattice-matched quantum wells, where the calculated absorption coefficient oscillates around the bulk form as we climb the staircase. For definiteness, we convert the units from the absorption per pass to the absorption coefficient per unit length using the nominal period that includes a 100-Å-wide barrier. This does not change the fact that the individual wells are essentially non-interacting in this calculation; i.e., the miniband width tends to zero.

The absorption between two subbands in an isolated quantum well in zero field is strong only when the electron and hole envelope functions have the same symmetry around the center of the well. This is because $|O_{eh}|^2 \approx |\int dz \sin(\pi\mu/t)\sin(\pi\nu/t)|^2 = 0$ unless $\mu + \nu$ is even. Thus the "allowed" transitions are CB1–HH1, CB3–HH1, . . . , CB2–HH2, CB4–HH2, . . . , CB1–HH3, CB3–HH3, . . . , etc. Because of HH–LH mixing, these "selection rules" become relaxed to some extent at higher $k_\parallel$.

To estimate the absorption between conduction and light-hole subbands, two approaches are at our disposal. A quick estimate uses the bulk wavefunctions in Eqs. (4.48c) and (4.48d), which tell us that the absorption is stronger for $z$-polarized light, unlike for HH subbands. Of course, if a plane wave is incident along the $z$ axis, it cannot be $z$ polarized because the light waves are transverse. For light incident at some angle $\theta$, the matrix element includes a factor of $\sin\theta$ to account for the dot product of the polarization direction and the growth axis:

$$|\langle c, 1|\nabla_{\boldsymbol{k}}H_e|v, l1\rangle|^2 = \frac{\hbar^2 E_P}{3m}|O_{el}|^2 \sin^2\theta \qquad (11.9)$$

by analogy with Eq. (11.4). The LH matrix element is a factor of 4/3 larger than that for HH from the squared pre-factors in Eq. (4.48a) and (4.48c). Assuming for simplicity that the electron and light-hole masses are the same, $m_r \approx m\frac{E_g}{2E_P}$, as before. As the light travels through the QW at angle $\theta$, the path length exceeds the well width by a factor of $1/\cos\theta$. Putting all the pieces together, we obtain

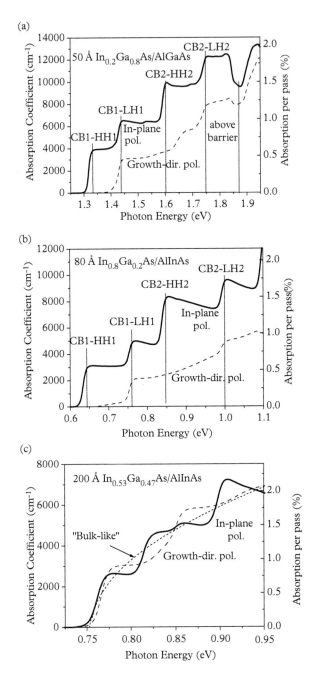

**Figure 11.1** *Absorption spectra for in-plane and growth-direction polarized light in (a) a 50-Å*
*In$_{0.2}$Ga$_{0.8}$As/Al$_{0.3}$Ga$_{0.7}$As quantum well with 1.4% compressive strain, (b) a 80-Å*
*In$_{0.8}$Ga$_{0.2}$As/Al$_{0.48}$In$_{0.52}$As quantum well with 1.8% compressive strain, and (c) a 200-Å*
*In$_{0.53}$Ga$_{0.47}$As/Al$_{0.48}$In$_{0.52}$As lattice-matched quantum well (all modeled as superlattices with*
*100-Å-wide barriers). The various transition energies and the onset of absorption in*
*quasi-continuous states are indicated. The approximate square-root bulk dependence is also*
*drawn for comparison in (c). The units are converted between absorption coefficient and*
*absorption per pass using the nominal superlattice period.*

$$\alpha_{QW,LH}(\hbar\omega) \approx \frac{2\pi\alpha_f}{3n_r}\frac{E_g}{\hbar\omega}|O_{el}|^2\frac{\sin^2\theta}{\cos\theta}H\left(\hbar\omega - E_{c,1} + E_{l,1}\right) \qquad (11.10)$$

Comparing this expression with Eq. (11.7), we see that the magnitudes of the CB–LH and CB–HH absorption are often similar, but much depends on the angle of incidence. The closer this angle is to 90° (i.e., to propagation in the plane), the stronger the CB–LH absorption. Of course, when the light actually propagates in the plane, we can no longer define the absorption per pass. In this case, we will need a different way to quantify the absorption, as discussed in Section 11.2.

We can also roughly estimate the CB–LH matrix element using the two-band Hamiltonian of Eq. (10.1) and neglecting the free-electron diagonal terms, as before:

$$\boldsymbol{H}_{2-band}(z) = \begin{bmatrix} E_c(z) & P\frac{\partial}{\partial z} \\ -P\frac{\partial}{\partial z} & E_v(z) \end{bmatrix} \qquad (11.11)$$

From Eqs. (10.3) and (10.4), we know that the $p_z$-like and $s$-like components of the CB wavefunction are related by

$$F_z^{CB} \approx \frac{P}{E_g}\frac{\partial}{\partial z}F_s^{CB} \qquad (11.12)$$

The same components of the LH wavefunction are also proportional:

$$F_s^{LH} \approx -\frac{P}{E_g}\frac{\partial}{\partial z}F_z^{LH} \qquad (11.13)$$

The LH subband is rarely at the top of the valence band, so we should replace $E_g$ in Eq. (11.13) with $E_{c,1} - E_{l,1} > E_g$ to be accurate. The operator for calculating the optical matrix elements in the case of Eq. (11.11) is

$$\nabla_{\boldsymbol{k}}H_{2-band} = \begin{bmatrix} 0 & iP \\ -iP & 0 \end{bmatrix}\hat{e}_z \qquad (11.14)$$

where $\hat{e}_z$ is the unit vector along the growth axis.

At this point, we can use Eq. (4.39) to show that

$$\langle c, 1|\nabla_{\boldsymbol{k}}H_{2-band}|v, l1\rangle = \begin{bmatrix} -iF_s^{CB*} & F_z^{CB*} \end{bmatrix}\begin{bmatrix} 0 & iP \\ -iP & 0 \end{bmatrix}\begin{bmatrix} iF_s^{LH} \\ F_z^{LH} \end{bmatrix}\hat{e}_z$$

$$= \hat{e}_z P \int dz\left(F_s^{CB*}(z)F_z^{LH}(z) + F_z^{CB*}(z)F_s^{LH}(z)\right) \qquad (11.15)$$

As usual in quantum mechanics, the overall phase does not matter, since only the square of the wavefunction enters the final expression. If both $F_s^{CB}$ and $F_z^{LH}$

are sine-like, with the other two components given by Eqs. (11.12) and (11.13), the second term under the integral is much smaller than the first unless the gap is narrow. Therefore, the square of the resulting matrix element is simply $\approx \frac{\hbar^2 E_P}{2m}|O_{el}|^2\sin^2\theta$. This differs a little from Eq. (11.9) because Eq.(11.11) neglects the spin–orbit interaction. What we have gained by returning to an explicit, albeit simplified, Hamiltonian is an expression for the square of the wavefunction overlap:

$$|O_{el}|^2 = \left| \int dz \left( F_s^{CB*}(z)F_z^{LH}(z) + F_z^{CB*}(z)F_s^{LH}(z) \right) \right|^2 \tag{11.16}$$

We will return to this form of the overlap in Section 11.2. Again, unless the gap happens to be particularly narrow, the selection rules are the same as for heavy-hole transitions, with LH replacing HH. On the other hand, if the gap *is* narrow, the second term in Eq. (11.16) may be comparable to the first. In this case, we should use a more realistic Hamiltonian, as discussed in Section 11.3.

## 11.2  Intersubband Transitions

In a quantum structure, wavevector is not necessarily conserved. Does that mean absorption that involves initial and final states in the same band is now possible? For example, can electrons make unassisted optical transitions within the conduction band? This is certainly not the case when it comes to transitions within the same subband! Here momentum is conserved in 2D rather than 3D. Only (unassisted) transitions *between different* subbands, illustrated in Fig. 11.2, are possible, and only because they are fundamentally *similar* to interband transitions considered in Section 11.1. This last point is not emphasized enough in much of the literature on intersubband transitions, which relies heavily on single-band models. Fortunately, it is quite easy to confirm using the full Hamiltonian.

Take a transition between a lower subband with index $\nu$ and a higher subband with index $\mu$ for the states of the two-band Hamiltonian in Eq. (10.1). The wavefunction components of each subband are

$$F_s^\nu = \sqrt{\frac{2}{t}}\sin\left(\frac{\nu\pi z}{t}\right) \tag{11.17a}$$

$$F_z^\nu = \frac{P}{E_g}\frac{\partial}{\partial z}F_s^\nu = \frac{P}{E_g}\frac{\nu\pi}{t}\sqrt{\frac{2}{t}}\cos\left(\frac{\nu\pi z}{t}\right) \tag{11.17b}$$

where for the final subband $\nu$ is replaced by $\mu$. To normalize the overall envelope function such that $\int dz\left[(F_s^\nu)^2 + (F_s^\nu)^2\right] = 1$, both components in Eq. (11.17) should be multiplied by $\left(1 + \frac{E_{\nu 0}E_p}{E_g^2}\right)^{-1}$. However, the normalization factor

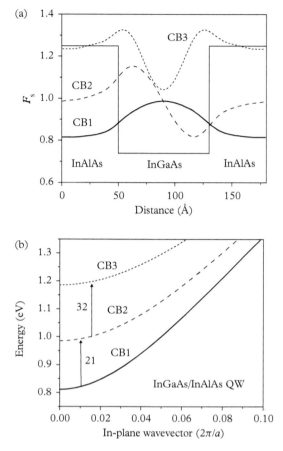

**Figure 11.2** *(a) S-like components of the envelope functions for three conduction subbands localized below potential barriers in a 80-Å In$_{0.53}$Ga$_{0.47}$As/In$_{0.52}$Al$_{0.48}$As quantum well, modeled as a superlattice. (b) Schematic of the possible intersubband transitions in this well with the in-plane CB1, CB2, and CB3 dispersions.*

$1 + \frac{E_{v0}E_p}{E_g^2}$ is close to unity when the subbands are not too high in the conduction band (or when the energy gap is not too narrow).

The CB–CB optical matrix element for the two-band Hamiltonian follows from Eq. (11.14) and bears more than a passing resemblance to the CB–LH transitions in Eq. (11.15):

$$\langle c, \mu | \nabla_{\boldsymbol{k}} H_{2-band} | c, v \rangle = [-i(F_s^\mu)^* \ (F_z^\mu)^*] \begin{bmatrix} 0 & iP \\ -iP & 0 \end{bmatrix} \begin{bmatrix} iF_s^v \\ F_z^v \end{bmatrix} \hat{e}_z$$

$$= \hat{e}_z P \int dz \left( (F_s^\mu)^*(z) F_z^v(z) + (F_z^\mu)^*(z) F_s^v(z) \right) \qquad (11.18)$$

Substituting the envelope-function components from Eq. (11.17), we see that the matrix element does not vanish only when $\mu$ and $\nu$ have different parities (so that $\mu - \nu$ is odd), e.g., $\nu = 1$ and $\mu = 2$ or $\nu = 4$ and $\mu = 7$. In other words, at zero field this requires one of the wavefunctions to be symmetric and the other antisymmetric about the center of the well at $z = t/2$. More generally, the matrix element is given by evaluating the integral:

$$\langle c, \mu | \nabla_{\boldsymbol{k}} H_{2-band} | c, \nu \rangle = \hat{e}_z \frac{P^2}{E_g} \int dz \left( (F_s^\mu)^* \frac{\partial F_s^\nu}{\partial z} + \frac{\partial (F_s^\mu)^*}{\partial z} F_s^\nu \right) \equiv \hat{e}_z \frac{P^2}{tE_g} \frac{8\mu\nu}{\mu^2 - \nu^2} O_{\mu\nu}$$

$$(11.19)$$

Here $\frac{P^2}{tE_g} = \frac{\hbar^2 E_P}{2mtE_g} \approx \frac{\hbar^2}{2m_e t}$, where $t$ is the integration range (well width) in Eq. (11.19), if we use the two-band value for the effective mass $m_e$. The overlap integral $O_{\mu\nu}$ is defined so that it is normalized to unity for an infinite square well. This overlap is in fact related to $O_{el}$ in Eq. (11.16) if we substitute the higher CB subband for the LH subband. With just a bit more work, we can derive the matrix element for the Kane Hamiltonian that includes spin–orbit coupling.

Notice again that the form of the matrix element in Eq. (11.18) is the same as for the CB–LH interband transitions in Eq. (11.15). Roughly speaking, the square of the optical matrix element is inversely proportional to the effective mass, which implies that both interband and intersubband transitions arise from the interaction with the valence bands. It is the band mixing due to quantum confinement along the growth direction that makes intersubband transitions possible. We could obtain essentially the same result from a one-band model with an effective mass, but the physical origin of the transitions then remains opaque!

If the envelope functions are close to the sine form in Eq. (11.17) and the energy gap is large compared to the subband energies, Eq. (11.19) becomes ($O_{\mu\nu} \approx 1$)

$$\langle c, \mu | \nabla_{\boldsymbol{k}} H | c, \nu \rangle \approx \frac{\hbar^2}{m_e t} \frac{4\mu\nu}{\mu^2 - \nu^2} \hat{e}_z$$

$$(11.20)$$

when $\mu - \nu$ is odd. The matrix element vanishes when $\mu - \nu$ is even. The most important intersubband transitions usually take place between neighboring subbands with $\mu = \nu + 1$, for which the matrix element in Eq. (11.20) becomes

$$\langle c, \nu + 1 | \nabla_{\boldsymbol{k}} H | c, \nu \rangle \approx \frac{\hbar^2}{m_e t} \frac{4\nu(\nu+1)}{2\nu+1} \hat{e}_z$$

$$(11.21)$$

The matrix elements drop off in magnitude as $\mu - \nu$ increases. For example, the ratios of the squared elements for the $\mu = 4 \rightarrow \nu = 1$ and $6 \rightarrow 1$ transitions are smaller than for the $2 \rightarrow 1$ transition by factors of 6.25 and 15.1, respectively.

How can we derive the same result from the one-band model of Eq. (8.23)? The matrix element we want here is $\nabla_{\boldsymbol{k}} H_{1-band} = -i\frac{\hbar^2}{m_e} \frac{\partial}{\partial z} \hat{e}_z$. We can also arrive at this answer by calculating the dipole matrix element: $\langle c, \mu | z | c, v \rangle = \int dz \, z \, (F_s^\mu)^* F_s^v$. For a free-particle-like Hamiltonian, the dipole is related to the general expression via Eq. (4.15) derived in Chapter 4:

$$\langle c, \mu | \nabla_{\boldsymbol{k}} H \cdot \hat{e}_z | c, v \rangle = i \langle c, \mu | z | c, v \rangle (E_\mu - E_v) \tag{11.22}$$

To relate better to the case in point, we show that this expression holds for the one-band Hamiltonian $H_{1-band} = -\frac{\hbar^2}{2m_e}\frac{\partial^2}{\partial z^2} + V(z)$ that is equivalent to Eq. (8.23) if all the materials in the structure have the same mass. If we evaluate the *commutator* of this Hamiltonian and the position operator $[H_{1-band}, z]$, we find that it is proportional to the operator we are interested in:

$$-\frac{\hbar^2}{2m_e}\left[\frac{\partial^2}{\partial z^2}(z\varphi) - z\frac{\partial^2\varphi}{\partial z^2}\right] = -\frac{\hbar^2}{m_e}\frac{\partial}{\partial z}\varphi = -i\frac{\partial H}{\partial k_z}\varphi \tag{11.23}$$

The potential terms vanish in the commutator.

Another way to approach the calculation is to apply the Hamiltonian in the commutator matrix element between states $\mu$ and $v$ to the state wavefunctions first. This generates the "free-particle" energies on the right-hand side of Eq. (11.22) multiplied by the dipole matrix element:

$$\langle c, \mu | [H, z] | c, v \rangle = \langle c, \mu | Hz - zH | c, v \rangle = (E_\mu - E_v) \langle c, \mu | z | c, v \rangle \tag{11.24}$$

Since the matrix element of the operator in Eq. (11.23) must be equal to that in Eq. (11.24), we obtain Eq. (11.22). For perfect confinement and parabolic dispersion, the dipole matrix element is

$$|\langle c, \mu | z | c, v \rangle| = \frac{8\mu v}{(\mu^2 - v^2)^2 \pi^2} t \, \delta_{\mu-v,odd} = \frac{4\sqrt{2}\mu v \hbar}{(\mu^2 - v^2)^{3/2}\pi (E_\mu - E_v)^{1/2} m_e^{1/2}} \delta_{\mu-v,odd} \tag{11.25}$$

where the Kronecker delta indicates a change in parity.

We are not through with commutators yet! Here is another that may not seem very useful at first glance: $[z, [H, z]]$. By analogy with Eq. (11.23)

$$-\frac{\hbar^2}{2m_e}\left[z, \frac{\partial^2}{\partial z^2}(z\varphi) - z\frac{\partial^2\varphi}{\partial z^2}\right] = -\frac{\hbar^2}{m_e}\left[z\frac{\partial\varphi}{\partial z} - \frac{\partial}{\partial z}(z\varphi)\right] = \frac{\hbar^2}{m_e}\varphi \tag{11.26}$$

We could also evaluate the matrix element of the commutator between states in the same subband:

$$\langle c, v | [z, [H, z]] | c, v \rangle = \langle c, v | 2zHz - z^2H - Hz^2 | c, v \rangle$$
$$= -2E_v \langle c, v | z^2 | c, v \rangle + 2 \langle c, v | zHz | c, v \rangle \tag{11.27}$$

If we sum over all the states in some basis, we see that $|c,v\rangle = \sum_\mu |c,\mu\rangle \langle c,\mu|c,v\rangle$ and $\langle c,v|AB|c,v\rangle = \sum_\mu \langle c,v|A|c,\mu\rangle \langle c,\mu|B|c,v\rangle$, where $A$ and $B$ are operators. This is the so-called *completeness relation* from quantum mechanics (and linear algebra). With this relation in mind, we write Eq. (11.27) as

$$-2E_v\langle c,v|z^2|c,v\rangle + 2 \langle c,v|zHz|c,v\rangle$$

$$= -2E_v \sum_\mu \langle c,v|z|c,\mu\rangle \langle c,\mu|z|c,v\rangle + 2 \sum_\mu \langle c,v|zH|c,\mu\rangle \langle c,\mu|z|c,v\rangle$$

$$= 2\sum_\mu \left(E_\mu - E_v\right) |\langle c,\mu|z|c,v\rangle|^2 \qquad (11.28)$$

Since Eqs. (11.26) and (11.28) describe the same commutator, we deduce that

$$\frac{2m_e}{\hbar^2} \sum_\mu \left(E_\mu - E_v\right) |\langle c,\mu|z|c,v\rangle|^2 = 1 \qquad (11.29)$$

Because we are summing over the square of the dipole, we can establish a relation between all the dipole elements that involve a single state. We do so by defining the dimensionless (normalized) *oscillator strength*: $f_{v\mu} \equiv \frac{2m}{\hbar^2}\left(E_\mu - E_v\right)|\langle c,\mu|z|c,v\rangle|^2$. With this definition, we have two equivalent statements of the *sum rule*:

$$\sum_\mu f_{v\mu} = \frac{m}{m_e} \qquad (11.30a)$$

$$\frac{m_e}{m} \sum_\mu f_{v\mu} = 1 \qquad (11.30b)$$

The total oscillator strength scales inversely with the electron effective mass. Alternatively, we could define the oscillator strength to include the inverse effective mass so that the oscillator strengths add up to 1. Either way, Eq. (11.30) gives us a measure of the relative strength of the transitions from one subband to any other. However, keep in mind that these expressions strictly apply only to the parabolic dispersions of a one-band "free-particle-like" Hamiltonian! They may not hold once we introduce nonparabolicity either as an add-on to the one-band model or via a more complex Hamiltonian.

In a parabolic well with perfect confinement, the dipole is given by Eq. (11.25), and we obtain for any subband $v$

$$\frac{64}{\pi^2} \sum_{\mu-v,odd} \frac{\mu^2 v^2}{\left(\mu^2 - v^2\right)^3} = 1 \qquad (11.31)$$

Again, we sum only over transitions with a parity change. Amazingly, the transitions between subbands 1 and 2 account for $\approx 96\%$ of the total oscillator strength for subband 1! This is an important result because most of the electrons often occupy CB1. The transitions from a higher subband can go both up (positive oscillator strengths) and down (negative oscillator strengths). Of course, this negative oscillator strength is just an accounting fiction, since $f_{\nu\mu} = -f_{\mu\nu}$. In general, the strongest transitions by far are those with $\mu = \nu - 1$ and $\mu = \nu + 1$. The oscillator strength normalized by the electron mass is plotted as a function of subband index in Fig. 11.3. We observe that $f_{\nu,\nu+1} \propto \nu$; i.e., the oscillator strength for a transition between successive subbands increases with subband index.

Does the concept of oscillator strength have any relevance for multiband Hamiltonians? Yes, but we must proceed with care! For the example of the two-band Hamiltonian of Eq. (11.11), we find by analogy with Eq. (11.23)

$$[H, Z] = -i\frac{\partial H}{\partial k_z} = \begin{bmatrix} 0 & P \\ -P & 0 \end{bmatrix} \qquad (11.32)$$

where $Z$ is the dipole matrix:

$$Z = \begin{bmatrix} z & 0 \\ 0 & z \end{bmatrix} \qquad (11.33)$$

By extending the analogy to Eq. (11.24), it is clear that

$$\langle c, \mu | [H, Z] | c, \nu \rangle = (E_\mu - E_\nu) \langle c, \mu | Z | c, \nu \rangle \qquad (11.34)$$

Now it immediately follows from Eqs. (11.32) and (11.34) that

$$\left\langle c, \mu \left| \frac{\partial H}{\partial k_z} \right| c, \nu \right\rangle = i \langle c, \mu | Z | c, \nu \rangle (E_\mu - E_\nu) \qquad (11.35)$$

Again, this is nothing more than a specific example of the more general result derived in Chapter 4 (Eq. (4.15)). Here we obtain Eq. (11.35) by explicit manipulations to illustrate how multiband Hamiltonians can be deployed in practical calculations.

We will now evaluate the $[z, [H, z]]$ commutator to find the oscillator strength. Somewhat unexpectedly, we find that

$$\langle c, \nu | [Z, [H, Z]] | c, \nu \rangle = i \left\langle c, \nu \left| \left[ Z, \frac{\partial H}{\partial k_z} \right] \right| c, \nu \right\rangle = 0 \qquad (11.36)$$

and $\sum_\mu f_{\nu\mu} = 0$. What is going on here is that the positive oscillator strengths for upward transitions are exactly canceled by the negative oscillator strengths for downward transitions (rather than leaving a residue of 1, as in the one-band model). It seems that in multiband models, we should always treat intersubband and interband transitions on the same footing! Of course, we are already aware of this from Eq. (11.18).

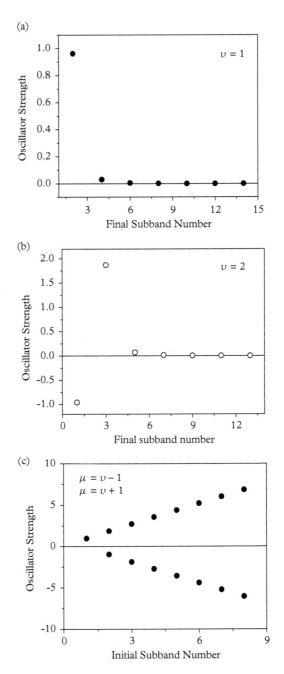

**Figure 11.3** *Oscillator strength normalized by the electron mass as a function of subband index in a square quantum well with infinite potential barriers: (a) for transitions from CB1, (b) from the second subband, and (c) for the strongest transitions in which the subband number changes by one (up or down).*

This discussion of the oscillator strengths is but a stepping-stone toward our ultimate goal, which is calculation of the intersubband absorption coefficient. For strictly parabolic bands, all subbands have the same (parallel) dispersion. Therefore, in the absence of broadening, intersubband transitions have a well-defined transition energy $\hbar\omega = E_{c,\mu} - E_{c,v}$. This makes for a strong contrast with interband transitions, for which there is a wide range of transition energies. Also, if the valence band is almost full, and the conduction band is almost empty, we can calculate the interband absorption coefficient over a wide range of energies by ignoring the carrier populations. This is because the populations achieved in practice are small enough that only transitions near the band edges are affected. However, intersubband absorption is only possible if there are electrons in the initial subband. Returning to the analogy with interband transitions, this is equivalent to having a (partially) full valence band. No wonder the intersubband absorption coefficient scales with carrier density, as we will confirm later.

We calculate the intersubband absorption per pass by inserting the occupation factors from Eq. (4.61) into Eq. (11.2):

$$
\alpha_{ib}(\hbar\omega) = \frac{\alpha_f}{\cos\theta \; n_r} \sum_{\mu,v} \int d^2k_\| \; |\langle c,\mu|\nabla_{\boldsymbol{k}} H|c,v\rangle|^2
$$

$$
\frac{\hbar\omega}{\left(E_{c,\mu} - E_{c,v}\right)^2} \; \sigma\left(E_{c,\mu} - E_{c,v} - \hbar\omega\right)\left[f_{c,v}\left(k_\|\right) - f_{c,\mu}\left(k_\|\right)\right] \qquad (11.37)
$$

Since there is no absorption at normal incidence, we need the angular factor from Eq. (11.10). We already know the origin of this factor from the discussion of CB–LH absorption. Note that in Eq. (11.37), the $\sin^2\theta$ factor is absorbed into the square of the matrix element.

Equation (11.37) looks suitable for numerical calculations, but does not lend itself readily to physical insight. To this end, we assume confinement in an infinite well, substitute the matrix element from Eq. (11.20), and then specialize to the absorption from subband $v$ to subband $\mu$ only. With the sum over spins doubling the total, the result is

$$
\alpha_{ib}(\hbar\omega) = \frac{2\alpha_f}{n_r} \frac{\sin^2\theta}{\cos\theta} \int d^2k_\| \left(\frac{\hbar^2}{m_e t} \frac{4\mu v}{\mu^2 - v^2}\right)^2
$$

$$
\frac{\hbar\omega}{\left(E_{c,\mu} - E_{c,v}\right)^2} \; \sigma\left(E_{c,\mu} - E_{c,v} - \hbar\omega\right)\left[f_{c,v}\left(k_\|\right) - f_{c,\mu}\left(k_\|\right)\right] \qquad (11.38)
$$

The carrier densities in the two subbands are simply

$$
n_\mu = \frac{2}{(2\pi)^2} \int d^2k_\| \; f_{c,\mu}\left(k_\|\right) \qquad (11.39)
$$

We can use Eq. (11.39) to rewrite Eq. (11.38) as

$$\alpha_{ib}(\hbar\omega) = \frac{4\pi^2\alpha_f}{n_r}\frac{\sin^2\theta}{\cos\theta}\left(\frac{\hbar^2}{m_e t}\frac{4\mu v}{\mu^2 - v^2}\right)^2\frac{\hbar\omega}{\left(E_{c,\mu} - E_{c,v}\right)^2}\sigma\left(E_{c,\mu} - E_{c,v} - \hbar\omega\right)\left(n_v - n_\mu\right)$$

(11.40)

This simplification is possible because the transition energy and matrix element are assumed to be independent of the in-plane wavevector $k_\parallel$. If we use the lineshape function from Eq. (4.56), $\sigma(0) = \frac{1}{\pi\Delta_l}$ at the peak energy $\hbar\omega = E_{c,\mu} - E_{c,v}$. Incidentally, the same is true of the Lorentzian function $\sigma(\delta) = \frac{\Delta_l}{\pi\left(\delta^2 + \Delta_l^2\right)}$ that is much more common in the literature. For other line shapes, the peak value may be different by a numerical factor of order unity. Therefore, the maximum absorption per pass becomes

$$\alpha_{ib,max} = \frac{4\pi\alpha_f}{\Delta_l\left(E_{c,\mu} - E_{c,v}\right)n_r}\frac{\sin^2\theta}{\cos\theta}\left(\frac{\hbar^2}{m_e t}\frac{4\mu v}{\mu^2 - v^2}\right)^2\left(n_v - n_\mu\right)$$

(11.41)

In practice, we often calculate the matrix element numerically, e.g. using Eq. (11.18) to replace the value in Eqs. (11.20) and (11.41). This is because few practical wells are deep enough for the infinite-well approximation to be quantitatively accurate. If the two subbands are mostly in different wells, i.e., the transition is *diagonal*, and $\left|O_{v\mu}\right|^2$ in Eq. (11.19) can be much smaller than one. If this is the case, the absorption is also much weaker, as we see from Eq. (11.19). It is common to express the matrix element in terms of the transition dipole in Eq. (11.22):

$$\alpha_{ib,max} = \frac{4\pi\alpha_f}{n_r}\frac{\sin^2\theta}{\cos\theta}\frac{E_{c,\mu} - E_{c,v}}{\Delta_l}|\langle c,\mu|z|c,v\rangle|^2\left(n_v - n_\mu\right)$$

(11.42)

Of course, the usual definition of the $z$ operator in a single-band model is only valid for parabolic dispersions. If nonparabolicity is significant, the dipole should be calculated from the general optical matrix element and Eq. (11.22). The somewhat ungainly $\frac{\sin^2\theta}{\cos\theta}$ factor happens to evaluate to one when the angle of incidence is at a realistic value, $\theta = 51.8°$.

Now we can put Eq. (11.41) to use in identifying several trends. Even though the absorption coefficient is inversely proportional to the subband separation $E_{c,\mu} - E_{c,v}$, this is compensated by the proportional change in $t^2$. Conversely, reducing $t$ does not enhance the absorption in general. In fact, the absorption is usually decreased because the dipole tends to drop by more than the change of the transition energy in a finite well (plus the wavefunction overlap tends to be smaller).

How does the absorption depend on effective mass? If all the parameters except $m_e$ and $t$ (which varies as $m_e^{-1/2}$ when the intersubband separation is constant) are fixed, we see that $\alpha_{ib,max}$ scales inversely with $m_e$. Therefore, the absorption at a given photon energy is stronger in a narrower-gap material with lower effective mass. However, we will see later that the stronger nonparabolicity in such materials largely neutralizes this effect when it comes to *peak* absorption.

Can we make the absorption stronger by populating $\nu > 1$ subbands and making upward transitions from them rather than the $\nu = 1$ subband, i.e., CB1? At first glance, the absorption is proportional to $(\nu/t)^2$, but the intersubband separation has the same dependence: $E_{c,\mu} - E_{c,\nu} \propto (\nu/t)^2$. In other words, transitions between higher subbands are no stronger even though the oscillator strength scales with subband index. Nor does it help for the electrons to be promoted beyond $\mu = \nu + 1$ because the matrix element falls off dramatically leading to $\alpha_{ib,max} \propto \left(\mu^2 - \nu^2\right)^{-2}$.

What is the absorption per pass for transitions between the two lowest subbands in an infinite quantum well ($\nu = 1$ and $\mu = 2$)? We estimate it using the following (reasonable) parameter values: a sheet density $n_1 = 10^{12}$ cm$^{-2}$ in the lower subband, an empty upper subband ($n_2 = 0$), a mass similar to InGaAs lattice-matched to InP ($m_e = 0.04m$), a refractive index $n_r = 3.3$, and a well thickness $t = 100$ Å, for which $E_2 - E_1 = 282$ meV. In a real quantum well with lattice-matched InAlAs barriers, $E_2 - E_1$ is somewhat smaller because the conduction-band offset is not much larger than $E_2 - E_1$. Finally, we take $\Delta_l = 0.05\,(E_2 - E_1) = 14$ meV and use the calculated matrix element $\langle c, 2 | \nabla_k H | c, 1 \rangle |= 5.1$ eV Å, which corresponds to a dipole matrix element of $|\langle z \rangle| = 18$ Å. This is similar to the matrix elements for interband transitions in Section 11.1. At an incidence angle of $\theta = 45°$, Eq. (11.41) gives an absorption per pass of 1.2%, which is not too far from the interband value in Section 11.1.

More generally, we compare the maximum intersubband and interband absorptions per pass in a type I quantum well with $|O_{eh}|^2 \approx 1$ and an intermediate angle using the ratio of Eqs. (11.7) and (11.41) (between subbands 1 and 2) with the matrix element in Eq. (11.25):

$$\frac{\alpha_{ib}}{\alpha_{QW}} \approx \frac{\hbar\omega}{\Delta_l} \frac{\hbar^2 n_1}{E_P m_e} \frac{m}{m_r} \approx \frac{\hbar^2 n_1}{m E_g} \frac{E_P}{E_g} \frac{E_g}{\Delta_l} \tag{11.43}$$

The last two factors are both of order 10–100, but the first is $\ll 1$ *percent* because real quantum well cannot be doped to levels higher than $\sim 10^{13}$ cm$^{-2}$. Overall, intersubband absorption is comparable to interband absorption at the same energy *only for maximum doping and minimum attainable linewidths*. The intersubband linewidth is again a combination of homogeneous broadening due to scattering within the initial and final subbands and inhomogeneous broadening due to well width variations in the growth plane.

Figure 11.4a plots the intersubband absorption between the lowest two subbands in 80-Å-wide GaAs/AlGaAs (on GaAs), In$_{0.53}$Ga$_{0.47}$As/InAlAs (on InP), and InAs/AlSb (on GaSb) quantum wells for an electron density of $10^{12}$ cm$^{-2}$. The results are expressed in terms of the absorption both per pass and per unit

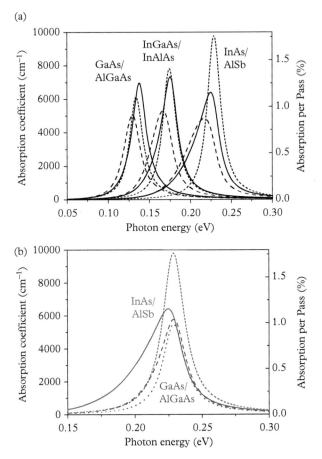

**Figure 11.4** *(a) Intersubband absorption spectra for the CB1–CB2 transition in 80-Å-wide GaAs/Al$_{0.3}$Ga$_{0.7}$As, In$_{0.53}$Ga$_{0.47}$As/In$_{0.52}$Al$_{0.48}$As, and InAs/AlSb quantum wells with sheet electron densities of $10^{12}$ cm$^{-2}$, assuming fixed broadening of $\Delta_l = 10$ meV, and Lorentzian line shape. The angle of incidence is taken to be $\theta = 45°$. The dotted lines represent the results for parallel dispersions, the dashed lines are for the full dispersions and matrix elements, and the solid line with many-body effects additionally included. (b) The red dotted and solid curves from (a) for the 80-Å-wide InAs/AlSb well along with results for a 47-Å-wide GaAs/Al$_{0.3}$Ga$_{0.7}$As well with the same intersubband transition energy (blue).*

length (after dividing by the well width). We compare the results for calculations using (1) parallel dispersions and Eq. (11.42); (2) full dispersions and Eq. (11.37); (3) full dispersions with many-body effects driven by the high electron densities, more about which is to follow. In all cases, only the lowest electron subband is occupied at room temperature, and the same Lorentzian broadening linewidth of $\Delta_l = 10$ meV is assumed. The matrix element at the zone center is always obtained from the full 8-band calculation. As the band-edge electron mass in bulk materials decreases, the intersubband transition blue-shifts (if the well width is held constant), and the absorption strength in the case of parallel dispersions goes up, as expected from the simple arguments that follow Eq. (11.42).

Unfortunately, nonparabocity broadens the intersubband transition in the InAs/AlSb much more than in the other two systems. The relative magnitude remains unchanged even when many-body effects are included. The careful reader may object that we should instead be comparing the *absorption per pass* for transitions with the same *energy* rather than in quantum wells with the same width. To do so, we plot in Fig. 11.4b the absorption coefficients for the case of parallel dispersions and the full calculation including many-body effects for the 47-Å-wide GaAs/AlGaAs well with the same intersubband separation as in the 80-Å-wide InAs/AlSb well (corresponding to a wavelength of 5.4 μm). The peak absorption per pass is now slightly stronger in the InAs well. The increased oscillator strength is instead transferred to the transitions at higher $k_{\parallel}$ that occur at lower photon energies. As a result, the intersubband transition is noticeably wider in the InAs/AlSb well *even for the same* $\Delta_l$.

Equation (11.19) shows that within the two-band model of Eq. (11.11), intersubband transitions couple only to light polarized along the growth direction $z$. Does this conclusion also hold for more accurate forms of the Hamiltonian? We know that the two-band model lacks the spin–orbit interaction, which we can restore by returning to the Kane Hamiltonian of Eq. (2.41). For simplicity, we disregard the free-electron diagonal terms and consider only one spin-degenerate set of states:

$$\boldsymbol{H}_{4\times4}\begin{bmatrix}iF_{s\uparrow}\\F_{xy\downarrow}\\F_{z\uparrow}\\F_{xy\uparrow}\end{bmatrix}=\begin{bmatrix}E_g&0&k_{z,eff}P&0\\0&-\frac{2\Delta}{3}&\frac{\sqrt{2}\Delta}{3}&0\\k_{z,eff}P&\frac{\sqrt{2}\Delta}{3}&-\frac{\Delta}{3}&0\\0&0&0&\frac{\hbar^2k^2}{2m_{HH}}\end{bmatrix}\begin{bmatrix}iF_{s\uparrow}\\F_{xy\downarrow}\\F_{z\uparrow}\\F_{xy\uparrow}\end{bmatrix} \tag{11.44}$$

where the wavefunction components are written as coefficients of the atomic-orbital functions. To make an approximate evaluation easier, we write Eq. (11.44) in terms of the "effective" wavevector $k_{z,eff}$ that follows from $E_{c,v}\approx\frac{\hbar^2k_{z,eff}^2}{2m_e}$ with $k_{z,eff}\leq\pi v/t$, where strict equality holds for infinite potential barriers.

It should be clear from Eq. (11.44) that the $s$-dominated envelope function couples not only to the $p_z$-like component, but also to the $p_x$ and $p_y$ components. To estimate the magnitude of these components, we use the second equation in the matrix of Eq. (11.44) and solve for $F_{xy}$ in terms of $F_z$ at an energy near the bottom of the electron subband $E_g+E_{c,v}$:

$$F_{xy}^v\approx\frac{\sqrt{2}}{3}\frac{\Delta}{E_g+\frac{2\Delta}{3}+E_{c,v}}F_z^v \tag{11.45}$$

What does this tell us about the transition strength? The full 8-band matrix has $s - x$ and $s - y$ terms proportional to $k_x$ and $k_y$, respectively, which produce matrix elements of the form:

$$\langle c, \mu | \nabla_{\boldsymbol{k}} H_{8-band} | c, \nu \rangle = \hat{e}_{x,y} P \int dz \left( \left( F_s^\mu \right)^* (z) F_{xy}^\nu (z) + \left( F_{xy}^\mu \right)^* (z) F_s^\nu (z) \right) \quad (11.46)$$

For a very rough estimate of the magnitude of the in-plane-polarized absorption, we take the ratio $\frac{F_{xy}^\nu}{F_z^\nu} \frac{F_{xy}^\mu}{F_z^\mu}$. This is far from a rigorous solution, but enough to show how the absorption for in-plane polarization scales with spin–orbit coupling. We immediately obtain

$$\frac{\alpha_{x,y}}{\alpha_z} \sim \frac{F_{xy}^\nu}{F_z^\nu} \frac{F_{xy}^\mu}{F_z^\mu} \approx \frac{2}{9} \frac{\Delta}{E_g + \frac{2\Delta}{3} + E_{c,\nu}} \frac{\Delta}{E_g + \frac{2\Delta}{3} + E_{c,\mu}} \quad (11.47)$$

When $E_g \gg \Delta, E_{c,\nu}$, e.g., for a GaAs-based quantum well, $F_{x,y}$ is essentially negligible. The interesting case is that of $\Delta \sim E_g$, e.g., in InAs-based wells. Equation (11.47) predicts a ratio of several percent. More generally, the scaling is close to the form $\Delta^2/\left(E_g + E_{c,\nu}\right)^2$. To check our naïve estimate, we calculate the eigenvalues and eigenfunctions of the Kane Hamiltonian and then apply Eq. (11.46). Figure 11.5 shows that the result of Eq. (11.47) is too high, and in practice the in-plane-polarized absorption remains below 1% of its $z$-polarized counterpart even for relatively narrow InAs/AlSb wells. Given how weak the absorption at normal incidence is, it has not found any practical application.

When electron densities become very high, the single-particle picture we have operated within so far must be revised. Repulsive interactions of the electrons shift the peak emission to $\hbar\omega > \Delta E \equiv E_{c,\mu} - E_{c,\nu}$. To find the shift, we start with a more general form of the response function that may be familiar from the theory of the dielectric function. This function reduces to the Lorentzian in the limit of $\Delta E \ll \left( E_{c,\mu} + E_{c,\nu} \right)/2$:

$$\sigma \left( \Delta E - \hbar\omega \right) \propto \mathrm{Im} \left( \frac{\hbar\omega}{\Delta E^2 - (\hbar\omega)^2 - i2\Delta_l \hbar\omega} \right) \quad (11.48)$$

There is still a peak at $\Delta E = \hbar\omega$, and the maximum amplitude is again inversely proportional to the linewidth. The factor of 2 multiplying $\Delta_l$ is needed so that the full width at half-maximum remains the same as for a Lorentzian.

Now we need to correct for the presence of all the other electrons. Their combined effect is represented by the total dielectric function

$$\varepsilon \left( \hbar\omega \right) = n_r^2 \left( 1 + \frac{\left( \hbar\omega_p \right)^2}{\Delta E^2 - (\hbar\omega)^2 - i2\Delta_l \hbar\omega} \right) \quad (11.49)$$

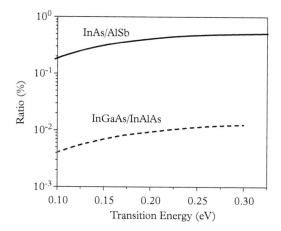

**Figure 11.5** *Ratio of the peak absorption coefficients for in-plane and out-of-plane polarizations in InGaAs/InAlAs and InAs/AlSb quantum wells as a function of transition energy determined by the well width. The in-plane polarized light is assumed to be at normal incidence, while $\theta = 45°$ for the out-of-plane polarization. At the highest transition energies, the second conduction subband is not confined to the well.*

which also peaks at the oscillator resonance $\Delta E = \hbar\omega$. Notice that the right-hand side of Eq. (11.48) is proportional to the imaginary part of Eq. (11.49), which will lead to considerable simplification. The plasma frequency $\omega_p$ in Eq. (11.49) scales with the difference between the electron densities in the two subbands:

$$\left(\hbar\omega_p\right)^2 = \frac{2e^2\left(n_\nu - n_\mu\right)S_{\nu\mu}\,\Delta E}{\varepsilon_0 n_r^2} \tag{11.50}$$

where $S_{\nu\mu}$ is an integral of the square of the wavefunction overlap over the well coordinate:

$$S_{\nu\mu} = \int_{-\infty}^{\infty}\left|\int_{-\infty}^{z}\left[F_s^\mu\left(z'\right)\right]^*F_s^\nu\left(z'\right)\,dz'\right|^2 dz \tag{11.51}$$

Using the envelope function for $0 < z < t$ from Eq. (11.17a) for a well with infinite barriers, we obtain

$$S_{\nu\mu} = \frac{\mu^2 + \nu^2}{\left(\mu^2 - \nu^2\right)^2\pi^2}\,t\,\delta_{\mu-\nu,odd} \tag{11.52}$$

This form is reminiscent of the expression for the dipole in Eq. (11.25). For $\mu = \nu + 1$, $S_{\nu\mu} \approx t/\left(2\pi^2\right)$, although $S_{\nu\mu}$ can be a little larger in practice because the electron wavefunctions extend slightly into the barriers. When $\mu \gg \nu$, $S_{\nu\mu} \approx$

$t/\left(\mu^2\pi^2\right)$, which means that the $S_{\nu\mu}\Delta E$ product remains approximately constant for large intersubband separations in an infinitely deep well. In a real well, $\Delta E$ does not increase as fast, and $S_{\nu\mu}\Delta E$ may drop off somewhat as $\mu - \nu$ increases.

The plasma frequency in Eq. (11.50) may seem unfamiliar, but it differs from the textbook value in parabolic systems only by a numerical factor of order 1. We can see this by substituting $\Delta E = \hbar^2\pi^2\left(\mu^2 - \nu^2\right)/\left(2m_e t^2\right)$ and the expression for $S_{\nu\mu}$ in Eq. (11.52), both of which are valid for the well with infinite barriers, into Eq. (11.50). The textbook value can in turn be derived by considering the motion of free electrons with mass $m_e$ in an electric field applied to a medium with permittivity $\varepsilon_r = n_r^2$.

To see how the response changes in the presence of other electrons, we calculate the "screened" function by taking a ratio of Eq. (11.48) and the dielectric function in Eq. (11.49):

$$\frac{\sigma\left(\Delta E - \hbar\omega\right)}{\varepsilon\left(\hbar\omega\right)} \propto \text{Im}\left(\frac{\hbar\omega}{\left[\Delta E^2 - (\hbar\omega)^2 - i2\Delta_l\hbar\omega\right]\left[1 + \frac{\left(\hbar\omega_p\right)^2}{\Delta E^2 - (\hbar\omega)^2 - i2\Delta_l\hbar\omega}\right]}\right)$$

$$= \text{Im}\left(\frac{\hbar\omega}{\Delta E^2 - (\hbar\omega)^2 + \left(\hbar\omega_p\right)^2 - i2\Delta_l\hbar\omega}\right) \qquad (11.53)$$

The screened response in Eq. (11.53) resembles the "bare" form in Eq. (11.48), but with the transition energy shifted from $\Delta E$ to $\Delta\tilde{E} = \sqrt{(\Delta E)^2 + \left(\hbar\omega_p\right)^2}$! This is known as the *depolarization* shift, and the hybrid excitation as the *intersubband plasmon*. In this simple example, the intersubband plasmon is just a blue-shifted intersubband resonance, by an amount proportional to the square of the electron density (for small shifts). As long as only one intersubband transition dominates, this is as far as we need to go to correct for many-body effects. Higher-order refinements are possible, but rarely essential. The depolarization shift calculated for an InGaAs/InAlAs quantum well is shown in Fig. 11.6.

If the electron density is high enough so that multiple subbands are occupied and multiple intersubband transitions are possible, the picture is more complicated because various intersubband plasmons interact with each other. The upshot is that most of the transition strength feeds into a single line at an energy that is a weighted average of the intersubband plasmon energies. This mechanism is analogous to the shift of the response maximum in Eq. (11.53) to an energy $\Delta\tilde{E}$, for which the real part of the expression in the brackets vanishes. Just as we introduced the intersubband plasmon to describe the peak in Eq. (11.53), we might as well think of the modified resonance as a new excitation, in this case, going by the catchy name of *multisubband plasmon*. As the well becomes wider, this plasmon transitions smoothly into a bulk-like plasma excitation associated with 3D electrons.

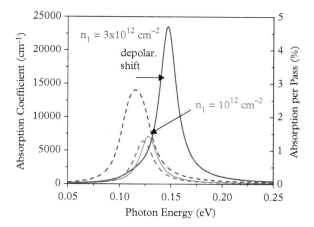

**Figure 11.6** *Effects of the depolarization shift on intersubband absorption in a 100-Å In$_{0.53}$Ga$_{0.47}$As/In$_{0.52}$Al$_{0.48}$As quantum well with electron densities of $10^{12}$ (red) and $3 \times 10^{12}$ cm$^{-2}$ (blue) in the lowest subband. The dashed (solid) curves do not (do) include many-body effects. To make the shift more obvious, the calculations are performed at $T = 10$ K with Lorentzian broadening and $\Delta_l = 10$ meV.*

For semiconductors with narrower gaps, the transition energy $\Delta E$ in Eq. (11.53) can vary strongly with $k_\parallel$, as illustrated in Fig. 11.7a for an 80-Å InGaAs/InAlAs quantum well. As opposed to interband transitions, the intersubband separation decreases at higher $k_\parallel$. This means that the intersubband absorption peak occurs at an energy below $\Delta E (k_\parallel = 0)$ and represents the "center-of-mass" of the oscillator strength. Do we expect the "smeared-out" oscillator strength to broaden the absorption lines? That depends on how high the plasma frequency and the electron density are. Figure 11.7b shows that many-body effects blue shift the line and narrow it down; i.e., they work against nonparabolicity. This is a $k_\parallel$-space analog of the narrowing observed for the multisubband plasmon. However, the effect can be masked by larger values of $\Delta_l$, which occur in even state-of-the-art material, as illustrated in Fig. 11.4.

So far we have treated only intersubband absorption in the conduction band, but we hasten to add that similar ideas apply to the transitions between valence subbands. For example, the structure of the two-band Hamiltonian tells us that LH–LH intersubband transitions at the zone center also couple to $z$-polarized light. We saw in Chapter 10 that the topmost valence subband is usually HH rather than LH, and that band mixing blurs the sharp distinction at higher $k_\parallel$. With this in mind, we formulate a few rules for intervalence transitions in 1D quantum structures:

(1) HH subbands do not couple strongly to $x - y$-polarized or $z$-polarized light, since they are composed of $p_x$ and $p_y$ components. Therefore,

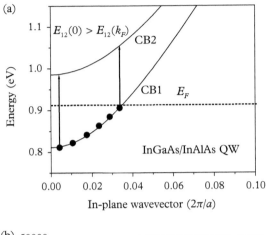

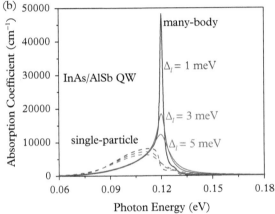

**Figure 11.7** *(a) Illustration of the effect of nonparabolicity on the intersubband transitions in the 80-Å $In_{0.53}Ga_{0.47}As/In_{0.52}Al_{0.48}As$ quantum well of Fig. 11.4. Nonparabolicity is more pronounced in wells based on bulk materials with narrow gaps. (b) Intersubband absorption spectra in a 150-Å InAs/AlSb quantum well with and without many-body effects for $\Delta_l = 1$ (blue), 3 (red), and 5 meV (green) for a sheet doping density of $10^{12}\ cm^{-2}$ at room temperature.*

the relevant $\nabla_k H_{8-band}$ matrix elements are proportional to the in-plane wavevector components.

(2) LH subbands couple to HH subbands under illumination by in-plane-polarized light, provided the parity of their $s$-like component matches that of the HH subband, as in the CB1–HH1, CB2–HH2, etc., transitions. The $s$-like component is proportional to the derivative of the $p_z$ component, which causes a flip from even to odd and vice versa; i.e., HH1 couples to LH2, LH4, etc., HH2 to LH1, LH3, etc. These intervalence transitions can be strong at normal incidence, just as the interband transitions are.

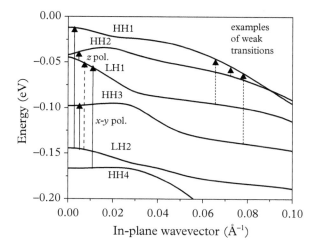

**Figure 11.8** *Allowed intervalence transitions in an 80-Å $In_{0.2}Ga_{0.8}As/Al_{0.3}Ga_{0.7}As$ quantum well. The transitions induced by in-plane-polarized (growth-direction-polarized) light are shown with solid (dashed) arrows. Some examples of weakly allowed transitions at large in-plane wavevectors are also shown.*

(3) Transitions between LH subbands with opposite symmetry, i.e., LH1–LH2, LH1–LH4, etc., are allowed for $z$-polarized light, by analogy with CB transitions.

(4) These selection rules also apply to transitions between valence minibands in a superlattice with $k_\parallel = 0$ and $0 < k_z < \frac{\pi}{d}$.

(5) All of the selection rules are gradually relaxed at higher $k_\parallel$, depending on the particular quantum-well/superlattice structure in question.

Figure 11.8 illustrates the allowed intervalence transitions in an InGaAs/AlGaAs quantum well. The strong allowed transitions are indicated with arrows near the zone center, and some examples of weak transitions due to band mixing at large $k_\parallel$ are also shown. How much absorption we see depends on the occupation of the initial states at the different $k_\parallel$, assuming final states are empty.

The character of interminiband transitions in a type I superlattice can be similar to intersubband transitions in an isolated well when the barriers are thick. For thin barriers and narrow minigaps, the absorption becomes more bulk-like. Figure 11.9 schematically illustrates the two limits. Remember that $k_z$ is conserved as long as a definite value of $k_z$ corresponds to both eigenstates involved in the transition, whether $k_z$ describes the entirety of the variation along the growth axis or the slow variation on the scale of multiple periods.

Intersubband transitions in quantum wires and interlevel transitions in quantum dots also occur when the lower states of these structures are populated. The selection rules and transition strengths depend on the relative dimensions of the

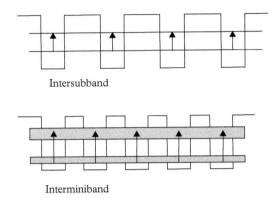

Intersubband

Interminiband

**Figure 11.9** *Schematic of intersubband absorption in multiple nearly isolated quantum wells versus interminiband absorption in a superlattice of strongly coupled wells.*

structures. The discrete transitions in a doped dot is usually between the lowest *s*-like state and the first excited *p*-like state as well as between the confined states and extended states beyond the potential barrier.

## 11.3  Interband Absorption in Quantum Wells and Superlattices

As in bulk semiconductors, the absorption in a quantum well is proportional to the product of the optical matrix element squared and the joint density of states. We saw in Section 11.2 that for intersubband absorption with parabolic dispersion, the joint density of states (JDOS) is a delta function at the transition energy. For interband transitions in an isolated quantum well with isotropic in-plane electron and hole dispersions, we obtain

$$\frac{\hbar^2}{2m}\mathcal{J}DOS^{QW} = \frac{\hbar^2}{2m}\frac{1}{(2\pi)^2}\left[\frac{2\pi k_\parallel}{\frac{dE_{ch}}{dk_\parallel}}\right] \tag{11.54}$$

where $E_{ch} = E_{ck,\mu} - E_{vk,v}$. If we once again assume parabolic bands, this expression becomes

$$\frac{\hbar^2}{2m}\mathcal{J}DOS^{QW} = \frac{\hbar^2}{2m}\frac{m_r}{2\pi\hbar^2} = \frac{1}{4\pi}\frac{m_r}{m}H\left(E_{ch}-E_g\right) \tag{11.55}$$

The normalized 2D JDOS is dimensionless and independent of energy when the bands are parabolic. This form is useful mainly for quick estimates, because the valence band is quite non-parabolic even for wider-gap materials.

We can rewrite the absorption between CB and HH subbands in Eq. (11.2) in a simplified form that should remind you of Eq. (4.47):

$$\alpha_{QW}(\hbar\omega) = \frac{8\pi^2 \alpha_f}{n_r} \frac{|M|^2}{\hbar\omega} \mathcal{J}DOS^{QW} \tag{11.56}$$

where the square of the optical matrix element $|M|^2 \equiv |\langle c, \mu | \nabla_k H | v, v \rangle|^2$ is assumed to be independent of $k_{\parallel}$. We have multiplied Eq. (11.56) by 2 to account for spin degeneracy. In real materials, the optical matrix element changes gradually with transition energy, but this dependence is mostly important for narrow-gap materials. As before, the 2D interband absorption is constant when the derivative of the transition energy depends linearly on wavevector.

What is the optical matrix element $|M|^2$ that enters Eq. (11.56)? Equation (4.49a) shows that the CB–HH matrix element vanishes when the wavevector and polarization are parallel, which means that we can think of the transition moment as a dipole perpendicular to $\boldsymbol{k}$. Taking the polarization to be in the $x - y$ plane of the quantum well (TE-polarized), we want to model the behavior of the optical matrix element. The first question is whether the wavevector in a quantum well lies entirely in the plane. This is not the case, because even at $k_{\parallel} = 0$ the electrons and holes have some energy due to confinement along the $z$ axis. The confinement can be included via the "effective" wavevector $k_{z,eff}$ that was defined in connection with Eq. (11.44). Now we add the in-plane motion and characterize the direction of $\boldsymbol{k}$ using the polar angle with respect to the growth direction:

$$\theta = \cos^{-1}\left(\frac{k_{z,eff}}{k}\right) \tag{11.57}$$

where $k = \sqrt{k_x^2 + k_y^2 + k_{z,eff}^2}$ is the total wavevector. Alternatively

$$\cos^2\theta = \frac{k_{z,eff}^2}{k_x^2 + k_y^2 + k_{z,eff}^2} \tag{11.58}$$

For the light polarized in the $x - y$ plane (corresponding to TE-polarized modes of a planar waveguide), the two possibilities illustrated in Fig. 11.10 are equally likely: (1) the polarization is in the *electron's* plane of incidence and (2) the polarization is perpendicular to that plane. Since the dipole is orthogonal to the wavevector and the matrix element is proportional to the dot product of the dipole and the polarization vector, we can immediately find the matrix elements. In the first case, the overall interaction (the square of the matrix element) is proportional to $\cos^2\theta$; i.e., it becomes stronger as the wavevector rotates away from the $x - y$ plane and no longer points along the light-polarization direction. In the second case, the matrix element always has its full value and is independent of $\theta$. Since both transition types contribute independently to the absorption, we take their average. Selecting

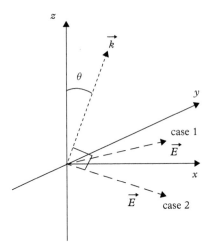

**Figure 11.10** *Directions of light polarization and electron wavevectors in Eqs. (11.57) and (11.58) for the two cases discussed in the text (polarization in and perpendicular to the electron's plane of incidence).*

a random polarization direction in the growth plane then halves the square of the matrix element for the overall result:

$$\left|M_{HH,x}^{QW}\right|^2 = \frac{1}{4}\left(1 + \cos^2\theta\right)P^2 \tag{11.59}$$

This matches the bulk result in Eq. (4.52a) only when the electron has no motion in the plane of the well ($\cos^2\theta = 1$). A plane wave at normal incidence can have two possible polarizations with the same matrix element.

For light polarized along the $z$ axis (this is TM polarization in the convention for planar waveguide modes), no transitions are possible unless the wavevector is rotated toward the growth plane by the in-plane motion. Another way to look at the problem is that the total interaction from three orthogonal polarization axes (2 in the $x-y$ plane and 1 along $z$) must equal the bulk value of $P^2$:

$$\left|M_{HH,z}^{QW}\right|^2 = P^2 - 2\left|M_{HH,x}^{QW}\right|^2 = P^2 - \frac{1}{2}\left(1 + \cos^2\theta\right)P^2 = \frac{1}{2}\sin^2\theta\,P^2 \tag{11.60}$$

This result is the same as Eq. (4.53a) if we set $\theta = \frac{\pi}{2} - \theta_i$, where $\theta_i$ is the angle between the polarization and the growth plane. This is because in Chapter 4 we allowed the polarization angle to vary, whereas here it is the wavevector direction that changes with respect to the polarization vector.

The polarization direction of the LH matrix element follows from the wavefunctions in Eqs. (4.48c) and (4.48d). But there is an easier way! The contributions from the HH and LH states at the same energy add up to twice the square of the bulk matrix element, which we know from Section 4.3 to be $\frac{2}{3}P^2$. Alternatively, we can start with $|M|^2 = P^2$ as the total interaction for the HH, LH, and SO bands in bulk. The SO matrix element is isotropic, so we can subtract one-third from the total immediately. This leads to the sum rule

$$\left|M_{HH}^{QW}\right|^2 + \left|M_{LH}^{QW}\right|^2 = P^2 - \left|M_{SO}^{QW}\right|^2 = \frac{2}{3}P^2 \tag{11.61}$$

By combining Eqs. (11.59), (11.60), and (11.61), we obtain

$$\left|M_{LH,x}^{QW}\right|^2 = \frac{2}{3}P^2 - \left|M_{HH,x}^{QW}\right|^2 = \frac{2}{3}P^2 - \frac{1}{4}\left(1 + \cos^2\theta\right)P^2 = \left(\frac{5}{12} - \frac{1}{4}\cos^2\theta\right)P^2 \tag{11.62a}$$

$$\left|M_{LH,z}^{QW}\right|^2 = \frac{2}{3}P^2 - \left|M_{HH,z}^{QW}\right|^2 = \frac{2}{3}P^2 - \frac{1}{2}\sin^2\theta P^2 = \left(\frac{1}{6} + \frac{1}{2}\cos^2\theta\right)P^2 \tag{11.62b}$$

We can check that the correct results are recovered at $k_\| = 0$, for which $\theta = 0$ and $\cos\theta = 1$:

$$\left|M_{HH,x}^{QW}\right|^2 (k_\| = 0) = \frac{1}{2}P^2 \tag{11.63a}$$

$$\left|M_{HH,z}^{QW}\right|^2 (k_\| = 0) = 0 \tag{11.63b}$$

$$\left|M_{LH,x}^{QW}\right|^2 (k_\| = 0) = \frac{1}{6}P^2 \tag{11.63c}$$

$$\left|M_{LH,z}^{QW}\right|^2 (k_\| = 0) = \frac{2}{3}P^2 \tag{11.63d}$$

These are exactly what we expect based on the wavefunctions in Eq. (4.48)! To complete our approximate calculation, we can now use the expressions for $|M|^2$ given by Eqs. (11.59), (11.60), (11.62a), and (11.62b) in Eq. (11.56).

For narrow quantum wells, the $p_z$ component of the CB envelope function is not negligible. To include the envelope-function character, we should replace $P^2$ in Eqs. (11.59), (11.60), (11.62a), and (11.62b) with $P^2|F_s|^2$, making sure that the envelope-function components are normalized so that $|F_s|^2 \leq 1$. Of course, these results are only approximate. For greater accuracy, we would need to calculate the matrix elements numerically from the Hamiltonian.

So far we have dealt exclusively with linearly polarized light, and assumed that the electron spin plays no role, apart from doubling the number of allowed transitions and the magnitude of the absorption coefficient. However, we can take a different tack and ask what polarization connects a particular CB spin state with one of the HH or LH spin states at $k_{\parallel} = 0$. Neglecting the $p_z$-like component of the electron state and using the HH wavefunction from Eq. (4.48a), the matrix element is maximum for

$$M_{+CB,HH} \rightarrow \frac{1}{\sqrt{2}} \left( e_x - ie_y \right) \tag{11.64}$$

because the intrinsic electron spin (but not the total angular momentum $\boldsymbol{J}$!) is conserved in the optical transition. This is a polarization vector characteristic of right circularly polarized light! Of course, we do not really expect to see circularly polarized emission because the other spin states for both electrons and holes are also occupied. The second spin transition contributes the following matrix element (from Eq. (4.48b)):

$$M_{-CB,HH} \rightarrow \frac{1}{\sqrt{2}} \left( e_x + ie_y \right) \tag{11.65}$$

which requires the emission of left-circularly polarized light. Another way to look at the problem is as a transition from a $| j = \frac{3}{2}, m_j = \pm\frac{3}{2} \rangle$ state to a $| j = \frac{1}{2}, m_j = \pm\frac{1}{2} \rangle$ state. The extra angular momentum is supplied by the intrinsic angular momentum of the photon of $\pm 1$, which is equivalent to the circular polarization.

The same analysis can be repeated for the CB–LH transitions, for which only the $| j = \frac{3}{2}, m_j = \pm\frac{1}{2} \rangle$ to $| j = \frac{1}{2}, m_j = \mp\frac{1}{2} \rangle$ transitions that involve right (left)-circularly polarized photons with angular momenta of $\pm 1$ are possible. Once again, the strict selection rules hold only at the zone center. As the states at higher in-plane wavevectors become occupied, band mixing relaxes the rules. In practice, circularly polarized emission can only be observed if one of the spin states is preferentially populated. This can happen, e.g., if we inject carriers from a contact with spin imbalance that occurs in a ferromagnetic material or by applying a strong magnetic field. Clearly, the lower the temperature, the fewer electronic states are occupied, and the stronger the effect is.

In fact, the CB–HH matrix elements discussed above decrease with $k_{\parallel}$ even when the HH–LH subband separation is large, as in narrow-gap wells. That does not necessarily mean that there is less absorption because the JDOS tends to be higher as the energy increases. Figure 11.11 shows the interband absorption spectra for several quantum wells. At low electron densities, the absorption coefficient in the InGaAs quantum well drops only slightly after the onset of the CB1–LH1 transition. The dip is more perceptible for the GaSb well. In both cases, the spectra vary non-monotonically only when the wells are quite narrow, so that the CB2–HH2 transition is well separated from CB1–HH1 and CB1–LH1.

We have not yet inserted the occupation factors. They are of little consequence for interband processes that occur at low carrier densities. Even if there are no electrons in the conduction band and no holes in the valence band, the many

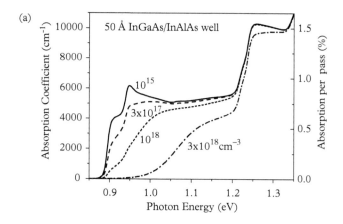

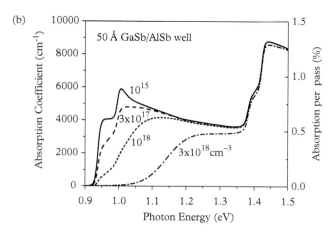

**Figure 11.11** *Interband absorption spectra for in-plane polarized light in 50-Å-wide (a) In$_{0.53}$Ga$_{0.47}$As/In$_{0.52}$Al$_{0.48}$As and (b) GaSb/AlSb quantum wells (modeled as superlattices with 100-Å-wide barriers) for several values of the electron density at room temperature. No holes are assumed to be present regardless of the electron density.*

electrons residing in the valence band in equilibrium are perfectly happy to make transitions to the conduction band. However, when one or both bands are heavily populated, we generalize Eq. (11.2) using what we learned from the discussion leading to Eq. (4.61):

$$\alpha_{QW}(\hbar\omega) = \frac{\alpha_f}{n_r} \sum_{\mu,v} \int d^2 k_\parallel \; |\langle c,\mu|\nabla_k H_e|v,v\rangle|^2$$

$$\frac{\hbar\omega}{\left(E_{ck,\mu} - E_{vk,v}\right)^2} \; \sigma\left(E_{ck,\mu} - E_{vk,v} - \hbar\omega\right) \left[f_v\left(E_{vk,v}\right) - f_c\left(E_{ck,\mu}\right)\right]$$

$$(11.66)$$

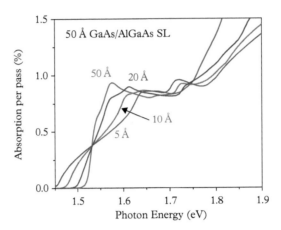

**Figure 11.12** *Absorption spectra for in-plane-polarized light in 50-Å-wide GaAs/Al$_{0.3}$Ga$_{0.7}$As superlattices with barrier thicknesses ranging from 50 to 5 Å. The spectra are expressed in units of absorbed fraction per pass for normally incident light to enable side-by-side comparison of the superlattices with different periods.*

At higher electron densities in Fig. 11.11, we see the Moss–Burstein effect that is a close analog of the bulk phenomenon illustrated in Fig. 4.9.

We already know from Section 4.4 that if $f_v\left(E_{vk,v}\right) - f_c\left(E_{ck,\mu}\right) < 0$ there is net photon gain rather than loss at the photon energy equal to the separation between the two states. In fact, many of the conclusions obtained in Chapter 4 apply to quantum structures, including the transparency condition $E_{Fc} - E_{Fv} = \hbar\omega$. The flat 2D JDOS for parabolic bands makes it easy to derive some trends, even if the conclusions are not very accurate for narrow-gap wells. In Section 11.4 we will develop the theory of optical gain in quantum structures.

In the limit of wide wells we already saw that there was a smooth crossover from bulk to quantum-well absorption spectra in Fig. 11.1c. A similar convergence occurs for superlattices with variable barrier thickness. Thinner barriers promote coupling between the neighboring wells, and the structures eventually turn into bona fide superlattices. As the minibands in the superlattices become wider, we recover the expected bulk absorption spectrum with its monotonic increase of the absorption coefficient with photon energy. Figure 11.12 shows an example of absorption spectra for a series of GaAs quantum wells with AlGaAs barriers of varying width.

We have already discussed type II superlattices such as the two-layer InAs/GaInSb structure in Chapter 10 (see Fig. 10.8). When the hole wells of such a structure are not too thick, the electron wavefunction extends throughout the SL, i.e., it is three-dimensional in nature, while the hole wavefunction remains confined to the hole well. When $k_z = \frac{\pi}{d}$, the electron envelope function flips its sign from one period to the next. This is because the phase factor $\exp\left(ik_z\Delta z\right)$

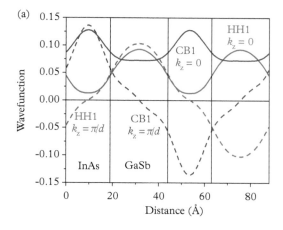

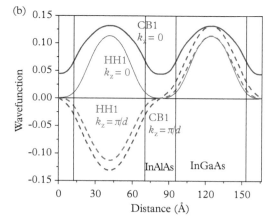

**Figure 11.13** *S-like component of the envelope function for the CB1 subband (blue) and the $p_x$-like component of the HH1 subband (red) at $k_z = 0$ (solid) and $k_z = \frac{\pi}{d}$ (dashed) in type II 19-Å InAs/25-Å GaSb (a) and type I 58-Å In$_{0.75}$Ga$_{0.25}$As/25-Å In$_{0.52}$Al$_{0.48}$As (b) superlattices. Two periods of each superlattice are shown.*

becomes $\exp(i\pi M)$, where $\Delta z = Md$, and $M$ is an integer. The phase factor is $+1$ when $M$ is even and $-1$ when $M$ is odd. Figure 11.13a illustrates this behavior of the envelope functions at $k_z = 0$ and $k_z = \frac{\pi}{d}$. Because the sign flip takes place in the middle of the hole well, which doubles as the electron barrier, the electron–hole overlap disappears as $k_z \to \frac{\pi}{d}$.

The growth-direction variation in Fig. 11.13a is very different from what happens in a type I superlattice, in which both envelope functions peak in the same layer and switch signs at the same time, as shown in Fig. 11.13b. Figure 11.14 compares the absorption coefficient at normal incidence for a 58-Å In$_{0.75}$Ga$_{0.25}$As/25-Å In$_{0.52}$Al$_{0.48}$As type I superlattice with energy gap

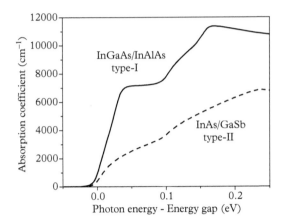

**Figure 11.14** *Absorption spectra at normal incidence (in-plane polarization) for type II 19-Å InAs/25-Å GaSb (on GaSb) and type I 58-Å $In_{0.75}Ga_{0.25}As$/25-Å $In_{0.52}Al_{0.48}As$ (on InP) superlattices at room temperature. To plot results on the same scale, the energy gap is subtracted from the horizontal axis.*

0.7 eV to that of a 19-Å InAs/25-Å GaSb type II superlattice with energy gap 0.27 eV. Apart from the differences in gaps and densities of states, the absorption mostly depends on the electron–hole overlap (squared) $|O_{eh}|^2$. It is close to one (92%) for the type I structure, but is a factor of 2 lower (45%) in the type II superlattice. Both superlattices feature nearly 2D confined HH subbands (with the LH1 subband ≈ 100 meV below HH1), while the CB1 miniband is 30 (230) meV wide in the type I (type II) superlattice. Because the electron wavefunction penetrates the barrier in both structures, the best comparison is in terms of the absorption per unit length rather than the absorbed fraction per pass.

We can see the effect of the growth-direction dispersion on the absorption more clearly in the "W" structure that was already discussed in Section 10.4. In this case, we will consider an 18-Å InAs/25-Å $Ga_{0.7}In_{0.3}$sb/18-Å InAs/AlSb four-layer superlattice structure. The InAs layers act as effective hole barriers and confine the HH subbands to the GaInSb layer. The width of the electron miniband is mostly controlled by the AlSb barrier. As the barrier thickness increases, the electron miniband shrinks, and the electrons become more two-dimensional. The absorption coefficient near the band edge increases and usually displays a peak or an inflection point. This is because the oscillator strength associated with the entire miniband becomes concentrated in a much narrower energy range. Figure 11.15 illustrates this behavior for the AlSb barrier thicknesses between 10 and 30 Å. The CB1 miniband width ranges from 105 meV for the 10-Å barrier to 31 meV for the 20-Å barrier, and to 5 meV for the 50-Å barrier. The second feature in the spectra is due to the CB1–LH1 transition and can be used to measure the HH1–LH1 splitting. The splitting is large when the GaInSb hole well is narrow

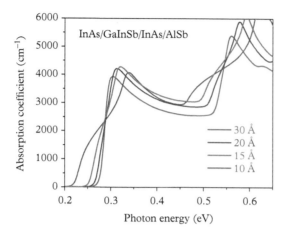

**Figure 11.15** *Absorption spectra at room temperature for type II 18-Å InAs/25-Å Ga$_{0.7}$In$_{0.3}$Sb/18-Å InAs/AlSb four-layer "W" superlattices with AlSb thicknesses between 10 and 30 Å. As the barrier becomes very thin, a transition from 2D to 3D density of states for electrons occurs. The holes remain 2D because they are confined by the InAs layers.*

and under heavy compressive strain (1.85%). On the other hand, it is not very sensitive to the AlSb thickness.

## 11.4 Optical Gain in Quantum Wells and Superlattices

We already know from Chapter 4 that absorption turns into gain when the conduction and valence subband populations are inverted, i.e. when there are enough electrons near the bottom of the conduction band and enough holes near the top of the valence band. This condition is stated mathematically as $f_v\left(E_{vk,v}\right) - f_c\left(E_{ck,\mu}\right) < 0$ in Eq. (11.66). The seeming simplicity conceals a subtlety: the occupation factors are applied to the quantum states before the broadening so that the inverted transitions can be swamped by the tails of the non-inverted higher-energy transitions for carrier densities close to transparency. Therefore, it appears that the transparency condition might be modified so the separation between the quasi-Fermi levels exceeds the energy gap by a small margin, approximately equal to the broadening linewidth. This may not be strictly valid from the thermodynamic standpoint, but the outcome of the calculation remains little changed. We will return to this point below when we compute the gain in real quantum wells.

More importantly, it turns out that in quasi-2D systems, i.e., quantum wells, we can write a simple expression for the optical gain useful for quick estimates of the transparency carrier density and differential gain. To derive this expression, we start with Eq. (1.43) and modify it to express the carrier density in a 2D parabolic, isotropic subband:

$$n = \frac{1}{(2\pi)^2} 2 \int_{E_{c1}}^{\infty} \frac{2\pi k f_c(E)}{dE/dk} dE = \frac{m_e}{\pi \hbar^2} \int_{E_{c1}}^{\infty} \frac{1}{1 + \exp\left[(E - E_F)/k_B T\right]} dE \quad (11.67)$$

where $E_{c1}$ is the energy minimum of the lowest conduction subband. Next, we carry out the integration

$$n = \frac{m_e k_B T}{\pi \hbar^2} \ln\left[1 + \exp\left(\frac{(E_F - E_{c1})}{k_B T}\right)\right] \equiv N_c \ln\left[1 + \exp\left(\frac{E_F - E_{c1}}{k_B T}\right)\right]$$

$$= -N_c \ln\left[1 - f_c(E_{c1})\right] \quad (11.68)$$

where $N_c$ is the electron effective density of states, defined by analogy with Eq. (1.50). We can invert the last expression to give the occupation factor at energy $E_{c1}$ in terms of $n$:

$$f_c(E_{c1}) = 1 - \exp\left(-n/N_c\right) \quad (11.69)$$

We could readily generalize this expression to multiple subbands, but we will not do so, bearing in mind that the result will in any event be approximate. Holes have a one-to-one correspondence to electrons with negative energies ($f_h = 1 - f_v$), so the occupation factor at the top of the valence band in the quantum well is

$$f_v(E_{v1}) = \exp\left(-p/N_v\right) \quad (11.70)$$

where $p$ is the hole density, the hole effective density of states is $N_v \equiv \frac{m_h k_B T}{\pi \hbar^2}$, and $E_{c1} - E_{v1} = E_g$.

Why do we wish to express the occupation factors for the band-edge states in terms of carrier densities? This is because the absorption coefficient (and gain) depend explicitly on the occupation factors. If we specialize Eq. (11.56) to the photon energy equal to the gap and multiply by the band-edge occupation factors from Eqs. (11.69) and (11.70), we obtain

$$g_{QW}(E_g) = -\alpha_{QW}(E_g) = \frac{8\pi^2 \alpha_f \, |M|^2}{n_r \quad E_g} \mathcal{J}DOS \left[f_c(E_{c1}) - f_v(E_{v1})\right]$$

$$\equiv g_s \left[1 - \exp\left(-\frac{n}{N_c}\right) - \exp\left(-\frac{p}{N_v}\right)\right] \quad (11.71)$$

where the factor $g_s$ is independent of the carrier densities. The gain in this expression is the maximum realized as a function of photon energy because $f_c(E_{ck}) - f_v(E_{vk})$ peaks at the band edges, while JDOS remains constant at all energies.

A key characteristic is how steep the increase of the gain with carrier density is. To this end, we assume that the electron and hole densities are nearly equal $n \approx p$, and differentiate Eq. (11.71) with respect to carrier density. We will confirm in

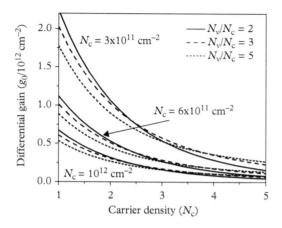

**Figure 11.16** *Differential gain in Eq. (11.72) for different values of $N_c$ and different $\frac{N_v}{N_c}$ ratios.*

Chapter 12 that the approximate equality is indeed a reasonable assumption. The resulting differential gain is

$$\frac{dg_{QW}}{dn}(E_g) = g_s\left[\frac{1}{N_c}\exp\left(-\frac{n}{N_c}\right) + \frac{1}{N_v}\exp\left(-\frac{n}{N_v}\right)\right] \tag{11.72}$$

The differential gain is maximized when the effective densities of states $N_c$ and $N_v$ are small, as shown in Fig. 11.16. In fact, this result holds even when the dispersion is nonparabolic.

The gain at threshold must be positive. Therefore, the threshold carrier density $n_{th}$ is bounded from below by the transparency carrier density $n_{tr}$, for which $g_{QW}(E_g) = 0$. The maximum differential gain with $g_{QW} > 0$ is observed just above $n_{tr}$ and drops off quite quickly with carrier density, as shown in Fig. 11.16. The behavior is similar in wider-gap quantum wells, as illustrated by the gain spectra of the InGaAs quantum well on GaAs for several carrier densities in Fig. 11.17a.

From Eq. (11.71), transparency occurs when

$$\exp\left(-\frac{n_{tr}}{N_c}\right) + \exp\left(-\frac{n_{tr}}{N_v}\right) = 1 \tag{11.73}$$

Unfortunately, there is no closed-form solution to this equation. To make further progress, we look at two important limits: (1) $N_c \approx N_v$ and (2) $N_c \ll N_v$, for which we can calculate the approximate results

$$n_{tr} \approx N_c \ln 2 \quad (N_c \approx N_v) \tag{11.74a}$$

$$n_{tr} \approx N_v \exp\left(-\frac{n_{tr}}{N_c}\right) \approx N_c\left[\frac{1}{3} + \frac{2}{3}\ln\left(\frac{N_v}{N_c}\right)\right] \quad (N_c \ll N_v) \tag{11.74b}$$

The last expression in Eq. (11.74b) is nothing more than a fit. We see that in most III–V semiconductor quantum wells with $N_c < N_v$, we have $n_{tr} \gtrsim N_c$. If we substitute $N_c$ from Eq. (11.68), the typical transparency sheet density becomes $n_{tr} \sim 5\text{–}15\times10^{11}$ cm$^{-3}$ ($5\text{–}15\times10^{17}$ cm$^{-3}$ if divided by typical well widths). This is close enough to the results of full calculations based on Eq. (11.66), as shown in Fig. 11.17b for quantum wells in several material systems. The transparency density in Eq. (11.74b) is a monotonically increasing function of $N_c$, and we know from Eq. (11.68) that $N_c$ scales approximately with energy gap. Therefore, $n_{tr}$ is noticeably lower in narrow-gap materials if we hold the other variables, such as well width and amount of strain, fixed. In fact, transparency in GaSb-based InGaAsSb wells emitting at 2–3 µm occurs at about half the carrier density of the InGaAs well emitting at 1 µm.

How does the transparency carrier density vary with temperature? The calculated values of $n_{tr}$ for three model quantum-well systems used in semiconductor lasers (see Chapter 12) are shown in Fig. 11.17c for the temperature range between 250 and 350 K. Because $n_{tr} \propto N_c \propto T$ in Eq. (11.74), we expect an approximate linear dependence when only one parabolic subband is occupied. Near room temperature, multiple valence subbands are usually occupied (with the possible exception of the mid-IR "W" structure), and the density of states increases faster than in the parabolic case even for a single subband in narrow-gap materials. Therefore, power-law fits ($n_{tr} \propto T^v$) in Fig. 11.17c yield $v = 1.4\text{–}1.5$ rather than 1. We will see how this result affects the performance of semiconductor lasers in Chapter 12.

To calculate an explicit expression for the differential gain in Eq. (11.72), we assume a CB1–HH1 transition, with $\theta \approx 0$ and TE-polarized light. The relevant matrix element is given by Eq. (11.63a), which we reproduce here, multiplied by the square of the envelope-function overlap:

$$|M_{HH}|^2 \approx \frac{\hbar^2}{2m}\frac{E_P}{2}|O_{eh}|^2 \qquad (11.75)$$

The maximum interband differential gain in Eq. (11.72) becomes

$$\left.\frac{dg_{QW}}{dn}\right|_{max} = \frac{4\pi^2\alpha_f}{n_r}\frac{E_P}{E_g}|O_{eh}|^2\left[\frac{\hbar^2}{2m}\mathscr{J}DOS\right]\left[\frac{1}{N_c}\exp\left(-\frac{n_{tr}}{N_c}\right)+\frac{1}{N_v}\exp\left(-\frac{n_{tr}}{N_v}\right)\right] \qquad (11.76)$$

Now we substitute the normalized JDOS from Eq. (11.55):

$$\left.\frac{dg_{QW}}{dn}\right|_{max} = \frac{\pi\alpha_f}{n_r}\frac{E_P}{E_g}\frac{m_r}{m}|O_{eh}|^2\left[\frac{1}{N_c}\exp\left(-\frac{n_{tr}}{N_c}\right)+\frac{1}{N_v}\exp\left(-\frac{n_{tr}}{N_v}\right)\right] \qquad (11.77)$$

If we also assume $N_c \ll N_v$, substitute the transparency density from Eq. (11.74b), and use the transparency condition Eq. (11.73), we obtain using the electron mass $m_e \approx mE_g/E_P$

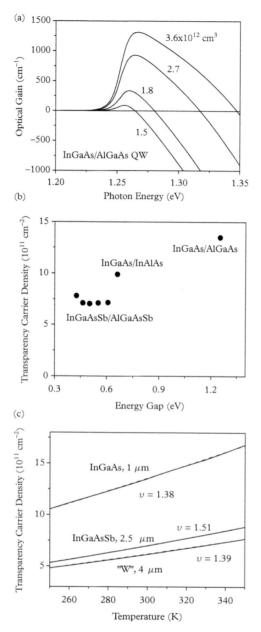

**Figure 11.17** *(a) Optical gain as a function of photon energy for an 80-Å-wide In$_{0.21}$Ga$_{0.79}$As/AlGaAs quantum well on GaAs (1.5% compressive strain) at several values of injected carrier density. (b) Transparency carrier densities for 80-Å-wide quantum wells in different material systems (InGaAs/Al$_{0.3}$Ga$_{0.7}$As on GaAs, InGaAs/InAlAs on InP, InGaAsSb/AlGaAsSb on GaSb with As fractions of 5, 10, 15, 20, and 25%) as a function of energy gap. All the wells are under 1.5% compressive strain, and all barriers are lattice matched. The AlGaAsSb barriers contain 25% Al for the InGaAsSb quantum wells with As fractions of 5, 10, and 15%, and 35% Al for the wells with As fractions of 20 and 25%. (c) Variation of the transparency carrier densities with temperature for the InGaAs/AlGaAs (E$_g$ = 1.2 eV, λ$_g$ = 1 μm), InGaAsSb/AlGaAsSb (E$_g$ = 0.5 eV, λ$_g$ = 2.5μm), and the "W" structure of Fig. 11.15 (E$_g$ = 0.3 eV, λ$_g$ = 4 μm) quantum wells at temperatures in the range 250–350 K. Fits to a power-law form are shown as dashed lines.*

$$\left.\frac{dg_{QW}}{dn}\right|_{max} \approx \frac{\pi^2 \alpha_f}{n_r} \frac{\hbar^2}{m_e k_B T} |O_{eh}|^2 \left(\frac{m_e}{m_h}\right)^{2/3} \tag{11.78}$$

Therefore, compressively strained hole wells with a small hole mass discussed in Section 10.1 are expected to have higher differential gain if the gap and electron mass are held approximately fixed.

How do the peak and differential gain for interband transitions compare with intersubband transitions studied in Section 11.2? To calculate the peak intersubband gain, we make use of Eqs. (11.19) and (11.41) for a transition between the two lowest subbands of a well with infinite barriers. Retaining the overlap integral and ignoring the angular factors for the moment, we obtain

$$g_{ib,max} \approx \frac{4\pi\alpha_f}{\Delta_l \left(E_{c,2} - E_{c,1}\right) n_r} \left(\frac{8\hbar^2}{3m_e t}\right)^2 |O_{12}|^2 \, (n_2 - n_1)$$

$$= \frac{512\alpha_f}{27\pi n_r} \frac{\hbar^2}{m_e \Delta_l} |O_{12}|^2 \, (n_2 - n_1) \tag{11.79}$$

where $O_{12} = \frac{3t}{16} \int dz \left( (F_s^2)^* \frac{\partial F_s^1}{\partial z} + \frac{\partial (F_s^2)^*}{\partial z} F_s^1 \right)$, as defined in Eq. (11.19), and $E_{c,2} - E_{c,1} = \frac{3\hbar^2}{2m_e} \frac{\pi^2}{t^2}$. If only the upper subband is occupied, and the barriers are much higher than the subband energies, Eq. (11.79) becomes

$$\left.\frac{dg_{ib}}{dn}\right|_{max} \approx \frac{512\alpha_f}{27\pi n_r} \frac{\hbar^2}{m_e \Delta_l} |O_{12}|^2 \tag{11.80}$$

Combining Eqs. (11.78) and (11.80), we set $E_{c,2} - E_{c,1} = E_g$ and calculate the ratio of the maximum differential gain for interband and intersubband transitions (electron masses are not necessarily the same):

$$\frac{\left.\frac{dg_{QW}}{dn}\right|_{max}}{\left.\frac{dg_{ib}}{dn}\right|_{max}} \approx 1.64 \left(\frac{m_{e,QW}}{m_{h,QW}}\right)^{2/3} \frac{m_{e,ib}}{m_{e,QW}} \frac{\Delta_l}{k_B T} \frac{|O_{eh}|^2}{|O_{12}|^2} \tag{11.81}$$

For quantum wells, we know that $m_h > m_e$, but not by a large margin, and for mid-infrared intersubband transitions we generally have $\Delta_l \lesssim k_B T$. Also, the electron mass in the quantum wells used for interband transitions (e.g., InAs) is usually somewhat smaller than in the wells used for intersubband transitions (e.g., InGaAs) at the same energy. With all these factors put together, we conclude the interband and intersubband differential gains are the same order of magnitude provided both transitions are spatially direct. We can tip the scales in either direction by using a type II interband or diagonal intersubband transition. In fact,

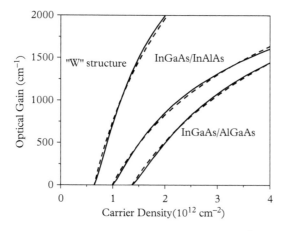

**Figure 11.18** *Optical gain as a function of carrier density for an 80-Å*
*$In_{0.21}Ga_{0.79}As/Al_{0.3}Ga_{0.7}As$ well (1.5% compressive strain, grown on GaAs), an 80-Å*
*$In_{0.75}Ga_{0.25}As/In_{0.52}Al_{0.48}As$ well (1.5% compressive strain, grown on InP), and an 18-Å*
*InAs/25-Å $Ga_{0.7}In_{0.3}Sb$/18-Å InAs/AlSb type II "W" structure (grown on GaSb). Fits to the*
*logarithmic functional form of Eq. (11.82) are also displayed, with the values of $g_0 = 3100$,*
*2700, 4000 $cm^{-1}$ for the InGaAs on GaAs, InGaAs on InP, and the "W" structure, respectively.*

these stand for the same physical configuration, in which the envelope functions
of the initial and final states are mostly contained in different layers.

The "natural" expressions for intersubband and quantum-well interband differential gain use units of area (or cross section). The units are the same as for
bulk differential gain, even though the natural units for the carrier density differ.
Of course, this is because the natural units for gain are also different ($cm^{-1}$ in
bulk versus no units in a quantum well). Semiconductor lasers in most cases use
optical waveguides that can be handled with a straightforward unit conversion, as
discussed below.

The optical gain as a function of carrier density for several different quantum
wells is shown in Fig. 11.18. The examples include an InGaAs/AlGaAs well grown
on GaAs ($E_g = 1.25$ eV), an InGaAs/InAlAs well grown on InP ($E_g = 0.67$ eV),
and an InAs/GaInSb/InAs/AlSb type II "W" structure with a thick AlSb barrier
grown on GaSb ($E_g = 0.29$ eV). The calculations assume equal electron and hole
densities. The transparency carrier density decreases as the gap narrows because
the JDOS is now lower, mostly because of the smaller $m_e$ and hence $N_c$ (see Fig.
11.17b). Even though $|O_{eh}|^2 \approx 0.5$ for the type II "W" structure, the differential
gain is higher than in type I wells! We see from Eq. (11.78) that this is due mostly to
the smaller $m_e$ in narrow-gap materials, i.e., InAs for the "W" structure in question.

Even though the magnitudes of the differential gain and transparency density
in Figs. 11.17 and 11.18 vary, the gain versus carrier density curve keeps the
same shape. Namely, it transitions from approximately linear slope near the

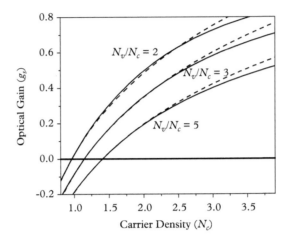

**Figure 11.19** *Comparison of the two different functional forms for the optical gain in quantum wells in Eqs. (11.71) (solid) and (11.82) (dashed lines) for three different values of $N_v/N_c$. The electron and hole densities are taken to be equal, and the slope near the transparency density is set to the same value.*

transparency point to sublinear, slowly saturating behavior at high carrier densities. Many analytical fits to this functional form are possible. Probably the simplest (but not necessarily the best!) is the logarithmic function:

$$g(n) = g_0 \ln \left( \frac{n}{n_{tr}} \right) \tag{11.82}$$

where in general the constant factor $g_0 \neq g_s$ from Eq. (11.71). Fits to the functional form of Eq. (11.82) are shown in Fig. 11.18. Equations (11.71) and (11.82) can be linearized near the transparency density, and to obtain the same slope we set $g_0 = x g_s$, where $x < 1$ is a function of $N_v/N_c$. Figure 11.19 shows that the two curves diverge only at high carrier densities. In addition, Fig. 11.18 indicates that these densities are high enough to be of little importance in the operation of photonic devices such as semiconductor lasers, as we will see in detail in Chapter 12.

How does the differential gain vary with temperature? Equation (11.78) tells us that in the simple case of a single parabolic subband, it scales inversely, or that the product of the transparency carrier density and differential gain is approximately independent of temperature. This implies a constant $g_0$ because the maximum differential gain for this form is simply $g_0/n_{tr}$, with values of $2–7 \times 10^{-9}$ cm in 2D units from the fits in Fig. 11.18. In reality, $g_0$ does vary with temperature, but the variation is quite slow.

Figure 11.18 implies that Eq. (11.82) is a reasonably accurate model of the gain curves in many different materials, including both bulk-like layers and quantum wells. To be sure, this fitted form makes no explicit reference to bulk band

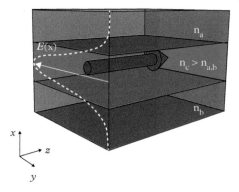

**Figure 11.20** *Schematic illustration of light propagating in a waveguide mode confined in a higher-index layer sandwiched between two lower-index layers (claddings).*

parameters, but its simplicity is a virtue when it comes to estimating the laser threshold current density, as we will see in Chapter 12. In principle, we could do a little better by introducing another parameter, as is sometimes done in the literature, e.g., $g(n) = g_0 \ln\left(\frac{n+n_1}{n_{tr}+n_1}\right)$. To keep matters simple, we will not do so in this book.

In the waveguide-based optical device illustrated in Fig. 11.20, light propagates in the plane of the quantum wells. The optical beam is confined primarily to the core that has a higher refractive index than the outer (cladding) layers. In a non-waveguide device, it makes a countable number of passes at some angle with respect to the plane. Most optical waveguides also feature at least some lateral confinement, but the growth-direction confinement is usually much stronger. Therefore, we need to calculate the absorption and gain per unit length in the plane of the structure. But we already mentioned in Chapter 4 that only a fraction of the optical mode interacts with the well/barrier region. What is this fraction?

The optical flux is given by Eq. (4.5), which we reproduce below:

$$\Phi = \frac{|\boldsymbol{E} \times \boldsymbol{H}|}{2\hbar\omega} \tag{11.83}$$

We can restate the problem as one of finding what fraction of the flux propagates in the QW medium. As long as the frequency dispersion is weak, this is equivalent to calculating what fraction of the electromagnetic energy given by Eq. (4.3) propagates in the wells. Since the optical flux propagates along the waveguide axis, $\boldsymbol{E} \propto e^{i\beta x}$ and the magnetic field is orthogonal to the electric field, $\boldsymbol{E} \cdot \boldsymbol{H} = 0$. Therefore, its magnitude is proportional to the square of the dominant electric-field component. The waveguide modes can be separated into two families: (1) with electric field in the plane $E_y$ (TE) or (2) with a component of the electric field along the growth direction (TM) $E_z$. For TM modes, there is also a component of the field along the propagation direction. In most of the cases discussed in this book, $E_x \ll E_z$.

The fraction of the light propagating in the wells is the *optical confinement factor*, which is defined a little differently for TE and TM modes:

$$\Gamma_{TE} = \frac{\int_0^{Mt} dz \, n_r(z) \left| E_y(z) \right|^2}{n_m \int_{-\infty}^{\infty} dz \, \left| E_y(z) \right|^2} \tag{11.84a}$$

$$\Gamma_{TM} = \frac{n_m \int_0^{Mt} dz \, n_r(z) \left( |E_x(z)|^2 + |E_z(z)|^2 \right)}{\int_{-\infty}^{\infty} dz \, n_r^2(z) |E_z(z)|^2} \tag{11.84b}$$

where $M$ is the number of wells, $n_m \equiv \beta/k_0$, $k_0 = 2\pi/\lambda$, and the origin is at one edge of the quantum-well region. Unless the refractive index is very different in different layers, Eqs. (11.84a) and (11.84b) give similar values. We can then use the simplified form

$$\Gamma = \frac{\int_0^{Mt} dz \, |\mathbf{E}(z)|^2}{\int_{-\infty}^{\infty} dz \, |\mathbf{E}(z)|^2} \tag{11.85}$$

This expression is typically within 10% of the values in Eq. (11.84).

In the calculations performed in Chapters 10 and 11, we often treat quantum wells as superlattices with barriers that are wide enough to prevent the envelope function from penetrating to neighboring wells, but otherwise chosen arbitrarily. In that case, optical gain and absorption are inversely proportional to the total normalization period that includes the well and barrier thickness. Therefore, in Eq. (11.85) we replace $t$ with period *provided* all optical properties are calculated consistently.

Provided $t \ll \lambda$, the optical flux within a given quantum well is nearly constant. The confinement factor for a single well becomes

$$\Gamma_{SQW} = \frac{t|\mathbf{E}(0)|^2}{\int_{-\infty}^{\infty} dz \, |\mathbf{E}(z)|^2} \equiv \frac{t}{d_{mode}} \tag{11.86}$$

If multiple identical wells are present, the common procedure is to multiply $\Gamma_{SQW}$ in Eq. (11.86) by the number of wells $M$. This is not applicable to widely separated wells, for which the electric field magnitude can be noticeably different, as observed for the cascade lasers discussed in Chapters 12 and 13. In that case, we can always beat a strategic retreat and use Eq. (11.84) to calculate the confinement factor.

Equation (11.86) casts the confinement factor as the ratio of the well width (or normalization period) to the modal extent $d_{mode}$. Therefore, to convert the

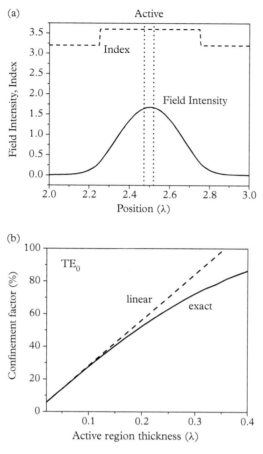

**Figure 11.21** *(a) Typical waveguide structure and mode profile and (b) optical confinement factor for the fundamental TE-polarized mode calculated numerically (solid) and using the linear form in Eq. (11.86) (dashed) versus active region thickness. The index of the separate-confinement region and the active region is 3.6, while the index of the claddings is 3.2. The thicknesses are normalized by wavelength, and the total thickness of the separate-confinement and active regions is assumed to be one-half the wavelength.*

absorption per pass in Eq. (11.66) into coefficient per unit length in a waveguide, we can simply divide by the width of the well (since the light no longer "passes through the well") and multiply by $\Gamma$ (to account for the fraction actually within the well). This is equivalent to dividing $\alpha_{QW}$ or $g_{QW}$ by the modal extent: $\alpha = \alpha_{QW}/d_{mode}$. It should make sense to the reader that the well width drops out here, since we have observed that the absorption per pass in a 2D system is universal (see Eq. (4.47)). Furthermore, if the optical gain or absorption is normalized to an arbitrarily chosen period, as discussed earlier, that period cannot enter observable characteristics. All this assumes that the absorption is not strong enough to alter the refractive index by much, which is true for most quantum wells with absorption per pass of order 1%.

The introduction of the confinement factor resolves the apparent tension between the definitions of gain for bulk and quantum-well materials. In a "bulk" material, the gain is proportional to the thickness of the active material (if it is not too thick), which is related to the conversion of the gain in an infinite bulk material to the modal gain in the core of an optical waveguide. For quantum wells, the confinement factor is proportional to the number of wells, but the well width in Eq. (11.86) merely cancels out $t$ that was introduced to convert from gain per pass to gain per unit length of the well. Ultimately, each well is treated as an indivisible (quasi-atomic) whole. This is true of interband and intersubband transitions alike. Of course, when the well becomes thick enough that multiple subbands are occupied and possibly contribute to gain, the distinction drawn here is blurred, which explains the conceptual confusion between 3D and 2D units you might find in the literature.

What happens if the waveguide is not planar, meaning that the mode extent is expressed in terms of area $A_{eff}$ rather than length $d_{mode}$? The integration in Eq. (11.85) should then be two-dimensional. The width of the well still drops out in the calculation, but the total absorption/gain may be lower if the mode spills out of the region containing the well(s) in the lateral direction. In most cases including the ubiquitous ridge waveguide, the lateral confinement factor does not become much less than one.

## 11.5    Absorption and Gain in Wurtzite Quantum Wells

As we learned in Section 3.5, the HH wavefunctions have the same form in the wurtzite and zinc-blende crystals. For this reason, Eq. (4.49a) applies to both crystal structures. On the other hand, the LH and CH wavefunctions in the wurtzite crystal contain $p_x, p_y$ as well as $p_z$ components. To find the form of the wavefunctions, we substitute the eigenenergies in Eq. (3.75) back into Eq. (3.74):

$$\psi^+_{LH} = -\sqrt{1 - r_z^2}\,(p_x + ip_y) \downarrow \, + r_z p_z \uparrow \qquad\qquad (11.87a)$$

$$\psi^+_{CH} = r_z\,(p_x + ip_y) \downarrow \, + \sqrt{1 - r_z^2}\, p_z \uparrow \qquad\qquad (11.87b)$$

where $r_z = \frac{\sqrt{2}\Delta_s}{E_{LH} - E_v}$. When the SO splitting is small, as it is in all of the interesting nitrides, the $p_x, p_y$ components dominate the LH wavefunction, while $p_z$ is more prominent in the CH wavefunction.

How do these results translate to quantum wells? The strength of the $p_z$ component in the top LH subband of a nitride well depends on the amount of strain and quantum confinement, and whether the strain is compressive or tensile. For most practical nitride wells, the top subband is still HH, so the CB–HH matrix elements are similar to Eqs. (11.59) and (11.60):

$$\left|M_{HH,x}^{QW}\right|^2 = \frac{1}{4}\left(1+\cos^2\theta\right)|O_{eh}|^2 P^2 \tag{11.88a}$$

$$\left|M_{HH,z}^{QW}\right|^2 = \frac{1}{2}\sin^2\theta\ |O_{eh}|^2 P^2 \tag{11.88b}$$

with $\left|M_{HH,x}^{QW}\right|^2 (k_\| = 0) = \frac{|O_{eh}|^2 P^2}{4}$ and $\left|M_{HH,z}^{QW}\right|^2 (k_\| = 0) = 0$ for the band-edge states. By analogy with Eqs. (11.63c) and (11.63d), the CB–LH transitions at the zone center have the matrix elements

$$\left|M_{LH,x}^{QW}\right|^2 (k_\| = 0) = 1 - r_z^2 |O_{el}|^2 P^2 \tag{11.89a}$$

$$\left|M_{LH,z}^{QW}\right|^2 (k_\| = 0) = r_z^2 |O_{el}|^2 P^2 \tag{11.89b}$$

To obtain the two CB–CH matrix elements at the zone center, we flip the two polarizations (see Eq. (11.87)). For wide wells, the overlap factor in Eqs. (11.88) and (11.89) can play an important role because of the internal fields. This is quite different from the zinc-blende materials, in which wide wells asymptotically approach the bulk behavior.

Since the $z$ axis is not equivalent to the in-plane axes, the momentum matrix element is in general different along the two directions. This means we should really use $P_x^2$ and $P_z^2$ instead of $P^2$ in Eqs. (11.88) and (11.89), depending on the polarization direction of the incident light. Even so, the two matrix elements are nearly the same in materials that are compositionally close to GaN, so we will elide that distinction here.

Using the full Hamiltonian of Eq. (3.76), we compute the polarization dependence of absorption in the 30-Å-wide GaN/AlGaN and InGaN/GaN quantum wells that we already considered in Section 10.5. The InGaN well is under compressive strain because both structures are assumed to be grown on GaN substrates or buffers. Figure 11.22 shows that as expected, the band-edge absorption is nearly all $x$-polarized and is due to the CB1–HH1 and CB1–LH1 transitions that merge with realistic broadening. In the GaN wells, the CB1–CH1 transition interacts mostly with the $z$-polarized light. But when we introduce compressive strain to the well (by adding In), the transition shifts to much higher energies.

Now we are ready to inject electrons and holes into the system and to compute the optical gain. Figure 11.23 shows the results for the InGaN/GaN well as a function of photon energy and for a few values of the carrier density. Not included into the calculations is that the electron and hole envelope functions do not have

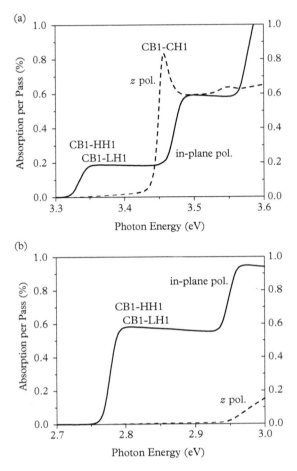

**Figure 11.22** *Absorption per pass for in-plane and out-of-plane-polarized light in 30-Å GaN/Al₀.₃Ga₀.₇N (a) and 20-Å In₀.₂Ga₀.₈N/GaN quantum wells. In both cases, the growth is assumed to be on a substrate or buffer with the GaN lattice constant.*

the same centroid position, and the internal field that develops due to the charge separation counteracts the built-in field to some extent, as already mentioned in Section 10.5. Because the densities of states are much larger in both the conduction and valence bands, the transparency carrier density is higher and the differential gain lower than in a zinc-blende well with moderate energy gaps (see Eqs. (11.74) and (11.78) in Section 11.4). However, take note that the calculations in Fig. 11.23 do not include self-consistent electrostatic and many-body effects, which increase the gain by boosting the electron–hole wavefunction overlap and the Coulomb interaction.

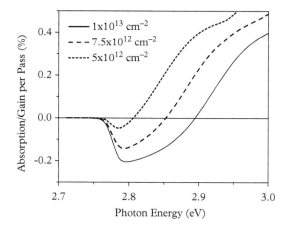

**Figure 11.23** *Optical gain spectra for a 20-Å $In_{0.2}Ga_{0.8}N/GaN$ quantum well for three sheet carrier densities above transparency. Negative absorption corresponds to optical gain.*

## 11.6   Radiative Recombination in Quantum Wells

The absorption and gain in a semiconductor medium are inextricably connected to how much light the medium emits. In general, a structure that hardly absorbs at a given wavelength also barely emits there when carriers are injected externally. Conversely, strong absorption leads to high optical gain and fast radiative recombination upon excitation. We already discovered the thermodynamic origin of the absorption/emission link for bulk semiconductors in Chapter 4. The story remains much the same for quantum structures.

To compute the radiative recombination rate in a quantum well, we generally follow the path charted for bulk materials that culminated in Eq. (4.71), but without a factor of $1/(2\pi)$ and one of the $k$ integrations:

$$R_{sp}^{QW,\perp,\parallel}(\hbar\omega) = \frac{\alpha_f n_r}{2\pi^2} \frac{1}{\hbar^3 c^2} \sum_{\mu,v} \int d^2 k_\parallel \, |\langle c, \mu | \nabla_{\boldsymbol{k}} H_{2-band} | v, v \rangle|^2$$

$$\times \frac{(\hbar\omega)^3}{\left(E_{ck,\mu} - E_{vk,v}\right)^2} \sigma\left(E_{ck,\mu} - E_{vk,v} - \hbar\omega\right) f_c\left(E_{ck,\mu}\right)\left[1 - f_v\left(E_{vk,v}\right)\right]$$

$$(11.90)$$

Since we have not yet included any polarization information, this expression differs from Eq. (4.71) by a factor of 2 (as well as the extra $k$ integration and the associated factor of $1/(2\pi)$).

The three spatial axes are no longer equivalent in a quantum well, so we cannot simply multiply by a factor of 2 to account for the polarization. Instead, we perform a more detailed averaging: $\frac{R_{sp}^x + R_{sp}^y + R_{sp}^z}{3}$, where $R_{sp}^x$ is the spontaneous emission rate for light propagating along the $x$ direction, etc. There are two possible polarizations for each in-plane propagation axis, in and out of the plane. Both polarizations for out-of-plane propagation are in the plane. Therefore, the average is $\frac{4R_{sp}^{\parallel}+2R_{sp}^{\perp}}{3}$, where $R_{sp}^{\parallel}$ ($R_{sp}^{\perp}$) is the spontaneous emission rate for light polarized in (out of) the plane. We can simplify somewhat by neglecting $R_{sp}^{\perp}$ for quantum wells with large compressive strain that pushes the light-hole subbands well down and out of the picture. In that case, the HH subbands near the top of the valence band in that case do not have substantial $p_z$ admixture, so that the right-hand side of Eq. (11.90) is replaced with $\frac{4}{3}R_{sp}^{\parallel}$.

This sets us up well for a numerical calculation, but let us estimate what to expect first. To this end, we reach into our customary bag of tricks: neglect the broadening, consider only transitions from one valence to one conduction subband, account for two sets of spin transitions by multiplying by a factor of 2, assume a constant optical matrix element, and approximate the joint density of states using the parabolic, isotropic form of the in-plane dispersion. The first two steps give the following:

$$
R_{sp}^{QW}(\hbar\omega) = \frac{2\alpha_f n_r}{3\pi^2} \frac{\hbar\omega}{\hbar^3 c^2} \int d^2 k_{\parallel} \, |\langle c, 1|\nabla_{\boldsymbol{k}} H_{2-band}|v, 1\rangle|^2
$$
$$
\times \, \delta\left(E_{ck,1} - E_{vk,1} - \hbar\omega\right) f_c\left(E_{ck,1}\right)\left[1 - f_v\left(E_{vk,1}\right)\right] \tag{11.91}
$$

With the factor of $1/(2\pi)^2$ subsumed into the JDOS, the other steps lead to

$$
R_{sp}^{QW}(\hbar\omega) = \frac{16\alpha_f n_r \hbar\omega}{3\hbar^3 c^2}|M|^2 \, JDOS \, f_c\left(\frac{m_r}{m_e}\left(\hbar\omega - E_g\right)\right)\left[1 - f_v\left(\frac{m_r}{m_h}\left(\hbar\omega - E_g\right)\right)\right]
$$
$$
\tag{11.92}
$$

For transitions between an electron subband and a heavy-hole subband, we can use the matrix element in Eq. (11.75), which we reproduce here:

$$
|M_{HH}|^2 \approx \frac{\hbar^2}{2m}\frac{E_P}{2}|O_{eh}|^2 \tag{11.93}
$$

Substituting into Eq. (11.92), we have

$$
R_{sp}^{QW}(\hbar\omega) = \frac{8\alpha_f n_r \hbar\omega E_P}{3\hbar^3 c^2}|O_{eh}|^2\left[\frac{\hbar^2}{2m} JDOS\right] f_c\left(1 - f_v\right) \tag{11.94}
$$

We can also insert the normalized JDOS from Eq. (11.55):

$$
R_{sp}^{QW}(\hbar\omega) = \frac{2\alpha_f n_r \hbar\omega E_P}{3\pi\hbar^3 c^2}\frac{m_r}{m}|O_{eh}|^2 f_c\left(1 - f_v\right) \tag{11.95}
$$

Now we just need to integrate over all photon energies to obtain the total radiative emission rate:

$$R_{sp}^{QW} = \frac{2\alpha_f n_r}{3\pi} \frac{E_P}{\hbar^3 c^2} \frac{m_r}{m} |O_{eh}|^2 \int_{E_g}^{\infty} \hbar\omega \, d(\hbar\omega) \, f_c(1 - f_v) \tag{11.96}$$

The integral above is straightforward if the carrier densities are non-degenerate and given by Eq. (4.76). In that case, we obtain

$$f_c(1 - f_v) \approx \exp\left(-\frac{\hbar\omega - E_g}{k_B T}\right) \frac{n}{N_c} \frac{p}{N_v} \tag{11.97}$$

where $n$ and $p$ are now 2D (sheet) rather than 3D densities. Note also that Eq. (11.97) follows directly from Eq. (11.68) by taking $E_{c1} \gg E_{Fc}$ and $E_{v1} \ll E_{Fv}$. By analogy with Eq. (4.82), we define

$$R_{sp}^{QW} = B_{QW}(E_g, T)\left(np - n_i^2\right) \tag{11.98}$$

The radiative recombination coefficient for quantum wells given by

$$B_{QW} = \frac{2\alpha_f n_r}{3\pi} \frac{E_P}{\hbar^3 c^2} \frac{m_r}{m} \frac{k_B T}{N_c N_v} |O_{eh}|^2 \left[E_g \int_0^{\infty} e^{-x} dx + k_B T \int_0^{\infty} x e^{-x} dx\right] \tag{11.99}$$

Assuming a unity wavefunction overlap in a type I well, we perform the integrals to simplify Eq. (11.99):

$$B_{QW} \approx \frac{2\pi \alpha_f n_r}{3} \frac{E_P}{mc^2} \frac{\hbar}{m_e + m_h} \left(\frac{E_g}{k_B T} + 1\right) \tag{11.100}$$

This is the analog of Eq. (4.84) for bulk semiconductors. We must not forget that Eqs. (11.100) and (4.84) apply strictly only to low carrier densities in a material with parabolic bands. For more accurate results, Eq. (11.90) should be evaluated numerically.

How high is the radiative rate in typical quantum wells? That depends on the electron and hole densities. If $n, p \gg n_i, N_A$, where $N_A$ is the doping density in the well, the recombination rate is proportional to $np$. If the electron and hole densities are equal, e.g., when charge neutrality is maintained under electrical injection or photoexcitation, $R_{sp}^{QW} \propto n^2$. Alternatively, if $n, p \gg n_i$ and $n, p \ll N_A$, $R_{sp}^{QW} \propto nN_A$ because the majority-carrier density approaches $N_A$. If the electron and hole densities are not much higher than $n_i$, we must evaluate the full expression.

Substituting typical values such as $T = 300$ K, $n_r = 3.5$, $E_g = 0.8$ eV, $E_P = 25$ eV, $m_e = 0.04m$, and $m_h = 0.15m$, we obtain $B_{QW} \approx 5 \times 10^{-4}$ cm$^2$/s. It may be easier and more useful to remember the recombination lifetime for an injected carrier: $\tau_r = 1/(nB_{QW})$ or

$$\tau_r = \frac{1}{N_A B_{QW}} \tag{11.101}$$

If $n$ and $N_A$ are not too high, e.g., $10^{11}$ cm$^{-2}$, the recombination lifetime is $\sim$20 ns. The lifetime becomes longer as the gap shrinks and as the wavefunction over-lap $|O_{eh}|^2$ is reduced in a type II well. The radiative emission rates for the In$_{0.75}$Ga$_{0.25}$As/In$_{0.52}$Al$_{0.48}$As type I and InAs/Ga$_{0.7}$In$_{0.3}$Sb/InAs/AlSb ("W") type II quantum wells calculated using Eq. (11.90) are plotted in Fig. 11.24. As in Section 11.4, we choose hole wells with $\approx$1.5% compressive strain because they are commonly used in the lasers and LEDs discussed in Chapter 12. They also provide the clearest illustration of quasi-2D behavior because the HH1–LH1 splitting is substantial. In contrast to the differential gain in Fig. 11.18, the smaller electron mass and the reduced wavefunction overlap lead to weaker radiative recombination and a longer radiative lifetime in the "W" structure. For low carrier densities, the radiative lifetime scales inversely with carrier density, as expected from Eq. (11.98). At higher carrier densities, the decrease slows down, with typical values of 30–50 ns for the "W" structure and 7–15 ns for the type I well.

To account for the growth-direction dispersion in a superlattice, we restore the integration over $k_z$. Figure 11.25 shows the radiative recombination coefficient $B$ for the InAs/GaInSb and InAs/InAsSb type II superlattices with an energy gap of 0.23 eV, and the longer-period InAs/InASb superlattice with an energy gap 0.11 eV as a function of temperature. The radiative recombination in a superlattice can be slower than in comparable quantum wells because the quasi-3D electrons in a superlattice spread out over a wider energy range. As expected from Eq. (11.100), the radiative coefficient is higher at low temperatures, where the carriers populate only states below or around the Fermi level. The $B$ values for these superlat-tices, which are commonly used in detector structures discussed in Chapter 14, range from $5\times10^{-11}$ to $1\times10^{-10}$ cm$^3$/s for operating temperatures in the range 50–150 K.

As discussed in Chapter 4, we *cannot* compare these estimates directly to the experimental results for structures with thicknesses comparable to the absorption depth. This is because some of the emitted photons are re-absorbed in the detector (or elsewhere) before they escape from the semiconductor sample. This effect is particularly salient in detector and solar-cell structures, which must be thick enough to absorb most of the incident radiation. For more on this *photon-recycling* effect, see Chapter 12.

Intersubband transitions can also produce photon emission. How do those processes compare with the interband rates in Fig. 11.24? To calculate the intersubband emission rate, we retrace our steps back to Eq. (11.90) and consider transitions between the two lowest subbands (CB1 and CB2). Since these transitions couple almost exclusively to in-plane-propagating (growth-direction-

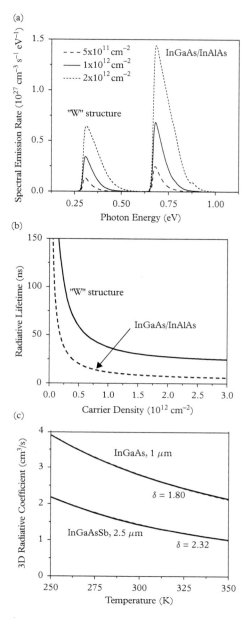

**Figure 11.24** *(a) Radiative emission spectra for a few carrier densities and (b) radiative lifetime as a function of carrier density for 80-Å In$_{0.75}$Ga$_{0.25}$As/In$_{0.52}$Al$_{0.48}$As type-I (1.5% compressive strain, grown on an InP substrate) and 18-Å InAs/25-Å Ga$_{0.65}$In$_{0.35}$Sb/18-Å InAs/50-Å AlSb type II "W" quantum wells. Emission in the "W" structure is at much lower photon energies (longer wavelengths) than in the type I well. (c) Temperature dependence of the radiative recombination coefficient at a sheet carrier density of $1 \times 10^{12}$ cm$^{-2}$ for the InGaAs/AlGaAs, InGaAsSb/AlGaAsSb, and the "W" quantum wells of Fig. 11.17c for temperatures in the range 250–350 K. Fits to a negative power-law functional form are shown as dashed lines.*

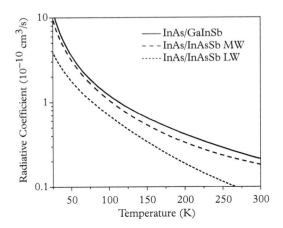

**Figure 11.25** *Radiative recombination coefficient as a function of temperature for 45-Å InAs/10 Å InAs$_{0.5}$Sb$_{0.5}$ and 18-Å InAs/25-Å Ga$_{0.7}$In$_{0.3}$Sb superlattices with the same gap of 0.23 eV at T = 150 K and the 100-Å InAs/22-Å InAs$_{0.5}$Sb$_{0.5}$ superlattice with a gap of 0.11 eV at T = 77 K. All of the structures are assumed to be grown on GaSb substrates. For simplicity, the energy gaps and electronic states are fixed at their values at T = 150 and 77 K, while the carrier statistics vary with temperature. The calculations also assume fully activated doping levels of $10^{15}$ cm$^{-3}$ (p-type for the InAs/GaInSb SL and n-type for the InAs/InAsSb SLs).*

polarized) light, the average reduces to $\frac{2R_{sp}^{\perp}}{3}$:

$$R_{sp}^{QW,isb}(\hbar\omega) = \frac{\alpha_f n_r}{3\pi^2} \frac{\hbar\omega}{\hbar^3 c^2} \int d^2 k_{\parallel} \, |\langle c, 2|\nabla_{\boldsymbol{k}} H_{Kane}|c, 1\rangle|^2$$
$$\times \sigma\left(E_{ck,2} - E_{ck,1} - \hbar\omega\right) f_c\left(E_{ck,2}\right)\left[1 - f_v\left(E_{ck,1}\right)\right] \qquad (11.102)$$

To make analytical progress, we assume that the matrix element is given by Eq. (11.21) and that the lower subband is completely empty:

$$R_{sp}^{QW,isb}(\hbar\omega) = \frac{\alpha_f n_r}{3\pi^2} \frac{\hbar\omega}{\hbar^3 c^2} \int d^2 k_{\parallel} \left(\frac{8\hbar^2}{3m_e t}\right)^2 \sigma\left(E_{ck,2} - E_{ck,1} - \hbar\omega\right) f_c\left(E_{ck,2}\right)$$
$$\qquad (11.103)$$

For two subbands with the same in-plane dispersion, the recombination rate in this expression is proportional to the carrier density in the upper subband: $n_2 = \frac{1}{2\pi^2} \int d^2 k_{\parallel} \, f_c\left(E_{ck,2}\right)$. Of course, we need to integrate over photon energy to obtain the overall recombination rate. As in the interband case, it is easier to think of the recombination lifetime for a given non-equilibrium electron than of the net radiative rate for the entire electron population in the upper subband. To convert between the two quantities, we use the expression

$$\frac{n_2}{\tau_{r,isb}} = \frac{2\alpha_f n_r}{3} \int_0^\infty d(\hbar\omega) \frac{\hbar\omega}{c^2} \frac{64\hbar}{9m_e^2 t^2} \left[ \frac{1}{2\pi^2} \int d^2 k_\parallel \, \sigma(\Delta E - \hbar\omega) f_c(E_{ck,2}) \right]$$

$$(11.104)$$

where $\Delta E \equiv E_{c,2} - E_{c,1}$. If the line shape function is narrow enough, we can move it out of the second integral and divide both sides by $n_2$:

$$\frac{1}{\tau_{r,isb}} = \frac{128\alpha_f n_r}{27} \int_0^\infty d(\hbar\omega) \frac{\hbar\omega}{c^2} \frac{\hbar}{m_e^2 t^2} \sigma(\Delta E - \hbar\omega)$$

$$(11.105)$$

In a well with infinite barriers, we have $\Delta E = \frac{3\hbar^2\pi^2}{2m_e t^2}$ and rewrite this expression as

$$\frac{1}{\tau_{r,isb}} = \frac{256\alpha_f n_r}{81\pi^2} \int_0^\infty d(\hbar\omega) \frac{\Delta E \, \hbar\omega}{\hbar m_e c^2} \sigma(\Delta E - \hbar\omega)$$

$$(11.106)$$

The broadening in a high-quality well is a small fraction of $\Delta E$. Once again, we assume that the photon energy remains approximately constant over the spectral bandwidth of the intersubband transition. This assumption allows us to take $\hbar\omega = \Delta E$ and to replace the broadening lineshape with a delta function:

$$\frac{1}{\tau_{r,isb}} \approx \frac{256\alpha_f n_r}{81\pi^2} \frac{\Delta E}{\hbar} \frac{\Delta E}{m_e c^2}$$

$$(11.107)$$

Finally, we are ready to substitute the typical parameters for intersubband transitions. For example, with $\Delta E = 0.2$ eV, $n_r = 3.4$, and $m_e = 0.04m$, we obtain $\tau_r \approx 40$ ns. Even though the expression for the interband radiative lifetime in Eq. (11.101) looks very different, the resulting values are comparable. To find out why, we consider the ratio of intersubband and interband lifetimes in Eqs. (11.100), (11.101), and (11.107):

$$\frac{\tau_{r,isb}}{\tau_{r,ib}} \approx \frac{27\pi^3}{128} \frac{m_e}{m_e + m_h} \frac{\hbar^2 N_A}{m\Delta E} \frac{E_P}{k_B T} \frac{E_g}{\Delta E}$$

$$(11.108)$$

This ratio is close to one for our choice of parameters, as expected from the same fundamental origin of interband and intersubband transitions. All the factors in Eq. (11.108) are of order unity (if we think of $\frac{\hbar^2 N_A}{m\Delta E} \frac{E_P}{k_B T}$ as a single factor). Figure 11.26 shows the calculated lifetime for the lowest intersubband transition in the InGaAs/InAlAs quantum well as a function of well width. The computation includes the numerically computed matrix elements and transition energies and assumes that the lower subband is completely empty. The estimate in Eq. (11.107) is quite close to the results in Fig. 11.26, although the agreement is not as good for wider wells and smaller transition energies; i.e., the lifetime does not really vary as $(\Delta E)^{-2}$. In some cases, we may also be interested in calculating the total radiative transition rate between subbands, which is an integral of the rate in

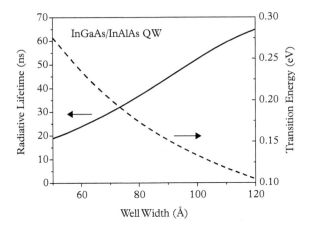

**Figure 11.26** *Radiative electron lifetimes and transition energies for the lowest intersubband transitions in $In_{0.53}Ga_{0.47}As/In_{0.52}Al_{0.48}As$ quantum wells as a function of well width. The lifetime is for an electron in the second subband, as calculated numerically for a completely full second subband with a density of $10^{11}$ $cm^{-2}$ and a completely empty first subband. Lorentzian broadening with a constant linewidth $\Delta_l = 10$ meV is also assumed.*

Eq. (11.102) over all occupied states. For incompletely inverted subbands, that total rate is proportional to the population inversion factor $(n_2 - n_1)/(n_2 + n_1)$.

The calculations in Fig. 11.26 assume single-particle intersubband transitions between inverted subbands. However, we already saw in Section 11.2 that many-body effects can modify these transitions considerably. The most dramatic change is the precipitous drop in the emission lifetime that occurs if the transitions are coherently coupled via intersubband or multisubband plasmons. This is in fact due to *superradiant* emission of an electron excited to the upper subband. The essence of the effect is that we know that one of the electrons is excited, but do not know which so that the excited state is a coherent superposition over all electrons present in the system rather than one particular electron as in Eq. (11.107) and Fig. 11.26. Superradiance requires the phase to remain constant; hence, it operates only within a wavelength-size area. As a result, the emission rate is enhanced by a factor of $n_1(\lambda/n_r)^2$. This factor can exceed a million for sheet doping densities of order $10^{13}$ $cm^{-2}$. Superradiant emission is a strong function of emission angle, as we might expect from the $\frac{\sin^2\theta}{\cos\theta}$ factor in Eq. (11.38). However, the emission maximum occurs at an intermediate angle rather than $\theta = 90°$ because the emission at large angles is "too strong" to couple effectively to free space. For details, see the references under "Suggested Reading."

In fact, intersubband and multisubband plasmons are just a few varieties of the more general class of excitations, known as *polaritons*. For example, an intersubband polariton is due to the coupling of the intersubband oscillator strength to a single mode of the electromagnetic field (provided the interaction

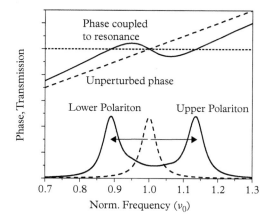

**Figure 11.27** *Schematic illustration of the strong coupling between a resonant excitation in the material (e.g., an intersubband transition) and a single cavity mode. The phase profiles corresponding to the left-hand side of Eq. (11.109a) are shown with and without material resonances. The transmission resonances with and without material excitations are also shown. The splitting between the polariton modes is symmetric in wavelength units or approximately symmetric if it is small on the scale of the center frequency.*

between the mode and the transition is large enough). How large must it be and how do we isolate a single mode from the continuum of such modes in free space with the density given by Eq. (4.68)? The interaction strength is expressed in terms of the loss rates for the intersubband excitation and the mode of the field. The former is the rate at which an electron excited to the upper subband relaxes to the lower subband (relaxation rate) or is scattered to a different state in the same subband (dephasing rate). The latter is the rate at which the photons in that mode escape to the outside or are absorbed within a lossy material. The interaction strength is essentially the rate at which an electronic excitation that corresponds to an electron in the upper subband and an empty state in the lower subband emits or absorbs a photon from the mode of the field. As discussed in Section 4.4, a single narrow mode of the field is available in a high-quality optical cavity and must be tuned very close to the intersubband line to improve the coupling.

If the interaction is larger than both loss rates, the line due to the intersubband absorption/emission splits into two, as shown schematically in Fig. 11.27. This regime is referred to as *strong coupling* (as opposed to weak coupling when the interaction is smaller than the loss). In this case, we can formally write the two absorption/emission peaks as being due to an in-phase and out-of-phase combination of the intersubband excitations and the cavity mode. When the interaction is so strong that it becomes comparable to the energy of the mode/excitation, we call that the *ultrastrong coupling* regime. It has been demonstrated experimentally for intersubband polaritons at long wavelengths.

The language of polariton physics may seem esoteric, but relevant to simple interpretations of many experiments, particularly with many material excitations coupled to a single mode. In such experiments, we observe only the transmission and/or reflection of the light entering and exiting the cavity. For these purposes, we assume a single longitudinal mode of the planar Fabry–Perot cavity, but we can readily generalize this discussion, even to include microcavities. Without the material absorption resonance, the cavity resonances correspond to standing waves fitting within the length of the cavity:

$$k_0 L_{cav} n_r = 2\pi \nu_l L_{cav} n_r = l\pi + \phi_0 \qquad (11.109a)$$

$$\nu_l = \frac{l\pi + \phi_0}{2\pi L_{cav} n_r} \qquad (11.109b)$$

where $\nu_l = 1/\lambda_l$ is the wavenumber, $\phi_0$ is the total phase shift at the two mirrors, and we assume for simplicity that the spectral dispersion of the refractive index is negligible in the absence of the resonance.

As we mentioned, the interaction reaches a maximum when the cavity mode frequency is tuned to the resonant material excitation. Because there is some absorption at the resonance, the real part of the refractive index changes in a well-defined way. This is because absorption appears in the imaginary part of the same complex quantity, which must be an analytic function of frequency. As a result, the two parts are connected via Kramers–Kronig relations. If the maximum absorption at the resonant wavevector $\nu_0 \approx \nu_l$ is $\alpha_0$, and the half-width in wavenumbers is $\delta_l$, the real part of the refractive index becomes approximately

$$n_r(\nu) \approx n_{r0} - \frac{1}{2}\frac{\alpha_0 \delta_l (\nu - \nu_0)}{(\nu - \nu_0)^2 + \delta_l^2} \qquad (11.110)$$

where $n_{r0}$ is the refractive index in the absence of the resonance. The refractive index becomes higher (lower) than $n_{r0}$ just below (above) the resonance, and remains unchanged when $\nu \approx \nu_0$. Therefore, as illustrated in Fig. 11.27, the original standing-wave condition is satisfied at frequencies $\nu_l$, $\nu_+$, and $\nu_-$, with

$$\nu_\pm \mp \nu_l \propto \alpha_0 \delta_l \qquad (11.111)$$

when the absorption is not too strong. However, when $\nu_0 \approx \nu_l$, the incident light does not couple well to the cavity because not much of it returns after a round trip through the absorbing medium. This prevents it from setting up the standing-wave interference near the absorption resonance. The result is only two resonances in the transmission and absorption, at nearly the same peak frequencies $\nu_\pm$, which correspond to the in-phase and out-of-phase polaritons. This discussion holds whenever the absorbing resonance has an approximately symmetric line shape (not necessarily Lorentzian), for any value of $l$, including

$l = 0$ in subwavelength cavities, as long as the cavity mode is determined by some kind of phase relationship that involves the complex permittivity modified by the resonant absorption. It also correctly predicts the loss and linewidth of the polariton resonances. This is the most natural description of strong-coupling effects when there are many material excitations coupled to the cavity mode. A quantum-optical description is more suitable when only one or a few emitters interact with a single mode.

## 11.7 Excitonic Effects in Quantum Wells

We already saw in Section 4.5 that excitons have higher binding energies in 2D than in 3D. Real quantum wells occupy a position intermediate between the 3D and 2D cases. This is because we cannot "squeeze" the well so it is narrow enough to disregard the vertical dimension entirely. Once the confinement energies reach the band offsets, narrow wells lose their confined states, and no excitons associated with the well can be observed. As a result, the modified Bohr radius in a narrow quantum well $a_{0,QW}$ is only about a third smaller than in bulk, and the binding energy for a GaAs/AlGaAs quantum well tops out at $\approx 9.5$ meV (as compared to $\approx 4$ meV in bulk GaAs). Figure 11.28 plots the typical dependence of the binding energy on the well width for a GaAs/Al$_{0.3}$Ga$_{0.7}$As quantum well.

Because the binding energy is higher in a quantum well, not only the HH exciton associated with CB1–HH1 transitions, but also the LH exciton due to CB1–LH1 transitions are experimentally accessible. To calculate the absorption per pass, we use the results of Sections 4.5 and 11.3. For the HH exciton, the absorption coefficient for $x$-polarized light at normal incidence is deduced by starting from Eq. (11.2) and following the same steps as for the bulk exciton:

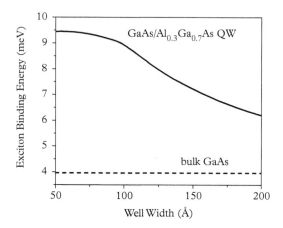

**Figure 11.28** *Variation of the binding energy of the fundamental CB–HH exciton with well width for an isolated GaAs/Al$_{0.3}$Ga$_{0.7}$As quantum well.*

$$\alpha_{QW}(\hbar\omega) = \frac{2\pi^2 \hbar \alpha_f E_P}{m \omega n_r A_{ex}} |O_{eh}|^2 \sigma (E_{ex} - \hbar\omega) \tag{11.112}$$

where $A_{ex}$ is the effective exciton area, and $|O_{eh}|^2$ is the square of the CB1–HH1 wavefunction overlap. To determine the effective exciton area, we again use the results from Section 4.5 to calculate

$$A_{ex} = 2\pi \int_0^\infty e^{-2r/a_{0,QW}} \, r \, dr = \frac{\pi a_{0,QW}^2}{2} \tag{11.113}$$

Substituting into Eq. (11.112), we obtain

$$\alpha_{QW}(\hbar\omega) = \frac{4\pi \alpha_f}{n_r} \frac{E_P}{\hbar\omega} \frac{\hbar^2}{m a_{0,QW}^2} |O_{eh}|^2 \sigma (E_{ex} - \hbar\omega) \tag{11.114}$$

We can compare the peak values of the absorption coefficients for bulk and quantum-well excitons in Eqs. (4.101) and (11.114) by multiplying the bulk absorption by the well width $t$ and assuming the same broadening and refractive index. We obtain

$$\frac{\alpha_{QW}}{\alpha_{bulk}} = \frac{2a_0^3}{t a_{0,QW}^2} |O_{eh}|^2 \tag{11.115}$$

For a well width of 50 Å, $a_0 = 130$ Å, $a_{0,QW} = 85$ Å, and $|O_{eh}|^2 \approx 1$, which means that exciton absorption in the well can be stronger by an order of magnitude for the same path length (for some number of passes or propagation in a waveguide). Unfortunately, the linewidths of the quantum-well excitons are often several times larger than in bulk because unavoidable interface fluctuations broaden their lines and decrease the peak absorption.

We see from Eq. (11.63) that the $x$-polarized (TE-polarized) absorption of the CB1–LH1 exciton is three times weaker than in Eq. (11.114), and $z$-polarized (TM-polarized) light couples only to the LH excitons. In addition to excitons associated with the lowest HH and LH transitions, higher-order excitons also exist. Instead of individual lines, they form peaks on the background that appears at the onset of the interband transitions.

As in bulk materials, as the energy gap of a quantum well becomes wider, the binding energy increases, and exciton transitions are more widely separated from the continuum interband spectrum. While they are nearly indiscernible in narrow-gap antimonide wells, even at low temperatures, they are much more pronounced in GaAs quantum wells. The binding energy reaches 30–40 meV in GaN/AlGaN wells and allows exciton transitions to be observed at room temperature. Because of the dependence on the electron–hole wavefunction overlap in Eq. (11.114), exciton transitions are much weaker in spatially indirect type II structures. This is a pity because the energy gap of type II wells can be tuned much more easily. Finally, the higher binding energy in a quantum well means that they are less susceptible to screening-induced bleaching at high carrier densities.

Because of their atomic-like nature, excitons are natural candidates for strong coupling and were in fact used for that purpose before intersubband transitions. The discussion of coupling to a single cavity mode in Section 11.6 applies to excitons in semiconductors (and other materials!) without any significant changes. The main difference is that excitons are well separated from band-to-band transitions at higher energies ($> 1$ eV), whereas intersubband transitions are most convenient when the energy is a fraction of an eV.

Excitons are hardly limited to III–V semiconductors. In fact, they occur in a vast range of materials that includes large and small organic molecules, wide-gap insulators, and monolayers of transition-metal dichalcogenides (TMDs) such as $MoS_2$, $WS_2$, $MoSe_2$, $MoTe_2$. They also come in several different varieties. For example, in some molecular solids, the Bohr radius is so small that the exciton is localized around a well-defined atomic position. However, the TMD excitons with transition energies in the visible mostly maintain their extended character despite the relatively large binding energies, as we will see below.

Like graphene, TMDs have the basic hexagonal 2D lattice structure, but with a different basis that includes two chalcogenide atoms (S, Se, or Te) for each transition-metal atom (Mo or W). Figure 11.29 shows in a plane view that this produces what appears to be a honeycomb lattice, but without a center of inversion symmetry, because the transition-metal atoms are displaced with respect to the lattice of the chalcogenide atoms. We already know from Chapters 2 and 8 that this means that the states are not spin degenerate, even at the Brillouin zone center. In fact, as in graphene, the lowest-energy states are located at the two $K$ points, but with opposite spins because of the differences between the transition-metal and chalcogenide atoms. In contrast to III–V semiconductors and graphene, the $K$-point states in TMDs have mostly $d$-orbital character. Whereas bulk TMD materials are indirect, TMD monolayers have a direct gap with large electron and hole effective masses that range from 0.3 to $0.7m$. This leads to exciton binding energies on the order of 0.5 eV, i.e. almost two orders of magnitude larger than in the III–V semiconductors! The corresponding Bohr radii are a few nm, i.e., smaller than in III–V materials, but still spanning a few lattice constants as do other spatially extended excitons. The large binding energies also mean that an entire series of exciton transitions may be observable, as shown in Fig. 11.30. However, in most practical cases, we are only interested in the ground-state exciton, which is by far the strongest and most distinct.

The spin splitting in TMD monolayers is much larger in the valence band than in the conduction band (a few hundred versus a few tens of meV), since the spin–orbit interaction is much stronger there. In fact, even the sign of the CB spin splitting can vary, depending on whether the transition metal is Mo or W. Of course, when electrons and holes are excited incoherently, both types of $K$ valleys are occupied at the same time, and both exciton transitions are seen

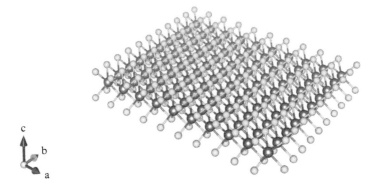

**Figure 11.29** *Lattice structure of TMD monolayers. The yellow atoms are chalcogenides (S, Se, or Te) and the darker atoms are transition metals (Mo or W).*

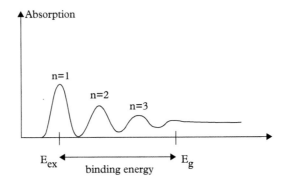

**Figure 11.30** *The series of transitions due to excitons with different quantum numbers in an ideal two-dimensional semiconductor. The magnitude of the band-to-band absorption, which sets in at energies greater than the energy gap, is enhanced by the Coulomb interaction.*

at any temperature. As discussed in Section 4.3, a change in the spin angular momentum by $m_s = \pm 1$ accompanies a definite circular polarization of the emitted photon. If we think of the photon as carrying an angular momentum $m_l = \mp 1$, the vertical angular momentum is conserved, as shown in Fig. 11.31. Because the spin states are flipped for inequivalent $K$ points, the spin and valley information are tightly correlated. As we already mentioned, the degeneracy of the $K$ points means that no overall spin or circular polarization is observed in equilibrium. Even if spin-polarized electrons are injected into the TMD monolayers, the circular polarization of the emitted photons is limited by valley and spin relaxation, as well as by the fact that the CB spin splitting is at best comparable to $k_B T$ at room temperature. Conversely, if we induce exciton transitions with circularly polarized light, the optical polarization determines both the spin state and the valley in

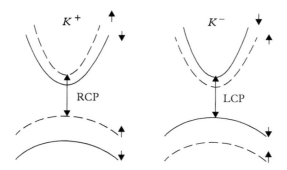

**Figure 11.31** *Illustration of two spin-selected transitions in a TMD monolayer that couple to right circularly polarized (RCP) or left circularly polarized (LCP) light. The two transitions occur at the same energies at inequivalent K points, which are indicated here as $K^+$ and $K^-$. The solid and dashed lines represent two different spin states, shown here as up and down.*

which the generated carriers reside. This remains true until the valley and spin populations equilibrate or the carriers recombine, whichever process occurs faster.

Analogs of type II structures can also occur in TMD monolayers. For example, when a monolayer of $WSe_2$ is stacked next to a monolayer of $MoSe_2$, the holes reside in $WSe_2$, while electrons mostly occupy the $MoSe_2$. The binding energy is reduced, and the recombination lifetime becomes longer because of the smaller electron–hole overlap. The emission dipole also acquires a strong out-of-plane character that contrasts with the 2D in-plane dipoles of the individual TMD monolayers. This is an example of how some of the concepts developed for III–V semiconductors can help us to understand novel materials!

## 11.8 Absorption, Gain, and Radiative Recombination in Quantum Wires and Dots

For most epitaxially grown structures, quantum confinement is along the vertical direction only. But this does not mean that multidimensional confinement is of no technological importance. The optical properties of quantum wires and dots are also well worth discussing because they have curious features that are not present in quantum wells and superlattices. To get some sense of the differences, we substitute the relevant JDOS and $|M|^2$ into Eq. (11.56). For the example of 1D, the JDOS for transitions between one pair of parabolic conduction and valence subbands is

$$\frac{\hbar^2}{2m}JDOS^{QWR} = \frac{\hbar^2}{2m}\frac{1}{2\pi}\left[\frac{dE_{ch}}{dk}\right]^{-1}_{E_{ch}=\hbar\omega} = \frac{1}{4\pi}\frac{m_r}{m}\frac{\hbar}{\sqrt{2m_r\left(\hbar\omega - E_g\right)}} \qquad (11.116)$$

where for simplicity the photon energy is taken to be the same as the transition energy $\hbar\omega = E_{ch}$. Another example might be a quantum wire made of a material with graphene-like linear dispersions $E_{ch} = 2\hbar v_F k$:

$$\frac{\hbar^2}{2m} \mathcal{J}DOS^{QWR} = \frac{1}{8\pi} \frac{\hbar}{m v_F} H\left(\hbar\omega - E_g\right) \qquad (11.117)$$

Figure 11.32 illustrates that Eqs. (11.116) and (11.117) have radically different forms. The JDOS in Eq. (11.116) diverges at the band edge $\hbar\omega = E_g$. On the other hand, Eq. (11.117) has the same functional form (the Heaviside step function at the energy gap) as the quantum-well JDOS in Eq. (11.55). In practice, any singularities are removed by broadening, and transitions between multiple subbands are present. But which JDOS is appropriate for a material with a narrow gap? To answer this question, we derive a more accurate JDOS from the expression for the transition energy $E_{ch} = \sqrt{E_g^2 + k^2 P^2}$ (compare with Eq. (4.59b)):

$$\frac{\hbar^2}{2m} \mathcal{J}DOS^{QWR} = \frac{\hbar^2}{4\pi m P} \frac{\hbar\omega}{\sqrt{(\hbar\omega)^2 - E_g^2}} \qquad (11.118)$$

Figure 11.32 shows that the joint density of states for the more accurate dispersion of Eq. (11.118) without any broadening diverges at $\hbar\omega = E_g$ just as the parabolic form. Of course, this is just because Eq. (11.118) reduces to Eq. (11.116) in that limit. An interesting observation is that the divergence is removed *even with no broadening* when the energy gap is precisely zero. In the limit of a very narrow gap, the divergence can be flattened with even a very small amount of broadening.

There is a progression from the JDOS for a bulk-like material (Eq. (4.28)), to a quantum well (Eq. (11.55)), and finally to a quantum wire (Eq. (11.116)). The natural units for the JDOS change from $J^{-1}cm^{-3}$ to $J^{-1}cm^{-2}$ to $J^{-1}cm^{-1}$. In tandem, the absorption (and normalized JDOS) units progress from per unit length to per pass (dimensionless), and finally to the absorption-length product $\alpha_{QWR}$. What are we to make of this behavior? The absorption per unit length enters Beer's law (that follows from Eq. (4.31)). For an optically thin quantum well, it is natural to measure how much is absorbed or gained per pass. We can relate this value to the case of waveguide propagation by normalizing by the mode extent in Eq. (11.86). If the light is focused on a quantum-well sample from an edge, and no waveguide is present, the beam spot size plays the role of mode extent.

How we use the absorption metric in quantum wires also depends on the direction of the light propagation, which can point along one of the confinement directions or along the wire axis. In practice, the light beam extends much farther laterally than the individual wires. If the light is incident on a wire array along, say the $x$ axis in Fig. 10.23, we divide the absorption-length product $\alpha_{QWR}$ by the period of the array along the $y$ axis. In this case, we recover the absorption per pass that is characteristic of a quantum well.

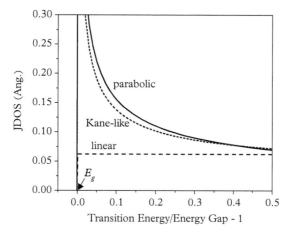

**Figure 11.32** *Joint density of states for the quantum wire with dispersion along the wire axis that follows the functional forms in Eqs. (11.116), (11.117), and (11.118). The assumed parameters are $E_g = 0.4$ eV, $m_r = 0.04m$, $P = \sqrt{\hbar^2 E_P/2m}$, $E_P = 25$ eV, and $v_F = 10^8$ cm/s, and no broadening is present.*

However, if the light is incident along the wire axis, it is likely propagating in an optical waveguide. In this case, we divide $\alpha_{QWR}$ by the modal area of the 2D (e.g., ridge) waveguide, or for a planar waveguide, by the product of the modal extent $d_{mode}$ and the array period along the orthogonal direction.

In much the same way we obtained Eq. (11.66), we write a general expression for the absorption in a quantum wire:

$$\alpha_{QWR,j}(\hbar\omega) = \frac{2\pi\alpha_f}{n_r} \sum_{\mu,\nu} \int dk \left| M_{\mu\nu,j}^{QWR} \right|^2 \frac{\hbar\omega}{\left(E_{ck,\mu} - E_{\nu k,\nu}\right)^2} 5em\sigma \left(E_{ck,\mu} - E_{\nu k,\nu} - \hbar\omega\right)$$

(11.119)

where $j$ refers to the polarization direction. We can include the occupation factors if the carrier densities are sufficiently high.

The relative dimensions of the wires determine the polarization dependence of the absorption. A simple example from Chapter 10 is a cylindrical quantum wire with the wavefunctions for the two top hole subbands given by Eqs. (10.46) and (10.47). The subbands are described by the total angular momentum $\mathbf{F} = \mathbf{J} + \mathbf{N}$, where $\mathbf{N}$ is the orbital angular momentum with respect to the wire ($z$) axis. We will only consider the absorption at its band-edge peak, which corresponds to $k = 0$.

Equation (10.46) shows that the top hole subband is dominated by the $p_z$-like component. The square of the matrix element is four times larger for $z$-polarized light than for the in-plane polarization. In contrast, transitions that involve the second hole subband absorb only $x$- and $y$-polarized light, because it has no $p_z$-like component. Therefore, for cylindrical wires, we can use Eqs. (11.63), (10.46), and (10.47) to write the matrix elements for band-edge transitions:

$$\left| M_{1/2,xy}^{QWR} \right|^2 (k=0) = \frac{1}{6} P^2 |O_{eh}|^2 = \frac{\hbar^2}{2m} \frac{E_P}{6} |O_{eh}|^2 \tag{11.120a}$$

$$\left| M_{1/2.z}^{QWR} \right|^2 (k=0) = \frac{2}{3} P^2 |O_{eh}|^2 = \frac{\hbar^2}{2m} \frac{2E_P}{3} |O_{eh}|^2 \tag{11.120b}$$

These equations include the envelope function overlap defined here as

$$|O_{eh}|^2 = \left| \iint dx \, dy \, \left( F_s^{CB} \right)^* (x,y) \, F_{x,y}^{HH} (x,y) \right|^2 \tag{11.121}$$

now with double integration. For a type I quantum wire, $|O_{eh}|^2 \approx 1$ unless at least one of the carriers is confined so weakly that the wavefunction spills out of the wire. The wavefunction for either electrons or holes may also peak outside the wire in a type II alignment. Most commonly, the same material surrounds the entire wire, when the wires are prepared by epitaxial overgrowth of an etched structure.

Does this result also hold for wires with square cross sections? To answer this question, we calculate the absorption spectrum for a 50-Å × 50-Å GaAs/AlGaAs quantum wire using the full band structure, matrix elements, and Eq. (11.119) and display the results in Fig. 11.33a. The $z/xy$ polarization ratio at $\hbar\omega \approx E_g$ is 4.5:1 rather than 4:1. If $t_x < t_y$, the band-edge absorption of the $x$-polarized light becomes stronger at the expense of $y$-polarized and, to some extent, $z$-polarized absorption, as illustrated in Fig. 11.33b. For transitions from CB1 to the second hole subband, the $x/y$ polarization ratio also departs from unity.

To extend these conclusions to wires with compressive and tensile strain, we need a detailed model of the strain tensor. The biaxial-strain model developed in Chapter 3 is no longer applicable here. The key is to consider whether the strain breaks the $x - y$ symmetry for wires with square cross sections.

Even though the joint density of states in quantum wires is sharply peaked, the states still form a quasi-continuum (unless the wire is very short). To obtain truly discrete states, we must look to isolated quantum dots. In that respect, these resemble atomic or defect states that are also tightly confined in space (of course, much more so than the dot states, which invariably extend over many atoms). Discreteness here has the working definition that the energy difference between the states be much larger than the broadening. The joint density of states then becomes a series of delta functions for each of the allowed transitions between the conduction and valence states. At higher energies, we observe transitions between the quasi-continuous, extended states above the confining potential. In practice, macroscopic samples often have a distribution of dot sizes, usually along the lateral directions, and transitions are broadened by that distribution, as illustrated in Fig. 11.34. We will focus on self-assembled quantum dots with planar density $N_{QD}$

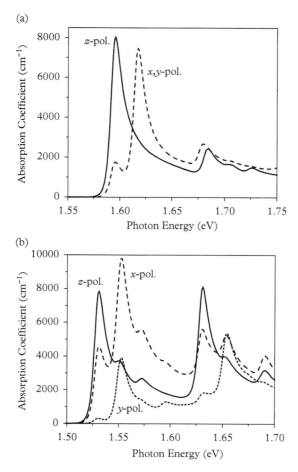

**Figure 11.33** *Absorption spectra for (a) 50-Å × 50-Å and (b) 50-Å × 100-Å GaAs/Al$_{0.3}$Ga$_{0.7}$As quantum wires calculated using the full band structure and optical matrix elements. We assume a broadening linewidth of $\Delta_l = 3$ meV to make the closely spaced transitions distinct and divide by the periods in both directions to obtain the absorption coefficient per unit length. It can be translated into other relevant quantities, as discussed in the text.*

grown by epitaxial techniques in one or more layers. In this case, the natural quantity for measuring the absorption is the cross section, which is converted to the absorbed fraction per pass through a layer of such dots $\alpha_{QD}$ by multiplying by $N_{QD}$. For propagation in a waveguide, the procedure is the same as for quantum wells discussed in Section 11.3.

As for quantum wires, the relative dot dimensions determine any polarization anisotropy. For example, the absorption in a cubic quantum dot is the same for light polarized along any of its confinement axes, but in general different for light polarized e.g. along the cube diagonal. Only a spherical dot has truly

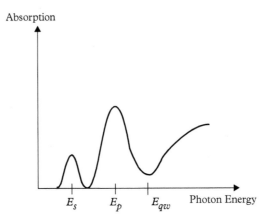

**Figure 11.34** *Illustration of the absorption spectrum due to broadened discrete transitions in shallow quantum dots (e.g., the self-assembled InAs/GaAs quantum dot) and the onset of transitions between extended states. Here the transitions between s-like and p-like shells appear as two distinct peaks. Note the similarity to the exciton transitions in Fig. 11.30.*

isotropic absorption. We mentioned in Section 10.6 that in a self-assembled dot, the envelope function is much more compressed vertically than laterally, i.e. it ise effectively a quantum well with some in-plane confinement. Typically, we may still classify the lowest-energy transition in such a dot as CB–HH based on the dominant wavefunction components. Then in-plane confinement enables the dot to absorb some of the light polarized along the $z$ axis. This is roughly equivalent to the CB–HH transition in a quantum well for states with $k_\parallel > 0$. The rough size scaling holds, i.e. if the in-plane dimensions are an order of magnitude larger than the vertical extent, the absorption of vertically polarized light is at least an order of magnitude weaker.

To derive the quantum-dot absorption cross section by analogy with Eqs. (11.66) and (11.119), we return to Eq. (11.1) and eliminate all three integrations over wavevector components (along with the corresponding factors of $\frac{1}{2\pi}$). For a plane wave propagating perpendicular to the growth direction, we can write the absorption per pass as

$$\alpha_{QD,xy}(\hbar\omega) = \frac{4\pi^2\alpha_f}{n_r}N_{QD}\sum_{\mu,v}\left|M^{QD}_{\mu v,xy}\right|^2\frac{\hbar\omega}{\left(E_{c,\mu}-E_{v,v}\right)^2}\sigma\left(E_{c,\mu}-E_{v,v}-\hbar\omega\right)$$

$$(11.122)$$

We have multiple ways to estimate the matrix element $M^{QD}_{\mu v,xy}$. In a numerical calculation, we can find it from the general Hamiltonian. If the dots are self-assembled, and their lateral sizes much larger than the vertical extent, the matrix elements are not much different from those in quantum wells. $M_{\mu\nu,xy}^{QD}$ includes an overlap integral of the form

$$|O_{eh}|^2 = \left| \iiint dx\, dy\, dz\, \left(F_s^{CB}\right)^* (x,y,z)\, F_{x,y}^{HH}(x,y,z) \right|^2 \tag{11.123}$$

We know that the transitions in quantum dots are between discrete states, so the absorption per pass at $\hbar\omega = E_{c,\mu} - E_{v,\nu}$ is inversely proportional to the broadening linewidth. The broadening is in part due to scattering as in bulk materials and quantum wells (see Eq. (4.56)), but, as mentioned earlier, is often dominated by the size distribution. In this case, the broadening is mostly *inhomogeneous* and conveniently modeled by a Gaussian (normal) energy profile:

$$\sigma(E) = \frac{1}{\sqrt{\pi}\,\Delta_l}\, \exp\left(-\frac{E^2}{\Delta_l^2}\right) \tag{11.124}$$

How does the absorption in a single layer of quantum dots compare to that in a single quantum well? To find out, we use Eqs. (11.7), (11.122), and (11.124) with $E_{c,1} - E_{v,1} = E_g = \hbar\omega$, and assume the same wavefunction overlap for both dots and wells. Finally, we also assume a single CB–HH transition in the QWs and a CB–HH-like transition in the dots. With these assumptions, we form the ratio of the peak dot and well absorption coefficients:

$$\frac{\alpha_{QD,max}}{\alpha_{QW,max}} \approx \frac{\sqrt{\pi}\,\hbar^2 N_{QD}}{m_r \Delta_l} \tag{11.125}$$

With order-of-magnitude estimates $\Delta_l \approx 30$ meV and $N_{QD} = 10^{11}$ cm$^{-2}$, $\frac{\alpha_{QD}}{\alpha_{QW}} \approx$ 10%. The lower absorption in quantum dots is partly because the dots fill only a fraction of the layer in which they reside, but mostly because of the wide distribution of transition energies. For example, if $\Delta_l$ could be lowered to the typical QW value of 5 meV, the QD and QW absorption coefficients would become comparable. However, it is difficult to reduce the spread of the size distribution because the self-assembled growth process only weakly selects for lateral dot size. In practice, multiple layers of dots are grown to boost the absorption and gain.

The areal carrier density in the dots under current injection is equal to the product of the dot density $N_{QD}$ and the occupation factor in each dot. By analogy with Eq. (11.78), we can write the maximum differential gain in quantum dots as

$$\left.\frac{dg_{QD}}{dn}\right|_{max} \approx \frac{\pi\sqrt{\pi}\,\alpha_f}{n_r} \frac{E_P}{E_g} \frac{\hbar^2}{m\Delta_l} |O_{eh}|^2 \tag{11.126}$$

Hence the ratio of the differential gains in Eqs. (11.78) and (11.126) is given by

$$
\frac{\left.\frac{dg_{QD}}{dn}\right|_{max}}{\left.\frac{dg_{QW}}{dn}\right|_{max}} \approx \frac{k_B T}{\sqrt{\pi}\,\Delta_l} \left(\frac{m_h}{m_e}\right)^{2/3}
\tag{11.127}
$$

This shows that the differential gain in state-of-the-art quantum dots is comparable to that in quantum wells; i.e., it is not down by an order of magnitude as the absorption strength from Eq. (11.125). The sharp JDOS is a big help, even though inhomogeneous broadening limits this effect via the denominator in Eq. (11.126). Again, with a lower inhomogeneous linewidth, the QD differential gain would be higher than in comparable QWs. Finally, the differential gain becomes lower as the required carrier density increases, and the QD lasers that require high gains operate on higher transitions.

The band alignment in a quantum dot can be either type I or type II along any of the three confinement axes. If the dots are close enough to each other either in the growth plane or vertically along the growth direction, we have a dot superlattice. Of course, above the quantum-dot confinement potential, the states are no longer confined. What happens at that point depends on how the dots are defined. As mentioned in Chapter 10, the so-called quantum-well-like InGaAs *wetting layer* forms in self-assembled dots, and the subbands of the wetting layer lie above the dot states.

The 3D confinement in quantum dots also leads to stronger absorption of in-plane-polarized light by interlevel transitions in the conduction band. Since the non-waveguide infrared detectors discussed in Chapter 14 must absorb light at normal incidence, this is a very useful feature indeed! If the in-plane confinement is not too strong, we can think of the dot envelope functions as having some admixture of $p_x$ and $p_y$, with the dominant $s$- and $p_z$-like components. It is the matrix elements connecting $s$ and $p_x/p_y$ that allow absorption at normal incidence. This is how confinement in extra dimensions blurs the distinction between interband and intersubband transitions.

What are the radiative recombination rates in quantum wires and dots? For wires, we return to Eq. (4.71) and eliminate two $k$-space integrations (with their $\frac{1}{2\pi}$ factors) as well as the factor of 2 from the two orthogonal polarizations:

$$
R_{sp}^{QWR,j}(\hbar\omega) = \frac{\alpha_f n_r}{\pi} \frac{1}{\hbar^3 c^2} \sum_{\mu,\nu} \int dk \, |\langle c|\nabla_k H_e|v\rangle|^2 \frac{(\hbar\omega)^3}{\left(E_{ck,\mu} - E_{vk,\nu}\right)^2}
$$
$$
\times \sigma\left(E_{ck,\mu} - E_{vk,\nu} - \hbar\omega\right) f_c\left(E_{ck,\mu}\right)\left[1 - f_v\left(E_{vk,\nu}\right)\right]
\tag{11.128}
$$

For simplicity, we also assume a transition between a pair of spin-degenerate conduction and valence subbands with parabolic dispersion and substitute the JDOS from Eq. (11.116), although the same approach works for any functional form of the JDOS. The matrix elements at the band edge for a cylindrical wire are given by Eq. (11.120). Next, we average over the three orthogonal propagation directions with their associated polarizations, as we did for quantum wells in Section 11.6. This procedure yields $2\frac{R^x_{sp}+R^y_{sp}+R^z_{sp}}{3} \equiv R^{QWR}_{sp}$. The total recombination rate that uses this weighted average becomes

$$R^{QWR}_{sp}(\hbar\omega) = \frac{\sqrt{2}\alpha_f n_r \sqrt{m_r} E_P}{3\pi m\hbar^2 c^2} |O_{eh}|^2 \frac{\hbar\omega}{\sqrt{(\hbar\omega - E_g)}} f_c(1-f_v) \qquad (11.129)$$

We can integrate this expression over all photon energies:

$$R^{QWR}_{sp} = \frac{\sqrt{2}\alpha_f n_r \sqrt{m_r} E_P}{3\pi m\hbar^2 c^2} |O_{eh}|^2 \int_{E_g}^{\infty} \frac{\hbar\omega}{\sqrt{(\hbar\omega - E_g)}} d(\hbar\omega) f_c(1-f_v) \qquad (11.130)$$

At low carrier densities, we can define a $B$ coefficient as in bulk materials and quantum wells, i.e., $R^{QWR}_{sp} \approx B_{QWR} np$. Using the integration variable $x \equiv (\hbar\omega - E_g)/k_B T$, we obtain

$$B_{QWR} = \frac{\sqrt{2}\alpha_f n_r E_P \sqrt{m_r k_B T}}{3\pi m\hbar^2 c^2} \frac{1}{N_c N_v} |O_{eh}|^2 \left[ E_g \int_0^{\infty} x^{-1/2} e^{-x} dx + k_B T \int_0^{\infty} x^{1/2} e^{-x} dx \right] \qquad (11.131)$$

The quantity in brackets can be expressed in terms of the gamma function $\Gamma(x)$:

$$E_g \int_0^{\infty} x^{-\frac{1}{2}} e^{-x} dx + k_B T \int_0^{\infty} x^{\frac{1}{2}} e^{-x} dx$$
$$= E_g \Gamma(1/2) + k_B T \Gamma(3/2) = \sqrt{\pi}\,(E_g + k_B T/2) \qquad (11.132)$$

When the gap is not too narrow, we can simplify by neglecting the second term in Eq. (11.131):

$$B_{QWR} \approx \frac{\alpha_f n_r}{3\sqrt{\pi}} \frac{\sqrt{2m_r k_B T}}{\hbar N_c N_v} \frac{E_P}{\hbar} \frac{E_g}{mc^2} |O_{eh}|^2 \qquad (11.133)$$

By analogy with quantum wells, we can define the effective electron and hole densities of states $N_c$ and $N_V$ assuming parabolic bands and spin degeneracy and using Eq. (11.116) as a model. We start from the densities of states in each band:

$$DOS^{e,QWR} = \frac{1}{\pi} \left[ \frac{dE_e}{dk} \right]^{-1} = \frac{1}{\pi} \frac{\sqrt{m_e}}{\hbar\sqrt{2(E_c - E_{c0})}} \qquad (11.134a)$$

$$DOS^{h,QWR} = \frac{1}{\pi} \left[ \frac{dE_h}{dk} \right]^{-1} = \frac{1}{\pi} \frac{\sqrt{m_h}}{\hbar\sqrt{2(E_v - E_{v0})}} \qquad (11.134b)$$

from which, by analogy with Eqs. (4.74) and (4.75), we have

$$n = N_c \exp\left(\frac{E_{Fc} - E_{c0}}{k_B T}\right) \qquad p = N_v \exp\left(-\frac{E_{Fv} - E_{v0}}{k_B T}\right) \tag{11.135a}$$

$$N_c \equiv \frac{\sqrt{k_B T m_e}}{\pi\sqrt{2}\hbar} \qquad N_v \equiv \frac{\sqrt{k_B T m_h}}{\pi\sqrt{2}\hbar} \tag{11.135b}$$

The definitions in Eq. (11.135b), as well as the usual expression $m_r = \frac{m_e m_h}{(m_e + m_h)}$ that holds for parabolic bands only, can be substituted into Eq. (11.130) to yield

$$B_{QWR} \approx \frac{\sqrt{(2\pi)^3}\alpha_f n_r}{3} \frac{E_P}{\sqrt{k_B T (m_e + m_h)}} \frac{E_g}{mc^2} |O_{eh}|^2 \tag{11.136}$$

Substituting the typical parameter values quoted in connection with Eqs. (11.100) and (11.101), we calculate $B_{QWR} \approx 3 \times 10^3$ cm/s. For a carrier density of $n = p = 10^5$ cm$^{-1}$, we estimate the radiative lifetime $\tau_r = \frac{1}{nB_{QWR}} \approx 3$ ns, which is somewhat shorter than in quantum wells.

To estimate the radiative recombination rate in a quantum dot, we perform similar, by now quite familiar, steps. For the last time we return to Eq. (4.71) and eliminate the integrations over all three wavevector components, along with a factor of 2 for the orthogonal polarizations in the density of optical modes:

$$R_{sp}^{QD,\perp,\|}(\hbar\omega) = \frac{2\alpha_f n_r}{\hbar^3 c^2} \sum_{\mu,\nu} \frac{|M_{c,\mu;v,\nu}|^2 (\hbar\omega)^3}{(E_{c,\mu} - E_{v,\nu})^2} \sigma\left(E_{c,\mu} - E_{v,\nu} - \hbar\omega\right) f_c (1 - f_v) \tag{11.137}$$

Using this expression, we can quickly estimate the rate for the transition between the lowest electron and highest hole states, both of them spin-degenerate. Confinement in the $x - y$ plane is usually weak, so we can use the matrix element from Eq. (11.93). As in the quantum-well case, the weighted average of the rates in Eq. (11.137) becomes $\frac{4}{3} R_{sp}^{\|}$:

$$R_{sp}^{QD}(\hbar\omega) = \frac{4\alpha_f n_r E_P}{3\hbar mc^2} |O_{eh}|^2 \frac{(\hbar\omega)^3}{(E_{c,1} - E_{v,1})^2} \sigma\left(E_{c,1} - E_{v,1} - \hbar\omega\right) f_c (1 - f_v) \tag{11.138}$$

We calculate the total transition rate by neglecting broadening and integrating over a delta function at $E_{c,1} - E_{v,1} = \hbar\omega = E_g$:

$$R_{sp}^{QD} \approx \frac{4\alpha_f n_r}{3} \frac{E_P}{\hbar} \frac{E_g}{mc^2} |O_{eh}|^2 f_c (1 - f_v) \tag{11.139}$$

We could now define the $B$ coefficient by factoring out the electron and hole occupation factors, $f_c$ and $1 - f_v$, but there is little reason to resort to this scheme here. If $f_c = f_v = \frac{1}{2}$, the radiative lifetime $\tau_r = 1/R_{sp}^{QD} \approx 2$ ns. Notice that the explicit temperature dependence of the radiative rate (for a fixed carrier density) is $(k_B T)^{-D/2}$, where $D$ is the number of dimensions (3 in bulk, 2 in quantum wells, 1 in quantum wires, and 0 in quantum dots).

Of course, Eqs. (11.136) and (11.139) are relatively crude estimates. For more accurate values, we might use expressions like Eqs. (11.128) and (11.137), average over all light propagation directions, and evaluate the formulas numerically. If the confinement potentials are not deep, and the electrons or holes occupy states that are not quasi-1D or 0D, we will need to include these extended states as well. Finally, if the density of modes is significantly different from a bulk semiconductor (Eq. (4.68)), we should use the modified expression in its place. Much faster emission is possible in a cavity with a large Purcell factor (see Eqs. (4.69) and (4.70)) so long as the dominant polarization of the cavity mode coincides with the largest matrix element of the active medium, be it a well, a wire, or a dot. Below we will return to this topic to estimate the Purcell effect in state-of-the-art cavities.

Just as in quantum wells, excitons exist in wires and dots and form distinctive features in the optical spectra at low temperatures if the gap is not too narrow. If strong inhomogeneous broadening due to a size distribution is present, excitons transitions are seen clearly when we isolate the emission from individual dots. In disk-like dots, excitons with four different total-spin states are possible, as shown in Fig. 11.35a. This is because we add HH states with $m_j = \pm 3/2$ to CB states with $m_s = \pm 1/2$, with all four sign combinations possible. Two of these four excitons have combined angular momentum projections of $m_j = \pm 2$, and do not couple to light (*dark excitons*). As we know from Chapter 4 and Section 11.7, this is because photons with circular polarization carry angular momentum $m_l = \pm 1$ along the vertical axis, so that angular momentum is not conserved. The two other excitons have $m_j = \pm 1$ and couple to circularly polarized photons (*bright excitons*).

Figure 11.35b shows that another transition is present in quantum dots charged with a single electron (or hole). In this case, there are again two optical transitions connected by photons with different circular polarizations. In the language of quantum information, we say that the dot's spin state is entangled with the emitted photon. Applying a magnetic field along the vertical axis, we split the spin states by an energy proportional to the field (*the Zeeman effect*), and all four states can be coupled by varying the (linear) polarization of the light. A charged exciton is also known as a *trion*.

Because any individual quantum dot has only a few discrete states, it resembles a model few-level system at low temperatures. Then, only a limited number of transitions, which can be distinguished in the emission spectra, are possible. The dots embedded in bulk-like barrier layers can be excited both optically and electrically. Unfortunately, as the temperature increases, additional states of the self-assembled dots fill up, and the picture becomes much more complicated. Semiconductor nanocrystals based on II–VI and lead-salt materials (CdSe, PbSe,

(a)

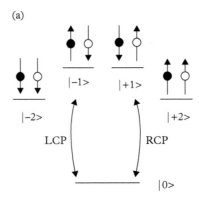

(b)

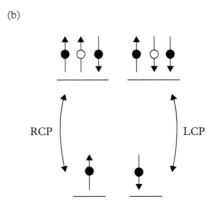

**Figure 11.35** *(a) Illustration of excitonic transitions in a single disk-like quantum dot. (b) Illustration of trion transitions in a single disk-like dot. The polarization selection rules for right-circularly polarized (RCP) and left-circularly polarized (LCP) light are indicated. The electrons are represented as filled circles, and holes as open circles. The up and down arrows stand for positive and negative projections of the angular momentum along the vertical axis.*

HgTe, etc.) have smaller lateral dimensions and can therefore maintain truly quantum features up to room temperature. Quantum dots and nanocrystals can be coupled to isolated modes of microcavities and subwavelength cavities, with strong coupling observed when the dot is positioned spatially near the antinode of the cavity field and spectrally with the exciton transition near the modal frequency. In this case, we can use Eqs. (4.69) and (4.70) to write

$$F_P = \frac{3}{4\pi^2} \frac{Q}{V_{eff}} \frac{\lambda_c^3}{n_r^2 n_g} \left( \frac{\boldsymbol{E} \cdot \boldsymbol{r}}{E_{max}r} \right)^2 \frac{1}{1 + 4Q^2 \left( \frac{v_0}{v_l} - 1 \right)^2} \tag{11.140}$$

where the penultimate factor is a simple consequence of the dipole Hamiltonian in Eq. (4.6), the last factor follows from Eq. (4.69) with the definitions in Eqs. (11.109) and (11.110), and $E_{max}$ is the maximum amplitude of the electric field of the cavity mode.

For a dot placed at the antinode of the field in an all-dielectric microcavity, e.g., based on photonic crystals with omitted holes or micropillars with Bragg mirrors, $V_{eff} \sim \frac{\lambda_c^3}{n_r^2 n_g}$. At low temperatures, the quantum-dot linewidth typically exceeds the cavity linewidth (the "good-cavity" limit of Section 4.4), with $Q \sim 10^4$. Therefore, for emitters spatially and spectrally matched to the cavity mode, the maximum Purcell factor is of order $10^3$, but decreases rapidly for any spectral or spatial mismatch. For subwavelength plasmonic cavities in the visible and near-IR, $Q \sim 20$, but this is compensated by the mode being at least an order of magnitude smaller than the wavelength in the material in each direction. Therefore, the theoretical maximum Purcell factor is comparable to or greater than that in all-dielectric cavities. While spectral mismatch is not a concern for such a low $Q$, plasmonic cavities are extremely sensitive to the emitter placement with respect to the field maximum. This makes it difficult to demonstrate $F_P \gg 100$ for any reasonable emitter distribution in the cavity.

So far we have assumed that the spin states of the dots are degenerate. In fact, the spin–spin interaction lowers the overall energy of dark excitons with parallel electron and hole spins relative to bright excitons with antiparallel spins. In self-assembled dots (as well as bulk semiconductors and quantum wells), the spin splitting is usually smaller than $k_B T$, even at cryogenic temperatures. The small splitting results in nearly equal populations of the two spin states in the conduction and valence bands. However, the splitting can be as large as several tens of meV in nanocrystals based on II–VI and III–V semiconductors, which makes the emission process via to the dark exciton very slow. The full description of these nanocrystals requires other higher-order corrections before it can be meaningfully compared to experimental data.

This completes our survey of the methods for calculating the quantum states and optical properties of semiconductor structures with confined electrons. Now we are ready to apply what we have learned to photonic devices!

## Suggested Reading

- A good overview of the optical transitions in zinc-blende and wurtzite quantum wells is available in S. L. Chuang, *Physics of Photonic Devices*, 2nd edn. (Wiley, New York, 2009).

- Calculations of the optical gain in zinc-blende wells are presented in a tutorial manner in L. A. Coldren, S. W. Corzine, and M. Mašanović, *Diode Lasers and Photonic Integrated Circuits*, 2nd edn. (Wiley, New York, 2012).

- To learn more about intersubband transitions in quantum wells and their applications to infrared lasers, see J. Faist, *Quantum Cascade Lasers* (Oxford University Press, Oxford, 2013). Quantum cascade lasers will also be covered in Part IV.

- For a discussion of multiband models for intersubband transitions, see C. Sirtori, F. Capasso, J. Faist, and S. Scandolo, "Nonparabolicity and a sum rule associated with bound-to-bound and bound-to-continuum intersubband transitions in quantum wells," *Physical Review B* **50**, 8663 (1994)

- J. Khurgin, "Comparative analysis of the intersubband versus band-to-band transitions in quantum wells," *Applied Physics Letters* **62**, 1390 (1993) emphasizes the fundamental similarity between intersubband and interband transitions when considered in the context of multiband models.

- For an introduction to including many-body effects into calculations of intersubband absorption, see R. J. Warburton, C. Gauer, A. Wixforth, J. P. Kotthaus, B. Brar, and H. Kroemer, "Intersubband resonances in InAs/AlSb quantum wells: Selection rules, matrix elements, and the depolarization field," *Physical Review B* **53**, 7903 (1996).

- A review of multisubband plasmons can be found in A. Vasanelli, Y. Todorov, and C. Sirtori, "Ultra-strong light-matter coupling and superradiance using dense electron gases," *Comptes Rendu Physique* **17**, 861 (2016).

- A simple model for the gain in parabolic, isotropic quantum wells was originally derived in K. J. Vahala and C. J. Zah, "Effect of doping on the optical gain and the spontaneous noise enhancement factor in quantum well amplifiers and lasers studied by simple analytical expressions," *Applied Physics Letters* **52**, 1945 (1988).

- The optical properties of cylindrical quantum wires are treated in P. C. Sercel and K. J. Vahala, "Polarization dependence of optical absorption and emission in quantum wires," *Physical Review B* **44**, 5681 (1991) using their ingenious formalism.

- For a review of excitons in TMD monolayers, see G. Wang, A. Chernikov, M. M. Glazov, T. F. Heinz, X. Marie, T. Amand, and B. Urbaszek, "*Colloquium*: Excitons in atomically thin transition metal dichalcogenides," *Reviews of Modern Physics* **90**, 021001 (2018).

- To learn more about excitons in quantum dots and dot molecules, see A. S. Bracker, D. Gammon, and V. L. Korenev, "Fine structure and optical pumping of spins in individual semiconductor quantum dots," *Semiconductor Science and Technology* **23**, 114004 (2008).

# Part IV

# Semiconductor Photonic Devices

# 12

# Interband Semiconductor Lasers and LEDs

In this chapter, we discuss the operation of conventional diode lasers based on quantum wells and quantum dots at different emission wavelengths. The recombination processes that control the threshold current density of the devices are described in detail, including recombination at defects, radiative and Auger recombination. The high-speed modulation and spectral characteristics of semiconductor lasers are also discussed. We continue by illustrating why interband cascade lasers can outperform diode lasers at mid-infrared wavelengths and describe their design and operating characteristics in detail. On the short-wavelength side of the spectrum, we discuss the nitride lasers and the factors that limit their performance. In addition to lasers, the principles underlying light-emitting diodes (LEDs) are outlined, and the proposed mechanisms for improving the extraction of the light from high-index semiconductor materials are described. The chapter concludes with a discussion of the performance of semiconductor optical amplifiers designed to amplify a weak input signal.

## 12.1   A Whirlwind Tour of Semiconductor Lasers

A laser converts (incoherent) electrical or optical input into a coherent optical beam. In semiconductor lasers, only electrical injection is of any practical interest. Most semiconductor lasers are based on quantum wells in the active gain medium, but the same ideas apply to, e.g., quantum-dot lasers discussed in Section 12.6. The electrical current injected into the laser creates excess electrons and holes in one or more active wells, and these carriers then recombine to emit light. The emitted photons are transmitted through the mirrors and may be absorbed somewhere in the cavity at some net rate per round trip, which represents the *cavity loss*. If the loss is higher than the optical gain from the active wells (calculated in Chapter 11), the emission remains weak because only a small fraction of the emitted radiation is observed, and only a fraction of the electron–hole recombination is radiative.

*Bands and Photons in III–V Semiconductor Quantum Structures.* Vurgaftman, Lumb, and Meyer,
Oxford University Press (2021). © Vurgaftman, Lumb, and Meyer.
DOI: 10.1093/oso/9780198767275.003.0012

However, if at some current the carrier density in the wells becomes high enough to compensate the loss, the cavity *taken as a whole* no longer absorbs the emitted photons. We saw in Chapter 4 that for this *threshold* condition to hold, we must inject enough electrons and holes to separate their quasi-Fermi levels by more than the energy gap. If we keep injecting carriers above the lasing threshold, the excess electron–hole pairs produce photons in one or few modes that are observed at the output. This is because the carrier density in the wells is pinned (or changes slowly with current), so nonradiative recombination processes continue at about the same rate as at threshold. The laser conversion from electrical input to optical output can be efficient provided (1) the threshold is not too high; and (2) the parasitic loss in the cavity is not much higher than the transmission through the mirror so emitted photons can find their way out of the cavity before they disappear. The emission is coherent, since the laser favors a limited set of cavity modes in both spectral and spatial domains. This happens because optical gain is spectrally narrow, and because the optical cavity provides feedback only for certain modal profiles, e.g., those that are guided in the cavity and reflected by the mirrors.

How does the injection of electrons and holes into the active wells of the semiconductor laser actually lead to optical gain? Think of a four-level scheme in which we pump carriers from the lowest level to the highest. The pumping mechanism here involves an entire electrical circuit, in which electrons removed from the valence band (injecting holes) travel around the circuit to enter the conduction band from the other side of the structure. The carriers in this four-level model can relax quite fast between the two top (conduction) and two bottom (valence) levels, but make only relatively slow (interband) transitions between the two lasing levels.

Of course, there are no discrete levels in quantum wells. Figure 12.1a shows that the role of the pump levels is played by higher-$k$ states as well as higher subbands in the conduction and valence bands. With scattering from phonons and other carriers, the electrons (holes) injected into those states relax on a very fast time scale ($\sim$1 ps) to the bottom of the conduction band and top of the valence band. Because the scattering is so fast, very few recombine before reaching the band edges. Now we can identify the bottom of the conduction band and top of the valence band with the two lasing levels. Their populations are inverted because electrons occupy the states at the bottom of the conduction band at the same time that holes fill the highest valence states. Of course, this is just a restatement of the condition for optical gain specified in Eq. (4.66): $E_{Fc} - E_{Fv} > E_g$.

For stimulated emission to take over above the lasing threshold, the light must not leave the cavity too fast through the mirrors. The semiconductor laser cavity typically consists of two mirrors separated by a distance that may be as short as the wavelength scale, but is typically many orders of magnitude longer. The light does not diffract as it travels between the mirrors because it is confined by an optical waveguide, as discussed in Chapter 11. Alternatively, we could say that reflections from the optical claddings cancel out the diffraction.

(a)

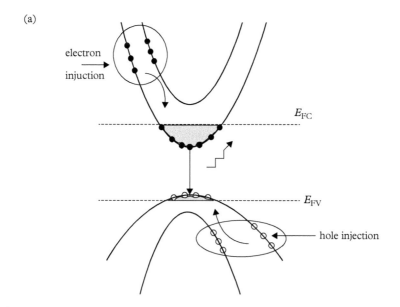

(b)

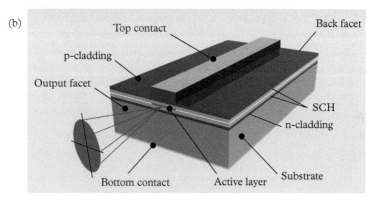

**Figure 12.1** *(a) Schematic of carrier injection into a semiconductor laser. The higher states in the conduction and valence bands act as the reservoir states of a four-level laser. (b) Schematic of an edge-emitting semiconductor laser structure with the main components identified. The most commonly employed shallow-ridge geometry is illustrated.*

Figure 12.1b shows a schematic of the semiconductor laser cavity. Depending on whether the long dimension of the active region is parallel or orthogonal to the cavity axis, this is an edge-emitting laser or a vertical-cavity surface-emitting laser (VCSEL). In edge-emitting lasers with typical cavity lengths of a few mm, thousands of wavelengths fit into one round trip between the mirrors. This often leads to multiple modes that share the same spatial profile along the directions orthogonal to the cavity axis.

How much optical gain does it take to reach the lasing threshold? From Beer's law in Chapter 4, we know that the optical intensity varies exponentially as the light propagates, with the exponent given by the difference between gain and loss. At each mirror, the reflected intensity is proportional to the incident intensity multiplied by the mirror reflectivity ($R_1$ or $R_2$). If the optical field reproduces itself after each round trip, the intensity at any given point in the cavity is the same before and after each round trip. Therefore, we can cancel it from both sides of the equation to obtain the threshold condition:

$$R_1 R_2 e^{2\Gamma_g g_{th} L_g - 2\Gamma_\alpha \alpha_i L_\alpha} = 1 \qquad (12.1)$$

Here $\Gamma_g$ is the gain confinement factor, $L_g$ is the length within the cavity over which the light experiences gain, $\Gamma_\alpha$ is the overlap of the distributed loss with the mode, $\alpha_i$ is the distributed loss in the cavity, e.g., due to parasitic absorption that does not generate electron–hole pairs or to optical scattering along the propagation path, and $L_\alpha$ is the length over which the light experiences the distributed loss. We could also include other sources of loss by analogy with Eq. (12.1).

We have already discussed the gain confinement factor in Chapter 4. It measures the overlap of the optical mode with the gain medium in the two directions perpendicular to the cavity axis. Along that axis, the light in the cavity forms standing waves that can enhance (by up to a factor of 2 compared to the average over the sinusoidal wave) or suppress the gain, depending on whether the gain region is at a peak or trough of the cavity optical field. In an edge-emitting laser, the gain material often occupies the entire cavity, which contains so many peaks and troughs that the average confinement factor $\Gamma_g$ applies when the carrier distribution is uniform.

Now we solve Eq. (12.1) to obtain the (average) threshold gain:

$$g_{th} = \frac{\Gamma_\alpha \alpha_i L_\alpha + \frac{1}{2} \ln \left( \frac{1}{R_1 R_2} \right)}{\Gamma_g L_g} \equiv \frac{\alpha_{cav} L_{cav}}{\Gamma_g L_g} \qquad (12.2)$$

We now know how much gain the active medium must provide to reach threshold, but how does the laser behave below and above that point? To answer this question, we construct a model for the photon density in the cavity. The rate at which photons are emitted into the cavity is the sum of the spontaneous emission rate calculated in Chapters 4 and 11 and the stimulated emission rate (proportional to the optical gain). Since a constant gain describes an exponential increase in the number of photons with distance, the stimulated emission rate is also proportional to the photon density itself and to the group velocity $v_g$ at which photons propagate through the cavity. The reverse of this process is the photon loss due to absorption, scattering, and leakage through the mirrors. Even though the mirror loss takes place at two planes only, we can distribute it over the entire cavity to avoid including the spatial dependence using Eq. (12.2). Averaging over the cavity axis, we can suggest a *rate equation* for the 2D photon density in one of the cavity modes:

$$\frac{dS}{dt} = v_g \left[ \Gamma_g g \frac{L_g}{L_{cav}} - \Gamma_\alpha \alpha_i \frac{L_\alpha}{L_{cav}} - \frac{1}{2L_{cav}} \ln\left(\frac{1}{R_1 R_2}\right) \right] S + r_{sp} \qquad (12.3)$$

The photon density is 2D (in plane) considering that the vertical extent of the mode $d_{mode}$ enters the confinement factors. This form is convenient for the dominant edge-emitting quantum-well lasers because when the carrier density is expressed in its most natural (2D) units, the photon density has the same dimensionality. A more general form of Eq. (12.3) would use a 3D photon density (or photon number) and then convert between the effective volumes for carriers and photons. In Eq. (12.3) the confinement factors take care of the conversion.

Clearly, we would like to maximize the first term and minimize the second term in the brackets of Eq. (12.3). Often, we cannot accomplish both of these two goals at the same time. For example, we might choose a cavity with large $d_{mode}$ to reduce $\Gamma_\alpha$, but in most cases this would also reduce $\Gamma_g$ if the number of wells $M$ is kept fixed. Nevertheless, such a trade-off may be worthwhile if the chief objective is to minimize internal loss. We will see shortly that low internal loss correlates with high efficiency.

The last term in Eq. (12.3) is typically a small fraction $\beta$ of the total spontaneous emission rate $R_{sp}(n)$ that was discussed in Chapters 4 and 11: $r_{sp}(n) = \beta R_{sp}(n)$. How small? A quick estimate might be to multiply the DOM in Eq. (4.68) by the spectral width of the emission ($\approx k_B T$) and by the modal cavity volume $V_{eff}$ from Eq. (4.69):

$$\beta \approx DOM\, k_B T\, V_{eff} = \frac{4k_B T n_r^2 n_g}{\lambda^2 \hbar c} V_{eff} = \frac{k_B T L_{cav}}{\hbar c} \frac{4n_r^2 n_g A_{eff}}{\lambda^2} \qquad (12.4)$$

where $A_{eff}$ is a 2D analog of $d_{mode}$ that was mentioned in Section 11.4. Keep in mind that the spontaneous emission factor $\beta$ has an upper limit of 1. It can approach this limit in very small cavities, in which the active medium interacts mostly with a single optical mode. This is often the same limit in which the Purcell factor from Eq. (4.70) is substantial (if $Q$ is high or the cavity is subwavelength).

If more than one mode experiences optical gain, Eq. (12.3) must be supplemented with a similar equation for each of the other modes. For the longitudinal modes of an edge-emitting laser, the only difference between the equations might be the magnitude of the optical gain. For other cavity configurations, the loss might also differ. For modes that are widely separated in wavelength (by an amount comparable to the spectral width of the spontaneous emission), $r_{sp}$ varies as well.

To compute the gain, we need to know the carrier densities. Assuming equal electron and hole densities, $n = p$, we recall that the creation of a photon requires the annihilation of an electron–hole pair; i.e., the carrier density decreases in the process of both spontaneous and stimulated emission. This is not the whole story, however, because electrons and holes can also disappear without producing a photon. In fact, in many kinds of lasers, nonradiative recombination dominates below the lasing threshold. We will learn more about the various (radiative and nonradiative) recombination mechanisms in Section 12.2, but for now we denote the *total* recombination rate as $R(n)$. That still does not account for the stimulated emission rate, which is the first term in the brackets in Eq. (12.3). To balance the disappearance of carriers from the active medium (in other words, electrons returning to the valence band), they must be pumped continuously into the system at a rate determined by the injected current density $\mathcal{J}$ (or the electron flux $\mathcal{J}/e$). The rate equation for carrier density then becomes

$$\frac{dn}{dt} = \frac{\mathcal{J}}{e} - R(n) - v_g \Gamma_g g(n) \frac{L_g}{L_{cav}} S \tag{12.5}$$

To solve the rate equations in steady state, we set the derivatives (left sides) to 0. The photon density in the cavity becomes

$$S_{ss}(n) = \frac{r_{sp}}{v_g \left( \Gamma_\alpha \alpha_i \frac{L_\alpha}{L_{cav}} + \frac{1}{2L_{cav}} \ln \left( \frac{1}{R_1 R_2} \right) - \Gamma_g g (n_{ss}) \frac{L_g}{L_{cav}} \right)} = \frac{r_{sp}}{v_g \left( \alpha_{cav} - \Gamma_g g (n_{ss}) \frac{L_g}{L_{cav}} \right)} \tag{12.6}$$

We assume that the absorption in the denominator is independent of carrier density. Figure 12.2 shows that the steady-state photon density starts out low, pumped by spontaneous emission into the mode. As the gain approaches its threshold value $g_{th}$ from Eq. (12.2), the denominator in Eq. (12.6) becomes

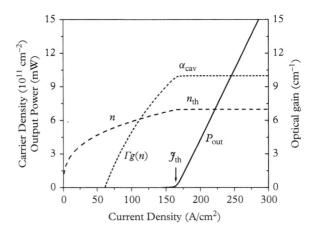

**Figure 12.2** *Output power, carrier density, and optical gain as a function of injected current density in a mid-infrared semiconductor laser with parameters similar to those of an interband cascade laser (see Section 12.4).*

very small, and the photon density increases much faster. In fact, Eq. (12.6) shows the lasing cavity never reaches the threshold gain. Instead, it approaches $g_{th}$ asymptotically as the photon density increases. For all practical purposes, we can say that the gain is *clamped* or *pinned* at its threshold value.

The steady-state solution of Eq. (12.5) is

$$\frac{\mathcal{J}}{e} = R(n_{ss}) + v_g \Gamma_g g(n_{ss}) \frac{L_g}{L_{cav}} S_{ss} \qquad (12.7)$$

Below the lasing threshold, the injected electrons go toward producing optical gain rather than cavity photons. In that regime, the second term is negligible, and to a good approximation the threshold current density is given by

$$\mathcal{J}_{th} = eR(n_{th}) \qquad (12.8)$$

where the threshold carrier density is defined such that $g(n_{th}) = g_{th} = \frac{\alpha_{cav}L_{cav}}{\Gamma_g L_g}$, with $g_{th}$ given by Eq. (12.2). We will see in Section 12.2 that when the active region has multiple wells, both the total recombination rate $R(n_{th})$ and $\mathcal{J}_{th}$ are proportional to the number of wells $M$ if the carriers are divided equally among the wells.

Where do the electrons go once the laser is above threshold and the gain becomes clamped? Since the first term and the $v_g \Gamma_g g(n_{ss}) \frac{L_g}{L_{cav}}$ proportionality factor in Eq. (12.7) are constant in this limit, only the photon density can increase in response to more current being injected. In fact, it increases linearly with the current in excess of the threshold. As the photon density grows, some of the photons leak out of the cavity through the mirrors, at a rate determined by the length of the cavity and the mirror reflectivity. This leakage represents the laser's output beam. How much power is available in that beam? To convert, we multiply by the modal volume divided by the vertical extent of the mode $\frac{V_{eff}}{d_{mode}} = \frac{L_{cav} A_{eff}}{d_{mode}} = L_{cav} w_{mode}$. We also multiply by the leakage rate and the energy of each photon $\hbar\omega$. The output is usually accessible from only one of the facets (the front facet). If both facets are identical, the photon leakage rate through one of them becomes $\frac{v_g}{2L_{cav}} \ln\left(\frac{1}{R}\right)$. As a result, we have

$$P_{out} = \frac{\hbar\omega V_{eff}}{d_{mode}} S_{ss} \frac{v_g}{2L_{cav}} \ln\left(\frac{1}{R}\right) \qquad V_{eff} = \hbar\omega \quad w_{mode} S_{ss} \frac{v_g}{2} \ln\left(\frac{1}{R}\right) \qquad (12.9)$$

To summarize, the light–current characteristics of a semiconductor laser with $\beta \ll 1$ can be written approximately in a piecewise form as

$$P_{out} \approx 0 \quad \mathcal{J} < \mathcal{J}_{th} \qquad (12.10a)$$

$$P_{out} \approx \frac{\hbar\omega}{e} \eta_e A(\mathcal{J} - \mathcal{J}_{th}) \quad \mathcal{J} > \mathcal{J}_{th} \qquad (12.10b)$$

where $P_{out}$ is the output power, $A$ is the volume of the active region, and $\eta_e$ is the laser's *external differential quantum efficiency* (EDQE) that will be described in more detail later. The steady-state solutions of the rate equations Eqs. (12.3) and (12.5) are shown in Fig. 12.2 using the parameters characteristic of a mid-infrared semiconductor laser.

We already mentioned in Chapter 4 that there is in fact less to the distinction between spontaneous and stimulated emission than meets the eye. For a laser cavity with a modal volume $V_{eff}$ much larger than $(\lambda/n_r)^3$, where $\lambda$ is the emission wavelength, the number of optical modes is so large that very few of them experience stimulated emission above threshold. A much larger fraction spectrally close to the energy gap receives photons from spontaneous emission, so that $\beta \ll 1$ according to Eq. (12.4). However, if we could somehow arrange for all of the spontaneous emission to go into a single mode with $\beta \sim 1$, we could distinguish it from stimulated emission only by measuring the statistics of the emitted photons. In fact, if $\beta = 1$ and all the recombination is radiative, the "knee" at the lasing threshold disappears from the light–current curve in Fig. 12.2 and Eq. (12.10). To see this mathematically, we set $R = r_{sp}$ and find from Eq. (12.6) that $R = S_{ss}v_g \left( \alpha_{cav} - \Gamma_g g\left(n_{ss}\right) \frac{L_g}{L_{cav}} \right)$. Substituting into Eq. (12.7), we see that $\frac{\mathcal{J}}{e} = v_g \Gamma_g \alpha_{cav} \frac{L_g}{L_{cav}} S_{ss}$; i.e., the emitted power is linear in $\mathcal{J}$ for *all current densities*. Of course, it is very difficult in practice to couple all (or nearly all) spontaneous emission into a single mode and to avoid any nonradiative recombination in the process.

How are electrons and holes injected efficiently into the active region of a conventional semiconductor laser? The electrical current serves both to populate the bottom of the conduction band and empty the top of the valence band (inject holes) of the active material. So it is natural to dope one side of the device $n$-type and the other $p$-type to position the respective quasi-Fermi levels near the conduction and valence bands. The opposite doping types meet at the active material, where a $p$–$n$ junction is formed, as illustrated schematically by the solid curves in Fig. 12.3a.

This was the approach adopted by early semiconductor laser designers, but it had an important flaw. Here it is: what prevents the injected electrons (holes) from overshooting the active region and recombining with majority holes (electrons) on the $p$ $(n)$ side of the device? To prevent this wasteful leakage, we can insert an electron barrier on the $p$ side and a hole barrier on the $n$ side of the active region. Ideally, these barriers should have heights much larger than $k_B T$. The dashed curves of Fig. 12.3a illustrate this *heterojunction laser* strategy. Figure 12.3b shows a more typical modern design, in which the $p$–$n$ junction is formed by doped cladding layers and lightly doped separate-confinement layers. The active region is either intrinsic or lightly doped. Because the loss is often low, the active region can be as thin as one or a few quantum wells. With this design change, the semiconductor lasers of the 1970s and 1980s were able to reach high performance levels. Another innovation that came in the 1980s was to strain quantum wells to

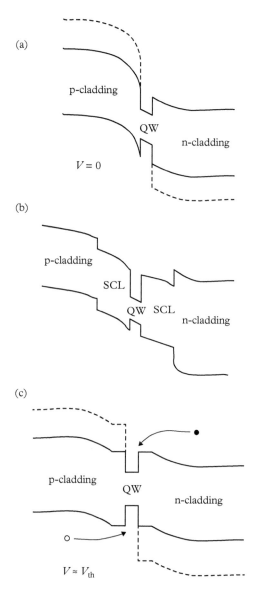

**Figure 12.3** *(a) A p–n junction at zero bias with one active quantum well that can inject carriers into a semiconductor laser. Higher band offsets (dashed curves) are desirable to prevent the injected electrons and holes from leaking into the opposite contact. (b) Heterojunction semiconductor laser with wider-gap materials used as claddings and narrower-gap materials used as separate-confinement layers (SCLs). (c) Semiconductor diode laser at the operating bias close to the lasing threshold.*

achieve higher differential gain (see Section 11.4). We will return to the question of how to design high-performance semiconductor lasers in Section 12.2.

Since the separation of the electron and hole quasi-Fermi levels at threshold must be larger than the energy gap, at least that much voltage must drop across the active region. This means that near threshold the bands near the active region become flatter, as shown in Fig. 12.3c. The full laser waveguide is reminiscent of Fig. 11.20, with the separate-confinement layers forming the high-index region and the claddings having a lower refractive index.

To find the external differential quantum efficiency ($\eta_e$) in Eq. (12.10b), we multiply the fraction of electrons injected above threshold that produce photons (defined as the *internal efficiency* $\eta_i$) by the fraction of photons generated in the cavity that actually contribute to emission in the output beam. To determine the second, we solve Eq. (12.7) for the photon density, and then differentiate with respect to the current density in the region above the lasing threshold:

$$\frac{dS_{ss}}{d\mathcal{J}} = \frac{1}{ev_g\alpha_{cav}} \tag{12.11}$$

The rate of photon emission from the cavity through the output facet with reflectivity $R_1$ is $R_{ph} \approx \frac{v_g}{2L_{cav}} \ln\left(\frac{1}{R_1}\right) \equiv v_g\alpha_{m1}$. This expression is exact if the two mirror reflectivity are equal, or if the back mirror is perfectly reflecting. Otherwise, different amounts of light are emitted through the two mirrors, but these cases are less important in practice. To convert to energy units, we multiply by the photon energy:

$$\frac{\hbar\omega}{e}\eta_e \approx \hbar\omega\frac{dS_{ss}}{d\mathcal{J}}R_{ph} \tag{12.12}$$

Substituting Eq. (12.11) into Eq. (12.12) and including the internal efficiency $\eta_i$, the EDQE becomes

$$\eta_e \approx \eta_i\frac{\frac{1}{2L_{cav}}\ln\left(\frac{1}{R_1}\right)}{\alpha_{cav}} = \eta_i\frac{\alpha_{m1}}{\alpha_{m1}+\alpha_{m2}+\alpha_i} \tag{12.13}$$

What causes the internal efficiency to drop below 1? One possibility is that some of the injected electrons and holes recombine outside the active region. This is equivalent to a parallel current path that produces no photons (or none at the right wavelength). Current leakage that involves no recombination is also possible, e.g., when some of the current flows along the sidewalls of the laser ridge. We can also have $\eta_i < 1$ when the gain does not pin perfectly above the lasing threshold (e.g., because of current-dependent heating and the resulting increase in $n_{th}$ with $\mathcal{J}$), so that additional current goes toward maintaining the threshold condition. If the internal efficiency remains the same below and above threshold, we can also revise the expression for threshold current density:

$$\mathcal{J}_{th} = \frac{eR(n_{th})}{\eta_i} \tag{12.14}$$

We will discuss the temperature dependence of the laser threshold in Section 12.2.

The EDQE is not a perfect measure of the device performance because it ignores the voltage drop over the device:

$$V = \hbar\omega/e + V_0 + \mathcal{J}\rho_s \tag{12.15}$$

where $\rho_s$ is the series resistance-area product, and $V_0$ is the component of the parasitic voltage that is independent of the injected current. The last two terms describe voltages that drop outside of the active region. These parasitic drops reduce the laser's overall power efficiency (power out divided by power in) of the laser, which is also known as the wall-plug efficiency (WPE). The WPE of a quantum-well laser is simply the product of the EDQE, the voltage efficiency $\frac{V}{\hbar\omega/e}$ and the threshold efficiency (ratio of the current above the lasing threshold to the total injected current):

$$\eta_{WPE} = \eta_e \frac{V}{\hbar\omega/e} \frac{\mathcal{J} - \mathcal{J}_{th}}{\mathcal{J}_{th}} = \eta_i \frac{\alpha_{m1}}{\alpha_{m1} + \alpha_{m2} + \alpha_i} \frac{\hbar\omega/e + V_0 + \mathcal{J}\rho_s}{\hbar\omega/e} \frac{\mathcal{J} - \mathcal{J}_{th}}{\mathcal{J}_{th}} \tag{12.16}$$

For operation far above threshold and a low parasitic voltage drop, the WPE is comparable to $\eta_e$. We will discuss the WPE of mid-infrared lasers in greater detail in Chapter 13.

Another signature of laser emission is a dramatic narrowing of the output spectrum. Well below threshold, the spectral width is given by the thermal spread of the radiating carriers, i.e. a few $k_B T$, as discussed in Section 4.4. In a typical edge-emitting laser, there is nothing to distinguish a large number (of order $\Delta\lambda L_{cav}/\lambda^2$, where $\Delta\lambda$ is the width of the optical gain) of longitudinal standing-wave modes from each other because $L_{cav} \gg \lambda$, and the mode order that determines the number of peaks is large. The photon density in each mode satisfies Eq. (12.6), with the gain now being a function of energy. At current densities well above threshold, the denominator of Eq. (12.6) becomes very small *only* for the mode(s) nearest to the gain peak. Therefore, if the gain spectrum is not too broad, most of the emission is directed into one (or at most a few) longitudinal modes. This is a dramatic change from the spontaneous emission spectrum.

Unfortunately, the single-mode emission of the Fabry–Perot laser in Fig. 12.1b is often unstable higher above threshold when the cavity is not too short. To understand why, recall that the lasing mode forms a standing wave of the form $\sin(m\pi z/L_{cav})$, with intensity nodes and antinodes that repeat every half a wavelength $[L_{cav}/m \approx \lambda/(2n_m)]$. The gain is supplied mostly at the antinodes of the cavity field, with no contribution from the nodes. As a result, the carrier density and gain near the nodes grow, while the gain at the antinodes remains pinned at threshold (see Fig. 12.4a). A neighboring mode has a standing wave shifted with

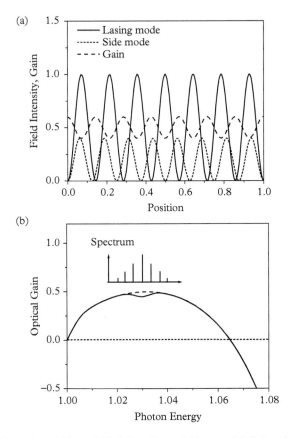

**Figure 12.4** *Illustration of (a) spatial hole burning and (b) spectral hole burning. In both cases, modes detuned from the gain peak can achieve the same gain as the lasing mode when the carriers that feed the lasing mode's gain become depleted.*

respect to the first, so that its nodes and antinodes do not coincide over much of the cavity. This means the neighboring mode can eventually have enough gain to lase. The reduction of gain available to the lasing mode is referred to as *spatial hole burning.*

The development of spatial hole burning is normally counteracted by carrier diffusion along the cavity axis. But far above threshold the lifetime shortens dramatically because of stimulated emission, and the carriers do not diffuse very far before they recombine. The distances between the nodes are also larger in longer-wavelength lasers, so spatial hole burning turns into a bigger problem. For lasers with wider ridges, lateral spatial hole burning can also occur. Both longitudinal and lateral hole burning can boost the internal loss somewhat above the simple average in Eq. (12.2).

As a side note, spatial hole burning is present only in cavities that support standing waves such as the Fabry–Perot resonator of Fig 12.1b. Now imagine a cavity that is formed from a ridge bent into a ring so that the mirrors are absent. If only clockwise (or counterclockwise) propagating modes lase in such a *ring cavity*, a standing wave does not become established, and spatial hole burning does not occur. It can reappear if both clockwise and counterclockwise modes lase and are phase locked. Ring cavities are much less popular than Fabry–Perot resonators because we must take active measures to extract the light from such a cavity, but they are gaining in currency.

In principle, single-mode emission can also be destabilized if the carriers contributing to the transition at the energy of the mode become depleted, by analogy with what happens in real space (see Fig. 12.4b). This scenario is known as *spectral hole burning*. However, because carriers scatter and replenish the gain on a sub-picosecond time scale, spectral hole burning is much less important than spatial hole burning in semiconductor lasers, even at currents well above threshold.

The conclusion is that to obtain robust single-mode emission, the cavity loss should be different for neighboring longitudinal modes, e.g., by taking advantage of their different frequencies. A well-known approach is the distributed-feedback (DFB) laser illustrated in Fig. 12.5. Here a refractive-index grating with a resonance tuned close to the gain peak generates spectrally selective feedback within the cavity. The grating can be placed within either the waveguide layers (Fig. 12.5a) or the top cladding (Fig. 12.5b). Of course, a single-mode laser should also operate in a single waveguide mode in the transverse and lateral directions. This is because a higher-order lateral mode can produce another longitudinal mode close

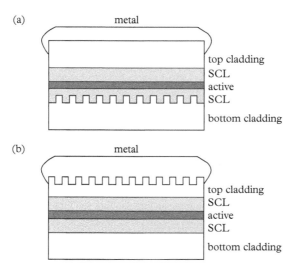

**Figure 12.5** *Schematic illustration of distributed-feedback semiconductor lasers with gratings below (a) and above (b) the gain region.*

to our target lasing mode. The strength of the index grating is adjusted so that the feedback over one cavity length is substantial.

If we can force the emission into a single spectral mode, the linewidth of that mode also narrows with increasing current density. When no current is injected, the linewidth is set by the sum of the rates at which the photons leak out of the cavity and at which they are lost within the cavity, which we define as the inverse of the photon lifetime $\tau_p$:

$$R_{ph,tot} \approx \frac{v_g}{2L_{cav}} \ln\left(\frac{1}{R_1 R_2}\right) + v_g \Gamma_\alpha \alpha_i \frac{L_\alpha}{L_{cav}} \equiv \frac{1}{\tau_p} \tag{12.17}$$

The denominator of Eq. (12.6) shows that above threshold almost all of the photon loss is compensated by gain. As a result, the effective photon escape rate slows dramatically:

$$\begin{aligned} R_{ph,tot} &\approx \frac{v_g}{2L_{cav}} \ln\left(\frac{1}{R_1 R_2}\right) + v_g \Gamma_\alpha \alpha_i \frac{L_\alpha}{L_{cav}} - v_g \Gamma_g g(\mathcal{J}) \frac{L_g}{L_{cav}} \\ &= \frac{r_{sp}}{S_{ss}} = \frac{r_{sp} \, \hbar \omega_{mode}}{P_{out}} \frac{v_g}{2} \ln\left(\frac{1}{R}\right) \end{aligned} \tag{12.18}$$

where we used Eq. (12.9) in the last equality. The single-mode laser linewidth is still equal to the effective photon decay rate, now given by Eq. (12.18). It turns out to be directly proportional to the spontaneous rate into the mode, and inversely proportional to the output power in that mode. Section 12.3 will discuss the semiconductor laser linewidth in greater detail.

If the laser emits into multiple modes, the "linewidth" depends on how many modes are excited and the free spectral range. The nonlinear dynamics discussed earlier make it difficult to derive the spectral envelope as a function of current and gain bandwidth. If multiple lateral or transverse modes are excited, there will be a mixture of different mode types in the envelope to boot.

Because the cavity in an edge-emitting laser is so long, the mirror loss is tolerable even if the mirrors are not very reflective. Since the cleaved facets of a semiconductor chip have $R_1, R_2 \approx 30$–$35\%$, we do not need to coat them for higher reflectivity to see lasing. To be sure, any light coming out the back facet is usually inaccessible, so to direct more power into the output beam, we should make that facet as reflective as possible.

Conversely, very short cavities such as those of a VCSEL require mirrors with $R_1, R_2 \gtrsim 99\%$. These mirrors often consist of quarter-wavelength stacks of semiconductor or dielectric materials with different refractive indices. Figure 12.6 shows a VCSEL geometry that emits light into a circularly symmetric beam from the top surface of the device. Because of its very short (wavelength-scale) cavity, the VCSEL emits in a single longitudinal mode. To ensure single-mode operation, the device must not lase in multiple lateral modes. One approach is

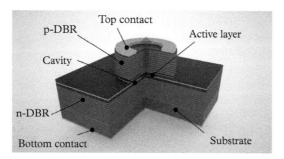

**Figure 12.6** *Schematic of a vertical-cavity surface-emitting semiconductor laser with emission from the top surface.*

to fabricate a skinny pillar that confines the modes via total internal reflection at the sidewalls. Another is to increase the loss of higher-order modes by cavity engineering. As in the case of the DFB laser, the threshold is at a minimum when the resonant wavelength matches the gain spectrum at the operating temperature. This is not necessarily easy because of imperfect fabrication, uncertainty in the modal index, and the fact that the gap shifts at a much higher rate with temperature than the modal index.

The theory developed here is equally applicable to both VCSELs and edge emitters. For an edge emitter, $L_g$ represents the active section of the cavity (the entire cavity length in most semiconductor lasers) and $\Gamma_g$ is the confinement factor calculated from Eq. (11.84) or (11.86). For a VCSEL, $L_g$ is the total active region thickness, e.g., for quantum wells modeled as superlattices, the normalization period multiplied by the number of wells $M$. Because the optical gain is inversely proportional to the period, it drops out of the rate equations, for both VCSELs and edge emitters.

As we saw in Section 4.4, a wavelength-size cavity with high $Q$ can modify the spontaneous emission rate by comparison with a bulk semiconductor. One way to do so is to make a VCSEL with highly reflective mirrors and pillars with a very small cross-sectional area. Another idea is to couple the active material to a subwavelength cavity supported by plasmon or phonon polariton resonances. While the $Q$ of these cavities is low ($\sim$10 for plasmons and $\sim$100 for phonons), they can enhance the emission via the Purcell effect from quantum wells or quantum dots.

Some wavelength-scale and subwavelength cavities can direct nearly all of the spontaneous emission into a single cavity mode. As discussed earlier, if we can also eliminate nonradiative recombination, the characteristic threshold behavior of Fig. 12.2 is suppressed, and the distinction between stimulated and spontaneous emission becomes blurred. But the absence of a threshold in the light–current curve does not guarantee a desirable laser. The wall-plug efficiency or the power output at a given current and voltage may be more important.

To make further progress, we must learn how to calculate the threshold current density of a particular semiconductor laser. This requires that we learn more about radiative and nonradiative recombination processes in semiconductors.

## 12.2   Radiative and Nonradiative Recombination Processes in Semiconductor Lasers

We will learn more about the carrier transport in *p–n* junctions in Chapter 14. For now we can simply think of the electrons and holes in a laser as flowing toward the active region from the opposite contacts, as shown in Fig. 12.2c. In this section, we assume that the current flow is limited by the recombination in the active region, as described by Eq. (12.8). The task is to understand what types of recombination mechanisms are important in different operating regimes.

In Chapter 11, we used Eq. (11.82) to approximate the optical gain in a quantum well as

$$g(n, T) \approx g_0 \ln \left( \frac{n}{n_{tr}(T)} \right) \tag{12.19}$$

where $n_{tr}$ is the temperature-dependent transparency carrier density and the coefficient $g_0$ is approximately independent of temperature. The maximum differential gain in the linear part of the gain curve is $g_0/n_{tr}(T)$. Although it is not difficult to compute the gain curve numerically, as in Chapter 11, we will continue to use the model of Eq. (12.19) for analytical calculations.

Inverting Eq. (12.19), we find an expression for the threshold carrier density $n_{th}$ in terms of the model's two parameters:

$$n_{th} = n_{tr}(T) \, \exp \left( \frac{\alpha_{cav}}{\Gamma_g g_0} \right) \tag{12.20}$$

where $\Gamma_g$ is the gain confinement factor defined in Eqs. (11.84) and (11.85). In edge-emitting lasers without a grating to provide spectrally selective feedback, lasing takes place in the mode nearest to the peak of the gain spectrum. In other device classes such as DFB lasers and VCSELs, any spectral misalignment between the gain peak and the loss minimum degrades the device performance because less gain is now available.

If the active region consists of $M$ identical wells, we can assume the injected current divides equally among them and parallel recombination occurs in each, as illustrated in Fig. 12.7. If the recombination rate in Eq. (12.14) is the same for each well, and the doping level is well below the injected carrier density ($n_{th} \approx p_{th}$), we obtain

electrons

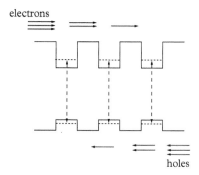

holes

**Figure 12.7** *Schematic of the current dividing equally (or nearly equally) among the active wells in a conventional semiconductor laser.*

$$\mathcal{J}_{th} = \frac{eMR(n_{th}, T)}{\eta_i} \qquad (12.21)$$

where the rate $R(n_{th}, T)$ represents the recombination in a single well only.

The temperature dependence of $\mathcal{J}_{th}$ follows its dominant recombination process and is often assumed to be exponential: $\mathcal{J}_{th} = \mathcal{J}_0 \exp\left(\frac{T}{T_0}\right)$, where $T_0$ is called the *characteristic temperature*. However, the functional form itself is no more than a fitting exercise, and in fact does not always lead to good fits over the entire range of temperatures. Quite often, the experimental variation of the threshold current density is slow at cryogenic temperatures, and then speeds up around room temperature. To cover this eventuality, we can say that the exponential dependence applies over a limited range $\Delta T_{max}$ around some $T_{ref}$ and the characteristic temperature as a fitting parameter in

$$\mathcal{J}_{th}\left(|T - T_{ref}| \lesssim \Delta T_{max}/2\right) \approx \mathcal{J}_{th}\left(T_{ref}\right) \exp\left(\frac{T - T_{ref}}{T_0}\right) \qquad (12.22)$$

We would like for $T_0$ to be as high as possible, but this cannot come at the expense of higher $\mathcal{J}_{th}(T_{ref})$. That is to say, a high characteristic temperature due to poor performance at lower temperatures is nothing to brag about.

We previously found in Chapters 4 and 11 that at low carrier densities, the spontaneous radiative recombination rate $R_{sp}$ scales as the product of the electron and hole densities $np$. However, we also saw that when the carrier populations become degenerate, the rate is no longer quadratic and asymptotically approaches a linear dependence in the limit $n$ (and $p) \to \infty$. The following approximate form spans the low- and high-density limits (see Fig. 4.14):

$$R_{sp}(n, p, T) = B(T)\frac{(np - n_0 p_0)}{1 + \frac{n}{n_s} + \frac{p}{p_s}} \qquad (12.23)$$

where the spontaneous radiative recombination coefficient $B(T)$ was defined in Chapters 4 and 11, $n_0$ and $p_0$ are the electron and hole densities in thermal equilibrium, and $n_s$ and $p_s$ are related to $N_c$ and $N_v$, respectively. In this expression we have written the equilibrium product of the carrier densities as $n_0 p_0$ rather than $n_i^2$ to make it applicable to degenerate populations. In that limit, $n_0 p_0 < n_i^2$ because the majority-carrier density no longer increases exponentially with Fermi level. Of course, when the injected carrier densities are much higher than the equilibrium densities, that term is unimportant anyway.

We also saw in Chapters 4 and 11 that for parabolic bands, the temperature dependence of the spontaneous radiative recombination coefficient $B$ follows a simple power law: $B(T) \propto T^{-\delta}$, with $\delta = 3/2$ for bulk materials and $\delta = 1$ for quantum wells with $E_g \gg k_B T$. The temperature dependence is stronger when the bands are not parabolic, as they are in real life, because the density of states increases more rapidly with energy, and the higher-energy states have a greater occupation probability at higher temperatures (see Fig. 11.24c). To keep the model simple, we retain the power-law dependence but broaden the potential parameter range to $1 \leq \delta \lesssim 2$.

As we mentioned in Section 12.1, electrons and holes can also recombine nonradiatively, i.e., without producing a photon. In *Auger recombination*, the energy of the recombining electron–hole pair is transferred to a third carrier, in either the conduction or valence band. That carrier jumps to an excited state, as illustrated in Fig. 12.8, with the total momentum (wavevector) conserved.

How does the Auger recombination rate depend on the electron and hole densities? We consider the case of an electron promoted to a higher state, as shown in Fig. 12.8a. The probabilities of finding the two electrons, with wavevectors $k_{i1}$ and $k_{i2}$, near the bottom of the conduction band and a hole, with wavevector $k_i$, near the top of the valence band are (see Eqs. (1.48)–(1.50))

$$f_c(k_{i1}) \approx \frac{n}{N_c} \exp\left(-\frac{E(k_{i1})}{k_B T}\right) \tag{12.24a}$$

$$f_c(k_{i2}) \approx \frac{n}{N_c} \exp\left(-\frac{E(k_{i2})}{k_B T}\right) \tag{12.24b}$$

$$1 - f_v(k_i) \approx \frac{p}{N_v} \exp\left(-\frac{E(k_i)}{k_B T}\right) \tag{12.24c}$$

The Auger rate is proportional to the probability $P_{A,e}$ that electrons occupy the first two states, while the third is vacant (meaning a hole occupies it):

$$P_{A,e} = f_c(k_{i1}) f_c(k_{i2}) [1 - f_v(k_i)] \approx \left(\frac{n}{N_c}\right)^2 \frac{p}{N_v} \exp\left(-\frac{E(k_{i1}) + E(k_{i2}) + E(k_i)}{k_B T}\right) \tag{12.25}$$

For an Auger process in which a hole jumps within the valence band, the probability is proportional to $P_{A,h} \propto np^2$.

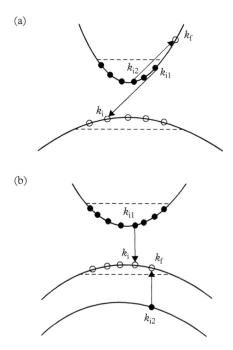

**Figure 12.8** *(a) Schematic of a multi-electron Auger recombination process. (b) Schematic of a multi-hole Auger recombination process in which a hole is promoted to a lower-lying band or subband. Dashed lines indicate quasi-Fermi levels for electrons and holes.*

Much like with radiative recombination, we expect the Auger rate to saturate at high carrier concentrations. How do we modify the probability in Eq. (12.25) to include the saturation? At low carrier densities, the recombination rate *per injected electron* is proportional to the density of holes in the valence band *and* the density of the third carrier species. However, at high (degenerate) densities the probability of finding both the recombining hole and the third carrier approaches unity, just as it does for radiative recombination. This leads to saturation of the Auger rate so that eventually it also varies linearly with carrier density. For practical materials, the densities are never high enough to occupy the final state. Therefore, we can model Auger recombination using

$$R_A(n,p,T) = C_e(T)\frac{\left(n^2 p - n_0^2 p_0\right)}{1 + \frac{np}{n_e p_e}} + C_h(T)\frac{\left(np^2 - n_0 p_0^2\right)}{1 + \frac{np}{n_h p_h}} \tag{12.26}$$

where $C_e$ and $C_h$ are the Auger coefficients for processes that involve two initial electrons and two initial holes, respectively, while $n_e$, $p_e$, $n_h$, and $p_h$ are related to $N_c$ and $N_v$. Once again, the negative terms ensure $R_A = 0$ in thermal equilibrium. We can apply Eq. (12.26) to either bulk materials, with carrier densities in units of cm$^{-3}$ and Auger coefficients in units of cm$^6$/s, or to quantum wells, with $n$ and $p$ in cm$^{-2}$ and $C_e$ and $C_h$ in cm$^4$/s. For superlattices, there is no single "correct" choice. As the minibands become wider, there is a continuous transition from 2D- to 3D-like behavior.

Unfortunately, the dimensional difference does not make it easy to compare bulk and quantum-well materials. For type I wells, we might simply scale by well width to put them on the same scale, but the choice is not black and white for type II wells. We will discuss this situation further in Section 12.4.

What is the temperature dependences of the Auger coefficients $C_e$ and $C_h$? To avoid detailed calculations, we assume parabolic bands in a bulk material and consider two limits: identical electron and hole masses ($m_e = m_h$) and $m_h \gg m_e$; i.e., the valence band is nearly flat compared to the conduction-band dispersion. Of course, the behavior in real materials is more complicated, but this is not a bad starting point.

From the energy conservation relation

$$\frac{\hbar^2 k_f^2}{2m_e} = E_g + \frac{\hbar^2 k_{i1}^2}{2m_e} + \frac{\hbar^2 k_{i2}^2}{2m_e} + \frac{\hbar^2 k_i^2}{2m_h} \tag{12.27}$$

we see that the final electron with energy $\frac{\hbar^2 k_f^2}{2m_e}$ must lie at least $E_g$ above the conduction-band minimum, which means $k_f$ is substantial (see Fig. 12.8a). The wavevector $k_f$ and the energy of the final electron are at a minimum when all of the wavevectors lie on the same line. The momentum-conservation condition then becomes

$$k_f = k_{i1} + k_{i2} + k_i \tag{12.28}$$

Since $k_f$ is large, one or more of the three initial carriers must populate a state with relatively high energy and low occupation probability. We conclude that to conserve both energy and momentum in simple bands, Auger recombination generally requires thermal activation. This is actually a good thing for the laser operation, insofar as the activation suppresses non-radiative decay. By contrast, the most likely radiative processes involve electrons and holes near the band extrema.

In the limit where the electron and hole masses are equal (e.g., in the lead salts), all three initial carriers carry the same penalty associated with low occupation probability at large wavevectors. In fact, the probability of a transition is maximized when all three have the same momentum, which justifies setting $k_{i1} = k_{i2} = k_i$ for a simple estimate. Substituting into Eqs. (12.27) and (12.28), we obtain

$$\frac{\hbar^2 k_f^2}{2m_e} = E_g + \frac{3\hbar^2 k_i^2}{2m_h} = E_g + 3E(k_i) \tag{12.29}$$

and $k_f = 3k_i$. Combining the two, we obtain

$$E(k_i) = \frac{E_g}{6} \qquad (12.30)$$

This considerably simplifies the overall Auger probability in Eq. (12.25):

$$P_{A.e} \approx \left(\frac{n}{N_c}\right)^2 \frac{p}{N_v} \exp\left(-\frac{3E(k_i)}{k_B T}\right) = \left(\frac{n}{N_c}\right)^2 \frac{p}{N_v} \exp\left(-\frac{E_g}{2k_B T}\right) \qquad (12.31)$$

Therefore, *in this particular limit*, the activation energy is $E_g/2$. For nondegenerate carriers in thermal equilibrium, the intrinsic $np$ product is itself proportional to $\exp\left(-\frac{E_g}{k_B T}\right)$, so the full dependence of the Auger recombination rate on energy gap scales as $\exp\left(-\frac{3E_g}{2k_B T}\right)$. This dependence is irrelevant for lasers, which operate far from thermal equilibrium, but it applies to detectors in which carrier densities are not very high. By symmetry, when the electron and hole masses are equal, Auger processes involving two initial holes have the same rate as those involving two initial electrons.

In the opposite extreme with a nearly flat valence band $\left(\frac{m_h}{m_e} \gg 1\right)$, the initial hole state can have a large wavevector without significantly increasing its energy and decreasing occupation probability. This means we can easily satisfy Eq. (12.28) by setting $k_f \approx k_i$. In effect, momentum conservation no longer matters, because the two initial electrons can have arbitrarily small wavevectors and energies. As a result, the Auger process is not activated:

$$P_{A.e} \approx \left(\frac{n}{N_c}\right)^2 \frac{p}{N_v} \qquad (12.32)$$

and typically occurs at a much higher rate. Non-activated Auger decay of this type makes carrier lifetimes quite short (and lasing thresholds rather high) in bulk narrow-gap III–V semiconductors.

This elementary picture, in which the Auger rate is governed by the hole-to-electron effective mass ratio, mostly applies to transitions within the same band or subband. Other Auger processes are also possible. For example, two-hole processes take place only when other valence bands or subbands are available for final transitions to conserve energy, as shown in Fig. 12.8b.

In Chapter 6, we saw that the SO band in many narrow-gap materials is separated from the top of the valence band by an energy close to the energy gap. When this happens, a hole near the top of the valence band can be promoted to the SO band in tandem with the disappearance of a precious electron–hole pair, as illustrated in Fig. 12.8b. Because states with small wavevectors are likely to be full, the Auger coefficient can become large and unactivated. The analog of this process in quantum wells occurs when the second hole is excited to a lower-lying valence subband (HH or LH). In some mid-IR and long-wave IR wells and superlattices, the multi-hole Auger coefficient associated with intervalence transitions is at least comparable to the multi-electron coefficient. By analogy with holes, a resonance with a higher-lying electron subband can also enhance the Auger rate. Finally, if the band offsets are limited, the transitions to unconfined states above the barriers also take place.

How are we to make sense of this wide variety of Auger mechanisms with different activation energies? First, note that the absence of activation is not equivalent to the lack of dependence on energy gap for a particular well or superlattice family. This is because additional transitions could come into play as the gap narrows. Regardless of the detailed mechanism, we are justified in coming to the following conclusions: (1) the Auger coefficient is higher (by orders of magnitude, because $E_g \gg k_B T$) in narrower-gap semiconductors and their quantum wells and (2) the Auger rate is low when the electron and hole effective masses are similar within a few $k_B T$ of the band extrema. In practice, we find that radiative recombination dominates in GaAs-based lasers for the near-IR and the visible, with Auger recombination paramount in GaSb-based interband lasers emitting at $\lambda \gtrsim 2$ μm. It follows that the strong compressive strain that reduces the hole mass and boosts the differential gain in quantum wells (see Chapters 10 and 11) should also act to suppress Auger recombination in the lasers emitting at longer wavelengths.

Recombination in most semiconductor active materials is dominated by either radiative or Auger processes. Nevertheless, recombination at crystal lattice defects becomes important when lattice-matched or coherently strained materials cannot be grown. This is the case for GaN-based blue-violet and ultraviolet lasers that have high dislocation densities. Such defects can spawn trap states in the energy gap, with the gap confining these states to the disturbance in the lattice. A carrier trapped by such a state is not available for transport in one of the bands. If both an electron and a hole can be trapped at the same defect, nonradiative recombination becomes possible.

The accepted model for defect-induced recombination was proposed by William Shockley and William Read, and, in a separate contribution, by Robert Hall. Here we calculate the Shockley–Read–Hall (SRH) rate for a single type of defect with volume density $N_t$ that creates discrete trap states having energy $E_t$. Recombination occurs when a given trap captures an electron and a hole in either order. When the recombination process is complete, the trap returns to its original charge state, and the combined energy of the electron–hole pair transfers to vibrations of the lattice ions that eventually dissipate as heat.

To calculate the net electron capture rate as simply as possible, we assume nondegenerate statistics. For the electron trap in Fig. 12.9, the capture rate is proportional to the electron density $n$, the trap density $N_t$, and the probability that the trap is empty $1 - f_t$. The electron can also escape from the trap before a hole is captured, which reduces the net capture rate. Since capture and escape can be thought of as the same mechanism run forward/backward in time, their

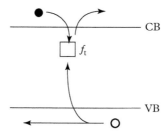

**Figure 12.9** *Schematic of an electron trap within the energy gap, which induces the Shockley–Read–Hall recombination of electron–hole pairs.*

rates share the same proportionality constant $\frac{1}{\tau_{n0}} \propto N_t$. This inverse lifetime is also proportional to the average velocity of the carriers. Since the thermal energy of a nondegenerate carrier is proportional to $k_B T$, both the average velocity and $\frac{1}{\tau_{n0}}$ are expected to scale as $T^{1/2}$.

The resulting electron capture rate becomes

$$R_{ce} = \frac{1}{\tau_{n0}} [(1 - f_t) n - f_t n_1] \tag{12.33}$$

The unknown parameter $n_1$ can be found by requiring the net capture rate to vanish, $R_{ce} = 0$, when no carriers are injected into the system, i.e. in equilibrium with $f_t$ given by the Fermi distribution (see Eq. (1.41)). For example, if the equilibrium Fermi level is equal to the trap energy $E_t$, $f_t = 1 - f_t = \frac{1}{2}$ and $n_1 = n_0 (E_F = E_t)$. In other words, $n_1$ is the equilibrium electron density when $E_F = E_t$. A more general expression is

$$n_1 = n_0 (1 - f_{t0}) / f_{t0} \tag{12.34}$$

where $f_{t0}$ is the value of $f_t$ under equilibrium conditions.

The rate for hole capture is proportional to $f_t$, because in this case recombination occurs only when an electron occupies the trap. By analogy with Eq. (12.33), we obtain

$$R_{ch} = \frac{1}{\tau_{p0}} [f_t p - (1 - f_t) p_1], \tag{12.35}$$

However, $\tau_{p0}$ is not necessarily equal to $\tau_{n0}$ because the conduction and valence bands in a III–V semiconductor are not symmetric. Following the same logic as above, $p_1 = p_0 (E_F = E_t)$, i.e., the hole density for $E_F = E_t$, and more generally

$$p_1 = p_0 f_{t0} / (1 - f_{t0}) \tag{12.36}$$

Combining Eqs. (12.34) and (12.36), we see that in this model with nondegenerate statistics, $n_1 p_1 = n_0 p_0 = n_i^2$, where $n_i$ is the intrinsic carrier concentration. Using Eq. (4.81), this leads to

$$n_1 p_1 = n_i^2 = N_C N_V \exp\left[\frac{(E_F - E_c)}{k_B T}\right] \exp\left[\frac{(E_v - E_F)}{k_B T}\right]$$
$$= N_C N_V \exp\left[-E_g / k_B T\right] \tag{12.37}$$

The rates $R_{ce}$ and $R_{ch}$, given in Eqs. (12.33) and (12.35), must be the same ($R_{ce} = R_{ch} \equiv R_{SRH}$) when the device is continuously pumped because carriers cannot accumulate without limit in steady state. This allows us to derive $f_t$ in terms of $n$ and $p$ so we can relate them to experimentally measurable quantities:

$$f_t = \left[\frac{n_1}{\tau_{n0}} + \frac{p_1}{\tau_{p0}}\right] / \left[\frac{n + n_1}{\tau_{n0}} + \frac{p + p_1}{\tau_{p0}}\right] \tag{12.38}$$

We now substitute Eq. (12.38) into Eq. (12.33) or Eq. (12.35), apply Eq. (12.37), and simplify to obtain:

$$R_{SRH} = \frac{(np - n_i^2)}{\tau_{n0} \tau_{p0}} / \left[\frac{n + n_1}{\tau_{n0}} + \frac{p + p_1}{\tau_{p0}}\right] \tag{12.39}$$

The second term in the numerator can be neglected when the injected carrier density is much higher than the intrinsic density. This is true for a semiconductor laser, but not for devices like detectors that operate near thermal equilibrium.

The SRH rate in Eq. (12.39) reaches a maximum when both $n_1$ and $p_1$ are small. This happens when the Fermi level is near mid-gap and $n_1 \approx p_1$. The physical reason is that a mid-gap electron (hole) trap is far more likely to be occupied (empty), and therefore available for hole (electron) capture. As the trap levels approach the band edges, $n_1$ and $p_1$ approach and exceed $n$ and $p$. If they dominate the denominator of Eq. (12.39), the SRH recombination rate becomes much lower.

When the doping density is much larger than the injected carrier density, e.g., $n \approx N_D \gg p, n_1, p \gg p_1$, and $np \gg n_i^2$, Eq. (12.39) simplifies to

$$R_{SRH} \approx \frac{p}{\tau_{p0}} \tag{12.40}$$

In the opposite limit of high non-equilibrium electron and hole densities, e.g., in a laser above threshold, the SRH rate is also linear in carrier density. For $n = p \gg n_i, n_1, p_1$ and taking $\tau_{n0} = \tau_{p0} \equiv \tau_0$, the recombination rate becomes

$$R_{SRH} \approx \frac{n}{2\tau_0} \tag{12.41}$$

We are now in a position to determine the temperature dependence of the threshold current density in Eq. (12.21) with the net recombination rate given by $R(n_{th}, T) = R_{sp}(n_{th}, T) + R_A(n_{th}, T) + R_{SRH}$. These recombination rates from Eqs. (12.23), (12.26), and (12.39) vary with temperature explicitly, as well as implicitly via threshold carrier density (decomposed into contributions from transparency carrier density, loss, and internal efficiency).

The fits to the transparency carrier density in Section 11.4 show that it varies as

$$n_{tr}(T) \propto T^{\upsilon} \qquad (12.42)$$

with $\upsilon = 1.4$–$1.5$ near room temperature, based on Fig. 11.17. If the loss increases gradually with temperature, we can assume a linear variation:

$$\alpha_{cav}(T) \approx \alpha_{cav}(T_{ref})\left(1 + \frac{T - T_{ref}}{\alpha'}\right) \qquad (12.43)$$

The same is true of the decrease in the internal efficiency: $\eta_i \approx \eta_i(T_{ref})$ $\left(1 - (T - T_{ref})/T_{\eta}\right)$, which if $T_{\eta} \gg T_{ref}$ we can also approximate as

$$\eta_i(T) = \eta_i(T_{ref}) \exp\left[-(T - T_{ref})/T_{\eta}\right] \qquad (12.44)$$

The radiative, Auger, and Shockley–Read–Hall recombination rates have different dependences on carrier density and temperature. To cover all possibilities, we write

$$R(n, T) \approx R_0 n^{\gamma} T^{-\delta} \exp\left(-\frac{T_{act}}{T}\right) \qquad (12.45)$$

where $1 \leq \gamma \leq 3$, $T_{act} = \frac{E_g}{ak_B}$ is the activation temperature, $2 \leq a < \infty$, $\delta \sim 0 - 2$, and the electron and hole densities are assumed to be equal. As we anticipated, a single characteristic temperature can be used only within a limited temperature range close to $T_{ref}$. The increase of the threshold current with temperature is not necessarily exponential, but it is at the very least superlinear.

Combining Eqs. (12.20), (12.21), (12.42), (12.43), (12.44), and (12.45), the net temperature dependence of the threshold current density has the form

$$\mathcal{J}_{th}(T) \approx \mathcal{J}_{th}(T_{ref})\left(\frac{T}{T_{ref}}\right)^{-\delta + \gamma\upsilon} e^{(T - T_{ref})/T_{\eta}} e^{\gamma(T - T_{ref})/T_{\alpha}} e^{T_{act}(T - T_{ref})/T_{ref}^2} \qquad (12.46)$$

where $T_{\alpha} \equiv \alpha' \Gamma g_{g0}/\alpha_{cav}(T_{ref})$. Typically, $T_{\alpha} \gg \alpha'$ because the threshold gain is lower than the maximum available (to minimize the threshold current density).

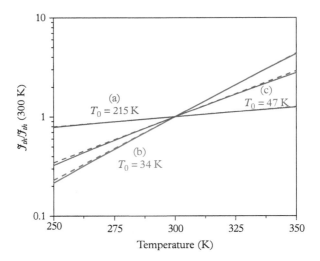

**Figure 12.10** *Model temperature dependences of the threshold current densities for several parameter sets chosen to approximate the behavior of a few representative semiconductor lasers at $T_{ref} = 300$ K: (a) a GaAs-based QW laser emitting near $\lambda = 1 \mu m$, with $\upsilon = 1.4$, $\delta = 1.8$, $\gamma = 1.7$, $T_{act} = 0$, $T_\alpha = 1000$ K, $T_\eta = 1000$ K (blue), (b) a notional GaSb-based QW laser emitting near $\lambda = 3 \mu m$, with $\upsilon = 1.5$, $\delta = 1$, $\gamma = 2.5$, $T_{act} = 1000$ K, $T_\alpha = 400$ K, $T_\eta = 300$ K (red), and (c) a mid-infrared laser emitting near $\lambda = 4 \mu m$, with $\upsilon = 1.4$, $\delta = 0$, $\gamma = 2.7$, $T_{act} = 0$, $T_\alpha = 200$ K, $T_\eta = 500$ K (magenta). The solid lines show the threshold current density according to Eq. (12.46), while the dashed lines are exponential fits with constant $T_0$.*

To calculate the value of $1/T_0$ at $T_{ref}$, we evaluate $\frac{d\ln(\mathcal{J}_{th}(T)/\mathcal{J}_{th}(T_{ref}))}{dT}$ at $T_{ref}$ and obtain

$$\left.\frac{1}{T_0}\right|_{T=T_{ref}} \approx \frac{\gamma}{T_\alpha} + \frac{1}{T_\eta} + \frac{T_{act}/T_{ref} - \delta + \gamma\upsilon}{T_{ref}} \tag{12.47}$$

where the last term represents the deviation from an exponential dependence. To keep it small, the temperature range should be within $\Delta T_{max} \ll T_{ref}/(T_{act}/T_{ref} - \delta + \gamma\upsilon)$. Figure 12.10 plots the temperature dependence of the threshold current density for several representative semiconductor lasers emitting at different wavelengths.

Equation (12.47) shows that the net temperature dependence combines contributions from the cavity loss (first term), internal efficiency (second term), recombination rate (third term), and transparency carrier density (last two terms). Unless $\mathcal{J}_{th}$ is quite high, we do not need to take into account the heating of the carriers in the active wells to a temperature above that of the lattice. The final consideration is that if the band offsets in the well are comparable to

$k_B T$, the carriers may "leak" out into the barriers. Most often, the characteristic temperature is dominated by the last ("transparency") term in Eq. (12.47).

How do the different laser classes (some of them illustrated in Fig. 12.10) fit into this model? SRH processes are mostly important in lasers based on nitrides and other novel materials. Otherwise, relatively few defects and dislocations are present in the active wells. In GaAs-based lasers emitting in the near-infrared and red ( 600–1100 nm), Auger recombination is weak because of the large activation energy that accompanies a wide gap (see Eq. (12.31)). Most of the threshold current density is due to radiative recombination. Section 11.4 showed that the typical threshold carrier density at room temperature in the InGaAs quantum-well laser on GaAs is $n_{th} \approx 1.5 \times 10^{12}$ cm$^{-2}$, with a radiative lifetime of $\approx$ 3–4 ns at this $n_{th}$. For a single well with $\eta_i \approx 1$ in a typical Fabry–Perot cavity, we estimate $\mathcal{J}_{th} \approx 70$ A/cm$^2$, which is close to the typical experimental results.

The internal loss and internal efficiency in near-IR and red lasers do not vary much with temperature, so $T_\alpha$ and $T_\eta$ are large and radiative emission is not activated ($T_a = 0$). Hence the characteristic temperatures for most GaAs-based QW lasers are dominated by the radiative recombination coefficient ($\delta \approx 1.8$) and transparency carrier density ($\gamma \upsilon \approx 2 - 3$) contributions to Eq. (12.47). Near room temperature, this implies $T_0 \approx 215$ K (see curve (a) in Fig. 12.10), which is broadly consistent with experimental data. The theoretical upper limit in this case is given by $T_{ref} / (\gamma \upsilon - \delta) \approx 500$ K.

As the wavelength becomes longer, Auger recombination comes to dominate the recombination rate. For InP-based lasers for the telecommunications range 1.3–1.6 $\mu$m, it is at least as important as the radiative process with coefficient $B_{QW}$ that decreases as the gap becomes narrower (see Eq. (11.100)). Does that mean that $\mathcal{J}_{th}$ is always higher for long-wavelength lasers? It turns out that the upward threshold trend is not as bad as might be expected because the electron and hole densities of states near the band edges are also lower when the gap is narrow. We will see in Section 12.4 that the threshold carrier densities in mid-IR ($\lambda = 3$–5 $\mu$m) GaSb-based type II QW lasers can be as low as 6–7$\times 10^{11}$ cm$^{-3}$, and the Auger lifetimes a little under 1 ns. With the lower $n_{th}$ roughly offsetting the shorter carrier lifetime, the threshold current density *per well* is often just a little higher than in a near-IR diode laser, i.e., only $\approx$ 100 A/cm$^2$.

However, the laser performance does suffer at longer wavelengths, where $T_0$ is typically lower and cavity losses higher. As a result, the efficiency and maximum output power are reduced in long-wavelength lasers. The Auger process has a strong dependence on carrier concentration ($\gamma \approx 2 - 3$), and $T_0$ depends greatly on whether the Auger process is activated. Curves (b) and (c) in Fig. 12.10 illustrate activated and non-activated results with parameters relevant to GaSb-based mid-IR lasers, with respective upper limits of $T_0 \approx 35$ and 50 K. Section 12.4 will consider an important practical example of a mid-IR laser. The interesting case of intersubband-based quantum cascade lasers that emit at wavelengths ranging from $\approx$ 3 $\mu$m to hundreds of $\mu$m is treated in Chapter 13.

Visible and ultraviolet GaN-based lasers are discussed in detail in Section 12.5. Relatively short Shockley–Read–Hall lifetimes along with a possible role for carrier leakage from the wells and Auger recombination lead to high threshold current densities of 3–5 kA/cm$^2$ at room temperature, with $T_0$ ranging widely between 100 and 200 K depending on the emission wavelength and barrier composition.

We already saw that while $T_0$ is a handy parameter for expressing the superlinear increase of the threshold current density with temperature, it is of limited predictive value. Whenever the temperature dependence of $\mathcal{J}_{th}$ is specified via $T_0$, the applicable range of temperatures should be included. Based on Eq. (12.47) we might expect $T_0$ to increase with reference temperature $T_{ref}$ when the $\frac{1}{T_0} \sim \frac{-\delta + \gamma \upsilon}{T_{ref}}$ term dominates, but this is rarely seen in practice. Possible reasons include (activated) carrier leakage, carrier heating, activated Auger mechanisms, a transition from the radiative-limited to Auger-limited regime, or a rapid increase of the internal loss with temperature.

Often we would like to optimize the number of wells $M$ for the lowest possible $\mathcal{J}_{th}$ given a particular value of $\alpha_{cav}$. Using Eqs. (12.20), (12.21), and (12.45), we obtain at $T = T_{ref}$

$$\mathcal{J}_{th} \approx M \mathcal{J}_{tr} e^{\gamma \alpha_{cav}/(M \Gamma_{SQW} g_0)} \tag{12.48}$$

where $\mathcal{J}_{tr}$ is the current density needed to reach transparency with a single active well, and $\Gamma_{SQW} = \frac{t}{d_{mode}}$ is the confinement factor *per well* from Eq. (11.86). Clearly, if the loss is very low, a single well achieves the lowest $\mathcal{J}_{th}$. In general, there is a balance between the extra current that supplies carriers to the other wells and the reduction in the threshold gain that additional wells provide. Figure 12.11 graphically illustrates the normalized threshold current density for $\gamma = 1.7$ (radiative limit) and $\gamma = 3$ (Auger limit) as a function of the normalized loss in the cavity $\alpha_{cav}/(\Gamma_{SQW} g_0)$ assuming equal electron and hole densities as before. In the radiative limit, a single well works well up to very high losses, at which point we may be better off using two wells. In contrast, in the Auger limit, the single-well region in Fig. 12.11b is compressed to very low loss values, with a transition to a broad minimum at 3–5 wells occurring for higher losses. While this illustrates the trend, in reality the laser designer may be well advised to use more than the bare minimum to account for effects that are omitted here, such as lattice and carrier heating, current leakage, etc.

Of course, the external differential quantum efficiency and the wall-plug efficiency given by Eqs. (12.13) and (12.16) are also important figures of merit. The WPE can be quite high (> 50%) for state-of-the-art lases emitting in the near-IR. Squeezing a few more percent from these devices is mostly an engineering challenge at this point. The WPE of mid-IR lasers is more interesting from the physics standpoint and will be discussed in Chapter 13. If the goal is to maximize output power and/or efficiency, in some cases we may use more wells than dictated by threshold considerations of Eq. (12.48).

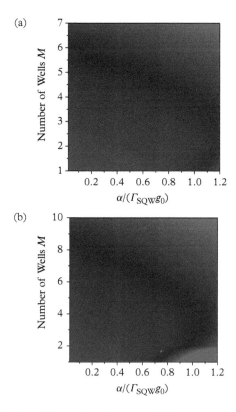

**Figure 12.11** *Optimization of the number of active wells in a semiconductor laser with threshold dominated by (a) radiative recombination with $\gamma = 1.7$ or (b) Auger recombination with $\gamma = 3$. The threshold current density varies from blue (low) to red (high), normalized to the transparency current density for a single well.*

## 12.3   High-Speed Modulation and Linewidth

Careful derivations of the high-speed modulation characteristics of quantum-well lasers are available from the sources listed under "Suggested Reading" at the end of this chapter. Instead of repeating those exercises, we will provide only an intuitive justification of the main results. We are interested in how fast the output power for a laser biased it well above the lasing threshold can be modulated by varying the injected current, as shown in Fig. 12.12. When the modulation is very slow, we expect the output to follow the CW values in Eq. (12.10b) and to remain in phase with the current:

$$P_{out}(t) \approx \frac{\hbar\omega}{e}\eta_e A(\mathcal{J}(t) - \mathcal{J}_{th}) \qquad (12.49)$$

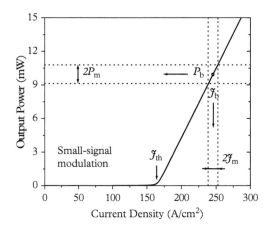

**Figure 12.12** *Illustration of the high-speed modulation of a semiconductor laser around a fixed set point well above the lasing threshold.*

For simplicity, we assume that the injected current is sinusoidal around some bias: $\mathcal{J}(t) = \mathcal{J}_b + \mathcal{J}_m \sin(\omega t)$, where $\omega$ is the modulation frequency. Of course, it is much easier to analyze the case of $\mathcal{J}_m \ll \mathcal{J}_b$, for which at low speeds we have $P_{out}(t) = P_b + P_{m0} \sin(\omega t)$, with $P_b = \frac{\hbar\omega}{e} \eta_e A (\mathcal{J}_b - \mathcal{J}_{th})$ and $P_{m0} = \frac{\hbar\omega}{e} \eta_e A (\mathcal{J}_m - \mathcal{J}_b)$. At what $\omega$ does the optical response become significantly weaker than the low-speed response? A useful figure of merit is the 3-dB point at which $P_m(\omega_{3dB}) = P_{m0}/2$.

To begin with, why is the optical response weaker at higher modulation frequencies? This is largely because the photon density in the cavity can no longer follow the change in the carrier density induced by the current modulation. To analyze the problem mathematically, we return to the rate equations Eqs. (12.3) and (12.5). While they look nonlinear, we can operate in a narrow range of currents and powers to linearize them. If we neglect the change in current density and spontaneously emitted photons, the linear forms become

$$\frac{dn_m}{dt} \approx -\frac{dR}{dn} n_m - \Gamma_g v_g \frac{dg}{dn} n_m S_b - \Gamma_g v_g g (n_b) S_m \approx -\left(\frac{dR}{dn} + \Gamma_g v_g \frac{dg}{dn} S_b \right) n_m - \frac{S_m}{\tau_p} \tag{12.50a}$$

$$\frac{dS_m}{dt} \approx \Gamma_g v_g \frac{dg}{dn} n_m S_b + \left[\Gamma_g v_g g (n_b) - \frac{1}{\tau_p}\right] S_m \approx \Gamma_g v_g \frac{dg}{dn} S_b n_m \tag{12.50b}$$

where $n_m$ and $S_m$ are the small amplitudes of the variation around the set point, the steady-state values indicated with the $b$ subscript are constants, the term in brackets in Eq. (12.50b) vanishes because the gain is pinned at its threshold value, and all the derivatives are taken at the bias point where $n = n_b$. For simplicity, we also assume here that the active medium fills the entire length of the cavity: $L_g = $

$L_{cav}$. Implicitly, we are neglecting any change in the gain induced by the photon density alone (e.g., due to hole burning). The quantity $\left(\frac{dR}{dn}\right)^{-1}$ is sometimes called the *differential carrier lifetime* and differs from our previous definition of carrier lifetime by a factor of $\gamma$, as should be clear from Eq. (12.45).

You may recall that two coupled first-order differential equations

$$\frac{dx}{dt} = -ay \tag{12.51a}$$

$$\frac{dy}{dt} = bx \tag{12.51b}$$

are equivalent to a harmonic oscillator at a frequency equal to the geometric mean of the two rate constants: $\omega_R = \sqrt{ab}$. To verify, differentiate one of the equations in Eq. (12.51) a second time and then substitute from the other equation to obtain: $\frac{d^2x}{dt^2} + \omega_R^2 x = 0$ with solutions $x, y \propto \sin(\omega_R t), \cos(\omega_R t)$.

The linearized rate equations for the semiconductor laser Eq. (12.50) are close to the form of Eq. (12.51), apart from the first term in Eq. (12.50a). The impact of this term will become clear in a moment. For now, defining $x \equiv n_m$ and $y \equiv S_m$, we see from Eq. (12.50) that $a = \Gamma_g v_g g(n_b) \approx \frac{1}{\tau_p}$ and $b = \Gamma_g v_g \frac{dg}{dn} S_b$. This means that the laser biased well above threshold tends to oscillate naturally at the *relaxation resonance frequency*:

$$\omega_R \approx \sqrt{\frac{\Gamma_g v_g \frac{dg}{dn} S_b}{\tau_p}} \tag{12.52}$$

The numerator under the square root is the differential change in the stimulated emission rate in response to a varying carrier density, while the denominator is how long photons remain in the cavity. These two major time scales determine the intrinsic small-signal modulation behavior of the laser, and we can think of the relaxation resonance frequency as their geometric mean. In physical terms, if the carrier density exceeds its steady-state level, the photon density responds by jumping above the steady state in tandem and forces the carrier density to fall below $n_b$. A decrease in the photon density and an increase in the carrier density will follow, after which the entire cycle is repeated. The period is $2\pi/\omega_R$, with $\omega_R$ given by Eq. (12.52). The laser emission behaves as a spring, with the *excess* photons and carriers acting as kinetic and elastic energy, respectively.

The form of Eq. (12.51) implies that these oscillations could go on forever, but they are in fact damped in real semiconductor lasers. The damping is due to the omitted $-\left(\frac{dR}{dn} + \Gamma_g v_g \frac{dg}{dn} S_b\right) n_m$ term in Eq. (12.50a), which describes the change in the recombination rate when the carriers are perturbed. We can think of it as a frictional term because $|n_m| > 0$ causes changes in the stimulated and other recombination rates that work against the perturbation; i.e., if $n_m > 0$ ($n_m < 0$), the recombination rates increase (decrease). Substituting $\frac{dx}{dt} = -cx - ay$ for Eq. (12.51a), we obtain the textbook equation for a damped oscillator.

The role of the injected current is to drive the modulation. When driven with a small-amplitude harmonic signal, the laser oscillates at the driving frequency $\omega$, but with an amplitude that depends on $\omega - \omega_R$. Putting all the pieces together, the response of the optical power to current modulation has the standard form for a damped, driven oscillator:

$$\frac{P_m}{\mathfrak{J}_m}(\omega) = \frac{H_0}{\omega_R^2 - \omega^2 - i\gamma\omega} \tag{12.53}$$

where $\gamma = \frac{dR}{dn} + \Gamma_g v_g \frac{dg}{dn} S_b \approx \frac{dR}{dn} + \omega_R^2 \tau_p$, and $H_0 = \frac{\hbar\omega}{e}\eta_e A \frac{\mathfrak{J}_m - \mathfrak{J}_b}{\mathfrak{J}_m}$ is the CW response that can be calculated from Eq. (12.49).

The amplitude of the typical response functions for semiconductor lasers is plotted in Fig. 12.13 for different values of the damping parameter $\gamma$. If the damping is not too strong, there is a peak at $\omega_R$ that tells us that the response is enhanced if the oscillator is driven near its natural frequency. If the damping is comparable to the relaxation resonance frequency, the frequency response is overdamped, and the peak at $\omega_R$ disappears. But even if $\gamma < \omega_R/2$, the 3-dB modulation frequency is only a little larger than $\omega_R$.

If we turn up the current passing through the laser, the carrier density and output power oscillate before settling down. Even though the change in carrier density leads that of the optical power, they are not completely out of phase so long as damping is present. In reality the delay is closer to a quarter of the oscillation ($\pi/2$). This is illustrated in Fig. 12.14a, with the parameters from

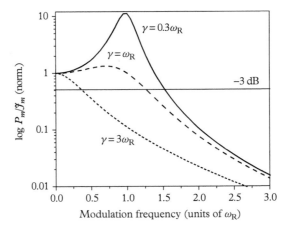

**Figure 12.13** *Small-signal modulation response of a typical semiconductor laser for three values of the damping rate.*

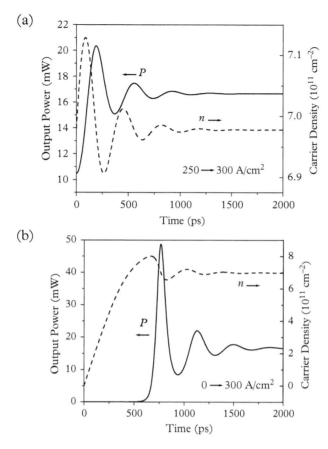

**Figure 12.14** *Time response of a mid-infrared semiconductor laser with the same parameters as in Fig. 12.2 to a small step increase in the drive current above the lasing threshold (from 250 to 300 A/cm²) and (b) to a large change from zero to a current well above the lasing threshold (300 A/cm²).*

Fig. 12.2 applicable to a mid-IR laser. If the dynamics are overdamped, the output power simply rises to the higher steady-state value without any oscillations at all.

In practical semiconductor lasers, other effects complicate this simple analysis. For example, hole burning compresses the gain at high photon densities and reduces the maximum modulation frequency. In quantum active structures such as quantum wells, wires, and dots, the carriers are typically captured from a quasi-bulk injection region into the structure itself. The capture time of at least 1 ps adds another limit on the modulation frequency. The final relaxation into a quantum dot may be particularly slow because the electrons must move down a ladder of discrete states. Often they can relax faster if they scatter off other electrons and holes in the same dot.

So far we have considered a laser biased well above threshold, and small changes of the current compared to the bias point. We have not addressed the question of how long it takes for the power to settle down when the laser is first turned on. To establish the steady-state carrier density takes approximately the carrier lifetime. If the lifetime varies strongly as a function of carrier density, which is often the case, the *large-signal* switching time is proportional to some average of the carrier lifetime over the relevant range of carrier densities. In general, the exact value can only be obtained numerically by integrating the rate equations Eqs. (12.3) and (12.5). For the example of the mid-IR semiconductor laser in Figs. 12.2 and 12.14a, the carrier lifetime at threshold is ≈ 700 ns, which results in a comparable delay time, as illustrated in Fig. 12.14b. Relaxation oscillations are also observed, but with a period that may be different from the small-signal response. The behavior of interband near-IR and mid-IR lasers is very different, apart from a longer carrier lifetime and somewhat higher carrier densities in the near-IR.

We saw from Eq. (12.18) that the laser linewidth should be inversely proportional to the output power. To make that expression quantitatively accurate, we must think deeper about the underlying physics. The linewidth does not really depend on how fast the intensity (or photon number) is lost from the cavity, but rather on how the phase $\phi$ of the output beam varies at a constant power level. The conversion factor turns out to be 1/2 because the phase change is proportional to half the intensity loss rate. The linewidth results from the random nature of photon emission, in the sense that we do not know exactly when a photon is emitted into the lasing mode rather than into one of many non-lasing modes. Sometimes it is claimed that spontaneous emission into the lasing mode is also random, but this is unjustified because, as we saw in Chapter 4, there is no fundamental difference between spontaneous and stimulated emission *into a given mode*. In any case, the random nature of the emission leads to fluctuations in the carrier density and available gain.

Any phase change due to the varying carrier density and optical gain affects (the real part of) the refractive index in the active medium because the gain is proportional to $\Delta n_i$, and the real and imaginary parts are physically connected. This means the laser linewidth is a sum of two contributions from the variation in (1) magnitude of the optical gain and (2) mode frequency due to index fluctuations $\Delta n_r$. Both are driven by fluctuations in the carrier density around its steady-state value, which are in turn caused by photon emission into non-lasing modes. The two contributions are out of phase because of the delay between the emission process and the corresponding change in carrier density and index obtained in connection with Fig. 12.14. The linewidth $\Delta \omega$ for this two-faceted random process is proportional to the phase variance $\langle \Delta \phi^2 \rangle$, which can be written

as $\Delta \omega \propto (\Delta n_i)^2 + (\Delta n_r)^2 = (\Delta n_i)^2 \left[ 1 + \left( \frac{\Delta n_r}{\Delta n_i} \right)^2 \right]$. If the first term is well described by one-half of Eq. (12.18), we can write

$$\Delta\omega = \frac{r_{sp}\,\hbar\omega\,w_{mode}}{2P_{out}}\,\frac{v_g}{2}\ln\left(\frac{1}{R}\right)\left[1+\left(\frac{\Delta n_r}{\Delta n_i}\right)^2\right] \qquad (12.54)$$

To make this expression more concise, we recast it in terms of the *linewidth enhancement factor*:

$$\alpha_e \equiv \frac{\Delta n_r}{\Delta n_i}\bigg|_{n_{th}} = \frac{\frac{dn_r}{dn}}{\frac{dn_i}{dn}}\bigg|_{n_{th}} = -\frac{4\pi}{\lambda}\,\frac{\frac{dn_r}{dn}\big|_{n_{th}}}{\frac{dg}{dn}\big|_{n_{th}}} \qquad (12.55)$$

where the negative sign makes the linewidth enhancement factor *positive* for most semiconductor lasers. Why are the signs of $\Delta n_r$ and $\Delta n_i$ different? To start with, the interband absorption is highly asymmetric with most of the oscillator strength concentrated well above the emission peak. Figures 12.15a and b illustrate that absorption induces a positive index change below the absorption peak. We can understand that behavior by thinking of a driving force with frequency below the natural oscillation frequency of the medium. Such a force always tends to lag the natural oscillation and hence is characterized by a lower phase velocity and higher index.

As the carrier density rises above transparency, the gain *reduces* that positive index change because it enters with the opposite sign, as shown in Fig. 12.15. In fact, a small change in the gain can produce a proportionally larger index variation. While $\alpha_e \sim 5-6$ for bulk materials, it drops to $\sim 2$ in quantum wells. By contrast, the emission typically takes place at the spectral peak of an atomic transition, where the differential index change vanishes and $\alpha_e = 0$. Intersubband and quantum-dot transitions in semiconductors can in principle approach that limit, but additional transitions typically make $\alpha_e$ non-zero. We emphasize that the linewidth enhancement factor is not fixed for any particular material, but depends on the wavelength of the mode, threshold gain, presence of nearby transitions, etc.

Apart from the effect of optical transitions, carriers shift the index directly because they oscillate in a collective plasmon mode at a frequency proportional to the square root of the density (in quasi-bulk materials). While for typical carrier densities in semiconductors, the frequency is quite low, in the THz or far-infrared, the change becomes more pronounced the longer the emission wavelength. The index change due to plasma excitations is negative because the gain peak is now *above* the fundamental oscillation (plasma) frequency.

Using the definition in Eq. (12.54) and converting from angular to linear frequency with $\Delta\nu = \Delta\omega/(2\pi)$, we rewrite Eq. (12.54) as

$$\Delta\nu = \frac{r_{sp}\,\hbar\omega\,w_{mode}}{8\pi P_{out}}\,v_g\ln\left(\frac{1}{R}\right)\left(1+\alpha_e^2\right) \qquad (12.56)$$

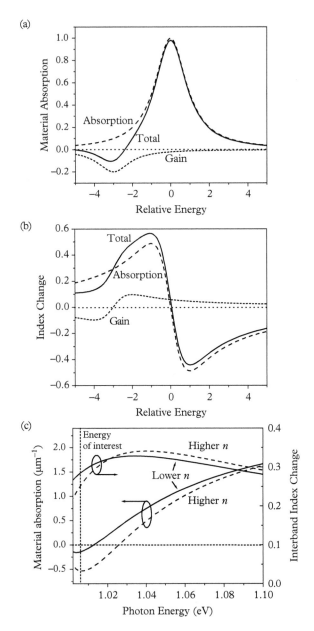

**Figure 12.15** *Illustration of the origin of the linewidth enhancement factor: (a) absorption and gain for two atomic-like transitions displaced by a few linewidths; (b) index change that corresponds to (a); (c) a more realistic representation of the interband absorption/gain and index change at two different carrier densities in a semiconductor material.*

Since $r_{sp}$ is not directly measurable, can we write this expression in an even more user-friendly form? To this end, we can use the relationship between the emission and absorption (gain) rates into a given mode (Eq. (4.86)):

$$r_{sp} = \frac{1}{w_{mode}L_{cav}} \frac{v_g \alpha_{cav}}{\exp\left[(\hbar\omega - eV)/k_B T\right] - 1} \tag{12.57}$$

where we multiplied $r'_{sp}$ by a factor of $\frac{1}{w_{mode}L_{cav}}$ to convert it into a rate per unit area $r_{sp}$. Substituting back into Eq. (12.56), we obtain

$$\Delta\nu = \frac{\hbar\omega}{8\pi P_{out}} \frac{v_g^2 \alpha_{cav}}{L_{cav}} \ln\left(\frac{1}{R}\right) \frac{1}{\exp\left[(\hbar\omega - eV)/k_B T\right] - 1} \left(1 + \alpha_e^2\right) \tag{12.58}$$

This expression is now suitable for practical estimates. For example, for $\alpha_e = 2$, $\alpha_{cav} = 10 \text{ cm}^{-1}$, $L_{cav} = 1$ mm, $R = 0.32$, $\frac{1}{\exp[(\hbar\omega - eV)/k_B T] - 1} \approx 2$, $n_g = 3.5$, $\hbar\omega = 1$ eV, we obtain $\Delta\nu \sim 1$ MHz/$P_{out}$ (mW).

Of course, the actual measured linewidth may well exceed the lower limit in Eq. (12.58). For example, the drive power and/or operating temperature of the laser are sometimes insufficiently stable. Low-frequency mechanical vibrations of the laser can also broaden the laser line. We can think of Eq. (12.58) as the fundamental quantum limit for the linewidth. We already know that many semiconductor lasers oscillate in multiple optical modes. The linewidth of the envelope of these modes is determined fundamentally by the width of the gain spectrum, but also by subtle nonlinear effects such as hole burning when the laser operates far above the lasing threshold. The linewidths of the individual modes may vary widely because the power each changes randomly even when the laser is driven CW.

## 12.4 Interband Cascade Lasers

Figure 12.11 shows that for sufficiently high loss, multiple active wells are needed to minimize the threshold current density. In most such lasers, the injected current splits (nearly) evenly among the wells. From the electrical point of view, the different wells then appear to be connected in parallel, as illustrated schematically in Fig. 12.16a. The required threshold voltage is given by

$$V_{th,p} = \hbar\omega/e + V_0 + \mathcal{J}_{th,p}\rho_s \tag{12.59}$$

where $\rho_s$ is the series resistivity in the circuit, and $V_0$ lumps together any voltage drop that is not proportional to the current density (including the voltage drop $\left(E_{F,n} - E_{F,p} - \hbar\omega\right)/e$ on the order of $k_B T/e$ needed to overcome the cavity loss). The objective of a good design is then to keep about the same current flowing in each parallel path through the individual wells.

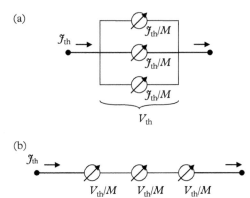

**Figure 12.16** *(a) Schematic of a quantum-well diode laser with injected current divided between the wells. (b) Schematic of a cascade laser with the same current passing through each active well.*

The total current in this parallel arrangement must be high enough to keep the carrier density in each well above transparency. But this is not the only possible configuration! The current can be much lower if the wells are connected in series rather than in parallel, as shown in Fig. 12.16b. Then it is the voltage drop (rather than current) over the active wells that scales with their number $M$:

$$V_{th,s} = M\left(\hbar\omega/e + V_0\right) + \mathcal{J}_{th,s}\rho_s \tag{12.60}$$

where $\mathcal{J}_{th,s} = \mathcal{J}_{th,p}/M$. The ratio of threshold power densities for the two cases becomes

$$\frac{\mathcal{J}_{th,p}V_{th,p}}{\mathcal{J}_{th,s}V_{th,s}} = \frac{M\mathcal{J}_{th,s}\left(\hbar\omega/e + V_0 + M\mathcal{J}_{th,s}\rho_s\right)}{\mathcal{J}_{th,s}\left[M\left(\hbar\omega/e + V_0\right) + \mathcal{J}_{th,s}\rho_s\right]} \tag{12.61}$$

Assuming the parameters $V_0, \rho_s, M$, and threshold current density per well $\mathcal{J}_{th,s}$ are the same in both cases, the first two terms in the numerator and denominator are equal. The difference is the last term proportional to $\rho_s$, which is a factor of $M^2$ smaller for the series connection. Since semiconductor lasers emitting at longer wavelengths benefit from as many as 5–10 wells, this factor is not negligible for typical values of $e\mathcal{J}_{th,s}\rho_s/\hbar\omega$. We conclude that the series scheme of Fig. 12.16b looks promising. But how do we implement it in practice?

To make the current flow successively through all the wells, we face the task of transferring those electrons that have already made a radiative (or nonradiative) transition from the conduction to the valence band back to the conduction band of the next active well. In principle, this can be done by inserting a tunnel diode into the current path, as shown in Fig. 12.17. The tunnel diode is doped heavily with opposite doping types on the two sides to create a strong band curvature near the junction. When a low reverse bias is applied to the junction, electrons

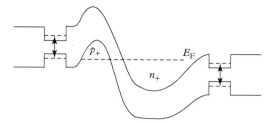

**Figure 12.17** *Active quantum wells connected by a tunnel junction.*

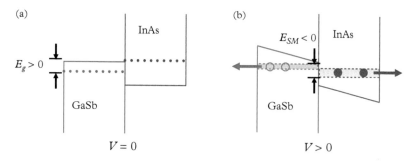

**Figure 12.18** *A type II quantum well consisting of a GaSb hole well and an InAs electron well, at zero bias (a) and under forward bias (b). An applied bias can turn the narrow gap at zero bias (a) into a semimetal with energy overlap between the conduction and valence bands (b).*

from the valence band on the *p* side can spill into unoccupied conduction-band states on the *n* side, with relatively low differential resistance. The catch is that at longer wavelengths, we may want to avoid heavy doping to minimize the internal loss due to free-carrier absorption.

Is there a better way? In the mid-1990s, Rui Yang proposed using the band overlap between InAs and GaSb to promote interband transfer, as shown in Fig. 12.18. While quantum confinement works against the band overlap, we can design the InAs and GaSb thicknesses so that a narrow gap exists (see Fig. 12.18a). When a forward bias is applied, as in Fig. 12.18b, the bands revert to a semimetallic lineup. Electrons and holes then coexist in quasi-thermal equilibrium, with electrons (holes) mostly in the InAs (GaSb) layer. The applied bias also leads to current flow, with holes being pulled away to the left (into the valence band of an active well), and electrons swept to the right (into the conduction band of the next active well), as shown in Fig. 12.19. The generated electron and hole densities must be equal to preserve overall charge neutrality, but they are different at any particular point along the current path. That charge imbalance generates electrostatic electric fields that oppose the applied field. The electrostatic internal fields are due to "zero-sum" carrier transfer, which means that their average over the entire active medium (or equivalently, within each of the periods) vanishes.

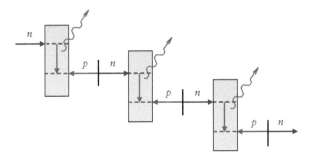

**Figure 12.19** *Schematic showing the internal generation of electrons and holes within each stage of an ICL.*

The total electric field at any point along the growth axis is a sum of the applied and internal fields.

If the type II GaSb–InAs structure in Fig. 12.18 is isolated and the electrons and holes have nowhere to flow, we can estimate their density and the magnitude of the internal electric field. The *semimetallic overlap* is defined as $\Delta_s(F_e) \equiv E_{v,GaSb} - E_{c,InAs}$, i.e., the difference between the top of the valence band in the GaSb well and the bottom of the conduction band in the InAs well. It depends explicitly on the applied electric field $F_e$ and the well widths. For a quick estimate, we take the bands to be parabolic and isotropic, with masses $m_h$ and $m_e$. If the electron and hole densities are equal ($n_{int} = p_{int}$), we can write at zero temperature

$$\frac{m_e}{\pi \hbar^2}(E_F - E_c) = \frac{m_h}{\pi \hbar^2}(E_v - E_F) \qquad (12.62)$$

where $E_F$ is the common quasi-Fermi level. This expression simplifies to

$$E_F = \frac{m_h}{m_e + m_h}\Delta_s(F_e) + E_c \qquad (12.63a)$$

or equivalently

$$E_F = E_v - \frac{m_e}{m_e + m_h}\Delta_s(F_e) \qquad (12.63b)$$

Since $m_h > m_e$, the quasi-Fermi level is closer to the top of the valence band than to the bottom of the conduction band. The same qualitative trends apply to higher temperatures and more accurate band dispersions, as long as $\Delta_s \gg k_B T$, which holds near and above the ICL threshold. The higher the external electric field $F_e$, the more carriers are generated on both sides of the semimetallic interface. The internally generated carrier density is

$$n_{int} \approx \frac{m_e m_h}{m_e + m_h}\frac{\Delta_s}{\pi \hbar^2} = \frac{m_r \Delta_s}{\pi \hbar^2} \qquad (12.64)$$

Now we calculate the internal field generated when carriers are separated on the two sides of the semimetallic interface. We assume the electrons and holes to be localized at distances $d_e$ and $d_h$ from the interface, which represent the centers of mass for the spatial distributions. This is a decent approximation even for realistic carrier distributions centered at the same positions. The problem then reduces to a sheet capacitor, with capacitance per unit area given by $\varepsilon_d/(d_e + d_h)$, where the zero-frequency permittivity $\varepsilon_d$ is a suitable average over the constituent layers. With this capacitance, the voltage induced between the two planes is

$$v_i \approx \frac{en_{int}}{\frac{\varepsilon_d}{d_e + d_h}} \approx \frac{m_r \Delta_s}{\pi \hbar^2} \frac{e(d_e + d_h)}{\varepsilon_d} \tag{12.65}$$

from which the average internal field is $F_i = \frac{m_r \Delta_s}{\pi \hbar^2} \frac{e}{\varepsilon_d}$. We see that the internal field increases with applied field $F_e$, via the field dependence of the semimetallic overlap.

To reuse electrons in the next ICL stage, which is connected in series, as shown in Fig. 12.19, an (external) voltage of at least $V_s = \hbar\omega/e$ must be dropped over the thickness of each stage. Since the electrons and holes in the ICL structure originate at the semimetallic interface, their density *at the applied field* $F_e$ must be high enough to reach the lasing threshold. If the density is too low because the overlap at zero field is too small, the threshold voltage exceeds its ideal value, and the wall-plug efficiency, maximum operating temperature, and maximum output power are reduced. At the other design extreme, there may be too many electrons and holes for $F_e$ that corresponds to $V_s = \hbar\omega/e$, which leads to higher free-carrier absorption and lower efficiency.

If we assume a threshold sheet density $n_{int} = 7 \times 10^{11}$ cm$^{-2}$ and a reduced mass close to $m_e = 0.023m$ in InAs (see Chapter 6), the required overlap is $\Delta_s \approx 100$ meV. The thinner the InAs and GaSb layers, the stronger the quantum confinement and the higher the applied field $F_e$ needed to generate $n_{int} \approx n_{th}$. Since the holes are much less sensitive to quantum confinement, we could fix the GaSb thickness at a reasonable value of 40–50 Å and vary only the InAs thickness. For most ICLs, that thickness is in the range 40–50 Å as well.

The basic ICL structure can accommodate a variety of different active regions. By far the most popular has been the InAs/GaInSb/InAs/AlSb "W" structure discussed in Chapter 10. In each stage of the full ICL structure, the single InAs and GaSb wells surrounding the semimetallic interface in Fig. 12.18 are replaced by multi-well electron and hole injectors as shown in Figs. 12.20 and 12.21. These play a dual role. First, they reduce the magnitude of the electric field $\sim \hbar\omega/(ed_s)$ at threshold, where $d_s$ is the total stage thickness (dominated by the electron and hole injectors). Second, the injectors with high barriers to the opposite carrier type block the electrons and holes from leaking out of the active "W" quantum wells before they have a chance to recombine radiatively. To illustrate these ideas, Fig. 12.21 shows the full band diagram and probability densities, i.e., squares of the envelope functions $|F_\mu|^2$, for selected electron and heavy-hole subbands and the dominant wavefunction components ($\mu = s$ for CB and $\mu = p_x, p_y$ for HH).

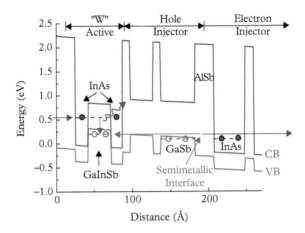

**Figure 12.20** *Semimetallic interface within each stage of an interband cascade laser structure. The directions of the electron and hole flow after they are generated at the interface at an applied bias are indicated.*

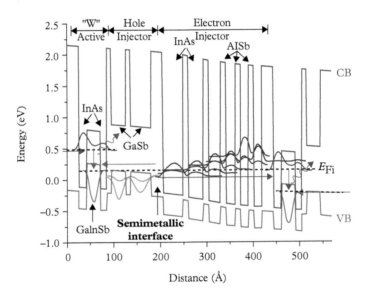

**Figure 12.21** *Band diagram and electron/hole probability densities in the full interband cascade laser structure. Zero probabilities are aligned with the corresponding subband extrema. The hole probabilities increase toward the bottom of the figure.*

To model the ICL of Fig. 12.21 efficiently, we take a number of shortcuts that shorten the computation time. For example, it is computationally intensive to solve the eight-band Hamiltonian for the full structure at many $k_\parallel$ points for each iteration of the internal-field solver (see Section 9.3). Instead, we solve the full Hamiltonian only at the zone center, which can be done using three-band (CB, LH, and SO) (or even two-band (CB and LH)) coupled equations, combined with a separate one-band solution for the HH band. This is equivalent to the quantum-confined Kane Hamiltonian of Eq. (10.13) (or the two-band Hamiltonian of Eq. (10.2)). To reduce the run time, we tabulate typical non-parabolic in-plane band dispersion relations for a four-layer "W" structure with similar quantum confinement using the full eight-band method. To compute the carrier densities, we use the subband energies given in Eq. (10.9):

$$E_v\left(k_\parallel\right) = E_v\left(k_\parallel = 0\right) + f\left(k_\parallel, E_v\left(k_\parallel = 0\right)\right) \qquad (12.66)$$

where the tabulated dispersion $f\left(k_\parallel\right)$ is substituted for each subband. We find that the error resulting from this procedure is smaller than the typical experimental uncertainties in layer thicknesses and compositions.

The general ICL design principles outlined here are well accepted by now, and have been verified and refined experimentally. However, that by no means implies there is nothing more to learn about ICL physics!

How are the electron and hole injectors constructed? We would like to block holes while permitting electrons to flow away from the semimetallic interface. To this end, the electron injector is typically a series of coupled InAs wells with large valence-band offsets with respect to GaSb. Coupled GaSb wells play the same role in the hole injector, as shown in Fig. 12.22. The barriers that separate the coupled wells in both injectors are usually AlSb, with thicknesses in the range 10–15 Å. Occasionally, AlGaSb and AlInSb are used to relieve the strain. To make a good barrier, the height must significantly exceed the subband energies in the wells, a condition that is easily fulfilled by AlSb or AlSb-rich alloys.

When an electric field is applied, the subband energies of isolated and coupled wells change in response (see Figs. 10.11 and 10.12). We already pointed out in Section 10.3 that a miniband can form in the presence of an applied field if the well thicknesses are chirped so that quantum confinement compensates the voltage drop (see Fig. 10.13). We can design the chirped superlattices for the electron and hole injectors by starting with the isolated-well subbands. For a target electric field of $F_e = 100$ kV/cm, the total stage thickness is approximately $\hbar\omega/eF_e$ (300–500 Å for wavelengths in the mid-IR). Excluding the active "W" thickness of ($\approx 80$ Å), how should this net thickness be subdivided between the two injectors? We prefer to make the electron injector much thicker than the hole injector because the lighter electrons are more easily shuttled across the injector. The hole injector can be as thin as a single GaSb well that feeds holes generated at the semimetallic interface into the GaInSb active hole well. However, most state-of-the-art ICL designs use two GaSb wells to thicken the barrier against electron tunneling from

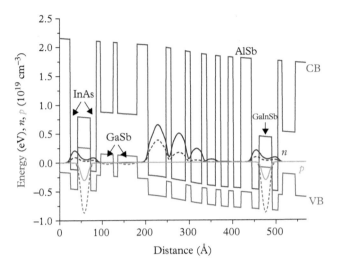

**Figure 12.22** *Position-dependent electron and hole densities that result from moderate (dashed lines) and high (solid lines) n-doping of the electron injector in an ICL.*

the active "W" QW region into the electron injector. Three to six InAs wells are common in the electron injector, with longer-wavelength ICLs generally having fewer because they operate at a lower applied bias per stage.

The final ingredients of the ICL layer structure in Fig. 12.21 are the AlSb barrier that separates the hole and electron injectors (at the semimetallic interface), and the AlSb barrier that separates the electron injector from the "W" active region. We need the first one to reduce parasitic interband absorption, but it cannot be so thick that transport would be impeded. This leads to thicknesses in the range 20–30 Å, with similar values used for the electron-injector/active AlSb barrier.

In modeling the ICL, we assume that the holes to the left of the semimetallic interface, as well as the electrons to the right, thermalize on a time scale much shorter than the carrier lifetime in the active region. For this reason, they support a common quasi-Fermi level that extends from the active hole well of any given stage to the active electron wells of the next stage. Because the highest valence subband in the active GaInSb hole well is typically 50–100 meV higher than in the hole injector, most of the generated holes occupy that well. Empirically, this design seems to have little or no negative impact on the hole transport, perhaps because the hole injector is very thin. Figure 12.22 illustrates the position-dependent electron and hole densities in an ICL stage.

Another wrinkle in the design is that we raise the subbands in the wells on the right side of the electron injector of Fig. 12.21 above the active wells of the next stage. This is done to reduce the electron density in the injector. Experiments have shown that as long as the difference is not too large, as in Figs. 12.21 and

12.22, working devices do not appear to feel any negative effects. The physical requirement is that the electron transit time must remain much shorter than the carrier lifetime in the active wells (several hundred ps).

We see from the density profiles in Fig. 12.22 that the InAs wells near the semimetallic interface are heavily populated. When the four middle wells of the electron injector is only moderately $n$-doped, this leads to electron densities in the active wells much lower than the hole densities. However, the difference disappears when the injector sheet doping reaches $5 \times 10^{12}$ cm$^{-2}$. In general, the electron injector is not doped uniformly, but the exact distribution of the charge in the injector is less important than the net sheet density.

Is it preferable to keep the electron and hole densities comparable? For a particular (multi-electron or multi-hole) Auger process dominating the lasing threshold, we may want to tilt the balance toward the carrier with the weaker Auger dependence (for the $n^2p$ Auger, keep $n \ll p$ and vice versa). To gain a quantitative understanding, we adopt the simple model of Eq. (11.71) applicable to quantum-well lasers in general. We assume that the cavity loss remains fixed, but this is not necessarily true when the hole density becomes high in long-wavelength lasers.

Since we want to maximize the gain at a given current density, we can fix the value of the general expression $(C_e/C_h)\,n^2p + np^2 \propto \mathcal{J}$. Casting the carrier densities in units of $N_c$ ($N \equiv n/N_c, P \equiv p/N_c$), this leaves two parameters: $N_{vc} \equiv N_v/N_c$ and $C_{eh} \equiv C_e/C_h$. Because the wells in the ICLs and other state-of-the-art lasers have significant compressive strain, the effective density of states ratio $N_{vc}$ is of order one, as discussed in Chapter 10. We now want to maximize the optical gain given by Eq. (11.71) as a function of the density ratio $N/P$.

We begin with the simpler task of deciding whether the $N/P$ ratio should differ from unity if the laser threshold is controlled by radiative recombination that scales as $NP$. We also omit the transition to a linear dependence at large densities in Eq. (12.23). If $N = P = N_0$ at some current density $\mathcal{J}_0$, the optical gain becomes

$$g\,(\mathcal{J}_0) = g_0 \left(1 - e^{-N_0} - e^{-N_0/N_{vc}}\right) \tag{12.67}$$

If $N_{vc} = 1$, this expression provides gain rather than loss only if $N_0 > \ln 2 \approx 0.693$ (or $n, p > 0.693 N_c$). The transparency carrier density $n_{tr}$ increases with $N_{vc}$. For example, if $N_{vc} = 2$ gain is possible when $N_0 > 2\ln\varphi \approx 0.962$, where $\varphi = \frac{1+\sqrt{5}}{2} \approx 1.618$ is the golden ratio. This result follows from converting Eq. (12.67) into a quadratic equation with the variable $e^{-N_0/2}$.

Now we perturb the original carrier densities by letting $N \to N_0 + \Delta N$ and $P \to N_0 - r\Delta N$, with $\Delta N \le N_0$. The condition of fixed $NP = N_0^2$ is equivalent to $\frac{\Delta N}{N_0} = \frac{1-r}{r}$, with $0 < r < 1$ for $\Delta N > 0$ and $1 < r < \infty$ for $\Delta N < 0$. This means that at a fixed current density $\mathcal{J}_0$, the difference in optical gain due to the perturbation becomes

$$\Delta g\left(\mathcal{J}_0\right) = g_0 \left[ e^{-N_0}\left(1 - e^{-\Delta N}\right) + e^{-\frac{N_0}{N_{vc}}}\left(1 - e^{\frac{r\Delta N}{N_{vc}}}\right) \right]$$

$$= g_0 \left[ e^{-N_0} - e^{-\frac{N_0}{r}} + e^{-\frac{N_0}{N_{vc}}} - e^{-\frac{N_0}{N_{vc}}r} \right] \tag{12.68}$$

We come out ahead by perturbing the densities only if $\Delta g\left(\mathcal{J}_0\right) > 0$. The gain versus current curve increases monotonically, so if we can obtain more gain at the same current, we can also reach $g_{th}$ at lower $\mathcal{J}$.

Figure 12.23 presents the results graphically as a function of $\Delta N/N_0$, for several values of $N_{vc}$ and gain at $\Delta N = 0$: $g = \frac{g_0}{10}, \frac{g_0}{4}, \frac{g_0}{2}$. Because of the compressive strain, we focus on the range $1 < N_{vc} < 5$. The gain is unchanged, $\Delta g = 0$, when $r = 1$ ($\Delta N = 0$) and $r = N_{vc}\left(\Delta N = -\frac{N_{vc}-1}{N_{vc}}\right)$. When the required gain is high, it helps to have a negative $\Delta N$; i.e., the hole density should be higher than the electron density. The increase in hole density is larger than the decrease in electron density in this scenario because $r > 1$. In contrast, when the gain is low, boosting the *electron* density makes it higher and potentially leads to a lower threshold. We can understand these results better by noticing that for high gains the $\exp\left(-\frac{n}{N_c}\right)$ term in Eq. (11.71) limits the gain, whereas in the low-gain limit, it is more sensitive to the $\exp\left(-\frac{p}{N_v}\right)$ term. The behavior is sometimes described as an increase in the differential gain accompanied by a higher transparency current density for $p$-doped structures (and the opposite for $n$-doped structures).

We also see that $N_{vc}$ must be large before it makes practical sense to disturb the electron–hole balance, and even then, the increase of the gain is limited. The improvement tends to decrease if we account for the transition to the linear carrier-density dependence and the full gain model. For this reason, the active wells of near-IR lasers are doped intentionally primarily to increase the differential gain and hence the modulation speed (see Eq. (12.52)), and less frequently to reduce the laser threshold. As we already mentioned, the differential gain can be higher in $p$-doped wells with $N_{vc} \gg 1$ because the first term in Eq. (11.72) is much larger. Even though the second term is then smaller, it contributes less to the differential gain when $N_{vc} \gg 1$.

Now we move on to the case of Auger recombination. If $N \rightarrow N_0 + \Delta N$ and $P \rightarrow N_0 - r\Delta N$, as before, the value of $C_{eh}N^2P + NP^2 = NP\left(C_{eh}N + P\right)$ remains fixed when $r = C_{eh}$ and $\frac{\Delta N}{N_0} = \left(1 - C_{eh}\right)/C_{eh}$. The change in gain becomes

$$\Delta g\left(\mathcal{J}_0\right) = g_0 \left[ e^{-N_0}\left(1 - e^{-N_0(1-C_{eh})/C_{eh}}\right) - e^{-N_0/N_{vc}}\left(e^{N_0(1-C_{eh})/N_{vc}} - 1\right) \right] \tag{12.69}$$

This expression is equivalent to Eq. (12.68) with $r$ replaced by $C_{eh}$. If $C_{eh} = 1$, no increase in the gain is possible, even if the densities of states in the conduction and valence bands are very different. But what if there is a distinct preference for either multi-electron or multi-hole processes? To estimate the effect, we can use the results of Fig. 12.23 with

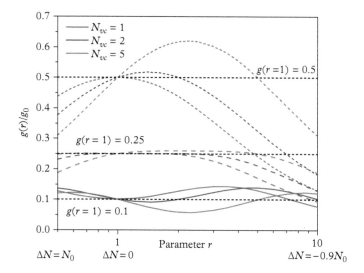

**Figure 12.23** *Optical gain as a function of relative electron and hole densities in the active wells, as expressed via the parameter r in Eq. (12.68), for several values of $N_{vc}$. The solid, dashed, and dotted lines are for $g/g_0$ values of 0.1, 0.25, and 0.5, respectively. The red, blue, and green lines correspond to relative valence band densities of states of $N_{vc} = \frac{N_v}{N_c} = 1, 2, 5$, respectively.*

$$r = C_{eh} = \frac{1}{1 + \Delta N/N_0} = \frac{N_0}{N_0 + \Delta N} \tag{12.70}$$

If $C_{eh} > 1$, we obtain $\Delta g > 0$ by reducing the electron density as long as the threshold gain is relatively high. By contrast, if multi-hole processes are stronger ($C_{eh} < 1$), increasing the electron density works only if the gain is low. At higher gains, the larger valence-band density of states suppresses the gain as the hole density drops. We could also say that for higher gains and $N_{vc} \gg 1$, a lower hole density reduces the separation between the electron and hole quasi-Fermi levels at the same injected current density.

What does the experimental evidence on the ICLs say? Figure 12.24 shows the thresholds of ICLs emitting near a wavelength of 3.8 μm as a function of sheet doping density in the electron injector, along with the calculated threshold electron and hole densities. The threshold has an apparent broad minimum that corresponds to roughly equal electron and hole densities (see also Fig. 12.22). Therefore, it is likely that $C_{eh} \sim 1$; i.e., neither the multi-electron nor multi-hole Auger mechanism dominates. Figure 12.25 shows the gain calculated for moderate and heavy doping of the electron injector, which supports the idea that the threshold decreases when the heavy doping of the electron injector raises the electron density in the active wells relative to the hole density. Such ICL designs are referred to as *carrier-rebalanced* structures.

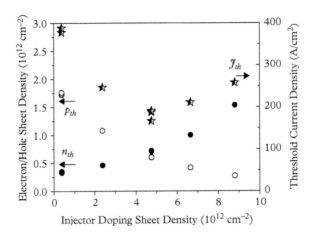

**Figure 12.24** *Experimental threshold current densities and calculated threshold carrier densities for different sheet doping densities in the electron injectors of ICLs emitting near a wavelength of 3.8 μm. The filled (open) points represent the threshold electron (hole densities) referring to the left axis. The stars represent the threshold current densities referring to the right axis.*

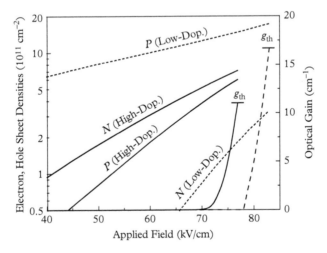

**Figure 12.25** *Calculated optical gain and sheet densities versus average applied electric field across a stage, for moderate (dashed) and heavy (solid) doping of the electron injectors.*

The ICL concept is not limited to the type II "W" InAs/GaInSb/InAs/AlSb active region shown in Fig. 12.20, which is currently its most mature variety. In principle, only a single type II interface may be enough to generate the optical gain we need, and the recombination rate should be lower in this structure because both nonradiative and radiative components scale with the electron–hole overlap in a similar way. In practice, it remains to be demonstrated that this design change benefits the laser threshold power. A more substantial modification is to use a more conventional type I well, e.g., a 100-Å-thick GaInAsSb in place of the "W" structure. Even though this requires a total revamping of the electron and hole injectors, the stronger gain of the type I cascade structure helps to minimize the number of stages in an optimal structure. Still, experimental work shows a clear benefit from using more than one stage, in particular when it comes to the maximum power and wall-plug efficiency, as discussed further in Section 13.3. Experimentally, type I cascade lasers with stages connected by tunnel junctions of Fig. 12.17 have been demonstrated.

## 12.5   Blue, Green, and Ultraviolet Nitride Lasers

The only direct-gap III–V materials that can emit on the blue side of the visible and into the ultraviolet are the nitrides. As we saw in Chapters 1 and 10, they have wurtzite crystal structure and strong built-in fields due to spontaneous polarization and the piezoelectric effect, if grown along a polar direction. From Chapter 6 we know that the nitrides form an isolated group of three binaries, with few possibilities for precise compensation of strain. For this reason, high dislocation densities and short Shockley–Read lifetimes are common and can dominate the threshold current density. Because the internal fields are large, and band offsets are often limited, carrier escape is another potential problem. Another practical difficulty, which is rarely encountered in other semiconductor lasers, is that the binding energies of acceptors such as Mg in AlGaN (the typical cladding material) are so high that few holes are available for electrical conduction at room temperature (with most of the holes remaining bound to the acceptors).

To determine the expected range of threshold current densities in the nitride lasers, we must combine the contributions due to Shockley–Read, radiative, and Auger recombination, because it is not immediately clear that any of them can be neglected completely. The widest variation has been reported for the Shockley–Read lifetime with values between 1 and 1000 ns, depending on growth conditions. The dislocations in nitride structures not only reduce the nonradiative lifetime, but can also lower the internal quantum efficiency via current leakage through the active region.

The radiative lifetimes for typical threshold carrier densities reaching $n_{th} \sim 10^{13}$ cm$^{-2}$ (see Section 11.5) should be shorter than in other III–V materials because of the wider gaps and higher density of states at the band edges, although the reduced wavefunction overlap brings them back into the range 3–10 ns for most nitride wells.

Because the gap is wide, we do not expect strong Auger recombination, but detailed calculations and measurements show that an indirect Auger process involving phonon and other scattering events can become significant at high carrier densities, with Auger coefficients of $C_e \sim C_h \sim 10^{-18} - 10^{-17}$ cm$^4$/s in 2D units. The Auger lifetimes are then 1–10 ns, and the Auger currents at threshold are 100–1000 A/cm$^2$. With the unavoidable contribution from radiative recombination, the nitride lasers should have room-temperature thresholds of at least 1 kA/cm$^2$, which is in agreement with the best experimental results.

To model optical gain and recombination in the InGaN and GaN quantum wells of Sections 10.5 and 11.5, we suggest a one-band Schrödinger equation for the conduction band and the six-band wurtzite Hamiltonian of Eq. (3.76) for the valence band, along with the polarization in Eq. (10.39). Because of the high carrier densities at threshold, we usually need to solve the Hamiltonians self-consistently with Poisson's equation, as described in Section 9.3. The reduced electron–hole overlap due to the polarization fields minus the opposing internal carrier-induced field affects both the optical gain and the radiative recombination rate.

Figure 12.26 shows a typical nitride laser structure, in which the claddings can be AlGaN, AlGaInN alloys or GaN/AlGaN superlattices, separate-confinement layers are GaN or InGaN with low doping, and the active wells are InGaN. The Al content in the cladding layers of blue and green lasers is typically quite small, ~5%, but sufficient to confine the mode without leading to an excessive binding energy. Because carriers can potentially leak out of the wells, some designs also insert an AlGaN current-blocking layer between the active wells and the separate-confinement layers. Many nitride lasers are grown on sapphire substrates using epitaxial laterally overgrown GaN buffers, but the recent availability of larger GaN substrates provides an attractive alternative.

The emission wavelength is determined by the alloy composition of the wells. For InGaN wells, the In fraction ranges from 5–10% for violet lasers ($\lambda \sim 410$ nm) to 15–20% for blue lasers ($\lambda \sim 450$ nm) to 30% for green lasers ($\lambda = 520$–30 nm). Moving into the ultraviolet (UV) requires that the InGaN or GaN wells be replaced by AlGaN; the Al fraction in the barriers and claddings is boosted to make sure carriers remain confined. As we saw in Section 11.5, all of these polar structures should have narrow well widths (2–3 nm) to keep electrons and holes close to each other. Typically only two or three wells are used, and good results are sometimes obtained with a single quantum well (see Section 12.1). It is also easier to inject carriers uniformly when there are only a few wells.

Figure 12.27 shows the variation of threshold current density at room temperature with emission wavelength, as compiled in a recent review paper. Notice first that the devices grown along semipolar and nonpolar directions currently do not perform as well as the more mature polar lasers, except in the green part of the spectrum. As we saw in Section 11.5, electrons and holes have stronger overlap

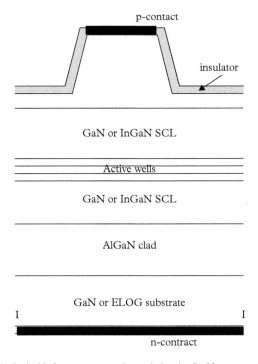

**Figure 12.26** *Typical nitride laser structure for emission in the blue part of the spectrum.*

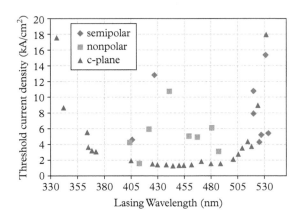

**Figure 12.27** *Threshold current density of the nitride lasers at room temperature as a function of emission wavelength. Reproduced with permission from M. T. Hardy et al., Materials Today* **14**, *408 (2011).*

in nonpolar nitride wells (similar to zinc-blende wells), so we expect these devices to improve as the growth technology matures. The performance also depends on the cavity axis because the gain in a nonpolar or semipolar well depends on the in-plane direction. The transparency carrier density should also be lower in wells grown at an angle to the $c$ axis.

Secondly, $\mathcal{J}_{th}$ has a broad minimum in the violet and blue, but increases rapidly on both sides of this sweet spot. What is the origin of this behavior? Wells with large In compositions needed to reach the green wavelengths are subject to heavy strain and large polarization fields of a few MV/cm. The strain can relax and create dislocations that act as Shockley–Read recombination centers. The large fields, which are approximately proportional to the In fraction, separate electrons and holes reducing gain and promoting carrier leakage. Currently, it remains unclear whether at least some of the increased threshold current density is due to stronger Auger recombination in narrower-gap green wells, as some theoretical models predict. The UV side suffers from analogous concerns about the strain and internal fields in AlGaN wells. Dislocation densities and hole activation also loom large.

A single blue and violet nitride laser ridge can produce 100 mW of CW output power. Optical loss is moderate in the best devices, and the high photon energy boosts the slope efficiency. The blue laser technology is quite mature, but many improvements in the design and performance of green and UV nitride devices may still be in the offing.

## 12.6 Quantum-Dot Lasers

We mentioned in Chapter 10 that the most popular variety of quantum dots is produced spontaneously during epitaxial growth. If a highly strained layer, in most cases InAs on GaAs, is grown beyond its critical thickness, 2D growth no longer takes place. Instead, InAs-rich regions prefer to "stay together" and form dots, usually embedded within an InGaAs quantum well with much smaller In fraction. While variation of the growth conditions can control the density, lateral size, and height of the dots to some extent, the spatial positions for dot formation within the growth plane appear to be completely random. Because of the randomness inherent in the self-assembly, quantum dots invariably come with a size distribution responsible for the inhomogeneous broadening discussed in Section 11.8.

Why should we bother with quantum dots considering the excellent performance of quantum-well lasers emitting at the same wavelengths? To answer this question, we examine Fig. 12.11 once again. When the cavity loss is low, $\mathcal{J}_{th}$ is bounded from below by the transparency current density of a single well. We cannot use a fraction of a well, so this is a fundamental limit. But quantum dots with a reduced fill factor *are* in some sense a fractional well, in that for typical dot densities $N_{QD}$, $\mathcal{J}_{tr,QD} \approx e\frac{N_{QD}}{\tau_{r,QD}} \ll e\frac{n_{tr,QW}}{\tau_{r,QW}}$. Here we assume similar radiative carrier lifetimes in wells and dots (see Section 11.8), while $N_{QD} \ll n_{tr,QW} \sim N_c \sim 10^{12}$ cm$^{-2}$. Originally, it was thought that the quasi-discrete density of states in a dot would supply high differential gain. Unfortunately, we saw in connection with Eq. (11.127) that the differential gain is comparable for self-assembled dots and wells. Therefore, quantum-dot lasers have lower threshold current densities than quantum-well lasers only when the cavity loss is quite low.

For a quantitative picture, we return to the typical quantum-dot experimental conditions considered briefly in Chapters 10 and 11. For InAs dots on GaAs substrates, typical areal densities $N_{QD}$ are in the range $10^{10}$–$10^{11}$ cm$^{-2}$. The dots used in lasers are often shaped like truncated cones or pyramids, with lateral extents of 10–30 nm and heights of 3–10 nm (corresponding to a $\approx$ 3:1 aspect ratio). To initiate the dot growth, a thin InAs wetting layer is formed on the GaAs surface, which after two monolayers transforms into dislocated-free islands that become the dots. Therefore, they typically end up on top of or embedded within an InGaAs quantum well with $\approx$ 20% In. It is important to remember that this well has electronic states close in energy to the confined dot states. Nevertheless, the dot potential is sufficient to confine multiple electron and hole levels (see Fig. 10.25(a)). Here we focus on relatively deep InAs dots, although shallow InGaAs dots as well as dots in other material systems such as InP and GaN can also be grown.

The differential gain in quantum dots shows a more dramatic decline than in a quantum well as the $s$ shell becomes full, and the gain peak shifts to higher shells and quasi-2D states. A simple model for the quantum-dot gain is an extension of the logarithmic form in Eq. (11.82):

$$g(n) = g_a \ln \left( \frac{n}{n_a} \right) + g_b \exp \left( -\frac{n}{n_b} \right) \qquad (12.71)$$

where the carrier density is given in 2D units, and $g_a$, $g_b$, $n_a$, and $n_b$ are fitting parameters. The $g_a$ and $g_b$ parameters scale with $N_{QD}$ and approximately inversely with the inhomogeneous broadening linewidth $\Delta_I$. By analogy with multiple wells, the quantum-dot gain is proportional to the number of dot layers $M$. Figure 12.28 plots typical quantum-well (see e.g. Fig. 11.18) and quantum-dot gain curves, in the modal gain units ($\Gamma g$) relevant to edge-emitting lasers. Section 11.8 showed that the carrier lifetimes in dots are only a little shorter than in wells, so the current density is approximately proportional to the carrier density in Fig. 12.28. Dots can readily produce modal gains of a few cm$^{-1}$ at current densities lower than in wells. While multiple layers of dots can achieve higher gains, the advantage over quantum wells disappears rapidly because all the layers need to be populated.

Experimental results largely agree with this simple model. For a single layer of dots in a low-loss cavity, the threshold current density can be as low as $\mathcal{J}_{th,QD} \approx e \frac{N_{QD}}{\tau_{r,QD}} \sim 10$ A/cm$^2$ at room temperature if we use $N_{QD} \sim 10^{11}$ cm$^{-2}$ and $\tau_{r,QD} \approx 2$ ns from Section 11.8. This is much lower than $\mathcal{J}_{th,QW} \approx 70$ A/cm$^2$

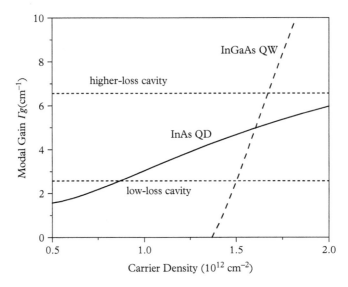

**Figure 12.28** *Illustration of the modal gain as a function of sheet carrier density (and hence current density) for a single InGaAs quantum well from Fig. 11.18 (dashed curve) and a single layer of InAs quantum dots with high $N_{QD}$. For a cavity with very low loss (lower dotted line), active regions based on quantum dots have a lower threshold current density, but when the loss is higher (upper dotted line), quantum wells are preferable.*

for a comparable quantum well from Section 12.1. Unfortunately, some laser classes such as VCSELs cannot achieve sufficiently low losses because the mirror loss per pass cannot be reduced arbitrarily. For quantum dots emitting at longer wavelengths, Auger recombination that potentially involves the quasi-2D states outside of the dot is a concern, but the details are complicated and structure-dependent.

Before the development of practical self-assembly techniques, it was thought that QD lasers would also exhibit low temperature sensitivity, i.e., a high value of $T_0$. In reality, undoped dots show characteristic temperatures similar to those of the quantum-well lasers from Section 12.1. Part of the problem is that at higher temperatures carriers occupy the higher-lying shells and quasi-2D states, which increases the threshold current density. One solution is to dope the dots $p$ type, so the hole density at threshold is much higher than the electron density. This actually increases the threshold current density at room temperature (see Section 12.4), but has the opposite effect on temperature sensitivity because the electron levels are more widely spaced and can produce high gain even when only the $s$ shell is occupied. Whether trading lower temperature sensitivity for a higher threshold is worthwhile depends on the application.

The gain from the fundamental $s$-shell transition in a quantum dot is quite symmetric, i.e., atomic-like, in stark contrast to the blue-weighted gain in a quantum well. As we discussed in Section 12.3, this means that the change in refractive index due to gain is odd about the transition wavelength and vanishes at the peak gain wavelength, i.e., that the linewidth enhancement factor $\alpha_e \approx 0$. Of course, we already know that higher-energy transitions are unavoidable and contribute to both absorption and gain (at high current densities). Neverthe- less, calculations show relatively low values of $\alpha_e \sim 1$ are possible in practical quantum dots. The low linewidth enhancement factor implies the potential for narrower lines and improved mode stability in single-mode lasers, as discussed in Section 12.3.

How do carriers injected from the contacts reach the localized states of the dots? In most semiconductor lasers, there is a continuum of states that extends from the wider-gap claddings to the active wells, and carriers can relax into the band- edge states by emitting lattice excitations called optical phonons. The emission is discussed in more detail in Chapter 13 in the context of quantum wells. This process is not possible in dots because the energy separations almost never match the phonon energies. However, this *phonon bottleneck* behavior is *not* observed experimentally in quantum dots used in semiconductor lasers. The culprits are apparently the phonon-transition broadening as well as electron–electron and electron–hole scattering. Even so, the carrier capture into the dots is a slow process that can limit the maximum modulation frequency discussed in Section 12.3. An interesting way to avoid this predicament is to add a quantum well separated from the dot by a thin barrier. The lowest subband of the well should be separated from the electron *s* shell in the dot by about one optical-phonon energy. This *tunneling injection* scheme illustrated in Fig. 12.29 avoids the need for injecting carriers from the top of the dot's potential barrier and speeds up the electron capture into the dot (the hole capture is generally much faster).

The strong inhomogeneous broadening hurts the threshold of quantum-dot lasers. Nevertheless, it is not without a silver lining for applications that require spectrally broad gain spectra, e.g., for wide tunability and short-pulse generation in mode-locked frequency combs discussed in more detail in Chapter 13. Another niche is the ability of quantum-dot lasers to operate at very long wavelengths when

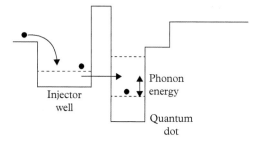

**Figure 12.29** *Tunneling injection into a quantum dot via a specially designed quantum well that is separated from the dots by a thin tunneling barrier.*

grown on GaAs substrates. Because GaAs–AlGaAs Bragg mirrors are easier to grow than all other III–V material pairs with a large refractive-index contrast, this is an important advantage for vertical emitters that use multiple dot layers.

Another interesting class of gain materials is quantum-dot-like nanocrystals formed from II–VI (CdSe, HgTe, etc.) and Pb-VI (PbSe, etc.) semiconductors. These nanocrystals are obtained by chemical synthesis methods in solution rather than embedding into another semiconductor like III–V dots. For this reason, carrier recombination at the surface of the nanocrystal looms large. To keep carriers away from the surface, the nanocrystal core is often enveloped by a wide-gap cap.

## 12.7 Light-Emitting Diodes and Devices

For light-emitting devices that never reach the lasing threshold, spontaneous emission continues to increase with current because the carrier density becomes higher. This emission is hardly negligible or useless. As we saw in Chapter 4, the spontaneous emission has a spectral width of $\sim k_B T$, while the emitted power depends on light collection and the presence of nonradiative recombination. Emission from the top or bottom of the sample fills a half space and is hence characterized by weak spatial and temporal coherence. Devices that rely on spontaneous emission from a semiconductor medium, such as a quantum well or superlattice, are termed *light-emitting diodes* (LEDs). Because some novel light-emitting structures (such as interband cascade regions from Section 12.4) do not employ $p$–$n$ junctions, we pull them under the umbrella by replacing the word "diode" in the LED acronym with "device."

What is the efficiency of the LED? Without a threshold, the probability of converting an electron–hole pair to a photon within the semiconductor is given by the ratio of the radiative recombination rate $R_{sp}$ to total recombination rate $R_{sp} + R_{nonrad}$ at the operating current density $\mathcal{J}$. Apart from that, some current never generates electron–hole pairs in the active medium, and some voltage drops outside of the active medium. The voltage efficiency of an LED is not much different from that of a laser if we ignore the distribution of photon energies. The efficiency observed from the outside also depends on how much of the emission can be extracted from the semiconductor and captured by an external optical system. Most III–V semiconductors have a refractive index of 3–4 and tend to trap much of the emitted light, as we will see below.

Putting these factors together, we can write the net WPE of an LED as

$$\eta_{WPE,LED}(\mathcal{J}) = \frac{R_{sp}}{R_{sp} + R_{nonrad}} \eta_{ex} \eta_I \eta_V \approx \frac{R_{sp}}{R_{sp} + R_{nonrad}} \frac{\hbar\omega/e + V_0 + \mathcal{J}\rho_s}{\hbar\omega/e} \eta_{ex} \eta_I$$

$$(12.72)$$

where $\eta_{ex}$ is the light extraction efficiency, $\eta_I$ is the current injection efficiency, and $\eta_V$ is the voltage efficiency. Optical loss usually has limited impact on the light extraction efficiency because the typical path length in the semiconductor is small.

Here we focus on the radiative and extraction efficiencies (the first and third factors in Eq. (12.72)). The radiative efficiency at a given current density depends on which of the nonradiative mechanisms (Shockley–Read–Hall and Auger, see Section 12.2) are present in the LED. Assuming a relatively large current that injects equal densities of electrons and holes, the SRH rate scales as $R_{SRH} \propto n$, according to Eq. (12.41), the Auger rate scales as $R_A \propto n^a$ (with $1 \le a \le 3$) from Eq. (12.26), and the radiative emission rate scales as $R_{sp} \propto n^b$, with $1 \le b \le 2$ from Eq. (12.23). When radiative recombination dominates (e.g. in the near-IR), the radiative efficiency can be high or even close to unity. The output power then increases linearly with current density, as in a laser. However, in mid-IR and visible/UV LEDs, it is far more common that $R_{nonrad} \gg R_{sp}$. Taking $R_{nonrad} \propto \mathcal{J} \propto n^\gamma$, with $1 \le \gamma \le 3$ to be fully general, we can write the dependence of the radiative efficiency on current density as

$$\frac{R_{sp}}{R_{sp} + R_{nonrad}} \propto n^{b-\gamma} \propto \left(\mathcal{J}^{\frac{1}{\gamma}}\right)^{b-\gamma} \propto \mathcal{J}^{\frac{b}{\gamma}-1} \qquad (12.73)$$

If the current injected into the LED is not too high, and the carriers are nondegenerate, $b \approx 2$ and $\gamma \approx 3$ in the Auger-dominated regime or $\gamma \approx 1$ when SRH processes dominate. This means the efficiency of Auger-dominated LEDs decreases with current. Can the LED efficiency become higher as more current is injected? This is only possible if the LED is SRH-dominated, but the injection is not so high that the carrier density becomes degenerate and the device heats up. Therefore, this behavior is rarely observed. More often, some current leaks and thereby limits the efficiency at low currents, e.g., when sidewalls of an LED mesa have lower resistance than bulk.

If the carrier density in long-wavelength LEDs approaches transparency ($\sim 10^{12}$ cm$^{-2}$), the carrier lifetime usually drops to a few hundred ps. On the other hand, the radiative lifetime remains $\sim 10$ ns, with some dependence on the active medium. This leads to radiative efficiencies ranging from 1 to 10%.

For many LEDs, the biggest hurdle is the low extraction efficiency. To understand why, we calculate the fraction of the light that escapes total internal reflection in a thick semiconductor slab. The light emerging from the surface must fall within a narrow cone defined by the emission angle $\theta_{TIR} = \sin^{-1}(1/n_r) \sim 14° - 19°$, which follows immediately from Snell's law when the maximum external angle is 90°, as shown in Fig. 12.30. We assume that both surfaces are flat, the bottom surface has reflectivity $R$, and there is no reabsorption of light in the active medium (or parasitic absorption elsewhere). We will see that the result depends on whether the emission is isotropic, as from a bulk-like semiconductor discussed in

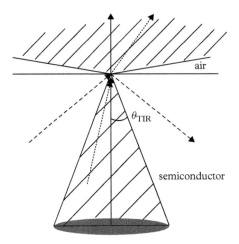

**Figure 12.30** *Extraction cone for light striking the surface of a semiconductor slab, as determined by the total internal reflection angle.*

Chapter 4, or polarized in the plane of the slab, as from a compressively strained well of Chapter 11.

For isotropic emission, the fraction of upward-propagating light within the cone bounded by $\theta_{TIR}$ becomes

$$\eta_{ex,iso} = \frac{1+R}{2} \frac{\int_0^{\theta_{TIR}} \sin\theta\, d\theta}{\int_0^{\pi/2} \sin\theta\, d\theta} = \frac{1+R}{2}(1-\cos\theta_{TIR}) = \frac{1+R}{2}\left(1-\sqrt{1-\frac{1}{n_r^2}}\right)$$

(12.74)

When $n_r$ is large, as in most III–V semiconductors (see Appendix D), this expression simplifies to $\eta_{ex,iso} \approx \frac{1+R}{4n_r^2}$, which is a small fraction ($< 5\%$). The reflection at the top surface does not appear in Eq. (12.74) because we assume the light can bounce from the top and bottom surfaces as many times as needed to get out, as long as it is within the escape cone.

A little more light can be extracted if the emission is polarized in the plane of the structure. Then the effective intensity varies with angle as $1+\cos^2\theta$ (see Eq. (11.59) for transitions to the heavy-hole subband at $k_{\|} = 0$). Vertical emission at $\theta = 0$ is twice as efficient as in-plane emission at $\theta = \pi/2$ because it can be directed into two different polarizations instead of one. If we integrate over $1 + \cos^2\theta$ (and normalize using the same weighting factor), we obtain

$$\eta_{ex,ip} = \frac{1+R}{2} \frac{\int_0^{\theta_{TIR}} \left(1+\cos^2\theta\right)\sin\theta\, d\theta}{\int_0^{\frac{\pi}{2}} \left(1+\cos^2\theta\right)\sin\theta\, d\theta} = \frac{1+R}{\frac{8}{3}}\left(\frac{4}{3} - \cos\theta_{TIR} - \frac{\cos^3\theta_{TIR}}{3}\right)$$

$$= \frac{1+R}{2}\left(1 - \frac{3}{4}\left[\sqrt{1-\frac{1}{n_r^2}} + \frac{1}{3}\left(1-\frac{1}{n_r^2}\right)^{3/2}\right]\right)$$

(12.75)

For large $n_r$, we obtain $\eta_{ex,ip} \approx \frac{3(1+R)}{8n_r^2}$, which is 50% higher than in the isotropic case. Of course, even for a compressively strained quantum well the band mixing away from $k_\parallel = 0$ makes the assumed $1 + \cos^2\theta$ dependence only approximate. Even so, we expect the extraction efficiency for this structure to be closer to Eq. (12.75) than to Eq. (12.74).

In any case, the extraction efficiency is low because the escape cone at the flat top surface is narrow. We cannot fix the problem by depositing an anti-reflection coating at the top surface because the path length becomes too large as the angle increases. In fact, with no reabsorption and a perfect reflector ($R = 1$) at the back, no improvement is possible.

Can we do better if the front surface is not flat? Wavelength-long corrugations on a sub-wavelength in-plane scale allow more light to escape by "softening" the index contrast between the semiconductor and the air. On the other hand, we can imagine corrugations on a wavelength as bending more of the light normal incidence, as shown in Fig. 12.31a. In either case, a larger fraction of the trapped light is released. To sum up, the surface texturing can be either random or periodic, at a scale below or at the wavelength in the material, i.e., $\leq \lambda/n_r$. Extraction efficiencies as high as 50%, i.e., an order of magnitude above the value for a flat, uncoated surface, have been obtained experimentally from textured surfaces. In addition, any photons absorbed in the active wells can be re-emitted if the radiative efficiency is high (photon recycling).

Another method of extracting more light is to pattern the front surface into a curved shape like a hemisphere or cone, as shown in Fig. 12.31b, with dimensions much larger than the emission wavelength. Once again, the curved top surface bends the light rays toward normal incidence (on a local scale), and thereby avoids trapping within the semiconductor.

Yet another approach is to reduce the thickness of the high-index slab to that of a single-mode waveguide ($\sim \lambda/2n_r$). This is experimentally challenging, but can allow most of the light to couple into the fundamental guided mode. Light propagating in the plane in that mode can then be coupled out using a grating with a period of $\lambda/n_m$, where $n_m$ is the modal index of the guided mode. In this case, the wavevector of the grating matches the propagation-direction wavevector of the light. If the radiative efficiency is not high, the grating must be strong enough to extract the light at a higher rate per unit length than the active-medium absorption coefficient. Figure 12.31c illustrates this geometry schematically.

The more reflective the bottom surface, the more light we expect to extract according to Eqs. (12.74) and (12.75). Unfortunately, the enhancement is limited to a factor of 2 when the emitter is far from the back surface. Another way to increase the extraction efficiency from a flat top surface suggests itself if we recall that nearly all the extracted light propagates close to normal incidence in the semiconductor, and that the spectral width on the order of $k_B T$ is usually much smaller than the photon energy. If the active medium is optically thin and can be placed close to a mirror, we can obtain a strong interference of the emission directed toward and away from the mirror. Because the electric field at a metallic surface nearly vanishes due to screening by charges in the metal, the field must rise

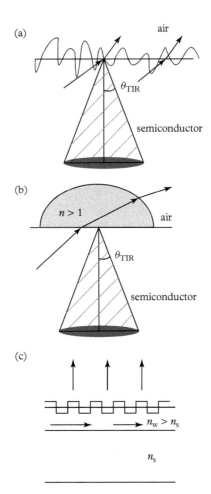

**Figure 12.31** *Various schemes for improving the extraction of photons from a semiconductor: (a) surface roughness, (b) a hemisphere at the emitting surface; (c) a thin single-mode waveguide patterned with a grating.*

back up to a peak a quarter-wavelength ($\approx \lambda/4n_r$) away from the mirror. Further peaks are also reproduced at half-wavelength spacings, as shown in Fig. 12.32. If the active medium is placed at one of these *antinodes* of the optical field, reflection from the mirror is coherently enhanced and the efficiency increases by up to a factor of 4 rather than 2. Accounting for the finite thickness ($t$) of the emitting material, the antinode enhancement factor is given by

$$\xi \approx \frac{2}{t} \int_{d-\frac{t}{2}}^{d+\frac{t}{2}} dx \, \cos^2 \left[ \frac{\pi n_r}{\lambda} (x - d) \right] = 1 + \frac{\sin\left( \frac{\pi n_r}{\lambda} t \right)}{\frac{\pi n_r}{\lambda} t} \qquad (12.76)$$

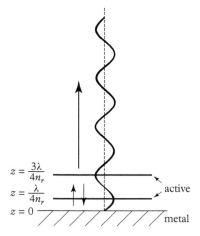

$z = \dfrac{3\lambda}{4n_r}$

$z = \dfrac{\lambda}{4n_r}$ active

$z = 0$ metal

**Figure 12.32** *Illustration of a coherent-interference scheme for improving photon extraction from a flat surface. Enhancement is obtained by placing active wells near a highly reflective backside mirror.*

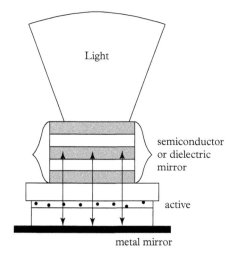

Light

semiconductor or dielectric mirror

active

metal mirror

**Figure 12.33** *Schematic of a resonant-cavity light-emitting diode.*

where $d$ is the active material position relative to the antinode. Clearly $\xi$ in Eq. (12.76) ranges from unity, for an active region that is half a wavelength long, to 2 for a vanishing active emitter thickness. This result holds as long as the emission is much narrower than the center wavelength.

This is promising, so can we do even better if we put the emitters at an antinode between two mirrors, as shown in Fig. 12.33? If their reflectivities, $R_t$ and $R_b$, are sufficiently high, and the cavity mode and emission wavelength are well matched, spontaneous emission *into the cavity mode* can be much faster than outside of the cavity. This follows from the Purcell factor in Eq. (4.70), which is proportional to $\frac{Q}{V_{eff}} \propto \frac{Q}{L_{cav}}$. The emission enhancement into the cavity mode is also proportional to the ratio between the cavity mode spacing (free spectral range) and the width of the cavity mode $\omega/Q$ (*finesse*). For a cavity with planar mirrors (Fabry–Perot), the emission enhancement and the finesse are proportional to $\frac{(R_t R_b)^{1/4}}{1-\sqrt{R_t R_b}}$. To see how much light emerges from the top mirror, we multiply by a factor of $2(1-R_t)/(2-R_t-R_b)$, i.e., the loss from the top mirror divided by the total mirror loss per pass. For $R_t, R_b \approx 1$, the enhancement is then close to the number of round trips in the cavity, $\frac{1-R_t}{(1-\sqrt{R_t R_b})^2}$. This evaluates to about an order of magnitude for realistic metallic mirrors with $R_b \sim 90\%$, which can be multiplied by the antinode enhancement factor in Eq. (12.76).

Unfortunately, spontaneous emission from a semiconductor is not very narrow. How does this affect the extraction efficiency from a resonant cavity (or a structure with a single mirror)? If we make the cavity $Q$ high by increasing the mirror reflectivities close to one, the width of the cavity mode shrinks with $1-\sqrt{R_t R_b}$ and can be much smaller than the emission bandwidth. For high-$Q$ cavities, the total enhancement over all wavelengths is then $\frac{1-R_t}{1-\sqrt{R_t R_b}}$, which is smaller and depends on the design of the mirrors. In fact, for maximum enhancement it pays to make the top mirror less reflective than the bottom mirror. For a single mirror, the large spectral spread means the enhancement near resonance is diluted, and the total extraction efficiency is noticeably lower.

Even though emission into the cavity mode is strongly enhanced, the *total* spontaneous emission may not be, e.g., because the emission bandwidth $\Delta\lambda \gg \lambda/Q$ or because many other modes are also present. If the spontaneous emission rate remains about the same, as it does in most semiconductor planar cavities, we can approximate the net extraction efficiency enhancement as

$$\frac{\eta_{ex,RCLED}}{\eta_{ex,0}} \approx \xi \frac{1-R_t}{1-\sqrt{R_t R_b}} \frac{\lambda}{\Delta\lambda} \frac{\lambda/2n_r}{L_{cav}} \tag{12.77}$$

For room-temperature operation, the resonant-cavity concept is promising for high-quality materials in the near-IR, but less attractive in the mid-IR where $\frac{\lambda}{\Delta\lambda} = \frac{\hbar\omega}{\Delta E}$ is less than an order of magnitude. While a resonant-cavity LED can narrow down spontaneous emission dramatically, that comes at the expense of ignoring transitions at other energies. The one exception is the true microcavity LED, in which most spontaneous emission goes into a single mode of the resonant cavity. Of course, this microcavity LED is simply a microcavity laser in which the cavity loss remains higher than the optical gain.

Current commercially available LEDs routinely take advantage of the encapsulation into a shaped low-index dielectric (Fig. 12.31b) as well as bottom reflectors to enhance the efficiency. Other approaches are less commonly used, with the surface texturing (Fig. 12.31a) being the easiest to implement in production.

## 12.8   Semiconductor Optical Amplifiers and High-Brightness Sources

Sometimes we already have a weak coherent source and would like to amplify its signal. Other times we may want to amplify spontaneously emitted radiation before making use of it. In both cases we need an *optical amplifier*, which has optical gain but does not lase. An amplifier can preserve most features of the input signal (e.g., bandwidth and coherence) and can be pumped electrically, just like a semiconductor laser.

An obvious way to prevent lasing is to spoil the reflection from the "mirrors." This turns out easier said than done because the lasing threshold scales logarithmically with $1/R$ (see Eq. (12.2)), while the gain continues to increase as we turn up the current density.

Figure 12.34 shows two of the geometries used in semiconductor optical amplifiers. The first involves tilting the amplifier ridge with respect to an anti-reflection-coated facet (Fig. 12.34a). Any light still reflected is then very unlikely to make it back into the ridge. To take full advantage of the available gain, we can also broaden the ridge as the light intensity becomes stronger in the process of amplification (Fig. 12.34b). Tapering spreads out the generated power and allows more light to be produced by a single ridge.

The rate equation in Eq. (12.3) specifies the time variation of the photon density assuming it is uniform throughout the cavity. A similar equation holds when light propagates with group velocity $v_g$ in the active material, but with time replaced by the spatial dimension along the axis of propagation. For simplicity, we assume that the entire region is filled with active material:

$$\frac{dS'}{dz} = \left(\Gamma_g g - \Gamma_\alpha \alpha_i\right) S' + \frac{r_{sp}}{v_g} \qquad (12.78)$$

Here the quantity $S'$ is the photon density per unit area perpendicular to the direction of propagation per unit time. To convert to power, we multiply by the

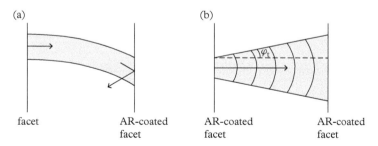

<table>
<tr><td>(a)</td><td></td><td>(b)</td><td></td></tr>
<tr><td>facet</td><td>AR-coated<br>facet</td><td>AR-coated<br>facet</td><td>AR-coated<br>facet</td></tr>
</table>

**Figure 12.34** *Some geometries for semiconductor optical amplifiers: (a) a tilted ridge waveguide that prevents back reflection and (b) a tapered ridge amplifier.*

cross-sectional area of the active material and the photon energy: $P \equiv \hbar\omega A S'$. We also denote $\gamma \equiv \Gamma_g g$, $\alpha \equiv \Gamma_\alpha \alpha_i$, $\rho \equiv \frac{r_{sp}}{v_g}$ to reduce the number of symbols:

$$\frac{dP}{dz} = (\gamma - \alpha)P + \rho \tag{12.79}$$

For full rigor, we could have derived Eq. (12.79) by starting from the wave equation interacting with the active medium. In any case, assuming that all the coefficients on the right-hand side are constants, we can immediately solve the resulting differential equation with the initial condition $P_i \equiv P(z = 0)$ to obtain the power at distance $L$ from the point at which the beam is injected:

$$P(L) = P_i e^{(\gamma - \alpha)L} + \frac{\rho}{\gamma - \alpha}\left(e^{(\gamma - \alpha)L} - 1\right) \tag{12.80}$$

The first term is the amplification of the signal injected at $z = 0$, while the second term is the *amplified spontaneous emission* (ASE) from the active material itself. When no beam is incident on the amplifier, it simply amplifies whatever light is emitted spontaneously inside. Because the LED emission is relatively weak, ASE can increase the power without making the emitted light spatially coherent. To be sure, the gain spectrum is sharper than the spontaneous emission from the LED, so the output from such a *superluminescent* device is narrower and less useful for applications that require broadband light.

Whenever we are interested in amplifying a weak input signal via the first term of Eq. (12.80), the second term acts as a noise source by adding light at other wavelengths. Therefore, ASE sets a lower limit on the input signal that can be rendered faithfully. Putting this issue aside, how much can we amplify the input? Equation (12.80) implies that the available gain for a given length of the active material is the only limit. But of course this is true only when the gain is independent of light intensity, which is not the case according to the rate equation for the carrier density [Eq. (12.5)]. In steady state

$$\Gamma_g g(n) S' = \frac{\mathcal{J}}{e} - R(n) \tag{12.81}$$

which we can rewrite as

$$\gamma(n) = \frac{\mathcal{J} A \hbar\omega}{eP} - \frac{R(n)A\hbar\omega}{P} \tag{12.82}$$

This expression for the gain is only implicit because both sides depend on the carrier density. To make analytical progress, we assume the gain to be linear in carrier density, i.e., $\gamma(n) = a(n - n_{tr})$, where $a = g_0/n_{tr}$ is the differential gain in the low-gain limit (see Eq. (11.82)), and that the same is true of the recombination rate: $R(n) = n/\tau$. Now we can find the carrier density as a function of light power:

$$n = \frac{\frac{\mathcal{J}A\hbar\omega}{eaP} + n_{tr}}{1 + \frac{A\hbar\omega}{\tau aP}} = \frac{\frac{\mathcal{J}\tau}{e} + \frac{n_{tr}\tau Pa}{A\hbar\omega}}{1 + \frac{\tau aP}{A\hbar\omega}} = \frac{\frac{\mathcal{J}\tau}{e} + \frac{n_{tr}P}{P_{sat}}}{1 + \frac{P}{P_{sat}}} \qquad (12.83)$$

where in the last step we defined the *saturation power* $P_{sat} \equiv \frac{A\hbar\omega}{\tau a}$. From Eq. (12.83), we can derive the desired expression for gain as a function of $P$:

$$\gamma = \frac{\frac{\mathcal{J}\tau}{e} - n_{tr}}{1 + \frac{P}{P_{sat}}} \equiv \frac{\gamma_0}{1 + \frac{P}{P_{sat}}} \qquad (12.84)$$

where $\gamma_0$ is the gain when the power is very small. More typically the assumptions of linear gain and recombination do not hold, and we must solve the more general form in Eq. (12.82). However, even then the conclusions reached below remain qualitatively valid.

If we substitute Eq. (12.84) into Eq. (12.79) and ignore the second term, we obtain

$$\frac{dP}{dz} = \left( \frac{\gamma_0}{1 + \frac{P}{P_{sat}}} - \alpha \right) P \qquad (12.85)$$

If $P(z) \ll P_{sat}$ for all $z$, the solution is exponential in distance, i.e., the first term in Eq. (12.80). However, in the opposite limit of $P(z) \gg P_{sat}$, the increase is not exponential! Even if the loss is very low, the power then increases only linearly with distance: $P(z) \approx \gamma_0 P_{sat} z + P_i$. We can obtain a better estimate for the increase in the power over length $L$ (assuming low loss) if we solve Eq. (12.85) by the separation of variables:

$$\int_0^L \left( 1 + \frac{1}{p} \right) dp = \int_0^L \gamma_0 \, dz \qquad (12.86)$$

where $p \equiv P/P_{sat}$. From this expression, we obtain

$$\ln\left( \frac{P(L)}{P_i} \right) + \frac{P(L)}{P_{sat}} - \frac{P_i}{P_{sat}} = \gamma_0 L \qquad (12.87)$$

Exponentiating both sides, we have

$$P(L) = P_i e^{\gamma_0 L} \, e^{-(P(L) - P_i)/P_{sat}} \qquad (12.88)$$

The second exponential factor reduces the unsaturated gain by an amount that depends on the ratio of propagating power to saturation power. Whenever the power exceeds $P_{sat}$ the gain drops and prevents us from increasing the power much further. Figure 12.35 plots the numerical solutions of Eq. (12.85) for a few parameter values.

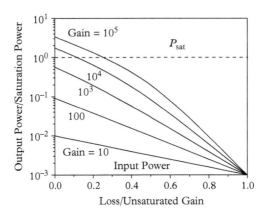

**Figure 12.35** *Solutions of Eq. (12.85) as a function of ratio between the loss and unsaturated gain $\alpha/\gamma_0$ for a few values of the overall unsaturated gain $e^{\gamma_0 L}$. The power is normalized to the saturation power, and the input power is taken to be $P_i = 10^{-3} P_{sat}$.*

Often we would like to amplify the input signal while maintaining its spatial and temporal coherence. This requires a single-mode amplifier waveguide, with cross-sectional area no greater than $\approx \lambda^2$. As a result, the saturation power $P_{sat} \equiv \frac{A \hbar \omega}{\tau a}$ and maximum output power are limited. We can overcome this limitation by adopting the tapered amplifier design from Fig. 12.34b. Here the input is amplified as it propagates along the radial direction within a full taper angle $2\varphi_t$. In cylindrical coordinates with azimuthal symmetry, Eq. (12.79) with the previous definitions and $\rho \rightarrow 0$ becomes

$$\frac{d\left(rS'\right)}{dr} = (\gamma - \alpha)\left(rS'\right) \tag{12.89}$$

Integrating across the vertical dimension and multiplying by the photon energy, we obtain

$$\frac{d(rI)}{dr} = (\gamma - \alpha)(rI) \tag{12.90}$$

where $I$ is the intensity in watts per centimeter. Substituting the total power $P = 2\varphi_t rI$ into Eq. (12.90), we see that Eq. (12.79) remains unchanged (with $z$ replaced by $r$). This means we can still use Eq. (12.85) within its realm of validity, if we use $P_{sat} = 2\varphi_t rMt\frac{\hbar\omega}{\tau a}$ . The saturation power increases the farther the beam propagates, which allows us to generate more power from any given amplifier. However, the taper angle must be small enough (5°–10°) to preserve the coherence of the input signal as it expands adiabatically.

The cylindrical symmetry approximation helps us understand the advantage of a tapered amplifier. But in fact the output facet is straight rather than curved, and the ridge expands from the wavelength scale rather than from zero. For a more accurate estimate, we must solve the full 2D propagation problem. The curvature of the phase front at the output facet means that the beam appears to originate from a point inside the amplifier, which requires care in evaluating the beam quality from the far-field pattern. The tapered ridge geometry can boost the output power for lasers with low reflection at the front facet as well as amplifiers.

Tapered amplifiers and lasers are not the only approaches available for enhancing the optical power emitted in a nearly diffraction-limited output beam. We can divide the others into two classes: (1) large-area single-mode waveguides achieved by coupling light generated in different regions of the waveguide; and (2) coherently or incoherently combining individual beams generated in separate single-mode waveguides.

The first class suffers from the need to dissipate heat from a large device area. This is difficult because almost all the heat must flow in a single (vertical) direction, while for a narrow ridge it can also flow sideways. In a wide ridge waveguide, regions that are far away laterally lase nearly independently of each other, so coherence can be destroyed by small changes in the intensity and phase. This is not the case in the so-called $\alpha$-DFB (a) and photonic-crystal DFB (b) approaches

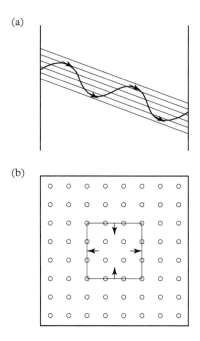

**Figure 12.36** *Illustration of the so-called $\alpha$-DFB and photonic-crystal DFB geometries for obtaining large-area single-mode emission from a semiconductor laser with lines and dots indicating grating structures.*

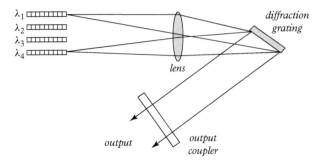

**Figure 12.37** *Typical geometry for spectral beam combining off the semiconductor chip using an adjustable diffraction grating.*

illustrated in Fig. 12.36. In both cases, the light can experience feedback along multiple non-parallel directions in the plane.

The second class of approaches is currently more popular. However, it is difficult to coherently combine the beams from many different ridges because the same phase must be maintained to high precision, which requires very fine control over it if the cavity is many wavelengths long. Nevertheless, there have been practical demonstrations of this *coherent beam-combining* scheme that use a common mirror. A technique that avoids this challenge at the expense of giving up some temporal coherence is *spectral beam combining*, in which each single-mode ridge emits at a slightly different wavelength. The beams are then combined, usually off the chip, using a diffraction grating with orientation such that each beam falls on the grating at a slightly different angle. Figure 12.37 illustrates the typical geometry for external spectral beam combining. In principle, spectral beam combining can also be deployed on the same chip, at the expense of greater complexity and potentially lower yield.

## Suggested Reading

- A few worthwhile and valuable book-length treatments of quantum-well interband semiconductor lasers are available: L. A. Coldren, S. W. Corzine, and M. L. Mašanović, *Diode Lasers and Photonic Integrated Circuits*, 2nd edn. (Wiley, New York, 2012), and S. L. Chuang, *Physics of Photonic Devices*, 2nd edn. (Wiley, New York, 2009). These references discuss the high-speed modulation characteristics of semiconductor lasers in detail, but their main focus is on the near-IR devices.

- A thorough description of the physics of interband cascade lasers can be found in I. Vurgaftman, W. W. Bewley, C. L. Canedy, C. S. Kim, M. Kim, J. R. Lindle, C. D. Merritt, J. Abell, and J. R. Meyer, "Mid-IR type-II interband cascade lasers," *IEEE Journal of Selected Topics in Quan-*

*tum Electronicsg* **17**, 1435 (2011), and I. Vurgaftman, R. Weih, M. Kamp, J. R. Meyer, C. L. Canedy, C. S. Kim, M. Kim, W. W. Bewley, C. D. Merritt, and J. Abell, "Interband cascade lasers," *Journal of Physics D* **48**, 123, 001 (2015).

- For review articles on nitride lasers, see M. T. Hardy, D. F. Feezell, S. P. DenBaars, and S. Nakamura, "Group-III nitride lasers: A materials perspective," *Materials Today* **14**, 408 (2011), and D. Sizov, R. Bhat, and C. E. Zah, "Gallium indium nitride-based green lasers," *Journal of Lightwave Technology* **30**, 679 (2012).

- The physics of quantum-dot lasers are discussed in W. W. Chow and F. Jahnke, "On the physics of semiconductor quantum dots for applications in lasers and quantum optics," *Progress in Quantum Electronics* **37**, 109 (2013).

- A good reference on the physics of the LEDs is E. F. Schubert, *Light-Emitting Diodes*, 2nd edn. (Cambridge University Press, Cambridge, UK, 2006).

- Tapered optical amplifiers and lasers are reviewed in J. N. Walpole, "Semiconductor amplifiers and lasers with tapered gain regions," *Optical and Quantum Electronics* **28**, 623 (1996).

- To learn more about different kinds of beam combining, see T. Y. Fan, "Laser beam combining for high-power, high-radiance sources," *IEEE Journal of Selected Topics in Quantum Electronics* **11**, 567 (2005).

# 13

# Quantum Cascade Lasers

This chapter describes the most commonly used approaches to computing the band structure of active materials with intersubband optical transitions. The physics of quantum cascade lasers (QCLs) is discussed in detail, including the mechanisms that limit the threshold current density, threshold voltage, wall-plug efficiency, and temperature sensitivity of state-of-the-art devices. The important roles of phonon and interface roughness scattering in determining threshold are emphasized. We also compare the performance of QCLs to other mid-IR lasers in considerable detail and make some conclusions as to which sources are preferred depending on the emission wavelength and application. Finally, we discuss the physical principles of laser-based frequency combs, including self-starting frequency-modulated QCL combs.

## 13.1   Hierarchy of Intersubband Models

As we saw in Chapters 10 and 11, optical transitions between the valence subbands can be quite complicated. Fortunately, all practical intersubband lasers employ transitions between the conduction subbands because they are easier to predict and are characterized by weaker intersubband scattering.

Many of the modeling approaches of Chapter 9 can be used to simulate these devices, most obviously by solving the 8-band $\mathbf{k} \cdot \mathbf{p}$ Hamiltonian. In practice, however, even an 8-band approach requires too much computing time to analyze the many-layered structures that form the active core of a QCL. So instead of applying a one-size-fits-all solution, we must decide which features of the band structure are essential, and which can be approximated without making the model too crude. It turns out it is rarely fatal to accept limited accuracy of the in-plane electron dispersion. This is because nonparabolicity is of secondary importance in many QCL materials such as GaAs and InGaAs lattice matched to InP. Furthermore, at low (nondegenerate) densities the electrons probe only the more nearly parabolic region near the conduction-band minimum. If push comes to shove, we can tabulate the in-plane dispersions for particular values of the subband minima (as we did for interband cascade laser in Section 12.4), and call on these

*Bands and Photons in III–V Semiconductor Quantum Structures.* Vurgaftman, Lumb, and Meyer,
Oxford University Press (2021). © Vurgaftman, Lumb, and Meyer.
DOI: 10.1093/oso/9780198767275.003.0013

dispersions whenever a more precise estimate becomes necessary. Otherwise, we can simply assume that the in-plane dispersion is parabolic.

How does this benign neglect of the exact in-plane dispersions help us? The band structure problem then reduces to computing energies at the $k_\parallel = 0$ point, where every state is spin degenerate. Instead of the 8-band Hamiltonian, we are left with the upper 3×3 block of the Kane Hamiltonian in Eq. (2.51) (or Eq. (2.41)), not counting the decoupled heavy-hole problem that is of little interest here:

$$
\mathbf{H}_{Kane} \begin{bmatrix} a_c \\ a_{LH} \\ a_{SO} \end{bmatrix} = \begin{bmatrix} E_g + (1+2F)\frac{\hbar^2 k^2}{2m} & \sqrt{\frac{2}{3}}\frac{\hbar}{m}kp_{cv} & -\sqrt{\frac{1}{3}}\frac{\hbar}{m}kp_{cv} \\ \sqrt{\frac{2}{3}}\frac{\hbar}{m}kp_{cv} & \frac{\hbar^2 k^2}{2m} & 0 \\ -\sqrt{\frac{1}{3}}\frac{\hbar}{m}kp_{cv} & 0 & -\Delta + \frac{\hbar^2 k^2}{2m} \end{bmatrix} \begin{bmatrix} a_c \\ a_{LH} \\ a_{SO} \end{bmatrix} \quad (13.1)
$$

Note that we now reference energies to the bottom of the lowest conduction band of any constituent materials (rather than the top of the valence band). Of course, we are always completely free to make that choice because absolute energies carry no physical meaning here.

We can go a step further by omitting the free-electron terms from the diagonal elements as well, on the grounds that $1 + 2F \approx 0$, and account for the other terms by correcting the interband coupling (see Chapter 9):

$$
\mathbf{H}_{Kane,s} \begin{bmatrix} a_c \\ a_{LH} \\ a_{SO} \end{bmatrix} = \begin{bmatrix} E_g & \sqrt{\frac{2}{3}}\frac{\hbar}{m}kp_{cv} & -\sqrt{\frac{1}{3}}\frac{\hbar}{m}kp_{cv} \\ \sqrt{\frac{2}{3}}\frac{\hbar}{m}kp_{cv} & 0 & 0 \\ -\sqrt{\frac{1}{3}}\frac{\hbar}{m}kp_{cv} & 0 & -\Delta \end{bmatrix} \begin{bmatrix} a_c \\ a_{LH} \\ a_{SO} \end{bmatrix} \quad (13.2)
$$

The quantum-confined version of Eq. (13.2) (with $-i\partial/\partial z$ replacing $k$) can then be solved as a straightforward eigenvalue problem, using one of the techniques (reciprocal space, finite difference, transfer matrix, etc.) described in Chapter 9. We only care about the solutions with energies above the conduction-band minimum ($E > E_g$) of the quantum-well material (or the material with the lowest conduction-band position in a multilayer structure).

Do we really need the 3-band Hamiltonian if we want to describe only the conduction-band states? In bulk, the solutions of Eq. (13.2) are given by Eq. (2.43):

$$
(E - E_g) E (E + \Delta) - \frac{\hbar^2 k^2}{2m} E_P \left( E + \frac{2\Delta}{3} \right) = 0 \quad (13.3)
$$

No one likes solving cubic equations, but we can approximate it with a quadratic equation provided $E_g \gg \Delta$. From Chapter 6, we know that this works well for

GaAs- and InP-based structures, but should give us pause when we are dealing with the antimonides (e.g., InAs wells and AlSb barriers). As we will see later, numerical results allay this concern even for the InAs/AlSb material system.

To convert Eq. (13.3) with $E > E_g$ to a quadratic equation, divide both terms by $\left(E + \frac{2\Delta}{3}\right)$ and approximate the denominator of the first term by

$$\frac{E(E+\Delta)}{\left(E + \frac{2\Delta}{3}\right)} \approx (E+\Delta)\left(1 - \frac{2\Delta}{3E}\right) \approx E + \frac{\Delta}{3} \tag{13.4}$$

Therefore, Eq. (13.3) simplifies to

$$(E - E_g)\left(E + \frac{\Delta}{3}\right) - \frac{\hbar^2 k^2}{2m} E_P \approx 0 \tag{13.5}$$

This happens to be the same as the eigenvalue equation for the 2-band bulk Hamiltonian:

$$\boldsymbol{H}_{Kane,2b}\begin{bmatrix} a_c \\ a_v \end{bmatrix} = \begin{bmatrix} E_g & \frac{\hbar}{m} k p_{cv} \\ \frac{\hbar}{m} k p_{cv} & -\frac{\Delta}{3} \end{bmatrix}\begin{bmatrix} a_c \\ a_v \end{bmatrix} \tag{13.6}$$

This is simply the 1D form of Eq. (8.24) (as well as Eq. (2.7) without the free-electron terms), except with the valence band shifted down by $\frac{\Delta}{3}$ so the "virtual" gap becomes $E_g + \frac{\Delta}{3}$. Of course, we would much prefer to work with a smaller matrix if the results are sufficiently close to Eq. (13.2).

Is the 2-band model the simplest that can account for nonparabolicity? In fact, we saw in Eq. (2.21) that we can do so with a single band by taking the effective mass to be energy dependent. Using Eq. (13.2), we replace Eq. (2.21) with

$$m_e(E) \approx m_e(0)\left(1 + \frac{2(E - E_g)}{3E_g} + \frac{E - E_g}{3(E_g + \Delta)}\right) \tag{13.7}$$

If we follow this route, we must solve the Schrödinger equation of the form

$$\left[-\frac{\hbar^2}{2}\frac{\partial}{\partial z}\frac{1}{m_e(E,z)}\frac{\partial}{\partial z} + V(z)\right]\Psi(z) = E\Psi(z) \tag{13.8}$$

We cannot solve Eq. (13.8) as a regular eigenvalue problem because the coefficients are not constant with eigenvalues. One straightforward method is to iterate a one-band solution, starting with a reasonable initial guess. The convergence can be quite fast, because of the linear dependence on energy in Eq. (13.7), but it is doubtful that this procedure can improve on the run time for the eigenvalue solution of the quantum-confined version of Eq. (13.6) (or even Eqs. (13.1) and

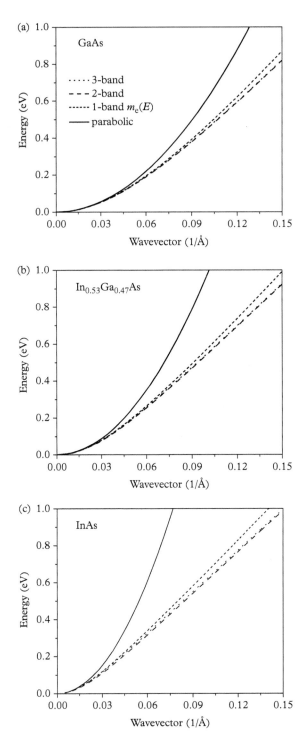

**Figure 13.1** *Electron dispersions as a function of k for the Hamiltonians of Eqs. (13.2) and (13.6), and a one-band model with the energy-dependent mass specified by Eq. (13.7), for GaAs (a), InGaAs lattice-matched to InP (b), and InAs (c). The solid curves represent the parabolic approximation, short-dashed curves are for the one-band model with an energy-dependent mass (Eq. (13.7)), dashed curves are for the two-band model of Eq. (13.5), and dotted curves are for the three-band model of Eq. (13.3) solved for their respective eigenvalues.*

(13.2)). Therefore, our preference is to model the conduction band of complicated multilayer structures by solving the two-band Hamiltonian of Eq. (13.6).

Figure 13.1 compares the electron dispersions based on Eqs. (13.2) and (13.6) for GaAs, $In_{0.53}Ga_{0.47}As$ lattice matched to InP, and InAs. The energies for the three- and two-band models are corrected by $\frac{\hbar^2 k^2}{2m}(1 + 2F)$ to make the dispersions agree with the recommended band parameters in Chapters 6 and 7. Somewhat surprisingly perhaps, the curves for the two- and three-band models are indistinguishable even for narrow-gap InAs. This is because when we have $E \gg E_g$, we also have $E \gg \Delta$, and the influence of spin–orbit splitting is weak. Figure 13.1 also plots the dispersions from Eq. (13.7) for GaAs, InGaAs, and InAs. The differences with the multiband models are minor and can be made even smaller if we adjust the $F$ parameter downward.

## 13.2   Quantum Cascade Lasers

The most obvious difference between the QCL and an interband laser is that the QCL generates gain by creating a population inversion between two conduction subbands, as discussed in Section 11.2. Current QCLs emit in the mid-infrared and far-infrared, because there are difficulties in designing structures with large intersubband transition energies $\Delta E \equiv E_{c,u} - E_{c,l}$. Many active transitions are generally needed to produce enough gain, so the QCL (like an ICL) benefits from a series electrical connection of the active wells shown in Fig. 13.2. The same current passes through each stage of the QCL successively, and the total voltage dropping over the active core scales with the number of stages. The cascaded arrangement in the QCL also helps to empty the lower lasing subband.

Figure 13.3 shows the subbands and corresponding envelope functions for a typical QCL designed for emission at $\lambda = 5$ µm, under an electric field close to the threshold value. The plot focuses on a single transition in active quantum wells surrounded by electron injectors that lead to and from the neighboring stages. The wavefunctions of states involved in the active transition are distributed across multiple wells. After an electron makes a radiative (or nonradiative) transition, it is extracted from the lower lasing subband via fast phonon scattering to the bottom of a different subband, separated from the first by a little more than one phonon energy. After the emission of the initial phonon, the electron may emit another one and then move along the injector miniband by other means. The injector miniband is formed by aligning the subbands of chirped wells in an applied field, as discussed in Section 10.3. As in the ICL case, the injector is long enough to be compatible with a threshold field of ~100 kV/cm and a total voltage drop that exceeds the photon energy, as discussed in the following.

Mid-IR QCLs generally employ InGaAs quantum wells and InAlAs barriers, while THz QCLs more commonly employ the GaAs/AlGaAs combination. What determines the particular well–barrier combination that is well suited for a given emission wavelength? If we are to prevent electron leakage from the upper

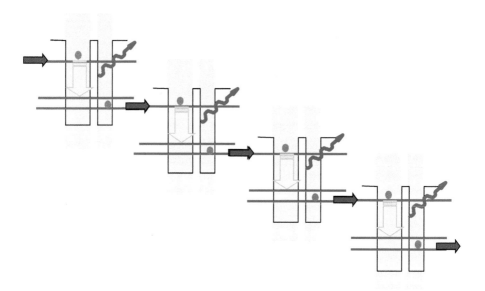

**Figure 13.2** *Schematic of the active intersubband transitions in quantum wells connected in series in a QCL.*

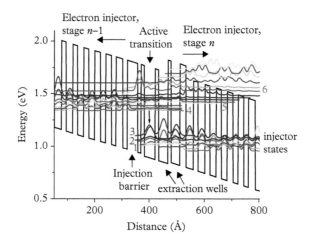

**Figure 13.3** *Subbands and corresponding wavefunctions in a biased QCL designed for emission at $\lambda = 5\,\mu m$. The wells are InGaAs, and the barriers InAlAs, with $\approx 1\%$ compressive strain in the wells and a compensating amount of tensile strain in the barriers. Subband 4 is the upper lasing subband, subband 3 is the lower lasing subband, subbands 2 and 1 are spaced by an optical-phonon energy, subbands 5 and 6 are potential leakage paths.*

lasing subband over the barriers, the conduction-band offset should be at least a factor of 2 larger than the photon energy. Since the offset for $In_{0.53}Ga_{0.47}As$ and $In_{0.52}Al_{0.48}As$ lattice-matched to InP is only $\approx 500$ meV (see Chapters 6 and 7), QCLs emitting at 4–8 μm wavelengths generally have wells with compressive strain (In compositions up to $\approx 70\%$) and barriers with tensile strain (In compositions as low as $\approx 25\%$). We will return to the question of electron leakage later in this section.

If we can empty the lower lasing subband much faster than the active transitions are made and if an electric current continuously pumps electrons into the upper lasing subband, we have a sort of quasi-equilibrium, in which the sheet densities in the two subbands satisfy the population-inversion condition $n_u \gg n_l$. This is the opposite of the relationship in thermal equilibrium: $n_l > n_u$, specifically $n_l = n_u e^{\Delta E / k_B T}$ using the Boltzmann distribution for non-degenerate carrier densities.

How much current must we inject to produce intersubband gain $g$? This occurs when the electron flux into the upper lasing subband $\mathcal{J}(g)/e$ equals the rate at which electrons make transitions between the upper and lower subbands $(n_u - n_l)/\tau_{ul} \approx n_u/\tau_{ul}$ (assuming there are no other processes that can circumvent both subbands). The gain is proportional to the carrier-density difference between the two subbands (see Eq. (11.79)). If we know the threshold gain $g_{th}$, the population inversion that corresponds to it is

$$n_u - n_l = g_{th} d / \left( \Gamma \frac{dg_{ib}}{dn} \right) \tag{13.9}$$

where $\Gamma$ is the optical confinement factor within the ridge waveguide, and $d$ is the thickness of each stage. We need to normalize by the stage thickness because the differential gain was defined as a cross section in Chapter 11, while the natural units for the carrier densities in Eq. (13.9) are 2D. We saw in Section 11.2 that intersubband gain is appreciable only for light polarized along the growth direction; therefore, QCLs cannot operate as VCSELs in Fig. 12.6 (in some cases, in-plane emission can be redirected vertically by diffraction).

If the lower lasing subband is nearly unpopulated, the current density required to produce this gain becomes

$$\mathcal{J}(g_{th}) \approx e g_{th} d / \left( \tau_{ul} \Gamma \frac{dg_{ib}}{dn} \right) \tag{13.10}$$

As before, $g_{th}$ is simply the cavity loss $\alpha_{cav}$, which includes both internal and mirror loss. Some current density $\mathcal{J}_{tr}$ is also needed to achieve transparency, which leads us to the final expression for the threshold current density:

$$\mathcal{J}_{th} \approx \frac{e \alpha_{cav} d}{\left( \tau_{ul} \Gamma \frac{dg_{ib}}{dn} \right)} + \mathcal{J}_{tr} \tag{13.11}$$

In fact, Eq. (13.11) holds for any semiconductor laser with gain that varies linearly with carrier density. The essence of the QCL then lies in the physical meaning of the various quantities. One implicit difference is that, as in an ICL, the QCL has multiple stages with the same current flowing through them. Finally, $\alpha_{cav}$ can be a function of current density via the dependence of the loss on the carrier densities $n_u$ and $n_l$.

As a first application of Eq. (13.11), we can explore whether it is preferable to have a "direct" active transition, in which the wavefunctions for both upper and lower subbands are localized in the same well, or an "indirect" one, where a barrier between two wells controls the wavefunction overlap. The two cases are illustrated in Fig. 13.4. In fact, the distinction is not hard and fast, since there is a continuous transition between fully isolated and fully hybridized subbands in different wells. The active transition for the QCL structure in Fig. 13.3 is mostly spatially direct, although the overlap is less than unity.

If both the scattering rate $1/\tau_{ul}$ and differential gain have the same dependence on the square of the overlap integral $|O_{12}|^2$ (e.g., linear), a nearly direct transition is preferred because high differential gain implies a lower value of $n_u$ via Eq. (13.9) and, hence, lower $\alpha_{cav}$ due to free-carrier absorption.

How much better can we do with a direct transition? To find out, we need to examine the physical mechanism that determines the scattering time $\tau_{ul}$ between the two subbands. Chapter 11 showed that radiative lifetime is relatively long, and in fact the scattering is almost always nonradiative. An important mechanism in polar III–V materials involves scattering from ionic vibrations (*optical* phonons) that produce a spatially varying potential. To avoid a tedious (and not particularly enlightening) derivation, we will use basic physics and dimensional analysis to estimate the strength and functional dependence of this scattering process.

Since the interaction between electrons and phonons is electromagnetic, the transition rate scales with the fine-structure constant $\alpha_f$, by analogy with absorption and emission of light. In fact, ionic potentials are short range, i.e., closer to the atomic than the optical scale, so the interaction is almost purely electrostatic and cannot depend on the speed of light. To eliminate it from the fine-structure constant, the scattering rate should be proportional to $\alpha_f c$.

Obviously, phonon energies and polarities vary, so we need to quantify the strength of the interaction. In fact, this information is already present in the measured permittivity. At very low frequencies, the ions respond to the driving fields, so the polarization includes the contribution from optical phonons. By contrast, the ions remain stationary at higher frequencies, so a "measure" of the interaction strength is obtained by subtracting the high-frequency polarization proportional to $1/\varepsilon(\infty)$ from the low-frequency polarization proportional to $1/\varepsilon(0)$. Using a refractive index of 3–4 at optical frequencies, we estimate $\frac{1}{\varepsilon(\infty)} \sim 0.1$. The phonon contribution is smaller: $\frac{1}{\varepsilon(0)} - \frac{1}{\varepsilon(\infty)} \approx 0.01$. The interaction is strongest in III–V materials with lighter atoms (such as Al and N), which are more polar and have larger phonon energies $\hbar\omega_0$.

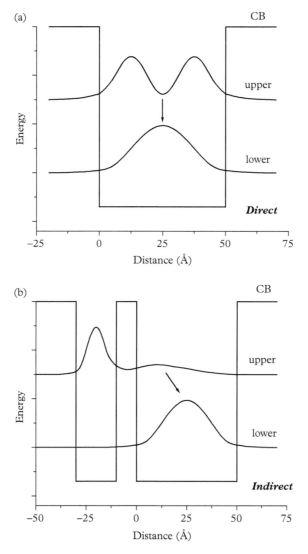

**Figure 13.4** *Schematics of spatially direct (a) and indirect (b) active intersubband transitions that can be used in QCLs.*

From our discussion of optical transitions in Chapter 4, we know that the absorption rate scales with (equilibrium) optical phonon number $n(\omega_0)$, while the emission rate scales as $n(\omega_0) + 1$. Because the phonons are simply vibration quanta, in equilibrium they follow Bose–Einstein statistics just like photons. Therefore, the phonon occupation is described by Eq. (4.91) with the chemical potential set to zero:

$$n(\omega_0) = \frac{1}{\exp(\hbar\omega_0/k_B T) - 1} \tag{13.12}$$

Of course, the phonons are not necessarily in equilibrium in a working laser. The electrons can be injected fast enough that $n(\omega_0)$ is much higher than the equilibrium value in Eq. (13.12), depending on how fast the optical phonons are removed from the system and eventually dissipated as heat. For now we assume that Eq. (13.12) holds in a QCL.

Optical phonon energies in the III–V materials currently used in QCLs fall into the relatively narrow range 30–40 meV. This is nearly an order of magnitude smaller than the intersubband separation $\Delta E$ used for the active transitions in mid-IR QCLs. If the energy carried off by the phonon is negligible, we can simply add the absorption and emission rates so that the total rate scales with $2n(\omega_0) + 1 \approx 1 + 2\exp(-\hbar\omega_o/k_B T) \approx 1.7$ at room temperature.

The "efficiency" with which an electron scatters from the optical phonon depends on $\Delta E$ and the momentum transferred to the phonon. We can approximate the momentum transfer by assuming parabolic dispersion and that $k_\parallel \approx 0$ for an electron in the upper subband, as in Fig. 13.5, to obtain

$$\Delta q = \frac{\sqrt{2m_e \Delta E}}{\hbar} \tag{13.13}$$

The scattering process is slow when the momentum transfer is large, which implies an inverse dependence. However, so far the only dimensional quantity in the derivation is the speed of light, so to convert the expression to a scattering

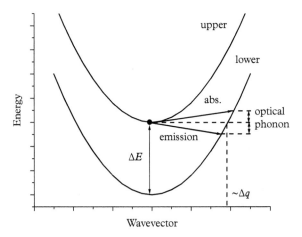

**Figure 13.5** *Illustration of momentum transfer due to phonon scattering in an intersubband transition.*

rate, we must multiply by a quantity with wavevector units. Therefore, we guess that we should multiply by $\Delta q$ and divide by $\Delta E \propto (\Delta q)^2$. To make the estimate dimensionally correct, we also multiply by the other important energy scale in the calculation, the optical-phonon energy $\hbar \omega_0$.

The total scattering rate is also proportional to the range of phonon wavevectors over which the wavefunction overlap $|I_{ul}|^2$ between the upper and lower subbands is close to its maximum value (of one for a direct transition). This range is some fraction of $\Delta q$ in Eq. (13.13), which we take to be one-half in this crude estimate. Now we multiply all the factors derived earlier:

$$\frac{1}{\tau_{ul}} \approx \alpha_f c \left[ \frac{1}{\varepsilon(0)} - \frac{1}{\varepsilon(\infty)} \right] [2n(\omega_0) + 1] \frac{\hbar \omega_0}{2\Delta E} \sqrt{\frac{2m_e \Delta E}{\hbar^2}} |I_{ul}|^2$$

$$\approx \alpha_f \left[ \frac{1}{\varepsilon(0)} - \frac{1}{\varepsilon(\infty)} \right] [2n(\omega_0) + 1] \omega_0 \sqrt{\frac{m_e c^2}{2\Delta E}} |I_{ul}|^2 \qquad (13.14)$$

This expression has the right units of phonon frequency! It combines several small dimensionless factors with the square root of the ratio of rest electron energy to the typical energy scale in the semiconductor, which is much larger than 1.

Is this estimate of the scattering rate reasonable? For a direct transition with $|I_{ul}|^2 \sim 1, \Delta E = 0.25$ eV, $\hbar \omega_0 = 0.030$ eV, and $m_e = 0.04m$, Eq. (13.14) yields $\tau_{ul} \sim 1$ ps at room temperature. Even though we obtained this estimate with relatively little work, it is close to a full calculation and displays the same trends. When the transition is indirect, the lifetime increases inversely with the wavefunction overlap between the two subbands. It becomes shorter as $\Delta E$ decreases, i.e., in a QCL emitting at longer wavelength. However, the variation is less than a factor of 2 over the wavelength range in which QCLs operate continuously at room temperature ($\approx 3.5$–12 μm or $\Delta E \approx 0.1$–0.35 eV). To be predictive, we should also include intersubband scattering mechanisms that do not involve phonons. For example, as the wells become narrower at shorter wavelengths, interface roughness scattering due to spatial fluctuations of the subband energies limits the upper-subband lifetime even further. Figure 13.6 shows the approximate variation of the phonon scattering rate between the first two subbands of $In_{0.53}Ga_{0.47}As$, GaAs, and InAs quantum wells with infinite barriers as a function of intersubband separation. Although you might think that a smaller effective mass should favor a lower scattering rate in InAs, the effects of nonparabolicity and a stronger interaction with optical phonons actually lead to slightly faster scattering than in InGaAs.

Now we use Eq. (13.11) to estimate the threshold current density $\mathcal{J}_{th}$ in a QCL with nearly direct active transitions and an intersubband lifetime of 1 ps. As a first step, we estimate the differential gain using Eq. (11.80) and multiply by the confinement factor of the QCL active core in a typical waveguide to obtain $\Gamma \frac{dg_{ib}}{dn}$.

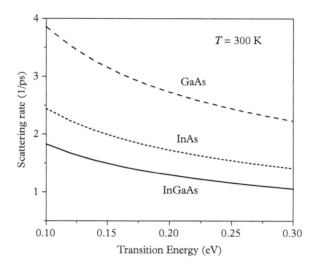

**Figure 13.6** *Approximate intersubband scattering rate at room temperature for the first two subbands of* $In_{0.53}Ga_{0.47}As$, *GaAs, and InAs quantum wells with infinite barriers as a function of intersubband separation. The scattering is assumed to take place from the bottom of the upper subband, and the rates for phonon absorption and emission are assumed to be the same, apart from their temperature dependence. Interface roughness scattering is not included, but can lead to a further increase in the scattering rate.*

In state-of-the-art QCLs, the confinement factor is large $\Gamma \approx 0.7$, while $d = 50$ nm. Continuing to ignore the lower-subband population, we obtain

$$\frac{\Gamma}{d} \frac{dg_{ib}}{dn}\bigg|_{max} \approx \frac{\Gamma}{d} \frac{512\alpha_f}{27\pi n_r} \frac{\hbar^2}{m_e \Delta_l} |O_{12}|^2 \qquad (13.15)$$

This is the differential gain at a photon energy equal to the intersubband separation $\Delta E$ (ignoring any plasma contribution). The broadening is dominated by roughness of the well interfaces (or equivalently, variations in the well width) and is taken to be $\Delta_l = 0.05\Delta E$. Figure 13.7 shows the variation of the calculated differential gain for a square quantum well with infinite barriers composed of one of the most important QCL well materials: GaAs, $In_{0.53}Ga_{0.47}As$, and InAs. With representative values of $m_e = 0.041\,m$, $\Delta E = 0.25$ eV, and $n_r = 3.2$, we calculate $\frac{\Gamma}{d}\frac{dg_{ib}}{dn}\big|_{max} \approx 2 \times 10^{-9}$ cm for a direct transition at $\lambda = 5.0$ μm.

Although the differential gain in Fig. 13.7 is not directly comparable to what we estimated in Chapter 11, we already know that the differences are small for spatially direct transitions. The actual thickness over which an optical transition occurs is a small fraction of the stage thickness. To convert to optical gain per unit length, we must divide by the thickness of each stage of the QCL (just as we might

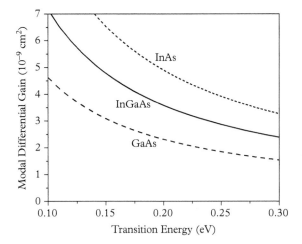

**Figure 13.7** *Modal differential gain at room temperature for transitions between the first two subbands of $In_{0.53}Ga_{0.47}As$, GaAs, and InAs quantum wells with infinite barriers as a function of intersubband separation. We assume that $\frac{\Delta_l}{\Delta E} = 0.05$, $Mt/d_{mode} \approx 0.7$, and $d = 50$ nm in Eq. (13.15).*

divide by the width of the quantum well before multiplying by the confinement factor to obtain the modal differential gain for an interband transition).

The final ingredient going into Eq. (13.11) is the optical loss in the QCL cavity. Because QCLs have only (very few) electrons and no holes, the loss due to free-carrier absorption in the active core and claddings tends to be lower. We assume a loss of 5 cm$^{-1}$ for mid-IR QCLs, evenly split between the mirror and internal loss.

Using $\frac{\Gamma}{d}\frac{dg_{ib}}{dn} \approx 2 \times 10^{-9}$ cm in Eq. (13.9), we obtain $(n_u - n_l)_{th}$ in the $10^9$ cm$^{-2}$ range. This is two orders of magnitude lower than in the ICL (and other interband lasers). It turns that the transparency current density $\mathcal{J}_{tr}$ (and carrier density) is comparable to that at threshold (rather than much higher). Equation (13.11) then indicates that $\mathcal{J}_{th}$ should be on the order 1 kA/cm$^2$, close to the values in state-of-the-art devices, but quite high in the context of other semiconductor lasers. The culprit is the much shorter (1 ps) upper-state lifetime determined by fast intersubband transitions. By contrast, interband transitions due to phonon scattering or any (nearly) elastic mechanism are not possible because the conduction and valence curve in the opposite directions. While the nitride lasers have similar threshold current densities, they are due mostly to high threshold carrier densities.

Why is the sheet carrier density at threshold of an interband laser ($n_{th} = 0.5$–$1 \times 10^{12}$ cm$^{-2}$ from Chapters 11 and 12) two orders of magnitude higher than in a QCL? Obviously, this is not due to a difference in the differential gain, as we learned from Eq. (11.81). In fact, the crucial point is that intersubband transitions do not absorb in the absence of carriers. By contrast, it is much more difficult to clear the top of the valence band of electrons to make an interband transition transparent. That takes a sheet density of $n_{tr} \sim N_c$ in quantum wells (see Eq. (11.74)). As a result, the QCLs operate even with their very short upper-state lifetime! Three orders of magnitude in the lifetime ($\sim 1$ ns versus $\sim 1$ ps) are partially counteracted by two orders of magnitude in $n_{th}$.

Is there any way to improve the intersubband lifetime to make it closer to the interband values? Even at low temperatures, the $2n(\omega_0) + 1$ factor remains substantial, so the only practical route is to remove the continuous distribution of energies. In principle, quantum dots offer discrete states within the conduction "band". However, many practical difficulties have been encountered in transferring the concepts developed for quantum wells to these devices, including efficient electron injection into and extraction out of quantum-dot states.

Figure 13.3 shows that the full QCL structure is relatively complicated, with most of the thickness of each stage taken up by the electron injector. A long stage is needed mostly to keep the electric field at a reasonable level ($< 100$ kV/cm). The QCL design also looks complex because electrons must be extracted as rapidly as possible from the lower lasing subband. The lower-state population is determined by the balance between the incoming transition rate from the upper subband to the net rate for all possible transitions out of the lower subband:

$$\frac{n_l}{n_u} = \frac{\tau_l}{\tau_{ul}} \tag{13.16}$$

where $\tau_l$ is the lower subband lifetime. For simplicity, we assume that electrons enter the lower subband only via transitions from the upper subband, even though other leakage paths exist in general.

How can we design a QCL with $\tau_l \ll \tau_{ul}$? According to Eq. (13.14), the phonon-limited lifetime is shorter when the intersubband separation is reduced. In fact, Eq. (13.14) is a simplified form that is valid for $\Delta E \gg \hbar\omega_0$. For $\Delta E < \hbar\omega_0$, emission from states near the bottom of the subband no longer conserves energy, even though phonon absorption and emission from higher-energy states can still take place. Therefore, the emission of optical phonons reaches a maximum when $\Delta E \approx \hbar\omega_0$ because the scattering involves phonons with nearly zero momentum. At the $\Delta E \approx \hbar\omega_0$ resonance in InGaAs, the calculated intersubband phonon scattering rate of $\approx 200$ fs sets a lower limit on $\tau_l$. In practice, exact resonances are rarely achieved (nor needed because phonon energies are different in different materials of the active core), and $\tau_l \approx 0.2 - 0.3\tau_{ul}$. Equation (13.16) indicates a corresponding increase in $n_u$ of 25–40%, needed to maintain the same population inversion $n_u - n_l$, and a similar increase in the threshold current density, not accounting for any extra free-carrier absorption.

Considering that the threshold carrier density is low, should we dope any layers of the QCL active core? In principle, it seems we could inject all the necessary electrons on one side of the device and extract them from the other. However, this approach fails miserably in practice because a non-uniform electric field develops across the active core. In state-of-the-art QCLs, the electron injector in each stage is doped to a level $n_D \gg (n_u + n_l)_{th}$. The doping is higher than expected because we need the extra carriers to support the current flow well above the lasing threshold (recall that $n_u$ and $n_l$ do not pin, only $n_u - n_l$ does) and because the injector subbands must also be occupied at threshold. Now we will consider the injector occupation in greater detail.

Figure 13.3 shows that state-of-the-art QCLs have multiple subbands in the electron injector, the lowest of which is separated from the lower lasing subband by $\delta E$. This arrangement helps to rapidly extract electrons from the lower lasing subband, but it also means that $\delta E$ must be at least $4 - 6k_B T$ (100–150 meV at room temperature) to prevent the electrons from "backing up" into that subband and destroying population inversion. The total population of the lower lasing subband is then

$$n_{l,th} = \frac{\tau_l}{\tau_{ul}} n_{u,th} + \exp\left(-\frac{\delta E}{k_B T}\right) n_D = \frac{\tau_l}{\tau_{ul}} \left(n_{th} + n_{l,th}\right) + \exp\left(-\frac{\delta E}{k_B T}\right) n_D \quad (13.17)$$

where $n_D$ is the sheet doping density per stage, and $n_{th} \equiv (n_u - n_l)_{th}$ needed to produce $g = \alpha_{cav}$. Rearranging terms, the population of the lower lasing subband at threshold becomes

$$n_{l,th} = \frac{\tau_l}{\tau_{ul} - \tau_l} n_{th} + \exp\left(-\frac{\delta E}{k_B T}\right) \frac{N_D}{1 - \tau_l/\tau_{ul}} = \frac{\tau_l}{\tau_{ul} - \tau_l} \frac{\alpha_{cav}}{\frac{\Gamma}{d} \frac{dg_{ib}}{dn}}$$

$$+ \exp\left(-\frac{\delta E}{k_B T}\right) \frac{n_D}{1 - \tau_l/\tau_{ul}} \quad (13.18)$$

Now we use this expression to estimate the transparency current density $\mathcal{J}_{tr}$ term in Eq. (13.11):

$$\mathcal{J}_{th} \approx \frac{e n_{u,th}}{\tau_{ul}} = \frac{e(n_{th} + n_{l}, th)}{\tau_{ul}} = \frac{e \alpha_{cav}}{(\tau_{ul} - \tau_l) \frac{\Gamma}{d} \frac{dg_{ib}}{dn}} + \exp\left(-\frac{\delta E}{k_B T}\right) \frac{e n_D}{(\tau_{ul} - \tau_l)} \quad (13.19)$$

In fact, electrons in the upper lasing subband do not scatter exclusively into the lower lasing subband. If $\tau_{ul} > \tau_u$, the current density is inversely proportional to $\tau_u$ rather than $\tau_{ul}$; we obtain

$$\mathcal{J}_{th} \approx \frac{e n_u}{\tau_u} = \frac{e \alpha_{cav}}{\tau_{eff} \frac{\Gamma}{d} \frac{dg_{ib}}{dn}} + \exp\left(-\frac{\delta E}{k_B T}\right) \frac{e n_D}{\tau_{eff}} \quad (13.20)$$

where $\tau_{eff} \equiv \tau_u (1 - \tau_l/\tau_{ul})$. In contrast to interband lasers, the transparency current density can become arbitrarily small when $\frac{\delta E}{k_B T} \gg 1$, as mentioned earlier.

If the lowest subband of the electron injector lines up with the upper lasing subband of the next stage, as shown in Fig. 13.3, the voltage drop per stage is $(\hbar\omega + \delta E)/e$, and the minimum thickness of each stage is $(\hbar\omega + \delta E)/(eF_{max})$, where $\hbar\omega \approx \Delta E$. For long emission wavelengths, most of the voltage goes to prevent the backfilling of the lower subband from carriers produced by the doping of each stage, i.e., to provide the required $\delta E$. This leads to a QCL stage thickness of 300–500 Å for QCLs emitting at $\lambda = 4.5 - 11\,\mu\text{m}$.

We cannot use the same well to contain both an active transition (0.1–0.3 eV) and a lower transition with $\Delta E \approx \hbar\omega_0$ (0.03–0.04 eV). The best option is to tailor a neighboring well coupled via a thin barrier to have a subband in the right place to extract electrons. State-of-the-art QCL designs do not stop here, but introduce another phonon-energy step down, which requires three tightly coupled wells that are more weakly coupled to the electron injector. The bottom of the injector miniband is designed to lie near the upper lasing subband and have little overlap with other subbands of the next stage at $k_\parallel = 0$. The upper lasing subband cannot empty into the injector because it faces an injector "*minigap*" rather than minibands (see Fig. 13.3).

How do we set the thickness of the "injection" barrier between two neighboring stages in Fig. 13.3? If the barrier is very thin, the lowest injector subband strongly couples to the upper lasing subband of the next stage. The model of Eqs. (1.24) and (1.25) tells us that the splitting between the two is $2V_0$, where $V_0$ is the interaction strength between the two subbands via the barrier. We can also express it using $V_0 \simeq \hbar/\tau_{tun}$, where $\tau_{tun}$ is the barrier tunneling time. If $V_0 < \Delta_l$, the two subbands cannot be resolved, as illustrated in Fig. 13.8. In this case, the oscillator strength between the upper and lower lasing subbands is insensitive to hybridization of the upper-subband envelope function between the injector and active wells. In practice, we estimate $V_0$ by searching for the minimum separation

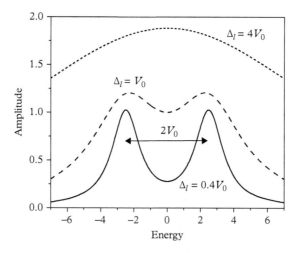

**Figure 13.8** *Illustration of "transmission resonances" for the lowest subband of the electron injector and the upper lasing subband, which are coupled via the injection barrier.*

between the two subbands as a function of applied field or barrier thickness (or some other suitable parameter) at $k_\parallel = 0$ and setting the splitting at the minimum to $2V_0$.

If the barrier thickness is so large that $\tau_{tun} > \tau_u$, it is tunneling through the barrier rather than the transitions between the subbands that limits the electron transport in a QCL stage. To avoid the barrier bottleneck, we design $\tau_{tun}$ to be $\leq 100$ fs ($V_0 > 6$ meV). We already saw that we would also like to have $V_0 < \Delta_l$. In the mid-IR, the typical broadening of $2\Delta_l \sim 20$–$30$ meV is dominated by interface roughness. This means the splitting should fall into the relatively narrow range of $10$–$15$ meV. As the line broadening decreases at longer wavelengths, the injection barrier should become thicker to make the splitting smaller, e.g., $5$–$10$ meV for a QCL emitting at $10\mu$m.

Electrons moving through the active stages of the QCL are driven by the applied electric field via phonon scattering and tunneling processes. Phonon scattering plays the dominant role in removing electrons from the lower lasing subband. The doping in each stage acts a reservoir, from which the electrons making the transitions are drawn. The rate of electron removal from the reservoir, given by the flux $\mathcal{J}/e$, cannot be greater than the rate at which they are replenished from the previous stage, $n_D/\tau_{trans}$, where $\tau_{trans}$ is the average time for an electron to transit one stage. Empirically, $\tau_{trans}$ is a few ps in most QCL designs, and typically $n_D \sim 10^{11}$ cm$^{-2} \gg n_{th}$. This is sufficient to support current densities as high as $\mathcal{J}_{max} \approx e n_D/\tau_{trans} \sim 5$ kA/cm$^2$. As discussed earlier, this "maximum" current density is achieved when the bottom of the injector aligns approximately with the upper lasing subband. We can imagine it as the point at which all the carriers in the injector are potentially available to participate in the transport.

The absorption loss in the active core generally scales with $n_D$, as does $\mathcal{J}_{max}$. To design a QCL with a minimum threshold (rather than high $\mathcal{J}_{max}$ and hence high powers), we may choose to reduce $n_D$. We can also optimize the number of stages in a QCL to minimize the threshold power for a given value of $n_D$, by adjusting the balance between a lower confinement factor and gain with fewer stages and higher voltage drop with more stages.

What is the temperature dependence of the QCL threshold and how does it compare with the interband semiconductor lasers analyzed in Section 12.2? We consider the following factors for operation near room temperature: (1) the phonon-limited lifetime of the upper lasing subband; (2) thermal backfilling of the lower lasing subband; (3) current leakage via subbands lying above the upper lasing subband; (4) any temperature variation of the optical transition broadening $\Delta_l$ that enters the differential gain (small if the broadening is dominated by interface roughness); and (5) any difference between electron and lattice temperatures.

The phonon-limited scattering rate between the upper and lower lasing subbands, $1/\tau_{u,LO}$, is proportional to $2n(\omega_0) + 1 \approx 1 + 2\exp(-\hbar\omega_0/k_B T)$. For simplicity, we assume that phonon scattering dominates the lifetime: $1/\tau_u \approx 1/\tau_{u,LO}$, although interface-roughness scattering can be comparable in the mid-IR. We take the ratio $\tau_l/\tau_{ul}$ to be independent of temperature because the occupation factor is present in both the numerator and denominator. Fixing all other parameters and finding the slope of $\mathcal{J}_{th}$ at $T_{ref}$, we obtain

$$T_\tau \approx T_{ref} \frac{k_B T_{ref}}{2\hbar\omega_0} \exp\left(\frac{\hbar\omega_0}{k_B T_{ref}}\right) \qquad (13.21)$$

which yields $T_\tau \approx 400$ K at $T_{ref} = 300$ K. More realistically, $T_\tau$ can be up to twice this value because interface-roughness scattering does not vary much with temperature.

To address the backfilling, we already know from Eq. (13.20) that we can reduce the transparency current density at the expense of a higher voltage drop. In a well-designed QCL, the backfilling should be weak, so the second term in Eq. (13.20) should be smaller than the first. The characteristic temperature due to backfilling then becomes

$$T_f \approx T_{ref} \frac{k_B T_{ref}}{\delta E} \frac{(n_u - n_l)_{th}}{n_D} \exp\left(\frac{\delta E}{k_B T}\right) \qquad (13.22)$$

While it is more difficult to model accurately the current leakage through non-lasing upper subbands in Fig. 13.3, we start by generalizing Eq. (13.20) as

$$\mathcal{J}_{th} = \mathcal{J}_{th,0} + \sum_i \mathcal{J}_{leak,i} e^{-(E_i - E_u)/k_B T} \equiv \mathcal{J}_{th,0} + \mathcal{J}_{leak} \qquad (13.23)$$

Here $E_i$ are the energy minima for subbands above the upper lasing subband, and $\mathcal{J}_{th,0}$ is the threshold current density computed from Eq. (13.20). Clearly, the higher these subbands are in energy, the lower this current leakage. The position of the subbands is in turn governed by the conduction-band offset, and we may even attribute the leakage problem to well-barrier combinations with insufficient offsets. In state-of-the-art mid-IR QCLs, we estimate the leakage-limited characteristic temperature $T_l$ is 500–1000 K, with the higher values obtained for devices with the most strain.

When all the main contributions to $T_0$ in a mid-IR QCL are assembled, we obtain

$$\frac{1}{T_0} \approx \frac{1}{T_\tau} + \frac{1}{T_f} + \frac{1}{T_l} + \frac{1}{T_b} \qquad (13.24)$$

where we introduce $T_b$ to cover the effects of electron heating and temperature-dependent linewidth broadening. Experimentally, the observed range of $T_0 \approx 150$–300 K is most likely dominated by a combination of leakage and phonon scattering. This is 3–5 times larger than the typical values for interband mid-IR lasers, a very important result for practical applications, as discussed in Section 13.3. At longer wavelengths, current leakage is less of a problem, but phonon scattering and broadening are worse, so that $T_0$ does not improve. Figure

13.9 shows the typical temperature dependence of $\mathcal{J}_{th}$ for notional mid-IR QCLs just above room temperature. Because the temperature range is smaller than $T_0$, the dependence appears nearly linear. Because of the slow temperature variation, even a small increase in $\mathcal{J}_{th}$ at room temperature can nullify the improvement in $T_0$, as illustrated by the two curves in Fig. 13.9.

While minimizing the threshold current and its temperature variation is very important in a QCL because of the short intersubband lifetime, we are also interested in how efficiently injected electrons are transformed into photons above the lasing threshold. To calculate the QCL EDQE, we write rate equations for the two lasing subbands and solve them in steady state. To do so, we add stimulated emission to Eqs. (13.16) and (13.20) and neglect the backfilling and leakage through higher subbands (compare to Eq. (12.5)):

$$\frac{dn_u}{dt} = \frac{\mathcal{J}}{e} - \frac{n_u}{\tau_u} - Sv_g \frac{\Gamma}{Md} \frac{dg_{ib}}{dn} (n_u - n_l) = 0 \qquad (13.25a)$$

$$\frac{dn_l}{dt} = \frac{n_u}{\tau_{ul}} - \frac{n_l}{\tau_l} + Sv_g \frac{\Gamma}{Md} \frac{dg_{ib}}{dn} (n_u - n_l) = 0 \qquad (13.25b)$$

The stimulated-emission term is proportional to the confinement factor per stage, which we approximate as the ratio of the total confinement factor and the number of stages $M$. If it varies from stage to stage, we need two equations like Eq. (13.25) for each stage. The rate equation for the photon density simply fixes the population inversion at the threshold level $(n_u - n_l)_{th}$ needed to produce modal gain $g_{th} = \alpha_{cav}$ according to Eq. (13.9).

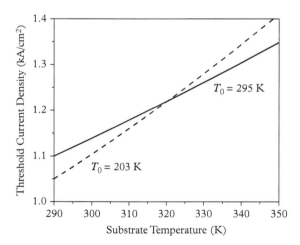

**Figure 13.9** *Typical temperature variation of the threshold current density in mid-IR QCLs near and above 300 K. Compare with curves (b) and (c) in Fig. 12.10 for interband mid-IR lasers.*

After some algebra, Eq. (13.25a) becomes

$$n_{u,th} = \tau_u \left( \frac{\mathcal{J}}{e} - S v_g \alpha_{cav}/M \right) \tag{13.26}$$

with the lower-subband density at and above threshold given by

$$n_{l,th} = n_{u,th} - (n_u - n_l)_{th} = n_u - \frac{\alpha_{cav} d}{\Gamma \frac{dg_{ib}}{dn}} \tag{13.27}$$

We now substitute the electron densities in the upper and lower subbands from these expressions into Eq. (13.25b):

$$\left( \frac{1}{\tau_{ul}} - \frac{1}{\tau_l} \right) \tau_u \left( \frac{\mathcal{J}}{e} - S v_g \alpha_{cav}/M \right) + \frac{\alpha_{cav} d}{\tau_l \Gamma \frac{dg_{ib}}{dn}} + S v_g \alpha_{cav}/M = 0 \tag{13.28}$$

in which we can use the previous definition $\tau_{eff} \equiv \tau_u (1 - \tau_l/\tau_{ul})$ to write

$$-\tau_{eff} \left( \frac{\mathcal{J}}{e} - S v_g \alpha_{cav}/M \right) + \frac{\alpha_{cav} d}{\Gamma \frac{dg_{ib}}{dn}} + \tau_l S v_g \alpha_{cav}/M = 0 \tag{13.29}$$

Collecting terms, we have

$$S v_g \alpha_{cav} \left( \tau_{eff} + \tau_l \right) /M + \frac{\alpha_{cav} d}{\Gamma \frac{dg_{ib}}{dn}} = \frac{\mathcal{J}}{e} \tau_{eff} \tag{13.30}$$

We can modify Eq. (12.12) to calculate the EDQE *per stage* for a cascade laser. The result is $\eta_e \approx \frac{e}{M} \frac{dS}{d\mathcal{J}} R_{ph}$, where the photon density is proportional to the output power. Using Eq. (13.30) in this expression, we obtain

$$\eta_e = \frac{e v_g \alpha_{m1}}{M} \frac{dS}{d\mathcal{J}} = \eta_i \frac{\tau_{eff}}{\tau_{eff} + \tau_l} \frac{\alpha_{m1}}{\alpha_{cav}} = \eta_i \frac{\tau_{eff}}{\tau_{eff} + \tau_l} \frac{\alpha_{m1}}{\alpha_{m1} + \alpha_{m2} + \alpha_i} \tag{13.31}$$

The internal efficiency $\eta_i$ specifies how much current actually flows through the active core. It can less than 1 if some of the current flows around the active transition either in the bulk of the device (e.g., leakage via the upper subbands) or at the sidewalls of the laser ridge. Equation (13.31) is the QCL analog of Eq. (12.13) for an interband laser, for which the time to inject and remove carriers is very short compared to carrier lifetime ($\tau_l \ll \tau_{eff}$).

Equation (12.16) shows that the wall-plug efficiency (WPE) is just the EDQE multiplied by the voltage efficiency and the threshold efficiency. The voltage efficiency is calculated from Eq. (12.60) for the cascaded (in-series) connection of the active wells with $V_0 = \delta E/e$:

$$\eta_v = \frac{\hbar\omega}{\hbar\omega + \delta E + e\mathcal{J}\rho_s/M} \tag{13.32}$$

Because $\delta E$ is comparable to $\hbar\omega$ in a QCL, the voltage efficiency is typically worse than in an interband laser even though $M \gg 1$.

The threshold efficiency remains the same as in Eq. (12.16), which gives the following result for the WPE:

$$WPE_{QCL} = \eta_e\eta_v\eta_{th} = \eta_i \frac{\tau_{eff}}{\tau_{eff}+\tau_l} \frac{\alpha_{m1}}{\alpha_{m1} + \alpha_{m2} + \alpha_i} \frac{\hbar\omega}{\hbar\omega + \delta E + e\mathcal{J}\rho_s/M} \frac{\mathcal{J} - \mathcal{J}_{th}}{\mathcal{J}} \tag{13.33}$$

We can use Eq. (13.33) to estimate the room-temperature WPE for notional QCLs emitting at 5 and 10 µm, with realistic input parameters summarized in Table 13.1. We also neglect loss from the back facet that can be coated for high reflection, i.e., $\alpha_{m2} \approx 0$, and take the threshold current density $\mathcal{J}_{th}$ to be 1.0 kA/cm$^2$ at 5 µm and 1.5 kA/cm$^2$ at 10 µm. This leads to a WPE of 15–20% at 5 µm and 5–10% at 10 µm at a current density of 3 kA/cm$^2$. The WPE grows linearly near the lasing threshold, where it is dominated by the current efficiency factor $\frac{\mathcal{J}-\mathcal{J}_{th}}{\mathcal{J}}$. At high currents, the WPE saturates when it approaches $\mathcal{J}_{max}$ or due to carrier and lattice heating (see Section 13.3). The actual WPE is a little lower than predicted by Eq. (13.33). The record reported WPE of 21% for a QCL emitting at 4.6 µm is close to our estimate.

How do we find the optimal number of stages $M$ in a QCL? First of all, we expect $\mathcal{J}_{th}$ to decrease with $M$ because the confinement factor $\Gamma$ and differential gain increase. Figure 13.10 shows the typical variation of $\Gamma$ with $M$ for a QCL emitting at 4.6 µm, assuming InP top and bottom cladding layers, a stage thickness of 500 Å, and several different thicknesses for the InGaAs separate-confinement layers between the claddings and the active core. The confinement factor starts out growing linearly with $M$, but then gradually saturates when the number of stages reaches $\approx 20$. If we push $M$ to very high values, the voltage drop increases proportionally, but the optical gain does not. Therefore, to maximize the WPE we must balance the improved voltage and threshold efficiency at higher $M$ with the accompanying reduction of EDQE (due to extra loss in the additional stages) and extra heating. If the QCL is doped more heavily to operate at higher

**Table 13.1** *Parameters used in the estimates of the wall-plug efficiencies for QCLs emitting at 5 and 10 µm.*

| $\lambda$ (µm) | $\eta_i$ | $\tau_{eff}$ (ps) | $\tau_l$ (ps) | $\alpha_{m1}$ (cm$^{-1}$) | $\alpha_i$ (cm$^{-1}$) | $\delta E$ (meV) | $\rho_s$ (m$\Omega$ cm$^2$) | $M$ | $J$ (kA/cm$^2$) |
|---|---|---|---|---|---|---|---|---|---|
| 5 | 0.95 | 1.1 | 0.25 | 1.6 | 1.5 | 100 | 0.7 | 30 | 3 |
| 10 | 0.95 | 0.7 | 0.25 | 1.6 | 2.5 | 100 | 0.7 | 40 | 3 |

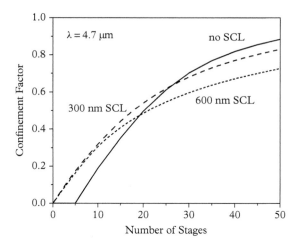

**Figure 13.10** *Typical variation of the confinement factor versus number of stages in a mid-IR QCL.*

$\mathcal{J}_{max}$, we often benefit from lower $M$ that minimizes the dissipated power density $\mathcal{J}V_{th} = M\mathcal{J}\,(\hbar\omega + \delta E) + \mathcal{J}^2\rho_s$ and also keeps the internal loss $\alpha_i \propto \Gamma$ at a manageable level. When the number of stages is relatively large ($> 20$), the InGaAs separate-confinement layers are not strictly necessary.

How much output power can be generated in a typical narrow-ridge mid-IR QCL? If we operate the device with widely spaced short pulses, the average heating is negligible. We use typical parameters ($w = 10\ \mu m$, $L_{cav} = 3\ mm$, $\mathcal{J}_{th} = 1\ kA/cm^2$, $M = 30$, and $\eta_e = 50\%$) in Eq. (12.10b) to obtain $P_{out} = wL_{cav}\,\hbar\omega\frac{M\eta_e(\mathcal{J}-\mathcal{J}_{th})}{e} > 2\ W$. This is high in the context of other semiconductor lasers owing to the combination of many stages, high operating current densities, and weak variation of $\mathcal{J}_{th}$ with temperature (see Section 13.3).

QCLs cover an exceptionally wide range of emission wavelengths, extending from $\approx 4\ \mu m$ to the THz. THz QCLs are the least developed because it is difficult to keep the upper-state lifetime higher than the intrasubband scattering time. Even though phonon emission cannot proceed when $\Delta E < \hbar\omega_o$, phonon absorption (from the bottom of the upper subband to a higher $k_\parallel$ state in the lower subband) becomes faster at higher temperature, being proportional to $n(\omega_0) \approx \exp\left(-\hbar\omega_o/k_B T\right)$. This is one factor that currently limits THz QCLs to temperatures well below ambient ($< 210\ K$). However, we will discuss a different approach to THz emission from the QCLs in Section 15.5.

Another advantage of the QCLs is their weak absorption at energies well below and above the inverted intersubband transition. Therefore, stages with several different intersubband transitions can be combined in the same waveguide to generate optical gain over a wide spectral range. This is not possible with interband transitions because of strong absorption above the quasi-Fermi level difference (see Eq. (4.64)). Lasers with broadband gain can be tuned over a wide range of emission wavelengths using spectrally selective external feedback.

Can we make LEDs based on intersubband transitions? Because the nonradiative lifetime due to phonon scattering is so short (ps) by comparison with the radiative lifetime, the radiative efficiency $\frac{R_{sp}}{R_{sp}+R_{nonrad}}$ in Eq. (12.72) is negligible. However, if we need incoherent radiation at very long wavelengths, we may amplify it using electrically pumped gain, as discussed in Section 12.8.

## 13.3 Comparing Different Classes of Mid-Infrared Lasers

At our disposal are several different classes of mid-IR lasers, namely, the conventional quantum-well laser, the ICL, and the QCL. How do we select the right device for a particular application in the mid-IR (3–5 μm)? We will use three figures of merit: the threshold power density ($P_{th} \equiv \mathcal{J}_{th} V_{th}$), the maximum WPE, and the maximum output power $P_{out}$, all of them evaluated in continuous-wave (cw) mode, which means we cannot neglect the heat generated by the device.

The active-core temperature at the input power density $P \equiv \mathcal{J}V$ is given by

$$T = T_s + R_{th}P = T_s + R_{th}\mathcal{J}V \tag{13.34}$$

where the thermal impedance-area product $R_{th}$ ranges from 1 to 10 K/(kW/cm$^2$), and $T_s$ is the temperature of the submount assembly (and most of the substrate). $T_s$ is also the temperature of the active core when $P \approx 0$ (for either near-zero current or pumping with short, widely spaced pulses). For simplicity, we assume a uniform temperature of the active core even if it is thick enough to support a thermal gradient.

As the current increases above $\mathcal{J}_{th}$, the laser begins to produce light. In the absence of heating, we assume an exponential increase in the threshold current density, as in Section 12.2 with constant $T_0$: $\mathcal{J}_{th} = \mathcal{J}_0 \exp\left(\frac{T}{T_0}\right)$. Inserting Eq. (13.34) into this exponential dependence, we obtain

$$\mathcal{J}_{th} = \mathcal{J}_0 \exp\left[(T_s + R_{th}\mathcal{J}_{th}V_{th})/T_0\right] \tag{13.35}$$

where $V_{th} = M(\hbar\omega + \delta E)/e + \mathcal{J}_{th}\rho_s$ (Eq. (12.60)). Using this form in Eq. (13.35) yields

$$\mathcal{J}_{th} = \mathcal{J}_0 \exp\left[(T_s + R_{th}\mathcal{J}_{th}(V_a + \mathcal{J}_{th}\rho_s))/T_0\right] \tag{13.36}$$

In general, Eq. (13.36) for $\mathcal{J}_{th}$ can only be solved numerically. Physically, it describes the increase in $\mathcal{J}_{th}$ that occurs because of heating *at that value* of $\mathcal{J}_{th}$. For a laser biased well above threshold, the extra heating of the active core means that a larger fraction of the injected current goes toward overcoming the cavity losses. We can also define the "effective" $\mathcal{J}'_{th}(\mathcal{J})$ for any injected current $\mathcal{J}$ at fixed $T_s$:

$$\mathcal{J}'_{th}(\mathcal{J}) = \mathcal{J}_0 \exp[T(\mathcal{J})/T_0] = \mathcal{J}_0 \exp\left[\left(T_s + R_{th}\mathcal{J}\left(V_a + \mathcal{J}\rho_s\right)\right)/T_0\right] \quad (13.37)$$

For a laser biased above the lasing threshold, $\mathcal{J}'_{th}(\mathcal{J}) < \mathcal{J}$. $\mathcal{J}'_{th}$ cannot be observed directly, but has important consequences for the maximum power and efficiency.

By analogy with the threshold current, we can also parameterize the dependence of EDQE on temperature. The literature often chooses an exponential dependence, but it is often indistinguishable experimentally from a linear function over the temperature range of interest:

$$\eta_e(T) = \eta_s\left(1 - \frac{\Delta T}{T_1}\right) \quad (13.38)$$

where $\eta_s \equiv \eta_e(T_s)$ and $\Delta T \equiv T - T_s$. The EDQE may decrease at higher temperatures owing to a higher internal loss or lower internal efficiency. We will use Eq. (13.38) without worrying about the physical origin of the decrease. For the estimates below to be valid, we ignore the effect of lattice heating in Eq. (13.38); i.e., $T_1$ should be measured in pulsed mode.

We would like to calculate the maximum WPE and output power and the current densities $\mathcal{J}_{wmax}$ and $\mathcal{J}_{pmax}$ at which they occur. Equation (12.10b) shows that the power at a fixed temperature is proportional to the product of $\eta_e$ and $\mathcal{J} - \mathcal{J}_{th}$. To account for heating, we replace $\mathcal{J} - \mathcal{J}_{th}$ with $\mathcal{J} - \mathcal{J}'_{th}$, which produces a maximum even when the EDQE decreases monotonically with temperature. If the temperature degradation of $\eta_e$ is not too rapid ($T_1 \gg \Delta T$ in Eq. (13.38)), the maximum of $\mathcal{J} - \mathcal{J}'_{th}$ nearly coincides with the maximum output power. Differentiating this term with respect to $\mathcal{J}$ and setting the result to zero, we find that (in the limit $T_1 \to \infty$) the power is approximately maximized at $\mathcal{J}_{pmax}$, for which

$$\left.\frac{d\mathcal{J}'_{th}}{d\mathcal{J}}\right|_{pmax} = 1 \quad (13.39)$$

i.e., when the effective threshold increases as fast as the current itself.

The WPE is proportional to the product of the EDQE, the voltage efficiency, and the threshold efficiency (see Eq. (13.33)). Again, the first two terms decrease slowly and monotonically as $\mathcal{J}$ increases, which means the maximum WPE should occur close to a peak in the threshold efficiency. Differentiating $\frac{\mathcal{J}-\mathcal{J}'_{th}}{\mathcal{J}}$ with respect to $\mathcal{J}$ and setting the result to zero, we obtain (in the limits $T_1 \to \infty$ and $\rho_s \to 0$)

$$\left.\frac{d\mathcal{J}'_{th}}{d\mathcal{J}}\right|_{wmax} = \frac{\mathcal{J}'_{th}(\mathcal{J})}{\mathcal{J}} \quad (13.40)$$

Insofar as $\mathcal{J}'_{th}(\mathcal{J}) < \mathcal{J}$, the maximum WPE is reached at $\mathcal{J}_{wmax} < \mathcal{J}_{pmax}$, for which there is less heating.

Now we can evaluate Eqs. (13.39) and (13.40) by substituting $\mathcal{J}'_{th}$ from Eq. (13.37). We start by assuming that either the series resistance or the current is low enough that $V_a \equiv M(\hbar\omega + \delta E)/e \gg \mathcal{J}\rho_s$. Then Eq. (13.40) for the maximum WPE immediately yields

$$\mathcal{J}_{wmax} \approx \frac{T_0}{R_{th}V_a} \tag{13.41}$$

This simple expression is worth remembering. Equation (13.34) shows that at this current density the active core temperature is $T_{wmax} \approx T_s + T_0$.

It is only a little more difficult to evaluate Eq. (13.39):

$$\mathcal{J}_{pmax} \approx \frac{1}{R_{th}V_a}\left[T_0\ln\left(\frac{T_0}{\mathcal{J}_0 R_{th}V_a}\right) - T_s\right] = \mathcal{J}_{wmax}\left[\ln\left(\frac{\mathcal{J}_{wmax}}{\mathcal{J}_0}\right) - \frac{T_s}{T_0}\right]$$

$$= \mathcal{J}_{wmax}\left[\ln\left(\frac{\mathcal{J}_{wmax}}{\mathcal{J}_0}\right) - \ln\left(\frac{\mathcal{J}_s}{\mathcal{J}_0}\right)\right] = \mathcal{J}_{wmax}\ln\left(\frac{\mathcal{J}_{wmax}}{\mathcal{J}_s}\right) \tag{13.42}$$

where $\mathcal{J}_s \equiv \mathcal{J}_{th}(T_s)$. Typically, $\mathcal{J}_{wmax} \gg \mathcal{J}_s$ and $\mathcal{J}_{pmax} > \mathcal{J}_{wmax}$. Once again, we calculate the active-core temperature when the output power is maximized: $T_{pmax} \approx T_s + T_0\ln\left(\frac{\mathcal{J}_{wmax}}{\mathcal{J}_s}\right)$. The projected output power and WPE can be calculated by substituting the parameter values for "state-of-the-art" QCLs and ICLs listed in Table 13.2.

At substrate temperature $T_{s,max}$ the threshold current density becomes too high for lasing to be possible. That happens when the reduction in the gain due to extra heating exceeds the increase in the gain from more carriers. In other words, we look for a point where the derivative of the threshold current density with respect to substrate temperature becomes infinite. Equivalently, we want to find $T_{s,max}$ at which $\frac{dT}{dT_s} \to \infty$, or $\frac{dT_s}{dT}\Big|_{T_{s,max}} = 0$, where $T_s = T - R_{th}\mathcal{J}_{th}V_a$

$= T - R_{th}V_a\mathcal{J}_0\exp\left(\frac{T}{T_0}\right)$ from Eq. (13.34):

Table 13.2 *Typical parameters for a QCL emitting at 5 μm and an ICL emitting at 3.5 μm.*

| Device | λ (μm) | $\eta_e$ | $T_0$ (K) | $T_1$ (K) | $R_{th}$ (Kcm²/kW) | $J_{th}$ (300 K) (kA/cm²) | $\rho_s$ (mΩcm²) | $M$ | $\delta E$ (meV) | $L_{cav}$ (mm) | $w$ (μm) |
|---|---|---|---|---|---|---|---|---|---|---|---|
| QCL | 5.0 | 0.75 | 200 | 350 | 2 | 1000 | 0.7 | 30 | 100 | 3 | 10 |
| ICL | 3.5 | 0.30 | 50 | 130 | 4 | 150 | 0.7 | 10 | 15 | 1 | 15 |

$$T_{s,max} = T_0 \ln \left( \frac{\mathcal{J}_{wmax}}{\mathcal{J}_0} \right) \qquad (13.43)$$

At $T_{s,max}$ both the WPE and output power decrease to zero (neglecting spontaneous emission). Equation (13.43) allows us to re-express $\mathcal{J}_{pmax}$ in Eq. (13.42) in terms of $T_{s,max}$:

$$\mathcal{J}_{pmax}(T_s) \approx \frac{T_{s,max} - T_s}{R_{th} V_a} = \mathcal{J}_{wmax} \frac{T_{s,max} - T_s}{T_0} \qquad (13.44)$$

Armed with these concepts and formulas, we can now estimate the temperature rise at the lasing threshold and how much the cw $\mathcal{J}_{th}$ differs from the pulsed value. If we assume that $\Delta T \ll T_0$ in a good laser at threshold, we can linearize the threshold change:

$$\Delta T \approx R_{th} V_a \mathcal{J}_s \left( 1 + \frac{\Delta T}{T_0} \right) \qquad (13.45)$$

From this expression, we find the temperature rise

$$\Delta T \approx \frac{T_0}{\frac{\mathcal{J}_{wmax}}{\mathcal{J}_s} - 1} \approx \frac{T_0 \mathcal{J}_s}{\mathcal{J}_{wmax}} \qquad (13.46)$$

and the change in threshold current density due to lattice heating

$$\mathcal{J}_{th}(T) - \mathcal{J}_s \approx \frac{\mathcal{J}_s^2}{\mathcal{J}_{wmax}} \qquad (13.47)$$

So far, we have simplified by neglecting the series resistance and temperature dependence of the EDQE. Returning to Eqs. (13.39) and (13.40), we can include higher-order terms at the expense of simplicity. For example, when the series resistance is non-zero, $\mathcal{J}_{wmax}$ is calculated from the quadratic equation

$$R_{th} (V_a + 2 \mathcal{J}_{wmax} \rho_s) \mathcal{J}_{wmax} = T_0 \qquad (13.48)$$

which yields

$$\mathcal{J}_{wmax} = -\frac{V_a}{4\rho_s} + \sqrt{\frac{V_a^2}{16\rho_s^2} + \frac{T_0}{2\rho_s R_{th}}} \qquad (13.49)$$

If $V_a \gg 8\mathcal{J}_{wmax} \rho_s$, we can expand Eq. (13.49) in a Taylor series to the third term:

$$\mathcal{J}_{wmax} \approx \frac{T_0}{R_{th} V_a} \left( 1 - \frac{2\rho_s T_0}{R_{th} V_a^2} \right) \qquad (13.50)$$

This represents the lowest-order correction to Eq. (13.41). We can also approximate the effect of finite $T_1$, but at this level of complexity it may be easier to maximize the WPE and output power numerically.

In a QCL, the increase of threshold current density with cavity loss is fast enough that the highest WPE and output power are obtained in 3- to 5-mm-long devices (with mirror loss no greater than 1–2 cm$^{-1}$). For many other semiconductor lasers, the threshold increase is much slower, so that the trade-off between higher threshold and higher $\alpha_m/\alpha_{cav}$ efficiency favors shorter devices. In particular, the maximum WPE in an ICL is achieved for a mirror loss as high as 15–20 cm$^{-1}$.

To illustrate these points mathematically, we calculate the optimum mirror loss assuming that $\mathcal{J}_{wmax}$ is fixed by lattice heating, e.g., via Eq. (13.50). We maximize the product of the EDQE and current efficiency:

$$\eta_e \eta_{th} \propto \frac{\alpha_m}{\alpha_m + \alpha_i} \frac{\mathcal{J}_{wmax} - \mathcal{J}_{tr} - b(\alpha_m + \alpha_i)}{\mathcal{J}_{wmax}} \tag{13.51}$$

where for simplicity we assume that the dependence of the threshold current density on cavity loss is linear with coefficient $b$. The peak efficiency then occurs at

$$\alpha_{m,opt} = \alpha_i + \sqrt{\alpha_i^2 + \alpha_i \left( \frac{\mathcal{J}_{wmax} - \mathcal{J}_{tr}}{b} - \alpha_i \right)} \approx 2\alpha_i \tag{13.52}$$

We can avoid making very short cavities by coating the output facet for low reflection to increase the mirror loss, subject to the limits discussed earlier. We could also optimize the ridge width for maximum output power. While the power may be expected to scale with ridge width, the higher threshold power and reduced lateral heat flow set a practical limit that depends on the operating wavelength, number of stages, and details of how the device is connected to the heat sink.

Next we will estimate the WPEs and output powers for the QCL operating at 5 μm and the ICL with type II "W" active region operating at 3.5 μm, using the parameters summarized in Table 13.2. The results represent "good" devices rather than the best ever reported. We do not include a conventional quantum-well laser emitting in the mid-IR because we already know the benefits of cascading. Type I ICLs emitting in the range 3–3.5 μm can have thresholds comparable to the type II devices, but require fewer stages owing to a higher gain.

We focus on a QCL with $M = 30$, even though there are reports of high WPE and output power from QCLs with more and fewer stages. For an ICL the threshold and power density are at a minimum with $M = 3 - 6$. To maximize the WPE and output power in an ICL, we would use 7–10 stages because the peak current is an order of magnitude beyond the threshold as well as a higher mirror loss, as discussed earlier. Finally, the QCL material system allows much better heat dissipation because the laser ridge can be overgrown with InP that has a low refractive index and high thermal conductivity. The rough rule of thumb is that the thermal conductivity of a binary semiconductor is a factor of 5 higher than in a ternary or quaternary alloy, and a factor of 5–10 higher than in a short-period superlattice like those used as cladding layers in the ICL. State-of-the-art thermal impedances are a factor of 2 lower for a QCL than for a ICL, if both devices are mounted epitaxial-side-down to reduce the thermal resistance between the active core and the heat sink.

Figure 13.11 shows the WPEs and output powers for devices with the parameters listed in Table 13.2. The corresponding threshold power density for a state-of-the-art QCL is of order 10 kW/cm². This is much higher than the typical value

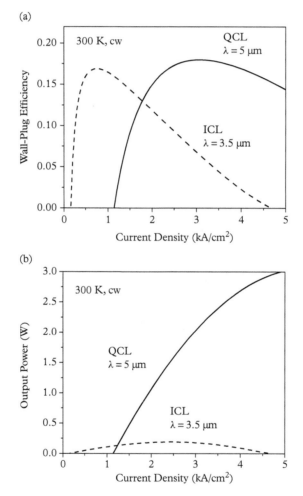

**Figure 13.11** *Wall-plug efficiency (a) and output power (b) versus current density for the QCL and ICL with the parameters listed in Table 13.2.*

of several hundred W/cm$^2$ for an ICL (which has much longer carrier lifetime), although we will see later that there is considerable variation as a function of wavelength.

Interestingly, the WPEs for both QCLs and ICLs have very similar maximum values close to 20%, despite numerous differences. Besides the very different input parameters, the maximum output power in Fig. 13.11b is much lower for the ICL (200 mW versus 2–3 W for the QCL). This is due to the much smaller $T_0$ (and to a lesser extent the smaller $T_1$), as well as the ICL's much higher internal loss. The ICL's stronger temperature dependence makes it difficult to add more stages, and its higher internal loss tips the balance toward a shorter cavity with smaller active area. In fact, the output power from an ICL (300–500 mW from a single ridge) is maximized for a cavity much longer than in Table 13.2, but at the expense of lower WPE. Figure 13.11a also shows that the QCL's weaker temperature dependence allows it to maintain high WPE up to the large currents that produce the most output power. This is much more difficult in the ICL (and most other interband lasers). On the other hand, Fig. 13.11 does not include the limited maximum current density determined by the doping in the injector. In many QCLs, as the current approaches $\mathcal{J}_{max} \sim 3 - 5$ kA/cm$^2$, the WPE and output power decrease, even though the lattice heating is still tolerable. Figure 13.12 illustrates the calculated dependence of the output power on operating temperature $T_s$ in a representative mid-IR QCL.

Figure 13.13 plots the experimental dependence of the ICL threshold current density on emission wavelength. The ICL performance drops off at $\lambda < 3.5$ μm, which is manifested in a gradual increase of the threshold current density as well as an efficiency "droop" at higher currents. The "droop" is not predicted by simple

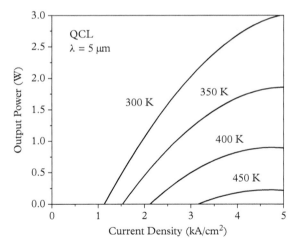

**Figure 13.12** *Output power in the QCL with the parameters listed in Table 13.2 versus current density at several different operating temperatures.*

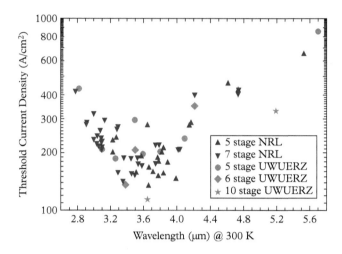

**Figure 13.13** *Pulsed threshold current densities at room temperature for broad-area ICLs emitting at different wavelengths in the mid-IR. Devices designed, grown, and fabricated by the groups at the Naval Research Laboratory and the University of Würzburg are shown.*

laser models and probably reflects an increase in the internal loss or a decrease in the internal efficiency at high currents. Some measurements imply that the carrier density does not pin at threshold, as shown in Fig. 12.2, but continues to rise at a reduced rate.

Figure 13.13 shows that ICLs have lower threshold current densities than QCLs (with typical $\mathcal{J}_{th} \geq 1$ kA/cm$^2$) to at least 6 μm. The advantage only grows when we take into account the different threshold voltages implied by the parameters in Table 13.2: > 10 V for the QCL versus 3–4 V for the ICL. At even longer wavelengths, QCLs are the coherent source of choice for both high-power and low-threshold operation. On the other hand, the QCLs operating in the spectral range 3–4 μm require highly strained layers to achieve a large conduction-band offset. For that reason, QCL performance currently trails off at λ < 4 μm, where ICLs boast both lower threshold and higher WPE.

As the wavelength decreases toward 2 μm, the need for cascading is less clear-cut because the parasitic voltage drop is a smaller fraction of the photon energy. For devices operating close to threshold, it may be hard to justify the added complexity of the cascade laser. To date, no QCLs emitting at room temperature at λ < 3 μm have been demonstrated, and there has been little recent progress in that direction despite the large offsets available in the InAs/AlSb and GaN/AlN systems.

## 13.4   Mode-Locked Lasers and Laser Frequency Combs

In Section 13.1, we mentioned that semiconductor lasers often operate on several sets of longitudinal modes, each set associated with a different lateral mode of the ridge waveguide. The emission of a ridge in which only a single lateral mode lases consists of nearly evenly spaced Fabry–Perot resonances with frequencies given by the product of the standing-wave number and the velocity at which light propagates in the cavity (see Fig. 13.14):

$$\omega_j = k_j \left[ \frac{dk}{d\omega} (\omega_j) \right]^{-1} = \frac{j\pi}{L_{cav}} v_g (\omega_j) \tag{13.53}$$

where the group velocity $v_g (\omega) = \left[ \frac{dk}{d\omega} \right]^{-1} = \left[ \frac{d(\omega n_m / c)}{d\omega} \right]^{-1} = \frac{c}{n_m + \omega dn_m / d\omega}$, where the index of the lateral mode $n_m$ is in general a function of frequency. Because the group velocity is not the same for different $j$, the longitudinal modes are not evenly spaced; i.e., it is *not* true in general that $\omega_j = \omega_0 + j\Delta\omega$. The spacing may look nearly uniform if $v_g$ does not vary fast with frequency, but it is not truly constant unless $v_g$ is a constant! We can cast the dependence of the group velocity on frequency in terms of the *group velocity dispersion* (GVD):

$$GVD = \frac{d}{d\omega} \frac{1}{v_g} = \frac{d^2 k}{d\omega^2} \tag{13.54}$$

In a cavity, a more common measure is the *group delay dispersion* (GDD), which is equal to the GVD multiplied by $2L_{cav}$. Of course, the GVD (or GDD) is not constant with frequency either, but in many cases we can assume it is flat over a certain frequency range, e.g., when the spectral dependence comes from the tail of a distant absorption line.

The frequency "comb" that consists of uncoupled longitudinal modes is not very useful because the individual lines are not precisely positioned with respect to each other. Also, the power in the individual lines can fluctuate strongly, even when the total output is nearly constant in cw mode. Despite these drawbacks, this multimode spectrum is tantalizingly close to an evenly spaced comb for modest GVD, as shown in Fig. 13.14. Is there a way to fix the frequencies with much greater precision, i.e., to "pull" the modes toward the desired frequencies? We would also like to impose some kind of phase relationship between them so the modes do not fluctuate dramatically in time. We can think of the problem in the following way: if the different modes are not connected by a tight phase relationship, we simply have a series of independent, random-phase emitters with line-to-line noise similar to that of a single emitter. In a true frequency comb, the line-to-line noise is much smaller because the same phase relationship holds over the entire comb spectrum that is much wider than the comb spacing $\Delta\omega$.

One approach to fixing the phases of the individual modes is to modulate the cavity loss (or gain) periodically at the cavity round-trip time $T = 2L_{cav}/v_g (\omega_0)$, where $\omega_0$ is the center frequency. Because of the modulation, each longitudinal mode acquires sidebands spaced by $\Delta\omega = \frac{2\pi}{T} = \frac{\pi}{L_{cav}} v_g (\omega_0)$. Then the difference between the positions of the sidebands and the unperturbed longitudinal modes in Eq. (13.53) is due solely to the GVD, which is typically small on the scale of two

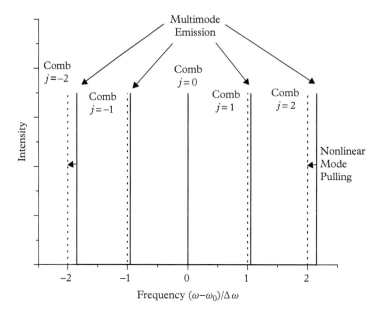

**Figure 13.14** *Schematic illustration of multimode lasing and frequency-comb formation in a cavity with group velocity dispersion.*

mode separations. The modes are pulled to the desired frequency-comb positions in Fig. 13.14, and in fact the structure favors the formation of short light pulses separated by the round-trip time (*active mode locking*). The classic example of this behavior is the coherent sum of $N$ phase factors that correspond to electric field amplitudes with the same phases (taken to be zero for simplicity):

$$e^{i\omega_0 t} \sum_{j=-N/2}^{N/2} e^{ij\Delta\omega t} = e^{i\omega_0 t} e^{-iN\Delta\omega t/2} \frac{1 - e^{iN\Delta\omega t}}{1 - e^{i\Delta\omega t}} = e^{i\omega_0 t} e^{-i\Delta\omega t/2} \frac{\sin(N\Delta\omega t)}{\sin(\Delta\omega t)}$$

$$(13.55)$$

The pulse described by Eq. (13.55) for large $N$ has a width of $T/N$ in the time domain. This is because for small $t$, the denominator of Eq. (13.55) is close to $\Delta\omega t$, and the expression transforms into a sinc function. On the other hand, the spectral width is $N\Delta\omega = \frac{2\pi N}{T}$ because $N$ equally spaced modes are locked together. If all the comb teeth are coherent, the total field is proportional to the number of teeth $N$, and the peak power is proportional to $N^2$. Because the pulse width is $T/N$, and the repetition rate is $T$, the *average* power scales with $N$ rather than $N^2$.

A more common approach to producing short-pulse frequency combs in interband semiconductor lasers is to introduce a saturable-absorber section into the laser cavity, as depicted schematically in Fig. 13.15a. The saturable absorber

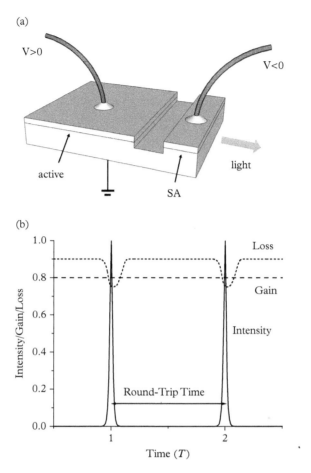

**Figure 13.15** *(a) Schematic of the typical cavity configuration for passively mode-locked semiconductor lasers with active gain and saturable absorber (SA) sections. (b) Illustration of the gain and loss in the saturable absorption as a function of time.*

typically consists of the active material under reverse bias. When the light intensity is low, it absorbs the incident light and does not allow it to propagate in the cavity. But at high intensity, the carrier density generated by the light can be high enough to make the absorber transparent by separating the electron and hole quasi-Fermi levels. The overall effect of the saturable absorber is to transmit the light near the peak of any pulse that propagates in the cavity while suppressing the tails of such a pulse, shortening it in the process, as shown in Fig. 13.15b. This approach is known as *passive mode locking*. Because carrier lifetimes in interband semiconductor lasers are quite long, saturable absorbers in these materials are "slow" (of the integrating type; i.e., it is the integral of the incident intensity up to a certain point in time that matters rather than the instantaneous power at that point). Despite that difference, the operation of passively mode-locked lasers largely follows the principle outlined above. To make sure that the saturable absorber is in the initial state before the next pulse arrives, it must recover on a time scale shorter than the cavity round-trip time. Applying reverse bias often sweeps the carriers out fast enough for this condition to hold. We can also assure a fast recovery by bombarding the saturable-absorption section with ions to create a high density of Shockley–Read–Hall-active traps and reduce the carrier lifetime, at the expense of a higher saturation intensity. The pulsing in a passively mode-locked laser can start spontaneously because random fluctuations are selected and amplified by the saturable absorber. The strongest such fluctuation can be transformed into a short pulse, while all other fluctuations decay.

As the carrier lifetime decreases, passive mode locking becomes more difficult because not only can the absorption recover fast, but so can the gain in the active section of Fig. 13.15(a). In fact, in the limit of very short carrier lifetime, the gain is out of phase with the light intensity in the pulse. That is, it is high when the intensity is low in the tails of the pulse and low near the peak. This lengthens any pulse that propagates in the cavity and in the end favors a nearly constant intracavity intensity. Therefore, passive mode locking is essentially impossible in a device with very short lifetimes like the QCL. Fortunately, this fast-carrier regime offers a completely different, yet equally practical route to a self-starting frequency comb.

We start by noticing that multimode operation (see Fig. 13.14) is driven by spatial hole burning discussed in connection with Fig. 12.4a. When spatial hole burning is active and the carrier density is periodically modulated along the cavity axis, the energy-conserving *four-wave mixing* (FWM) processes illustrated in Fig. 13.16 can generate a frequency comb. FWM occurs when a change in the intensity of one mode affects its neighbors because they share the same pool of carriers, which creates an effective optical nonlinearity. FWM conserves energy and hence demands a constant separation between the lines (as in a frequency comb). The continuous interplay between hole burning and FWM processes in

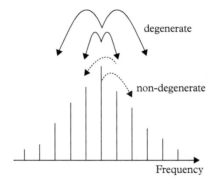

**Figure 13.16** *Frequency-domain illustration of the four-wave mixing processes that help to establish FM frequency-comb operation in a QCL.*

the material with short carrier lifetime leads to a quasi-cw comb that can extend over a large fraction of the gain spectrum of the laser (positive gain minus cavity loss is still needed to lase in the first place). Of course, FWM takes place to some extent in any semiconductor laser, but strong carrier diffusion typically suppresses the population modulation due to spatial hole burning, so frequency combs are rarely observed in normal laser operation.

How can a comb develop without being associated with a short pulse via Eq. (13.55)? Think of it as analogous to a frequency-modulated (FM) signal with a constant intensity and a range of sidebands separated by the modulation frequency (in this case, $\Delta\omega = \frac{2\pi}{T}$). Written as the real part of the electric field at some point in the cavity, its general form is similar to

$$\text{Re}\,(E(z)) = E_m(z)\cos\left(\omega_0 t + \varphi(t)\right) = E_m(z)\cos\left(\omega_0 t + \frac{A_m}{\Delta\omega}\cos\left(\Delta\omega t\right)\right) \quad (13.56)$$

with sidebands at $\omega_0 + j\Delta\omega$, where $j$ is an integer. Of course, the exact spectrum produced by the comb is not necessarily that of a typical FM signal, so the second equality in Eq. (13.56) is for illustration purposes only. This is because the shape of the gain spectrum tends to favor more emission near its peak and less near its tails. The *instantaneous frequency* $\omega_{inst} = \omega_0 + \frac{d\varphi}{dt}$ varies over a round trip because the device then takes full advantage of the entire available gain spectrum rather than oscillating at one or a few frequencies near the gain peak. The variation of $\omega_{inst}$ in a QCL comb is illustrated in Fig. 13.17, with the parameter space that leads to this dependence still under investigation. A similar linear change has been observed (without a saturable absorber) in other semiconductor lasers that do

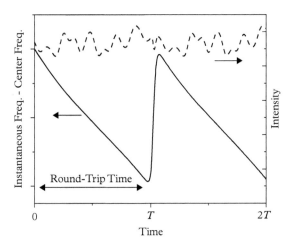

**Figure 13.17** *Schematic illustration of the typical phase and intensity variation in a QCL frequency-comb cavity as a function of time.*

not have a very short carrier lifetime including the ICL. Again, much remains to be understood about the conditions that are conducive to FM combs in different types of lasers, but strong spatial hole burning appears to be an essential ingredient.

In some QCL combs the teeth are not separated by $\Delta\omega$, but by some multiple of the free spectral range $N\Delta\omega$. These *harmonic* frequency combs are produced by the interplay of dispersion and the effective nonlinearity that peaks at $N\Delta\omega$. On the other hand, if a harmonic comb results from passive mode locking, the time-domain picture corresponds to multiple pulses propagating in the cavity at the same time.

Laser combs remain an active research field. As mentioned earlier, it appears that some active media support both short pulses via passive mode locking (in a cavity with a well-designed saturable absorber) and a quasi-cw FM laser comb (without a saturable absorber). Whatever the mechanism of frequency-comb formation, the total comb bandwidth is often determined by the GVD in Eq. (13.54) rather than the width of the gain spectrum unless specific measures are taken to minimize the GVD. This is because mode locking and FWM can only do so much to oppose the tendency of the GVD to place modes in the "wrong" places, as shown in Fig. 13.14. To improve the bandwidth, we might want to adjust the waveguide dispersion of the lasing mode, or introduce dispersive elements that cancel the GVD in the cavity.

## Suggested Reading

- Quantum cascade lasers are described in more detail in a book by J. Faist, *Quantum Cascade Lasers* (Oxford University Press, Oxford, 2013).

- The main ideas for mode locking can be found in A. E. Siegman, *Lasers* (University Science Books, 1986).

- QCL frequency combs are a quickly developing field, so be forewarned that the information available at the time of publication is subject to rapid obsolescence! One review article is J. Faist, G. Villares, G. Scalari, M. Rösch, C. Bonzon, A. Hugi, and M. Beck, "Quantum cascade laser frequency combs," *Nanophotonics* **5**, 272 (2016). The linear chirp is discussed in M. Singleton, P. Jouy, M. Beck, and J. Faist, "Evidence of linear chirp in mid-infrared quantum cascade lasers," *Optica* **5**, 948 (2018).

- No comprehensive theoretical model that explains all the observations exists at the time of writing, but some of the physics is beginning to be understood. FM combs in quantum-well lasers are treated in M. Dong, S. T. Cundiff, and H. G. Winful, "Physics of frequency-modulated comb generation in quantum-well diode lasers," *Physical Review A* **97**, 053822 (2018) and QCL combs in N. Opačak and B. Schwarz, "Theory of frequency-modulated combs in lasers with spatial hole burning, dispersion, and Kerr nonlinearity," *Physical Review Letters* **123**, 243902 (2019).

# 14

# Semiconductor Photodetectors

In this chapter, we describe the operating principles of photoconductive and photovoltaic detectors based on III–V semiconductors. The electrical characteristics of both photodiodes and majority-carrier barrier structures are discussed starting with the diffusion equation. We outline the figures of merit used to evaluate the performance of infrared photodetectors including the responsivity, dark-current density, and normalized detectivity. We discuss bulk-like and type II superlattice photodetectors and how the multistage arrangement of interband cascade detectors (ICDs) can reduce the dark-current density at the expense of a lower responsivity. Detectors that employ intersubband optical transitions, namely, quantum-well infrared photodetectors (QWIPs) and quantum cascade detectors (QCDs) are also discussed. We consider how the dark-current density can be suppressed in resonant-cavity and thin waveguide-based detectors. We conclude with a discussion of the requirements for high-speed operation and an overview of novel types of detectors that draw their inspiration from III–V semiconductor devices.

## 14.1 Photoconductive Detectors

A photodetector is a device that responds to light illumination by changing the current that flows through the device. Semiconductor photodetectors fall into two broad categories: *photoconductive* and *photovoltaic*. A photoconductive detector relies on a change in the current flow that takes place when it is exposed to light. Essentially, it is a light-sensitive resistor with an ohmic $[V = IR(P)]$ behavior. In contrast, a photovoltaic detector can operate at zero bias without any current flowing in the dark. Illumination then leads to current flow, even at zero bias. In this chapter, we will focus on detectors for the mid-wave (3–8 μm) and long-wave (8–12 μm) IR, with similar concepts often applying to other wavelength ranges.

How do we characterize the performance of a photodetector? It responds to individual photons by producing electrons (in contrast, *bolometers* change conductance in response to the total amount of generated heat). Therefore, the signal in a photoconductive detector corresponds to the change in electron flux upon illumination. In relative terms, it is the ratio of that flux change to the

*Bands and Photons in III–V Semiconductor Quantum Structures.* Vurgaftman, Lumb, and Meyer,
Oxford University Press (2021). © Vurgaftman, Lumb, and Meyer.
DOI: 10.1093/oso/9780198767275.003.0014

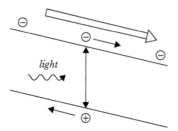

**Figure 14.1** *Schematic illustration of an interband photoconductive detector.*

photon flux incident on the detector. If that was the end of the story, such a detector would be sensitive even to single photons. In practice, we can only see signals that are sufficiently above the noise level. The relevant noise here comes from thermal fluctuations that generate electrons, which then flow through the device and are indistinguishable from the light-induced current flow. To estimate detectable signals, we calculate the signal-to-noise ratio.

Figure 14.1 schematically illustrates an interband version of such a detector. This optically sensitive device can be as simple as a rectangular piece of doped semiconductor with electrical contacts (that transmit at least some of the incident light) on top and bottom. Light shining from one direction, with a photon energy larger than the material's energy gap generates electrons and holes along the device thickness. The absorbed fraction is proportional to $1 - \exp(-\alpha L)$, where $\alpha$ is the absorption coefficient of the absorbing material, and $L$ is its thickness. More generally, we call the total fraction of incident photons that produce electron–hole pairs the *quantum efficiency* $\eta_a$. The generation rate per unit volume in the semiconductor is $\eta_a I_\omega / (\hbar \omega L)$, where the incident light intensity $I_\omega$ (in units of W/cm$^2$) is divided by $\hbar \omega$ to convert to the photon flux per unit area. When a vertical bias is applied, any photogenerated carriers that do not recombine along the way are collected at the two contacts.

What is the resulting photocurrent density? In equilibrium, the generation rate $\eta_a I_\omega / (\hbar \omega L)$ must equal the recombination rate $n_e / \tau_e$, where $n_e$ is the optically generated carrier density, and $\tau_e$ is the carrier lifetime, which is potentially a function of $n_e$ (see Section 12.2). The photoexcited carrier density becomes

$$n_e = \eta_a \frac{I_\omega}{L \, \hbar \omega} \tau_e \tag{14.1}$$

These carriers are driven along the photoconductor by the bias. The electron photocurrent is given by

$$\mathcal{J}_p = e n_e \mu_e F_e = e n_e v_e \tag{14.2}$$

where $F_e$ is the applied electric field, and the electron mobility $\mu_e$ quantifies the linear relationship between electron velocity and applied field, $v_e = \mu_e F_e$, which is valid at low fields. Holes are usually much less mobile, so we may omit them here. At room temperature, typical electron mobility in III–V semiconductors is a few thousand cm$^2$/Vs. However, at low temperatures the mobility can depend strongly on the impurity density and quantum confinement.

Combining Eqs. (14.1) and (14.2), we obtain

$$\mathcal{J}_p = e\eta_a \frac{I_\omega}{L}\frac{1}{\hbar\omega}\tau_e\mu_e F_e = \frac{e\eta_a I_\omega}{\hbar\omega}\frac{v_e}{L}\tau_e = \frac{e\eta_a I_\omega}{\hbar\omega}g_p \tag{14.3}$$

The factor $e\eta_a \frac{I_\omega}{\hbar\omega}$ is simply the absorbed photon flux times the electron charge. We can also define the *responsivity*: $R_p \equiv \frac{\mathcal{J}_p}{I_\omega} = \frac{e\eta_a g_p}{\hbar\omega}$. It turns out the total change in current is larger by a factor known as the *photoconductive gain*: $g_p = v_e\tau_e/L$, which we can rewrite as

$$g_p = \frac{\tau_e}{\tau_{trans}} \tag{14.4}$$

Here $\tau_{trans} \equiv L/v_e$ is the total transit time through the device. If the device consists of $M$ repeating units of length $d = L/M$, the denominator in Eq. (14.4) is proportional to $M$. The photoconductive gain can be of order 100–1000, since typical carrier lifetimes are in the ns range and much longer than the transit time. For a sample thickness of a few μm, which corresponds to the typical absorption depth, the transit time is in the ps range.

We can understand photoconductive gain better by noticing that the electrons collected at the contacts are not necessarily the same as those generated by the incident light. This is because the photoexcited electrons do not simply stay put, waiting to recombine with holes. Instead, the applied field shuttles them through the sample between the two contacts, which replenish them on the ps transit time scale. What we observe is the electron flow at the contacts, at a rate proportional to the light intensity shining on the sample. The photon flux simply "gates" how much current can flow, rather than directly supplying electrons for this purpose. We will see in Section 14.2 that the photon flux does set an upper limit on the photocurrent in a photovoltaic detector, but even then the collected carriers are not necessarily those excited by the light.

How does this signal (photocurrent) compare with the noise due to random fluctuations of the current in the device? The material's electrical resistance under bias and in the dark is finite. Therefore, some *dark current* will flow between the contacts, with wide fluctuations around a mean value that result, e.g., from electrons escaping into the conduction band from traps. To characterize the fluctuations, think of each individual electron arriving at the contacts as discrete independent events. The number of such events $n$ in a given time interval $\tau_{int}$ is described by the Poisson distribution. For our purposes, the only important feature of the Poisson statistics is that the standard deviation is the square root of the mean, or equivalently, the variance and the mean are the same, i.e., $\langle n^2 \rangle - \langle n \rangle^2 = \langle n \rangle$. Here the quantities in angle brackets represent random variables. The mean current is given by the electron charge times the mean number of collection events divided by the observation time, i.e., $\langle i_d \rangle = e \langle n \rangle / \tau_{int}$. Using the relationship between the mean and variance for Poisson statistics, we write $i_n^2 \equiv \langle i_d^2 \rangle - \langle i_d \rangle^2 = e \langle i_d \rangle / \tau_{int}$. We can express this in terms of frequency bandwidth by noticing that we must sample twice in a time interval $\tau_{int}$ to capture the maximum frequency variation of $\Delta f = \frac{1}{2\tau_{int}}$. This is the famous *Nyquist–Shannon sampling theorem* (see also Section 9.2). Finally, the cross-sectional area of the photoconductor $A$ is arbitrary, so we can work with current densities with a view of normalizing it out of the final expression. Putting together all the factors, we obtain the variance of the dark-current density:

$$j_n^2 = \frac{2e\mathcal{J}_d \Delta f}{A} \tag{14.5}$$

where $\mathcal{J}_d = \langle j_d \rangle$ is the mean dark-current density.

Matters are in fact a little more complicated for a photoconductive detector because the actual recombination time is a random variable rather than the fixed value $\tau_e$. This phenomenon turns out to contribute as much noise as the discrete nature of the electrons. In many photoconductive detectors, the current noise is subject to much the same gain as the photocurrent. This is because the carrier number fluctuates due to generation or recombination, and the rate of these fluctuations is amplified to parallel with the optical generation rate. With this in mind, we modify Eq. (14.5) to read

$$j_n^2 = \frac{4eg_p\mathcal{J}_d \Delta f}{A} \tag{14.6}$$

To reiterate, Eq. (14.6) is applicable to *photoconductive detectors only*, as well as to devices in which the dark current and photocurrent are subject to the same gain. On the other hand, we will see in Section 14.2 that Eq. (14.5) remains correct for photovoltaic detectors.

The shot noise mechanism that leads to Eq. (14.5) is not the only one present in photodetectors. Another important source of noise, known as $1/f$ noise, has a spectral density that increases inversely with frequency (of the changes in the photocurrent) and is particularly pronounced for frequencies below 10–100 kHz. Contrast this behavior with the "white," i.e., frequency-independent, noise in Eqs. (14.5) and (14.6), which is simply proportional to the total bandwidth $\Delta f$. Another contribution (thermal *Johnson noise*) is important for operation at low voltages. In this case, the variance is inversely proportional to the resistance of the device at zero bias. We will learn how to calculate its impact in Section 14.2. In what follows, we assume that white (shot or Johnson) noise dominates and note that the physical origin of $1/f$ noise is often unclear, but correlates with defect densities and current leakage.

To construct a figure of merit for the minimum detectable signal subject to dark-current noise in the photodetector, we divide the responsivity, which measures the photocurrent per incident intensity by the standard deviation of the dark current:

$$D = \frac{R_p}{j_n A} = \frac{\eta a}{2\hbar\omega}\sqrt{\frac{g_p e}{\mathscr{J}_d \Delta f A}} \tag{14.7}$$

This expression includes the bandwidth and cross-sectional area, which vary from one detector to the next. Of course, we would much rather work with a quantity that does not depend on these device-specific quantities. For this reason, we define the *specific detectivity $D^*$* as

$$D^* \equiv D\sqrt{\Delta f A} = \frac{\eta a}{2\hbar\omega}\sqrt{\frac{g_p e}{\mathscr{J}_d}} \tag{14.8}$$

The units here may seem unpalatable, cm Hz$^{1/2}$/W, but this is a small price to pay for pinning down the fundamental limit on photodetector operation. These units are also named *Jones* in honor of R. Clark Jones, an American physicist who proposed this useful definition in 1957.

Even though it looks like the detectivity is proportional to the square root of the gain, this is not necessarily the case. The dark-current density is also inversely proportional to the transit time, so we cannot raise the gain without degrading the dark-current noise. In fact, some of the best available detectors operate with no gain at all, as we will see in Section 14.2.

If we replace $\mathscr{J}_d/e$ in Eq. (14.8) with the flux of background thermal radiation falling on the device, we can calculate the detectivity that corresponds to the environmental noise. The background radiation closely follows the blackbody form, which was first derived by Max Planck. Its average photon number is given by Eq. (4.91) with $V = 0$. If the device has a lower detectivity in the dark, it is said to be *background limited*. The exact value of the background limit depends not only on the temperature of the environment seen by the detector, but also on its field of view and its spectral response. If the detector senses a signal on top of the background radiation, the latter may be viewed as another shot-noise source with variance proportional to the flux rather than $\mathscr{J}_d/e$ in Eq. (14.6).

How much dark current should we expect for a photoconductive detector with a bulk absorber? At the operating temperature, there will be some carriers in the material with density $n_{th}$. For example, in an *n*-doped material at higher temperatures, $n_{th}$ is approximately equal to the donor density $N_d$, so the dark-current density is

$$\mathscr{J}_d = eN_d\mu_e F_e = eN_d v_e \tag{14.9}$$

The same reasoning applies to $p$-doped materials. Even if the material is not doped, we can lower the dark current only down to the lower limit set by the intrinsic carrier density $n_i$ in Eq. (4.81). Assuming $\mu_e \gg \mu_h$, we obtain

$$\mathcal{J}_d \approx e n_i \mu_e F_e = e \mu_e F_e \sqrt{N_c N_v} \, \exp\left(-\frac{E_g}{2 k_B T}\right) \tag{14.10}$$

The intrinsic carrier density is substantial in narrow-gap materials at higher temperatures.

For the example of a doped material, we can use Eqs. (14.2) and (14.9) to find

$$\frac{\mathcal{J}_d}{\mathcal{J}_p} = \frac{N_d}{n_e} \tag{14.11}$$

where we have substituted the photoexcited carrier density $n_e$ from Eq. (14.1). In an undoped material, we would replace $N_d$ with $n_i$ in Eq. (14.11). In other words, in a photoconductive device the ratio of dark to photogenerated current density is the same as the ratio of equilibrium to photoexcited density. Usually this ratio is much larger than 1, and the photocurrent is a small fraction of the total current. If the device consists of a number of repeating units, the same current must flow through all of them, so that the photocurrent is independent of the number of periods $M$.

Because the dark current is quite high in photoconductive detectors, we need a large gain in Eq. (14.4) to improve $D^*$. This requires either a long carrier lifetime $\tau_e$ or short transit time $\tau_{trans}$. We can shorten the transit time by thinning the absorber, but then the quantum efficiency $\eta_a$ decreases proportionally. Hence the design options in a photoconductive detector are limited. Fortunately, we will see in Section 14.2 that we do not have to settle for high dark current in a semiconductor detector.

## 14.2  Photovoltaic Detectors

Can a detector be designed to suppress the dark current? In fact, such designs are possible, but usually at the price of a gain $g_p$ close to 1. Instead of increasing the flow of majority carriers, the strategy is to block it to the extent possible and instead collect minority carriers on the opposite sides of the device. These structures, known as *photovoltaic* detectors, operate at zero or low bias.

One straightforward way to block the flow of majority carriers (here, electrons) is an *nBn* structure (for $n$-type/Barrier/$n$-type) that introduces a large barrier in the conduction band and no barrier in the valence band, as illustrated in Fig. 14.2. As long as the carrier density in the barrier is low, most of the applied voltage drops there. At relatively low applied bias, majority electrons to the left of the barrier cannot get through to the right side and into the contact. Electrons on the right side

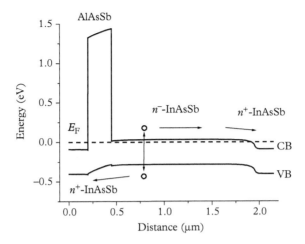

**Figure 14.2** *Band diagram for a unipolar nBn interband photovoltaic detector with an 1.5-μm-thick InAsSb absorber, thinner heavily doped InAsSb left and right contacts, and an AlAsSb majority-carrier barrier that blocks majority-carrier transport. The InAsSb and AlAsSb lattice constants approximately match the GaSb substrate. Barrier and absorber compositions are adjusted to match the lattice constants to that of the GaSb substrate. The doping is set to $10^{15}$ cm$^{-3}$ n-type in the InAsSb absorber and $10^{15}$ cm$^{-3}$ p-type in the AlAsSb barrier at a temperature of 130 K.*

of the barrier cannot flow either, because no compensating current comes from the left to replenish them and balance the charge. Of course, *some* dark current still flows in a photovoltaic detector, which will be estimated later. However, it is much lower than in a photoconductive detector, which by design must have unimpeded majority-carrier transport, with or without photoexcitation.

When light falls on the photovoltaic structure of Fig. 14.2, excess holes generated by an incident flux can flow across the "barrier" layer because it presents no barrier to minority holes! They can then continue to the contact on the right side, or recombine with an electron that is immediately replaced by an electron from the contact, without violating either charge balance or current continuity. Because the rate at which carriers are generated in any part of the structure is bounded by the incident flux, the photocurrent cannot exceed the flux, i.e., $g_p \leq 1$. Otherwise, some charge would inevitably build up in some region of the device. Therefore, the device responsivity is characterized by the quantum efficiency $\eta_a$ alone.

A photovoltaic detector commonly operates at zero or low bias. Can we extend the expression in Eq. (14.8) to zero bias? Even when the mean dark current is zero, the current still fluctuates with a standard deviation proportional to the "thermal voltage," $k_B T/e$, and inversely proportional to the differential device resistance–area product at zero bias $R_0 A \equiv \frac{\partial V}{\partial \mathcal{J}_d}$. This is the Johnson noise mentioned in Section 14.1. One way to think about this noise is to imagine random voltage swings, on the scale of the thermal voltage, driving current fluctuations around zero that limit the signal-to-noise ratio. Therefore, when $V = 0$, we replace $\mathcal{J}_d \rightarrow \frac{2k_BT}{eR_0A}$ and Eq. (14.5) with

$$j_n^2 = \frac{4k_BT\Delta f}{R_0A^2} \tag{14.12}$$

It may not look it, but $R_0A$ is a fundamental quantity derivable from the basic device and material characteristics. We can also generalize Eq. (14.8) to the case of zero bias by neglecting the photoconductive gain:

$$D^* = \frac{e\eta_a}{2\hbar\omega}\sqrt{\frac{R_0A}{k_BT}} \tag{14.13a}$$

If the bias *is not* zero, we must sum the contributions from the differential resistance and the dark-current density in the shot-noise variance, by using Eqs. (14.5) and (14.12):

$$D^* = \frac{e\eta_a}{\hbar\omega}\frac{1}{\sqrt{\frac{4k_BT}{R_0A} + 2e\mathcal{J}_d}} \tag{14.13b}$$

One of the terms in the denominator of Eq. (14.13b) often dominates the other.

What happens as the dark current becomes very small and the differential resistance very large? We saw in Section 14.1 that the current noise is then dominated by the absorbed background flux due to blackbody radiation. To obtain the so-called background-limited detectivity (BLIP), we substitute $\mathcal{J}_d \rightarrow e\Phi_B$ in Eq. (14.5), where $\Phi_B$ is the background photon flux in units of $s^{-1}$ cm$^{-2}$:

$$D^*_{BLIP} = \frac{e\eta_a}{\hbar\omega}\frac{1}{\sqrt{2e^2\Phi_B}} \tag{14.13c}$$

To be very general, we might combine Eqs. (14.13b) and (14.13c) as follows:

$$D^*_{BLIP} = \frac{e\eta_a}{\hbar\omega}\frac{1}{\sqrt{\frac{4k_BT}{R_0A} + 2e\mathcal{J}_d + 2e^2\Phi_B}} \tag{14.13d}$$

The background flux falls on the detector within the limited field of view of its detector optics, which can be much smaller than the full $2\pi$ solid angle. In many cases, the detector is at a lower temperature than the background, so radiation coming from the detector itself is negligible compared to the outside flux.

We learned in Chapters 6 and 7 that we cannot always find bulk III–V materials with the desired band offsets, gaps, and lattice constants. These practical limits meant that before the advent of superlattices and other advances in epitaxial structures, photovoltaic detectors blocked majority carriers using exclusively the *built-in potential* of a *p–n* junction, as shown in Fig. 14.3. How does the junction accomplish this task? The Fermi level in an *n*-type (*p*-type) material is in or near the conduction (valence) band. This means the bands on opposite sides of the junction must be separated by an energy close to $E_g$. We can also imagine electrons (holes) crossing the junction to the opposite side, where the Fermi level is lower (higher). This leaves behind oppositely charged bare ions in the *depletion region* that straddles both sides of the junction. The charge separation stops when an internal field opposing any further flow of electrons (holes) across the junction becomes high enough.

What is the magnitude of the built-in potential for a *homojunction* formed with the same material on both sides (see Fig. 14.3)? It is simply the difference between the band edges on both sides of the junction: $eV_{bi} = E_{c,p} - E_{c,n} = E_g + E_{v,p} - E_{c,n}$. Here $E_{c,p}$ is the energy of the conduction-band minimum on the *p*-side, $E_{c,n}$ is its energy on the *n*-side, etc., when the band edges are measured in the flat-band regions far away from the junction. For example, if both sides are doped non-degenerately, we can use the Boltzmann carrier statistics of Eq. (1.50): $n = N_c \exp\left[(E_F - E_{c,n})/k_B T\right]$ and $p = N_v \exp\left[(E_{v,p} - E_F)/k_B T\right]$, with $E_{c,n} \gg E_F$ and $E_{v,p} \ll E_F$. The built-in voltage is then

$$eV_{bi} = E_g + k_B T \left[ \ln\left(\frac{n}{N_c}\right) + \ln\left(\frac{p}{N_v}\right) \right] \qquad (14.14)$$

As expected, as long as $E_g \gg k_B T$ and the doping is significant, it is close to the energy gap. For very narrow gaps on both sides of the junction and/or very low doping, the junction is not sufficient to block the flow of majority carriers.

We can use the fact that the built-in voltage for an unbiased device is the potential difference between the *n* and *p* sides in equilibrium, with the Fermi level flat across the junction, to connect carrier densities of the same type on the two sides. For example, $n_n = N_c \exp\left[(E_F - E_{c,n})/k_B T\right]$ and $n_p = N_c \exp\left[(E_F - E_{c,p})/k_B T\right] = N_c \exp\left[(E_F - E_{c,n} - eV_{bi})/k_B T\right] = n_n \exp(-eV_{bi}/k_B T)$. Here we assume that all dopants are ionized, and $N_D, N_A \gg n_i$ so that $n_n \approx N_D$ and $p_p \approx N_A$ from charge neutrality.

Figure 14.3 shows that joining the two regions with opposite doping types causes the Fermi level to cross the middle of the energy gap, where the carrier densities are very low when $E_g \gg k_B T$. Therefore, the built-in electrostatic potential is created by immobile charged ions. The electric field due to this potential cannot be uniform across the depletion region and abruptly drops to zero at its boundaries. Instead, it peaks near the middle of the depletion region and decreases gradually in both directions. We can define the "depletion width" in terms of the average electric field and built-in voltage: $W \equiv V_{bi}/F_{i,av}$.

Now we will find the actual field profile. The total uncompensated charge must add up to zero when integrated over both sides of the junction. This means that

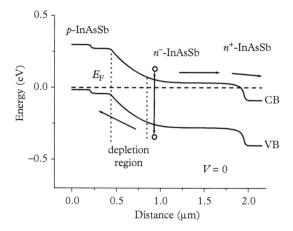

**Figure 14.3** *Band diagram for an InAsSb homojunction photovoltaic detector with an InAsSb absorber. The absorber thickness is taken to be 1.5 μm, the temperature is 130 K, the doping is set to $10^{15}$ $cm^{-3}$n-type in the InAsSb absorber, and $3 \times 10^{16}$ $cm^{-3}$p-type across the junction. The contacts are assumed to be doped more heavily.*

the ionization extends proportionally deeper into the lower-doped material. For a "one-sided" junction with lower doping density $N_b$, the sheet-charge density on the low-doped side is $\rho W \equiv eN_b W$, where $\rho$ is the 3D charge density. Gauss's law links the electric field $F_i$ to $\rho$ in the following way:

$$\frac{\partial F_i}{\partial x} = \frac{\rho}{\varepsilon_s} \tag{14.15}$$

where $\varepsilon_s$ is the static dielectric constant. For an abrupt junction, the charge on either side is constant, and the field varies linearly from the middle of the depletion region. Therefore, the average field is half its maximum value:

$$F_{i,av} = \frac{\rho W}{2\varepsilon_s} = \frac{eN_B W}{2\varepsilon_s} \tag{14.16}$$

Using this expression, we compute the depletion width $W$ as

$$W = V_{bi}/F_{i,av} = \sqrt{\frac{2\varepsilon_s V_{bi}}{eN_B}} \tag{14.17}$$

Apart from the factor of $\sqrt{2}$, we could have guessed this expression using dimensional analysis. The actual depletion width is slightly smaller than in Eq. (14.17) because the carrier densities have exponential tails rather than a top-hat profile.

Because the lower-doped side of the junction accommodates most of the depletion region, that is where most of the applied and built-in voltages drop. The voltage and carrier density in Eq. (14.17) are exactly proportional, which means that the relative voltage drop on the two sides scales inversely with the ratio of the doping densities: $V_n N_D = V_p N_A$ or

$$V_n/N_A = V_p/N_D \tag{14.18}$$

where $V_n + V_p = V_{bi}$.

The built-in potential presents an intrinsic barrier to current flow, which becomes higher (lower) when a reverse (forward) bias is applied to the junction. At sufficiently high forward bias, the junction essentially disappears, and the device acts as a resistor. If we consider the built-in voltage to be fully analogous to an applied voltage $V$, we can simply replace $V_{bi}$ in Eq. (14.17) and elsewhere with $V_{bi} - V$. For example, Eq. (14.17) becomes

$$W = V_{bi}/F_{i,av} = \sqrt{\frac{2\varepsilon_s (V_{bi} - V)}{eN_B}} \tag{14.19}$$

How do the carrier densities vary near the edges of the depletion region? Just beyond the edge, the majority-carrier density nearly equals the doping density with no appreciable change in the Fermi level. The minority-carrier density is then nearly the same as in the absence of a junction. But applying a reverse bias of only a few $\frac{k_B T}{e}$ changes the situation dramatically, as shown in Fig. 14.4. Why does such a small voltage $(eV \ll E_g)$ have a large effect on the minority-carrier density? This is because the density depends exponentially on the minority-carrier quasi-Fermi level. A reverse bias pulls minority electrons out of the $p$-doped region and minority holes out of the $n$-doped region. Conversely, a forward bias injects extra minority carriers into those regions. As expected, a small bias does not have as much effect on the majority carriers because they must still balance the static dopant charge.

Now we can derive the spatial variation of the minority-carrier densities outside the depletion region under an applied bias. For simplicity, we consider the case in which one side of the junction is doped much more heavily than the other, so that only what happens on the lower-doped side matters. This is because the minority-carrier density on the heavily doped side is negligible. Once we are beyond the depletion region, there is hardly any voltage drop, and the minority electrons (holes) must *diffuse* through the nearly flat-band portion of the $p(n)$ layer to the edge of the depletion region, where they are swept across the junction by the electrostatic field. We understand diffusion to be a random process, with the carriers migrating from a region with relatively high minority-carrier density (the interior of a doped layer) to a region with lower minority density (the edge of the depletion region). In the diffusion process, each carrier is buffeted by multiple

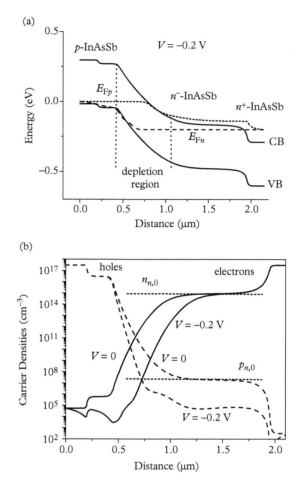

**Figure 14.4** *Band diagram (a) and carrier concentrations (b) for the homojunction photovoltaic detector of Fig. 14.3 under a reverse bias of $-0.2$ V and operating at 130 K.*

collisions that erase all memory of its history. For this reason, the *ratio* of minority-carrier densities at any two points depends only on their separation, but *not* on the distance from the depletion region. This is only possible when the carrier density varies exponentially with position. If we assume the junction is at $x = 0$, the edge of the depletion region is at $x = W$. The minority-carrier density $n_{min}$ ($n_p$ for electrons in a $p$-type region or $p_n$ for holes in an $n$-type region) far away from the junction ($x \gg W$) must revert to its equilibrium value $n_{min,0}$. If the reverse bias is high enough to remove all the carriers that reach the edge of the depletion region, $n_{min}(x)$ has the form

$$n_{min} = n_{min,0} \left[ 1 - \exp\left( -\frac{x - W}{L_d} \right) \right] \tag{14.20}$$

where $L_d$ is the *diffusion length*; i.e., the characteristic distance over which the minority carriers can diffuse.

The diffusion length increases when the carriers stick around longer (longer lifetime $\tau$) and move faster despite collisions (higher mobility $\mu$). To make the dimensions work out, we multiply $\tau\mu$ by a quantity with units of voltage, and then take the square root. The obvious choice for the multiplicative factor is the thermal voltage $k_B T/e$, which turns out to be correct for non-degenerate carriers. Therefore, the diffusion length in Eq. (14.20) is given by

$$L_d = \sqrt{\frac{\mu \tau k_B T}{e}} \tag{14.21}$$

Since a non-degenerate carrier density depends exponentially on the quasi-Fermi level, which shifts with applied voltage, we can guess that $n_{min}(V) - n_{min}(V = 0)$ should be proportional to $\exp(eV/k_B T)$. By definition, we should also have $n_{min}(V = 0) = n_{min,0}$. Using Eq. (14.20), the full expression for minority-carrier density outside the depletion region $(x > W)$ becomes

$$n_{min}(V, x) = n_{min,0} \left[ 1 + \left( \exp\left( \frac{eV}{k_B T} \right) - 1 \right) \exp\left( -\frac{x - W}{L_d} \right) \right] \tag{14.22}$$

We obtained this standard result with minimal effort. The standard route to Eq. (14.22) is to solve a second-order differential equation that represents pure carrier diffusion in the absence of an electric field or a carrier-generation source:

$$L_d^2 \frac{\partial^2 \Delta n}{\partial x^2} = \Delta n \tag{14.23}$$

where $\Delta n(x) \equiv n_{min}(x) - n_{min,0}$, with the boundary condition $n_{min}(V, W) = \exp\left( \frac{eV}{k_B T} \right) n_{min,0}$ at $x = W$. We notice that an applied voltage of only $V \approx -3k_B T/e$ saturates the dark current. Operation at a low reverse bias is consistent with the photovoltaic mode of detector operation.

This brief introduction to $p$–$n$ junctions should be sufficient to understand the basic operating characteristics of photovoltaic detectors. As discussed earlier, these detectors rely on either a majority-carrier barrier or the built-in voltage of a $p$–$n$ junction. We will use the junction as our example, but similar principles apply to both. The absorbed optical flux generates excess carriers (well below the doping levels), which in turn produce a flow of photocurrent. The two are not necessarily equal because carriers generated far away from the junction $(x > L_d)$ recombine before they can diffuse far enough to be collected across the junction. Any minority carrier that does reach the junction (or is photoexcited in the depletion region) is swept across for easy collection. Once in the region where it is the majority carrier, it can "nudge" other carriers of the same type, so recombination no longer prevents the photocurrent from flowing.

Of course, the absorbed flux is not uniform, but drops off exponentially with distance in the absorber according to Beer's law. Therefore, the optical flux $\Phi$ absorbed in a small slice at position $x$ is given by $\Phi \alpha \exp(-\alpha x)$. In quasi-equilibrium, this flux is equal to the excess carrier density divided by the carrier lifetime. Equivalently, the excess carrier density due to the absorbed light is equal to the absorbed flux times the minority-carrier lifetime. Therefore, in the presence of illumination, we can replace Eq. (14.23) with

$$L_d^2 \frac{\partial^2 \Delta n}{\partial x^2} + \Phi \alpha \tau \ \exp(-\alpha x) = \Delta n \tag{14.24}$$

For simplicity, we ignore the boundary condition at large $x$ by assuming that the absorber extends infinitely far from the junction. We guess that the full solution under illumination includes the terms

$$n_{min}(x) = n_{min,0} \left[ 1 - \exp\left( -\frac{x - W}{L_d} \right) \right] + C_1 \exp\left( -\frac{x - W}{L_d} \right) + C_2 \exp(-\alpha x) \tag{14.25}$$

where the constants $C_1$ and $C_2$ are determined by the boundary conditions and Eq. (14.22). The first two terms have the exponential dependence required by Eq. (14.23), and the last term follows the variation of the light intensity with distance, i.e., the inhomogeneous term in Eq. (14.24). We separate the first and second terms so that Eq. (14.25) reduces to Eq. (14.22) when the illumination is turned off and $V \ll -k_B T/e$.

The boundary condition at the edge of the depletion region is $n_{min}(x = W, V \ll -k_B T/e) \approx 0$. Substituting into Eq. (14.25), we express $C_1$ in terms of $C_2$ via $C_1 = -C_2 \exp(-\alpha W)$:

$$n_{min}(x) = n_{min,0} \left[ 1 - \exp\left( -\frac{x - W}{L_d} \right) \right] - C_2 e^{-\alpha W} \exp\left( -\frac{x - W}{L_d} \right) + C_2 \exp(-\alpha x) \tag{14.26}$$

Inserting this expression into Eq. (14.24), the first two homogeneous terms disappear, and we obtain an expression for $C_2$:

$$C_2 \left( 1 - \alpha^2 L_d^2 \right) \exp(-\alpha x) = \Phi \alpha \tau \exp(-\alpha x) \tag{14.27}$$

which leads to

$$C_2 = \frac{\Phi \alpha \tau}{1 - \alpha^2 L_d^2} \tag{14.28}$$

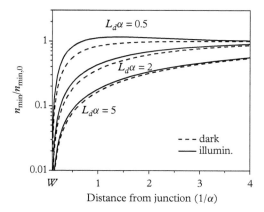

**Figure 14.5** *Spatial variation of the minority-carrier density in quasi-neutral regions with low fields and diffusion-dominated transport in the dark and under illumination. We assume* $W = 0.1/\alpha$, *and* $\Phi\alpha\tau e^{-\alpha W} = n_{min,0}$ *for illustration purposes for several values of* $L_d$. *Distances are normalized to the absorption depth.*

Upon substitution into Eq. (14.26), we finally arrive at

$$n_{min}(x) = n_{min,0}\left[1 - \exp\left(-\frac{x-W}{L_d}\right)\right] + \frac{\Phi\alpha\tau}{\alpha^2 L_d^2 - 1}e^{-\alpha W}$$
$$\times \left[\exp\left(-\frac{x-W}{L_d}\right) - \exp\left(-\alpha\left(x-W\right)\right)\right] \qquad (14.29)$$

Figure 14.5 illustrates typical spatial dependences of the minority-carrier density in the dark and under illumination.

Using Eq. (14.22), we can derive both the *diffusion-limited* photocurrent and the *diffusion-limited* dark current. Since diffusion is driven by a concentration gradient, it is proportional to the derivative of the carrier density along the photodetector axis (or its gradient if the variation is along more than one direction). Assuming that the junction is along the $x$ axis and that carriers are uniform in the other two dimensions, the quantity of interest is the current density per unit area orthogonal to the junction. The dimensionally correct form consistent with Eq. (14.23) is simply

$$\mathcal{J}_d = -k_B T\mu\frac{\partial n_{min}}{\partial x} \qquad (14.30)$$

where we assume that the current is due to holes, with the negative sign indicating flow in the direction of a decreasing concentration gradient. We could also have derived Eq. (14.29) by making the current continuity condition with recombination, $\frac{\partial \mathcal{J}_d}{\partial x} = \frac{\Delta n}{\tau}$, consistent with Eq. (14.23). Here the relevant mobility is that for *minority carriers*, and the negative sign corresponding to the flow of negatively charged electrons is flipped for a minority hole current. Since the current is continuous, we can evaluate it at the edge of the depletion region, where only minority carriers contribute. Substituting Eq. (14.29) into Eq. (14.30) and evaluating at $x = W$, we obtain for $V \ll -k_B T/e$

$$\mathcal{J}_d|_{x=W} = \frac{k_B T \mu \, n_{min,0}}{L_d} + e\Phi \frac{\alpha L_d}{1 + \alpha L_d} e^{-\alpha W} \qquad (14.31)$$

When the applied bias does not satisfy the condition $V \ll -k_B T/e$, the first term is proportional to $n_{min}(V, W) - n_{min,0}$, as given by Eq. (14.22). If the junction is not very asymmetric, we should also add a term originating from the other side of the junction, which is completely analogous to the first term of Eq. (14.31). Note that the currents from the opposite sides of the junction are due to different minority-carrier types: electrons from the $p$ side and holes from the $n$ side, and that the diffusion length can be different in distinct regions of the device.

Within the approximation of a thick absorber $L_a \gg L_d$, which we made in deriving Eq. (14.29), the first term of Eq. (14.31) represents the diffusion current density in the absence of illumination ("in the dark") $\mathcal{J}_{d,dark}$. The second term is the diffusion-limited photocurrent density, which represents the absorbed flux that can be collected at the junction. To make Eq. (14.31) suitable for comparison to experimental data, we recall that the equilibrium minority-carrier density is close to the density of ionized dopants. For non-degenerate statistics, the product of the equilibrium electron and hole densities is independent of the Fermi level position and equal to $n_i^2$, which means that $n_{min,0} \approx n_i^2/N_D$. When the region from which minority carriers are collected is heavily doped, this expression needs to account for saturation of the increase in the majority-carrier density with Fermi level that is the hallmark of degeneracy.

In the opposite limit of an absorber much thinner than the diffusion length ($L_a \ll L_d$), the diffusion current depends on what happens on the other side of the absorber. According to Eq. (14.30), the diffusion current scales with the change in the minority-carrier density between the edge of the depletion region ($x = W$) and the other side of the absorber ($x = W + L_a$). If $n_{min}(x = W + L_a)$ is fixed, $n_{min}(x = W) \to 0$ in reverse bias, and the gradient (and diffusion current) scale with $1/L_a$ for $L_a \ll L_d$. However, if we can continue to extract minority carriers at $x = W + L_a$, the gradient need not become large. If we also arrange for the minority-carrier density in the region with $x > W + L_a$ to be much smaller than in the absorber ($W < x < W + L_a$), we can neglect the former contribution to the diffusion current. As a result, the diffusion current decreases by a factor of $L_a/L_d$ compared to the first term in Eq. (14.31). This means we can collect nearly all the carriers, but with the absorber thickness rather than the diffusion length limiting how many. The dark current asymptotically approaches the two forms

$$|\mathcal{J}_{d,dark}| = \frac{k_B T \mu}{L_d} \frac{n_i^2}{N_D}, \quad L_a \gg L_d \qquad (14.32a)$$

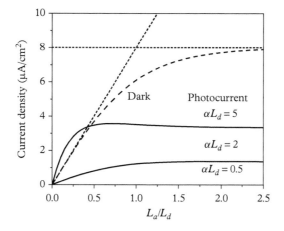

**Figure 14.6** *Dark-current density and photocurrent density versus absorber thickness normalized to the diffusion length, along with the limiting forms for the dark-current density. We assume $W = 0.1/\alpha$, $n_{min,0} = 10^{10}$ cm$^{-3}$, and $e\Phi e^{-\alpha W} = \frac{k_B T \mu}{2 L_d} \frac{n_i^2}{N_D}$ for illustration purposes. The photocurrent curves are shown for several values of the product of absorption coefficient and diffusion depth. Full expressions for the photocurrent (not shown in the text) are used. All of the photocurrent and dark-current generation is assumed to take place in the quasi-neutral region; i.e., the G–R contribution is set to zero. Short-dashed curves correspond to the limiting forms in Eq. (14.32).*

$$\left| \mathcal{J}_{d,dark} \right| = \frac{e L_a}{\tau} \frac{n_i^2}{N_D}, \quad L_a \ll L_d \tag{14.32b}$$

With the dark current varying as $\tanh(L_a/L_d)$, the two forms in Eq. (14.32) are collectively accurate to within $\approx 25\%$ for any absorber thickness. Figure 14.6 plots the resulting variation of dark-current density with relative absorber thickness, along with the estimates from Eq. (14.32) for comparison. The photocurrent follows a similar dependence, which is also plotted in Fig. 14.6.

In practical devices, the dark current does not come solely from diffusion. However, this is the *only fundamental contribution*, because the diffusion of thermally generated carriers is physically indistinguishable from the diffusion of photogenerated carriers. In other words, the two terms in Eq. (14.31) cannot be separated, hence we cannot suppress the diffusion dark current and maintain high quantum efficiency. Dark current can also arise from thermal generation in the depletion region (*generation–recombination current*) or on exposed surfaces. In a narrow-gap or heavily doped device, carriers can even quantum-mechanically tunnel from one side of the junction to the other. Much of the day-to-day research on IR photodetectors aims at minimizing these non-fundamental contributions.

The photocurrent in Eq. (14.31) approaches the absorbed flux $\Phi e^{-\alpha W}$ when the diffusion length significantly exceeds the absorption depth. This term should be multiplied by $1 - R$ if the device's top surface reflects some of the incident beam. In general, the flux in the absorber can be derived more rigorously from an electromagnetic model. If the depletion region has the same gap as the absorber, it can contribute an additional photocurrent of up to $1 - e^{-\alpha W}$ of the incident flux. However, this extra absorption comes at the price of rapid thermal generation in the depletion region. If the equilibrium minority-carrier density is given by $n_{min,0} = \frac{n_i^2}{N_D} \ll n_i$ because $N_D \gg n_i$, the extra generation–recombination (G–R) dark current becomes

$$\left| \mathcal{J}_{gr,dark} \right| \approx \frac{e n_i W}{\tau} \tag{14.33}$$

and can easily dominate the diffusion dark current of Eq. (14.32).

We can extend Eq. (14.31) to an absorber with finite thickness, but we will not reproduce the derivation here. For a thick absorber, the quantum efficiency due to absorption in the quasi-neutral region only becomes

$$\eta_a = \frac{\alpha L_d}{1 + \alpha L_d} e^{-\alpha W} \tag{14.34}$$

In the opposite limit of $L_a \ll L_d$, we multiply by $L_a/L_d$ as before, to account for the limited collection. If $\alpha L_d \ll 1$ as well, the quantum efficiency has a particularly simple form:

$$\eta_a \approx \alpha L_a e^{-\alpha W} \tag{14.35}$$

If the top surface does not transmit all the light, we include an additional factor of $1 - R$. Most practical photodetectors are anti-reflection coated to ensure $R \approx 0$.

Figure 14.6 compares these simple expressions to a full solution for different absorber thicknesses (relative to $L_d$ and $\alpha$). It is not necessary to increase the absorber thickness much beyond $L_a \approx L_d$ because the quantum efficiency saturates. The dark-current density from Eq. (14.32b) is then

$$\left| \mathcal{J}_{d,dark} \right| \approx \frac{e L_d}{\tau} \frac{n_i^2}{N_D} \tag{14.36}$$

If the experimental dark current is higher, it is likely that generation–recombination in Eq. (14.33) or tunneling contributes (see Section 14.3 for a fuller discussion).

So far, we have assumed that the carrier lifetime $\tau$ is a constant. In fact, we know from Section 12.2 that this is only true when it is controlled by Shockley–Read–Hall processes from defects that are not affected by the doping. If radiative recombination is weak, the Auger lifetime varies approximately inversely with the square of the carrier density. Therefore, instead of the inverse dependence of $\mathcal{J}_{d,dark}$ on $N_D$, it may vary linearly. In some devices, there is an optimum value of the doping density that minimizes the diffusion dark current. If Auger recombination dominates, the G–R dark current in Eq. (14.33) is not necessarily higher than the diffusion dark current $|\mathcal{J}_{d,dark}| \propto N_D$.

Even though the photovoltaic detector can operate at zero bias (or at a bias of a few $k_B T/e$), this is not always possible in practice. The problem is that in attempting to block the majority carriers, we may inadvertently impede the transport of minority carriers. One example is a structure similar to Fig. 14.2, but with a small valence band offset (as well as a large conduction band offset). To collect minority carriers, we may then need to apply a much higher reverse bias to overcome the minority-carrier barrier. This situation is described quantitatively in Section 14.3.

In contrast to semiconductor lasers, the carrier densities in photodetectors are quite low, so Auger recombination dominates only when there are few effective traps, or the bandgap is narrow and the operating temperature relatively high. Radiative recombination may also be weak if the absorber is thick for high quantum efficiency, which causes most of the emitted light to be re-absorbed before it reaches the surface. Even if not absorbed in the first pass, it may hit the surface at an angle larger than that for total internal reflection (see Section 12.7), and make additional passes through the absorber. Overall, radiative recombination rarely limits the carrier lifetime in a photodiode with a thick absorber. It follows that the main battle in developing a mid-wave ($\lambda$ = 3–5 µm) or long-wave (8–12 µm) IR detector with high quantum efficiency and low dark current is to suppress Shockley–Read–Hall recombination as well as to obtain a long carrier lifetime $\tau$.

With $\tau_{n0} \approx \tau_{p0} \equiv \tau_0$, Eq. (12.39) for the SRH recombination rate simplifies to

$$R_{SRH} \approx \frac{\left(np - n_i^2\right)}{(n+p+n_1+p_1)\tau_0} \tag{14.37}$$

This process is most effective when $n_1 \approx p_1 \approx n_i$, i.e., when the traps are near the middle of the gap. The rate is then

$$R_{SRH} \approx \frac{\left(np - n_i^2\right)}{(n+p+2n_i)\tau_0} \tag{14.38}$$

The electron and hole densities at the junction vary with position, and so does the recombination rate in Eq. (14.37). The net generation rate under reverse bias is determined by an integral over position and is dominated by the depletion region, where carrier densities are lower than the intrinsic density: $\int R_{SRH} dx \approx n_i W(V)/(2\tau_0)$. The voltage-dependent $W$ can be approximated using Eq. (14.19), but also could be *defined* to be proportional to this integral. It could then be used in Eq. (14.33) with $\tau = 2\tau_0$. On the other hand, for diffusion current in the quasi-neutral region, the majority-carrier density is much larger than the minority-carrier density, so the relevant lifetime is $\tau = \tau_0$.

## 14.3   Majority-Carrier Barrier Structures

We already know from Section 14.2 that a photodetector that blocks the flow of majority carriers is not expected to have any gain. Both the photocurrent and dark current are due to the generation of minority carriers in the absorber, by optical means for one and by thermal excitation for the other. The *nBn* detector of Fig. 14.2 is the best-known example of a majority-carrier barrier structure. It uses bulk-like layers of InAsSb as the absorber and contact layers, and a 250-nm-thick layer of AlAsSb as the barrier that blocks the flow of majority electrons (but not minority holes). All of these layers can be lattice-matched to GaSb (see Chapter 7). The doping is higher in the contact regions at the edges of the devices and low in the barrier, so that ideally any applied bias falls over the barrier. In fact, we will see later with Fig. 14.8 that we prefer the barrier to be lightly *p*-type (rather than *n*-type) because the large conduction-band offset combined with a significant electron density can bend the bands downward. Though no junction is present, the diffusion currents are analogous to a photodiode and given by Eqs. (14.31) and (14.32). This is because minority carriers are depleted at the edge of the absorber in reverse bias. We may also flip all the polarities and band alignments, so the barrier blocks holes rather than electrons, and construct a *pBp* structure instead of *nBn*. As shown in Chapters 6 and 7, the selection of bulk materials with large valence-band offsets and narrow gaps is very limited, so the *pBp* structure is most often used with superlattices (see Section 14.4).

Why might the majority-carrier barrier structure be preferable to a *p–n* junction? A comparison of Figs. 14.2 and 14.3 shows that a good *nBn* design has no depletion region with very low carrier densities. The advantage is that the G–R dark current in the ideal *nBn* structure is suppressed. To illustrate, we calculate the dark-current–voltage characteristics for the structures of Figs. 14.2 and 14.3 using the standard drift–diffusion model that reduces to the analytical expressions in Section 14.2 in the appropriate limits. The model includes the diffusion current similar to Eq. (14.32) as well as the "drift" current proportional to the carrier density and electric field as in Eq. (14.2). The electric field is computed self-consistently from the Poisson equation, Eq. (14.15). For simplicity, we also assume negligible Auger recombination so that a temperature-independent lifetime of $\tau_0 = 100$ ns is due only to the Shockley–Read–Hall mechanism. The results of this reduced model are displayed in Fig. 14.7.

The dark-current density is much lower in the *nBn* structure at low temperatures and low voltages. This is because the factor $n_i/N_D \propto \exp\left(-E_{gB}/2k_B T\right)$ increases strongly with temperature, and we assume that the density of ionized dopants that contribute carriers remains the same at all temperatures (this may be true down to temperatures below 77 K). At higher applied biases, the depletion region begins to appear even in the *nBn* structure and extends deeper into the absorber, so the dark-current density rises with voltage due to the G–R contribution. Even though the fundamental dark-current limit is the same for both structures, the results

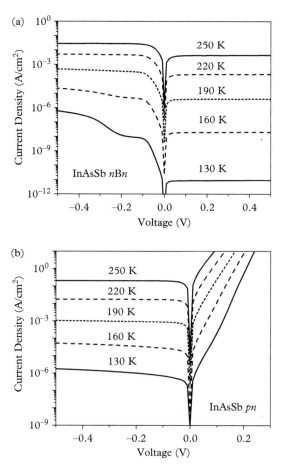

**Figure 14.7** *Dark-current–voltage characteristics for the majority-carrier barrier (a, with the band diagram shown in Fig. 14.2) and photodiode (b, with the band diagram shown in Fig. 14.3) devices at several operating temperatures. The tunneling currents are not taken into account in the drift–diffusion model used to calculate these curves. We also double the absorber thicknesses in Figs. 14.2 and 14.3 to make them closer to commercially useful photodetectors.*

do not converge even at high temperatures, in part because the intrinsic carrier density approaches $N_D = 10^{15}$ cm$^{-3}$ in the absorber, and the *p–n* junction no longer blocks majority-carrier transport well enough.

The structures also look very different in forward bias. The barriers in the *p–n* junction are gradually reduced, and the current flow increases exponentially with voltage, as expected from Eqs. (14.22) and (14.31). However, the majority carriers are still impeded in the *nBn* structure. In fact, the dark current actually drops because it becomes difficult to collect thermally generated carriers from the low-doped absorber.

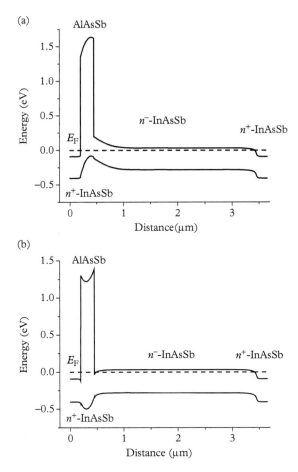

**Figure 14.8** *Band diagrams for the InAsSb/AlAsSb nBn structure with 3-μm-thick absorbers at zero bias and barrier doping levels that correspond to (a) a hole density of $1 \times 10^{16}$ cm$^{-3}$ and (b) an electron density of $1 \times 10^{16}$ cm$^{-3}$.*

The *nBn* structure may seem an ideal photovoltaic detector, but there are practical complications. For example, if the barrier doping is not under control and becomes too heavily *p*-type, the structure comes to resemble two back-to-back *p–n* junctions (see Fig. 14.8a). Nor do we want an *n*-type barrier, which blocks the flow of minority holes (see Fig. 14.8b). Neither situation is acceptable in a good photodetector, which means that the doping in the barrier needs to be just right.

As long as the depletion region is mostly in the barrier, the extra dark G–R current is proportional to $n_i \propto \exp\left(-E_{gB}/2k_B T\right)$, where $E_{gB}$ is the barrier (wider) gap. The key to good performance is to have so little depletion in the narrow-gap absorber that the diffusion current dominates. To diagnose the dominant dark-current component, we can plot the dark current versus $1/T$ (see Fig. 14.9)

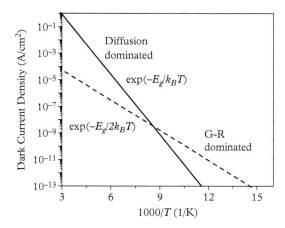

**Figure 14.9** *Typical variation of dark-current density with inverse temperature. The regions where diffusion and G–R currents dominate are indicated.*

and extract the activation energy $E_a$. For a diffusion-controlled device with dark-current scaling as $n_i^2 \propto \exp\left(-E_g/k_B T\right)$, as in Eq. (14.32), $E_a \approx E_g\,(T = 0)$. However, if the G–R current from the depletion region dominates, the dark current scales as $n_i$, and $E_a \approx \frac{E_g}{2}$ for a homojunction and $\frac{E_g}{2} < E_a < \frac{E_{gB}}{2}$ when the depletion region is mostly confined to the wide-gap barrier. Because of the faster increase in $n_i^2$, the diffusion (G–R) current dominates at higher (lower) temperatures (and tunneling at very low temperatures).

The *nBn* structure of Fig. 14.2 has another basic defect in that some of the incident light will be absorbed in the narrow-gap contact layer before it reaches the other side of the barrier. Of course, we could illuminate from the other side of the device, but this is not always convenient. So can we make a contact with a wider gap? Unfortunately, that tends to block the flow of minority carriers because the conduction bands will be nearly aligned at zero bias.

By doping the barrier *p*-type, we have now come nearly full circle between the *p–n* junction and *nBn* architectures. Can we combine the best features of both structures to overcome the problems with unipolar majority-carrier barriers (*nBn* and *pBp*)? As before, we can attempt to confine the depletion region to a wide-gap layer next to the absorber. But instead of transitioning to the narrower-gap *n*-type contact on the other side, we can continue with a wider-gap *p* region. Of course, the barrier should only block majority carriers and leave minority carriers unimpeded. This means that for an *n*-type absorber the entire discontinuity should occur in the conduction band rather than valence band. This ideal alignment of a *heterojunction* photodiode is illustrated in Fig. 14.10.

Unfortunately, we do not have a very wide palette of bulk materials to choose from, as mentioned earlier. To cut the knot, we can take advantage of superlattices as effective absorbing and minority-carrier-blocking materials with tailored band

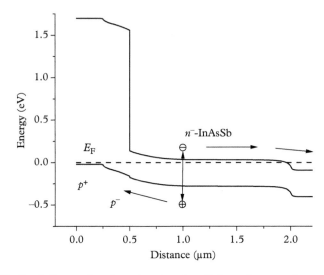

**Figure 14.10** *Band diagram for a heterojunction photodiode analogous to the majority-carrier-barrier structure of Fig. 14.2. The valence-band offset is zero to allow unimpeded hole transport. This cannot be achieved using lattice-matched AlAsSb, but potentially possible using superlattice barrier p regions. Alternatively, superlattices with the desired gaps can be used in both the absorber and the p region.*

offsets and energy gaps (see Section 14.4). For now we outline the main features of the heterojunction device.

For the structure of Fig. 14.10, we would like to dope the wide-gap region next to the absorber much more lightly than the absorber so that most of the depletion region sees a wider gap (by analogy with the *nBn* structure). What is the optimal position for the junction? If the wide-gap region is doped *p*-type, the depletion region is more likely to extend into the absorber than if the junction is shifted entirely to the wide-gap region. That means the wide-gap region next to the absorber should be *n*-type, so we can call the structure $pB_nn$. The $pB_nn$ and $pB_pn$ dark-current characteristics are compared in Fig. 14.11. Clearly, the $pB_nn$ structure has lower G–R currents for all absolute doping levels in the barrier. This is because the effective junction is shifted away from the narrow-gap absorber, and nearly all the voltage falls in the wide-gap barrier, where G–R is reduced exponentially with gap. The higher the doping in the barrier relative to the doping in the absorber, the wider the region with suppressed G–R dark current under reverse bias. Of course, when the doping in the barrier is very low, it does not matter as much whether it is *p*- or *n*-type.

If the absorber is *p*-type, all the polarities can be reversed without changing the basic physics; i.e., the $pB_nn$ and $pB_pn$ structures are transformed into $nB_pp$ and $nB_np$, respectively. We may prefer a *p*-type absorber because the electron mobility is much higher than the hole mobility (see also Section 14.4). We could also grade

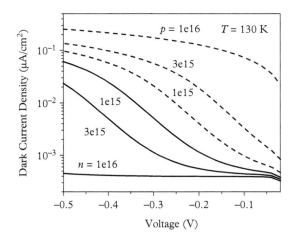

**Figure 14.11** *Dark-current density as a function of reverse bias for pB$_n$n (solid curves) and pB$_p$n (dashed curves) heterojunction photodiodes with various barrier doping levels. The absorber doping is $1 \times 10^{16}$ cm$^{-3}$, minority-carrier lifetime is 100 ns, and all dopants are assumed to be activated so that the doping density is equal to the carrier density.*

the gap of the $B_n$ region, starting from the edge of the absorber, on the assumption that a series of small barriers is not as effective at blocking minority carriers as a single, much larger barrier.

What happens when we do have a minority-carrier barrier? We consider the $nB_np$ structure with a $p$-doped absorber with an effective conduction density of states $N_{c+}$, and a wider-gap $B_n$ region with an effective conduction density of states $N_{c-}$ (here and below, the plus sign stands for the narrow-gap side of the interface, and the minus sign for the wide-gap side). Because the $B_n$ region is very lightly doped, according to Fig. 14.11 it does not matter much whether it is $p$-type or $n$-type. Again, the arguments apply with equal force to an $n$-type narrow-gap absorber once we interchange the symbols referring to electrons and holes (donors and acceptors). At the interface between the absorber and barrier, the conduction-band discontinuity $\Delta E_c$ blocks the flow of minority electrons. Because the Fermi level is the same on both sides of the interface *at zero bias*, the electron density on the narrow-gap side $n^+$ is connected to the density on the wider-gap side $n^-$ via the relationship $n^- = n^+ \frac{N_{c-}}{N_{c+}} e^{-\Delta E_c / k_B T}$. We neglect the fact that the $B_n$ region is usually followed by a wider-gap $n$ region (see Fig. 14.12a), because it does not affect the calculated dark-current density.

The current density under bias is proportional to the imbalance between the two electron densities. It is also proportional to the average electron velocity normal to the interface, which we might approximate using Boltzmann statistics for a parabolic band:

$$v_n = \frac{\int v_z e^{-m_e v_z^2/(2k_B T)}\, dv_z}{\int e^{-m_e v_z^2/(2k_B T)}\, dv_z} = \left( \frac{k_B T}{m_e} \right)^{1/2} \left( \frac{1}{2\pi} \right)^{1/2} \tag{14.39}$$

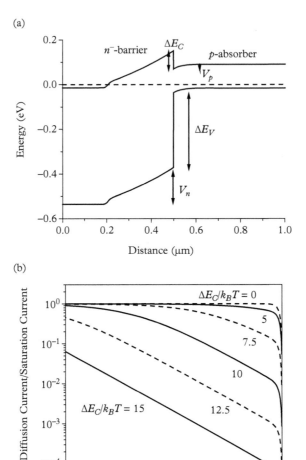

**Figure 14.12** *(a) Band diagram and (b) "diffusion" current normalized to the saturation current versus reverse bias voltage, as calculated from Eq. (14.46) for a $nB_pp$ heterojunction diode with a small conduction-band barrier between the wide-gap and narrow-gap regions. We take $r\frac{N_{c+}}{N_{c-}} = 0.1$, $\frac{N_D}{N_A+N_D} = 0.1$, $eV_{bi}/k_BT = 30$, and show the results in (b) for several relative barrier heights $\Delta E_c/k_BT$.*

The current density across the interface then takes the form

$$\mathcal{J}_{int} = ev_n\left(n^- - n^+\frac{N_{c-}}{N_{c+}}e^{-\Delta E_c/k_BT}\right) \tag{14.40}$$

Now we would like to express the electron densities at the interface in terms of the equilibrium electron densities on the two sides of the junction. If the $B_p$ and

$n$ regions are the same material with the same energy gap, we can relate all the densities to the electron density in that layer, which is approximately equal to the density of ionized dopants $N_D$. Figure 14.12a shows that the difference in the band-edge positions of the $n$ region and at the $B_p/p$ interface is equal to $V_n$, which is the built-in voltage minus any applied voltage dropped across the wider-gap layers. Therefore, for non-degenerate electrons we can write $n^- \approx N_D e^{-eV_n/k_B T}$. Similarly, on the $p$ side we have $n^+ \approx n_p e^{eV_p/k_B T}$, where $n_p(V)$ is the (minority) electron density at the edge of the depletion region in the $p$-type absorber. Because the conduction-band edge shifts by $V_{bi}$ in the transition from $n$ region to $p$ region, the equilibrium electron density $n_{p,0} = n_p(V=0)$ is

$$n_{p,0} \approx n_i^2/N_A \approx \frac{N_{c+}}{N_{c-}} N_D e^{-(eV_{bi}-\Delta E_c)/k_B T} \tag{14.41}$$

The electron density at the edge of the depletion region on the $p$ side follows from Eq. (14.22). As before, the total potential dropping over the device is the sum of the built-in and applied voltages, i.e., $V_n + V_p = V_{bi} - V$, with the $V_n/V_p$ ratio given by Eq. (14.18), and $n^+ \approx n_i^2/N_A e^{eV_p/k_B T}$. Equation (14.40) then becomes

$$\mathcal{J}_{int} = ev_n \left( N_D e^{-eV_n/k_B T} - n_p \frac{N_{c-}}{N_{c+}} e^{(eV_p-\Delta E_c)/k_B T} \right) \tag{14.42}$$

Since the current must be continuous throughout the structure, this is also the diffusion-limited value at the edge of the depletion region on the $p$ side. We can write it as a version of Eq. (14.36) generalized to any bias:

$$\mathcal{J}_d = \frac{eL_d \left( n_p - n_i^2/N_A \right)}{\tau} \tag{14.43}$$

As required, this expression reduces to zero in equilibrium where $n_p = n_{p,0} = n_i^2/N_A$, and saturates in reverse bias where $n_p \approx 0$.

Now we solve for $n_p(V)$ by setting $\mathcal{J}_d = \mathcal{J}_{int}$ and defining the ratio of the "diffusion" velocity to the average thermal velocity $r \equiv L_d/(\tau v_n)$:

$$n_p \left( r + \frac{N_{c-}}{N_{c+}} e^{(eV_p-\Delta E_c)/k_B T} \right) = N_D e^{-eV_n/k_B T} \tag{14.44}$$

We multiply both sides by $\frac{N_{c+}}{N_{c-}} e^{-(eV_p-\Delta E_c)/k_B T}$ and use Eq. (14.41) to express all the electron densities in terms of $n_i^2/N_A$:

$$n_p = \frac{n_i^2/N_A \left( e^{eV/k_B T} + r \frac{N_{c+}}{N_{c-}} e^{(-eV_p+\Delta E_c)/k_B T} \right)}{1 + r \frac{N_{c+}}{N_{c-}} e^{(-eV_p+\Delta E_c)/k_B T}} \tag{14.45}$$

Substituting into Eq. (14.43), we finally obtain the desired expression for the current density:

$$\mathcal{J}_d = \frac{e\left(L_d/\tau\right)\left(n_i^2/N_A\right)\left(e^{eV/k_BT} - 1\right)}{1 + r\frac{N_{c+}}{N_{c-}}e^{(-eV_p+\Delta E_c)/k_BT}} \tag{14.46}$$

This looks like the diffusion-limited current density in a *p–n* junction divided by a correction term. The correction disappears when $\Delta E_c = 0$ and $r \ll 1$:

$$\mathcal{J}_d = e\left(L_d/\tau\right)\left(n_i^2/N_A\right)\left(e^{eV/k_BT} - 1\right) \tag{14.47}$$

The voltage dependence here reduces to the so-called "ideal diode" equation, which is a good approximation for the diffusion current in a diode based on a homojunction.

Equation (14.18) shows that $V_p = N_D/(N_A + N_D)(V_{bi} - V)$ based on the voltage division between the two sides. In this case, we must maintain $N_A \gg N_D$ in order to keep most of the depletion region on the wide-gap $B_n$ side of the junction and to suppress the G–R current, which means the voltage fraction on the narrow-gap *p* side of the junction is $V_p \ll V_{bi} - V$. In many photodetectors, it is also true that $r \ll 1$; i.e., diffusion is slow compared to the transport across the heterointerfaces. For a barrier small on the scale of $k_BT$, $\Delta E_c \lesssim k_BT$, the denominator reduces to unity, and we revert to Eq. (14.47). However, if $\Delta E_c \gg k_BT$ and $V = 0$, $r\frac{N_{c+}}{N_{c-}}e^{(-eV_p+\Delta E_c)/k_BT} \gg 1$, and the minority current is suppressed. Figure 14.12b illustrates that it takes a substantial reverse bias before this term becomes less than 1. Neglecting the small built-in voltage dropping over the *p* side, the voltage required for the current to converge to the saturation limit is

$$|V| \approx \frac{N_D + N_A}{N_D}\left[\frac{\Delta E_c}{e} - \frac{k_BT}{e}\ln\left(\frac{N_{c-}}{rN_{c+}}\right)\right] - V_{bi} \tag{14.48}$$

Remember that this expression is valid only when $\Delta E_c \gg k_BT$. Otherwise, the barrier is of little consequence to the device performance.

To summarize this discussion, if the ratio of doping densities on the two sides of the junction satisfies $\frac{N_A}{N_D} \gg 1$, only a small fraction of the applied voltage drops on the "right" side of the junction, i.e., before the band discontinuity is reached. Figure 14.12b shows that at $T = 77$ K, $\Delta E_c \lesssim 30$ meV does not require much extra voltage. However, with $\Delta E_c \sim 100$ meV the bias needed for minority carriers to move freely can be as high as $-0.5$ V. If the doping densities on the two sides of the junction are more balanced, minority carriers are not blocked as effectively. But then the G–R current looms as a potential problem if the temperature is not high enough (see Section 14.2). Of course, in practice we always wish to have $\Delta E_c \lesssim k_BT$, but the imperfect knowledge of band offsets and practical design limitations often conspire against us.

What if $\Delta E_c$ in Fig. 14.12a is negative, i.e., the conduction-band edge in the wide-gap material is *lower* than in the narrow-gap absorber? Obviously, there will be no barrier to the leftward minority-carrier flow, but that does not mean we are out of the woods yet. For long-wave detectors, the gap is small enough that the band discontinuity can be similar. In this case, we may observe strong interband tunneling (see Section 12.4) even with a low reverse bias, which leads to high dark current. Therefore, $|\Delta E_c|$ must lie within a narrow range of a few $k_B T$, particularly for long-wave detectors with absorber gaps of only 0.1–0.15 eV.

As mentioned earlier, these results apply (with a suitable swapping of symbols) to conduction-band offsets in $nB_np$, $nB_pp$, and $nB_pn$ devices, as well as to valence-band offsets in $pB_pn$, $pB_nn$, and $pB_np$ devices. Both the diffusion current and photocurrent are blocked by minority-carrier barriers, so Eq. (14.48) specifies the *operating bias* of the heterojunction photodiode (defined as the voltage needed to reach, say, 90% of the maximum possible quantum efficiency).

## 14.4   Comparison of Bulk and Type II Superlattice IR Detectors

For the past few decades, high-sensitivity photodetectors in the mid-wave ($\lambda = 3$–5 μm) and long-wave (8–12 μm) IR have been almost exclusively HgCdTe devices. As mentioned in Chapter 4, in terms of their band structure the HgCdTe and InAsSb alloys with the same gap are very nearly identical twins. The copious work on HgCdTe is due to a quirk of fate that nearly fixes its lattice constant over a wide range of gaps, all of which can be grown on the same CdZnTe substrates. In stark contrast, the InAsSb gap covers the 4–12 μm range, but the GaSb lattice constant is matched only at a single composition in the mid-wave.

One potential solution is to grow InAsSb on a metamorphic buffer with the desired lattice constant, to reduce the density of dislocations that would otherwise be too high for a good detector. We could even avoid a random alloy by growing a "digital" structure with the same gap, namely, a superlattice with layers so thin they do not confine the electrons and holes. These possibilities remain under study, but neither is currently used in practical detectors.

Of course, we do not need to grow or otherwise emulate a bulk alloy to have a very narrow energy gap. We saw in Chapters 10 and 11 that a type II superlattice such as InAs/Ga(In)Sb or InAs/InAsSb can do that perfectly well, while balancing the strain in the two layers and allowing epitaxial growth on GaSb with a minimal density of dislocations and other defects. We can view them as *effective materials*, with adjustable energy gaps and, to some extent, positions of the conduction- and valence-band edges. We can then slide the band edges to desired positions that prevent minority carriers from being blocked (see Fig. 14.12).

Section 10.2 already discussed the band structure properties of type II super-lattices with gaps in the range 0.1–0.4 eV. Here we will also use the 8-band **k·p** method with periodic boundary conditions with the interface terms set to 0 (see Fig. 8.6 for an estimate of the error incurred in this procedure). Recall that in a type II superlattice such as InAs/GaSb, the holes are confined to the material with the highest valence-band maximum ("hole well") and the electrons are mostly contained in the material with the lowest conduction band minimum ("electron well"), but penetrating into the other layers (see Fig. 10.8b). Therefore, the holes in a type II SL are essentially 2D, but the electrons retain most of their 3D nature.

The design goals for type II SLs suitable for photodetectors are: (1) to target a particular energy gap; (2) to maintain strain balance with respect to the GaSb substrate; (3) to maximize the absorption coefficient near the band edge; and (4) efficient minority-carrier transport along the growth direction, i.e., decent growth-direction mobility for the minority carriers. To achieve these goals, we must work with a few degrees of freedom, namely: (1) the width of the electron well with fixed composition (most commonly InAs); (2) the width of the hole well; (3) the composition of the hole well, e.g., $x$ in $Ga_{1-x}In_xSb$. For simplicity, we do not take advantage of variable interface bond type, e.g., InSb or GaAs-like bonds (see Fig. 8.5) because the hole-well composition can serve the same function.

Are these knobs sufficient to design a serviceable structure? We will see below that to maximize the absorption coefficient, the hole-well width must fall within a narrow range. Then we can adjust the InAs thickness to arrive at the desired gap. To compensate the tensile strain in that layer, we vary the In fraction in the GaInSb. The last knob is not available in the so-called "Ga-free" InAs/InAsSb type II SL. There the Sb fraction in the hole well is selected to maximize the absorption, and strain is compensated via the ratio between the InAs and InAsSb thicknesses. Unfortunately, we will see below that this approach does not guarantee strong absorption in the long-wave IR.

Why should the hole well in a type II SL be much thinner than the electron well? To answer this question, we plot in Fig. 14.13 the electron–hole overlap squared $|O_{eh}|^2$ for the CB1 and HH1 subbands for InAs/GaInSb SLs with energy gaps of 0.106 and 0.23 eV as a function of GaInSb thickness. We saw in Section 11.1 that the absorption coefficient near the band edge is proportional to $|O_{eh}|^2$. Fig. 14.13 also plots the ratio of the GaInSb thickness to the total period of the SL, $t_h/d$ at 77 K.

A thin GaInSb layer does not block electrons, so in that limit the CB1 envelope function remains nearly the same throughout the superlattice. But when the hole well becomes thick, the electron penetration is much less effective (see Fig. 10.8b for an intermediate case). On the other hand, the HH1 envelope function remains confined to the hole well for most detector structures of interest. Obviously, we need electrons and holes to overlap to have any interband transitions at all, so the strongest optical processes nearly always occur in or near the hole well. Because the electrons do not penetrate a thick layer as well, the hole well should remain thin. Even though the calculations in Fig. 14.13 imply that the GaInSb layers should be 15–20 Å thick, it turns out that the full evaluation leads to similar band-edge absorption coefficients as long as the hole well is *no wider* than 25–30 Å.

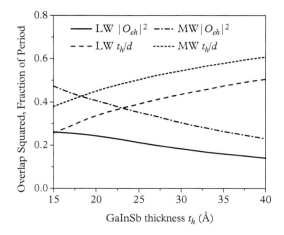

**Figure 14.13** *Electron–hole wavefunction overlap squared $|O_{eh}|^2$ (solid and dash-dotted curves) and the ratio of the Hole-well width to the total superlattice period $t_h/d$ (dashed and dotted curves) for InAs/GaInSb SLs with gaps of 0.106 and 0.23 eV at $T = 77$ K as a function of hole well width $t_h$. The In fraction in the hole well is adjusted to maintain strain balance to the GaSb substrate.*

This is because the wavefunction overlap is not the only quantity that enters the absorption coefficient. Equation (11.56) shows it is also proportional to the joint density of states, which is in turn dominated by the electron mass. In other words, we want the (in-plane) electron mass to be as large possible for the particular gap. Here the type II SLs shine by comparison with bulk HgCdTe and InAsSb with narrow gaps in the range 0.1–0.15 eV. Whereas the bulk electron mass scales with energy gap, the mass in the type II SL is determined by the InAs electron well, which has a much wider gap. Unfortunately, the type II SL's JDOS advantage peaks for long-wave absorbers with electron mass roughly twice that of a bulk alloy with the same gap, which also have a relatively small wavefunction overlap.

The final piece of the puzzle is that when the electron–hole overlaps and joint densities of states in two superlattices are similar, the absorption coefficient *per unit length* of the light at normal incidence, and polarized in the plane of the structure, is inversely proportional to the period (the absorption per period being approximately the same). Figure 14.14 shows absorption spectra that were calculated with full in-plane and out-of-plane dispersions and matrix elements for type II SLs with gaps of 0.106 and 0.27 eV, and bulk InAsSb with the same gaps. The trends agree well with our reasoning above. The absorption in the long-wave region is stronger in bulk InAsSb, despite its lower joint density of states, because $|O_{eh}|^2 \approx 0.20 - 0.25$ in the type II SLs.

The InAs/Ga(In)Sb superlattice can span a wide spectral range via simultaneous adjustments of the InAs thickness and the GaInSb composition, as discussed earlier. Unfortunately, in practice it suffers from a relatively short minority-carrier

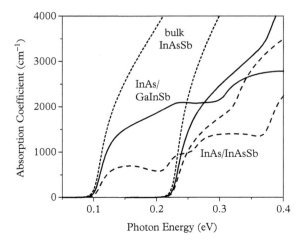

**Figure 14.14** *Absorption spectra for InAs/GaInSb (solid curves) and InAs/InAsSb (dashed curves) type II SLs with energy gaps of 0.106 and 0.23 eV for in-plane-polarized light at normal incidence. For comparison, the dotted curves show the results for bulk InAsSb alloys with the same energy gap. The absorption of out-of-plane polarized light is very weak near the band edge in type II SLs.*

lifetime dominated by SRH recombination at defects (rather than Auger or radiative processes that impose a more fundamental limit). For the relevant operating range of 60–140 K, the typical lifetime in a long-wave InAs/GaInSb SL is ~30 ns. This is much lower than the lifetime of a few μs in state-of-the-art epitaxially grown HgCdTe, and leads to higher dark-current densities. The shortfall is not for the lack of trying. Recently, interest shifted to the "Ga-free" InAs/InAsSb structures (see Section 10.2). At the low temperatures relevant to the photodetector operation, they boast lifetimes comparable to HgCdTe (several μs) for mid-wave absorbers and a few hundred ns for long-wave absorbers. At room temperature, Auger recombination dominates and the lifetime becomes much shorter. But of course, minority-carrier lifetime is not the only factor determining the dark-current density (see Eq. (14.32b)):

$$\left| \mathcal{J}_{d,dark} \right| = \frac{eL_a}{\tau} \frac{n_i^2}{N_D} \tag{14.49}$$

Therefore, the figure of merit should be the product of the minority-carrier lifetime and the doping density in the absorber (more precisely, the majority-carrier density, which is nearly the same at the relevant operating temperatures), $N_D\tau$. We can benefit from heavier doping as long as the lifetime does not vary more rapidly than the doping. However, data appears to indicate that the $N_D\tau$ product in InAs/Ga(In)Sb superlattices does, in fact, decrease for doping levels $> 10^{16}$ cm$^{-3}$ (and possibly for densities as low as $1–3\times10^{15}$ cm$^{-3}$ in InAs/InAsSb SLs).

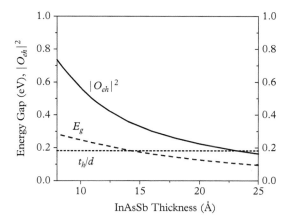

**Figure 14.15** *Electron–hole wavefunction overlap squared $|O_{eh}|^2$ and energy gap as a function of hole-well width $t_h$ for InAs/InAs$_{0.5}$Sb$_{0.5}$ SLs with a constant ratio of hole-well width to SL period $t_h/d$ (needed to compensate the strain) at $T = 77$ K.*

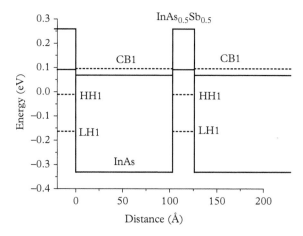

**Figure 14.16** *Band diagram for a 101 Å/25 Å long-wave InAs/InAs$_{0.5}$Sb$_{0.5}$ SL at $T = 77$ K.*

Given this lifetime advantage, why do we even bother discussing the InAs/Ga(In)Sb SL? We start with Fig. 14.15 that plots $|O_{eh}|^2$ as a function of energy gap for a long-wave InAs/InAs$_{0.5}$Sb$_{0.5}$ SL (see Fig. 14.16 for band profiles and subband energies of a representative structure). We assume an average Sb fraction of 50% in the hole well, which is near the maximum demonstrated experimentally, although the Sb profile appears to deviate substantially from a square wave. The high Sb fraction in the hole well leads to semimetallic overlap of the conduction band in InAs and the valence band in InAsSb, which maximizes the wavefunction overlap and absorption coefficient. However, we see that even with 50% Sb the maximum $|O_{eh}|^2$ in Fig. 14.15 is somewhat smaller than in Fig. 14.13 for a long-wave InAs/GaInSb SL.

The absorption coefficients for InAs/GaInSb and InAs/InAsSb SLs, plotted in Fig. 14.14, are inversely proportional to the period. That is, the absorption per unit length is stronger in a superlattice with shorter period when the gap and overlap are fixed. Therefore, the InAs/GaInSb SL (with a period of 64 Å) has nearly a factor of 2 advantage over the InAs/InAsSb SL (with a period of 126 Å). The difference is smaller in the mid-wave, but the InAs/GaInSb SL is still slightly thinner and more absorbing near the band edge.

Lest the reader think that lifetime and absorption coefficient are the only stories, consider that the longest lifetimes occur in unintentionally doped Ga-free superlattices, which turn out to be *n*-type because donor-like native defects dominate. Because we must rely on hole diffusion, thick InAs layers become a real burden. Not only do they reduce the electron–hole overlap, but they also block the hole transport. This is a concern because Eq. (14.34) tells us that the minority-carrier diffusion length must exceed the absorption depth, or the generated carriers end up recombining before they can produce photocurrent. Fortunately, mid-wave InAs/InAsSb superlattice absorbers have much thinner InAs layers.

To quantify the problem, we estimate the hole mass along the growth direction. A heavy mass usually leads to a poor mobility in the growth direction (see the brief mention of the conductivity mass in Section 10.2). It turns out we cannot accurately estimate the width of the HH1 miniband by looking only at $k_\parallel = 0$, because it broadens substantially for $k_\parallel > 0$. To fix this defect, we average over the in-plane dispersion, weighted by the Boltzmann factor $e^{-E(k_\parallel)/k_B T}$ relevant to the non-degenerate statistics of minority carriers, and plot the resulting conductivity mass in Fig. 14.17 for both electrons and holes in the InAs/GaInSb and InAs/InAsSb SLs. We find that the hole masses in the two types of superlattices are quite similar in the mid-wave, but that the Ga-free devices are at a marked disadvantage from a transport point of view in the long wave. It seems the electron well should be as thin as possible to maximize *both* the hole mobility and the absorption coefficient near the band edge.

Another consideration is that the unintentionally doped InAs/Ga(In)Sb SL tends to be *p*-type. Electrons as minority carriers are quite convenient because they are less confined and generally have much higher growth-direction mobility (see Fig. 14.17). Unfortunately, *p*-type type II SLs suffer from severe leakage at the mesa sidewalls that become inverted, i.e., *n*-type. Although this affliction is not incurable, a definitive, universal remedy has yet to emerge.

Because the hole mobility along the growth direction in a long-wave Ga-free superlattice is so low ($< 10$ cm$^2$/Vs is common) by semiconductor standards, it becomes even more important to maximize the lifetime so the photogenerated holes can diffuse farther than the absorber thickness ($\alpha L_d \geq 1$) and maintain a high quantum efficiency. Here the diffusion-limited dark-current density is given by

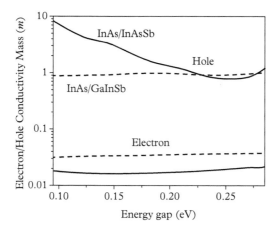

**Figure 14.17** *Growth-direction hole and electron conductivity masses for strain-balanced InAs/InAs$_{0.5}$Sb$_{0.5}$ and InAs/GaInSb type II SLs at T = 77 K as a function of energy gap. For the InAs/InAsSb SL, the ratio of layer thicknesses is kept the same, while the hole-well width in the InAs/GaInSb SL is kept at 25 Å, and the In composition is varied to compensate for the strain.*

Eq. (14.49), and we can reduce the G–R current using an $nB_np$ or $pB_pn$ configuration, depending on the absorber doping. For a $p$-type InAs/GaInSb SL, we need a hole barrier with near-zero conduction-band discontinuity that is also strain-balanced to the GaSb substrate. These requirements are nicely filled by an InAs/AlInSb SL, in which the In fraction in the AlInSb barrier controls the overall strain balance. Figure 14.18a shows the resulting band diagram.

It is less obvious how to find a good electron barrier with near-zero valence-band offset to complement an $n$-type Ga-free absorber, but band-structure calculations indicate that a simple downscaling of the absorber layers works well (e.g., reducing both layer thicknesses by a factor of 2 maintains the strain balance assuming the Sb composition does not change). However, in practice the unintentionally doped InAs/InAsSb is lightly $n$-type, which means we need to look for a good electron barrier if we do not wish to counter-dope and possibly sacrifice the carrier lifetime. Figure 14.18b shows the coupling of an $n$-type InAs/InAsSb SL absorber with an AlGaAsSb barrier, with quaternary composition that is lattice matched to GaSb and also chosen to match its valence band maximum to that of the SL.

Figure 14.19 shows calculations of the dark-current versus voltage character-istics at 77 K for both of the structures from Fig. 14.18. Assuming a minority electron mobility of 1200 cm$^2$/Vs and lifetime of 30 ns, the diffusion length in the InAs/GaInSb SL from Fig. 14.18a is approximately 5 µm. This implies it should be possible to extract nearly all the photoexcited minority electrons. For the InAs/InAsSb SL of Fig. 14.18b, we assume a much longer hole lifetime of 300 ns, but also a hole mobility of only 30 cm$^2$/Vs. Even though these values are again close to the maximum, they result in a hole diffusion length of only

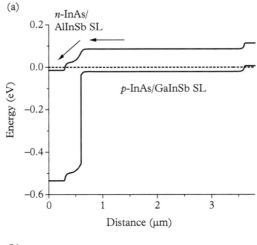

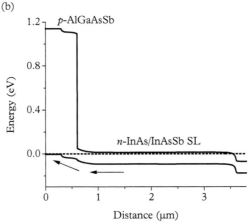

**Figure 14.18** *Band diagrams for (a) an nB$_n$p detector with an InAs/GaInSb SL absorber and InAs/AlInSb SL hole barrier and (b) a pB$_p$n detector with an InAs/InAsSb SL absorber and an Al$_{0.6}$Ga$_{0.4}$As$_{0.06}$Sb$_{0.94}$ electron barrier at T = 77 K. Both detectors have energy gaps of 0.106 eV. We assume that the absorber in (a) is doped p-type to $1 \times 10^{16}$ cm$^{-3}$ and in (b) n-type to $1 \times 10^{15}$ cm$^{-3}$, which happen to be representative background doping levels. The unintentionally doped barriers next to the absorber are assumed to have $3 \times 10^{15}$ holes (a) and electrons (b) in the limit of a thick layer in equilibrium. Both absorbers are 3 μm thick, which is less than the electron diffusion length in (a), but slightly exceeds the estimated hole diffusion length in (b), as discussed in the text.*

2.4 μm. We see from Fig. 14.19 that the dark-current density is a little lower in the Ga-free SL, but because we have trouble extracting minority carriers, this difference is not dispositive. The dark-current densities for both of these SLs are only a few times higher than state-of-the-art HgCdTe at low voltages. They are

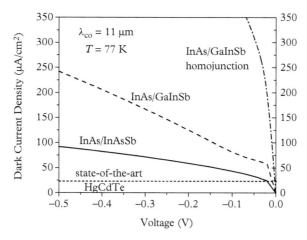

**Figure 14.19** *Dark-current density versus voltage for the two detector structures in Fig. 14.18 at T = 77 K. Shown for comparison is the state-of-the-art dark-current density in HgCdTe photodetectors with the same cut-off wavelength (gap) (horizontal dotted line) and the dark-current density versus voltage for a InAs/GaInSb SL homojunction (dash-dotted line). The calculations assume a minority-carrier lifetime of 30 ns and electron mobility of 1200 cm²/Vs for the InAs/GaInSb SL of Fig. 14.18a, and a hole lifetime of 300 ns and mobility of 30 cm²/Vs for the InAs/InAsSb SL of Fig. 14.18b.*

also much lower than in homojunction devices, even though G–R current comes to dominate at higher voltages (which we know because the current–voltage curves are not very flat). Residual G–R current is present because near the junction we cannot prevent some electric field from penetrating into the absorber, even when the barrier doping is less than in the absorber (see Section 14.3).

Are these two type II SLs the only game in town when it comes to long-wave III–V-based detectors? Apart from metamorphic designs, we can take advantage of the stronger band-edge absorption in the "W" structure (see Fig. 10.16), and similar ideas can be carried over to the Ga-free structures. However, the stronger absorption of a "W" design should be balanced against *even poorer* hole transport for an *n*-type absorber. It is unknown at this time whether a switch to a *p*-type Ga-free absorber can be successfully pulled off, e.g., by retaining a long minority-carrier lifetime and avoiding sidewall leakage.

So how does the performance of type II superlattice detectors compare with state-of-the-art HgCdTe photodiodes? As we already mentioned, the dark-current density in Fig. 14.19 approaches what HgCdTe can deliver today, with values of order 10 µA/cm² at 77 K in the long-wave (and similar results in the mid-wave at higher operating temperatures of 140–150 K). Because of poor hole transport, long-wave superlattices have not yet demonstrated high quantum efficiency, comparable to HgCdTe (>70%), in the long wave. The best reports of 30–35% correspond to $D^*$ of up to $10^{12}$ cm Hz$^{1/2}$/W, with lower values at longer wavelengths and higher temperatures, as expected from Eqs. (14.13b) and (14.32b), and the temperature dependence of the intrinsic carrier density in Eq. (4.81).

The HgCdTe technology has not stood still either, with reports of much lower dark-current densities achieved with thick, fully depleted absorbers. In practical terms, being able to grow type II SLs on much larger GaSb wafers than the CdZnTe substrates available for HgCdTe is a big selling point in its own right. Other advantages include lower cost and higher "operability"; i.e., a larger fraction of array elements operate successfully within a narrow range of performance metrics.

## 14.5   Interband Cascade Detectors

Have we overlooked any approaches to maximizing the detectivity $D^*$ in Eq. (14.13b)? The detectivity of a device operating at zero bias is proportional to $\sqrt{R_0 A}$. We also know from the discussion of ICLs that the voltage increases while the current decreases when multiple stages are stacked so the same current passes through the entire staircase (see Section 12.4), which should lead to a higher differential resistance. It seems that we can convert the ICL to an interband cascade detector (ICD), by operating the same structure at zero or reverse bias. A further wrinkle illustrated in Fig. 14.20 would be to replace the "W" quantum wells in each stage of the ICL with a type II InAs/GaInSb superlattice absorber (see Section 14.4). We may consider this path if the quantum efficiency of the conventional detector is low because the carriers cannot all be collected (see Eq. (14.34)), so we can come out ahead by splitting the absorber with complete carrier collection in each stage. Our goal here is to quantify any possible improvement.

As in the ICL, both contacts of the ICD are $n$-type, and holes are present only internally. For the ICD operated at zero bias, the conduction-band minimum in the chirped electron injector rises from left to right. On the left, the minimum is aligned with the top of the valence band in the absorbing region of the given stage, and connected to it via a thin hole injector. On the right, it is matched to the bottom of the conduction band in the absorbing region of the next stage. Electrons generated in the absorber face a large barrier on the right, but are free to move to the neighboring stage of the device on the left. Therefore, the electron-injection barrier in the ICD functions much like the barrier in the $nBn$ structure of Fig. 14.2. An electron that makes it to the valence band of the neighboring stage, via interband tunneling, can be excited back to the conduction band by absorbing an additional photon. In operating the ICD, we may sometimes want to apply a low reverse bias to promote carrier transport in the desired direction (see Section 14.3). The characteristics of an ICD can be calculated using the same

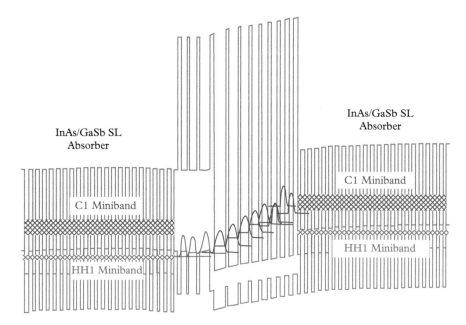

**Figure 14.20** *Schematic of the ICD structure operated close to zero bias. In this structure, the absorber in each stage is an InAs/GaSb type II superlattice. Reproduced from R. T. Hinkey and R. Q. Yang, J. Appl. Phys. **114**, 104,506 (2013), with the permission of AIP Publishing.*

principles as for an ICL, i.e., tabulating the in-plane dispersions and making assumptions about transport being faster than the recombination time within each stage.

The ICD is designed so that current can flow within a given stage only when a new electron–hole pair is created by photoexcitation or thermal generation. But because current cannot simply disappear in the middle of the structure, the same amount must flow through all the stages. The incident beam depletes as it passes through the structure, which means the stages farther away from the light source absorb fewer photons over the same thickness (assuming normal incidence). For this reason, we want thicker absorbers farther from the plane of incidence, to compensate for the reduced optical flux. In principle, we can always make the photocurrents in the different stages equal if we know the absorbed fraction per period of the superlattice (or per well) accurately. In practice, this can be done only approximately, and the observed current becomes a balancing act between the different optical fluxes and thermal generation rates in the various stages.

The ICD differs from conventional photovoltaic detectors in that multiple absorption events are needed to sustain an observable photocurrent. This leads to a low quantum efficiency for the ICD, but does not preclude achieving a higher detectivity than for a "single-stage" device with the same absorber, as we will see in the following. We might also add that the ICD is a perfect example of how the particular electrons generated by incident light are not necessarily the same ones that reach the contacts. In fact, the electrons generated in interior stages cannot possibly reach the contacts, because they undergo continual recycling and interband transitions within each successive stage.

What is the ideal absorbing region in an ICD? It is not a single "W" quantum well, since it absorbs < 1% of the incident light for a very low quantum efficiency. Nor do we want *too much* absorption within each stage, which would severely deplete the optical beam after propagation through only a few stages and remove the rationale for a cascade detector. In any event, the ICD is only attractive when $\alpha L_d \ll 1$, since it offers no advantage when carriers generated in a region thick enough to absorb most of the light are easily collected. Between those two limits, a good target might be a lightly doped type II SL (either two-layer or "W") that absorbs several percent of the light per stage. In some cases, the ICD must be processed from the same wafer material as the ICL because we would like for them to be a part of the same photonic integrated circuit. Because the light then propagates in the growth plane, nearly all the light can be absorbed even with a few stages.

What compensates for the comparatively low quantum efficiency of an ICD? The key is that while it needs multiple absorption events to create the photocurrent normally generated by a single absorption, multiple consecutive thermal gener-ation processes must also occur for any dark current to flow. The probability of a distinct thermal event in each stage decreases linearly with the number of stages. Hence the variance of the fluctuating dark current is $i_n^2 \propto 1/M$, where $M$ is the number of stages. Assuming an absorber thickness in each stage close to that for an ideal single-stage detector, the detectivity given by Eq. (14.8) varies inversely with the standard deviation of the dark-current noise, which means it is proportional to $\sqrt{M}$. However, we mentioned earlier that the entire concept is only worthwhile when $\eta_a \ll 1$ for the best single-stage device we can make (subject to some constraints). In that case the ICD's quantum efficiency is not much worse than in a single-stage detector, but the noise can be lower.

To express these ideas mathematically, we assume $\alpha L_d \ll 1$ and the quantum efficiency of an ICD is given by Eq. (14.35). The photocurrent can be estimated from the absorption in the *last* stage of the device, at which point the beam is the mostly depleted. Hence we replace $W$ with $(M-1)L_a$, where $L_a$ is now the length of each absorbing section (assumed to be constant for now):

$$\eta_M \approx \alpha L_a e^{-\alpha(M-1)L_a} \tag{14.50}$$

The dark-current density for a thin absorber in Eq. (14.32b) scales with $L_a$, but the noise scales with $\sqrt{L_a}/\sqrt{M}$, as we noted earlier. The ICD detectivity with $M$ stages is then proportional to

$$D_M^* \propto \frac{\alpha L_a}{\sqrt{L_a}} \sqrt{M} e^{-\alpha(M-1)L_a} \tag{14.51}$$

For $M \gg 1$, the exponent simplifies slightly to

$$D_M^* \propto \alpha \sqrt{L_a} \sqrt{M} e^{-\alpha M L_a} \tag{14.52}$$

The detectivity in Eq. (14.52) has a maximum as a function of $M$. Differentiating with respect to $M$ and setting the derivative to 0, the optimal number of stages is

$$M_{opt} = \frac{1}{2\alpha L_a} > \frac{1}{2\alpha L_d} \tag{14.53}$$

Substituting this result back into Eq. (14.52) yields

$$D_{M,opt}^* \propto \sqrt{\frac{\alpha}{2}} e^{-1/2} \tag{14.54}$$

On the other hand, the detectivity for the best possible single-stage device with the same absorber, doping, and other characteristics follows from Eq. (14.51) with $M = 1$ and $L_a \approx L_d$:

$$D_1^* \propto \frac{\alpha L_d}{\sqrt{L_d}} = \alpha \sqrt{L_d} \tag{14.55}$$

Now all that remains is to take the ratio of Eqs. (14.54) and (14.55):

$$\frac{D_M^*}{D_1^*} = \frac{e^{-1/2}}{\sqrt{2\alpha L_d}} \approx \frac{0.43}{\sqrt{\alpha L_d}} \tag{14.56}$$

As expected, the ICD detectivity is higher only when $\alpha L_d \ll 1$, and each stage is much thinner that the diffusion length. Figure 14.21 illustrates the behavior graphically, with the more general expressions for quantum efficiency and dark-current density, but still taking $L_a \ll L_d$. If $\alpha L_d = 0.1$, the detectivity for a multi-stage ICD is $\approx 40\%$ higher than for the simpler device. This is probably not worth the trouble, in view of the inevitable imperfections of practical ICDs. The scheme could become more attractive when $\alpha L_d$ is even smaller, but then requires 20–100 stages to converge to the maximum theoretical improvement in Eq. (14.56).

However, this is not yet the ultimate limit because we assumed that the photocurrent is limited by the absorption in the last stage. What if we follow an earlier suggestion and adjust the absorber thickness so that each stage absorbs the same amount of light? Working along similar lines, we derive

$$D_M^* \propto \frac{1 - e^{-\alpha \sum_m L_m}}{M} \sqrt{M} \sqrt{\sum_m L_m^{-1}} = \frac{1 - e^{-\alpha \sum_m L_m}}{\sqrt{M}} \sqrt{\sum_m L_m^{-1}} \tag{14.57}$$

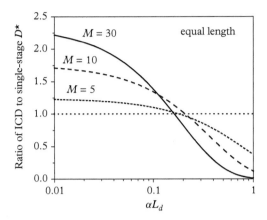

**Figure 14.21** *Ratio of optimal equal-length multistage and single-stage (dotted curve) detectivities for ICDs as a function of $\alpha L_d$, for designs with 5, 10, and 30 stages.*

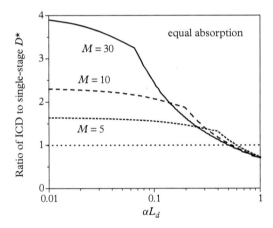

**Figure 14.22** *Ratio of optimal variable-length multistage and single-stage detectivities for ICDs as a function of $\alpha L_d$ for designs with different numbers of stages.*

where $L_m < L_d$ are the lengths of the $m = 1, \ldots, M$ individual stages. The relative improvement is

$$\frac{D_M^*}{D_1^*} \approx \frac{1 - e^{-\alpha \sum_m L_m}}{\alpha L_d \sqrt{M}} \sqrt{\sum_m \frac{L_d}{L_m}} \tag{14.58}$$

Figure 14.22 shows the ratio of multistage to single-stage detectivities, when the ICD stage lengths are adjusted for optimal sensitivity. With a maximum improvement larger than that for equal-length absorbers, the ICD may have a substantial edge when $\alpha L_d = 0.1$. This value is potentially relevant to the Ga-free type II SL detectors of Section 14.4. It is a pity then that quite a few stages are required and the structure is rather complex. And we should not forget that if the diffusion length can be improved by other means, the entire ICD concept becomes moot! As discussed earlier, this result does not apply to the ICL–ICD combination in an integrated design.

## 14.6   Quantum-Well Infrared Photodetectors and Quantum Cascade Detectors

Stepping back another pace, we can ask whether only interband transitions can be used for photodetectors. We already saw that QCLs have many advantages over interband lasers for wavelengths longer than 4 μm. Might we also enlist intersubband transitions to make photodetectors with long cut-off wavelengths? A nice side benefit of the switch would be the use of more mature material systems based on GaAs and InP, with the attendant benefits of manufacturability and yield.

   Although we might imagine many varieties of intersubband photodetectors, it is perhaps simplest to emulate the photoconductor structure of Fig. 14.1. If we nearly align the second subband in a heavily doped quantum well with the top of its barrier, as shown in Fig. 14.23, electrons excited to that subband by the incident light can escape, and flow in the continuum above the barriers when driven by an applied electric field. This concept gained currency in the 1990s under the name *quantum-well infrared photodetector* (QWIP). If the intersubband energy difference is much larger than $k_B T$ and the doping is not too high, the upper subband is nearly empty in thermal equilibrium and the intersubband absorption coefficient is proportional to the sheet doping density $n_d$ (see Eq. (11.42)).

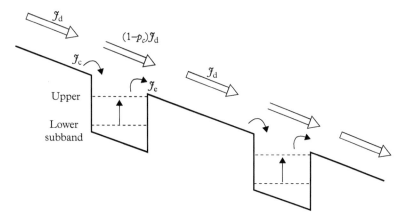

**Figure 14.23** *Schematic of a QWIP in which electrons in a lower subband can be excited to the continuum by incident light.*

When the wells are almost isolated, as in Fig. 14.23, some of the electrons flowing above a given well are captured, balanced by the emission of a few electrons in the well into barrier states. The escape rate must equal the capture rate if the current is to remain continuous throughout the QWIP structure. The escape/capture current should also be a small fraction of the total flowing between the top and bottom contacts. We can write it as

$$\mathcal{J}_e = \mathcal{J}_c = \frac{e n_d}{\tau_{ul}} \tag{14.59}$$

where $\tau_{ul}$ is the scattering time from the lower to the quasi-bound upper subband, via states at larger $k_{\parallel}$.

The time of a few ps needed to capture a carrier into the well, $\tau_c$, is usually much longer than the time it takes to traverse one period of a QWIP with $d \sim 10$ nm: $\tau_{trans} = d/v_e \sim 0.1$ ps, where $v_e$ is the electron velocity $\leq 10^7$ cm/s. The capture probability $p_c$ is approximately equal to the ratio of these time scales:

$$p_c = \frac{\tau_{trans}}{\tau_c + \tau_{trans}} \approx \frac{\tau_{trans}}{\tau_c} \tag{14.60}$$

Therefore, we expect the capture probability to be small, e.g., $p_c \sim 0.1$. The current captured by the well is the product of the total dark current and the capture probability: $\mathcal{J}_c = \mathcal{J}_e = \mathcal{J}_{d,QWIP} p_c$.

Now we combine Eqs. (14.59) and (14.60) to find the dark current in the QWIP:

$$\mathcal{J}_{d,QWIP} = \frac{\mathcal{J}_e}{p_c} \approx \frac{e n_d}{\tau_{ul}} \frac{\tau_c}{\tau_{trans}} = \frac{e n_d}{d} \frac{\tau_c}{\tau_{ul}} v_e \tag{14.61}$$

This is the QWIP form of the photoconductive dark current discussed in Section 14.1. If the capture and escape rates are equal in steady state so that carriers neither accumulate nor disappear from the well, we can write

$$\frac{n_d}{\tau_{ul}} = \frac{d\, N_{3d}}{\tau_c} \tag{14.62}$$

where $N_{3d}$ is the volume electron density above the top of the barrier. Substituting Eq. (14.62) into Eq. (14.61), we obtain Eq. (14.9), with $N_d$ replaced by the equivalent value $N_{3d}$.

We can use the photocurrent in Eq. (14.3) for the QWIP, but a few important corrections are needed. First, by analogy with the dark current, the *observed* photocurrent is given by the ratio of the escape current and the capture probability $\mathcal{J}_{p,QWIP} \propto \frac{\mathcal{J}_{p,e}}{p_c}$, where $\mathcal{J}_{p,e}$ is given by Eq. (14.3). Because any photogenerated carriers must also escape from the well, the photocurrent is also proportional to the escape probability $p_e$. The photoexcited electrons in a well-designed QWIP should be able to escape well before they relax back into the lowest subband of the well. For this reason, $p_e$ approaches unity in state-of-the-art devices. As in the case of the ICD of Section 14.5, the same current flows sequentially through all the wells, which implies that $\mathcal{J}_{p,QWIP} = \frac{p_e \mathcal{J}_{p,e}}{M p_c}$.

Now compare the photocurrent expression $\mathcal{J}_p = \frac{\mathcal{J}_{p,e} p_e}{M p_c}$ to the dark current $\mathcal{J}_{d,QWIP} = \mathcal{J}_e / p_c$ flowing above the barriers. As in the case of the ICD, increasing the number of wells does not affect the photocurrent, so when $\eta_a \ll 1$ the product of the gain and the absorption per well becomes

$$\eta_a g_{p,QWIP} = \frac{p_e}{M p_c} \approx \frac{p_e \tau_c}{M \tau_{trans}} \tag{14.63}$$

As discussed above, $\frac{p_e}{p_c} \sim 10$ in the QWIP, so Eq. (14.63) this still leads to photoconductive gains below unity. The QWIP photocurrent is proportional to the time needed to capture an electron into the well, rather than the electron–hole recombination time (see Eq. (14.4)), because $\tau_c$ now governs the rate at which excess carriers disappear.

Clearly, the QWIP is still subject to the main limitation of photoconductive detectors: a significant dark current. In the QWIP it scales with $n_d$, which must be high enough to absorb much of the incident light. By analogy with the ICD, the QWIP noise current is continuous through the entire structure, which means the detectivity is given by a slightly modified version of Eq. (14.8):

$$D^*_{QWIP} = \frac{\eta_a}{2\hbar\omega} \sqrt{M} \sqrt{\frac{e g_{p,QWIP}}{\mathcal{J}_{d,QWIP}}} \tag{14.64}$$

where the $\eta_a g_p$ product is given by Eq. (14.63), and the dark-current density by Eq. (14.61). $D^*_{QWIP}$ increases with doping when the absorption efficiency is low, then once the absorbed fraction begins to saturate, the dark-current dependence on $n_d$ takes over. Therefore, there is an optimal value of the doping density in a QWIP that corresponds to a Fermi level a few $k_B T$ above the bottom of the lowest electron subband.

We learned in Section 14.2 that intersubband transitions couple predominantly to light polarized along the growth direction, particularly in wider-gap semiconductors such as GaAs that are commonly used for QWIPs. On the other hand, normally incident light propagates along the growth axis and is polarized in the growth plane. How can we couple the light passing through the quantum wells to intersubband transitions? Several approaches can deal with this "selection-rule" mismatch. Section 14.7 will discuss the use of a grating to redirect the light into the growth plane. We can also improve the absorption at normal incidence by confining electrons in the plane as well as the growth direction, e.g., using interlevel transitions in quantum dots. This version of the QWIP is called the *quantum-dot infrared photodetector* (QDIP). One limitation of the QDIP is the broad distribution

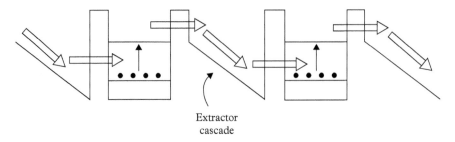

Extractor
cascade

**Figure 14.24** *Schematic of a quantum cascade detector structure.*

of dot sizes in practical samples, which affects the amount of absorption and device design.

Of course, we already know that the photovoltaic configuration has an advantage in terms of dark current. In fact, the QWIP can be re-engineered into a photovoltaic detector. A particularly relevant version is the *quantum cascade detector* (QCD), which is illustrated in Fig. 14.24. As the name implies, the QCD is based on a modification of the quantum cascade laser structure for operation close to zero bias. Instead of exciting carriers optically to a subband close to the barrier states, the intersubband transitions in a QCD take place deep in the wells. This means the current flows almost entirely from well to well, via tunneling or phonon-assisted processes in the chirped injectors. In other words, transport in a QCD closely resembles that in a QCL, and in fact the basic QCL design is a good starting point, albeit with a small red shift of the center absorption with respect to the QCL emission. Since the QCD operates at zero bias, the injector subbands do not align into a miniband as in the QCL. Electrons that are excited to a higher subband decay and move in real space through the QCD injector until they reach the next active well, where another photon must be absorbed and the entire sequence repeats. The QCD resembles the ICD in this respect, and was in fact the first to be demonstrated experimentally.

Because there is no need to capture electrons in a QCD, the quantum efficiency is proportional to the escape probability from the well to the injector $p_e$ (see Eq. (14.63)). As in the QWIP and ICD, the photocurrent does not scale with the number of stages, and a photon must be absorbed in each stage for the photocurrent to flow. We can write the escape probability as

$$p_e = \frac{1/\tau_{ui}}{1/\tau_{ul} + 1/\tau_{ui}} = \frac{\tau_{ul}}{\tau_{ul} + \tau_{ui}} \tag{14.65}$$

where $\tau_{ul}$ is the scattering time between the upper and lower active subbands and $\tau_{ui}$ is the scattering time between the upper subband and the closest injector subband. Obviously, $\tau_{ul} \gg \tau_{ui}$ in a good QCD design; i.e., electrons escape mostly through the injector.

The dark-current density in a QCD is given by

$$\mathcal{J}_{d,QCD} = \frac{e n_d}{\tau_{trans}} \tag{14.66}$$

where $\tau_{trans}$ is the transit time through one stage. Despite its apparent simplicity, Eq. (14.66) conceals some complicated physical mechanisms that govern the transport between stages. These are subsumed into $\tau_{trans}$, which has a typical value of several ps.

By analogy with the already discussed ICL/ICD pair, the main attraction of the QCD is that a photodetector can be built on the same basic platform as the emitter (QCL). Even though the standard QCL structure does not function particularly well as a QCD, it can be modified to perform both roles adequately. When a QCD is paired with a QCL, high sensitivity is usually not required, and the dual functionality makes integrated designs possible.

## 14.7 Resonant-Cavity and Waveguide-Enhanced Detectors

Section 12.7 discussed different ways to improve the extraction of radiation from a semiconductor slab. Here we will deal with an inverse problem: how to couple radiation into a slab to produce photocurrent. Can we apply what we have learned about emitters here? For example, we could constructively interfere light at normal incidence with a mirror (Fig. 12.32), form a resonant cavity that induces multiple passes through the absorber (Figs. 12.33 and 14.25), or couple the light to an in-plane waveguide for increased effective propagation length in the absorber (Figs. 12.31(c) and 14.26). Of course, this analogy does not mean that the design constraints are the same for emission and absorption processes. For one thing, the spectrum of spontaneously generated light is usually a few $k_B T$ wide (apart from the quasi-atomic lines emitted by quantum dots at low temperatures), whereas the light incident on a detector could come either from a narrow-line external laser or at the other extreme from very broad thermal radiation. The potential for enhancement is much stronger when the radiation to be detected has a very narrow linewidth, so we can design a wavelength-selective cavity or grating to transform the incident beam.

We begin with the vertical resonant-cavity detector shown in Fig. 14.25. While the front mirror with reflectivity $R_t$ needs to transmit a fraction of the radiation, the back mirror ($R_b$) can absorb the light that is not reflected (i.e., it can be a metal mirror). The Fabry–Perot cavity formed by the mirrors has an effective length $L_{cav} = L_m + L_{eff,t} + L_{eff,b}$, where $L_{eff,t}$ and $L_{eff,b}$ are the effective penetration depths into the two mirrors. As in Section 12.7, the quantum efficiency boost is proportional to the number of passes through the absorber, and also to the antinode enhancement factor $\xi$ given by Eq. (12.76). Assuming that only a small

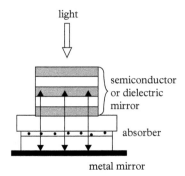

**Figure 14.25** *Schematic of a resonant-cavity detector.*

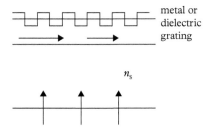

**Figure 14.26** *Schematic of a waveguide-coupled detector, in which a grating diffracts light at normal incidence into the waveguide mode.*

fraction of the light $\alpha L_a \ll 1$ is absorbed in a single pass through the absorber with thickness $L_a$, the quantum efficiency at the cavity resonance becomes

$$\eta_a = \frac{[1 + R_b (1 - \alpha L_a)] (1 - R_t)\, \xi \alpha L_a}{\left(1 - \sqrt{R_t R_b}\, (1 - \alpha L_a)\right)^2} \qquad (14.67)$$

by analogy with the arguments in Section 12.7.

Even when the top mirror is highly reflective, Eq. (14.67) shows that much of the light can be absorbed as long as $R_b$ is also close to one. Mathematically, this is because $R_t$ appears in both in the numerator and denominator. Physically, what happens is that the light reflected from the back mirror interferes with the incident light, so that the reflected field is canceled when the wavelength matches the cavity resonance.

Is there any downside to this enhancement of the quantum efficiency? Equation (14.67) strictly applies only on resonance, which has a finite linewidth given by

$$\Delta\lambda_{FWHM} = FSR \frac{1 - \sqrt{R_t R_b}\, (1 - \alpha L_a)}{\pi\, (R_t R_b)^{1/4}\, (1 - \alpha L_a/2)} \qquad (14.68)$$

where the free spectral range (mode separation) is given by

$$FSR = \frac{\lambda_{res}^2}{2n_{eff}L_{cav}} \tag{14.69}$$

where $\lambda_{res}$ is the resonant wavelength, and $n_{eff}$ is the effective index for light propagating in the cavity. Therefore, high quantum efficiency from a weak absorber is possible, but only in a very narrow spectral range. If we would like to detect the light from a narrowband source, e.g., a single-mode laser, this is a worthwhile trade-off. The narrow line can also reduce the noise away from the resonance, within the combined stopband of the two mirrors. Of course, when the detected signal is spectrally broad (e.g., blackbody radiation), the resonant cavity is of no help.

Consider two limits of the expressions in Eq. (14.67): (1) $R_b \approx 1$ and (2) $R_t \approx R_b \equiv R$. In the first limit, we have

$$\eta_a = \frac{(2 - \alpha L_a)(1 - R_t)\xi \alpha L_a}{\left(1 - \sqrt{R_t}(1 - \alpha L_a)\right)^2} \tag{14.70a}$$

$$\Delta\lambda_{FWHM} = FSR\frac{1 - \sqrt{R_t}(1 - \alpha L_a)}{\pi (R_t)^{1/4}(1 - \alpha L_a/2)} \tag{14.70b}$$

If we further neglect $\alpha L_a$ and use $\xi \approx 2$, we obtain

$$\eta_a \approx \frac{4\left(1 + \sqrt{R_t}\right)}{1 - \sqrt{R_t}}\alpha L_a \approx \frac{FSR}{\Delta\lambda_{FWHM}}\frac{4\left(1 + \sqrt{R_t}\right)}{\pi (R_t)^{1/4}}\alpha L_a \tag{14.71}$$

When $R_t$ is close to 1, as needed to maximize the enhancement in Eq. (14.70a), the product of $\eta_a$ and $\Delta\lambda_{FWHM}$ is essentially independent of cavity design. As $\eta_a$ rises, the linewidth shrinks. For $R_t = 90\%$ $\eta_a$ is enhanced over the single-pass value $\alpha L_a$ by two orders of magnitude.

In the second limit, we have

$$\eta_a = \frac{[1 + R(1 - \alpha L_a)](1 - R)\xi \alpha L_a}{(1 - R(1 - \alpha L_a))^2} \tag{14.72a}$$

$$\Delta\lambda_{FWHM} = FSR\frac{1 - R(1 - \alpha L_a)}{\pi \sqrt{R}(1 - \alpha L_a/2)} \tag{14.72b}$$

This is a more realistic scenario. If we once again neglect $\alpha L_a$ and use $\xi \approx 2$, we have

$$\eta_a = \frac{2(1 + R)}{1 - R}\alpha L_a = \frac{FSR}{\Delta\lambda_{FWHM}}\frac{2(1 + R)}{\pi \sqrt{R}}\alpha L_a \tag{14.73}$$

Again, when the mirrors are highly reflective there is a clear connection between quantum efficiency and bandwidth. We also observe from Eq. (14.73) that in the limit $1 - R \ll 1$, the enhancement approaches $4/(1 - R)$ while the linewidth is approximately $FSR\,(1 - R)/\pi$. To boost the quantum efficiency by two orders of magnitude now requires $R \geq 0.96$ for both mirrors. If the back mirror is already very reflective, as in Eqs. (14.70) and (14.71), the quantum efficiency for the same top mirror increases by a factor of 2.

Figure 14.27 plots the quantum-efficiency enhancement and cavity $Q\left(= \frac{\lambda_{res}}{\Delta\lambda_{FWHM}}\right)$ for a wider range of $R_b$ and $R_t$. Clearly, we should not make the front mirror more reflective than the back mirror because the cancellation mentioned above becomes less effective. Also, the transmission through the

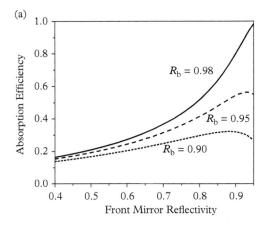

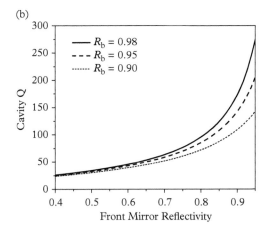

**Figure 14.27** *Quantum-efficiency enhancement (a) and cavity Q (b) for resonant-cavity detectors with variable mirror reflectivities, assuming $L_{cav} = 2\lambda_{res}/n_{eff}$, $\alpha L_a = 1\%$, and $\xi = 2$.*

mirrors should not drop below the single-pass absorption $\alpha L_a$. In fact, Eqs. (14.67)–(14.73) and Fig. 14.27 neglect parasitic loss in the cavity. To include it, we replace $1 - R_t$ in Eqs. (14.67) and (14.70a) by the transmission of the front mirror $T_t \neq 1 - R_t$, and $(1 - \alpha L_a)$ in the denominator by $(1 - \alpha L_a - \alpha_i L_{cav})$. This assumes that the total loss is small and distributed over the entire cavity length. For example, Eq. (14.67) becomes

$$\eta_a = \frac{[1 + R_b (1 - \alpha L_a)] \, T_t \xi \alpha L_a}{\left(1 - \sqrt{R_t R_b} \, (1 - \alpha L_a - \alpha_i L_{cav})\right)^2} \qquad (14.74)$$

When parasitic loss is included, $1 - R$ should not be lower than the total absorption in the cavity (both useful and parasitic). With a low enough parasitic loss, we can obtain very high quantum efficiency with linewidths equivalent to $Q \sim 10^4$, even when the single-pass absorption is only 1%.

We already saw that a small single-pass absorption combined with high $\eta_a$ can open a few doors (see Sections 14.4 and 14.5). One simple example is a diffusion-limited photodetector with dark-current density scaling with $L_a$ in the $L_a \ll L_d$ limit, as in described by Eq. (14.32b). In a conventional photodetector, high quantum efficiency requires $\alpha L_a = 2$–3. According to Eq. (14.8), reducing the absorber thickness by two orders of magnitude would translate to an order-of-magnitude improvement in $D^*$. Of course, once the absorber is much thinner there is no guarantee that the device remains diffusion limited, with dark current described by Eq. (14.32b). In particular, it can be difficult to keep the G–R and other dark current small when the absorber is thin. This means that the resonant-cavity detector makes the most sense at higher temperatures and longer wavelengths (narrower absorber gaps), where the intrinsic carrier density is high.

We can also benefit from a thin absorber if the structures cannot be grown to the required thickness, e.g., because they are not fully strain balanced. In Section 14.4, we discussed how strain balancing can limit the absorption strength. When Auger recombination dominates, we may even come out ahead by fitting the entire absorber in the depletion region.

The resonant cavity does not necessarily need two separate mirrors that create a vertical Fabry–Perot geometry. For example, we can use a subwavelength resonator supported by plasmonic and other polaritonic modes. If it is sufficient to enhance the quantum efficiency by a factor of 4 over the single-pass value, we can impose constructive interference with a single metal mirror (see Fig. 12.32). We can also create a cavity in the plane of the absorber, if the incident light is redirected to an in-plane waveguide mode. Of course, the in-plane dimensions are typically much longer than the vertical dimensions, so diffraction by a suitable grating to the in-plane mode may be enough by itself (see Fig. 14.26). If it is inconvenient to define a dielectric waveguide in the grown detector structure, the surface plasmon polariton mode of the metal grating can guide the light. The grating does not even need to couple to an in-plane waveguide mode. Sometimes diffracting the light

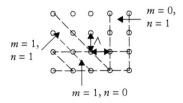

**Figure 14.28** *Schematic of a grating that leads to two-dimensional diffraction processes.*

into a large angle with respect to normal increases the path length enough to make the quantum efficiency high.

What should this grating look like? A one-dimensional line grating, with period $\Lambda = \lambda/n_{eff}$ and made from a noble metal that supports a plasmon mode at the desired wavelength, works well for incident radiation polarized perpendicular to the grating. This is because the grating reciprocal lattice vector of magnitude $2\pi n_{eff}/\lambda$ matches the propagation-direction wavevector for light traveling in the waveguide mode, and the incoming light does not need any wavevector component in the propagation direction. However, a line grating does not interact with light polarized along the grating lines. To couple an arbitrary polarization, we can use a two-dimensional grating such as that shown in Fig. 14.28. The resonant wavelengths for a square grating at normal incidence become

$$\lambda_{m,n} = \frac{\Lambda}{n_{eff}} \frac{1}{\sqrt{m^2 + n^2}} \qquad (14.75)$$

where $m$ and $n$ are integers. If either $m = 0$ or $n = 0$, the diffraction is exactly analogous to a line grating. When both are non-zero, the light diffracts from both rows of features at the same time. More complicated grating geometries are also possible. For example, a two-dimensional grating based on a hexagonal lattice has a higher degree of symmetry than the square lattice. The period of the grating can be varied, either to increase the bandwidth over which the light couples to the grating or to improve the trapping within the sample when the grating does not couple to a waveguide mode.

How strong should diffraction from the grating be? If the coupling of incident light is much stronger than the absorption, most of the light will couple back out before it is absorbed. On the other hand, if it is much weaker than the absorption, most of the incident light will be reflected from the metal grating rather than absorbed. Therefore, the coupling coefficient for the diffraction process should be similar to the waveguide mode's absorption coefficient per unit length, taking into account the overlap of the mode with the absorber region. This description can be formalized using coupled-mode theory, but an intuitive understanding is sufficient for our purposes.

A grating that redirects the beam at normal incidence is also very useful for QWIPs, which absorb in-plane-polarized light very weakly. In this case, the improvement due to diffraction can be quite dramatic, because the baseline quantum efficiency is so low. The situation is very different for interband transitions, whether in bulk or quantum wells. There the light at normal incidence sees at least the same absorption, so the enhancement is more subtle unless the absorber layer is very thin indeed.

How broad are the resonances associated with a waveguide-enhanced detector? For a metal grating, the plasmon resonance in the IR can be on the order of a few hundred nm, much broader than any resonant-cavity scheme, because the substantial metal loss puts a ceiling on $Q$. We should be aware of these differences in practical designs.

## 14.8   Novel Photodetector Structures and High-Speed Operation

The concepts discussed in Chapter 14 apply not only to the mature III–V and II–VI semiconductor detectors, but also to new classes of devices. To see how they compare to the better-established detectors, we can use the concept of the normalized detectivity $D*$ specified in Eqs. (14.8) and (14.13).

An important device with applications to both established III–V and novel materials is the *phototransistor* illustrated in Fig. 14.29. In this photoconductive detector, current flow in the dark is blocked by a barrier created by either opposite-type doping (*npn* or *pnp*) or a heterojunction (by analogy with the photovoltaic

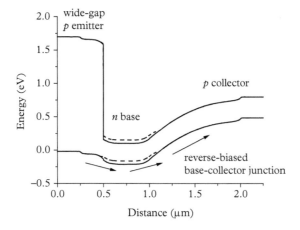

**Figure 14.29** *Schematic of a pnp phototransistor with absorbing base based on the InAsSb/AlAsSb structure in Fig. 14.10. Here the collector region is also absorbing, whereas the emitter has a wider gap with the valence band offset that prevents hole injection from the absorber. When light shines on the device, electrons are trapped in the n-doped base, which lowers the potential barrier to holes (dashed line) and increases the current flow to the collector.*

nBn detector). The absorption of light takes place in the barrier or its immediate vicinity, which we can term the *base* by analogy with a three-terminal electrically controlled transistor. In the *pnp* structure shown in Fig. 14.29, the photogenerated carriers are quickly separated by the potential barriers, with electrons swept out of the base, while the holes are trapped there. The resulting charge separation lowers the potential barrier for the electron flow between the contacts by an amount (see Eq. (12.65))

$$eV_e \approx \frac{e^2 n_e}{\frac{\varepsilon_d}{d}} \approx \frac{e^2}{\varepsilon_d} \eta_a \frac{I_\omega}{\hbar\omega} \tau_e \tag{14.76}$$

where $d$ is the average distance between the separated photogenerated carriers, which is taken to be close to the base width. The lowering of the barrier allows additional current to flow, as expressed by $\mathcal{J}_p \propto \exp(eV_e/k_B T)$. This implies large photoconductive gain is possible if the photocurrent is properly balanced against the dark current flowing when $V_e = 0$.

We can also think of the phototransistor in terms of a bipolar (*pnp* or *npn*) transistor with two terminals (emitter and collector) and a floating base. The photon flux absorbed in the base plays the role of the base current. The result is much the same: strong amplification of the current for base widths much smaller than the minority-carrier diffusion length in the base (and relatively low base doping). This is because the electrons in the base must diffuse to the reverse-biased base-collector junction before they are separated from the holes.

What distinguishes a photoconductive detector from a phototransistor? In a photoconductor, the extra current is directly proportional to the carrier density generated by the incident light. That current continues to flow until the additional carriers recombine. By contrast, the photogenerated carriers in a phototransistor modulate the current by lowering a potential barrier, which can produce a much stronger dependence on photoexcited carrier density. Both devices can exhibit a large gain, as quantified by the additional current density divided by the incident optical flux (divided by the electron charge).

To see how a novel material can be configured as a phototransistor, consider Fig. 14.30 with the conductive layer composed of a monolayer of graphene deposited on a suitable substrate, typically Si with a thin $SiO_2$ insulating layer on top. The substrate can be biased to gate the Fermi level in the graphene. Of course, graphene by itself is a poor detector material because it has no energy gap (see Section 5.1), and hence the photoexcited carriers scatter back to the Fermi level within $\approx 100$ fs. Graphene also absorbs very little light in a single monolayer (see Section 4.2), and it is cumbersome to contact multiple monolayers. To make the approach more attractive, the graphene surface between the source and drain contacts can be *sensitized* with nanoparticles or quantum dots that absorb at the desired wavelengths. Following absorption of an incident photon, one carrier type rapidly transfers to the graphene (by hopping transport, tunneling, or dipole–dipole interactions). The opposite carrier is left behind in the nanoparticle, which

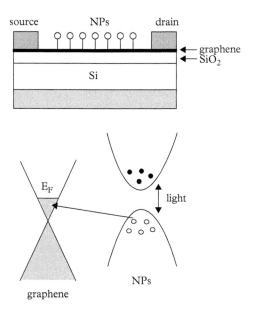

**Figure 14.30** *Schematic of a phototransistor based on graphene sensitized with nanoparticles or quantum dots.*

raises the Fermi level in the graphene via capacitive coupling (see Eq. (14.76)). This stimulates additional current flow in the graphene.

The device in Fig. 14.30 was found experimentally to have a large gain. Of course, as in any photoconductive structure the challenge is to suppress the current flow in the dark. To do so, the energy shift in Eq. (14.76) should be as large as possible. In the structure of Fig. 14.30, this implies separating the charges as far as possible, a strategy that runs into the difficulty of transferring carriers efficiently from the nanoparticles to the graphene at large distances. Even with this complication, this alternative to semiconductor-based detectors is worth further study because it avoids complicated and expensive epitaxial growth.

Sometimes we would like to detect an optical signal that varies more rapidly than the typical time scales discussed earlier. For example, we may want to communicate optically and encode information in the modulation of the optical power level. Such variation can also indicate spectral changes in the input signal, e.g., after it passes through a medium we want to sense and undergoes beating with another beam that is closely matched in frequency (*heterodyning*). Whatever the origin of the rapid variation, can it be sensed by the photodetectors discussed here? If they respond too slowly, the high-frequency component of the incoming modulated signal would be lost when the optical signal is transformed into an electrical one; i.e., it would function as a *low-pass filter*.

The relevant time scales for high-speed operation depend on whether the detector is photoconductive or photovoltaic, as well as whether the detector "stores" the light in a resonant-cavity structure to enhance the absorption (see Section 14.7). It also depends on purely electrical properties, such as how much charge the detector's junction can store during regular operation. We start with the fundamental limitations.

The photoexcited carriers in a photoconductor can contribute additional current for as long as they are present in the channel, i.e., until they recombine in an interband photoconductor or are captured by the well in a QWIP. If the gain is high and the transit time is short, we can think of them as recirculating many times in the circuit until this happens. Therefore, in this case the relevant time scale is the carrier recombination time $\tau_e$ or capture time $\tau_{ul}$, i.e., the average time for a photogenerated carrier to disappear from the conducting channel after cessation of the optical signal. The same time scale applies to phototransistors, which differ only in the mechanism by which the channel current is modulated by the generated carriers.

In a photovoltaic device, the transit time comes to the fore. Here we define it as the time needed for a minority electron or hole to move from where it is generated to a contact. However, in practice it only needs to reach the heavily doped region next to the contact, because beyond that numerous other carriers are available to complete the external circuit. So the transit time in a photovoltaic device is the time needed for a minority carrier in a doped region to diffuse to the junction, plus the time required to sweep it across the depletion region:

$$\tau_{transit} = \frac{\langle d \rangle^2}{2D_{e,h}} + \frac{W}{v_{e,h}} = \frac{e\langle d \rangle^2}{2\mu_{e,h}k_B T} + \frac{W}{v_{e,h}} \tag{14.77}$$

where $\langle d \rangle$ is the average distance from the point at which a minority carrier is generated to the edge of the depletion region; $W$ is the width of the depletion region; and $D_{e,h}$, $\mu_{e,h}$, and $v_{e,h}$ are the minority-carrier diffusion coefficient, mobility, and velocity (not necessarily proportional to the mobility in the depletion region, where the field is high).

One approach to reducing the transit time is to place the entire absorber within the depletion region, which eliminates the first term in Eq. (14.77). This is a popular design in high-speed near-IR and visible detectors, but runs into difficulties at longer wavelengths, where the absorption length is often quite large: $\frac{1}{\alpha} \gg W$. It is easier to fit the thinner absorber within the depletion region of resonant-cavity structures, or if the G–R dark current is too high, we can generate all the carriers near the edge of the depletion region so $\langle d \rangle$ is minimized.

How is the transit time in Eq. (14.77) related to the maximum frequency at which the detector responds to the optical signal? The conventional figure of merit is the frequency at which the square of the response falls by half (3 dB, see Section 12.3):

$$f_{3dB} = \frac{\omega_{3dB}}{2\pi} \approx \frac{2.4}{2\pi\tau_{transit}} \tag{14.78}$$

As expected, it is roughly the reciprocal of the transit time. For a depletion width of 100 nm that contains the entire absorber and a maximum electron velocity of $10^7$ cm/s, $f_{3dB}$ can reach several hundred GHz. The polar opposite is the typical mid-wave or long-wave structure, where carriers must diffuse several microns with mobilities of only a few hundred at cryogenic temperatures. This leads to $f_{3dB}$ of only tens of MHz (see Eqs. (14.77) and (14.78)).

Clearly, conventional photovoltaic structures do not work well for high-speed detection at longer wavelengths. But all is not lost because the absorber thickness can be reduced to make them faster, either by sacrificing quantum efficiency or by making the structure more complicated, e.g., using a resonant cavity from Section 14.7. On the other hand, photoconductive QWIPs are relatively fast because the capture time is at most a few ps and potentially allows $f_{3dB} \sim 100$ GHz (ignoring a slow photocurrent component associated with the recharging of the wells).

Of course, it is not enough for the detector to have a fast transit time. For example, the intrinsic capacitance of a $p$–$n$ junction is inversely proportional to the width of the depletion region $W$:

$$C_j = \frac{\varepsilon_d}{W}A \qquad (14.79)$$

where $A$ is the device area. In combination with the circuit's load resistance of $R_L = 50 \ \Omega$, the junction capacitance acts as a low-pass filter with 3 dB frequency:

$$f_{3dB} = \frac{1}{2\pi R_L C_j} = \frac{W}{2\pi R_L \varepsilon_d A} \qquad (14.80)$$

If the depletion width is 100 nm, and the device area is 100 μm × 100 μm, this leads to $f_{3dB} \sim 300$ MHz, which can be much lower than the transit-time limitation! The solution is to extend the depletion width and make the device area as small as possible. It is also fortunate that photoconductive devices do not necessarily exhibit significant capacitances.

Either the transit time or the $R_L C_j$ time constant dominates in many practical devices, but in general both limit the detector speed. To complete the survey, a resonant cavity enhances the quantum efficiency of a thin absorber at the price of effectively storing the light in the cavity. The relevant time scale is the round-trip time in the cavity, $\tau_{rt} = 2L_{cav}n_{eff}/c$, divided by the total loss over the round trip, $\alpha_t \approx 2 - R_b - R_t + 2\alpha L_a$ (for high-reflectivity mirrors and thin absorbers, see Section 14.7):

$$\tau_{light} = \frac{\tau_{rt}}{\alpha_t} = \frac{2L_{cav}n_{eff}/c}{2 - R_b - R_t + 2\alpha L_a} \qquad (14.81)$$

In a typical resonant cavity with $L_{cav}$ of a few μm, the round-trip time $\tau_{rt}$ is a few tens of fs, while the total loss per round trip is at least a few percent. Under these conditions, the 3-dB frequency due to stored light is at least 100 GHz, which is high enough for most applications, assuming the extrinsic capacitance in Eq. (14.79) does not dominate.

## Suggested Reading

- A general reference for the physics of *p–n* junctions and other basic semiconductor devices is S. M. Sze, *Physics of Semiconductor Devices*, 2nd edn. (Wiley, New York, 1981).

- A detailed, but somewhat dated description of IR photodetectors is available in A. Rogalski and J. Piotrowski, "Intrinsic infrared detectors," *Progress in Quantum Electronics* **12**, 87 (1988). This reference is useful for its description of the physics rather than the details of the device performance available at the time of its publication over thirty years ago.

- The physics of QWIPs and a few other IR detectors is reviewed in H. Schneider and H. C. Liu, *Quantum Well Photodetectors* (Springer, Berlin, 2007).

- A recent description of the work on type II superlattice detectors can be found in D. Z. Ting, A. Soibel, A. Khoshakhlagh, S. A. Keo, S. B. Rafol, A. M. Fisher, B. J. Pepper, E. M. Luong, C. J. Hill, and S. D. Gunapala, "Advances in III–V semiconductor infrared absorbers and detectors," *Infrared Physics and Technology* **97**, 210 (2019).

- The theory of interband cascade detectors is covered in R. T. Hinkey and R. Q. Yang, "Theory of multiple stage interband cascade photovoltaic devices and ultimate performance level comparison of multiple-stage and single-stage interband infrared detectors," *Journal of Applied Physics* **114**, 104,506 (2013).

- Resonant-cavity photodetectors are reviewed in M. S. Ünlü and S. Strite, "Resonant cavity enhanced photonic devices," *Journal of Applied Physics* **78**, 607 (1995).

# 15

# Solar Cells, Thermophotovoltaics, and Nonlinear Devices Based on Quantum Wells

In this chapter, we describe the basic principles behind the solar-cell operation using both an empirical picture and fundamental thermodynamic relationships. We consider how semiconductor materials are selected for use in solar cells and why materials with different gaps need to be stacked to improve the conversion efficiency. We also discuss advanced solar-cell concepts such as quantum-well, intermediate-band, and hot-carrier solar cells. Thermophotovoltaic devices that are similar to solar cells, but designed for emission peaks at much lower effective temperatures than the surface of the sun (and narrower gaps), are also discussed, and multistage thermophotovoltaic devices are described in detail. We conclude the book by presenting the basic nonlinear physics of intersubband transitions in quantum wells, and how we can take advantage of these physical principles for second-harmonic generation and difference-frequency mixing. The important application of generating THz emission from mid-IR quantum cascade lasers using difference-frequency mixing is emphasized.

## 15.1 Basics of Solar Cells

In Chapter 14, we saw that an optical flux incident on a photovoltaic detector produces electrical current through its terminals. Most photodetectors are designed for low photon fluxes, but what happens if we shine a high-power laser with photons of energy $\hbar\omega_p > E_g$ on a photovoltaic structure? As expected, we observe a "reverse" photocurrent proportional to the incident flux. When the device is short-circuited ($V = 0$), only this photocurrent flows in the circuit. On the other hand, if we disconnect the terminals from the external circuit to prevent current flow, some voltage develops across the terminals. This is the forward bias that drives just enough current to balance the photocurrent produced by the laser. For an operating bias between zero and the open-circuit voltage, the product of the

*Bands and Photons in III–V Semiconductor Quantum Structures.* Vurgaftman, Lumb, and Meyer,
Oxford University Press (2021). © Vurgaftman, Lumb, and Meyer.
DOI: 10.1093/oso/9780198767275.003.0015

voltage and current is negative. This means we are generating electrical power by illuminating the photovoltaic structure with a laser!

To describe these ideas mathematically, we assume the photovoltaic device is a *p–n* junction with one side much thicker and more lightly doped than the other. We further assume that the G–R current from the depletion region is much smaller than the diffusion current from the thick absorber (e.g., this is an optimized heterojunction), so that we can describe the dark current using Eq. (14.47). Then to account for optical excitation, we add the second term from Eq. (14.31) in the limits $\alpha L_d \gg 1$ and $\alpha W \ll 1$:

$$\mathcal{J}_{tot} = e(L_d/\tau)\left(n_i^2/N_A\right)\left(e^{eV/k_B T_c} - 1\right) - e\Phi \equiv \mathcal{J}_0\left(e^{eV/k_B T_c} - 1\right) - e\Phi \quad (15.1)$$

where $\Phi$ is the absorbed optical flux, and $T_c$ is the cell's operating temperature. The incident and absorbed fluxes are the same if we neglect reflection and scattering losses from the front surface and continue with the assumptions of a thick absorber and transparent junction: $\alpha L_d \gg 1$ and $\alpha W \ll 1$.

The short-circuit current density is given by

$$\mathcal{J}_{sc} = \mathcal{J}_{tot}(V = 0) = -e\Phi \quad (15.2)$$

and the open-circuit voltage follows from the condition $\mathcal{J}_{tot} = 0$:

$$V_{oc} = \frac{k_B T_c}{e}\ln\left(1 + \frac{e\Phi}{\mathcal{J}_0}\right) \approx \frac{k_B T_c}{e}\ln\left(\frac{e\Phi}{\mathcal{J}_0}\right) \quad (15.3)$$

The energy-gap dependence of $n_i^2 \propto \exp\left(-E_g/k_B T_c\right)$ then leads to

$$V_{oc} \approx \frac{k_B T_c}{e}\ln\left(\frac{\tau \Phi}{L_d}\frac{N_A}{N_c N_v}\right) + \frac{E_g}{e} \quad (15.4)$$

Assuming that the incident flux is not high enough to generate degenerate carrier densities, we estimate

$$V_{oc} \approx \frac{E_g - A k_B T_c}{e} \quad (15.5)$$

where $A$ is of order 1–10. The higher the flux from the laser, the more closely the open-circuit voltage comes to matching the energy gap. When the incident flux is too high, the device does not behave as a junction.

The typical current–voltage characteristics of an illuminated photovoltaic device are shown in Fig. 15.1. Under a load in the circuit, it operates at a small forward bias $V < V_{oc}$, and a reverse current flows through the terminals (see Eq. (15.1)). This means the device acts as a current source powered by the external laser, as shown in the equivalent-circuit diagram at the right of Fig. 15.1. In other

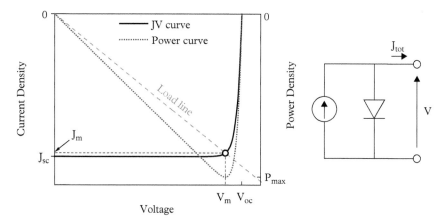

**Figure 15.1** *Typical current–voltage characteristics of a solar cell (left) and the equivalent circuit for an ideal cell (right). The maximum power point occurs at a voltage slightly less than the open-circuit voltage $V_m < V_{oc}$, where the current density to the load is slightly less than the short-circuit value: $\mathfrak{J}_m < \mathfrak{J}_{sc}$.*

words, it converts the external optical power into electrical power in the circuit. Is that interesting enough for practical applications? That depends on how efficiently we can perform the power conversion, i.e., on the *conversion efficiency*.

To estimate the conversion efficiency, we maximize the electrical output power, as illustrated in Fig 15.1. With no series resistance, this occurs at a voltage $V_m$ that is lower than $V_{oc}$ by a few $k_B T_c/e$, owing the exponential dependence of dark current on voltage. Maximizing the product of dark-current density and voltage in Eq. (15.1), we obtain

$$V_{oc} - V_m \approx \frac{k_B T_c}{e} \ln\left(\frac{eV_m}{k_B T}\right) \tag{15.6}$$

For example, the difference between energy gap and open-circuit voltage in a practical solar cell at room temperature is 0.3–0.4 eV (depending on the incident flux). If we assume $V_{oc} \sim 1$ V, $V_{oc} - V_m \approx 0.1$ V according to Eq. (15.6).

The current density $\mathfrak{J}_m$ that corresponds to the maximum output power is lower than the short-circuit photocurrent, but not by much because the current remains saturated until we approach within a few $\frac{k_B T_c}{e}$ of $V_{oc}$. Ideally, this current is close to $e\Phi\left(1 - \frac{k_B T_c}{eV_m}\right)$. From Eq. (15.6), we see that high conversion efficiency requires that the energy gap be much larger than $k_B T_c$. The conversion efficiency $\eta_c$ becomes

$$\eta_c = \frac{\mathfrak{J}_m V_m}{\Phi \hbar \omega_p} = \frac{E_g - k_B T_c \left[A' + \ln\left(\frac{E_g}{k_B T_c}\right) + 1\right]}{\hbar \omega_p} \tag{15.7}$$

where $\hbar\omega_p$ is the pump photon energy, and $A' \sim 1$–$10$, depending on the pump intensity. Since only photons with $\hbar\omega_p > E_g$ are absorbed in the active material, the maximum conversion efficiency in the high-intensity limit is given by

$$\eta_{c,\text{max}} \approx \frac{E_g - k_B T \ln\left(\frac{E_g}{k_B T_c}\right)}{\hbar\omega_p} \tag{15.8}$$

Assuming $E_g \sim 1.4$ eV and $\hbar\omega_p = E_g + 0.1$ eV, $\eta_{c,\text{max}}$ can be as high as 85% for an intense monochromatic pump beam that is almost completely absorbed. In fact, this is not a strict upper limit because Eq. (15.8) shows that we can make $E_g$ arbitrarily close to $\hbar\omega_p$ in an ideal device with a very long diffusion length.

Of course, laser pumping is rarely a practical solution to the generation of electricity. Fortunately, we have access to an alternative source that does not ultimately draw its power from the grid. It is powered by nuclear fusion, with no danger of running out of fuel within the next several billion years. It is also widely available during the daytime, when the weather is clear, comes free to the end user, and is safely situated 93 million miles away. To be sure, solar radiation is hardly monochromatic. Instead, its spectrum closely approximates the distribution of a $\approx 5800$ K blackbody, near the temperature of its outer surface. The intensity of the radiation reaching the outer surface of the Earth's atmosphere is roughly constant, at 1366 W/m². The source is well collimated, with a divergence half-angle of only $\theta_d = 0.26°$, due to the immense distance. We can also focus the sunlight, to *concentrate* the intensity falling on the device area. As the solar radiation traverses the atmosphere, its spectrum acquires dips at several wavelengths, due to atmospheric scattering and absorption by the constituent gases, as shown in Fig. 15.2a. While the total flux making its way to the Earth's surface depends on the time of day, spatial location, and weather, the total (i.e., direct and diffuse) global average impinging on a surface with normal pointing directly at the sun is nearly 1000 W/m². To account for the differences, it is common to use three standard benchmark solar spectra: AM0, AM1.5D, and AM1.5G. Here "AM" stands for "air-mass," or the equivalent number of atmospheric thicknesses traversed for a given sun elevation angle. AM1 corresponds to the solar spectrum when the sun is directly overhead, e.g., at solar noon on the equator on the vernal or autumnal equinox, while AM0 is the spectrum hitting the top of the atmosphere. The D and G denote the direct and "global" (direct and diffuse) light. For convenience, the AM1.5G spectrum rounds to exactly 1000 W/m².

What conversion efficiency can we expect when we generate electricity from solar radiation? It is equal to the efficiency of transferring radiative power from a blackbody source at $T_s = 5800$ K to a non-equilibrium blackbody cell at $T_c = 300$ K, under an applied bias $V = V_m$ when we apply Eqs. (4.91) and (4.93). The ultimate limit is the Carnot efficiency that follows directly from the laws of thermodynamics, which we can derive by equating the emitted and absorbed energies, as well as the entropy differences of the hot and cold objects. This best possible (reversible) scenario yields

$$\eta_C = 1 - \frac{T_c}{T_s} \approx 95\% \tag{15.9}$$

As expected, the Carnot efficiency vanishes when $T_c = T_s$ because we can only move heat from a warmer object to a cooler object. To move it in the opposite direction, we need to expend energy rather than produce it. In practice, the entropy change of the cold object is larger, so Eq. (15.9) is only an upper limit.

The solar cell under forward bias $V = V_m$ also emits radiation and, for this reason, it cannot capture the entire incident power of the sunlight. This is because emission counteracts the photocurrent and can reduce $\mathcal{J}_m$, $V_m$, and $V_{oc}$. We might like to engineer the density of modes in the medium surrounding the solar cell to minimize this emission, but this is not easy. Assuming the "free-space" density of modes in Eq. (4.68), how much does emission reduce the efficiency?

The emission loss is irreversible because only a tiny fraction of the solar radiation intercepts the Earth in a narrow-divergence beam ($\Omega_{abs} = 2\pi \int_0^{\theta_d} \cos\theta \sin\theta d\theta = \pi \sin^2\theta_d$), whereas the solar cell emits in all directions ($\Omega_{ems} = 2\pi \int_0^{\pi/2} \cos\theta \sin\theta d\theta = \pi$). Here the integration is over the directional factor $\cos\theta$ that applies to a thin and flat emitter/absorber. In other words, an increase in entropy occurs because the emitted radiation is lost irretrievably to the cool surroundings.

If we recall that the solid angle enters the emission rate in the pre-factor of Eq. (4.93), the open-circuit voltage is reduced by

$$\Delta V_{oc} = \frac{k_B T_c}{e} \ln\left(\frac{\Omega_{ems}}{\Omega_{abs}}\right) = \frac{k_B T_c}{e} \ln\left(\frac{\pi}{\pi \sin^2\theta_d}\right) \approx 0.28 \text{ V} \tag{15.10}$$

which is similar to what we found in Eq. (15.5). We can make up some ground by increasing the divergence angle via concentration, but then the optics must cover an area much larger than the device itself. The theoretical maximum concentration achievable in this manner is $C = 1/\sin^2\theta_d \approx 46,000$. The practical maximum of $C \approx 100 - 1000$ is limited by the optics, device heating, and series resistance. Using the relation $V_{oc}(C) = V_{oc}(C = 1) + \frac{k_B T}{e} \ln C$ that follows from Eq. (15.10), we can claw back a little more than half of the reduction in the open-circuit voltage due to the solid-angle mismatch. The emission from the solar cell also directly reduces the operating current, but this is a relatively small effect.

The simplest solar-cell design is a single junction with fixed energy gap. What should this gap be? Are we better off with the strong photocurrent from a narrow-gap material that captures more of the sunlight, or with the high open-circuit voltage that comes with a wide gap according to Eq. (15.5)? In this case, moderation is a virtue. With a narrow gap, hardly any voltage develops across the

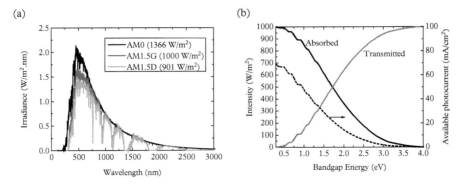

**Figure 15.2** *(a) Spectral irradiance for the three common standard reference solar spectra and their integrated power densities. (b) Absorbed and transmitted power density versus absorber bandgap. The two curves intersect at ≈ 1.7 eV. The available photocurrent versus bandgap energy is also shown.*

terminals. But if the gap is too wide, the photocurrent is low because only a small fraction of the photons is absorbed. A "moderate" gap allows photon energies in the "middle" of the solar spectrum to be absorbed. Using the data in Fig. 15.2a, we can estimate the optimal gap from the incident irradiance, i.e., the distribution of solar photons per unit energy $n_{ph}(\hbar\omega)$, weighted by the photon energy. The distribution's center of mass is found from

$$\int_0^{E_{g,opt}} \hbar\omega\, n_{ph}(\hbar\omega)\, d(\hbar\omega) = \int_{E_{g,opt}}^{\infty} \hbar\omega\, n_{ph}(\hbar\omega)\, d(\hbar\omega) \qquad (15.11)$$

This exercise leads to an optimal gap of ≈ 1.7 eV, as determined graphically in Fig. 15.2b for the AM1.5G spectrum. It is more accurate to maximize the output power $\mathcal{J}_m V_m$ because the open-circuit voltage increases approximately linearly with gap, whereas the available photocurrent decreases much faster. We account for this effect by shifting the gap slightly below 1.7 eV and make it even narrower for higher sunlight concentrations.

To estimate the operating bias and current, we again assume that the only (fundamental) loss is due to the emitted radiation. Using Eq. (4.93) for the radiative flux emerging from the semiconductor toward the top surface with an absorbing substrate and $E_g \gg k_B T_c$, we write

$$R_{sp,areal} \approx \frac{R_{sp,app}}{\alpha} \approx \frac{n_r^2}{2\pi^2} \frac{E_g^2 k_B T_c}{\hbar^3 c^2} \exp\left[\frac{eV - E_g}{k_B T}\right] \qquad (15.12)$$

where the radiation is assumed to come from a layer with thickness equal to the absorption depth $1/\alpha$.

Now we subtract the photocurrent from the dark current due to radiative recombination $eR_{sp,areal}$, which leads to the current–voltage relation

$$\mathcal{J}(V) = \frac{en_r^2}{2\pi^2} \frac{E_g^2 k_B T_c}{\hbar^3 c^2} \exp\left[\frac{eV - E_g}{k_B T_c}\right] - en_{ph} \tag{15.13}$$

Extracting the efficiency from this expression gives us the already familiar result for a single junction solar cell, shown in Fig. 15.3. For the AM1.5G spectrum, the peak efficiency is $\approx$ 33% for an energy gap of $\approx$ 1.3 eV. Setting the total current to 0, we can find the open-circuit voltage

$$V_{oc} = \frac{E_g}{e} - \frac{k_B T_c}{e} \ln\left[\frac{n_r^2}{2\pi^2} \frac{E_g^2 k_B T_c}{\hbar^3 c^2 n_{ph}}\right] \approx 1.0 \text{ V} \tag{15.14}$$

The subtracted term follows from the thermodynamic argument discussed earlier. The operating bias is lower still, by a few $\frac{k_B T_c}{e}$ (see Eq. (15.6)). In practice, the conversion efficiency never reaches this upper limit even in the most mature solar-cell materials such as single-crystal Si and GaAs. For example, the best result reported for a GaAs photovoltaic device is $\approx$ 29% for unconcentrated sunlight. The record conversion efficiency for Si solar cells is currently 26.7%, with very minor increases for over a decade. This is because the dark current is never purely radiative, owing to the Shockley–Read–Hall and Auger recombination mechanisms, the absorption of the sunlight is incomplete, and parasitic resistances are invariably present in the circuit. Figure 15.3 shows a few recent efficiency records for single-junction solar cells under the AM1.5G spectrum.

To summarize, we have identified four main mechanisms that limit the efficiency of solar cells:

(1) the Carnot efficiency in Eq. (15.9), which quantifies the average emission energy of the source in terms of the thermal energy in the cell;

(2) the mismatch between the solid angles of the source and cell in Eq. (15.10), which governs how much radiative emission is irretrievably lost from the cell;

(3) the fraction of radiation not absorbed because it falls below the energy gap of the absorber; and

(4) the energy dissipated when photoexcited electrons and holes relax to their respective band edges before they are extracted from the cell. This mechanism makes the gap the upper limit on the open-circuit voltage.

We cannot do anything about the Carnot efficiency, which in any case is quite high. It is indeed possible to gain back some of the inefficiency by concentrating the sunlight, i.e., narrowing the ratio of solid angles. We also know from the

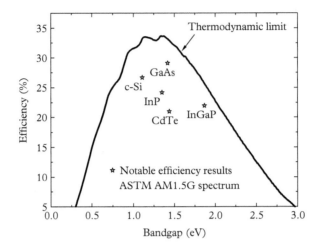

**Figure 15.3** *Limiting efficiency versus energy gap for a single-junction solar cell under AM1.5G illumination with only radiative recombination in the cell. The stars show experimentally demonstrated efficiencies from single-junction solar cells without concentration based on different semiconductor materials.*

monochromatic pump example that if we fully overcome the final two limitations (no below-gap absorption and energy relaxation), we can at least double the conversion efficiency without concentration. This list is not exhaustive, and real devices also suffer from non-fundamental limitations such as nonradiative recombination, parasitic series resistance, metal contacts shading some of the absorbing area, etc.

Is there a practical way to reduce the below-gap and energy relaxation losses? We can imagine the broad solar spectrum as a sum of many spectrally narrow sources. Figure 15.4a illustrates that if we split the radiation into multiple spectral channels and let each pass through a photovoltaic element designed for absorption at that channel's wavelength, we approach the case with a single-wavelength pump laser analyzed earlier. More practically, we could build our device in layers. Figure 15.4b shows that the topmost layer should have the widest gap to make it transparent to all but the shortest-wavelength radiation, the second layer has a narrower gap that absorbs somewhat longer wavelengths, and so on down to the narrowest-gap junction. We also need to connect all the junctions electrically in series. Since wider-gap junctions contribute the most to the open-circuit voltage, we can stop adding channels when the gap approaches a certain multiple of $k_B T_c$. Because of the series connection, we also want to match the current densities generated by all the junctions, by analogy with equal photocurrents generated by individual stages of an ICD (see Section 14.5).

To convert the scheme in Fig. 15.4 into a practical device, we stack multiple junctions with different energy gaps and maximize the product of the

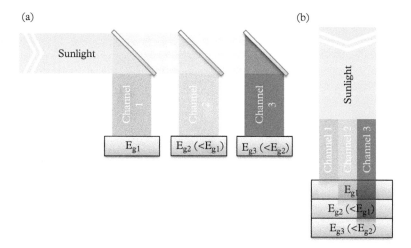

**Figure 15.4** *Illustration of the spectral-splitting approach to realizing higher conversion efficiency in a solar cell: (a) conceptual scheme, and (b) the most common experimental realization (here with three different channels/junctions).*

photocurrent and total voltage (equal to the sum of the voltages over all the junctions). There are practical limits to how many junctions it makes sense to include in a single solar cell. Under AM1.5G illumination, 37.8% efficiency has been achieved using four junctions, and for highly concentrated sunlight the current record stands at 46%. Commercial devices use three or four *p–n* junctions, while research-scale demonstrations of five- and six-junction devices are now commonplace. The returns from using even more junctions diminish rapidly, and multi-junction devices must contend with non-idealities and the unavoidable subcell current mismatches that result from diurnal and seasonal spectral variations.

The subcells of multi-junction devices are connected by tunnel junctions (see the discussion in connection with Fig. 12.17). As an example, Fig. 15.5 shows the equilibrium band diagram for a two-junction solar cell. The cell consists of an InGaP *n–p* junction and a GaAs *n–p* junction separated by a thin, highly doped GaAs *p–n* tunnel junction. Many solar cells add wider-gap heterostructures to suppress nonradiative interface recombination at the outer extremities of the subcell junctions. These are known as window (or front surface field) layers at the side of the cell facing the sun, and as back surface field layers at the rear. The choice of energy gap and doping for these layers is based on trade-offs of the surface recombination velocity at the heterointerface, series resistance for majority carriers traversing the interface, and, for window layers in particular, transparency to the photons absorbed in the device.

Better known is a multi-junction solar cell based on the Ge substrate, devised for applications in terrestrial concentrator systems and currently the industry

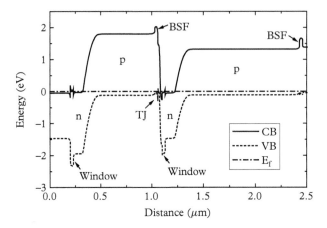

**Figure 15.5** *Equilibrium band diagram for a simple InGaP/GaAs two-junction (2J) solar cell connected via a heavily doped tunnel junction (TJ). Back surface field (BSF) and window layers are indicated.*

workhorse for space power conversion. The narrowest-gap junction is in the Ge, which has an indirect gap of 0.66 eV and direct gap at the Γ point of 0.8 eV. InGaAs and GaInP junctions, with gaps of 1.4 and 1.85 eV, respectively, are grown on top of the Ge substrate.

Because any epitaxial layer with low defect density must be lattice matched to the substrate, the menu of available options for multi-junction cells is somewhat limited. In fact, the theoretical optimal gaps almost never correspond to materials with the same lattice constant. An example is the putative three-junction device on a GaAs substrate shown in Fig. 15.6. The main difference from the Ge-based solar cell is that the Ge junction is replaced with a subcell having a theoretical optimal gap of ≈ 1.0 eV (see Fig. 15.7 for the calculation). On a GaAs substrate, the lower junction can be composed of InGaAs quantum wells with strain-compensating barriers, an alternative bulk material such as InGaAsN, or even multiple layers of InAs quantum dots. We already saw that the energy gaps are chosen to match the photocurrents in each junction, because contacting each junction individually is too cumbersome. Many ideas attempt to further expand the playing field, including metamorphic growth and wafer bonding to bring together materials grown on dissimilar substrates, e.g., GaAs and InP, or GaAs and GaSb.

As in a single-junction solar cell, the efficiency of a multi-junction device increases with concentration, as shown in Fig. 15.8 for two- and three-junction devices. While the higher efficiency is a nice bonus, in this case the main motivation for concentrating optics is a smaller solar-cell area. The space occupied by these expensive, epitaxially grown materials can then be replaced with much cheaper glass or plastic optics.

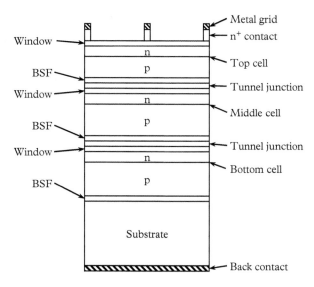

**Figure 15.6** *Layer structure for a three-junction solar cell on a GaAs substrate.*

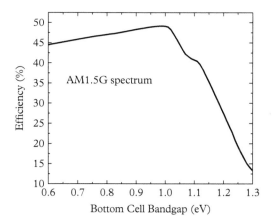

**Figure 15.7** *Limiting efficiency of a three-junction solar cell consisting of InGaP and GaAs top junctions as a function of the energy gap of the bottom subcell.*

## 15.2   Advanced Solar-Cell Concepts

Short of splitting the solar spectrum into distinct wavelength bands or using a multi-junction cell, can we improve the conversion efficiency simply by inserting some material with narrower gap into the basic *p–n* junction? The idea is to capture some of the photons with energy below the gap of the original junction,

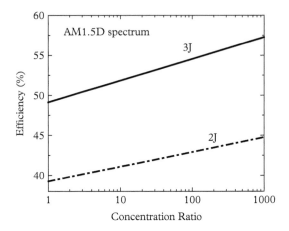

**Figure 15.8** *Limiting efficiency of two-junction and three-junction solar cells versus concentration ratio. In practice, heating, series resistance, and acceptance angle limit the maximum concentration ratio to $C \approx 100 - 1000$.*

for conversion to additional current. But the problem is that if the same electron and hole quasi-Fermi levels extend over the entire device, any increase in the photocurrent is accompanied by a decrease of the open-circuit voltage. This is a simple consequence of the reduced separation between the two quasi-Fermi levels. If the extra material is very thick, the conversion efficiency remains the same as for an entire junction made of the wider-gap material, whereas if it is very thin the efficiency hardly changes.

Do the quasi-Fermi levels always remain flat over the entire absorber? It has been reported that narrower-gap quantum wells may be inserted into the junction without shifting the open-circuit voltage by the full difference in the gaps. Whether or not this surprising finding withstands the test of time, the concept of a quantum-well solar cell is well worth discussing.

Figure 15.9a is the schematic band diagram of a solar cell with 10 quantum wells. Here InGaAs wells with gap 50–300 meV lower than GaAs (1.42 eV) are inserted into the depletion region of what is otherwise a single GaAs homojunction. The GaAsP barriers between the wells induce tensile strain that compensates the compressive strain of the wells. As we saw in Chapter 11, a single pass through a quantum well absorbs 0.6–0.7% of the incident radiation at a wavelength close to the absorption edge, via CB1–HH1 transitions (see Fig. 11.1). There-fore, we need at least 20–40 such wells to significantly boost the photocurrent. Figure 15.9b shows the absorption spectrum for this quantum well and barrier system, compared to that of bulk GaAs. Ideally, all the wells should fit inside the depletion region of the junction, so the built-in electric field can extract the photogenerated electrons and holes via thermal emission from the wells (with the escape time much shorter than the recombination time). An alternate design

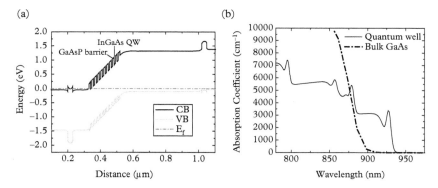

**Figure 15.9** *(a) Schematic band diagram of a single-junction solar cell with narrower-gap 80 Å In$_{0.12}$Ga$_{0.88}$As quantum wells separated by 100 Å GaAs$_{0.8}$P$_{0.2}$ barriers, inserted in the depletion region to improve the short-circuit current. On the p side, the structure includes an additional Al$_{0.2}$Ga$_{0.8}$As back surface field barrier, while on the n side, it contains an In$_{0.49}$Ga$_{0.51}$P window layer. (b) Absorption spectrum of the InGaAs/GaAsP strain-compensated well–barrier pair from (a), compared to that of bulk GaAs. Exciton resonances near the band-edge transitions are included, in contrast to Fig. 11.1.*

choice may be to couple the wells via thin barriers, and then extract the carriers at the ends of the depletion region. In any case, this means the background doping should be low enough to accommodate the combined well and barrier thicknesses in the depletion region (see Eq. (14.17)).

Even if the quasi-Fermi levels shift by the full difference in gap when the wells are inserted into the junction, the concept may still be useful in multi-junction cells. For example, in a two-junction GaAs/InGaP cell, the InGaP subcell generates more photocurrent, which limits the overall conversion efficiency. GaAs-based quantum wells can boost the photocurrent produced there to closely match the InGaP subcell. In other multi-junction cells, we may want to insert InGaAsP quantum wells into the In$_{0.49}$Ga$_{0.51}$P junction. These wells with an In fraction higher than 49% are also compressively strained and their gap is 100–300 meV narrower than bulk InGaP. We can compensate the strain with a higher Ga fraction in the InGaP barriers, by analogy with the InGaAs/GaAsP well–barrier pair. As discussed in Section 15.1, we can also use an effective quantum-well "material" with a gap of $\approx$ 1.0 eV, most often strain-compensated InGaAs/GaAsP, for the entire bottom junction of a three-junction solar cell.

It seems we can generalize as follows: The extra quantum-well absorption may be viewed as being due to a band of states within the energy gap of the bulk-like semiconductor that forms the junction, as shown in Fig. 15.10a. This generalization led to proposals of *intermediate-band* solar cells that do not necessarily use quantum wells. The materials can be as varied as InAs/GaAs quantum dots or highly mismatched alloys with gaps below that of the main junction. The key requirement is that the intermediate band should conduct within

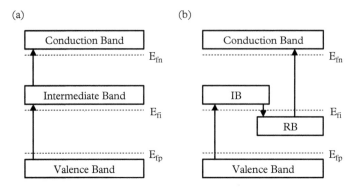

**Figure 15.10** *(a) Schematic of the general intermediate-band solar-cell concept. (b) Schematic of an intermediate-band (IB) solar cell with a ratchet band (RB).*

the cell, without being intimately connected to the conduction or valence bands in Fig. 15.10a. This means the entire structure does not share the same quasi-Fermi levels. It follows that the ideal intermediate-band cell has three quasi-Fermi levels: for the primary conduction and valence bands, $E_{Fn}$ and $E_{Fp}$, and the quasi-Fermi level of the isolated intermediate band, $E_{Fi}$.

The intermediate band supports an additional absorption channel between the conduction and valence bands, which boosts the short-circuit current with below-gap photons. Since the photoexcited carriers are extracted only from the conventional bands with no direct electrical contact to the intermediate-band states, the operating voltage remains high. Carriers can only transition to and from the intermediate band by emitting or absorbing a photon. To promote transitions from the intermediate band to the conduction band, we want to fill the former, either by doping or at high intensities by pumping carriers from the valence band.

One problem with the intermediate band is that after a large number of electrons have been promoted to it, further transitions are blocked via the final-state occupation factor. Can we circumvent this blocking? The proposal illustrated in Fig. 15.10b involves splitting off a so-called *ratchet band*, which provides a reservoir for carriers that rapidly relax from the intermediate band. The energy difference between the intermediate and ratchet bands should be much smaller than the gap, but also $> k_B T_c$ to prevent thermal backfilling that returns carriers to the intermediate band. Two kinds of optical transitions are possible: (1) from the valence band to the intermediate band; and (2) from the ratchet band to the conduction band. The rapid relaxation means that the intermediate band remains nearly empty, while the ratchet band is mostly full. It is not easy to find appropriate material combinations and optical properties, but research in this area continues.

Another rather speculative advanced solar-cell concept is the *hot-carrier* solar cell illustrated in Fig. 15.11. The idea is to prevent the loss of efficiency due to carrier thermalization following the absorption of photons with energies well above the energy gap. If these carriers can be collected by contacts that only

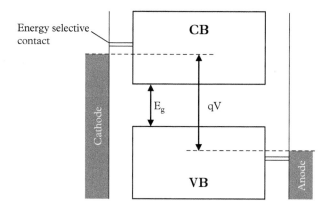

**Figure 15.11** *Schematic band diagram for a notional hot-carrier solar cell with energy-selective contacts.*

accept carriers in a specific narrow energy range, e.g., well above the conduction-band maximum, the conversion efficiency can be enhanced without resorting to multiple junctions. Even though carriers only at specific energies are extracted, the photocurrent can presumably be almost as high as without energy-selective contacts if a thermal distribution is established on a very fast time scale. Any improvement would be observed as an increase in the open-circuit voltage over the value from Eq. (15.5). In practice, it is difficult to prevent hot carriers in a semiconductor from relaxing very rapidly. Materials and structures with slow relaxation must be found or developed for this purpose.

Even if these more exotic concepts never become practical, there are other ways to improve on the state of the art. For example, we can raise the open-circuit voltage for a solar cell with strong nonradiative recombination by making the active material thinner. To avoid a reduction in photocurrent, we can apply the light-trapping concepts that were explored for photodetectors in Section 14.7. We could even incorporate small particles that scatter light in all directions, to increase the effective path length through the absorbing material. These should be positioned near the back surface, so the light is not scattered before it reaches the absorber. Of course, none of these potential improvements can overcome the fundamental radiative emission limit (see Section 15.1). But they may still be valuable in practice, since even a small boost to the performance of these mature devices counts.

## 15.3   Thermophotovoltaic Devices

We already mentioned in Section 15.1 that the solar spectrum, as observed from the ground, is not much different from a blackbody emitting at 5800 K, the surface temperature of the Sun. More generally, any hot object that does not perfectly

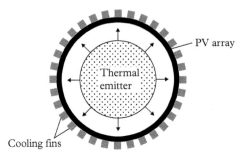

**Figure 15.12** *Top view schematic of a thermophotovoltaic device for converting heat to electricity that uses a concentric cylinder geometry.*

trap all its internal radiation emits as a blackbody at its surface temperature. We certainly would not want to heat an object just to collect this radiation, but energy production and other industrial processes often generate enough waste heat that recycling it into electricity may be worth the trouble. An alien may even perceive all human activity as a giant waste heat source that radiates into the cold space surrounding us. Why not try to recover at least some fraction of that radiation?

By analogy with the solar cell, the gap in a power-generating cell is tuned to the center-of-mass photon energy of the blackbody radiation, which corresponds to a temperature much lower than the surface of the Sun. Ideally, the surface of the cell surrounds the hot object, as exemplified by the concentric cylinder geometry shown schematically in Fig. 15.12 (one of the many configurations adopted in practice). We can apply the approach used in Fig. 15.2 to estimate the optimal energy gap for such a *thermophotovoltaic* (TPV) device. Figure 15.13 shows resulting optimal gap and maximum power conversion efficiency for a single-junction TPV structure, as a function of the effective blackbody emitter temperature (in the range 700–2000°C). These results represent the fundamental limits because we neglect nonradiative recombination and series resistance as well as assume a view factor of unity; i.e., all photons leaving the emitter strike the surface of the device. We see from the figure that to optimally convert the radiation from a 700–2000°C emitter, the energy gap should range in the mid-IR to near-IR (0.2–0.7 eV).

What materials are used in practical TPV devices? $In_{0.53}Ga_{0.47}As$ (with gap 0.74 eV) is quite common for higher emitter temperatures, although its conversion efficiency is limited by the range of photon energies that can be absorbed. Metamorphic InGaAs, with In fraction > 53% and gap down to 0.6 eV, is a better fit to the common blackbody spectra. We can also switch to a GaSb substrate, on which InGaAsSb quaternaries with gap near 0.5 eV offer a better match to an emitter with effective temperature ~1500°C. While we might want an even narrower gap, miscibility issues with the quaternary alloy make this a difficult proposition. Multi-junction TPV devices are also on the rise, to provide higher

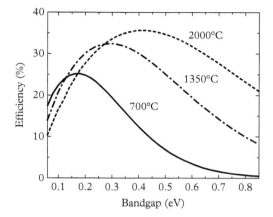

**Figure 15.13** *Optimal TPV energy gap and maximum conversion efficiency as a function of effective emitter temperature. The simulations assume operation at 300 K, and carefully account for the thermal energy's much stronger influence on the narrow-gap TPV cell's performance. Nonradiative recombination and series resistance are neglected, and all photons leaving the emitter strike the surface of the device.*

power conversion efficiencies by capturing a wider range of photon energies as in multi-junction solar cells.

At the narrow-gap end, a single junction has low conversion efficiency because the open-circuit voltage is small, especially compared to the total series voltage drop. We can increase $V_{oc}$ by connecting multiple junctions in series, as in a multi-junction TPV or solar cell. However, we also know that the connecting tunnel junctions in an antimonide device can be replaced with semimetallic interfaces based on type II InAs/GaSb quantum wells (see Section 12.4). Then the gap in the different stages can be the same, or decreasing from top to bottom as in a multi-junction solar cell. This multistage interband cascade TPV device is a close cousin of the ICD (see the band diagram in Fig. 14.20) with an InAs/GaInSb superlattice absorber, but operating in forward bias. Once again, we would adjust the stage lengths to generate the same photocurrent in each stage.

An interband cascade TPV device with absorber gap near 0.25 eV and seven active stages achieved an open-circuit voltage of 0.65 V at room temperature. The dark-current–voltage curve for a diffusion-limited interband cascade TPV device with the same energy gap and applied voltage in every stage follows from Eq. (15.1):

$$\mathcal{J}_d = \mathcal{J}_0 \left( e^{eV/Mk_B T_c} - 1 \right) \tag{15.15}$$

where $M$ is the number of stages, and $\mathcal{J}_0 \approx e(L_d/\tau)\left(n_i^2/N_A\right)$ is inversely proportional to the minority-carrier lifetime in the absorber. As we saw in Section 12.2,

Shockley–Read–Hall and Auger nonradiative processes often dominate the lifetime in a bulk or superlattice narrow-gap material. If the dark current is *not* diffusion limited, the curves can still be fitted using Eq. (15.15) with the parameter $M_{eff} \geq M$. The open-circuit voltage is then analogous to Eq. (15.3):

$$V_{oc} = \frac{M_{eff} k_B T_c}{e} \ln \left( 1 + \frac{e\Phi}{\mathcal{J}_0} \right) \tag{15.16}$$

where $\Phi$ is the absorbed blackbody flux. As in an ICD, $M_{eff}$ should approach $M$ at higher operating temperatures, where diffusion is more likely to dominate. By analogy with Eqs. (15.4) and (15.5), the maximum open-circuit voltage is bounded from above by the product of the number of stages and the energy gap of each stage, $ME_g/e$. However, the actual value is usually much lower because we are far away from the radiative limit.

Another pathway to improving the TPV performance is to keep the gap wider than optimal (say, 0.5–0.6 eV for an InP- or GaSb-based bulk TPV cells), and modify the *emission* of the hot object. We discussed in Section 4.4 that this can be done by changing the density of optical modes in Eq. (4.68). While the blackbody mode density in a homogeneous environment increases as $(\hbar\omega)^2$, the density for an emitter with wavelength-scale patterning can be below the energy gap and, much higher, just above it. A hot object with this density of states is called a *selective emitter*. We can also boost the selectivity by inserting a filter that reflects below-gap photons back into the emitter. Another approach to enhancing the radiative transfer is to reduce the spatial gap separating the TPV device from the emitter, to a value much smaller than the peak emission wavelength. Near-field coupling then speeds up the radiative rate, although thermal insulation of the TPV device becomes difficult. Whatever the configuration, most practical applications require a conversion efficiency $\geq 20\%$ from a hot object with temperature $\geq 1000°C$. Such an ambitious goal remains elusive at this time.

## 15.4   Basics of Nonlinear Optics in Quantum Wells

All the optical effects discussed so far have been *linear*, including interband and intersubband absorption, gain, and radiative recombination. The transition strength of a linear process is independent of optical intensity; e.g., the absorption coefficient is the same regardless of the beam strength. On the other hand, in a *nonlinear* process the nature of the interaction depends on the light intensity because multiple photons interact with electronic transitions at the same time. The most important nonlinear processes are of second and third order, with higher-order effects becoming progressively weaker. We will limit this discussion to second-order nonlinear processes in quantum wells. While bulk III–V semiconductors also experience nonlinear effects, they are not very different physically from the same phenomena in other nonlinear crystals.

The overall effect of a linear optical process scales is proportional to the optical electric field. Traditionally, the effect is described by the *polarization* of the semiconductor medium (not to be confused with the electric field polarization!) In a second-order process, the material polarization is proportional to the square of the electric field for a single beam, or to the product of the fields from two different beams. The proportionality factor is styled as the *susceptibility*. The second-order susceptibility is $\chi_{ijk}^{(2)}$, where $i,j,k$ refer to the directions of the material polarization and the polarizations of the two electric fields. Therefore, the material polarization is defined

$$P_i = \varepsilon_0 \sum_{jk} \sum_{pq} \chi_{ijk}^{(2)} \left(\omega_p + \omega_q, \omega_p, \omega_q\right) E_j\left(\omega_q\right) E_k\left(\omega_p\right) \tag{15.17}$$

where $\omega_p$ and $\omega_q$ are the frequencies of the two electric fields, which can be the same.

For a satisfying description of nonlinear susceptibility, we return to the perturbation solution for electronic transitions from Section 4.1. We expanded the time-dependent wavefunction in terms of stationary electronic eigenfunctions $\varphi_l(\mathbf{r})$ at energies $\hbar\omega_l$:

$$\psi\left(\mathbf{r}, t\right) = \sum_l a_l(t)\, \varphi_l\left(\mathbf{r}\right) e^{-i\omega_l t} \tag{15.18}$$

We assume that the expansion produces time-dependent amplitudes $a_l^0(t)$ in the absence of an incident field. Time-dependent perturbation theory tells us that the new amplitudes in an applied field are obtained by integrating over the original $a_l^0(t)$ and the elements of the perturbing potential expanded in terms of the same eigenfunctions:

$$a_m^1(t) = \frac{1}{i\hbar} \sum_l \int_{-\infty}^{t} dt'\, a_l^0\left(t'\right) V_{ml}\left(t'\right) e^{i\omega_{ml} t} \tag{15.19}$$

where $\omega_{ml} = \omega_m - \omega_l$. We are simply integrating over the perturbing potential that was not included in the original wavefunction (recall the time-dependent form of the Schrödinger equation $\frac{\partial \psi}{\partial t} = \frac{1}{i\hbar}\left(H_0 + V\right)\psi$).

The perturbing potential is the dipole–field product, as described in Chapter 4, but now expanded over all the field components and with the time dependence explicitly included:

$$V_{ml}(t) = e \sum_p \mathbf{E}\left(\omega_p\right) \cdot \mathbf{r}_{ml} e^{-i\omega_p t} \tag{15.20}$$

We substitute Eq. (15.20) into Eq. (15.19) and assume that initially $a_s^0 \approx 1$ and $a_l^0 \approx 0$ for all $l \neq s$. The integration yields the linear contribution due to optical transitions:

$$a_m^1(t) = -\frac{e}{\hbar} \sum_p \frac{\boldsymbol{E}\left(\omega_p\right) \cdot \boldsymbol{r}_{ms}}{\omega_{ms} - \omega_p} e^{i\left(\omega_{ms} - \omega_p\right)t} \tag{15.21}$$

Unless $\omega_{ms} = \omega_p$ the phase factor averages to zero for $t \to \infty$. In other words, we recover the Dirac delta function $\delta\left(\omega_{ms} - \omega_p\right)$, similar to Eq. (4.8), but obtained more rigorously than in Section 4.1.

Substituting $a_m^1$ in Eq. (15.21) in place of $a_l^0$ in Eq. (15.19), we can calculate the *second-order* perturbative contribution

$$a_n^2(t) = \frac{1}{i\hbar} \sum_m \int_{-\infty}^t dt' a_m^1\left(t'\right) V_{nm}\left(t'\right) e^{i\omega_{nm}t}$$

$$= \frac{e}{i\hbar} \sum_m \int_{-\infty}^t dt' a_m^1\left(t'\right) \sum_q \boldsymbol{E}\left(\omega_q\right) \cdot \boldsymbol{r}_{nm} e^{-i\omega_q t} e^{i\omega_{nm}t} \tag{15.22}$$

Integrating this expression, we obtain

$$a_n^2(t) = \frac{e^2}{\hbar^2} \sum_{pq} \sum_m \frac{\left[\boldsymbol{E}\left(\omega_q\right) \cdot \boldsymbol{r}_{nm}\right]\left[\boldsymbol{E}\left(\omega_p\right) \cdot \boldsymbol{r}_{ms}\right]}{\left(\omega_{ns} - \omega_p - \omega_q\right)\left(\omega_{ms} - \omega_p\right)} e^{i\left(\omega_{ns} - \omega_p - \omega_q\right)t} \tag{15.23}$$

The material polarization has several terms. The most relevant to us is $\boldsymbol{P} = eN\langle\psi^0|\boldsymbol{r}|\psi^2\rangle + \ldots$, where $N$ is the volume density of possible transitions in the material. Using this form and Eq. (15.23), we can write the polarization as

$$\boldsymbol{P} = \frac{Ne^3}{\hbar^2} \sum_{pq} \sum_{mn} \frac{\boldsymbol{r}_{sn}\left[\boldsymbol{E}\left(\omega_q\right) \cdot \boldsymbol{r}_{nm}\right]\left[\boldsymbol{E}\left(\omega_p\right) \cdot \boldsymbol{r}_{ms}\right]}{\left(\omega_{ns} - \omega_p - \omega_q\right)\left(\omega_{ms} - \omega_p\right)} e^{-i\left(\omega_p + \omega_q\right)t} + \ldots \tag{15.24}$$

which along with Eq. (15.17) leads to the second-order susceptibility

$$\chi_{ijk}^{(2)}\left(\omega_p + \omega_q, \omega_p, \omega_q\right) = \frac{Ne^3}{\varepsilon_0 \hbar^2} \sum_{mn} \frac{r_{sn}^i r_{nm}^j r_{ms}^k}{\left(\omega_{ns} - \omega_p - \omega_q\right)\left(\omega_{ms} - \omega_p\right)} + \ldots \tag{15.25}$$

This nonlinear coefficient involves contributions from three different transitions ($s \to m, m \to n, n \to s$). A good illustration of applying Eq. (15.25) is the quantum-well structure with three subbands shown in Fig. 15.14, for which $s \to 1$, $m \to 2, n \to 3$. Since we already know that the intersubband dipoles point along the growth axis, we can assume all the fields are $z$-polarized, i.e., $i = z, j = z, k = z$. For an optical beam propagating at some angle $\theta$ with respect to the growth axis, each dipole scales by a factor $\sin\theta$ to account for the dot product between the field and the dipole, by analogy with Eq. (11.38). This means we can rewrite Eq. (15.25) for the structure of Fig. 15.14 as

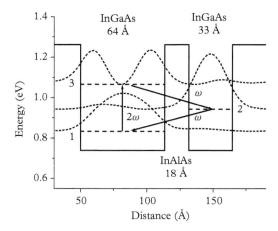

**Figure 15.14** *Coupled quantum-well structure with three almost equally spaced subbands for second-harmonic generation at 5 μm with pumping at 10 μm.*

$$\chi^{(2)}_{zzz}\left(\omega_p+\omega_q,\omega_p,\omega_q\right)=\frac{Ne^3z_{13}z_{12}z_{23}\sin^3\theta}{\varepsilon_0\left(E_{13}-\hbar\omega_p-\hbar\omega_q\right)\left(E_{12}-\hbar\omega_p\right)} \tag{15.26}$$

This expression uses the notation of Chapter 11, e.g., $z_{\mu\nu}=\langle c,\mu|z|c,\nu\rangle$. Here $N$ is the three-dimensional electron density per period of the structure, i.e., including both wells and barriers. If the envelope functions do not extend significantly into the barriers, the barriers do not contribute much to the nonlinear process.

Missing from Eqs. (15.18)–(15.26) is the electron's finite lifetime in a subband state, i.e., the effect of rapid intersubband and intrasubband scattering discussed in Chapter 13. We can correct for that effect by adding an imaginary component to the intersubband energy separations. This is similar to what we did in Section 4.3, apart from the use of a hyperbolic-secant line shape instead of a Lorentzian. In this manner, we generalize Eq. (15.26) to read

$$\chi^{(2)}_{zzz}\left(\omega_p+\omega_q,\omega_p,\omega_q\right)=\frac{Ne^3z_{13}z_{12}z_{23}\sin^3\theta}{\varepsilon_0\left(E_{13}+i\hbar\Gamma_{13}-\hbar\omega_p-\hbar\omega_q\right)\left(E_{12}+i\hbar\Gamma_{12}-\hbar\omega_p\right)}$$
$$\tag{15.27}$$

The transition linewidths $\Gamma_{12}$ and $\Gamma_{13}$ may scale with the corresponding transition energies $E_{12}$ and $E_{13}$, or have some other dependence based on the nature of the transition. Here we assume $\Gamma_{12}=\Gamma_{13}\equiv\Delta_l/\hbar$ for simplicity, where we can think of $\Delta_l/\hbar$ as a scattering rate (due to phonons or some other mechanism). We will focus on the case of *second-harmonic generation* with a single pump beam ($\hbar\omega_p=\hbar\omega_q\equiv\hbar\omega$), but the physics are quite similar when there are two pump beams with different frequencies, which produce *sum-frequency* (and difference-frequency) *generation*. For second-harmonic generation, Eq. (15.27) becomes

$$\chi^{(2)}(2\omega,\omega,\omega) = \frac{Ne^3 z_{13}z_{12}z_{23}\sin^3\theta}{\varepsilon_0(2\hbar\omega - E_{13} - i\Delta_l)(\hbar\omega - E_{12} - i\Delta_l)} \tag{15.28}$$

where for simplicity we have started omitting the polarization subscripts in the susceptibility. The second-order susceptibility in Eq. (15.28) peaks at *double resonance*: $E_{12} = \hbar\omega$, $E_{13} = 2\hbar\omega$, $E_{23} = \hbar\omega$. The magnitude of the susceptibility at resonance scales inversely with $\Delta_l^2$. The assumption is that all the electrons initially occupy the lowest subband, or that the quantum well is doped to a level for which electrons occupy only a single subband. Otherwise, Eq. (15.28) should be corrected by an occupation factor. Even though the nonlinear susceptibility in Eq. (15.28) seems independent of growth direction; in fact, its sign flips when we reverse the sequence of the wells.

As the pump and second-harmonic beams propagate, e.g., along the $x$ axis, they experience both loss and nonlinear conversion. They will also diffract if there is no waveguide to confine the beams, but the diffraction is negligible if the spot sizes are large enough. We write the fields as $\boldsymbol{E}(\omega) = \boldsymbol{z}E_\omega e^{i(\omega t - k_\omega x)}$ and $\boldsymbol{E}(2\omega) = \boldsymbol{z}E_{2\omega}e^{i(2\omega t - k_{2\omega}x)}$, and derive a beam propagation equation starting from the one-dimensional wave equation and the assumption of slow field variation on the scale of a wavelength. Collecting all the phase factors in the nonlinear term, we obtain

$$\frac{dE_\omega}{dx} = -\frac{\alpha_\omega}{2}E_\omega - i\frac{k_0}{n_\omega}S_2\chi^{(2)}E_{2\omega}E_\omega^* e^{-i(k_{2\omega}-2k_\omega)x} \tag{15.29a}$$

$$\frac{dE_{2\omega}}{dx} = -\frac{\alpha_{2\omega}}{2}E_{2\omega} - i\frac{k_0}{n_{2\omega}}S_2\chi^{(2)}E_\omega E_\omega e^{-i(2k_\omega-k_{2\omega})x} \tag{15.29b}$$

where $\alpha_\omega$ and $\alpha_{2\omega}$ ($n_\omega$ and $n_{2\omega}$) are the absorption coefficients (refractive indices) at the pump and second-harmonic frequencies, $k_0 = \frac{\omega}{c} = \frac{2\pi}{\lambda}$, $S_2 \approx (N_1 - 2N_2 + N_3)/N$ is a saturation factor that accounts for depopulation of the lowest subband under strong pumping, and $N_1$, $N_2$, and $N_3$ are the carrier densities in the three subbands. For propagation at an angle with respect to normal, $x$ is replaced with $\frac{x}{\cos\theta}$. For the sake of simplicity, we assume in what follows that Eq. (15.29) describes a waveguide-like structure with overlap factors close to unity.

We see from Eq. (15.29) that second-harmonic generation competes with intersubband and free-carrier absorption. Instead of jumping immediately into the full problem, we start by estimating the maximum possible conversion efficiency in a transparent material, i.e., with $\alpha_\omega = 0$ and $\alpha_{2\omega} = 0$. Assuming also that only the lowest subband remains populated at any intensity ($S_2 \approx 1$), Eq. (15.29b) becomes

$$\frac{dE_{2\omega}}{dx} = -i\frac{k_0}{n_{2\omega}}\chi^{(2)}E_\omega E_\omega e^{-i\Delta kx} \tag{15.30}$$

where $\Delta k \equiv k_{2\omega} - 2k_\omega$, with $k_{2\omega} = 2\omega n_{2\omega}/c$ and $k_\omega = \omega n_\omega/c$.

The phase term in Eq. (15.30) causes the amplitude of the second-harmonic beam to oscillate with period $2L_{coh} \equiv \frac{2\pi}{\Delta k} = \frac{c\pi}{\omega(n_{2\omega}-n_\omega)} = \frac{\lambda}{2(n_{2\omega}-n_\omega)}$ as it propagates, as illustrated in Fig. 15.15. This is known as the *coherence length*. The oscillations occur because in the absence of loss, the conversion is reversible. The pump power can feed the second harmonic or vice versa, with the direction of power flow flipping every coherence length. Of course, the reversibility is of no help if we wish to generate as much power in the second harmonic as possible, which may require the beam to propagate over a distance much larger than the coherence length. Therefore, ideally we prefer to be very close to the *phase-matching* condition $\Delta k = 0$, which is equivalent to $n_{2\omega} = n_\omega$ and nominally leads to an infinite coherence length.

If phase matching is satisfied and $\alpha_{2\omega} = 0$, $E_{2\omega}$ grows linearly with distance $L$ according to Eq. (15.29b):

$$E_{2\omega}(x=L) = -i\frac{k_0 L}{n_\omega}\chi^{(2)}E_\omega E_\omega \tag{15.31}$$

The measurable quantity is the field intensity rather than amplitude. To convert, we recall Eq. (4.5) and write

$$I_{2\omega} = \frac{cn_{2\omega}\varepsilon_0}{2}|E_{2\omega}|^2 \tag{15.32a}$$

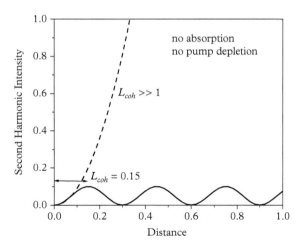

**Figure 15.15** *Variation of the second-harmonic intensity with propagation distance for* $\Delta k = 0$ *and* $\Delta k > 0$, *no absorption, and no pump depletion.*

$$I_\omega = \frac{cn_\omega \varepsilon_0}{2}|E_\omega|^2 \tag{15.32b}$$

The intensity of the second-harmonic beam at position $L$ is then

$$I_{2\omega}(L) = \frac{2k_0^2}{n_\omega^3 c\varepsilon_0}\left(\chi^{(2)}L\right)^2 I_\omega^2(0) \tag{15.33}$$

At double resonance, the square of the second-order susceptibility from Eq. (15.28) is

$$\left(\chi^{(2)}\right)^2 = \frac{N^2 e^6 z_{12}^2 z_{13}^2 z_{23}^2 \sin^6\theta}{\varepsilon_0^2 \Delta_I^4} = \frac{\left(4\pi\alpha_f\right)^3 \varepsilon_0 c^3 \hbar^3 N^2 z_{12}^2 z_{13}^2 z_{23}^2 \sin^6\theta}{\Delta_I^4} \tag{15.34}$$

The corresponding power conversion efficiency is

$$\eta_c\left(I_\omega\right) \equiv \frac{I_{2\omega}(L)}{I_\omega(0)} = \frac{2\omega^2}{n_\omega^3 c^3 \varepsilon_0}\left(\chi^{(2)}L\right)^2 I_\omega(0) \tag{15.35}$$

To summarize, the second-harmonic intensity increases as the square of the pump intensity, while the conversion efficiency increases linearly.

We already know that the output from a laser increases with pump intensity above the lasing threshold. It is certainly quite helpful that not only the power, but also the second-harmonic conversion efficiency go up if we pump harder. Of course, we arrived at this result following a number of simplifying assumptions, namely transparency, phase matching, and no depletion of the pump. The last condition is particularly worrisome, because it cannot remain satisfied as the conversion continues if we wish to convert a large fraction of the pump to the second harmonic.

More generally, what is the optimal propagation length $L$ in Eq. (15.35)? Because we do not wish deplete the pump to no useful purpose, the propagation length should not exceed the absorption depth at the pump wavelength, i.e., $L_{max} \approx \cos\theta/\alpha_\omega$. Taking the maximum intersubband absorption coefficient at resonance from Eq. (11.42) and assuming that all the electrons occupy the lowest subband, we obtain

$$\alpha_\omega = \alpha_{12} = \frac{N4\pi\alpha_f\hbar\omega z_{12}^2 \sin^2\theta}{\Delta_I n_\omega} \tag{15.36}$$

where for simplicity we take the absorption $\alpha_\omega$ to consist entirely of the intersubband component $\alpha_{12}$.

What is the maximum conversion efficiency at $L = L_{max}$? The absorbed flux is given by $\alpha_{12}I_\omega/\hbar\omega$. To avoid depopulating the lowest subband, the maximum pump intensity $I_{max}$ should excite fewer than $N/2$ electrons to the second subband.

We also assume that the lifetime in the second subband is dominated by scattering to the first subband; i.e., it is equal to $\tau_{12}$. These assumptions lead to the maximum pump intensity that saturates the intersubband transition:

$$I_{max} = \frac{N\hbar\omega}{2\tau_{12}\alpha_{12}} \tag{15.37}$$

Substituting $I_{max}$ and $L_{max}$ for $I_\omega$ and $L$, and using Eq. (15.34) in Eq. (15.35), we obtain

$$\eta_{max} = \frac{N^3(\hbar\omega)^3}{\tau_{12}n_\omega{}^3\alpha_{12}{}^3} \frac{(4\pi\alpha_f)^3 z_{12}^2 z_{13}^2 z_{23}^2 \sin^6\theta}{\Delta_l{}^4} \tag{15.38}$$

Also substituting Eq. (15.38) into Eq. (15.37), we find a simple expression for the conversion efficiency under the ideal conditions

$$\eta_{max} = \frac{\hbar}{\tau_{12}\Delta_l}\left(\frac{z_{13}z_{23}}{z_{12}^2}\right)^2 \tag{15.39}$$

The nonlinear susceptibility in Eq. (15.34) peaks at double resonance, but this is also where intersubband absorption is at its strongest. If much of the light is simply absorbed, high conversion efficiency is unlikely. For this reason, it often helps to detune from double resonance by $\delta_{12} = E_{12} - \hbar\omega$ and $\delta_{13} = E_{13} - 2\hbar\omega$ to reduce the absorption (see Section 15.5). Assuming Lorentzian broadening, we can generalize Eq. (15.39) for arbitrary detuning:

$$\eta_{max} = \frac{\hbar}{\tau_{12}\Delta_l}\left(\frac{z_{13}z_{23}}{z_{12}^2}\right)^2 \frac{(\delta_{12}^2 + \Delta_l^2)^2}{(\delta_{13}^2 + \Delta_l^2)\Delta_l^2} \tag{15.40}$$

The broadening in a practical well may not be Lorentzian. For example, fluctuations in the well width at different positions in the plane of the structure lead to a Gaussian line shape. In general, the Lorentzian shape overestimates the magnitude of the absorption and index change when the detuning is large.

Equations (15.39) and (15.40) are good only for rough estimates of the potential conversion efficiency. For greater realism, we should solve Eq. (15.29) along with the rate equations for the carrier populations in the three subbands as a function of pump intensity and position. The typical values $N = 10^{18}$ cm$^{-3}$, $\hbar\omega = 0.12$ eV, $\tau_{12} = 1$ ps, and $\alpha_{12} = 1000$ cm$^{-1}$ yield $I_{max} \approx 10$ MW/cm$^2$, which causes too much lattice heating in cw operation. If we use more practical cw intensities of tens of kW/cm$^2$, the conversion efficiency is much lower. While higher efficiencies are possible at short pulses with low duty cycles, the range of practical applications then narrows dramatically.

## 15.5  Quantum-Well and Quantum-Cascade Nonlinear Devices

The *asymmetric double quantum-well* (ADQW) of Fig. 15.14 is the most flexible intersubband scheme for second-harmonic generation. Typically, many repeats of the ADQWs are needed to achieve high efficiencies. In the absence of a waveguide, the light may be coupled from an angled polished facet and make multiple bounces in the wells, possibly using reflections from a mirror deposited on top of the sample.

InGaAs/InAlAs ADQWs grown on InP (see Fig. 15.14) are the preferred material system for second-harmonic generation. The other logical alternative, GaAs/AlGaAs, runs into a limited conduction-band offset that makes it unsuitable for generating mid-IR light. If we pump in the 8-to 10-μm band for output at 4–5 μm, the less mature InAs/AlSb wells grown on GaSb are not required, although we may resort to that system at even shorter wavelengths.

How do we design the ADQW structure? One strategy is to position the second subband in the narrower well. Then the first and third subbands correspond to the two lowest subbands of the wider well. As a result, $z_{13}$ is relatively large because the transition is direct, wheras $z_{12}$ is small because it is indirect. The barrier thickness in Fig. 15.14 determines just how spatially indirect this transition is. Although the condition $\frac{z_{13}}{z_{12}} \gg 1$ maximizes the figure of merit $\eta_{max}$ in Eq. (15.39), this is not the end of the story. To begin with, $z_{23}$ is reduced in tandem with $z_{12}$. Also, a large value of $z_{13}$ means that absorption of the second harmonic (rather than the pump beam) dominates the dynamics. In this case, we might revise the figure of merit using

$$L_{max} \approx \frac{\cos\theta}{\alpha_{2\omega}} \tag{15.41}$$

which leads to

$$\eta_{max,s} = \frac{\hbar}{\tau_{12}\Delta_l} \left( \frac{z_{12}z_{23}}{2z_{13}^2} \right)^2 \tag{15.42}$$

In this limit, we are hardly better off with $1 \to 3$ being a direct transition! Nor does it help to have a long $\tau_{12}$ that goes with an indirect $1 \to 2$ transition because then the saturation intensity in Eq. (15.37) becomes low. The only way out of this labyrinth is to balance the matrix elements such that $\left( \frac{z_{13}z_{23}}{z_{12}^2} \right)^2$ is on the order of 1. In fact, it rarely exceeds 2–3 even in a well-designed device. With $\tau_{12} \sim 1$ ps and $\Delta_l \sim 10$ meV (see Chapter 13), we estimate a maximum second-harmonic generation efficiency of 20% at the saturation intensity from Eq. (15.39). Practical values at double resonance are much lower. How can we do better?

We have not yet tried to take advantage of the detuning as described in connection with Eq. (15.40). If we detune both the pump and second harmonic from resonance by similar amounts $\delta \equiv \frac{\delta_{12}}{\Delta_l} \approx \frac{\delta_{13}}{\Delta_l}$, the figure of merit $\eta_{max}$ increases by a factor of $(\delta^2 + 1)$. Even though the Lorentzian line shape does not really hold for large detuning, this suggests a promising direction. Can the conversion efficiency be much higher for $\delta \gg 1$?

This is in fact a bridge too far. What we overlooked is that once intersubband absorption becomes very weak away from resonance, the total absorption comes to be dominated by other loss mechanisms, e.g., intraband free-electron absorption. This means that if we overdo the detuning, $\chi^{(2)}$ drops without any tangible benefit in the loss. In practice, it turns out that $\delta$ of order 1 is about the best we can do. It also pays to set $|\delta_{12}| > |\delta_{13}|$ by a slight margin because the second harmonic builds up gradually.

Figure 15.16 plots the second-harmonic conversion efficiency $\eta_c$ as a function of propagation distance for three different values of $\delta$, obtained numerically from the beam propagation equations (Eq. (15.29)) and the rate equations for carrier populations in the three subbands. The calculations assume typical parameters for InGaAs/InAlAs ADQWs, and ignore heating at the optimal value of $I_\omega$ in the tens of MW/cm$^2$ range. The best practical conversion efficiency, which tops out near 15%, is obtained for negative detuning, i.e., with subband energies below the photon energy. At lower pump intensities that are more practical for cw operation, the maximum conversion efficiency remains well below 10%.

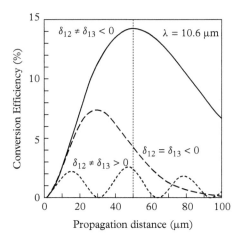

**Figure 15.16** *Solutions of Eq. (15.29) for the ADQW structure similar to those of Fig. 15.14 with various detuning parameters including optimized unequal detuning $\delta_{12} \neq \delta_{23}$ (sold curve), optimized equal detuning $\delta_{12} = \delta_{23} < 0$ (dashed), and identical detuning in the opposite direction (dotted). Reproduced with permission from I. Vurgaftman et al., IEEE J. Quantum Electron.* **32**, *1334 (1996).*

The results in Fig. 15.16 show that we would like to propagate over a significant distance in the sample. How do we make the coherence length long enough without modifying the rest of the structure? One strategy is to insert a separate *phase-matching* quantum-well region, possibly with one well per period, with an intersubband transition tuned for exact compensation of the mismatch in $n_{2\omega}$ and $n_{\omega}$. For Lorentzian broadening, the intersubband contribution to the refractive index is

$$n(\omega) = -\frac{\hbar c \alpha_{\omega}}{2\Delta E} \frac{(\hbar\omega - \Delta E)\,\Delta_l}{(\hbar\omega - \Delta E)^2 + \Delta_l^2} \tag{15.43a}$$

$$\Delta n(2\omega) = -\frac{\hbar c \alpha_{\omega}}{2\Delta E} \frac{(2\hbar\omega - \Delta E)\,\Delta_l}{(2\hbar\omega - \Delta E)^2 + \Delta_l^2} \tag{15.43b}$$

where $\alpha_{\omega}$ is the on-resonance absorption coefficient from Eq. (15.36) for an intersubband transition with energy $\Delta E$. With normalized detuning $\delta_{pm} \equiv (\hbar\omega - \Delta E)/\Delta_l$, the absorption coefficient decreases as $\delta_{pm}^{-2}$ when $\delta \gg 1$, but the index change drops only as $\delta_{pm}^{-1}$. This allows us to nudge the index just enough for phase matching with hardly any parasitic absorption. For the more common case of $n_{2\omega} > n_{\omega}$, we would like the index change to have opposite signs at the pump and second harmonic: $\Delta n(\omega) > 0$ and $\Delta n(2\omega) < 0$. To this end, we set $\Delta E \approx 3\hbar\omega/2$ and adjust the doping until $n_{2\omega} \approx n_{\omega}$. The extra absorption should be weak as long as $\hbar\omega \gg \Delta_l$, which definitely holds for state-of-the-art mid-IR materials.

We can also fix the phase mismatch problem by reversing the sign of the nonlinear susceptibility every coherence length (but keeping its magnitude the same!), a method known as *quasi-phase matching*. Because power begins flowing back to the pump beam after propagation over a distance of $L_{coh}$, changing the sign of $\chi^{(2)}$ reverses the direction of the power flow (see the second term in Eq. (15.29)). The most common quasi-phase-matching scheme is to periodically modulate the sign of the polarization field applied to a ferroelectric bulk crystal (*periodic poling*). While most III–V semiconductors are not ferroelectric, we can flip the direction of the crystal plane instead, e.g., by growing materials such as GaAs and GaN on a seeded template. However, periodic poling of intersubband devices becomes quite difficult to carry off, and to date no one has attempted it experimentally.

Why does anyone want nonlinear devices when we already have excellent mid-IR lasers based on the cascade configurations discussed in Chapters 12 and 13? Actually, the development of high-power QCLs for the mid-IR has led to something of a renaissance in quantum-well nonlinear optics, particularly when it comes to THz generation at room temperature. We pointed out in Section 13.2 that it is intrinsically challenging for a QCL (or any other gain medium) to produce direct THz generation at a temperature high enough that the thermal energy is comparable to or larger than the energy separation between the upper and lower lasing levels. However, this fundamental limit goes away if we have two mid-IR laser lines that create THz photons via a close cousin of second-harmonic and sum-frequency generation, the nonlinear process called *difference-frequency mixing*. This second-order process can mix two closely spaced pump beams, with frequencies $\omega_1$ and $\omega_2$ (typically in the mid-IR), to produce an output beam at the THz difference frequency $\Delta\omega = \omega_1 - \omega_2 \ll \omega_1, \omega_2$. We can even create THz photons *inside* the laser cavity rather than with external pumping! Neglecting population saturation and specifics of the in-cavity process, the propagation of the three beams can be described by

$$\frac{dE_{\omega 1}}{dx} = -\frac{\alpha_{\omega 1}}{2} E_{\omega 1} - i\frac{k_0}{n_{\omega 1}}\, \chi^{(2)}\left(\Delta\omega, \omega_1, \omega_2\right) E_{\Delta\omega} E_{\omega 2} e^{-i(k_{\Delta\omega} - k_{\omega 1} + k_{\omega 2})x} \quad (15.44a)$$

$$\frac{dE_{\omega 2}}{dx} = -\frac{\alpha_{\omega 2}}{2} E_{\omega 2} - i\frac{k_0}{n_{\omega 2}}\, \chi^{(2)}\left(\Delta\omega, \omega_1, \omega_2\right) E_{\Delta\omega} E_{\omega 1}^* e^{-i(k_{\Delta\omega} - k_{\omega 1} + k_{\omega 2})x} \quad (15.44b)$$

$$\frac{dE_{\Delta\omega}}{dx} = -\frac{\alpha_{\Delta\omega}}{2} E_{\Delta\omega} - i\frac{k_0}{n_{\Delta\omega}}\, \chi^{(2)}\left(\Delta\omega, \omega_1, \omega_2\right) E_{\omega 1} E_{\omega 2}^* e^{i(k_{\Delta\omega} - k_{\omega 1} + k_{\omega 2})x} \quad (15.44c)$$

Again, we do not include the modal overlaps of the three beams propagating in the laser waveguides.

Difference-frequency mixing requires a large $\chi^{(2)}\left(\Delta\omega, \omega_1, \omega_2\right)$ combined with low absorption at all three frequencies. We can detune the resonances, much as we did with second-harmonic generation, to reduce the absorption faster than $\chi^{(2)}$. In this case, phase matching requires $k_{\Delta\omega} = k_{\omega 1} - k_{\omega 2}$, which is equivalent to

$$\frac{\Delta\omega}{c} n_{\Delta\omega} \cos\theta = \frac{\omega_1}{c} n_{\omega 1} - \frac{\omega_2}{c} n_{\omega 2} \quad (15.45)$$

Here $\theta$ is the angle at which the THz radiation propagates with respect to the cavity axis (the angle is zero for two mid-IR laser beams). For $\omega_1 \approx \omega_2$, Eq. (15.45) simplifies to

$$n_{\Delta\omega} \cos\theta = n_{\omega 1} + \omega_1 \frac{\partial n_\omega}{\partial \omega}\bigg|_{\omega=\omega_1} \quad (15.46)$$

If $\theta = 0$, we are usually left with no prospects for phase matching. However, if $\theta > 0$ and $n_{\Delta\omega} > n_{\omega 1}, n_{\omega 2}$ in the substrate, it becomes a simple matter of picking the right angle. Of course, $n_{\Delta\omega} > n_{\omega 1}, n_{\omega 2}$ is equivalent to anomalous dispersion, which is available near intersubband resonances (see Eq. (15.43)).

How can we generate QCL beams at two closely spaced frequencies? All it takes is a stage design with two slightly different transition energies in the same active core! Since the absorption in a QCL is weak on both sides of the gain line, powers up to 1 W can be produced at both frequencies. Alternatively, we can make do with the same stage design if two different intersubband transitions

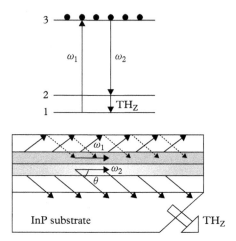

**Figure 15.17** *Illustration of THz difference-frequency mixing in a mid-IR QCL laser core, with the THz beam emerging from the cavity at an angle.*

simultaneously provide lasing at photon energies having the desired separation. Because the difference-frequency mixing takes place within an optical cavity, the QCL pump beams, which make multiple passes through the active material, can lase in a single mode if the cavity has distributed-feedback gratings.

Figure 15.17 illustrates that the THz beam emerging from the device may not align with the cavity axis. The typical THz propagation angle is $\sim 20 - 30° \gg \theta_{TIR} = \sin^{-1}(1/n_r) \sim 15°$, where $\theta_{TIR}$ is the total internal reflection angle (see Section 12.7). The light emitted upward is not necessarily lost, because it can be reflected downward by a top metal mirror as shown in Fig. 15.17. To collect the THz output, the output facet can be polished at an angle close to $\theta$.

Currently, difference-frequency THz QCLs emit only a few µW of power, i.e., six orders of magnitude lower than for a state-of-the-art mid-IR device. The conversion efficiencies also lag one to two orders of magnitude below simple projections. Most of the shortfall may be due to strong free-carrier absorption at THz frequencies. Nonetheless, the ability to produce coherent cw THz radiation at room temperature goes a long way.

## Suggested Reading

- A solid book-length overview of solar-cell physics may be found in J. A. Nelson, *The Physics of Solar Cells* (Imperial College Press, London, 2003).
- Fundamental limits are treated empirically by C. H. Henry, "Limiting efficiencies of ideal single and multiple energy gap terrestrial solar cells," *Journal of Applied Physics* **51**, 4494 (1980) and from a thermodynamic point

of view by L. C. Hirst and N. J. Ekins-Daukes, "Fundamental losses in solar cells," *Progress in Photovoltaics* **19**, 286 (2011).

- The theory of interband cascade photovoltaic devices is covered in R. T. Hinkey and R. Q. Yang, "Theory of multiple stage interband cascade photovoltaic devices and ultimate performance level comparison of multiple-stage and single-stage interband infrared detectors," *Journal of Applied Physics* **114**, 104,506 (2013).

- The basics of nonlinear optics are reviewed in A. Yariv, *Quantum Electronics*, 3rd edn. (Wiley, New York, 1988) and R. Boyd, *Nonlinear Optics* (Academic Press, San Diego, CA, 1992).

- The figure of merit for intersubband second-harmonic generation discussed here first appeared in J. R. Meyer, C. A. Hoffman, F. J. Bartoli, E. R. Youngdale, and L. R. Ram-Mohan, "Momentum-space reservoir for enhancement of intersubband second-harmonic generation," *IEEE Journal of Quantum Electronics* **31**, 706 (1995).

- THz difference-frequency generation using mid-IR QCLs is discussed in greater detail by K. Fujita, S. Jung, Y. Jiang, J. H. Kim, A. Nakanishi, A. Ito, M. Hitaka, T. Edamura, and M. A. Belkin, "Recent progress in terahertz difference-frequency quantum cascade laser sources," *Nanophotonics* **7**, 1795 (2018).

# Appendix A
# Physical Constants, Units, and Other Useful Physical Relations

This appendix contains a compendium of the physical constants, the units in which the most important quantities are expressed, as well as some of the key physical relations from the text. Not all of the quantities given here are explicitly used by the text, but they should still be useful to the readers. Four significant figures are usually given, even though the calculations in this field rarely need to be so precise (and are often dominated by uncertainties in the band parameters discussed in Chapters 6 and 7). See the text for the definition of symbols.

## A.1 Physical Constants

Speed of light in vacuum: $c = 2.998 \times 10^{10} \frac{\text{cm}}{\text{s}}$

in a semiconductor: $\frac{c}{n_r} = \frac{299.8 \times 10^{10}}{n_r} \frac{\mu\text{m}}{\text{ps}}$, where $n_r$ is the refractive index

Elementary charge: $e = 1.602 \times 10^{-19}$ C

Free electron mass: $m = 5.686 \times 10^{-16} \frac{\text{eV s}^2}{\text{cm}^2} = 9.109 \times 10^{-31}$ kg

Electron or hole effective mass in a semiconductor: $m_{e,h} = 5.686 \times 10^{-16} \frac{\text{eV s}^2}{\text{cm}^2} \left( \frac{m_{e,h}}{m} \right)$

Permittivity of vacuum: $\varepsilon_0 = 8.854 \times 10^{-14} \frac{\text{F}}{\text{cm}} = 8.854 \times 10^{-14} \frac{\text{C}}{\text{V cm}}$

Permitivity in a semiconductor: $\varepsilon = \varepsilon_r \varepsilon_0 = n_r^2 \varepsilon_0$ when the imaginary part is small or zero

Permeability of vacuum: $\mu_0 = 1.257 \times 10^{-8} \frac{\text{H}}{\text{cm}} = 1.257 \times 10^{-4} \frac{\text{kg cm}}{\text{C}^2}$

Impedance of free space: $Z_0 = \sqrt{\frac{\mu_0}{\varepsilon_0}} = \mu_0 c = \frac{1}{\varepsilon_0 c} = 376.7\ \Omega$

Reduced Planck constant: $\hbar \equiv \frac{h}{2\pi} = 6.582 \times 10^{-16}$ eV s

$\Rightarrow \frac{\hbar^2}{2m_{e,h}} = 3.810 \times 10^{-16}\ \text{eV cm}^2 \left( \frac{m_{e,h}}{m_0} \right)^{-1}$

Boltzmann constant: $k_B = 8.617 \times 10^{-5} \frac{\text{eV}}{\text{K}}$

$\Rightarrow k_B T = 25.85\ \text{meV} \left( \frac{T}{300\text{K}} \right)$

## A.2  Units

$1 \text{ kg} = 10^{-4} \frac{\text{J s}^2}{\text{cm}^2} = 6.242 \times 10^{14} \frac{\text{eV s}^2}{\text{cm}^2}$

$1 \text{ J (joule)} = 6.242 \times 10^{18} \text{ eV}$

$1 \text{ GPa (gigapascal)} \equiv 10^9 \frac{\text{J}}{\text{m}^3} = 6.242 \times 10^{21} \frac{\text{eV}}{\text{cm}^3}$

$1 \text{ G (gauss)} \equiv 10^{-4} \text{ T (tesla)} = 10^{-8} \frac{\text{V s}}{\text{cm}^2}$

$1 \text{ F (farad)} \equiv 1 \frac{\text{C}}{\text{V}}$

$1 \text{ A} \equiv 1 \frac{\text{C}}{\text{s}}$

$1 \text{ } \Omega \equiv 1 \frac{\text{V}}{\text{A}}$

$\alpha L = \frac{\zeta}{10} \ln(10) = 0.2303 \, \zeta$, where $\zeta$ is attenuation in dB

$\nu = \frac{c}{\lambda} = \frac{2.998 \times 10^{14}}{\text{s}} \left( \frac{\lambda}{1 \,\mu\text{m}} \right)^{-1}$, where $\nu$ is optical frequency and $\lambda$ is wavelength

$\hbar\omega = h\nu = \frac{2\pi c\hbar}{\lambda} = 1.240 \text{ eV} \left( \frac{\lambda}{1 \,\mu\text{m}} \right)^{-1}$

$100 \text{ THz} = (2.998/\mu\text{m})$

$\text{Wavenumbers} = 10^4 \text{ cm}^{-1} \left( \frac{\lambda}{1 \,\mu\text{m}} \right)^{-1} = 8065 \text{ cm}^{-1} \left( \frac{\hbar\omega}{1 \text{ eV}} \right)$

$\text{Jones} = \text{cm Hz}^{1/2}/\text{W}$

## A.3  Other Useful Physical Relations

Normalized band and joint density of states:

$$\frac{\hbar^2}{2m} DOS = \frac{\hbar^2}{2m} \frac{1}{(2\pi)^3} \left[ \frac{4\pi k^2}{dE/dk} \right] = \frac{\sqrt{2Em_r^3}}{4\pi^2 m\hbar} = 4.104 \times 10^5 \text{cm}^{-1} \left[ \frac{E}{100 \text{ meV}} \right]^{\frac{1}{2}} \left( \frac{m_r}{m} \right)^{3/2}$$

[Eqs. (1.46) and (4.34)]

Carrier density in 3D with no degeneracy and parabolic bands:

$$n = N_c \exp\left( \frac{E_F}{k_B T} \right) = \frac{\sqrt{\frac{(k_B T m_e)^3}{2\pi^3}}}{\hbar^3} \exp\left( \frac{E_F}{k_B T} \right)$$

$$= 2.510 \times 10^{19} \text{ cm}^{-3} \left[ \frac{T}{300 \text{ K}} \right]^{3/2} \left( \frac{m_e}{m} \right)^{3/2} \exp\left( \frac{E_F}{k_B T} \right)$$

[Eqs. (1.50), (4.76), and (4.77)]

in 2D:

$$n = N_c \exp\left(\frac{E_F}{k_B T}\right) = \frac{k_B T m_e}{\pi \hbar^2} \exp\left(\frac{E_F}{k_B T}\right) = 1.079 \times 10^{13} \text{cm}^{-2} \left[\frac{T}{300 \text{ K}}\right] \left(\frac{m_e}{m}\right) \exp\left(\frac{E_F}{k_B T}\right)$$

Fermi level for degenerate carrier density in 2D and parabolic bands:

$$E_F = \frac{\pi \hbar^2}{m_e} n = 2.396 \text{ meV} \left(\frac{n}{10^{12} \text{ cm}^{-2}}\right) \left(\frac{m_e}{m}\right)^{-1}$$

Energy-dependent electron mass:

$$m_e(\Delta E) \approx m \frac{\Delta E + E_g}{2\Delta E + E_g + E_P} \approx m_e(0) \left(1 + \frac{\Delta E}{E_g}\right) \qquad \text{[Eqs. (2.20) and (2.21)]}$$

Electron mass near the bottom of the band:

$$\frac{1}{m_e} \approx \frac{1}{m} \left[1 + 2F + \frac{E_P\left(E_g + {}^{2\Delta}\!/_3\right)}{E_g\left(E_g + \Delta\right)}\right] \qquad \text{[Eq. (3.2)]}$$

Heavy- and light-hole masses:

$$m_{HH}[001] = \frac{m}{\gamma_1 - 2\gamma_2}, \quad m_{LH}[001] = \frac{m}{\gamma_1 + 2\gamma_2}, \quad m_{HH}[111]$$

$$= \frac{m}{\gamma_1 - 2\gamma_3}, \quad m_{LH}[111] = \frac{m}{\gamma_1 + 2\gamma_3} \qquad \text{[Eq. (3.36)]}$$

Split-off mass:

$$\frac{m}{m_{SO}} \approx \gamma_1 - \frac{E_P}{3E_g} \frac{\Delta}{E_g + \Delta} \qquad \text{[Eq. (3.37)]}$$

Relation between in-plane and out-of-plane biaxial strain:

$$\epsilon_{zz} = -\frac{2c_{12}}{c_{11}} \epsilon_{xx} = -\frac{2c_{12}}{c_{11}} \frac{a_0 - a}{a},$$

where $a$ is the lattice constant of the layer, and $a_0$ is the lattice constant of the substrate [Eqs. (3.57) and (3.58)]

Optical intensity:

$$\hbar \omega \Phi = \frac{c n_r \varepsilon_0}{2} |E|^2 = 4.645 \times 10^{-3} \text{W/cm}^2 \left(\frac{n_r}{3.5}\right) \left(\frac{|E|^2}{1 \text{ V}^2/\text{cm}^2}\right) \qquad \text{[Eq. (4.5)]}$$

Fine structure constant:

$$\alpha_f = \frac{e^2}{4\pi c \varepsilon_0 \hbar} = 7.297 \times 10^{-3} \approx \frac{1}{137} \qquad \text{[Eq. (4.33)]}$$

Absorption coefficient in 3D parabolic system with constant matrix element:

$$\alpha\left(\hbar\omega\right) = \frac{4\pi\,k_r}{\lambda} = \frac{2\alpha_f\,E_P}{3n_r\,\hbar\omega}\frac{\sqrt{2\left(\hbar\omega - E_g\right)m_r{}^3}}{\hbar m}$$

$$\approx 1.4 \times 10^4\;\mathrm{cm}^{-1}\sqrt{\frac{\hbar\omega - E_g}{E_g}} \qquad \text{[Eqs. (4.34) and (4.35)]}$$

Absorption per pass in 2D parabolic system with constant matrix element:

$$\alpha_{QW}\left(\hbar\omega\right) = \frac{\pi\,\alpha_f}{n_r}\frac{E_P}{\hbar\omega}\frac{m_r}{m}|O_{eh}|^2 H\left(\hbar\omega - E_g\right)$$

$$= 0.655\%\left(\frac{n_r}{3.5}\right)^{-1}\left(\frac{E_P}{\hbar\omega}\right)\left(\frac{m_r}{m}\right)|O_{eh}|^2 H\left(\hbar\omega - E_g\right) \qquad \text{[Eq. (11.7)]}$$

Density of optical modes in 3D:

$$DOM = \frac{1}{\pi^2}\frac{(\hbar\omega)^2}{\hbar^3 c^3}n_r^2 n_g = 5.654 \times 10^{14}\frac{1}{\mathrm{eV\,cm}^3}\left(\frac{\hbar\omega}{1\;\mathrm{eV}}\right)^2\left(\frac{n_r^2 n_g}{3.5^3}\right) \qquad \text{[Eq. (4.68)]}$$

Purcell factor at field and spectral maximum:

$$F_P = \frac{3}{4\pi^2}\frac{Q}{V_{eff}}\frac{\lambda_c^3}{n_r^2 n_g} = 607.9\left(\frac{Q}{10^3}\right)\left(\frac{V_{eff}}{(\lambda_c/(2n_r))^3}\right)^{-1}\left(\frac{n_r}{n_g}\right) \qquad \text{[Eq. (4.70)]}$$

Intrinsic carrier density for parabolic bands:

$$n_i = (N_c N_v)^{\frac{1}{2}}\exp\left(-\frac{E_{c0} - E_{v0}}{2k_B T}\right) = \left(\frac{k_B T}{2^{\frac{1}{3}}\pi\hbar^2}\right)^{\frac{3}{2}}(m_e m_h)^{\frac{3}{4}}\exp\left(-\frac{E_g}{2k_B T}\right)$$

$$= 2.510 \times 10^{19}\mathrm{cm}^{-3}\left(\frac{m_e m_h}{m^2}\right)^{3/4}\left(\frac{T}{300\;\mathrm{K}}\right)^{3/2}\exp\left(-\frac{E_g}{2k_B T}\right) \qquad \text{[Eq. (4.81)]}$$

Radiative emission rate in 3D parabolic system:

$$R_{sp,net} = B\left(np - n_i^2\right)$$

$$B = \frac{2\sqrt{2}\pi^{\frac{3}{2}}\alpha_f n_r}{3}\frac{E_P}{mc^2}\left(\frac{\hbar^2}{(m_e + m_h)k_B T}\right)^{\frac{3}{2}}\frac{1}{\hbar}\left(E_g + \frac{3}{2}k_B T\right)$$

$$= 1.304 \times 10^{-12}\frac{\mathrm{cm}^3}{\mathrm{s}}\left(\frac{n_r}{3.5}\right)\left(\frac{E_P}{25\;\mathrm{eV}}\right)\left(\frac{T}{300\;\mathrm{K}}\right)^{-\frac{1}{2}}\left(\frac{m_e + m_h}{m}\right)^{-3/2}\left(\frac{E_g}{k_B T} + \frac{3}{2}\right)$$

$$\text{[Eqs. (4.82) and (4.84)]}$$

2D parabolic system:

$$B_{QW} \approx \frac{2\pi \alpha_f n_r}{3} \frac{E_P}{mc^2} \frac{\hbar}{m_e + m_h} \left( \frac{E_g}{k_B T} + 1 \right)$$

$$= 3.029 \times 10^{-6} \frac{cm^2}{s} \left( \frac{n_r}{3.5} \right) \left( \frac{E_P}{25 \text{ eV}} \right) \left( \frac{m_e + m_h}{m} \right)^{-1} \left( \frac{E_g}{k_B T} + 1 \right) \qquad \text{[Eq. (11.100)]}$$

Modified Bohr radius:

$$a_0 = \frac{4\pi \varepsilon_s \hbar^2}{m_r e^2} = \frac{\hbar (\varepsilon_s/\varepsilon_0)}{\alpha_f m_r c} = 6.483 \text{ Å} \left( \frac{\varepsilon_s}{12.25\varepsilon_0} \right) \left( \frac{m_r}{m} \right)^{-1} \qquad \text{[Eq. (4.97a)]}$$

Coulomb binding energy:

$$E_B = \frac{m_r e^4}{2\hbar^2 (4\pi \varepsilon_s)^2} = \frac{m_r c^2 \alpha_f^2}{2(\varepsilon_s/\varepsilon_0)^2} = 90.68 \text{ meV} \left( \frac{m_r}{m} \right) \left( \frac{\varepsilon_s}{12.25\varepsilon_0} \right)^{-2} \qquad \text{[Eq. (4.97b)]}$$

Subband energies in a quantum well with parabolic dispersions and infinite barriers:

$$E_v = \frac{\hbar^2}{2m_e} \frac{v^2 \pi^2}{t^2} = 3.764 \text{ meV } v^2 \left( \frac{m_e}{m} \right)^{-1} \left( \frac{t}{100\text{Å}} \right)^{-2}$$

Peak intersubband absorption coefficient per pass:

$$\alpha_{ib,\max} = \frac{4\pi \alpha_f}{\Delta_l (E_\mu - E_{,v}) n_r} \frac{\sin^2 \theta}{\cos \theta} \left( \frac{\hbar^2}{m_{et}} \frac{4\mu v}{\mu^2 - v^2} \right)^2 (n_v - n_\mu)$$

$$= 1.521\% \left[ \frac{\mu v}{\mu^2 - v^2} \right]^2 \left( \frac{n_r}{3.5} \right)^{-1} \left( \frac{m_e}{0.04m_0} \right)^{-2} \left( \frac{t}{100 \text{ Å}} \right)^{-2} \left( \frac{E_\mu - E_{,v}}{100 \text{ meV}} \right)^{-1}$$

$$\left( \frac{\Delta_l}{10 \text{ meV}} \right)^{-1} \left( \frac{n_v - n_\mu}{10^{11} \text{cm}^{-2}} \right) \left( \frac{\sin^2 \theta}{\cos \theta} \right) \qquad \text{[Eq. (11.41)]}$$

Approximate maximum gain in quantum wells:

$$g_{QW} (E_g) = g_s \left[ 1 - \exp \left( -\frac{n}{N_c} \right) - \exp \left( -\frac{p}{N_v} \right) \right] \qquad \text{[Eq. (11.71)]}$$

Capacitance per unit area $A$:

$$C = \frac{Q}{VA} = \frac{e\Delta N}{V} = \frac{\varepsilon_s}{d},$$

where $\varepsilon_s$ is the zero-frequency permittivity and $d$ is the distance between layers of charge:

$$\Rightarrow V = 0.1477 \text{ V} \left( \frac{\Delta N}{10^{12} \text{cm}^{-2}} \right) \left( \frac{d}{10 \text{ nm}} \right) \left( \frac{\varepsilon_s}{12.25} \right)^{-1}$$

Wall-plug efficiency of a semiconductor laser:

$$\eta_{WPE} = \eta_i \frac{\alpha_{m1}}{\alpha_{m1} + \alpha_{m2} + \alpha_i} \frac{\hbar\omega/e + V_0 + \mathcal{J}\rho_s}{\hbar\omega/e} \frac{\mathcal{J} - \mathcal{J}_{th}}{\mathcal{J}_{th}} \qquad \text{[Eq. (12.16)]}$$

LED:

$$\eta_{WPE,LED}(\mathcal{J}) \approx \frac{R_{sp}}{R_{sp} + R_{nonrad}} \frac{\hbar\omega/e + V_0 + \mathcal{J}\rho_s}{\hbar\omega/e} \eta_{ex}\eta_I \qquad \text{[Eq. (12.72)]}$$

Linewidth of a semiconductor laser:

$$\Delta\nu = \frac{\hbar\omega}{8\pi P_{out}} \frac{v_g^2 \alpha_{cav}}{L_{cav}} \ln\left(\frac{1}{R}\right) \frac{1}{\exp\left[\frac{(\hbar\omega - eV)}{k_B T}\right] - 1} \left(1 + \alpha_e^2\right)$$

$$= 2.327 \text{ kHz} \left(\frac{\lambda}{3 \text{ μm}}\right)^{-1} \left(\frac{P_{out}}{10 \text{ mW}}\right)^{-1} \left(\frac{n_g}{3.5}\right)^{-2} \left(\frac{\alpha_{cav}}{10 \text{ cm}^{-1}}\right) \left(\frac{L_{cav}}{1 \text{ mm}}\right)^{-1}$$

$$\times \ln\left(\frac{0.3}{R}\right) \frac{1}{\exp\left[(\hbar\omega - eV)/k_B T\right] - 1} \left(1 + \alpha_e^2\right) \qquad \text{[Eq. (12.58)]}$$

Threshold current density of a QCL:

$$\mathcal{J}_{th} \approx \frac{e\alpha_{cav}}{\tau_{eff} \frac{\Gamma}{d} \frac{dg_{ib}}{dn}} + \exp\left(-\frac{\delta E}{k_B T}\right) \frac{en_D}{\tau_{eff}} \approx \frac{e\alpha_{cav}}{\tau_{eff} \frac{\Gamma}{d} \frac{dg_{ib}}{dn}}$$

$$= 640.8 \text{ A/cm}^2 \left(\frac{\alpha_{cav}}{2 \text{ cm}^{-1}}\right) \left(\frac{\tau_{eff}}{1 \text{ ps}}\right)^{-1} \left(\frac{1}{5 \times 10^{-10} \text{ cm}} \frac{\Gamma}{d} \frac{dg_{ib}}{dn}\right)^{-1} \qquad \text{[Eq. (13.20)]}$$

where the effective carrier lifetime is $\tau_{eff} \equiv \tau_u (1 - \tau_l/\tau_{ul})$

temperature dependence:

$$\mathcal{J}_{th} = \mathcal{J}_0 \exp\left[(T_s + R_{th}\mathcal{J}_{th}(V_a + \mathcal{J}_{th}\rho_s))/T_0\right] \qquad \text{[Eq. (13.37)]}$$

Detectivity:

$$D^* = \frac{e\eta_a}{\hbar\omega} \frac{1}{\sqrt{\frac{4k_B T}{R_0 A} + 2e\mathcal{J}_d + 2e^2 \Phi_B}} \approx \frac{e\eta_a}{\hbar\omega} \frac{1}{\sqrt{2e\mathcal{J}_d}},$$

when operating at significant bias and not background limited    [Eq. (14.13d)]

Depletion width of a one-sided p-n junction:

$$W = V_{bi}/F_{i,av} = \sqrt{\frac{2\varepsilon_s V_{bi}}{eN_B}} = 116.4 \text{ nm} \left(\frac{\varepsilon_s}{12.25}\right)^{\frac{1}{2}} \left(\frac{V_{bi}}{1 \text{ eV}}\right) \left(\frac{N_B}{10^{17}\text{cm}^{-3}}\right)^{-\frac{1}{2}}$$

$$\text{[Eq. (14.17)]}$$

Carrier mobility:

$$\mu = \frac{e\tau_m}{m_{e,h}}, \text{ where } \tau_m \text{ is momentum relaxation time}$$

$$\Rightarrow \tau_m = \frac{m_{e,h}\mu}{e} = 0.5686 \text{ ps} \left(\frac{\mu}{10^3 \text{cm}^2/\text{Vs}}\right)\left(\frac{m_{e,h}}{m}\right)$$

Minority carrier diffusion coefficient:

$$D = \frac{\mu k_B T}{e} = 25.85 \frac{\text{cm}^2}{\text{s}} \left(\frac{T}{300\text{K}}\right)\left(\frac{\mu}{10^3 \text{cm}^2/\text{Vs}}\right)$$

Diffusion length:

$$L_d = \sqrt{D\tau} = \sqrt{\frac{\mu \tau k_B T}{e}} = 8.145 \mu\text{m} \left(\frac{T}{77 \text{ K}}\right)^{\frac{1}{2}} \left(\frac{\mu}{10^3 \text{cm}^2/\text{Vs}}\right)^{\frac{1}{2}} \left(\frac{\tau}{100 \text{ ns}}\right)^{\frac{1}{2}} \quad \text{[Eq. 14.21]}$$

Diffusion-limited dark current density:

$$|\mathcal{J}_{d,dark}| = \frac{eL_a}{\tau} \frac{n_i^2}{N_D} = 1.602 \ \mu\text{A/cm}^2 \left(\frac{L_a}{1 \ \mu\text{m}}\right)\left(\frac{\tau}{100 \text{ ns}}\right)\left(\frac{n_i}{10^{13}\text{cm}^{-3}}\right)^2 \left(\frac{N_D}{10^{16}\text{cm}^{-3}}\right)$$
$$\text{[Eq. (14.32b)]}$$

Plasma energy:

$$\hbar\omega_p = \left(\frac{Ne^2}{\varepsilon m_{e,h}}\right)^{1/2} = 10.61 \text{ meV}\left(\frac{N}{10^{18}\text{cm}^{-3}}\right)^{1/2} \left(\frac{\varepsilon}{12.25\varepsilon_0}\right)^{-1/2} \left(\frac{m_{e,h}}{m}\right)^{-1/2}$$

# Appendix B
# Hole Effective Masses for Wurtzite Materials

The approximate hole effective masses near $k = 0$ for the wurtzite materials for wavevector directions along the $c$ axis and along the $x$ or $y$ axis can be derived from Eqs. (3.76) and (3.79). They are expressed in terms of the $A_1, A_2, A_3, A_4, A_5, A_6$ parameters as well as the energies of the LH and CH bands $E_{LH}$ and $E_{CH}$ at $k = 0$ given by Eq. (3.75) with $E_v = 0$ (and are shown as negative here):

$$m_{HH}^{\parallel} \approx -\frac{m}{A_1 + A_3} \tag{B.1}$$

$$m_{HH}^{\perp} \approx -\frac{m}{A_2 + A_4} \tag{B.2}$$

$$m_{LH}^{\parallel} \approx -\frac{m}{A_1 + \frac{E_{LH}}{E_{LH} - E_{CH}} A_3} \tag{B.3}$$

$$m_{LH}^{\perp} \approx -\frac{m}{A_2 + \frac{E_{LH}}{E_{LH} - E_{CH}} A_4} \tag{B.4}$$

$$m_{CH}^{\parallel} \approx -\frac{m}{A_1 + \frac{E_{CH}}{E_{CH} - E_{LH}} A_3} \tag{B.5}$$

$$m_{CH}^{\perp} \approx -\frac{m}{A_2 + \frac{E_{CH}}{E_{CH} - E_{LH}} A_4} \tag{B.6}$$

# Appendix C
# Loehr's Parametrization of the Second-Nearest-Neighbor EBOM

We saw in Section 5.2 that the second-nearest-neighbor effective bond-orbital method (EBOM) with fcc lattice symmetry can be reduced to the 8-band $\mathbf{k} \cdot \mathbf{p}$ method, including second-order terms due to the interactions with remote bands. Since the 8-band $\mathbf{k} \cdot \mathbf{p}$ approach takes experimental band energies and curvatures (effective masses) at the $\Gamma$ and $X$ points as its parameters, we can do the same for this version of the EBOM. The detailed derivation may be found in John Loehr's book cited under "Suggested Reading" for Chapter 3, and here we summarize some approximate relations between the matrix elements in Table 5.1 and the experimental parameters.

A few definitions will simplify the expressions given below:

$$E_{v,s} \equiv E_v - \frac{\Delta}{3} \tag{C.1}$$

$$r \equiv \frac{E_g}{E_g + \Delta} \tag{C.2}$$

$$h = \frac{\hbar^2}{2ma^2} \tag{C.3}$$

Here $E_v$ is the material's valence-band maximum, which is conventionally set to 0 in a bulk material, but becomes important when heterointerfaces are considered. We also define the conduction-band minimum as $E_c \equiv E_v + E_g$.

With these definitions, the parameters have the values

$$V_{ss0} = \frac{5E_c + 3E_{X1c}}{8} + \frac{6h}{m_e} + (2+r)\left(\frac{E_{X5v} - E_{X3v}}{4} - 12h\gamma_3\right) \tag{C.4}$$

$$V_{ss1} = \frac{E_c - E_{X1c}}{16} \tag{C.5}$$

$$V_{ss2} = \frac{E_{X1c} - E_c}{16} - \frac{h}{m_e} + (2+r)\left(\frac{E_{X3v} - E_{X5v}}{24} + 2h\gamma_3\right) \tag{C.6}$$

$$V_{pp\sigma 0} = \frac{5E_{v,s} + 3E_{X3v}}{8} - 6h(\gamma_1 - 2\gamma_3) \tag{C.7}$$

$$V_{pp\sigma 1} = \frac{E_{v,s} - 3E_{X3v} + 2E_{X5v}}{16} \tag{C.8}$$

$$V_{pp\sigma 2} = \frac{2E_{X5v} - E_{X3v} - E_{v,s}}{16} + h(\gamma_1 + 4\gamma_2 - 6\gamma_3) \tag{C.9}$$

$$V_{pp\pi 1} = \frac{E_{v,s} + E_{X3v} - 2E_{X5v}}{16} \tag{C.10}$$

$$V_{pp\pi 2} = \frac{E_{X5v} - E_{v,s}}{16} + h(\gamma_1 - 2\gamma_2) \tag{C.11}$$

$$V_{sp\sigma 1} = \frac{1}{4}\sqrt{\frac{E_g(E_{X3v} - E_{X5v})}{4} + 12E_g h\gamma_3} \tag{C.12}$$

Chapters 6 and 7 tabulate the parameters entering these expressions for various III–V semiconductors.

# Appendix D
# Table of Optical Parameters for Bulk III–V Semiconductors

Table D1 *Recommended real and imaginary parts of the permittivities of III–V zinc-blende binary compounds GaAs, AlAs, InAs, GaP, and InP.*

| Energy (eV) | GaAs | | AlAs | | InAs | | GaP | | AlP | |
|---|---|---|---|---|---|---|---|---|---|---|
| | $\varepsilon_r$ | $\varepsilon_i$ | $\varepsilon_r$ | $\varepsilon_i$ | $\varepsilon_r$ | $\varepsilon_i$ | $\varepsilon_r$ | $\varepsilon_i$ | $\varepsilon_r$ | $\varepsilon_i$ |
| 6 | −4.8914 | 6.6574 | −6.8583 | 8.0036 | −2.7481 | 5.6169 | −6.0583 | 7.4507 | −7.69643 | 10.05499 |
| 5.98 | −4.9291 | 6.703 | −6.878 | 8.1072 | −2.7412 | 5.6166 | −6.151 | 7.5376 | −7.8105 | 10.25527 |
| 5.96 | −4.9718 | 6.7492 | −6.903 | 8.2122 | −2.7399 | 5.6161 | −6.2501 | 7.6266 | −7.92457 | 10.45554 |
| 5.94 | −5.0195 | 6.7962 | −6.9329 | 8.319 | −2.7438 | 5.6157 | −6.3558 | 7.7181 | −8.12182 | 10.68421 |
| 5.92 | −5.0718 | 6.8444 | −6.9676 | 8.4281 | −2.7528 | 5.6157 | −6.4679 | 7.8126 | −8.33041 | 10.91675 |
| 5.9 | −5.1285 | 6.8943 | −7.0067 | 8.5399 | −2.7664 | 5.6163 | −6.5865 | 7.9106 | −8.50069 | 11.16979 |
| 5.88 | −5.1896 | 6.9459 | −7.0502 | 8.6548 | −2.7846 | 5.6178 | −6.7118 | 8.0126 | −8.64748 | 11.43539 |
| 5.86 | −5.2549 | 6.9996 | −7.0977 | 8.773 | −2.8071 | 5.6203 | −6.8439 | 8.1189 | −8.79512 | 11.70959 |
| 5.84 | −5.3243 | 7.0555 | −7.1493 | 8.8951 | −2.8338 | 5.6239 | −6.9832 | 8.23 | −8.94408 | 11.99697 |
| 5.82 | −5.3977 | 7.114 | −7.2045 | 9.0212 | −2.8647 | 5.6289 | −7.1298 | 8.3463 | −9.08656 | 12.29263 |
| 5.8 | −5.4752 | 7.1752 | −7.2635 | 9.1518 | −2.8997 | 5.6353 | −7.2843 | 8.4683 | −9.2032 | 12.62145 |
| 5.78 | −5.5565 | 7.2394 | −7.326 | 9.2872 | −2.9386 | 5.6434 | −7.4473 | 8.5967 | −9.32137 | 12.95028 |

*(continued)*

**Table D1** *Continued*

| Energy (eV) | GaAs | | AlAs | | InAs | | GaP | | AlP | |
|---|---|---|---|---|---|---|---|---|---|---|
| | $\varepsilon_r$ | $\varepsilon_i$ | $\varepsilon_r$ | $\varepsilon_i$ | $\varepsilon_r$ | $\varepsilon_i$ | $\varepsilon_r$ | $\varepsilon_i$ | $\varepsilon_r$ | $\varepsilon_i$ |
| 5.76 | −5.6418 | 7.3068 | −7.3918 | 9.4278 | −2.9816 | 5.6533 | −7.6193 | 8.732 | −9.47646 | 13.27945 |
| 5.74 | −5.7309 | 7.3775 | −7.4609 | 9.5739 | −3.0285 | 5.6651 | −7.8012 | 8.8752 | −9.63156 | 13.60863 |
| 5.72 | −5.824 | 7.4519 | −7.5331 | 9.7258 | −3.0796 | 5.6791 | −7.9939 | 9.0272 | −9.76118 | 13.99133 |
| 5.7 | −5.921 | 7.5301 | −7.6084 | 9.8841 | −3.1347 | 5.6956 | −8.1984 | 9.1892 | −9.88828 | 14.37934 |
| 5.68 | −6.022 | 7.6125 | −7.6864 | 10.049 | −3.1939 | 5.7147 | −8.4161 | 9.3629 | −10.0009 | 14.81096 |
| 5.66 | −6.127 | 7.6992 | −7.7672 | 10.221 | −3.2574 | 5.7369 | −8.6481 | 9.5503 | −10.1102 | 15.25248 |
| 5.64 | −6.2359 | 7.7905 | −7.8505 | 10.401 | −3.3251 | 5.7626 | −8.8957 | 9.7539 | −10.1971 | 15.73712 |
| 5.62 | −6.349 | 7.8868 | −7.9361 | 10.588 | −3.3971 | 5.7922 | −9.1603 | 9.977 | −10.2764 | 16.23613 |
| 5.6 | −6.4662 | 7.9882 | −8.024 | 10.784 | −3.4733 | 5.8263 | −9.4427 | 10.223 | −10.3471 | 16.7692 |
| 5.58 | −6.5876 | 8.095 | −8.1138 | 10.989 | −3.5535 | 5.8654 | −9.7437 | 10.498 | −10.4142 | 17.3162 |
| 5.56 | −6.7134 | 8.2075 | −8.2054 | 11.203 | −3.6376 | 5.9103 | −10.063 | 10.806 | −10.4404 | 17.89778 |
| 5.54 | −6.8436 | 8.3261 | −8.2987 | 11.427 | −3.7251 | 5.9615 | −10.4 | 11.154 | −10.4491 | 18.49418 |
| 5.52 | −6.9786 | 8.4511 | −8.3933 | 11.661 | −3.8155 | 6.0198 | −10.752 | 11.55 | −10.3604 | 19.11552 |
| 5.5 | −7.1185 | 8.5828 | −8.4891 | 11.906 | −3.908 | 6.0857 | −11.115 | 12.001 | −10.2339 | 19.74656 |
| 5.48 | −7.2636 | 8.7215 | −8.5858 | 12.163 | −4.0018 | 6.1597 | −11.484 | 12.516 | −10.1029 | 20.39381 |
| 5.46 | −7.4143 | 8.8677 | −8.6832 | 12.431 | −4.0957 | 6.2422 | −11.85 | 13.102 | −9.97067 | 21.0459 |
| 5.44 | −7.5712 | 9.0219 | −8.7813 | 12.712 | −4.1885 | 6.333 | −12.201 | 13.765 | −9.7815 | 21.71351 |

| | | | | | | | | | | |
|---|---|---|---|---|---|---|---|---|---|---|
| 5.42 | -7.7348 | 9.1847 | -8.88 | 13.006 | -4.2791 | 6.4321 | -12.526 | 14.513 | -9.58148 | 22.38409 |
| 5.4 | -7.9059 | 9.3567 | -8.9792 | 13.314 | -4.3663 | 6.5387 | -12.807 | 15.348 | -9.25793 | 23.06017 |
| 5.38 | -8.0854 | 9.539 | -9.079 | 13.637 | -4.4489 | 6.652 | -13.027 | 16.269 | -8.92608 | 23.73677 |
| 5.36 | -8.2746 | 9.7327 | -9.1797 | 13.975 | -4.5261 | 6.7706 | -13.167 | 17.274 | -8.57943 | 24.41601 |
| 5.34 | -8.4748 | 9.9395 | -9.2817 | 14.331 | -4.5977 | 6.8932 | -13.207 | 18.352 | -8.20141 | 25.09198 |
| 5.32 | -8.6872 | 10.162 | -9.3856 | 14.706 | -4.6634 | 7.0181 | -13.127 | 19.492 | -7.71534 | 25.75666 |
| 5.3 | -8.9135 | 10.402 | -9.4919 | 15.102 | -4.7239 | 7.1438 | -12.912 | 20.673 | -7.20194 | 26.41302 |
| 5.28 | -9.1549 | 10.665 | -9.6013 | 15.523 | -4.7801 | 7.2686 | -12.548 | 21.871 | -6.64754 | 27.0569 |
| 5.26 | -9.4124 | 10.954 | -9.7136 | 15.974 | -4.8334 | 7.3915 | -12.028 | 23.061 | -6.06368 | 27.67634 |
| 5.24 | -9.6862 | 11.276 | -9.8279 | 16.461 | -4.8857 | 7.5116 | -11.349 | 24.21 | -5.46059 | 28.27984 |
| 5.22 | -9.9756 | 11.638 | -9.9417 | 16.99 | -4.9388 | 7.6288 | -10.519 | 25.289 | -4.76307 | 28.83633 |
| 5.2 | -10.278 | 12.047 | -10.05 | 17.572 | -4.9949 | 7.7432 | -9.5518 | 26.267 | -4.04333 | 29.37727 |
| 5.18 | -10.589 | 12.514 | -10.143 | 18.213 | -5.056 | 7.8557 | -8.4681 | 27.119 | -3.29038 | 29.85021 |
| 5.16 | -10.901 | 13.046 | -10.21 | 18.921 | -5.1239 | 7.9678 | -7.2963 | 27.824 | -2.52167 | 30.29452 |
| 5.14 | -11.204 | 13.653 | -10.233 | 19.702 | -5.2001 | 8.0811 | -6.0689 | 28.369 | -1.72366 | 30.68567 |
| 5.12 | -11.484 | 14.341 | -10.191 | 20.555 | -5.2855 | 8.1976 | -4.821 | 28.751 | -0.85368 | 30.9974 |
| 5.1 | -11.724 | 15.115 | -10.062 | 21.475 | -5.3808 | 8.3197 | -3.5868 | 28.977 | 0.05505 | 31.26636 |
| 5.08 | -11.905 | 15.977 | -9.8235 | 22.445 | -5.4862 | 8.4494 | -2.3969 | 29.061 | 0.85766 | 31.44891 |
| 5.06 | -12.004 | 16.922 | -9.457 | 23.444 | -5.6015 | 8.5889 | -1.2754 | 29.027 | 1.64211 | 31.60156 |

(continued)

**Table D1** *Continued*

| Energy (eV) | GaAs $\varepsilon_r$ | GaAs $\varepsilon_i$ | AlAs $\varepsilon_r$ | AlAs $\varepsilon_i$ | InAs $\varepsilon_r$ | InAs $\varepsilon_i$ | GaP $\varepsilon_r$ | GaP $\varepsilon_i$ | AlP $\varepsilon_r$ | AlP $\varepsilon_i$ |
|---|---|---|---|---|---|---|---|---|---|---|
| 5.04 | −12.001 | 17.94 | −8.9511 | 24.441 | −5.7263 | 8.7401 | −0.23792 | 28.902 | 2.39394 | 31.6932 |
| 5.02 | −11.874 | 19.015 | −8.3049 | 25.401 | −5.8604 | 8.9047 | 0.70967 | 28.716 | 3.11353 | 31.74684 |
| 5 | −11.605 | 20.124 | −7.5299 | 26.292 | −6.0032 | 9.0847 | 1.5707 | 28.495 | 3.80831 | 31.78242 |
| 4.98 | −11.183 | 21.24 | −6.6488 | 27.083 | −6.1547 | 9.2816 | 2.3562 | 28.264 | 4.3401 | 31.83883 |
| 4.96 | −10.603 | 22.331 | −5.6938 | 27.758 | −6.3148 | 9.4977 | 3.0827 | 28.04 | 4.83849 | 31.97603 |
| 4.94 | −9.8667 | 23.363 | −4.7008 | 28.311 | −6.4837 | 9.7357 | 3.7696 | 27.832 | 5.29608 | 32.21195 |
| 4.92 | −8.9881 | 24.304 | −3.7034 | 28.751 | −6.6615 | 9.999 | 4.4372 | 27.642 | 5.82537 | 32.65513 |
| 4.9 | −7.9879 | 25.124 | −2.7272 | 29.1 | −6.8477 | 10.292 | 5.1056 | 27.468 | 6.41027 | 33.18232 |
| 4.88 | −6.8946 | 25.802 | −1.7848 | 29.388 | −7.041 | 10.621 | 5.7923 | 27.296 | 7.09386 | 33.82781 |
| 4.86 | −5.7416 | 26.323 | −0.87345 | 29.644 | −7.2382 | 10.993 | 6.5086 | 27.108 | 8.04568 | 34.50282 |
| 4.84 | −4.5647 | 26.684 | 0.023299 | 29.893 | −7.4338 | 11.415 | 7.2563 | 26.879 | 9.18022 | 35.1516 |
| 4.82 | −3.3984 | 26.89 | 0.93242 | 30.148 | −7.6191 | 11.895 | 8.0255 | 26.586 | 10.61313 | 35.7013 |
| 4.8 | −2.2729 | 26.956 | 1.885 | 30.406 | −7.7816 | 12.438 | 8.7964 | 26.212 | 12.17003 | 36.06452 |
| 4.78 | −1.2121 | 26.903 | 2.9094 | 30.648 | −7.9059 | 13.048 | 9.5438 | 25.75 | 13.77085 | 36.29482 |
| 4.76 | −0.2313 | 26.758 | 4.0237 | 30.842 | −7.9739 | 13.722 | 10.242 | 25.203 | 15.38135 | 36.33423 |
| 4.74 | 0.66262 | 26.547 | 5.2307 | 30.946 | −7.9668 | 14.452 | 10.869 | 24.585 | 17.0877 | 36.1947 |
| 4.72 | 1.4705 | 26.296 | 6.5147 | 30.92 | −7.868 | 15.225 | 11.41 | 23.912 | 18.84686 | 35.95953 |
| 4.7 | 2.1988 | 26.026 | 7.8424 | 30.727 | −7.6656 | 16.018 | 11.856 | 23.205 | 20.65878 | 35.631 |

| | | | | | | | | | | |
|---|---|---|---|---|---|---|---|---|---|---|
| 4.68 | 2.8574 | 25.755 | 9.1678 | 30.349 | −7.3555 | 16.807 | 12.204 | 22.485 | 22.49246 | 35.13723 |
| 4.66 | 3.4606 | 25.5 | 10.439 | 29.781 | −6.943 | 17.569 | 12.459 | 21.773 | 24.45795 | 34.47474 |
| 4.64 | 4.0303 | 25.272 | 11.607 | 29.042 | −6.4418 | 18.283 | 12.627 | 21.082 | 26.54023 | 33.60536 |
| 4.62 | 4.5948 | 25.072 | 12.63 | 28.161 | −5.8717 | 18.938 | 12.72 | 20.428 | 28.61527 | 32.30339 |
| 4.6 | 5.1771 | 24.882 | 13.483 | 27.182 | −5.2527 | 19.534 | 12.75 | 19.817 | 30.60111 | 30.53082 |
| 4.58 | 5.7849 | 24.678 | 14.152 | 26.149 | −4.5988 | 20.081 | 12.728 | 19.255 | 32.3437 | 28.40103 |
| 4.56 | 6.4119 | 24.437 | 14.639 | 25.106 | −3.9128 | 20.593 | 12.667 | 18.744 | 33.7198 | 25.94304 |
| 4.54 | 7.0454 | 24.146 | 14.959 | 24.092 | −3.1838 | 21.084 | 12.576 | 18.285 | 34.4862 | 23.28964 |
| 4.52 | 7.6703 | 23.799 | 15.133 | 23.137 | −2.3905 | 21.556 | 12.464 | 17.875 | 34.76038 | 20.7195 |
| 4.5 | 8.2717 | 23.393 | 15.185 | 22.262 | −1.5085 | 21.993 | 12.338 | 17.511 | 34.6241 | 18.29782 |
| 4.48 | 8.8353 | 22.932 | 15.143 | 21.479 | −0.52089 | 22.363 | 12.205 | 17.191 | 34.14214 | 16.08746 |
| 4.46 | 9.3493 | 22.424 | 15.031 | 20.793 | 0.57289 | 22.622 | 12.069 | 16.91 | 33.47647 | 14.20347 |
| 4.44 | 9.8045 | 21.878 | 14.871 | 20.205 | 1.7503 | 22.724 | 11.932 | 16.666 | 32.70214 | 12.58657 |
| 4.42 | 10.195 | 21.305 | 14.681 | 19.71 | 2.9662 | 22.632 | 11.799 | 16.454 | 31.87468 | 11.18213 |
| 4.4 | 10.518 | 20.718 | 14.474 | 19.303 | 4.1602 | 22.331 | 11.67 | 16.272 | 31.0282 | 9.96913 |
| 4.38 | 10.775 | 20.128 | 14.262 | 18.976 | 5.2696 | 21.834 | 11.547 | 16.116 | 30.19207 | 8.92419 |
| 4.36 | 10.967 | 19.546 | 14.052 | 18.725 | 6.243 | 21.176 | 11.43 | 15.984 | 29.33808 | 8.04642 |
| 4.34 | 11.101 | 18.981 | 13.85 | 18.547 | 7.0488 | 20.408 | 11.32 | 15.874 | 28.57741 | 7.27828 |
| 4.32 | 11.182 | 18.44 | 13.66 | 18.437 | 7.6779 | 19.586 | 11.218 | 15.785 | 27.83431 | 6.59998 |

(continued)

Table D1 Continued

| Energy (eV) | GaAs | | AlAs | | InAs | | GaP | | AlP | |
|---|---|---|---|---|---|---|---|---|---|---|
| | $\varepsilon_r$ | $\varepsilon_i$ | $\varepsilon_r$ | $\varepsilon_i$ | $\varepsilon_r$ | $\varepsilon_i$ | $\varepsilon_r$ | $\varepsilon_i$ | $\varepsilon_r$ | $\varepsilon_i$ |
| 4.3 | 11.218 | 17.927 | 13.488 | 18.397 | 8.14 | 18.755 | 11.123 | 15.714 | 27.1214 | 5.99946 |
| 4.28 | 11.214 | 17.448 | 13.341 | 18.429 | 8.4576 | 17.953 | 11.035 | 15.66 | 26.46651 | 5.46468 |
| 4.26 | 11.179 | 17.002 | 13.228 | 18.534 | 8.6583 | 17.203 | 10.954 | 15.623 | 25.8333 | 4.9826 |
| 4.24 | 11.118 | 16.592 | 13.163 | 18.715 | 8.7697 | 16.515 | 10.88 | 15.602 | 25.22321 | 4.54764 |
| 4.22 | 11.037 | 16.216 | 13.167 | 18.973 | 8.8157 | 15.892 | 10.812 | 15.597 | 24.68505 | 4.13946 |
| 4.2 | 10.941 | 15.874 | 13.262 | 19.299 | 8.8153 | 15.331 | 10.752 | 15.607 | 24.14725 | 3.75202 |
| 4.18 | 10.834 | 15.564 | 13.474 | 19.677 | 8.7825 | 14.826 | 10.697 | 15.632 | 23.62177 | 3.38274 |
| 4.16 | 10.719 | 15.285 | 13.827 | 20.077 | 8.7271 | 14.372 | 10.649 | 15.674 | 23.09118 | 3.03472 |
| 4.14 | 10.601 | 15.035 | 14.333 | 20.46 | 8.6559 | 13.962 | 10.608 | 15.731 | 22.5531 | 2.71441 |
| 4.12 | 10.481 | 14.811 | 14.985 | 20.776 | 8.5732 | 13.59 | 10.575 | 15.806 | 22.04617 | 2.42811 |
| 4.1 | 10.36 | 14.611 | 15.759 | 20.982 | 8.4824 | 13.253 | 10.548 | 15.898 | 21.54937 | 2.17996 |
| 4.08 | 10.242 | 14.434 | 16.608 | 21.052 | 8.3855 | 12.946 | 10.53 | 16.009 | 21.09469 | 1.96743 |
| 4.06 | 10.126 | 14.278 | 17.481 | 20.984 | 8.2843 | 12.667 | 10.521 | 16.139 | 20.64547 | 1.78403 |
| 4.04 | 10.014 | 14.141 | 18.338 | 20.807 | 8.18 | 12.413 | 10.522 | 16.291 | 20.17415 | 1.62247 |
| 4.02 | 9.9073 | 14.021 | 19.169 | 20.563 | 8.0738 | 12.182 | 10.533 | 16.465 | 19.72264 | 1.4782 |
| 4 | 9.8053 | 13.916 | 19.999 | 20.291 | 7.9668 | 11.973 | 10.556 | 16.664 | 19.32996 | 1.35021 |
| 3.98 | 9.7089 | 13.826 | 20.883 | 19.995 | 7.8596 | 11.783 | 10.592 | 16.889 | 18.95979 | 1.23864 |
| 3.96 | 9.6181 | 13.75 | 21.875 | 19.627 | 7.7531 | 11.612 | 10.641 | 17.143 | 18.64299 | 1.14328 |

| | | | | | | | | | | |
|---|---|---|---|---|---|---|---|---|---|---|
| 3.94 | 9.5332 | 13.685 | 22.986 | 19.095 | 7.648 | 11.458 | 10.703 | 17.432 | 18.25703 | 1.0629 |
| 3.92 | 9.454 | 13.631 | 24.162 | 18.299 | 7.5449 | 11.319 | 10.778 | 17.762 | 17.91679 | 0.99529 |
| 3.9 | 9.3806 | 13.587 | 25.28 | 17.177 | 7.4442 | 11.195 | 10.868 | 18.146 | 17.62357 | 0.93781 |
| 3.88 | 9.3127 | 13.553 | 26.195 | 15.75 | 7.3464 | 11.085 | 10.977 | 18.604 | 17.32009 | 0.88793 |
| 3.86 | 9.2503 | 13.527 | 26.788 | 14.109 | 7.2519 | 10.987 | 11.12 | 19.163 | 17.05581 | 0.8434 |
| 3.84 | 9.1931 | 13.509 | 27.001 | 12.396 | 7.1608 | 10.901 | 11.327 | 19.858 | 16.82394 | 0.80342 |
| 3.82 | 9.1408 | 13.499 | 26.852 | 10.753 | 7.0735 | 10.825 | 11.664 | 20.728 | 16.53692 | 0.76671 |
| 3.8 | 9.0934 | 13.496 | 26.418 | 9.2878 | 6.9901 | 10.76 | 12.24 | 21.789 | 16.25535 | 0.73346 |
| 3.78 | 9.0505 | 13.5 | 25.8 | 8.0514 | 6.9106 | 10.703 | 13.209 | 22.996 | 16.01229 | 0.70392 |
| 3.76 | 9.0121 | 13.511 | 25.092 | 7.0472 | 6.8352 | 10.655 | 14.728 | 24.192 | 15.77697 | 0.67794 |
| 3.74 | 8.9779 | 13.529 | 24.364 | 6.248 | 6.7638 | 10.614 | 16.875 | 25.099 | 15.5971 | 0.65527 |
| 3.72 | 8.9478 | 13.554 | 23.659 | 5.6148 | 6.6963 | 10.581 | 19.541 | 25.376 | 15.40182 | 0.63437 |
| 3.7 | 8.9218 | 13.586 | 22.999 | 5.1083 | 6.6328 | 10.554 | 22.408 | 24.763 | 15.22082 | 0.61301 |
| 3.68 | 8.8999 | 13.624 | 22.392 | 4.6957 | 6.573 | 10.533 | 25.033 | 23.227 | 15.0045 | 0.58889 |
| 3.66 | 8.8821 | 13.67 | 21.837 | 4.3514 | 6.5168 | 10.517 | 27.032 | 20.993 | 14.84244 | 0.56101 |
| 3.64 | 8.8685 | 13.724 | 21.329 | 4.0573 | 6.4641 | 10.507 | 28.225 | 18.437 | 14.67627 | 0.53023 |
| 3.62 | 8.8592 | 13.785 | 20.863 | 3.8007 | 6.4146 | 10.501 | 28.657 | 15.922 | 14.5141 | 0.49862 |

(*continued*)

**Table D1** *Continued*

| Energy (eV) | GaAs | | AlAs | | InAs | | GaP | | AlP | |
|---|---|---|---|---|---|---|---|---|---|---|
| | $\varepsilon_r$ | $\varepsilon_i$ | $\varepsilon_r$ | $\varepsilon_i$ | $\varepsilon_r$ | $\varepsilon_i$ | $\varepsilon_r$ | $\varepsilon_i$ | $\varepsilon_r$ | $\varepsilon_i$ |
| 3.6 | 8.8545 | 13.854 | 20.431 | 3.5734 | 6.3683 | 10.501 | 28.52 | 13.689 | 14.37338 | 0.46858 |
| 3.58 | 8.8545 | 13.931 | 20.031 | 3.3696 | 6.3249 | 10.504 | 28.036 | 11.826 | 14.19914 | 0.44112 |
| 3.56 | 8.8597 | 14.017 | 19.657 | 3.1855 | 6.2841 | 10.512 | 27.381 | 10.317 | 14.02847 | 0.41649 |
| 3.54 | 8.8703 | 14.112 | 19.306 | 3.0182 | 6.246 | 10.524 | 26.667 | 9.1034 | 13.89562 | 0.39447 |
| 3.52 | 8.8869 | 14.216 | 18.977 | 2.8655 | 6.2102 | 10.54 | 25.956 | 8.1214 | 13.76782 | 0.37464 |
| 3.5 | 8.9099 | 14.331 | 18.668 | 2.7253 | 6.1766 | 10.56 | 25.275 | 7.315 | 13.62161 | 0.35635 |
| 3.48 | 8.9401 | 14.456 | 18.377 | 2.5958 | 6.1451 | 10.584 | 24.637 | 6.6407 | 13.50829 | 0.33924 |
| 3.46 | 8.978 | 14.592 | 18.103 | 2.4756 | 6.1156 | 10.613 | 24.043 | 6.066 | 13.31567 | 0.32321 |
| 3.44 | 9.0245 | 14.741 | 17.844 | 2.363 | 6.088 | 10.645 | 23.489 | 5.5676 | 13.22818 | 0.30808 |
| 3.42 | 9.0807 | 14.902 | 17.601 | 2.2566 | 6.0623 | 10.682 | 22.97 | 5.1296 | 13.08444 | 0.29364 |
| 3.4 | 9.1477 | 15.077 | 17.372 | 2.1549 | 6.0384 | 10.724 | 22.481 | 4.7412 | 13.00259 | 0.27987 |
| 3.38 | 9.227 | 15.267 | 17.157 | 2.0559 | 6.0164 | 10.77 | 22.019 | 4.3951 | 12.81632 | 0.26683 |
| 3.36 | 9.3204 | 15.473 | 16.954 | 1.9565 | 5.9965 | 10.821 | 21.582 | 4.0858 | 12.70122 | 0.25432 |
| 3.34 | 9.4298 | 15.697 | 16.76 | 1.8533 | 5.9785 | 10.877 | 21.166 | 3.8092 | 12.62472 | 0.24239 |
| 3.32 | 9.5576 | 15.939 | 16.568 | 1.7459 | 5.9629 | 10.939 | 20.771 | 3.5615 | 12.51336 | 0.23102 |
| 3.3 | 9.706 | 16.203 | 16.373 | 1.6393 | 5.9496 | 11.006 | 20.396 | 3.3395 | 12.40251 | 0.22008 |
| 3.28 | 9.8768 | 16.491 | 16.175 | 1.5414 | 5.9389 | 11.079 | 20.041 | 3.1402 | 12.28686 | 0.20967 |
| 3.26 | 10.073 | 16.813 | 15.98 | 1.4557 | 5.931 | 11.158 | 19.704 | 2.9609 | 12.18763 | 0.19968 |

| | | | | | | | | | | |
|---|---|---|---|---|---|---|---|---|---|---|
| 3.24 | 10.305 | 17.186 | 15.793 | 1.3802 | 5.9263 | 11.243 | 19.385 | 2.799 | 12.09863 | 0.19007 |
| 3.22 | 10.608 | 17.632 | 15.617 | 1.3106 | 5.925 | 11.335 | 19.083 | 2.6522 | 12.01241 | 0.18084 |
| 3.2 | 11.055 | 18.146 | 15.449 | 1.2441 | 5.9274 | 11.433 | 18.798 | 2.5184 | 11.91155 | 0.17195 |
| 3.18 | 11.719 | 18.652 | 15.29 | 1.1792 | 5.934 | 11.539 | 18.529 | 2.3958 | 11.81144 | 0.16338 |
| 3.16 | 12.609 | 19.013 | 15.137 | 1.1154 | 5.9452 | 11.652 | 18.274 | 2.2824 | 11.72115 | 0.15512 |
| 3.14 | 13.619 | 19.106 | 14.991 | 1.0524 | 5.9614 | 11.772 | 18.033 | 2.177 | 11.64234 | 0.14711 |
| 3.12 | 14.582 | 18.915 | 14.85 | 0.99003 | 5.9832 | 11.9 | 17.805 | 2.0781 | 11.53727 | 0.13937 |
| 3.1 | 15.37 | 18.535 | 14.714 | 0.928 | 6.011 | 12.036 | 17.588 | 1.9847 | 11.45146 | 0.1318 |
| 3.08 | 15.952 | 18.099 | 14.583 | 0.86622 | 6.0454 | 12.181 | 17.383 | 1.8957 | 11.39777 | 0.12448 |
| 3.06 | 16.367 | 17.708 | 14.457 | 0.80466 | 6.0871 | 12.334 | 17.188 | 1.8105 | 11.30918 | 0.11733 |
| 3.04 | 16.678 | 17.422 | 14.339 | 0.74324 | 6.1366 | 12.496 | 17.003 | 1.7285 | 11.21958 | 0.11034 |
| 3.02 | 16.95 | 17.271 | 14.233 | 0.68056 | 6.1946 | 12.669 | 16.826 | 1.6489 | 11.15862 | 0.10349 |
| 3 | 17.253 | 17.277 | 14.15 | 0.5993 | 6.2615 | 12.851 | 16.658 | 1.5715 | 11.0732 | 0.09675 |
| 2.98 | 17.682 | 17.454 | 14.059 | 0.45684 | 6.3377 | 13.044 | 16.497 | 1.4958 | 11.00319 | 0.09012 |
| 2.96 | 18.392 | 17.775 | 13.882 | 0.30107 | 6.4235 | 13.251 | 16.344 | 1.4213 | 10.9468 | 0.0836 |
| 2.94 | 19.586 | 18.061 | 13.669 | 0.21389 | 6.5193 | 13.473 | 16.198 | 1.3475 | 10.87053 | 0.07717 |
| 2.92 | 21.301 | 17.877 | 13.477 | 0.16626 | 6.6271 | 13.718 | 16.06 | 1.2736 | 10.81832 | 0.0708 |
| 2.9 | 23.097 | 16.809 | 13.302 | 0.13182 | 6.7532 | 13.994 | 15.93 | 1.1983 | 10.73231 | 0.0645 |

(continued)

Table D1 *Continued*

| Energy (eV) | GaAs | | AlAs | | InAs | | GaP | | AlP | |
|---|---|---|---|---|---|---|---|---|---|---|
| | $\varepsilon_r$ | $\varepsilon_i$ | $\varepsilon_r$ | $\varepsilon_i$ | $\varepsilon_r$ | $\varepsilon_i$ | $\varepsilon_r$ | $\varepsilon_i$ | $\varepsilon_r$ | $\varepsilon_i$ |
| 2.88 | 24.288 | 15.018 | 13.137 | 0.10582 | 6.9116 | 14.308 | 15.806 | 1.1191 | 10.6766 | 0.05823 |
| 2.86 | 24.66 | 13.137 | 12.982 | 0.086449 | 7.1257 | 14.663 | 15.687 | 1.034 | 10.61189 | 0.052 |
| 2.84 | 24.529 | 11.575 | 12.836 | 0.072228 | 7.4253 | 15.046 | 15.568 | 0.94428 | 10.51991 | 0.04581 |
| 2.82 | 24.217 | 10.333 | 12.697 | 0.061937 | 7.8347 | 15.424 | 15.451 | 0.85482 | 10.47181 | 0.03965 |
| 2.8 | 23.841 | 9.3029 | 12.566 | 0.054589 | 8.3587 | 15.749 | 15.346 | 0.76819 | 10.4232 | 0.03351 |
| 2.78 | 23.428 | 8.4255 | 12.441 | 0.049392 | 8.9729 | 15.97 | 15.278 | 0.65319 | 10.35811 | 0.02743 |
| 2.76 | 22.993 | 7.6741 | 12.322 | 0.045714 | 9.6277 | 16.058 | 15.182 | 0.44034 | 10.31459 | 0.02149 |
| 2.74 | 22.548 | 7.0313 | 12.209 | 0.043053 | 10.265 | 16.014 | 14.942 | 0.22981 | 10.22772 | 0.01588 |
| 2.72 | 22.107 | 6.4819 | 12.101 | 0.04102 | 10.837 | 15.865 | 14.665 | 0.14295 | 10.19853 | 0.01086 |
| 2.7 | 21.677 | 6.0119 | 11.997 | 0.03933 | 11.32 | 15.657 | 14.438 | 0.11117 | 10.12338 | 0.00674 |
| 2.68 | 21.265 | 5.6087 | 11.898 | 0.037791 | 11.714 | 15.434 | 14.244 | 0.090046 | 10.07617 | 0.00373 |
| 2.66 | 20.875 | 5.2612 | 11.803 | 0.036299 | 12.033 | 15.231 | 14.067 | 0.073535 | 10.00125 | 0.00181 |
| 2.64 | 20.508 | 4.9592 | 11.71 | 0.034817 | 12.3 | 15.073 | 13.903 | 0.060323 | 9.95037 | 7.56E − 04 |
| 2.62 | 20.164 | 4.6941 | 11.622 | 0.033335 | 12.542 | 14.976 | 13.75 | 0.049643 | 9.91694 | 2.65E − 04 |
| 2.6 | 19.844 | 4.4585 | 11.535 | 0.031854 | 12.789 | 14.95 | 13.605 | 0.040999 | 9.84623 | 7.72E − 05 |
| 2.58 | 19.545 | 4.2462 | 11.452 | 0.030372 | 13.081 | 15.002 | 13.468 | 0.034033 | 9.79526 | 1.96E − 05 |

| 2.56 | 19.265 | 4.0526 | 11.371 | 0.028891 | 13.47 | 15.125 | 13.337 | 0.028453 | 9.74989 | 4.04E−06 |
|------|--------|--------|--------|----------|-------|--------|--------|----------|---------|----------|
| 2.54 | 19.001 | 3.8738 | 11.293 | 0.027409 | 14.023 | 15.275 | 13.213 | 0.024006 | 9.69613 | 6.52E−07 |
| 2.52 | 18.752 | 3.7076 | 11.216 | 0.025927 | 14.789 | 15.352 | 13.094 | 0.020466 | 9.66812 | 9.54E−08 |
| 2.5 | 18.515 | 3.5524 | 11.142 | 0.024446 | 15.741 | 15.211 | 12.98 | 0.017633 | 9.61193 | 1.16E−09 |
| 2.48 | 18.289 | 3.4072 | 11.07 | 0.022964 | 16.737 | 14.736 | 12.87 | 0.015333 | 9.55969 | 0 |
| 2.46 | 18.073 | 3.2712 | 11 | 0.021483 | 17.576 | 13.939 | 12.765 | 0.013417 | 9.49943 | 0 |
| 2.44 | 17.865 | 3.1439 | 10.932 | 0.020001 | 18.122 | 12.966 | 12.664 | 0.01176 | 9.48548 | 0 |
| 2.42 | 17.665 | 3.0247 | 10.865 | 0.01852 | 18.373 | 11.995 | 12.567 | 0.010264 | 9.43556 | 0 |
| 2.4 | 17.473 | 2.9131 | 10.8 | 0.017038 | 18.422 | 11.132 | 12.473 | 0.008854 | 9.39831 | 0 |
| 2.38 | 17.288 | 2.8085 | 10.737 | 0.015556 | 18.369 | 10.393 | 12.382 | 0.007481 | 9.353 | 0 |
| 2.36 | 17.111 | 2.7104 | 10.675 | 0.014075 | 18.268 | 9.7531 | 12.295 | 0.006119 | 9.31419 | 0 |
| 2.34 | 16.94 | 2.6185 | 10.615 | 0.012593 | 18.141 | 9.1855 | 12.209 | 0.004759 | 9.26549 | 0 |
| 2.32 | 16.775 | 2.5321 | 10.556 | 0.011112 | 17.994 | 8.6749 | 12.127 | 0.0034 | 9.23131 | 0 |
| 2.3 | 16.617 | 2.451 | 10.498 | 0.00963 | 17.832 | 8.2135 | 12.046 | 0.002057 | 9.19687 | 0 |
| 2.28 | 16.465 | 2.3746 | 10.441 | 0.008149 | 17.658 | 7.7962 | 11.968 | 0.000878 | 9.17589 | 0 |
| 2.26 | 16.318 | 2.3025 | 10.386 | 0.006667 | 17.478 | 7.4189 | 11.892 | 0.000198 | 9.1343 | 0 |
| 2.24 | 16.178 | 2.2345 | 10.332 | 0.005186 | 17.295 | 7.0777 | 11.818 | 1.79E−05 | 9.08721 | 0 |

(continued)

**Table D1** *Continued*

| Energy (eV) | GaAs | | AlAs | | InAs | | GaP | | AlP | |
|---|---|---|---|---|---|---|---|---|---|---|
| | $\varepsilon_r$ | $\varepsilon_i$ | $\varepsilon_r$ | $\varepsilon_i$ | $\varepsilon_r$ | $\varepsilon_i$ | $\varepsilon_r$ | $\varepsilon_i$ | $\varepsilon_r$ | $\varepsilon_i$ |
| 2.22 | 16.042 | 2.17 | 10.279 | 0.003705 | 17.11 | 6.769 | 11.746 | 0 | 9.06797 | 0 |
| 2.2 | 15.912 | 2.1089 | 10.227 | 0.002242 | 16.928 | 6.4894 | 11.676 | 0 | 9.02378 | 0 |
| 2.18 | 15.787 | 2.0507 | 10.176 | 0.000957 | 16.749 | 6.2358 | 11.609 | 0 | 8.98578 | 0 |
| 2.16 | 15.667 | 1.9952 | 10.126 | 0.000216 | 16.574 | 6.0053 | 11.543 | 0 | 8.95804 | 0 |
| 2.14 | 15.552 | 1.9421 | 10.076 | 1.95E − 05 | 16.406 | 5.7951 | 11.479 | 0 | 8.91766 | 0 |
| 2.12 | 15.441 | 1.891 | 10.029 | 0 | 16.245 | 5.6028 | 11.417 | 0 | 8.88243 | 0 |
| 2.1 | 15.335 | 1.8419 | 9.9823 | 0 | 16.09 | 5.426 | 11.356 | 0 | 8.8724 | 0 |
| 2.08 | 15.233 | 1.7943 | 9.937 | 0 | 15.942 | 5.2627 | 11.297 | 0 | 8.84004 | 0 |
| 2.06 | 15.135 | 1.7481 | 9.8927 | 0 | 15.802 | 5.1112 | 11.239 | 0 | 8.79862 | 0 |
| 2.04 | 15.04 | 1.7032 | 9.8495 | 0 | 15.668 | 4.9697 | 11.183 | 0 | 8.78338 | 0 |
| 2.02 | 14.95 | 1.6592 | 9.8072 | 0 | 15.541 | 4.8369 | 11.128 | 0 | 8.7511 | 0 |
| 2 | 14.862 | 1.6161 | 9.7658 | 0 | 15.419 | 4.7116 | 11.075 | 0 | 8.72517 | 0 |
| 1.98 | 14.778 | 1.5738 | 9.7254 | 0 | 15.303 | 4.5929 | 11.023 | 0 | 8.68974 | 0 |
| 1.96 | 14.698 | 1.5321 | 9.6858 | 0 | 15.192 | 4.4801 | 10.972 | 0 | 8.65655 | 0 |
| 1.94 | 14.62 | 1.4908 | 9.6471 | 0 | 15.085 | 4.3724 | 10.922 | 0 | 8.6389 | 0 |
| 1.92 | 14.545 | 1.4499 | 9.6092 | 0 | 14.983 | 4.2697 | 10.873 | 0 | 8.60726 | 0 |
| 1.9 | 14.473 | 1.4093 | 9.5721 | 0 | 14.883 | 4.1714 | 10.826 | 0 | 8.59178 | 0 |

| 1.88 | 14.405 | 1.3689 | 9.5358 | 0 | 14.788 | 4.0775 | 10.779 | 0 | 8.54925 | 0 |
| 1.86 | 14.339 | 1.3284 | 9.5003 | 0 | 14.695 | 3.9876 | 10.734 | 0 | 8.53628 | 0 |
| 1.84 | 14.278 | 1.2871 | 9.4654 | 0 | 14.605 | 3.9016 | 10.689 | 0 | 8.50653 | 0 |
| 1.82 | 14.22 | 1.2432 | 9.4312 | 0 | 14.518 | 3.8194 | 10.646 | 0 | 8.48635 | 0 |
| 1.8 | 14.165 | 1.1939 | 9.3978 | 0 | 14.434 | 3.7407 | 10.604 | 0 | 8.46593 | 0 |
| 1.78 | 14.108 | 1.1379 | 9.3651 | 0 | 14.353 | 3.6653 | 10.562 | 0 | 8.44213 | 0 |
| 1.76 | 14.043 | 1.0787 | 9.3331 | 0 | 14.274 | 3.5932 | 10.522 | 0 | 8.40845 | 0 |
| 1.74 | 13.971 | 1.023 | 9.3017 | 0 | 14.198 | 3.5241 | 10.482 | 0 | 8.39204 | 0 |
| 1.72 | 13.895 | 0.97552 | 9.271 | 0 | 14.124 | 3.458 | 10.443 | 0 | 8.37489 | 0 |
| 1.7 | 13.822 | 0.9356 | 9.2409 | 0 | 14.053 | 3.3946 | 10.405 | 0 | 8.34548 | 0 |
| 1.68 | 13.754 | 0.89997 | 9.2114 | 0 | 13.984 | 3.3337 | 10.368 | 0 | 8.32631 | 0 |
| 1.66 | 13.69 | 0.86596 | 9.1825 | 0 | 13.917 | 3.2754 | 10.332 | 0 | 8.30926 | 0 |
| 1.64 | 13.63 | 0.83237 | 9.1542 | 0 | 13.853 | 3.2194 | 10.296 | 0 | 8.30466 | 0 |
| 1.62 | 13.573 | 0.79879 | 9.1265 | 0 | 13.792 | 3.1655 | 10.261 | 0 | 8.26591 | 0 |
| 1.6 | 13.519 | 0.76509 | 9.0993 | 0 | 13.732 | 3.1138 | 10.227 | 0 | 8.25146 | 0 |
| 1.58 | 13.468 | 0.73119 | 9.0727 | 0 | 13.675 | 3.0639 | 10.194 | 0 | 8.23119 | 0 |
| 1.56 | 13.419 | 0.69699 | 9.0466 | 0 | 13.62 | 3.0159 | 10.161 | 0 | 8.2183 | 0 |

*(continued)*

**Table D1** *Continued*

| Energy (eV) | GaAs | | AlAs | | InAs | | GaP | | AlP | |
|---|---|---|---|---|---|---|---|---|---|---|
| | $\varepsilon_r$ | $\varepsilon_i$ | $\varepsilon_r$ | $\varepsilon_i$ | $\varepsilon_r$ | $\varepsilon_i$ | $\varepsilon_r$ | $\varepsilon_i$ | $\varepsilon_r$ | $\varepsilon_i$ |
| 1.54 | 13.373 | 0.66243 | 9.0209 | 0 | 13.567 | 2.9696 | 10.129 | 0 | 8.20728 | 0 |
| 1.52 | 13.33 | 0.62742 | 8.9958 | 0 | 13.517 | 2.9249 | 10.098 | 0 | 8.18606 | 0 |
| 1.5 | 13.292 | 0.59195 | 8.9713 | 0 | 13.469 | 2.8816 | 10.068 | 0 | 8.16202 | 0 |
| 1.48 | 13.261 | 0.55612 | 8.9472 | 0 | 13.422 | 2.8397 | 10.038 | 0 | 8.14413 | 0 |
| 1.46 | 13.245 | 0.51896 | 8.9236 | 0 | 13.378 | 2.7991 | 10.008 | 0 | 8.13222 | 0 |
| 1.44 | 13.265 | 0.4588 | 8.9006 | 0 | 13.336 | 2.7596 | 9.9798 | 0 | 8.10614 | 0 |
| 1.42 | 13.285 | 0.30084 | 8.878 | 0 | 13.296 | 2.7212 | 9.9518 | 0 | 8.09077 | 0 |
| 1.4 | 13.165 | 0.097602 | 8.8558 | 0 | 13.258 | 2.6838 | 9.9244 | 0 | 8.08298 | 0 |
| 1.38 | 12.967 | 0.013088 | 8.8342 | 0 | 13.221 | 2.6472 | 9.8976 | 0 | 8.05734 | 0 |
| 1.36 | 12.815 | 0.001468 | 8.813 | 0 | 13.187 | 2.6115 | 9.8714 | 0 | 8.04935 | 0 |
| 1.34 | 12.7 | 0.00033 | 8.7922 | 0 | 13.155 | 2.5764 | 9.8457 | 0 | 8.02013 | 0 |
| 1.32 | 12.603 | 7.47E−05 | 8.7719 | 0 | 13.124 | 2.542 | 9.8206 | 0 | 8.00763 | 0 |
| 1.3 | 12.517 | 1.35E−05 | 8.752 | 0 | 13.095 | 2.5081 | 9.796 | 0 | 8.0011 | 0 |
| 1.28 | 12.438 | 1.94E−06 | 8.7325 | 0 | 13.068 | 2.4746 | 9.772 | 0 | 7.98263 | 0 |
| 1.26 | 12.366 | 0 | 8.7135 | 0 | 13.043 | 2.4416 | 9.7485 | 0 | 7.97431 | 0 |
| 1.24 | 12.297 | 0 | 8.6948 | 0 | 13.019 | 2.4089 | 9.7256 | 0 | 7.96623 | 0 |
| 1.22 | 12.233 | 0 | 8.6766 | 0 | 12.997 | 2.38E+00 | 9.7031 | 0 | 7.94973 | 0 |

| | | | | | | | | | | |
|---|---|---|---|---|---|---|---|---|---|---|
| 1.2 | 12.173 | 0 | 8.6587 | 0 | 12.977 | 2.34E+00 | 9.6811 | 0 | 7.93703 | 0 |
| 1.18 | 12.115 | 0 | 8.6413 | 0 | 12.958 | 2.3123 | 9.6597 | 0 | 7.91644 | 0 |
| 1.16 | 12.06 | 0 | 8.6242 | 0 | 12.941 | 2.2803 | 9.6387 | 0 | 7.90181 | 0 |
| 1.14 | 12.008 | 0 | 8.6075 | 0 | 12.925 | 2.2484 | 9.6182 | 0 | 7.89306 | 0 |
| 1.12 | 11.957 | 0 | 8.5912 | 0 | 12.91 | 2.2165 | 9.5982 | 0 | 7.87692 | 0 |
| 1.1 | 11.909 | 0 | 8.5752 | 0 | 12.897 | 2.1846 | 9.5786 | 0 | 7.86795 | 0 |
| 1.08 | 11.863 | 0 | 8.5596 | 0 | 12.886 | 2.1526 | 9.5595 | 0 | 7.86374 | 0 |
| 1.06 | 11.818 | 0 | 8.5444 | 0 | 12.875 | 2.1204 | 9.5409 | 0 | 7.84287 | 0 |
| 1.04 | 11.775 | 0 | 8.5296 | 0 | 12.866 | 2.0881 | 9.5227 | 0 | 7.83767 | 0 |
| 1.02 | 11.734 | 0 | 8.515 | 0 | 12.859 | 2.0556 | 9.5049 | 0 | 7.82204 | 0 |
| 1 | 11.694 | 0 | 8.5009 | 0 | 12.852 | 2.0228 | 9.4876 | 0 | 7.80844 | 0 |
| 0.98 | 11.655 | 0 | 8.487 | 0 | 12.847 | 1.9897 | 9.4706 | 0 | 7.80641 | 0 |
| 0.96 | 11.618 | 0 | 8.4736 | 0 | 12.844 | 1.9563 | 9.4542 | 0 | 7.78867 | 0 |
| 0.94 | 11.582 | 0 | 8.4604 | 0 | 12.841 | 1.9224 | 9.4381 | 0 | 7.77438 | 0 |
| 0.92 | 11.547 | 0 | 8.4476 | 0 | 12.84 | 1.8882 | 9.4224 | 0 | 7.7712 | 0 |
| 0.9 | 11.514 | 0 | 8.4351 | 0 | 12.841 | 1.8534 | 9.4071 | 0 | 7.76558 | 0 |

*(continued)*

**Table D1** *Continued*

| Energy (eV) | GaAs $\varepsilon_r$ | GaAs $\varepsilon_i$ | AlAs $\varepsilon_r$ | AlAs $\varepsilon_i$ | InAs $\varepsilon_r$ | InAs $\varepsilon_i$ | GaP $\varepsilon_r$ | GaP $\varepsilon_i$ | AlP $\varepsilon_r$ | AlP $\varepsilon_i$ |
|---|---|---|---|---|---|---|---|---|---|---|
| 0.88 | 11.481 | 0 | 8.4229 | 0 | 12.843 | 1.8181 | 9.3923 | 0 | 7.75156 | 0 |
| 0.86 | 11.45  | 0 | 8.411  | 0 | 12.848 | 1.7821 | 9.3778 | 0 | 7.74667 | 0 |
| 0.84 | 11.42  | 0 | 8.3995 | 0 | 12.855 | 1.7451 | 9.3637 | 0 | 7.73353 | 0 |
| 0.82 | 11.391 | 0 | 8.3883 | 0 | 12.866 | 1.7057 | 9.35   | 0 | 7.72611 | 0 |
| 0.8  | 11.363 | 0 | 8.3773 | 0 | 12.882 | 1.6608 | 9.3367 | 0 | 7.72189 | 0 |
| 0.78 | 11.335 | 0 | 8.3667 | 0 | 12.898 | 1.606  | 9.3238 | 0 | 7.71195 | 0 |
| 0.76 | 11.309 | 0 | 8.3564 | 0 | 12.908 | 1.5404 | 9.3112 | 0 | 7.70293 | 0 |
| 0.74 | 11.284 | 0 | 8.3464 | 0 | 12.904 | 1.4704 | 9.299  | 0 | 7.69611 | 0 |
| 0.72 | 11.259 | 0 | 8.3367 | 0 | 12.888 | 1.4063 | 9.2872 | 0 | | |
| 0.7  | 11.236 | 0 | 8.3273 | 0 | 12.866 | 1.3539 | 9.2757 | 0 | | |
| 0.68 | 11.213 | 0 | 8.3181 | 0 | 12.847 | 1.3111 | 9.2646 | 0 | | |
| 0.66 | 11.191 | 0 | 8.3093 | 0 | 12.834 | 1.273  | 9.2539 | 0 | | |
| 0.64 | 11.17  | 0 | 8.3007 | 0 | 12.826 | 1.2361 | 9.2435 | 0 | | |
| 0.62 | 11.15  | 0 | 8.2925 | 0 | 12.822 | 1.199  | 9.2334 | 0 | | |
| 0.6  | 11.13  | 0 | 8.2845 | 0 | 12.82  | 1.1613 | 9.2237 | 0 | | |
| 0.58 | 11.112 | 0 | 8.2768 | 0 | 12.82  | 1.1229 | 9.2143 | 0 | | |
| 0.56 | 11.094 | 0 | 8.2694 | 0 | 12.823 | 1.0838 | 9.2052 | 0 | | |
| 0.54 | 11.077 | 0 | 8.2622 | 0 | 12.828 | 1.0439 | 9.1964 | 0 | | |

| | | | | | | | | |
|---|---|---|---|---|---|---|---|---|
| 0.52 | 11.06 | 0 | 8.2553 | 0 | 12.835 | 1.0032 | 9.188 | 0 |
| 0.5 | 11.044 | 0 | 8.2487 | 0 | 12.845 | 0.96167 | 9.1799 | 0 |
| 0.48 | 11.029 | 0 | 8.2424 | 0 | 12.858 | 0.91938 | 9.1722 | 0 |
| 0.46 | 11.015 | 0 | 8.2364 | 0 | 12.877 | 0.87653 | 9.1648 | 0 |
| 0.44 | 11.001 | 0 | 8.2306 | 0 | 12.903 | 0.83347 | 9.1578 | 0 |
| 0.42 | 10.988 | 0 | 8.225 | 0 | 12.947 | 0.79039 | 9.1511 | 0 |
| 0.4 | 10.975 | 0 | 8.2198 | 0 | 13.035 | 0.73431 | 9.1447 | 0 |
| 0.38 | 10.964 | 0 | 8.2148 | 0 | 13.178 | 0.5666 | 9.1387 | 0 |
| 0.36 | 10.952 | 0 | 8.2101 | 0 | 13.164 | 0.23991 | 9.133 | 0 |
| 0.34 | 10.942 | 0 | 8.2056 | 0 | 12.941 | 0.043707 | 9.1276 | 0 |
| 0.32 | 10.932 | 0 | 8.2014 | 0 | 12.759 | 0.008108 | 9.1225 | 0 |
| 0.3 | 10.923 | 0 | 8.1975 | 0 | 12.644 | 0.00335 | 9.1177 | 0 |
| 0.28 | 10.914 | 0 | 8.1938 | 0 | 12.56 | 0.001454 | 9.1133 | 0 |
| 0.26 | 10.906 | 0 | 8.1903 | 0 | 12.493 | 0.000569 | 9.1091 | 0 |
| 0.24 | 10.899 | 0 | 8.1872 | 0 | 12.437 | 0.000199 | 9.1053 | 0 |
| 0.22 | 10.892 | 0 | 8.1843 | 0 | 12.391 | $6.17E-05$ | 9.1018 | 0 |
| 0.2 | 10.886 | 0 | 8.1816 | 0 | 12.352 | $1.70E-05$ | 9.0985 | 0 |

Table D2 Recommended real and imaginary parts of the permittivities of III–V zinc-blende binary compounds InP, GaSb, AlSb, and InSb.

| Energy (eV) | InP | | GaSb | | AlSb | | InSb | |
|---|---|---|---|---|---|---|---|---|
| | $\varepsilon_r$ | $\varepsilon_i$ | $\varepsilon_r$ | $\varepsilon_i$ | $\varepsilon_r$ | $\varepsilon_i$ | $\varepsilon_r$ | $\varepsilon_i$ |
| 6 | −3.457 | 5.8761 | −4.8667 | 4.5624 | −5.2483 | 4.3235 | −3.8894 | 3.796 |
| 5.98 | −3.4541 | 5.8904 | −4.8824 | 4.6098 | −5.247 | 4.3777 | −3.8923 | 3.839 |
| 5.96 | −3.4587 | 5.9053 | −4.9007 | 4.6572 | −5.2486 | 4.4315 | −3.8966 | 3.8818 |
| 5.94 | −3.4704 | 5.9213 | −4.9216 | 4.7049 | −5.2521 | 4.485 | −3.9023 | 3.9247 |
| 5.92 | −3.4892 | 5.9389 | −4.945 | 4.7534 | −5.2572 | 4.5384 | −3.9091 | 3.9677 |
| 5.9 | −3.515 | 5.9587 | −4.971 | 4.8027 | −5.2645 | 4.5917 | −3.9169 | 4.0109 |
| 5.88 | −3.5475 | 5.9812 | −4.9994 | 4.8533 | −5.2735 | 4.6451 | −3.9257 | 4.0542 |
| 5.86 | −3.5868 | 6.007 | −5.0301 | 4.9056 | −5.2843 | 4.6985 | −3.9354 | 4.0976 |
| 5.84 | −3.6326 | 6.0369 | −5.063 | 4.9598 | −5.2962 | 4.752 | −3.946 | 4.1413 |
| 5.82 | −3.6847 | 6.0716 | −5.0981 | 5.0164 | −5.3097 | 4.8057 | −3.9574 | 4.1851 |
| 5.8 | −3.7426 | 6.1121 | −5.1351 | 5.0758 | −5.3255 | 4.8596 | −3.9696 | 4.2291 |
| 5.78 | −3.8059 | 6.1593 | −5.1736 | 5.1384 | −5.3433 | 4.9138 | −3.9826 | 4.2732 |
| 5.76 | −3.8735 | 6.214 | −5.2134 | 5.2048 | −5.3625 | 4.9683 | −3.9965 | 4.3176 |
| 5.74 | −3.9443 | 6.2769 | −5.2539 | 5.2752 | −5.3844 | 5.0234 | −4.0112 | 4.3621 |
| 5.72 | −4.017 | 6.3486 | −5.2945 | 5.35 | −5.4087 | 5.0791 | −4.0269 | 4.4069 |

| | | | | | | | | |
|---|---|---|---|---|---|---|---|---|
| 5.7 | −4.0898 | 6.4291 | −5.3346 | 5.4295 | −5.4346 | 5.1358 | −4.0435 | 4.4519 |
| 5.68 | −4.1613 | 6.5181 | −5.3733 | 5.5139 | −5.4632 | 5.1937 | −4.0612 | 4.4972 |
| 5.66 | −4.2296 | 6.6147 | −5.4097 | 5.6032 | −5.4946 | 5.2533 | −4.0802 | 4.5428 |
| 5.64 | −4.2934 | 6.7175 | −5.4428 | 5.6971 | −5.5293 | 5.315 | −4.1005 | 4.5889 |
| 5.62 | −4.352 | 6.8245 | −5.4718 | 5.7953 | −5.566 | 5.3796 | −4.1224 | 4.6356 |
| 5.6 | −4.405 | 6.9335 | −5.4957 | 5.8973 | −5.6062 | 5.4478 | −4.146 | 4.6832 |
| 5.58 | −4.4531 | 7.0424 | −5.5136 | 6.0021 | −5.6483 | 5.5203 | −4.1714 | 4.7319 |
| 5.56 | −4.4975 | 7.149 | −5.5249 | 6.1088 | −5.6929 | 5.5982 | −4.1989 | 4.7821 |
| 5.54 | −4.5403 | 7.2516 | −5.5291 | 6.2161 | −5.7381 | 5.6822 | −4.2285 | 4.8345 |
| 5.52 | −4.5842 | 7.3493 | −5.5259 | 6.3227 | −5.7829 | 5.773 | −4.2601 | 4.8895 |
| 5.5 | −4.6321 | 7.4421 | −5.5153 | 6.4272 | −5.8271 | 5.8712 | −4.2936 | 4.948 |
| 5.48 | −4.6869 | 7.5307 | −5.4975 | 6.5279 | −5.8673 | 5.9769 | −4.3286 | 5.0107 |
| 5.46 | −4.7512 | 7.6168 | −5.4733 | 6.6236 | −5.902 | 6.0898 | −4.3644 | 5.0784 |
| 5.44 | −4.8273 | 7.7028 | −5.4437 | 6.7128 | −5.9302 | 6.209 | −4.4 | 5.1519 |
| 5.42 | −4.9167 | 7.7914 | −5.41 | 6.7946 | −5.9497 | 6.3331 | −4.4341 | 5.2315 |
| 5.4 | −5.02 | 7.8857 | −5.3736 | 6.8679 | −5.9576 | 6.4601 | −4.4652 | 5.3175 |
| 5.38 | −5.1375 | 7.9888 | −5.3361 | 6.9323 | −5.9544 | 6.5874 | −4.4914 | 5.4096 |
| 5.36 | −5.2686 | 8.1035 | −5.2993 | 6.9876 | −5.9393 | 6.712 | −4.5111 | 5.507 |

(continued)

Table D2 *Continued*

| Energy (eV) | InP | | GaSb | | AlSb | | InSb | |
|---|---|---|---|---|---|---|---|---|
| | $\varepsilon_r$ | $\varepsilon_i$ | $\varepsilon_r$ | $\varepsilon_i$ | $\varepsilon_r$ | $\varepsilon_i$ | $\varepsilon_r$ | $\varepsilon_i$ |
| 5.34 | −5.4128 | 8.2322 | −5.2651 | 7.0339 | −5.9123 | 6.8307 | −4.5225 | 5.6083 |
| 5.32 | −5.569 | 8.3774 | −5.2351 | 7.0716 | −5.875 | 6.9406 | −4.5242 | 5.7115 |
| 5.3 | −5.7366 | 8.5412 | −5.2109 | 7.1017 | −5.8303 | 7.0388 | −4.5154 | 5.8143 |
| 5.28 | −5.9149 | 8.7258 | −5.1938 | 7.1253 | −5.7805 | 7.1233 | −4.496 | 5.9137 |
| 5.26 | −6.1031 | 8.9337 | −5.185 | 7.1438 | −5.7286 | 7.1928 | −4.4668 | 6.0069 |
| 5.24 | −6.3007 | 9.1682 | −5.1853 | 7.1587 | −5.6789 | 7.247 | −4.4292 | 6.0912 |
| 5.22 | −6.5067 | 9.433 | −5.195 | 7.1716 | −5.635 | 7.2864 | −4.3857 | 6.1644 |
| 5.2 | −6.7192 | 9.733 | −5.2144 | 7.1839 | −5.6006 | 7.3126 | −4.3393 | 6.2251 |
| 5.18 | −6.935 | 10.074 | −5.2432 | 7.1973 | −5.5777 | 7.328 | −4.2932 | 6.2725 |
| 5.16 | −7.149 | 10.461 | −5.2812 | 7.2131 | −5.5692 | 7.3354 | −4.2508 | 6.3071 |
| 5.14 | −7.3537 | 10.9 | −5.3277 | 7.2324 | −5.5756 | 7.3381 | −4.2154 | 6.3302 |
| 5.12 | −7.5392 | 11.396 | −5.3819 | 7.2563 | −5.5974 | 7.3392 | −4.1894 | 6.3439 |
| 5.1 | −7.6929 | 11.95 | −5.443 | 7.2856 | −5.6351 | 7.3417 | −4.1747 | 6.3508 |
| 5.08 | −7.8004 | 12.561 | −5.5099 | 7.3209 | −5.6866 | 7.3481 | −4.1722 | 6.3538 |
| 5.06 | −7.8469 | 13.222 | −5.5818 | 7.3624 | −5.7502 | 7.3605 | −4.1822 | 6.3557 |
| 5.04 | −7.8187 | 13.923 | −5.6579 | 7.4104 | −5.8253 | 7.3803 | −4.204 | 6.3594 |
| 5.02 | −7.7058 | 14.646 | −5.7374 | 7.4648 | −5.9089 | 7.4081 | −4.2365 | 6.3669 |

| | | | | | | | |
|---|---|---|---|---|---|---|---|
| 5 | −7.504 | 15.373 | −5.8196 | 7.5254 | −5.9999 | 7.4444 | −4.2781 | 6.3798 |
| 4.98 | −7.2168 | 16.081 | −5.904 | 7.5921 | −6.0968 | 7.489 | −4.3269 | 6.3993 |
| 4.96 | −6.8563 | 16.753 | −5.9903 | 7.6646 | −6.199 | 7.5414 | −4.3814 | 6.4259 |
| 4.94 | −6.441 | 17.379 | −6.078 | 7.7424 | −6.3051 | 7.601 | −4.4399 | 6.4594 |
| 4.92 | −5.9931 | 17.956 | −6.1672 | 7.8254 | −6.4151 | 7.6672 | −4.5011 | 6.4996 |
| 4.9 | −5.5321 | 18.496 | −6.2576 | 7.9132 | −6.5292 | 7.7394 | −4.5641 | 6.5458 |
| 4.88 | −5.068 | 19.022 | −6.3495 | 8.0054 | −6.6468 | 7.8171 | −4.6283 | 6.5975 |
| 4.86 | −4.5954 | 19.562 | −6.443 | 8.102 | −6.7696 | 7.9 | −4.6932 | 6.6539 |
| 4.84 | −4.091 | 20.137 | −6.5385 | 8.2028 | −6.8974 | 7.9881 | −4.7587 | 6.7143 |
| 4.82 | −3.5158 | 20.757 | −6.6363 | 8.3076 | −7.0309 | 8.0815 | −4.825 | 6.7783 |
| 4.8 | −2.8244 | 21.408 | −6.7367 | 8.4166 | −7.1716 | 8.1806 | −4.8924 | 6.8454 |
| 4.78 | −1.9772 | 22.051 | −6.8402 | 8.5298 | −7.3195 | 8.2863 | −4.961 | 6.9155 |
| 4.76 | −0.95414 | 22.625 | −6.9475 | 8.6474 | −7.4758 | 8.3992 | −5.0312 | 6.9883 |
| 4.74 | 0.23544 | 23.059 | −7.059 | 8.7697 | −7.6404 | 8.5205 | −5.1035 | 7.0638 |
| 4.72 | 1.55 | 23.291 | −7.1755 | 8.8972 | −7.8141 | 8.6514 | −5.1783 | 7.1422 |
| 4.7 | 2.922 | 23.277 | −7.2975 | 9.0306 | −7.9966 | 8.7931 | −5.2559 | 7.2237 |
| 4.68 | 4.2714 | 23.005 | −7.4257 | 9.1704 | −8.1888 | 8.9469 | −5.3367 | 7.3084 |

*(continued)*

Table D2 *Continued*

| Energy (eV) | InP | | GaSb | | AlSb | | InSb | |
|---|---|---|---|---|---|---|---|---|
| | $\varepsilon_r$ | $\varepsilon_i$ | $\varepsilon_r$ | $\varepsilon_i$ | $\varepsilon_r$ | $\varepsilon_i$ | $\varepsilon_r$ | $\varepsilon_i$ |
| 4.66 | 5.5218 | 22.499 | −7.5609 | 9.3177 | −8.3897 | 9.1138 | −5.421 | 7.3969 |
| 4.64 | 6.6146 | 21.805 | −7.7037 | 9.4736 | −8.5992 | 9.2949 | −5.5092 | 7.4894 |
| 4.62 | 7.5163 | 20.986 | −7.8548 | 9.6394 | −8.8173 | 9.4908 | −5.6015 | 7.5864 |
| 4.6 | 8.2194 | 20.103 | −8.0147 | 9.8167 | −9.0442 | 9.702 | −5.6981 | 7.6885 |
| 4.58 | 8.7368 | 19.209 | −8.1841 | 10.007 | −9.2809 | 9.9289 | −5.7994 | 7.7963 |
| 4.56 | 9.094 | 18.345 | −8.3633 | 10.214 | −9.5288 | 10.172 | −5.9055 | 7.9104 |
| 4.54 | 9.3214 | 17.536 | −8.5527 | 10.438 | −9.7908 | 10.433 | −6.0166 | 8.0316 |
| 4.52 | 9.4488 | 16.792 | −8.7523 | 10.684 | −10.07 | 10.713 | −6.1329 | 8.1605 |
| 4.5 | 9.5019 | 16.118 | −8.9617 | 10.954 | −10.371 | 11.018 | −6.2545 | 8.2981 |
| 4.48 | 9.5012 | 15.51 | −9.1803 | 11.253 | −10.699 | 11.355 | −6.3814 | 8.4452 |
| 4.46 | 9.4619 | 14.965 | −9.4068 | 11.584 | −11.057 | 11.734 | −6.5138 | 8.6029 |
| 4.44 | 9.3953 | 14.476 | −9.639 | 11.954 | −11.446 | 12.171 | −6.6517 | 8.7722 |
| 4.42 | 9.3093 | 14.037 | −9.8733 | 12.367 | −11.863 | 12.682 | −6.7952 | 8.9544 |
| 4.4 | 9.2097 | 13.643 | −10.105 | 12.829 | −12.296 | 13.288 | −6.9444 | 9.151 |
| 4.38 | 9.1006 | 13.288 | −10.326 | 13.347 | −12.728 | 14.008 | −7.0991 | 9.3636 |
| 4.36 | 8.9853 | 12.97 | −10.529 | 13.926 | −13.132 | 14.855 | −7.2593 | 9.5943 |
| 4.34 | 8.8661 | 12.684 | −10.7 | 14.569 | −13.473 | 15.839 | −7.4247 | 9.8453 |
| 4.32 | 8.745 | 12.428 | −10.826 | 15.279 | −13.713 | 16.955 | −7.5947 | 10.119 |

| | | | | | | | | |
|---|---|---|---|---|---|---|---|---|
| 4.3 | 8.6235 | 12.199 | −10.89 | 16.053 | −13.812 | 18.186 | −7.7682 | 10.42 |
| 4.28 | 8.5031 | 11.994 | −10.874 | 16.884 | −13.733 | 19.504 | −7.9434 | 10.751 |
| 4.26 | 8.3848 | 11.811 | −10.761 | 17.761 | −13.451 | 20.864 | −8.1177 | 11.116 |
| 4.24 | 8.2695 | 11.649 | −10.538 | 18.667 | −12.951 | 22.217 | −8.2872 | 11.52 |
| 4.22 | 8.1578 | 11.506 | −10.191 | 19.578 | −12.241 | 23.507 | −8.4463 | 11.966 |
| 4.2 | 8.0503 | 11.379 | −9.7178 | 20.469 | −11.341 | 24.685 | −8.5881 | 12.459 |
| 4.18 | 7.9475 | 11.268 | −9.1206 | 21.311 | −10.289 | 25.712 | −8.7037 | 13.001 |
| 4.16 | 7.8496 | 11.171 | −8.4105 | 22.079 | −9.1296 | 26.565 | −8.7826 | 13.593 |
| 4.14 | 7.7568 | 11.086 | −7.6063 | 22.747 | −7.9086 | 27.236 | −8.813 | 14.234 |
| 4.12 | 7.6693 | 11.013 | −6.7321 | 23.301 | −6.6688 | 27.73 | −8.7823 | 14.919 |
| 4.1 | 7.5868 | 10.951 | −5.8157 | 23.731 | −5.4451 | 28.064 | −8.6781 | 15.64 |
| 4.08 | 7.5094 | 10.898 | −4.8853 | 24.035 | −4.2615 | 28.26 | −8.4891 | 16.386 |
| 4.06 | 7.437 | 10.853 | −3.967 | 24.221 | −3.1338 | 28.341 | −8.2061 | 17.141 |
| 4.04 | 7.3692 | 10.816 | −3.0825 | 24.303 | −2.0702 | 28.329 | −7.8239 | 17.887 |
| 4.02 | 7.3058 | 10.787 | −2.2472 | 24.296 | −1.0731 | 28.243 | −7.3413 | 18.605 |
| 4 | 7.2465 | 10.764 | −1.471 | 24.221 | −0.14141 | 28.1 | −6.7628 | 19.273 |

(continued)

**Table D2** *Continued*

| Energy (eV) | InP $\varepsilon_r$ | InP $\varepsilon_i$ | GaSb $\varepsilon_r$ | GaSb $\varepsilon_i$ | AlSb $\varepsilon_r$ | AlSb $\varepsilon_i$ | InSb $\varepsilon_r$ | InSb $\varepsilon_i$ |
|---|---|---|---|---|---|---|---|---|
| 3.98 | 7.1911 | 10.748 | −0.75743 | 24.095 | 0.72719 | 27.91 | −6.0985 | 19.872 |
| 3.96 | 7.1393 | 10.737 | −0.10512 | 23.936 | 1.5362 | 27.685 | −5.3638 | 20.384 |
| 3.94 | 7.0906 | 10.732 | 0.49116 | 23.758 | 2.2892 | 27.431 | −4.5785 | 20.798 |
| 3.92 | 7.0449 | 10.733 | 1.0391 | 23.57 | 2.9886 | 27.154 | −3.7651 | 21.108 |
| 3.9 | 7.0019 | 10.739 | 1.547 | 23.378 | 3.6368 | 26.86 | −2.9459 | 21.315 |
| 3.88 | 6.9614 | 10.751 | 2.0227 | 23.186 | 4.2365 | 26.555 | −2.1407 | 21.426 |
| 3.86 | 6.9232 | 10.768 | 2.4727 | 22.994 | 4.7896 | 26.242 | −1.3647 | 21.452 |
| 3.84 | 6.8873 | 10.791 | 2.902 | 22.8 | 5.2981 | 25.927 | −0.6281 | 21.405 |
| 3.82 | 6.8535 | 10.821 | 3.3135 | 22.603 | 5.7662 | 25.614 | 0.062898 | 21.3 |
| 3.8 | 6.8219 | 10.856 | 3.7081 | 22.401 | 6.1969 | 25.307 | 0.70479 | 21.146 |
| 3.78 | 6.7926 | 10.899 | 4.0856 | 22.191 | 6.5947 | 25.011 | 1.2957 | 20.955 |
| 3.76 | 6.7656 | 10.948 | 4.4446 | 21.974 | 6.9652 | 24.726 | 1.8355 | 20.737 |
| 3.74 | 6.7412 | 11.004 | 4.7838 | 21.749 | 7.3136 | 24.455 | 2.3258 | 20.501 |
| 3.72 | 6.7196 | 11.068 | 5.1023 | 21.519 | 7.6442 | 24.196 | 2.7701 | 20.257 |
| 3.7 | 6.7012 | 11.14 | 5.3988 | 21.285 | 7.9618 | 23.946 | 3.1735 | 20.01 |
| 3.68 | 6.6861 | 11.22 | 5.6728 | 21.051 | 8.2686 | 23.704 | 3.5419 | 19.767 |
| 3.66 | 6.6748 | 11.309 | 5.9248 | 20.818 | 8.5648 | 23.464 | 3.8814 | 19.528 |

| | | | | | | | | |
|---|---|---|---|---|---|---|---|---|
| 3.64 | 6.6678 | 11.407 | 6.1562 | 20.59 | 8.8487 | 23.225 | 4.1973 | 19.295 |
| 3.62 | 6.6656 | 11.514 | 6.3693 | 20.368 | 9.1176 | 22.984 | 4.4939 | 19.068 |
| 3.6 | 6.6685 | 11.631 | 6.5661 | 20.156 | 9.3668 | 22.742 | 4.7744 | 18.844 |
| 3.58 | 6.6773 | 11.758 | 6.7491 | 19.954 | 9.5917 | 22.502 | 5.0405 | 18.623 |
| 3.56 | 6.6924 | 11.896 | 6.9214 | 19.762 | 9.7887 | 22.267 | 5.293 | 18.403 |
| 3.54 | 6.7145 | 12.046 | 7.0859 | 19.581 | 9.9563 | 22.046 | 5.5319 | 18.182 |
| 3.52 | 6.744 | 12.207 | 7.2456 | 19.409 | 10.094 | 21.847 | 5.7569 | 17.962 |
| 3.5 | 6.7814 | 12.381 | 7.4021 | 19.245 | 10.205 | 21.677 | 5.9676 | 17.742 |
| 3.48 | 6.8269 | 12.57 | 7.5571 | 19.087 | 10.293 | 21.545 | 6.164 | 17.523 |
| 3.46 | 6.8806 | 12.774 | 7.7116 | 18.934 | 10.367 | 21.458 | 6.3467 | 17.307 |
| 3.44 | 6.9423 | 12.999 | 7.8661 | 18.783 | 10.434 | 21.424 | 6.517 | 17.095 |
| 3.42 | 7.0127 | 13.25 | 8.0205 | 18.631 | 10.506 | 21.449 | 6.6765 | 16.887 |
| 3.4 | 7.0958 | 13.536 | 8.1734 | 18.478 | 10.598 | 21.539 | 6.8272 | 16.683 |
| 3.38 | 7.2033 | 13.872 | 8.3237 | 18.321 | 10.728 | 21.695 | 6.9706 | 16.481 |
| 3.36 | 7.3587 | 14.266 | 8.4695 | 18.16 | 10.92 | 21.917 | 7.1078 | 16.281 |
| 3.34 | 7.5969 | 14.714 | 8.609 | 17.995 | 11.202 | 22.194 | 7.2383 | 16.078 |

(continued)

Table D2 *Continued*

| Energy (eV) | InP | | GaSb | | AlSb | | InSb | |
|---|---|---|---|---|---|---|---|---|
| | $\varepsilon_r$ | $\varepsilon_i$ | $\varepsilon_r$ | $\varepsilon_i$ | $\varepsilon_r$ | $\varepsilon_i$ | $\varepsilon_r$ | $\varepsilon_i$ |
| 3.32 | 7.9514 | 15.188 | 8.74 | 17.825 | 11.604 | 22.501 | 7.3608 | 15.873 |
| 3.3 | 8.4353 | 15.64 | 8.8609 | 17.652 | 12.146 | 22.8 | 7.4725 | 15.662 |
| 3.28 | 9.0298 | 16.024 | 8.9701 | 17.476 | 12.835 | 23.038 | 7.5702 | 15.447 |
| 3.26 | 9.6948 | 16.32 | 9.0665 | 17.301 | 13.651 | 23.157 | 7.6505 | 15.228 |
| 3.24 | 10.398 | 16.541 | 9.1491 | 17.127 | 14.551 | 23.11 | 7.7108 | 15.01 |
| 3.22 | 11.141 | 16.713 | 9.2175 | 16.956 | 15.469 | 22.873 | 7.7497 | 14.795 |
| 3.2 | 11.956 | 16.842 | 9.2724 | 16.792 | 16.338 | 22.449 | 7.7673 | 14.589 |
| 3.18 | 12.879 | 16.89 | 9.3144 | 16.635 | 17.097 | 21.871 | 7.7651 | 14.396 |
| 3.16 | 13.913 | 16.783 | 9.3445 | 16.488 | 17.707 | 21.191 | 7.7458 | 14.218 |
| 3.14 | 15.001 | 16.443 | 9.3639 | 16.352 | 18.155 | 20.465 | 7.7128 | 14.058 |
| 3.12 | 16.035 | 15.839 | 9.3739 | 16.229 | 18.447 | 19.745 | 7.6699 | 13.917 |
| 3.1 | 16.907 | 15.009 | 9.3766 | 16.118 | 18.602 | 19.07 | 7.6204 | 13.796 |
| 3.08 | 17.548 | 14.04 | 9.3736 | 16.022 | 18.644 | 18.47 | 7.5671 | 13.693 |
| 3.06 | 17.943 | 13.035 | 9.3665 | 15.939 | 18.597 | 17.965 | 7.5125 | 13.608 |
| 3.04 | 18.13 | 12.078 | 9.3564 | 15.87 | 18.485 | 17.574 | 7.4582 | 13.538 |
| 3.02 | 18.168 | 11.213 | 9.3446 | 15.815 | 18.334 | 17.315 | 7.4053 | 13.483 |
| 3 | 18.113 | 10.45 | 9.3327 | 15.774 | 18.174 | 17.209 | 7.3545 | 13.442 |

| 2.98 | 18.002 | 9.7773 | 9.3216 | 15.746 | 18.052 | 17.276 | 7.3063 | 13.413 |
|------|--------|--------|--------|--------|--------|--------|--------|--------|
| 2.96 | 17.86  | 9.182  | 9.312  | 15.731 | 18.035 | 17.528 | 7.2611 | 13.396 |
| 2.94 | 17.697 | 8.651  | 9.3044 | 15.729 | 18.211 | 17.95  | 7.2191 | 13.389 |
| 2.92 | 17.522 | 8.1749 | 9.2996 | 15.739 | 18.68  | 18.483 | 7.1804 | 13.394 |
| 2.9  | 17.339 | 7.7463 | 9.2983 | 15.762 | 19.528 | 19.001 | 7.1454 | 13.408 |
| 2.88 | 17.153 | 7.3596 | 9.301  | 15.798 | 20.775 | 19.32  | 7.1143 | 13.433 |
| 2.86 | 16.967 | 7.0095 | 9.3083 | 15.846 | 22.338 | 19.238 | 7.0872 | 13.467 |
| 2.84 | 16.783 | 6.6915 | 9.3206 | 15.908 | 24.017 | 18.605 | 7.0644 | 13.511 |
| 2.82 | 16.603 | 6.4014 | 9.3389 | 15.983 | 25.542 | 17.402 | 7.0462 | 13.564 |
| 2.8  | 16.428 | 6.1356 | 9.3641 | 16.073 | 26.678 | 15.758 | 7.033  | 13.627 |
| 2.78 | 16.259 | 5.8908 | 9.3973 | 16.177 | 27.295 | 13.889 | 7.0249 | 13.699 |
| 2.76 | 16.096 | 5.6641 | 9.4399 | 16.297 | 27.384 | 12.036 | 7.0225 | 13.781 |
| 2.74 | 15.939 | 5.453  | 9.4933 | 16.433 | 27.055 | 10.39  | 7.0261 | 13.873 |
| 2.72 | 15.788 | 5.2554 | 9.5595 | 16.588 | 26.468 | 9.0447 | 7.0362 | 13.974 |
| 2.7  | 15.642 | 5.0697 | 9.641  | 16.761 | 25.777 | 8.0024 | 7.0534 | 14.085 |

(continued)

**Table D2** *Continued*

| Energy (eV) | InP | | GaSb | | AlSb | | InSb | |
|---|---|---|---|---|---|---|---|---|
| | $\varepsilon_r$ | $\varepsilon_i$ | $\varepsilon_r$ | $\varepsilon_i$ | $\varepsilon_r$ | $\varepsilon_i$ | $\varepsilon_r$ | $\varepsilon_i$ |
| 2.68 | 15.5 | 4.8945 | 9.7413 | 16.955 | 25.088 | 7.2098 | 7.0784 | 14.207 |
| 2.66 | 15.363 | 4.7287 | 9.8651 | 17.172 | 24.453 | 6.5974 | 7.1118 | 14.339 |
| 2.64 | 15.23 | 4.5717 | 10.019 | 17.414 | 23.89 | 6.1034 | 7.1546 | 14.482 |
| 2.62 | 15.1 | 4.4227 | 10.214 | 17.681 | 23.393 | 5.6836 | 7.2076 | 14.637 |
| 2.6 | 14.973 | 4.2812 | 10.463 | 17.969 | 22.95 | 5.3104 | 7.2721 | 14.803 |
| 2.58 | 14.849 | 4.1469 | 10.781 | 18.27 | 22.547 | 4.9687 | 7.3491 | 14.981 |
| 2.56 | 14.728 | 4.0193 | 11.183 | 18.565 | 22.174 | 4.6505 | 7.4398 | 15.171 |
| 2.54 | 14.61 | 3.8981 | 11.676 | 18.825 | 21.824 | 4.3522 | 7.5454 | 15.376 |
| 2.52 | 14.495 | 3.7831 | 12.254 | 19.019 | 21.491 | 4.072 | 7.6671 | 15.597 |
| 2.5 | 14.382 | 3.6738 | 12.898 | 19.116 | 21.172 | 3.8086 | 7.8076 | 15.84 |
| 2.48 | 14.272 | 3.5701 | 13.574 | 19.092 | 20.866 | 3.5613 | 7.9741 | 16.112 |
| 2.46 | 14.165 | 3.4715 | 14.24 | 18.944 | 20.571 | 3.329 | 8.1834 | 16.421 |
| 2.44 | 14.06 | 3.3779 | 14.857 | 18.683 | 20.287 | 3.1108 | 8.4646 | 16.765 |
| 2.42 | 13.959 | 3.2889 | 15.394 | 18.335 | 20.014 | 2.9056 | 8.8518 | 17.122 |
| 2.4 | 13.86 | 3.2043 | 15.835 | 17.932 | 19.751 | 2.7121 | 9.3666 | 17.443 |
| 2.38 | 13.763 | 3.1239 | 16.176 | 17.507 | 19.498 | 2.5288 | 9.9973 | 17.665 |

| | | | | | | | |
|---|---|---|---|---|---|---|---|
| 2.36 | 13.67 | 3.0474 | 16.424 | 17.09 | 19.255 | 2.3543 | 10.694 | 17.736 |
| 2.34 | 13.579 | 2.9745 | 16.591 | 16.701 | 19.021 | 2.1869 | 11.386 | 17.645 |
| 2.32 | 13.491 | 2.9052 | 16.691 | 16.357 | 18.797 | 2.0251 | 12.01 | 17.418 |
| 2.3 | 13.406 | 2.839 | 16.739 | 16.07 | 18.58 | 1.8672 | 12.535 | 17.103 |
| 2.28 | 13.323 | 2.7759 | 16.75 | 15.849 | 18.371 | 1.7115 | 12.954 | 16.748 |
| 2.26 | 13.243 | 2.7156 | 16.739 | 15.705 | 18.168 | 1.5559 | 13.282 | 16.385 |
| 2.24 | 13.166 | 2.6579 | 16.724 | 15.644 | 17.97 | 1.3977 | 13.535 | 16.034 |
| 2.22 | 13.091 | 2.6027 | 16.729 | 15.675 | 17.775 | 1.2297 | 13.73 | 15.703 |
| 2.2 | 13.019 | 2.5498 | 16.781 | 15.8 | 17.562 | 1.0468 | 13.875 | 15.397 |
| 2.18 | 12.949 | 2.4989 | 16.914 | 16.014 | 17.316 | 0.87232 | 13.979 | 15.118 |
| 2.16 | 12.882 | 2.45 | 17.168 | 16.301 | 17.051 | 0.73051 | 14.045 | 14.869 |
| 2.14 | 12.817 | 2.4028 | 17.581 | 16.627 | 16.787 | 0.61984 | 14.079 | 14.657 |
| 2.12 | 12.755 | 2.3572 | 18.181 | 16.94 | 16.531 | 0.53365 | 14.085 | 14.485 |
| 2.1 | 12.695 | 2.3132 | 18.979 | 17.169 | 16.287 | 0.46665 | 14.067 | 14.361 |

(continued)

**Table D2** *Continued*

| Energy (eV) | InP | | GaSb | | AlSb | | InSb | |
|---|---|---|---|---|---|---|---|---|
| | $\varepsilon_r$ | $\varepsilon_i$ | $\varepsilon_r$ | $\varepsilon_i$ | $\varepsilon_r$ | $\varepsilon_i$ | $\varepsilon_r$ | $\varepsilon_i$ |
| 2.08 | 12.637 | 2.2704 | 19.957 | 17.238 | 16.056 | 0.41448 | 14.032 | 14.294 |
| 2.06 | 12.582 | 2.2289 | 21.059 | 17.073 | 15.839 | 0.37365 | 13.99 | 14.294 |
| 2.04 | 12.529 | 2.1885 | 22.202 | 16.629 | 15.634 | 0.34135 | 13.954 | 14.371 |
| 2.02 | 12.478 | 2.1491 | 23.282 | 15.897 | 15.441 | 0.31536 | 13.944 | 14.536 |
| 2 | 12.429 | 2.1106 | 24.206 | 14.915 | 15.26 | 0.29392 | 13.99 | 14.797 |
| 1.98 | 12.382 | 2.0729 | 24.904 | 13.751 | 15.09 | 0.2756 | 14.13 | 15.155 |
| 1.96 | 12.337 | 2.0359 | 25.339 | 12.489 | 14.929 | 0.25928 | 14.414 | 15.594 |
| 1.94 | 12.294 | 1.9995 | 25.508 | 11.22 | 14.777 | 0.24412 | 14.894 | 16.073 |
| 1.92 | 12.253 | 1.9636 | 25.443 | 10.018 | 14.632 | 0.22952 | 15.613 | 16.518 |
| 1.9 | 12.213 | 1.9282 | 25.195 | 8.9357 | 14.495 | 0.21513 | 16.577 | 16.825 |
| 1.88 | 12.176 | 1.8932 | 24.821 | 7.9992 | 14.363 | 0.20078 | 17.736 | 16.883 |
| 1.86 | 12.141 | 1.8585 | 24.375 | 7.2114 | 14.236 | 0.18644 | 18.984 | 16.608 |
| 1.84 | 12.107 | 1.8241 | 23.898 | 6.5614 | 14.114 | 0.1721 | 20.175 | 15.978 |
| 1.82 | 12.075 | 1.7899 | 23.421 | 6.0303 | 13.996 | 0.15775 | 21.173 | 15.049 |
| 1.8 | 12.045 | 1.7557 | 22.963 | 5.597 | 13.883 | 0.14341 | 21.891 | 13.93 |
| 1.78 | 12.016 | 1.7216 | 22.535 | 5.2415 | 13.772 | 0.12907 | 22.304 | 12.747 |

| | | | | | | | | |
|---|---|---|---|---|---|---|---|---|
| 1.76 | 11.989 | 1.6875 | 22.143 | 4.9457 | 13.665 | 0.11473 | 22.448 | 11.609 |
| 1.74 | 11.965 | 1.6532 | 21.788 | 4.6942 | 13.561 | 0.10039 | 22.39 | 10.582 |
| 1.72 | 11.942 | 1.6186 | 21.469 | 4.4738 | 13.46 | 0.086048 | 22.202 | 9.6916 |
| 1.7 | 11.92 | 1.5838 | 21.184 | 4.2722 | 13.361 | 0.071707 | 21.94 | 8.9332 |
| 1.68 | 11.901 | 1.5485 | 20.926 | 4.0784 | 13.263 | 0.057365 | 21.642 | 8.2898 |
| 1.66 | 11.884 | 1.5127 | 20.684 | 3.8853 | 13.168 | 0.043025 | 21.331 | 7.7424 |
| 1.64 | 11.868 | 1.4764 | 20.445 | 3.695 | 13.073 | 0.02872 | 21.019 | 7.2745 |
| 1.62 | 11.856 | 1.4393 | 20.203 | 3.5176 | 12.978 | 0.015055 | 20.715 | 6.8727 |
| 1.6 | 11.845 | 1.4014 | 19.965 | 3.361 | 12.882 | 0.004855 | 20.423 | 6.5262 |
| 1.58 | 11.838 | 1.3626 | 19.738 | 3.2248 | 12.788 | 0.000713 | 20.146 | 6.2258 |
| 1.56 | 11.835 | 1.3223 | 19.526 | $3.10E+00$ | 12.7 | $3.77E-05$ | 19.887 | 5.9637 |
| 1.54 | 11.838 | 1.2794 | 19.329 | $2.99E+00$ | 12.617 | 0 | 19.646 | 5.733 |
| 1.52 | 11.849 | 1.2309 | 19.144 | $2.88E+00$ | 12.537 | 0 | 19.423 | 5.5276 |
| 1.5 | 11.866 | 1.1713 | 18.968 | $2.78E+00$ | 12.461 | 0 | 19.218 | 5.3424 |
| 1.48 | 11.885 | 1.0942 | 18.801 | $2.68E+00$ | 12.388 | 0 | 19.029 | 5.1732 |
| 1.46 | 11.897 | 0.99727 | 18.641 | 2.5869 | 12.317 | 0 | 18.856 | 5.0166 |
| 1.44 | 11.888 | 0.88767 | 18.487 | 2.4967 | 12.248 | 0 | 18.696 | 4.8699 |

*(continued)*

Table D2 *Continued*

| Energy (eV) | InP | | GaSb | | AlSb | | InSb | |
|---|---|---|---|---|---|---|---|---|
| | $\varepsilon_r$ | $\varepsilon_i$ | $\varepsilon_r$ | $\varepsilon_i$ | $\varepsilon_r$ | $\varepsilon_i$ | $\varepsilon_r$ | $\varepsilon_i$ |
| 1.42 | 11.856 | 0.78051 | 18.34 | 2.4103 | 12.182 | 0 | 18.547 | 4.7311 |
| 1.4 | 11.813 | 0.69034 | 18.198 | 2.3274 | 12.118 | 0 | 18.409 | 4.5987 |
| 1.38 | 11.78 | 0.62252 | 18.061 | 2.2478 | 12.056 | 0 | 18.28 | 4.4718 |
| 1.36 | 11.807 | 0.5532 | 17.929 | 2.1715 | 11.995 | 0 | 18.158 | 4.3496 |
| 1.34 | 11.863 | 0.32526 | 17.802 | 2.0983 | 11.937 | 0 | 18.044 | 4.2316 |
| 1.32 | 11.66 | 0.056809 | 17.679 | 2.028 | 11.88 | 0 | 17.935 | 4.1175 |
| 1.3 | 11.423 | 0.005023 | 17.56 | 1.9605 | 11.825 | 0 | 17.832 | 4.007 |
| 1.28 | 11.272 | 0.00172 | 17.445 | 1.8957 | 11.772 | 0 | 17.733 | 3.8998 |
| 1.26 | 11.155 | 0.000675 | 17.334 | 1.8335 | 11.72 | 0 | 17.64 | 3.7956 |
| 1.24 | 11.056 | 0.000233 | 17.227 | 1.7738 | 11.67 | 0 | 17.55 | 3.6944 |
| 1.22 | 10.97 | 7.02E − 05 | 17.123 | 1.7165 | 11.621 | 0 | 17.465 | 3.5957 |
| 1.2 | 10.892 | 1.83E − 05 | 17.023 | 1.6614 | 11.573 | 0 | 17.384 | 3.4995 |
| 1.18 | 10.821 | 4.14E − 06 | 16.926 | 1.6086 | 11.527 | 0 | 17.306 | 3.4054 |
| 1.16 | 10.756 | 0 | 16.833 | 1.5579 | 11.482 | 0 | 17.232 | 3.3133 |
| 1.14 | 10.695 | 0 | 16.743 | 1.5091 | 11.439 | 0 | 17.162 | 3.2228 |

| | | | | | | | |
|---|---|---|---|---|---|---|---|
| 1.12 | 10.638 | 0 | 16.657 | 1.4623 | 11.396 | 0 | 17.095 | 3.1339 |
| 1.1 | 10.584 | 0 | 16.573 | 1.4172 | 11.355 | 0 | 17.032 | 3.046 |
| 1.08 | 10.534 | 0 | 16.494 | 1.3738 | 11.315 | 0 | 16.973 | 2.9584 |
| 1.06 | 10.486 | 0 | 16.417 | 1.3319 | 11.277 | 0 | 16.919 | 2.8689 |
| 1.04 | 10.44 | 0 | 16.344 | 1.2914 | 11.239 | 0 | 16.869 | 2.7739 |
| 1.02 | 10.396 | 0 | 16.274 | 1.2522 | 11.202 | 0 | 16.816 | 2.6707 |
| 1 | 10.355 | 0 | 16.208 | 1.2141 | 11.167 | 0 | 16.753 | 2.5624 |
| 0.98 | 10.315 | 0 | 16.146 | 1.177 | 11.132 | 0 | 16.678 | 2.4579 |
| 0.96 | 10.277 | 0 | 16.086 | 1.1406 | 11.099 | 0 | 16.595 | 2.3655 |
| 0.94 | 10.241 | 0 | 16.031 | 1.105 | 11.066 | 0 | 16.512 | 2.2866 |
| 0.92 | 10.206 | 0 | 15.979 | 1.0698 | 11.034 | 0 | 16.435 | 2.2168 |
| 0.9 | 10.173 | 0 | 15.932 | 1.0349 | 11.004 | 0 | 16.364 | 2.1518 |
| 0.88 | 10.141 | 0 | 15.889 | 1.0001 | 10.974 | 0 | 16.298 | 2.0894 |
| 0.86 | 10.11 | 0 | 15.85 | 0.9653 | 10.945 | 0 | 16.236 | 2.0288 |
| 0.84 | 10.08 | 0 | 15.817 | 0.93015 | 10.917 | 0 | 16.176 | 1.9699 |
| 0.82 | 10.052 | 0 | 15.791 | 0.89446 | 10.89 | 0 | 16.119 | 1.9126 |
| 0.8 | 10.024 | 0 | 15.774 | 0.85793 | 10.864 | 0 | 16.065 | 1.8568 |

(continued)

Table D2 *Continued*

| Energy (eV) | InP | | GaSb | | AlSb | | InSb | |
|---|---|---|---|---|---|---|---|---|
| | $\varepsilon_r$ | $\varepsilon_i$ | $\varepsilon_r$ | $\varepsilon_i$ | $\varepsilon_r$ | $\varepsilon_i$ | $\varepsilon_r$ | $\varepsilon_i$ |
| 0.78 | 9.9979 | 0 | 15.77 | 0.8202 | 10.838 | 0 | 16.013 | 1.8025 |
| 0.76 | 9.9726 | 0 | 15.794 | 0.77835 | 10.814 | 0 | 15.962 | 1.7497 |
| 0.74 | 9.9482 | 0 | 15.878 | 0.69337 | 10.79 | 0 | 15.914 | 1.6984 |
| 0.72 | 9.9248 | 0 | 15.955 | 0.44634 | 10.767 | 0 | 15.868 | 1.6484 |
| 0.7 | 9.9023 | 0 | 15.802 | 0.13817 | 10.744 | 0 | 15.824 | 1.5997 |
| 0.68 | 9.8807 | 0 | 15.541 | 0.017956 | 10.723 | 0 | 15.782 | 1.5523 |
| 0.66 | 9.8599 | 0 | 15.352 | 0.003 | 10.702 | 0 | 15.742 | 1.5062 |
| 0.64 | 9.8399 | 0 | 15.216 | 0.001429 | 10.682 | 0 | 15.704 | 1.4612 |
| 0.62 | 9.8208 | 0 | 15.106 | 0.000741 | 10.662 | 0 | 15.667 | 1.4174 |
| 0.6 | 9.8024 | 0 | 15.011 | 0.000364 | 10.644 | 0 | 15.633 | 1.3746 |
| 0.58 | 9.7848 | 0 | 14.927 | 0.000168 | 10.626 | 0 | 15.601 | 1.3329 |
| 0.56 | 9.7679 | 0 | 14.851 | 7.32E − 05 | 10.608 | 0 | 15.57 | 1.2921 |
| 0.54 | 9.7517 | 0 | 14.781 | 2.98E − 05 | 10.592 | 0 | 15.542 | 1.2522 |
| 0.52 | 9.7363 | 0 | 14.718 | 1.14E − 05 | 10.576 | 0 | 15.516 | 1.2131 |
| 0.5 | 9.7215 | 0 | 14.659 | 4.09E − 06 | 10.561 | 0 | 15.492 | 1.1748 |
| 0.48 | 9.7074 | 0 | 14.605 | 1.37E − 06 | 10.546 | 0 | 15.47 | 1.1372 |
| 0.46 | 9.6939 | 0 | 14.554 | 0 | 10.532 | 0 | 15.45 | 1.1002 |
| 0.44 | 9.6812 | 0 | 14.507 | 0 | 10.519 | 0 | 15.433 | 1.0637 |

| | | | | | | | | |
|---|---|---|---|---|---|---|---|---|
| 0.42 | 9.669 | 0 | 14.463 | 0 | 10.506 | 0 | 15.418 | 1.0277 |
| 0.4 | 9.6575 | 0 | 14.422 | 0 | 10.494 | 0 | 15.405 | 0.99212 |
| 0.38 | 9.6466 | 0 | 14.383 | 0 | 10.483 | 0 | 15.395 | 0.95683 |
| 0.36 | 9.6363 | 0 | 14.348 | 0 | 10.472 | 0 | 15.388 | 0.92178 |
| 0.34 | 9.6266 | 0 | 14.315 | 0 | 10.462 | 0 | 15.383 | 0.88688 |
| 0.32 | 9.6175 | 0 | 14.284 | 0 | 10.452 | 0 | 15.383 | 0.85204 |
| 0.3 | 9.609 | 0 | 14.255 | 0 | 10.443 | 0 | 15.386 | 0.81718 |
| 0.28 | 9.6011 | 0 | 14.229 | 0 | 10.435 | 0 | 15.394 | 0.78225 |
| 0.26 | 9.5937 | 0 | 14.205 | 0 | 10.427 | 0 | 15.411 | 0.7472 |
| 0.24 | 9.587 | 0 | 14.182 | 0 | 10.42 | 0 | 15.44 | 0.7119 |
| 0.22 | 9.5807 | 0 | 14.162 | 0 | 10.413 | 0 | 15.501 | 0.66976 |
| 0.2 | 9.5751 | 0 | 14.144 | 0 | 10.407 | 0 | 15.614 | 0.55884 |
| 0.18 | | | | | | | 15.65 | 0.29374 |
| 0.16 | | | | | | | 15.469 | 0.067049 |
| 0.14 | | | | | | | 15.257 | 0.007363 |
| 0.12 | | | | | | | 15.093 | 0.001516 |
| 0.1 | | | | | | | 14.925 | 0.000577 |
| 0.08 | | | | | | | 14.682 | 0.000207 |
| 0.06 | | | | | | | 14.183 | 6.61E − 05 |
| 0.04 | | | | | | | 12.379 | 1.89E − 05 |

# Index

**Igor Vurgaftman** received the Ph.D. degree in electrical engineering from the University of Michigan, Ann Arbor, MI, in 1995. Since then, he has been with the Optical Sciences Division of the Naval Research Laboratory (NRL), Washington, DC, where he is currently Head of the Quantum Optoelectronics Section. At NRL, he has investigated mid-infrared lasers based on interband and intersubband transitions, type-II superlattice photodetectors, as well as the physics of various plasmonic and polaritonic devices. He is the author of more than 285 refereed articles in technical journals, cited more than 14,000 times (H-index of 46), as well as more than 25 patents. Dr. Vurgaftman is a recipient of the IEEE Photonics Society Engineering Achievement Award (2012), Dr. Dolores M. Etter Top Navy Scientists and Engineers of the Year Award (2015), NRL Sigma Xi Pure Science (2012) and Young Investigator (2005) Awards. He is a Fellow of the Optical Society and the American Physical Society.

**Matthew Lumb** gained his Ph.D in 2009 from the Department of Physics at Imperial College London in the field of InAs quantum dot saturable absorbers. In 2009, Dr. Lumb joined UK-based startup QuantaSol Ltd., developing state-of-the-art solar cells using strain-balanced quantum wells. In 2011, Dr. Lumb transitioned to the Electrical and Computer Engineering department at George Washington University, based fulltime as a contractor at the Naval Research Laboratory, in Washington DC. His research activities involve numerous aspects of optoelectronic device R&D, with a particular emphasis on high efficiency, multi-junction photovoltaics and related devices. In 2018, Dr. Lumb founded Polaris Semiconductor LLC, and is currently developing a novel DC power management technology funded by the National Science Foundation. Dr. Lumb has coauthored more than 100 refereed journal articles and conference proceedings, has given numerous invited talks including being a plenary speaker at the 2017 CPV-13 conference in Ottawa, has a H-index of 18 and holds 2 patents.

**Jerry R. Meyer** received a Ph.D. in physics from Brown University in 1977. Since then he has carried out basic and applied research at the Naval Research Laboratory in Washington DC, where he is the Navy Senior Scientist for Quantum Electronics (ST). His research has focused on semiconductor optoelectronic materials and devices, especially new classes of lasers and detectors for the infrared. Dr. Meyer is a Fellow of the Optical Society of America (OSA), the American Physical Society (APS), the Institute of Physics (IOP), the Institute of Electrical and Electronics Engineers (IEEE), and SPIE. He is a recipient of the Presidential Rank Award (2016), ONR's Captain Robert Dexter Conrad Award for Scientific Achievement (2015), NRL's E. O. Hulbert Annual Science Award (2012), the IEEE Photonics Society Engineering Achievement Award (2012), the Dr. Dolores M. Etter Top Navy Scientists and Engineers of the Year Award (2008), and the NRL Edison Chapter Sigma Xi Award for Pure Science (2003). He has coauthored more than 390 refereed journal articles that have been cited more than 27,000 times (H-Index of 65), 39 patents granted, and more than 190 Invited, Plenary, Keynote, and Tutorial conference presentations.